DIGITAL ELECTRONICS: Fundamental Concepts And Applications

DIGITAL ELECTRONICS: Fundamental Concepts And Applications

CHRISTOPHER E. STRANGIO

Lecturer,
Lowell Institute School,
Massachusetts Institute of Technology

PRENTICE-HALL, INC., *Englewood Cliffs, New Jersey 07632*

Library of Congress Cataloging in Publication Data
Strangio, Christopher E
Digital electronics.

Includes index.
1. Digital electronics. I. Title.
TK7868.D5S77 621.3815 79-23183
ISBN 0-13-212100-X

Editorial production supervision
and interior design by: JAMES M. CHEGE

Page layout by: RITA K. SCHWARTZ

Manufacturing buyer: GORDON OSBOURNE

Printed in the United States of America

10 9 8 7 6 5 4 3 2

PRENTICE-HALL INTERNATIONAL, INC., *London*
PRENTICE-HALL OF AUSTRALIA PTY. LIMITED, *Sydney*
PRENTICE-HALL OF CANADA, LTD., *Toronto*
PRENTICE-HALL OF INDIA PRIVATE LIMITED, *New Delhi*
PRENTICE-HALL OF JAPAN, INC., *Tokyo*
PRENTICE-HALL OF SOUTHEAST ASIA PTE. LTD., *Singapore*
WHITEHALL BOOKS LIMITED, *Wellington, New Zealand*

Dedicated to

Dr. Bruce D. Wedlock

CONTENTS

PREFACE

A *signal* is a means used to convey information from one point to another. It may be embodied as a visual, audible, electrical, or other variation that can be transmitted in an appropriate medium through space. Traditional electrical signals have been analog in nature, in which case an informational quantity, such as temperature, is proportionally related to an electrical parameter, such as voltage. *Analog signals* smoothly span a continuous range of voltage (or current), and are restricted only in the absolute limits of the range. In contrast, a *digital signal* is a numerically coded representation of an informational quantity, and is confined to a small number of discrete levels of voltage (or current). The majority of digital signals are binary coded, thereby limited to exactly two discrete voltage (or current) levels, and appear as a sequence of 'on'-'off' conditions (high or low voltages) on an electrical conductor. *Digital Electronics* concerns the generation, processing, and storage of digital signals, and establishes a framework for the analysis, design, and construction of digital systems.

The intent of this book is to convey the principle concepts of digital electronics, and offer a thoughtful, practical approach to digital circuit analysis and design. Digital computers comprise one specific branch of the digital systems tree, and are also discussed. Since this book is more particularly focused on the hardware issues, computers are introduced from this vantage point. As prerequisites, students should have had only a basic course in algebra, and some background in simple electrical circuits as might be found in an electric shop course or in introductory physics. Semiconductor circuit analysis is not involved. The expected audience is upper level technical school students, and college undergraduates specializing in any of the branches of engineering and science.

A teacher's greatest responsibility is to cultivate the student's ability to think. That philosophy is the theme of this book. Rather than present a catalog of techniques, a fewer number of carefully selected techniques are examined in detail, and applied to typical analysis or design problems. Answers to "Why?" and "How?" are deemed as important as those for "How to . . . ?" Above all, it is important that the student be equipped to assimilate new developments as advancing technology produces them. In this, the ability to think clearly about digital concepts is invaluable, and is strongly emphasized in the presentation.

Chapter One introduces the student to the essential concepts in digital electronics, and tries to establish a framework of thinking that will facilitate comprehension of the topics to follow. Although relays are used very rarely in modern digital systems, their basic simplicity is relied on here to permit a clear, understandable description of logic functions without forcing the student to understand semiconductor circuits, and without using the unconvincing "black box" approach to describe logic functions.

Chapter Two addresses Boolean algebra from the switching logic viewpoint introduced in Chapter One, and develops a useful set of Boolean properties. Some educators feel that the usefulness of Boolean algebra and Karnaugh map techniques is minimal at the present and will decline in the future. In this text, it is felt that the *method of thinking* developed in a study of Boolean algebra and related mathematical techniques is crucially important in perceiving efficient digital systems and in evolving simple, minimal hardware solutions to design problems, even though certain specific techniques may not themselves be used frequently.

Combinational logic is the basis of all further chapters in this book, and is introduced in Chapter Three. A realistic, practical approach is integrated with the Boolean algebra concepts of Chapter Two to validate by use those symbolic techniques that have been presented. Karnaugh maps build further on the mathematical framework started with Boolean algebra, and are substantiated by application to problems with solutions that are not obvious.

In Chapter Four, binary digital computation is explained emphasizing the natural relationship between decimal arithmetic and binary arithmetic. The number circle is introduced as an adaptation of the number line, thereby accounting for the limited numerical capacity of all digital hardware. Octal and hexadecimal codes are explained to benefit those students intending subsequent study in the area of digital computers and microprocessors.

As the first formal presentation of memory elements, the flip-flop is carefully described in Chapter Five, using as the chief subject a latch. Building around the latch, clocked flip-flops, monostables, and astables are described. Most importantly, the sections on applications bring together the key ideas behind flip-flops. The concept of a digital "module," such as a shift register or counter, becomes apparent, and their use as subsystems in an integrated digital system is explained.

In Chapter Six, a detailed examination of counters offers the first opportunity to examine the intricacies of timing and synchronization in digital circuits. The race problem encountered in decoding output states of a ripple counter motivates the synchronous counter. Generalizing the natural binary count sequence to an arbitrarily permuted one, programmed counters ease the student into the concept of a sequential circuit—a difficult topic thoroughly covered in the following chapter. Special purpose counters for timing and clock generation demonstrate how counters can be used for purposes other than just counting.

One of the most intriguing subjects in this book concerns sequential circuits, and is covered in Chapter Seven. The operation of a sequential circuit is not obvious from the logic diagram, but slowly unfolds as succeeding stages of analysis progress. The lack of obviousness in a sequential circuit's operation gives the designer the feeling that some primitive form of life has been created, and drives the curiousity to understand. Asynchronous sequential circuits are presented first primarily because they were historically developed first. The natural process of evolution in engineering design is taken advantage of by solving the race/hazard problem in asynchronous circuits with synchronous circuit design. Since the process of synchronous circuit analysis is virtually identical to that of asynchronous circuits, time is saved in presenting synchronous circuits, and the concepts of asynchronous circuits are reinforced.

Digital circuit fault analysis is discussed in Chapter Eight, and is a subject that has been, surprisingly, ignored in many texts. Experience has shown that students interested in other than a strictly theoretical knowledge of digital electronics, want and need an understanding of how a digital circuit can malfunction. Some basic concepts are presented that explain the effect of various circuit faults, and propose corrective actions.

As the cost of digital integrated circuits continues to fall, many electronic systems once relying exclusively on analog circuitry will be supplemented or replaced by digital circuitry. That transition, however, does not change the analog nature of many real-world signals. Consequently, the use of analog-digital conversion circuits to make actual signals amenable to digital processing must increase. Chapter Nine is one of the most important in preparing students for the future, as indicated by present trends. The principles of digital-to-analog and analog-to-digital conversion are presented, followed by detailed examples that illustrate some important applications.

Chapter Ten very carefully introduces the student to the concept of programmed logic. The idea of a computer is developed from the digital electronics basics presented in Chapters One through Seven, and is motivated by the need for complex and flexible digital processing. Simple program examples illustrate how the basic components of the computer's CPU interact to provide a functional processing unit. Algorithms are described, and some practical program examples demonstrate the usefulness of computers. Microprocessors are described as a specialized subset of the general purpose digital computer. The factors that have influenced microprocessor development are emphasized.

An appendix is provided with the company names and addresses of most of the major semiconductor manufacturers. This will assist those interested in obtaining data books and applications manuals on the latest technological innovations and logic families.

Several individuals have made invaluable contributions to this book, and have my deepest gratitude. I am particularly indebted to Rick Bahr and Eliot Moss for their consistent and devoted effort in technically reviewing the entire manuscript. These are two rare and talented engineer/scientists who have the unique combination of technical expertise and supurb command of the English language. In his reviewing the manuscript, Professor Tsute Yang of Villanova University has provided some fundamental suggestions that have greatly enhanced the pedagogical value of this book. For these suggestions, and for his time and patience, I am very grateful. My thanks also go to Ken Wacks for his review and comments in the final stages of this project.

CHRISTOPHER E. STRANGIO

Cambridge, Massachusetts

DIGITAL ELECTRONICS: Fundamental Concepts And Applications

1 LOGIC CONCEPTS

1.1 INTRODUCTION

1.1.1 Prelude

People make simple decisions every day. When to arise in the morning, when to eat, and whether to take an umbrella to work are thoughts that cross our minds frequently. Whenever decisions like this are made, one or more different factors must be accounted for, such as the time of day or the condition of the weather, before action is taken. Perhaps a number of conditions must exist to justify a decision, or perhaps a single condition might result in several decisions. Our upbringing has trained us to respond in particular ways to conditions of our environment, and we can do this now without conscious thought. Sometimes complex decisions must be made, such as in selecting an occupation or choosing a place to live. Such complex decisions involve many factors and must be carefully considered before a course of action is selected. In making important decisions that may have a permanent effect on our lives, we rely much less on our childhood training, and instead try to anticipate and understand the results a decision may bring. Our attempt to understand the effect of a decision requires that we analyze the situation and draw some kind of logical conclusion from the information available to us. Common to both simple and complex decisions is the act of drawing on an available body of knowledge, deliberating alternative courses of action, and finally selecting a course of action that will achieve a desired result.

As mankind has come to rely increasingly on machines to do work, certain elementary decision operations have become mechanized. A decision machine will accept inputs from human beings or other machines and respond according to a predetermined plan that is specified by the machine's design. A simple combination lock is an example of a decision machine. Appropriately turning the dial to the proper settings provides input information that results in the opening of a lock. If a setting is incorrect, the machine's design prevents the lock from opening. In the sense that the combination lock permits or prevents action based on input information, it is a decision machine.

Electrical circuits may also be used to mechanize decision operations. In an electrical circuit, though, the condition of switches, sensors, or other components modifies the flow of current in the circuit. Decision circuits in which only a limited number of fixed current or voltage values are possible, determined by the setting of switches or the condition of other components, are referred to as ***logic circuits***. The simple two-switch hall light is an example of an elementary logic circuit. In the same sense that we account for various factors when we make a decision, a logic circuit responds to the condition of switches or other components to cause a specified current or voltage in the circuit.

Our basic aim is to study logic circuits. In doing so, we will gain an understanding of their operation and be able to apply our knowledge to the practical solution of real problems. Many examples will be encountered to illustrate the principles and to reinforce important concepts in logic circuit analysis. We will see how simple electrical and electronic circuits can make decisions for us to provide services that would otherwise be difficult or impossible to obtain.

1.1.2 Historical Background

Logic circuits had their beginning well over one hundred years ago with the development of the railway signal interlock system in 1855. The increasing volume of rail traffic at this time required rail lines to be shared by a number of different trains, placing a heavy burden on dispatchers in scheduling track usage. Railway signal interlock switches prevented the dispatcher from incorrectly issuing a "route clear" message for a track that was presently in use. This early system did not of itself initiate action, but simply prevented erroneous action by the dispatcher based on the rule that no two trains should run on the same track at the same time.

Although refinements and sophistication were made in railway signal and control equipment, the great revolution in logic circuitry began in 1879, when the first patent on an automatic telephone switching system was issued to Connolly, Connolly, and McTighe. Because of increased telephone usage that overtaxed the capability of human operators to make circuit connections for each call, automatic systems were developed. This was particularly true in urban areas, where the number of telephones and constant business activity required many daily telephone connections.

Perhaps as significant as the invention of the telephone itself was the installation of the first completely automatic telephone exchange at LaPorte, Indiana, in 1892. The heart of this system was the Strowger step-by-step switch, an electromechanical device in which a movable arm advances on the command of dial pulses to one of many switch contacts. This selection process would proceed as each digit was dialed, energizing at each level another step-by-step switch, until connection with the called party was achieved. The Strowger switch is still used today in many small exchanges.

Logic circuitry, such as might have been found in early automatic telephone switching equipment, was developed long before the invention of the transistor. This early circuitry was built with various complex forms of electrically controlled switches, of which the Strowger switch was one. The modern relay is a descendant of the original electrically controlled switches and is widely used in high-current switching networks, such as in elevator control circuits. It is important to note, though, that for almost fifty years, logic circuit design techniques relied almost exclusively on intuition and experience, with no mathematical tools available to analyze circuit operation. Thus, with the exception of the simplest circuits, it was not possible to know with certainty whether an optimum design had been achieved.

A great stride was made in 1938, when Shannon presented a means of symbolically analyzing the behavior of switching circuitry. Shannon's work was closely related to a system of symbolic logic developed by George Boole in 1854. Although it was Shannon who revived the seemingly forgotten work of Boole, it was Boole's symbolic algebra that has survived improvement and refinement to become known today as "Boolean algebra." Indeed, even with the introduction of the transistor in the late 1950's for use as a switching agent, the representation of the new "electronic" logic circuits with Boolean techniques was still perfectly valid. This is because the Boolean representation of a logical function is independent of the physical means by which that function is made operational. Thus, we might have a transistor logic circuit that is faster, cheaper, quieter, and cooler than its corresponding relay logic circuit, but both will be symbolically represented in the same way with Boolean algebra.

1.1.3 Definition of Digital Systems

The word "information" can be defined in several ways. One definition that will be useful to us has "information" meaning "knowledge pertaining to something we wish to describe." For example, information about a particular radio receiver might indicate its frequency range, power requirements, and size. Information about the present environment might include temperature, humidity, ambient light, and air circulation. Electronic devices are almost exclusively used for *information processing*, as opposed to electrical devices, which are often used for energy conversion. Information processing means different things to different people; however, we will consider information processing to include the following operations:

1. ***Information translation.***
2. ***Information manipulation.***
3. ***Information storage.***

Traditional electronic devices, and historically the first, relied on ***analog information processing***. Specifically, a transducer would convert some quality of the environment, such as temperature, into an *analogous* corresponding voltage or current. This operation is the *translation* of environmental information into a proportional electrical quantity. The resulting electrical signal might then be converted to another electrical signal related in a well-defined way to the first. The new signal may, for instance, be an amplified version of the original signal. Operating on one electrical signal to produce another is information *manipulation* since the signal that is modified was initially made to correspond to information by use of the transducer. Finally, it may be necessary to store the resulting signal for future reference. To do so is information *storage*.

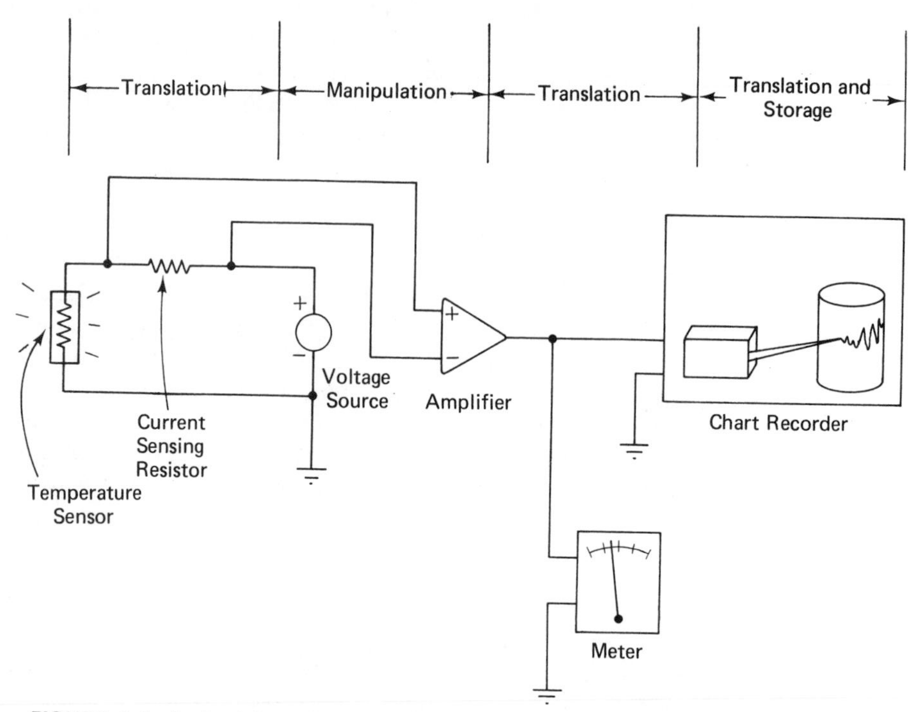

FIGURE 1-1 Analog information processing system that measures and records temperature.

As an example of an analog information processing system, consider an electronic temperature measurement device. A temperature sensor is used that is a two-terminal electrical component whose resistance decreases as the temperature increases (a silicon chip, for example). By connecting this sensor in a circuit with a battery, we would find that the current flow through the circuit increased as the temperature increased. Temperature, thus, will have been translated into a corresponding current flow. In order to drive a display device, the current flow must be amplified. Manipulation of the signal in this manner is achieved with a linear amplifier. A display device, such as a d'Arsonval meter, is used to translate the electrical output of the amplifier into a needle position on the meter that is easily read by an observer. Finally, a record of the temperature variation over a period of time is made with a chart recorder, enabling the storage of the temperature information indicated on the meter. This system is illustrated in Figure 1-1.

Analog information processing systems have been used for many years in radio and television, audio recording, and radar, to name only a few applications. Consequently, the technology associated with analog systems has been developed to a very advanced state. Relatively new techniques for information processing have undergone tremendous development only over the last generation, and rely on a numerical representation of information. Known as ***digital information processing***, these techniques offer certain capabilities that would be achieved only with great difficulty, if at all, using analog systems. In particular, digital information processing systems provide the following capabilities:

1. Complex mathematical functions can be evaluated, such as the trigonometric and logarithmic functions found on many hand-held calculators. Virtually any degree of accuracy desired can be obtained by providing sufficient digital hardware and allowing suitable computation time.
2. Sequential switching operations involved in the control of electrical equipment are possible. Elevator control and telephone switching are examples.
3. Numerical models of engineering designs can be used to test the properties of a design before actually constructing it.
4. Simple decision functions, such as those found in burglar alarm systems or electronic code locks, are easily and inexpensively built.
5. Using certain digital codes, many errors resulting from electrical noise during the long-distance transmission of information can be detected and corrected.
6. Advanced semiconductor technology permits the construction of digital circuits on microscopically small chips of silicon material, offering great savings in space, power, and cost.
7. Pertinent to all of the capabilities mentioned thus far is the simplicity, low cost, and speed of digital circuits and systems in performing a specified function.

An extremely important capability to the scientific and engineering community is that many technological problems can be numerically modeled and tested in a digital information processing system, while a much smaller number of problems can be modeled adequately with an analog electrical quantity for processing on an analog system. Testing a numerical model of a device or process can save untold hours and great expense over actual construction.

Digital information processing involves the same three basic operations as analog processing: information translation, manipulation, and storage. The operations take different forms, however. Translation requires the conversion of information into a suitable numerical code. Manipulation results in the formation of new numbers related to the original numbers by some mathematical operation. Storage requires the recording of numbers on some storage medium.

Although telephone switching techniques are perhaps the oldest form of electrical digital information processing, they nevertheless provide a good example of the digital system concept.* The telephone instrument present at a subscriber's residence serves as an information translation device in two ways. First, acoustical excitations at the mouthpiece are converted to corresponding electrical oscillations. Second, and relevant to the example, numbers printed on the dial are converted to a sequence of pulses when the dial mechanism is operated. The number of pulses generated equals the numerical value of the dialed digit (except 0, when ten pulses are provided), and in this sense, printed information is translated into a numerically coded electrical

* The example that is given pertains to the interconnection of two telephone circuits by dialing and does not concern the transmission of voice signals over wire, since such transmission is an analog process.

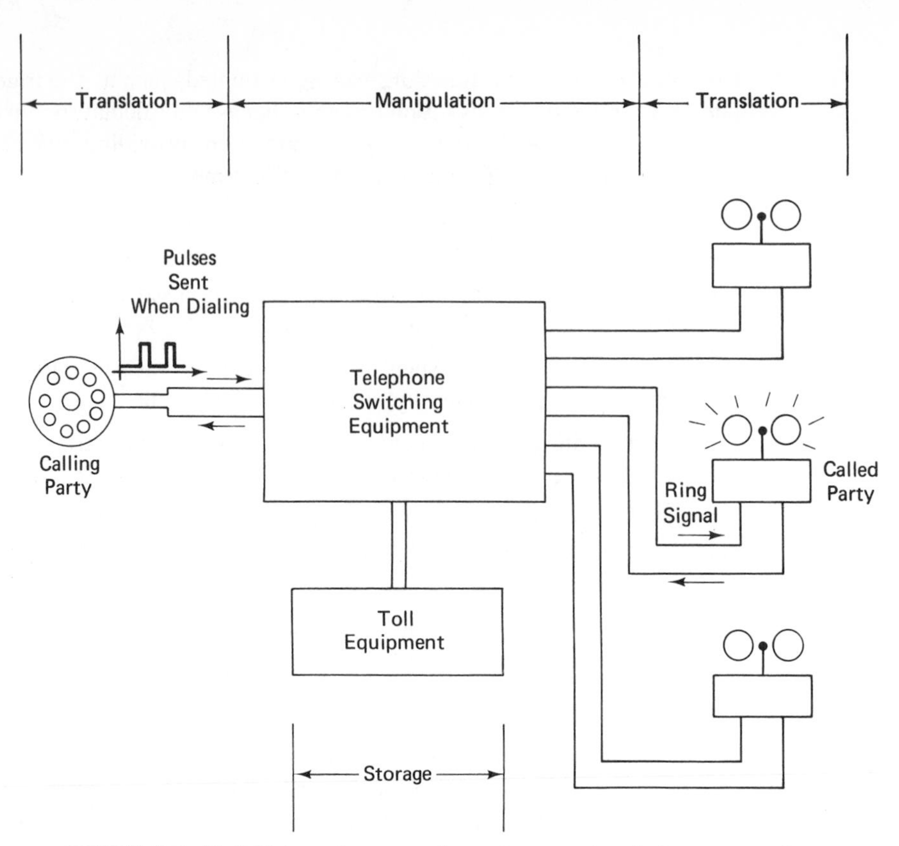

FIGURE 1-2 Digital information processing system to make telephone connections.

signal. Switching equipment in the telephone exchange is actuated by the numerically coded electrical signal to produce a new signal at the telephone instrument of the called party, namely a ring signal, and to electrically connect the two telephones. Recording equipment at the telephone exchange stores the number dialed and connection time to facilitate billing. Figure 1-2 illustrates this system.

A telephone exchange is a very complex form of digital information processing system. Prior to understanding the details of any complex digital system, it is first necessary to understand the simpler building blocks from which complex systems are constructed. To do this, we will examine the simplest functional elements and explain their interconnection to form basic digital operations. A digital system, then, will be considered to be a collection of interconnected digital circuits. As a whole, the system may perform complex tasks at high speed, while each digital circuit will individually operate in a simple and comprehensible way.

Our studies will involve two aspects of digital circuitry:

1. The *symbolic description* of digital logic, relying on Boolean algebra and similar basic mathematical techniques.
2. The *electronic implementation* of digital logic, using available electronic components.

In the same way that Boolean algebra survived the transition from relay logic in the 1950's to transistor logic in the 1960's, the symbolic techniques for describing digital logic will survive technological advances in electronics that may be unforeseen today. The method by which we symbolically describe a certain logic function, therefore, is independent of the means used to construct it. Indeed, the means used to construct a logic function need not necessarily be electrical at all! A small field computer was built for the U.S. Army some years ago whose logic functions were implemented with pneumatic valves, needing only compressed air to operate.

Topics related to the symbolic description of digital logic include

1. Properties of numbers.
2. Numerical codes.
3. Boolean algebra.
4. Logic functions.

Topics related to the electronic implementation of digital logic include

1. The switching speed of transistor circuits.
2. Transistor circuit power requirements.
3. Electrical interference between nearby wires in a digital circuit.
4. The voltage levels associated with logic values.

A digital circuit cannot be realized without some knowledge of both the symbolic and electronic techniques that are involved. We will proceed in our studies with this in mind, frequently demonstrating the usefulness of particular techniques by presenting practical examples.

1.2 SWITCH NETWORKS

1.2.1 Electrical Current Flow in Circuits

Switch circuits possessing some form of logic structure first appeared in the railway signal interlock system. Particular combinations of switch settings would cause appropriate signal lamps on the rail line to luminesce, thereby indicating the traffic condition of the unseen path ahead. We will begin our investigation of logic circuits by examining switch circuits, since switch circuits contain no complex components and are easy to understand.

The most elementary switch circuit that can be described is shown in Figure 1-3.

FIGURE 1-3 *Simple switch circuit.*

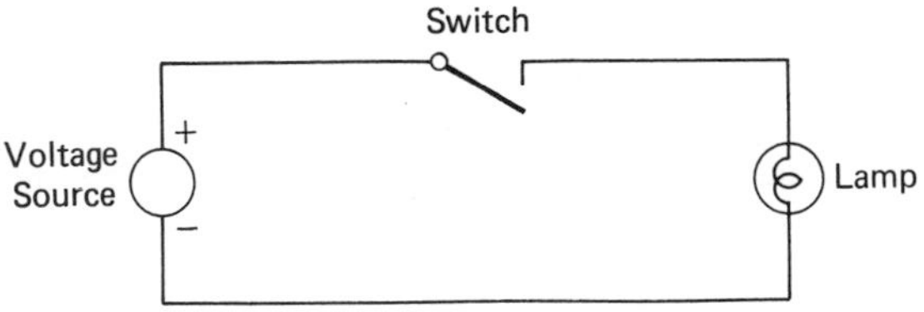

Condition Table

Condition of Switch	Condition of Lamp
Open	Off
Closed	On

Connected in series are a voltage source (which could be a battery or a power supply), a single-pole single-throw switch, and an indicator lamp. With the switch closed, a path for current flow through the lamp is established and the lamp glows. With the switch open, no current can flow in the circuit and the lamp does not glow. This operation can be symbolically represented by labeling the switch 'S', the lamp 'L', and indicating their condition by the numerical values '0' and '1'. A '0' would show an open switch or a dark lamp, while a '1' would show a closed switch or a glowing lamp. The circuit and condition table of Figure 1-3 are redrawn in Figure 1-4 to illustrate this symbolic description. The symbol '0' will always be associated with the *off* state, while the symbol '1' will always be associated with the *on* state.

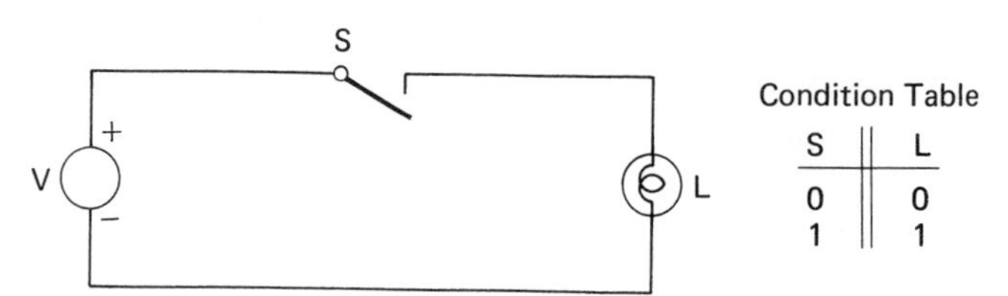

Condition Table

S	L
0	0
1	1

FIGURE 1-4 *Possible conditions in a simple switch circuit, described symbolically.*

1.2.2 Logic Functions with Switches

Consider the circuit of Figure 1-5. This circuit is different than that of Figure 1-4 only in that two switches are present instead of one. In this case, two paths of current flow are possible. With only switch 1 closed, current flows exclusively through the upper branch, and the lamp glows. With only switch 2 closed, current flows exculsively through the lower branch, and the lamp also glows. With both switches closed, the current divides equally between both branches, still permitting the lamp to glow.

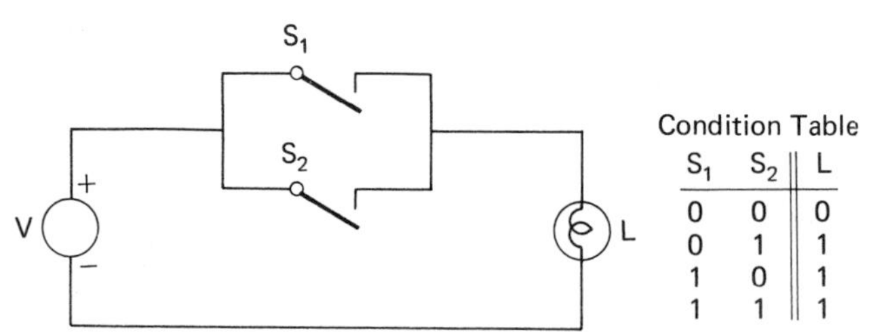

Condition Table

S_1	S_2	L
0	0	0
0	1	1
1	0	1
1	1	1

FIGURE 1-5 *Switch circuit utilizing two switches to create an OR function.*

However, if both switches are open, an open circuit results and the lamp is dark. This operation is specified in the condition table. By connecting the two switches in parallel as in Figure 1-5, an OR function is obtained in that the lamp will glow when either S_1 *or* S_2 is closed. Figure 1-6 illustrates a rearrangement in the connection of the two switches so that a series circuit is formed. Now, an AND function is obtained since the only condition that permits the lamp to glow is one where both S_1 *and* S_2 are closed. All other combinations result in an open circuit with the lamp dark.

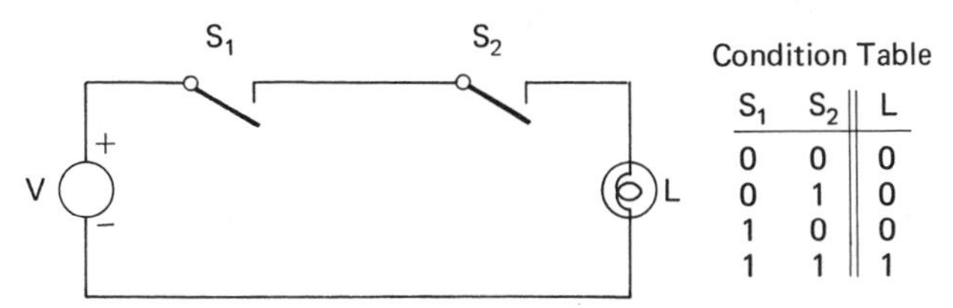

S_1	S_2	L
0	0	0
0	1	0
1	0	0
1	1	1

FIGURE 1-6 *Switch circuit utilizing two switches to create an AND function.*

Note that in both the OR and the AND functions, a condition table requiring *four* entries was needed to account for all possible combinations of 'on' and 'off' in the two switches. Testing each of the four possible combinations of the two switches to determine the condition of the light exhausts all permissible states a two-switch circuit may have. The condition table is more commonly referred to as the ***truth table***, a term of historical origin found in writings on mathematical logic. We will adopt this term to be consistent with most texts describing digital logic.

The OR and AND functions that have been discussed utilized two switches in a parallel or series connection. It is not necessary, however, that we be limited to only two switches. It is possible, for example, to build an AND function having three or more switches connected in series. A three-switch AND function is shown in Figure 1-7. Because there are three switches in this circuit instead of two, the truth table must contain more entries to account for all possible combinations of 'on' and 'off' in the switches. Specifically, every time a new switch is added, the number of truth table entries must double, permitting both the 'on' and 'off' conditions of the added switch to be combined with all 'on' and 'off' combinations of the other switches. In the same manner, a three-switch OR function can be constructed, but in this case the three switches would be connected in parallel.

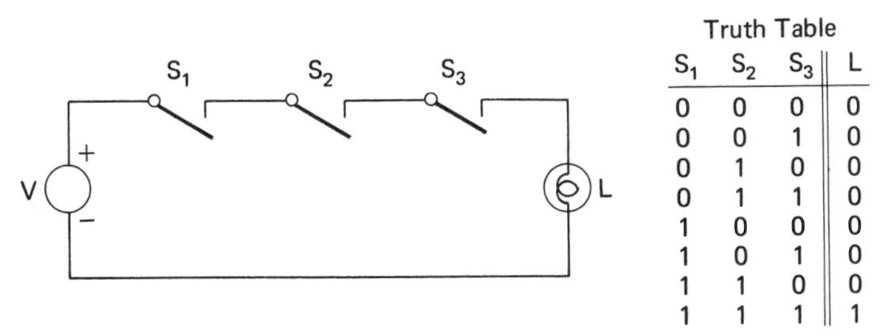

S_1	S_2	S_3	L
0	0	0	0
0	0	1	0
0	1	0	0
0	1	1	0
1	0	0	0
1	0	1	0
1	1	0	0
1	1	1	1

FIGURE 1-7 *Three-switch AND function. The truth table must contain eight entries to account for all possible switch settings.*

It is important to note several things at this time. First, the number of entries in a truth table depends only on the number of switches in a switch circuit, not on how the switches are connected. Second, the progression of the switch settings as one moves down the truth table is always the same, regardless of the manner in which the switches are connected. Although a different progression than the one used would have been acceptable (as long as all input combinations were accounted for), the sequence that was used in the OR and AND truth tables has a special meaning. Namely, it is a ***binary number*** counting sequence. Decimal numbers, which we all are familiar with, have a natural counting sequence: 0, 1, 2, 3, 4, 5, 6, 7, 8, 9, 10, ... Binary numbers use only the symbols '0' and '1,' and also have a natural counting sequence: 0, 1, 10, 11, 100, 101, 110, 111, 1000, 1001, 1010, ... This natural binary counting sequence will be explained more fully in a later chapter. Listing the switch settings in the natural binary progression has the additional advantage of allowing us to quickly see whether a certain logic function is represented in a truth table. With the switch settings following a natural binary progression, we can readily recognize an AND function by noting that only the last entry (all switches closed) results in the lamp glowing. A final item to note is that the particular connection of the switches in a switch circuit *is* reflected by the values entered in the 'L' column of the truth table. A different connection of the switches may result in alternative paths for current flow and thus cause the lamp to glow for different switch settings. The sequence of switch settings, then, can be considered an *input* sequence, while the corresponding lamp conditions will be an *output* sequence and will reveal the way the switches are connected.

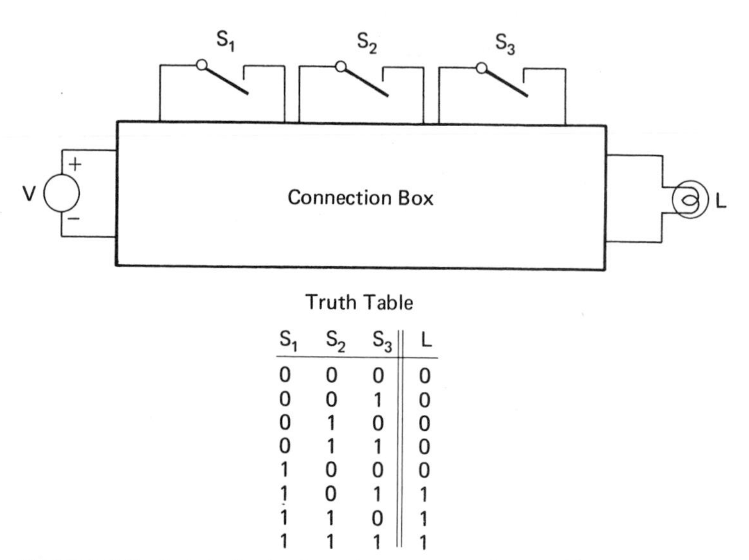

Truth Table

S_1	S_2	S_3	L
0	0	0	0
0	0	1	0
0	1	0	0
0	1	1	0
1	0	0	0
1	0	1	1
1	1	0	1
1	1	1	1

FIGURE 1-8 *Switch circuit with unknown connections, and its experimentally determined truth table.*

To demonstrate these ideas, consider the following example. A switch circuit is constructed that contains a voltage source, three switches interconnected in an unknown manner, and a lamp, as shown in Figure 1-8. The circuit is tested by setting the switches in a natural binary progression and recording the condition of the lamp for each setting. In this way, a truth table is developed. Can the connection of the switches be inferred from the truth table? Since the truth table contains all necessary information about the logic function of the circuit, the switch connections can be determined. We note from the truth table that with S_1 off, L is always off, and with

S_1 on, L is on only if S_2 or S_3 is on. This indicates that S_1 is connected in series with the parallel combination of S_2 and S_3, one arrangement of which is shown in Figure 1-9. Note that although the truth table tells us that S_1 is connected in series with the parallel combination of S_2 and S_3, it does not indicate the exact order: the parallel combination of S_2 and S_3 could have come first, instead of as shown in the circuit of Figure 1-9, where S_1 is first. Therefore, we cannot tell the exact circuit connections, only the series-parallel switch connections that are necessary.

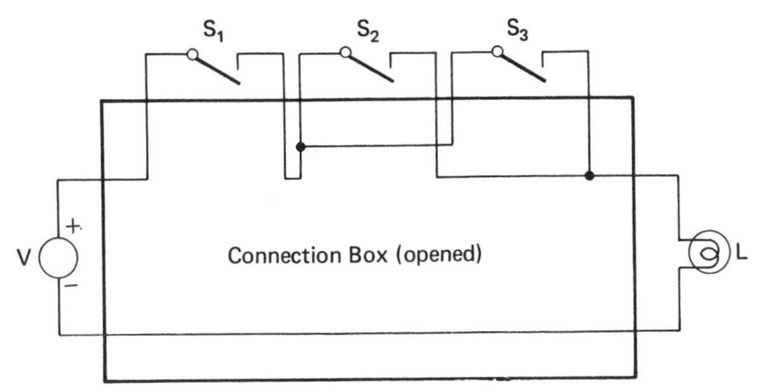

FIGURE 1-9 Connection that satisfies the requirements of the truth table in Figure 1-8.

The switch circuits that we have examined thus far have been quite simple in structure, but have allowed us to see that different connections of switches produce different effects. For a particular connection of switches, we have seen that a truth table states exactly the relationship between switch settings and lamp conditions. By doing so, the truth table specifies the logical function of the circuit. We will say that a logic circuit possesses a ***logic function*** that is completely specified by its truth table.

An interesting and practical example utilizing switches in a logic circuit concerns a burglar alarm system. In this system, three doors are connected in a sensor circuit such that if any door is opened with the system activated, an alarm bell will sound. This is accomplished by mounting a pushbutton in each door frame and connecting the pushbuttons in a series circuit. With a door closed, its pushbutton switch is closed, and with a door open, its pushbutton switch is open. Figure 1-10 illustrates the burglar

FIGURE 1-10 Simple burglar alarm circuit. An AND logic function ensures that the alarm bell is off only when all door-frame switches are on, indicating that all doors are closed.

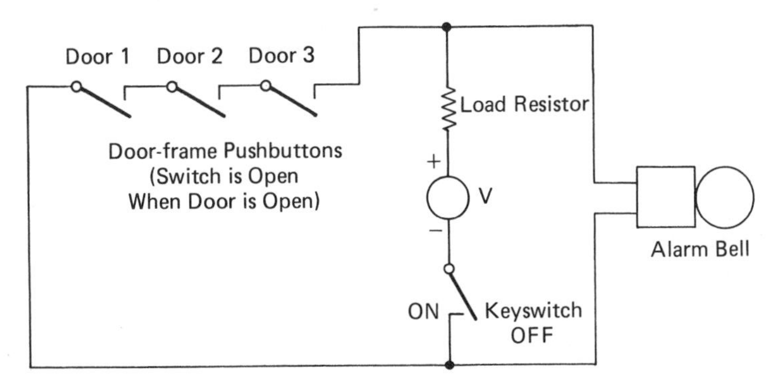

alarm circuit. With the present circuit, the alarm bell will ring with a door open, but the bell would be silent if the door were then closed. This would not be satisfactory in a normal system since it is desirable that the alarm bell continue to ring after unauthorized entry has occurred. To achieve a held alarm signal, the circuit requires some form of memory. Although we are not ready at this time to add a memory element to the circuit, we will return to this example later, after memory elements have been described.

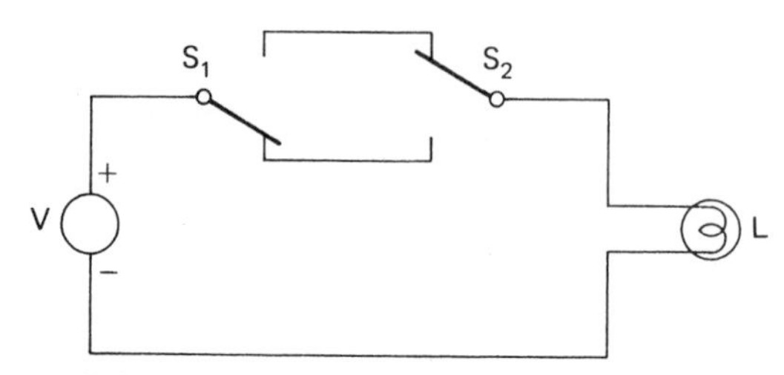

FIGURE 1-11 *Typical two-switch light circuit.*

As a second example, consider a typical two-switch light circuit that permits the on-off control of a single lamp from either of two switches. Such a circuit is shown in Figure 1-11. An important point to note in this circuit is that the switches possess a different contact arrangement than those that have been discussed previously. In particular, each switch has two positions, a different circuit being made for each position. This type of switch is referred to as a *single-pole double-throw* switch, abbreviated SPDT; a valid electrical connection is made whether the switch is set in one position or the other. Figure 1-12 compares an SPST switch with an SPDT switch. It seems that a

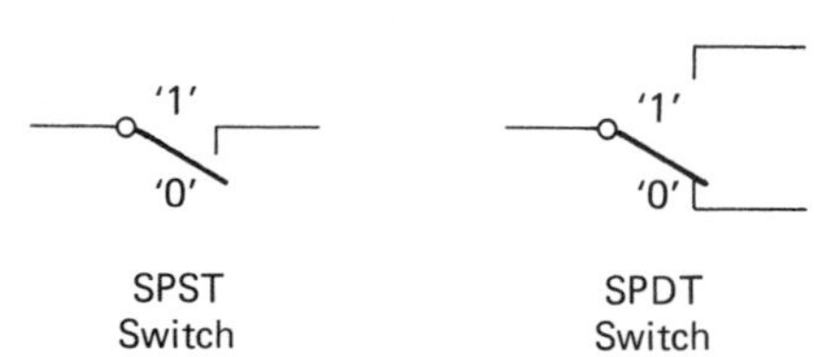

FIGURE 1-12 *Comparison of two different switch types.*

problem is created by the SPDT switch when we attempt to construct a truth table to describe a circuit's operation. Earlier, a value of '1' was associated with a closed switch and a value of '0' was associated with an open switch. However, with an SPDT switch, either position can be considered to be a closed circuit, depending on how the switch is connected. To resolve this difficulty, the actual position of the switch is used instead of an electrical condition to define its logic state. That is, a value of '1' is associated with a switch whose toggle is thrown in an "up" position, and a value of '0' is associated with a switch whose toggle is thrown in a "down" position. Of course, "up" and "down" are relative to the observer's view of a page. In this book, though, "up" will always refer to the top of a page, and all switches will be drawn

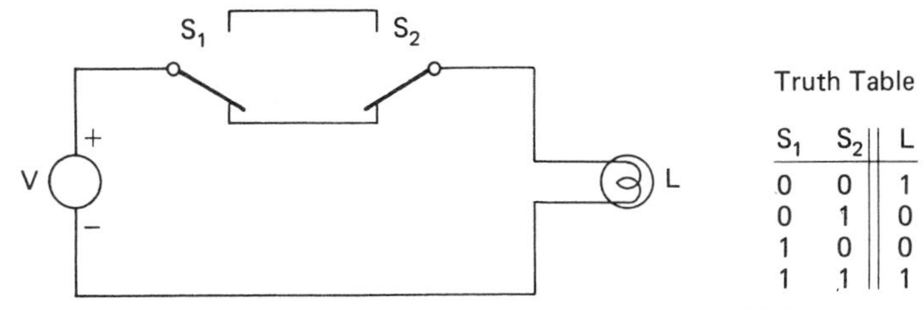

Truth Table

S_1	S_2	L
0	0	1
0	1	0
1	0	0
1	1	1

FIGURE 1-13 *Two-switch light control circuit with its corresponding truth table.*

accordingly. Using this convention, the two-switch light circuit can be defined with a truth table as shown in Figure 1-13. Note that four input combinations are possible because there are two switches, each of which has two positions; the electrical nature of the switch contacts *does not* affect the possible input combinations in the truth table. The added contacts in the switch, and the nature of the wiring, however, do affect the logical operation of the circuit and are reflected in the output column, labeled 'L', of the truth table.

1.2.3 Relay Logic

Thus far, all switches involved in example circuits have been manual switches, in that some external physical motion was required to cause a change in switch position and the related change in the electrical contacts within the switch. It is possible to design an *electrically controlled* switch so that an electrical current can induce a change in switch position. Such a switch is referred to as a ***relay*** and operates when current is passed through an electromagnet in the relay. This is illustrated in Figure 1-14. The energized electromagnet attracts a metallic leaf connected to which is one or more toggles of a switch mechanism. The particular mechanical configuration of

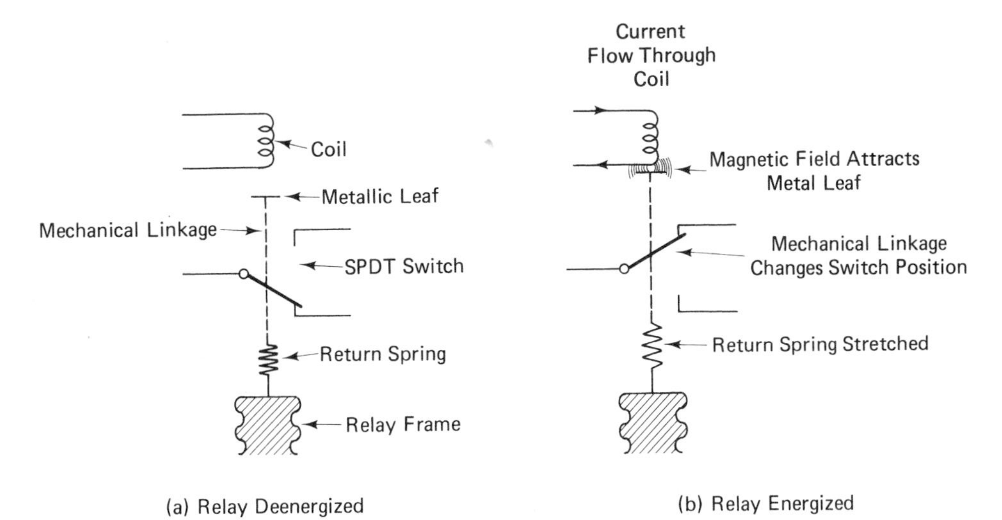

FIGURE 1-14 *Operation of a relay. The return spring is not normally shown in a relay symbol.*

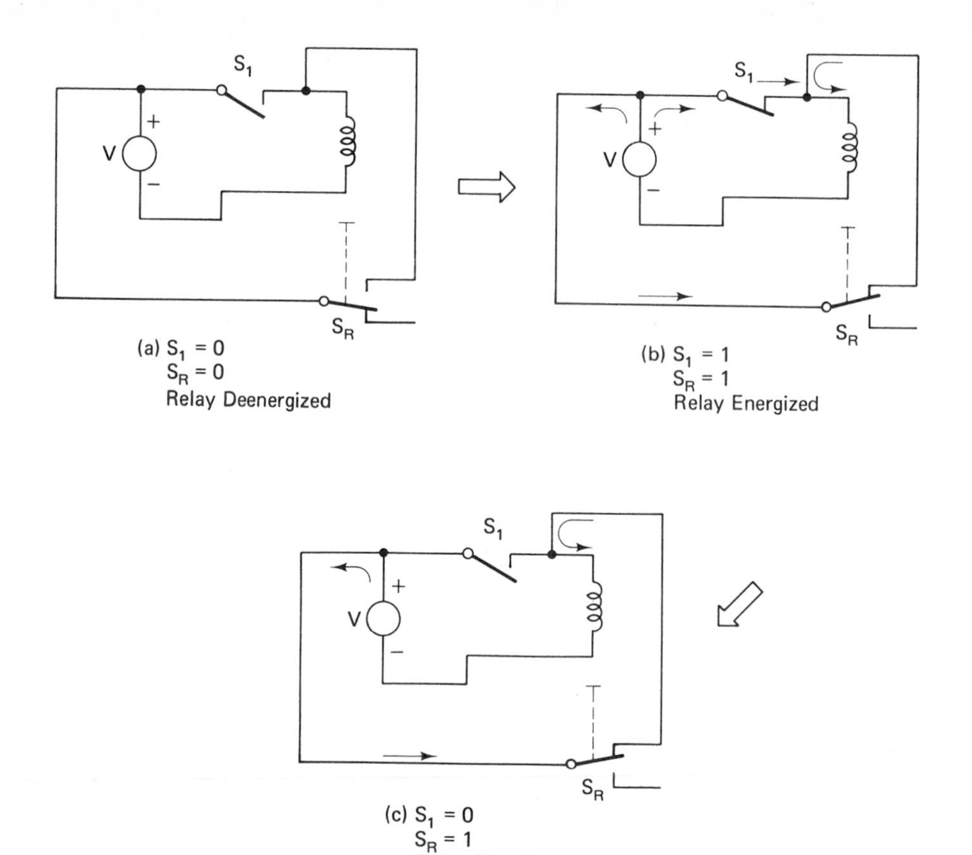

FIGURE 1-15 *Sequence of changes illustrating how a relay circuit can "remember" the depression of switch S_1.*

the coil, metallic leaf, mechanical linkage, and switch contacts varies from one type of relay to another, and depends on what application the manufacturer had in mind when the relay was designed. We will represent the relay schematically as shown in Figure 1-15; the return spring is not normally drawn.

The primary advantage of relays in logic circuits is their ability to be energized via the contacts of *their own* switch network. Such potential offers the feature of memory in a relay circuit. This important idea is illustrated in Figure 1-15. Switch S_1 is initially open and the relay is deenergized. As S_1 is closed, current flow through the coil energizes the relay and operates S_R. With S_R operated, a second path is provided for the coil current through switch S_R. Now, when S_1 is opened, only one path for the coil current is removed, the other being maintained by the relay's energized state. Thus, the relay is holding itself 'on' when S_1 is opened, and thereby preserves an indication of an event that occurred earlier, namely the closing of S_1. Only when the power is removed will the return spring allow the relay to assume its deenergized position.

Addition of another switch to the circuit as shown in Figure 1-16 permits removal of the coil's hold current. This important addition provides the capability of energizing or deenergizing the relay, and hence changing the position of all switches con-

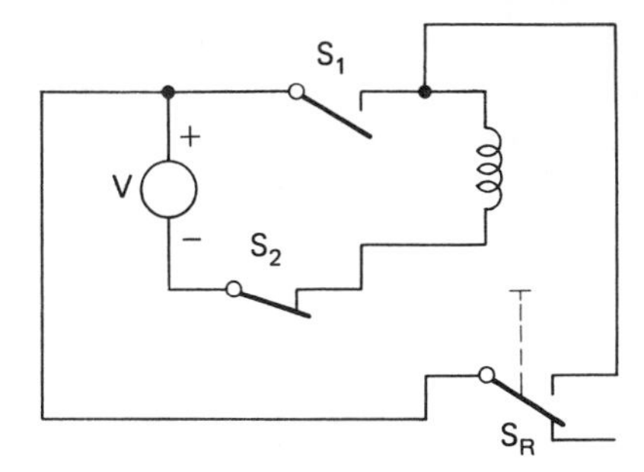

Truth Table

S_1	S_2	Coil (Energized = 1)
0	0	0
0	1	Z
1	0	0
1	1	1

FIGURE 1-16 *The addition of switch S_2 in the coil's current return path allows the relay to be deenergized. Since we cannot directly control S_R, it will not appear in the truth table.*

nected to its mechanical linkage, on our command. The truth table for this circuit, also shown in Figure 1-16, explicitly indicates the state of the coil excitation for three of the four possible combinations of S_1 and S_2. One combination, however, will not actively cause a specific change in the circuit, but rather enables the coil to maintain either an energized or deenergized state, depending on the prior setting of the switches. Specifically, if the switches were operated in the order (S_1, S_2): (1, 1) ⟶ (0, 1), we know that the coil would be energized. If the switches were operated in the order (S_1, S_2): (0, 0) ⟶ (0, 1), though, the coil would be deenergized. Since under the switch setting (S_1, S_2): (0, 1) the coil may be either steadily energized or steadily deenergized, it is necessary to use a special symbol, '*Z*' in this case, to show in the truth table that *two* conditions are possible. '*Z*' will be referred to as a ***binary variable*** because it may assume either one of two possible values, '0' or '1'.

The relay circuit of Figure 1-16 is frequently employed in the on-off control of high-power electrical machinery. With switches S_1 and S_2 replaced by pushbuttons, S_1 normally open and S_2 normally closed, a high-power control circuit would take the form shown in Figure 1-17. Another switch is added to the relay's mechanical

FIGURE 1-17 *High-power control circuit utilizing a relay and two pushbutton switches S_1 and S_2. S_1 is normally open and S_2 is normally closed.*

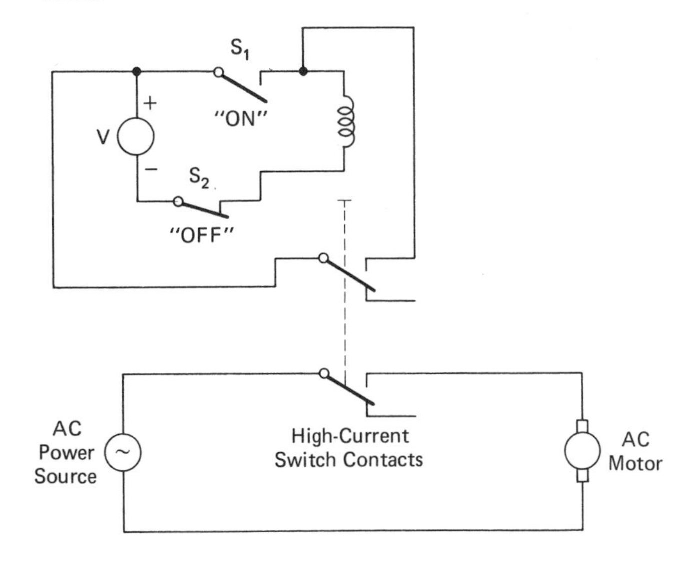

linkage to provide the desired high-power electrical control. Operating pushbutton S_1 turns the controller on, while operating pushbutton S_2 turns the controller off. Although the switches used in the relay hold circuit may be of the low-current, low-voltage type, the power control switch must have a suitable electrical rating.

1.2.4 Combinational Logic versus Sequential Logic

Several important observations can be made relating to the logic circuits that have been discussed. First, we examined the interconnection of single-pole single-throw switches to form the basic logic functions of OR and AND. These simple functions are completely and explicitly specified by their respective truth tables, which, when stated, exhaustively define a single output condition for each possible input condition. Furthermore, any circuit that is a composite of OR's and AND's utilizing more than two switches has an equally well defined truth table that also gives a fixed relationship between input and output. The second type of circuit we examined made use of an electrically controlled switch known as a relay. In this circuit, a feedback path was provided so that the relay's coil current could be controlled by its own switch elements. In writing a truth table for the relay circuit, one output entry could not be explicitly stated in terms of only the present inputs, but rather had to be represented by the binary variable 'Z'. It was noted that Z could assume the values of '1' or '0', depending upon the prior sequence of changes that the input variables had undergone.

> Whenever a logic circuit is explicitly defined by its truth table to provide a fixed, invariant relationship between input and output, the circuit is referred to as a ***combinational circuit***. A combinational circuit contains no memory or feedback paths, and will always operate in accordance with its truth table regardless of any prior input sequences to which the circuit may have been exposed.
>
> Whenever a logic circuit cannot be explicitly defined by its truth table but instead requires the entry of a binary variable for one or more of its output conditions, the circuit is referred to as a ***sequential circuit***. A sequential circuit possesses memory as the result of feedback paths, and may operate differently for a given input condition, depending upon the prior input sequence applied to the circuit.

Combinational circuits and sequential circuits make up two important branches of study in the field of digital electronics. Because a sequential circuit can be considered to be composed of a combinational circuit with feedback paths, a study of combinational logic must precede a study of sequential logic so that combinational logic concepts are well understood. Both branches of study are interesting, although sequential logic is often intriguing as well since the operation of a sequential circuit may involve very complex interdependencies among the components from which it is made.

With some knowledge of sequential relay circuits, let us return to the burglar alarm example of Section 1.2.2. An undesirable feature of that system had the alarm signal being turned off when an illegally opened door was then closed by the inturder. The system should operate in such a manner that the alarm signal is maintained after a door has been opened, regardless of any subsequent action by the intruder, whether

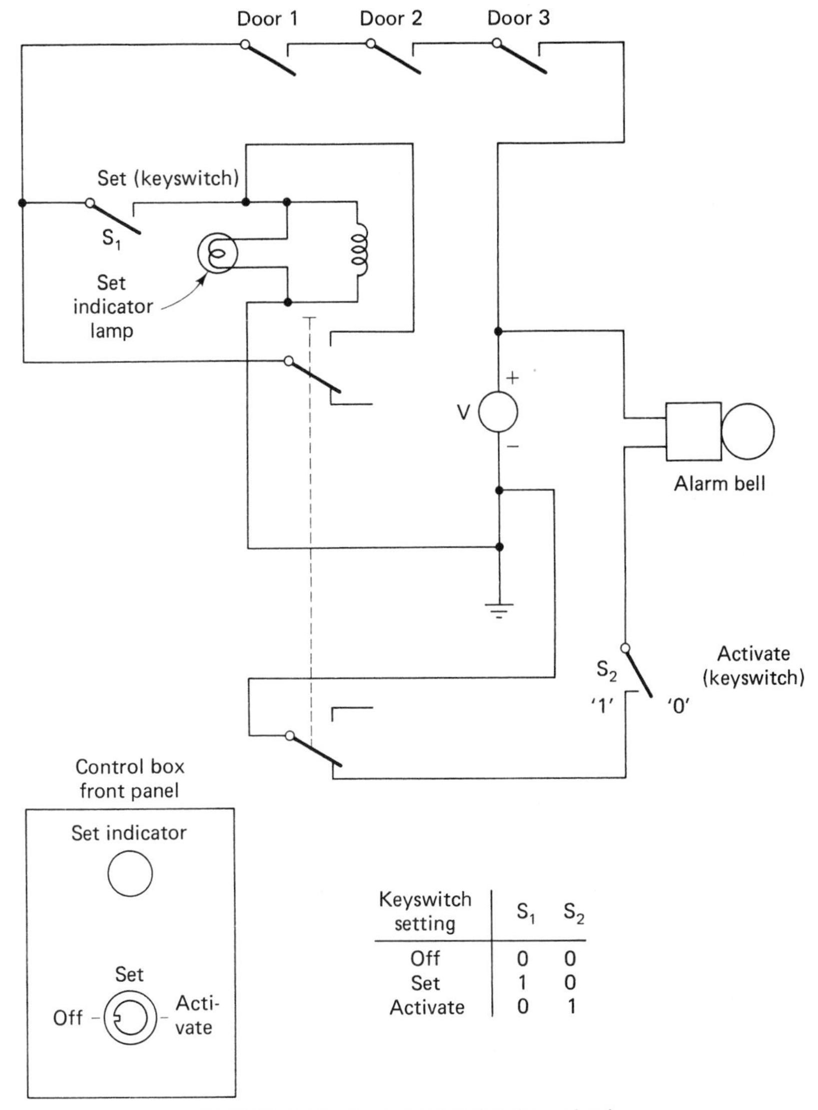

Keyswitch setting	S_1	S_2
Off	0	0
Set	1	0
Activate	0	1

FIGURE 1-18 *Improved burglar alarm circuit.*

that be closing the door, cutting the sensor wire, or damaging the control box. What is needed, then, is a circuit whereby a memory element is caused to change state if any interruption occurs in the flow of current through the sensor circuit. The only permissible action taken to silence the alarm signal would be the authorized deactivation of the alarm system. A circuit that accomplishes this is shown in Figure 1-18. A keyswitch and indicator lamp are located outside the protected area, as shown in the figure. Several features should be noted in the improved burglar alarm circuit. First, the relay cannot be energized with the "set" button until all doors are closed; the logic AND condition imposed by the wiring of the door-frame switches and the set pushbutton must be met. The operator, therefore, must ensure that all doors are closed before activating the alarm. With the door conditions met and the "set" switch closed, the "set" indicator will glow, showing that the relay is energized and that the alarm may now be activated. Second, note that after the alarm is activated, any disruption in current flow through the sensor circuit will deenergize the relay. However, simply reestablishing the closed sensor circuit by closing a door will not energize the relay again. Thus, momentarily breaking the flow in the sensor circuit will sound and hold an alarm signal until the keyswitch is turned off. Finally, since the sensor circuit is a series circuit that is normally closed when the burglar alarm is activated, the alarm signal will sound if the wire interconnecting the doors is cut. Furthermore, the sensor circuit could be made to include the keyswitch mounting box so that any tampering with the keyswitch would generate an alarm signal. The important issues to be realized from this example are as follows:

1. A given set of conditions must be met and maintained in the sensor circuit in order to prevent generation of an alarm signal. That only one of a multitude of door conditions is acceptable is implemented using the logic AND function in the form of a series electrical circuit.
2. A memory element must be used to record an interruption in current flow through the sensor circuit. The memory element used is a relay connected so that it may be electrically latched. The memory element can be reset only by use of the keyswitch, ensuring a continuous alarm signal should unauthorized entry occur.

1.2.5 Limitations of Switch Networks

Although the utility of switch and relay networks is evident from previous examples, a number of limitations inherent in the mechanical nature of these devices restrict or exclude their use in many present-day applications. During the late 1950's, switch circuits began to utilize transistors as replacements for relays and other electromechanical switching devices. This was possible because a transistor can be made to operate as an electrically controlled switch. Today, the only logic and switching circuits that *do not* utilize transistors are those involved in the switching of high current loads, or those used in an extremely high electrical noise environment. Soon, semiconductor engineering will reach such a state of advancement that even high-current and high-noise environment switching may be economically accomplished with solid-state devices.

An abundance of advantages make transistors more desirable than relays for switching purposes. The important differences are explained:

1. Transistors can be switched from the conductive to the nonconductive state many times faster than relays. The fastest relay requires at least 1 ms (10^{-3} second) to switch, while the best transistor can be switched in less than 1 ns (10^{-9} second), at least 1 million times faster.
2. Transistors have no moving parts and will never naturally wear out. Relays, however, have a limited life because of friction and arcing at the switch contact points. A good relay may be guaranteed for several million energize/

deenergize operations. By comparison, a transistor used in a high-speed digital circuit may switch states several million times *per second.*

3. Transistor circuits can be designed to consume extremely little power, while a constant flow of considerable current must be supplied to a relay coil to maintain it in its energized state. A complete transistor circuit may draw only a few microamps of current, while a single relay may require at least several milliamps in order to become energized.
4. Entire transistor circuits can be fabricated on a silicon chip $\frac{1}{10}$ in. square, while relays, at their smallest, are many times larger.
5. Transistors are largely insensitive to mechanical vibration, while relays may falsely switch under conditions of vibration.
6. Both transistors and relays can be difficult to manufacture; however, their respective manufacturing processes are so different as to be not comparable.
7. Because of the increased switching speed and sensitivity of transistors over relays, transistors are much more susceptible to electromagnetic interference and noise than are relays.
8. The understanding of transistor operation is not obvious and requires special training, while relay operation is easy to understand and apply.

With this comparison, it can be understood why switch networks using relays have become unacceptable. Of course, the inexpensiveness of transistors had a great deal to do with their acceptance as switching elements, although with such a list of overwhelming advantages, the cost of a transistor could have been considerably higher and it still would have enjoyed great popularity.

Thus far, some basic concepts have been presented relating to switch circuits that operate in accordance with specified logic criteria. With the advent of transistors, though, simple switch circuits have become outmoded and are no longer acceptable as a means of implementing logic operations. With this motivation, we begin our study of electronic logic. It is interesting and useful to note that a basic understanding of transistor operation will not be needed until Chapter 8. This is because our primary concern initially is with the analysis and application of logic circuit principles, not involving the particular technology with which these principles are made operational. Logic functions will be represented symbolically and their interconnection studied in detail.

1.3 LOGIC GATES

1.3.1 Logic State Representations

In the switch circuits that we discussed, the flow of current through a circuit was governed by a switch network. When the setting of the switches was such that current could flow through the switch network, a lamp connected in series with the network would glow, indicating that a positive state of conduction existed. The state of con-

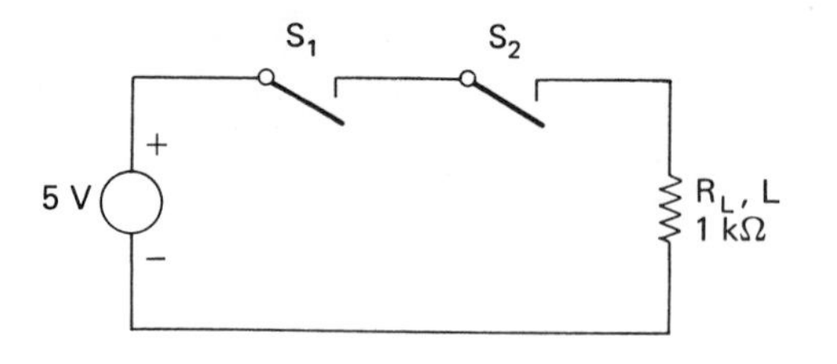

Conduction State S_1	S_2	Circuit Current Flow
Open	Open	0 mA
Open	Closed	0
Closed	Open	0
Closed	Closed	5 mA

(a) Current Mode AND

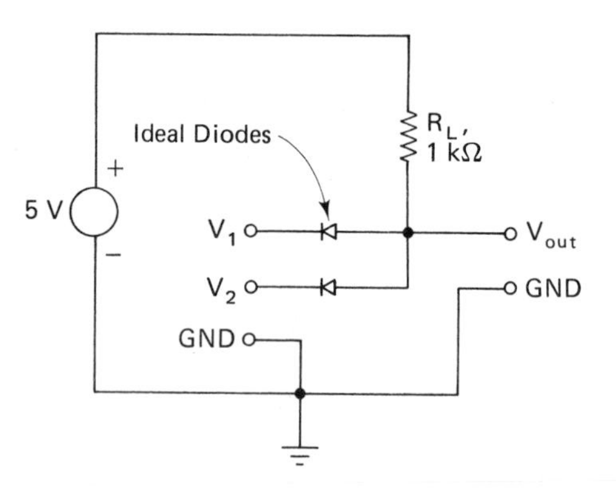

Input Voltage Level V_1	V_2	Output Voltage Level V_{out}
0 V	0 V	0 V
0	+5	0
+5	0	0
+5	+5	+5

(b) Voltage Mode AND

FIGURE 1-19 *Implementing the AND logic function with a current mode circuit* (a), *and a voltage mode circuit* (b).

duction in which current flows was symbolically represented by a '1' and that where no current flows was represented by '0'.

Instead of utilizing current flow, electronic logic makes use of *voltage levels* to represent logic '1' and logic '0' conditions. A switch circuit is compared to a simple diode circuit in Figure 1-19 to illustrate how a logic function can be implemented using current *or* voltage as the measured parameter. In Figure 1-19(a), a two-input AND function is shown constructed with switches. In Figure 1-19(b), the ideal diodes permit unimpeded current flow only in the direction of the arrow in the diode symbol. Thus, when V_1 or V_2 is at 0 V, the current that flows through R_L will be drawn through the corresponding diode and the output voltage will be 0 V. Only when both V_1 and V_2 are at +5 V is it impossible for current to flow through R_L. With no current flow through R_L, there can be no voltage drop across it by Ohm's law, and hence the output voltage will be +5 V. If '0' is associated with the voltage value of 0 V, and '1' is associated with the voltage value of +5 V, the voltage table of Figure 1-19(b) can be rewritten in the form of a truth table, as shown in Figure 1-20, and an AND function becomes evident. A logic circuit in which logic values are represented by voltage levels is referred to as a ***voltage mode*** circuit. Similarly, a logic circuit in which logic values are represented by current levels is referred to as a ***current mode*** circuit. Logic functions may be constructed with voltage mode circuits as well as with current mode circuits.

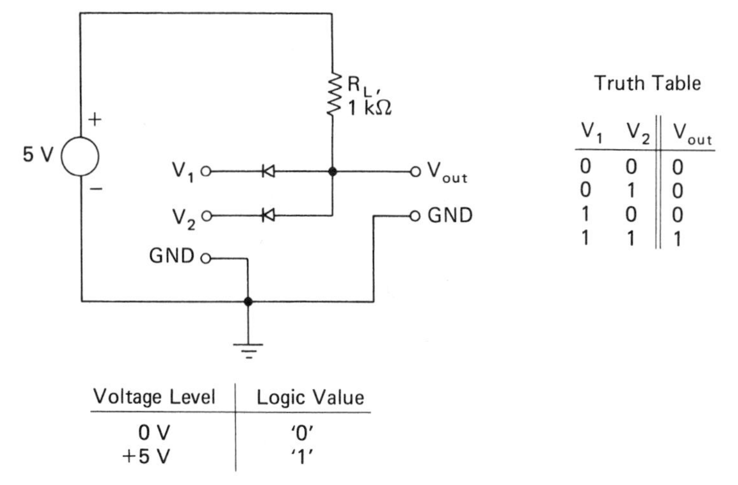

V_1	V_2	V_{out}
0	0	0
0	1	0
1	0	0
1	1	1

Voltage Level	Logic Value
0 V	'0'
+5 V	'1'

FIGURE 1-20 *Representation of voltage levels by logic values to form a truth table.*

Consider the voltage mode AND circuit of Figure 1-20 to be represented schematically as shown in Figure 1-21. In this representation, the exact electrical components used to construct the function are not specified, thus generalizing the AND function to simply the input/output relationship stated by the truth table. This is a crucially important generalization because it enables us to represent a logic function in a way that is independent of distinct electrical components, and thereby *immune* to technological advances. Furthermore, we are usually interested in only the basic functional operation of a logic element such as the AND box of Figure 1-21, not the specific voltage or current levels inside the device. The generalization of a logic function to a "black box" that operates on inputs to produce an output provides a shorthand way of describing the logic function. Even when working in the laboratory with voltage mode logic circuits, one often needs to know only the voltage levels that correspond to logic values in order to verify proper operation of a circuit.

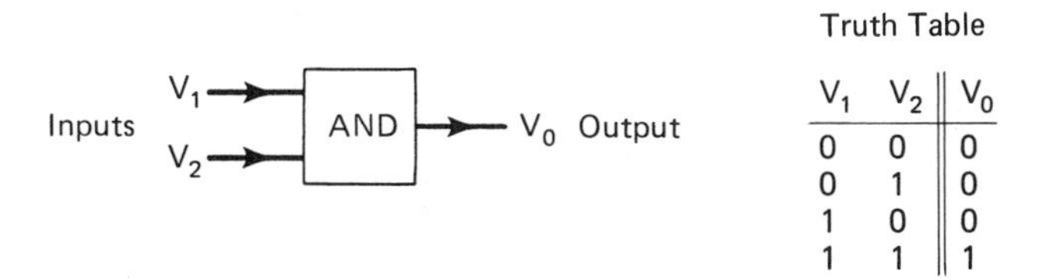

V_1	V_2	V_0
0	0	0
0	1	0
1	0	0
1	1	1

FIGURE 1-21 *Symbolic representation of the AND function that does not specify the electrical components from which it is constructed.*

1.3.2 Logic Gate Symbols and Functions

To continue our generalization of logic circuits as functional units that do not specify electrical components, some new symbols are introduced. A number of conventions have been adopted in the schematic representation of basic logic functions.

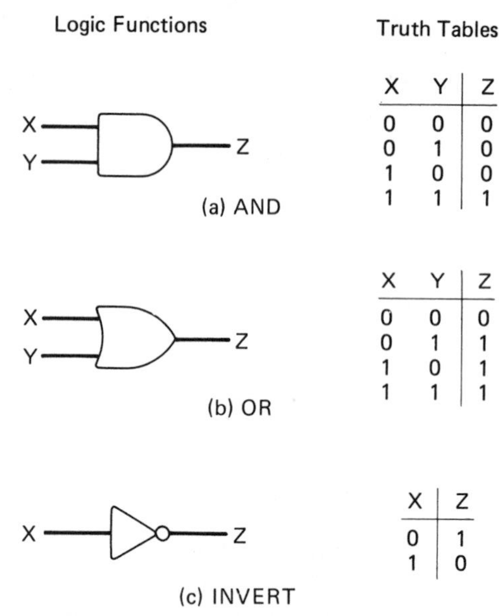

X	Y	Z
0	0	0
0	1	0
1	0	0
1	1	1

X	Y	Z
0	0	0
0	1	1
1	0	1
1	1	1

X	Z
0	1
1	0

FIGURE 1-22 *Graphical symbols adopted to represent the three elementary logic functions AND, OR, and INVERT.*

The conventions relate to specific shapes that are to be associated with logic functions when drawing logic circuit diagrams. The three elementary logic functions of AND, OR, and INVERT are depicted in Figure 1-22, relating the adopted symbol, the English-given name, and the truth table. Each of the symbols in Figure 1-22 is referred to as a ***logic gate***. Figure 1-22(a), for instance, relates the AND gate to the AND function truth table. "Logic gate" may also refer to the physical entity, made up of electrical components, that realizes a logic function.

The input variables 'X' and 'Y' each assume logic values of '1' or '0' and will cause the output variable of a specified logic gate to assume a value consistent with the gate's truth table. The voltage level corresponding to a certain logic value may be applied directly from a voltage source, or may be derived from the output of another gate. Thus, the ability to interconnect gates becomes evident. As an example, consider the interconnection of an AND gate and an INVERT gate as shown in Figure 1-23. A truth table is shown that relates the input variables 'X' and 'Y' to the output variable 'W'. 'Z' is considered an output of the AND gate and asserts logic values in

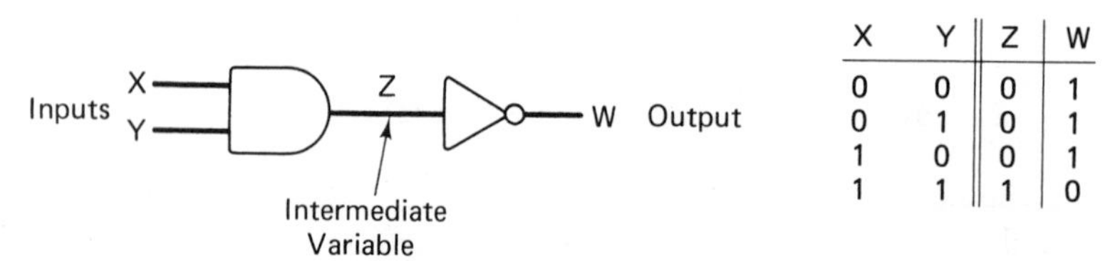

X	Y	Z	W
0	0	0	1
0	1	0	1
1	0	0	1
1	1	1	0

FIGURE 1-23 *Interconnection of an AND and INVERT gate to produce 'W.'*

accordance with the AND truth table. However, 'Z' is also an input to the INVERT gate. The purpose of an INVERT gate is to ***complement*** (not compliment) its input value, that is, produce the binary value at its output that is opposite to that existing at its input. Thus, for each value of 'Z' in the truth table of Figure 1-23, a value of '*W*' will be produced that is the complement of the value of '*Z*'. Whenever gates are interconnected, the resulting logic function can always be found by considering each possible combination of input values, recording the values of ***intermediate variables*** in a truth table, and deducing the final output.

Several combinations of logic functions are so frequently used that special symbols have been adopted for them. These composite functions are illustrated in Figure 1-24. In the first case, an AND function is complemented to produce a NAND function. Placing an N in front of AND to produce NAND is a contraction of *N*ot *AND*.

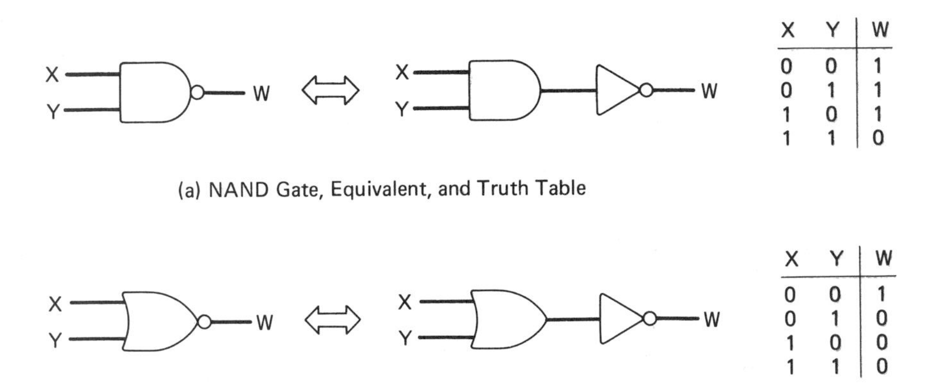

X	Y	W
0	0	1
0	1	1
1	0	1
1	1	0

(a) NAND Gate, Equivalent, and Truth Table

X	Y	W
0	0	1
0	1	0
1	0	0
1	1	0

(b) NOR Gate, Equivalent, and Truth Table

FIGURE 1-24 *Two composite functions: NAND and NOR. NAND comes from Not AND, and NOR comes from Not OR.*

In the second case, an OR function is complemented to produce a NOR function. A circle separating a logic gate from an input or output lead indicates that inversion occurs as a signal is transmitted through that circle. In addition to using an ***inversion circle*** to create a NAND from an AND, and a NOR from an OR, Figure 1-25 illustrates other possibilities.

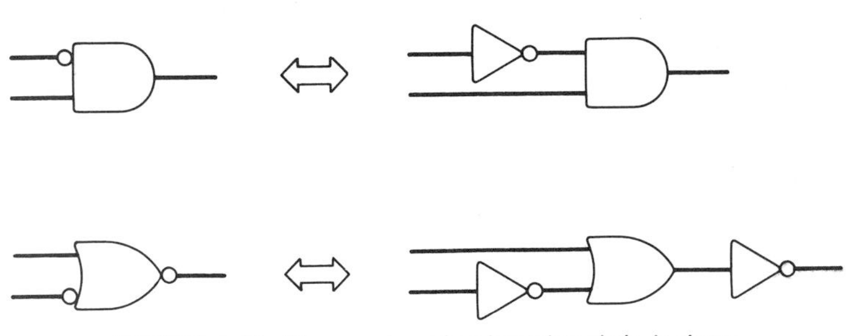

FIGURE 1-25 *The meaning of an* inversion circle *is shown with equivalent gate combinations.*

Typically, the electronics involved in the construction of NAND and NOR functions is less complex than that necessary for just an AND or OR function. Less complex electronics means fewer components per gate, and higher switching speed. These factors make NAND and NOR gates somewhat more desirable than their AND and OR counterparts.

1.3.3 Equivalent Functions

The three basic functions of AND, OR, and INVERT are the "atomic" operations of digital electronics. All digital electronic circuits from the simplest detection circuit to powerful digital computers can be expressed in terms of only AND, OR, and INVERT gates. Sequential and memory circuits require only the addition of feedback lines to basic AND, OR, and INVERT logic in order to be realized. Of course, the representation of a digital circuit in a logic diagram will probably not use only the basic functions, but instead will have large-scale functions represented by other symbols in combination with the basic functions. The object of a logic diagram is to communicate the operation of a digital circuit to a human being, and anything that lends itself to clarity and simplicity will be desired. So, although one could always express a complex circuit in terms of only these basic functions, it is not usually done to avoid an unnecessarily detailed logic diagram.

An ***equivalent function*** possesses the same truth table as a given function, but is represented by a different gate or connection of gates. One logic function that is frequently used is shown in Figure 1-26, with a special symbol representing its equivalent function. The Exclusive OR function will assert a logic '1' whenever either one *but not both* of its inputs are logic '1'. The Exclusive OR can be represented by a connection of the basic logic functions as shown in Figure 1-26(a) (realize that the NAND can be considered as an AND followed by an INVERT), but is not a basic logic function itself.

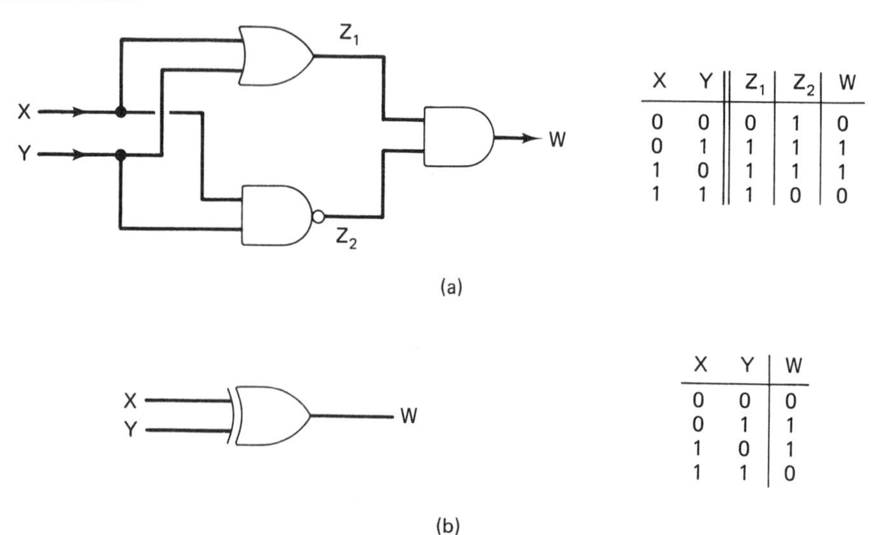

X	Y	Z_1	Z_2	W
0	0	0	1	0
0	1	1	1	1
1	0	1	1	1
1	1	1	0	0

X	Y	W
0	0	0
0	1	1
1	0	1
1	1	0

FIGURE 1-26 *A frequently used composite function is the Exclusive OR, shown with its schematic representation and truth table.*

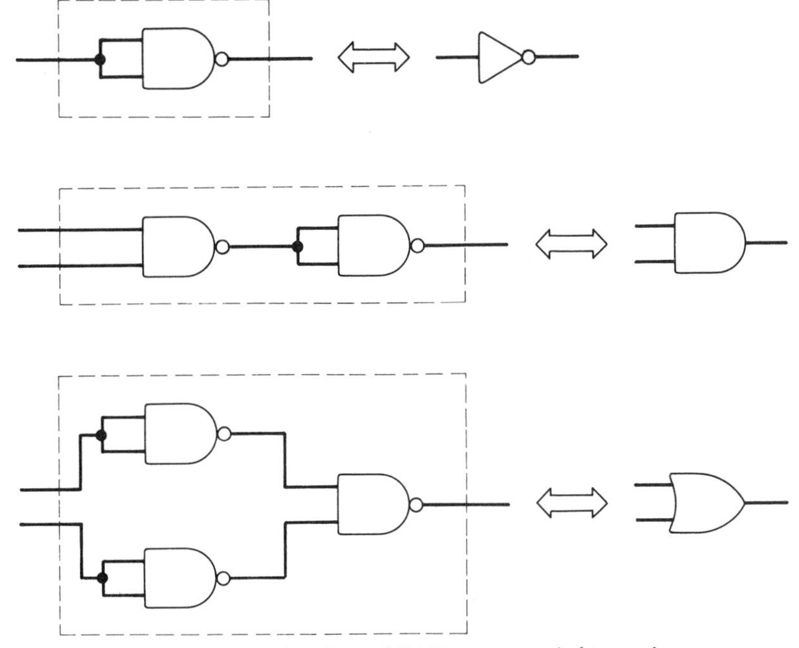

FIGURE 1-27 The combinations of NAND gates needed to produce any one of the three basic functions. The NAND is, therefore, a universal gate.

Figure 1-27 shows how different combinations of the NAND gate can be used to create any one of the three basic functions of AND, OR, and INVERT. This capability of the NAND gate makes it a ***universal gate***, since any conceivable logic function can be made from a suitable interconnection of NAND gates.

With the exception of the INVERT gate, which has only one input, we have discussed exclusively gates with two inputs. It is possible and sometimes desirable to have gates with multiple inputs, an example of which is shown in Figure 1-28. A three-input AND, for instance, behaves in the same sense as a two-input AND in that an output of '1' is asserted only when all inputs are '1'. However, the third input doubles the number of possible input combinations that must be accounted for in the truth table. Multiple-input AND gates are particularly useful in detecting the simul-

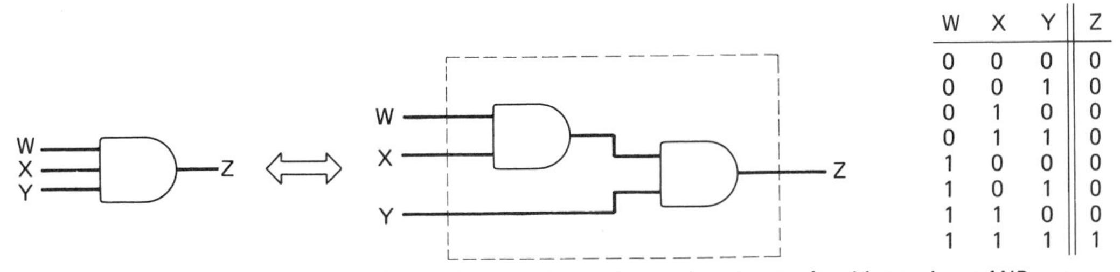

W	X	Y	Z
0	0	0	0
0	0	1	0
0	1	0	0
0	1	1	0
1	0	0	0
1	0	1	0
1	1	0	0
1	1	1	1

FIGURE 1-28 A three-input AND gate, its equivalent function made with two-input AND gates, and its truth table.

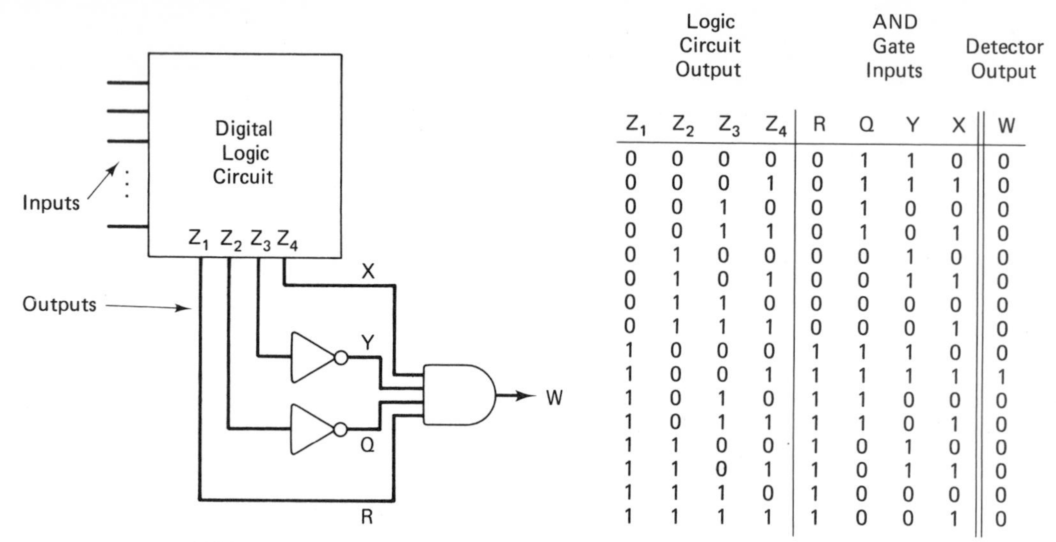

Logic Circuit Output				AND Gate Inputs				Detector Output
Z_1	Z_2	Z_3	Z_4	R	Q	Y	X	W
0	0	0	0	0	1	1	0	0
0	0	0	1	0	1	1	1	0
0	0	1	0	0	1	0	0	0
0	0	1	1	0	1	0	1	0
0	1	0	0	0	0	1	0	0
0	1	0	1	0	0	1	1	0
0	1	1	0	0	0	0	0	0
0	1	1	1	0	0	0	1	0
1	0	0	0	1	1	1	0	0
1	0	0	1	1	1	1	1	1
1	0	1	0	1	1	0	0	0
1	0	1	1	1	1	0	1	0
1	1	0	0	1	0	1	0	0
1	1	0	1	1	0	1	1	0
1	1	1	0	1	0	0	0	0
1	1	1	1	1	0	0	1	0

FIGURE 1-29 *Detection of the output state '1 0 0 1' using a four-input AND gate and two INVERT gates.*

taneous occurrence of specified binary values on a set of wires. Consider a large-scale function to be represented by a box as shown in Figure 1-29. It is desired to detect only the condition '1 0 0 1' on the four output lines; the occurrence of these four values is to cause a '1' to be generated, while all other combinations should result in a '0'. A four-input AND gate and two INVERT gates are used to detect this condition. In order that the AND produce a '1' output, all of its inputs must be '1'. However, since two '0's are in the set of values to be detected, the '0's must be complemented before being sensed by the AND gate so that the desired output will be achieved. The logical functions of typical three-input gates are shown in Figure 1-30.

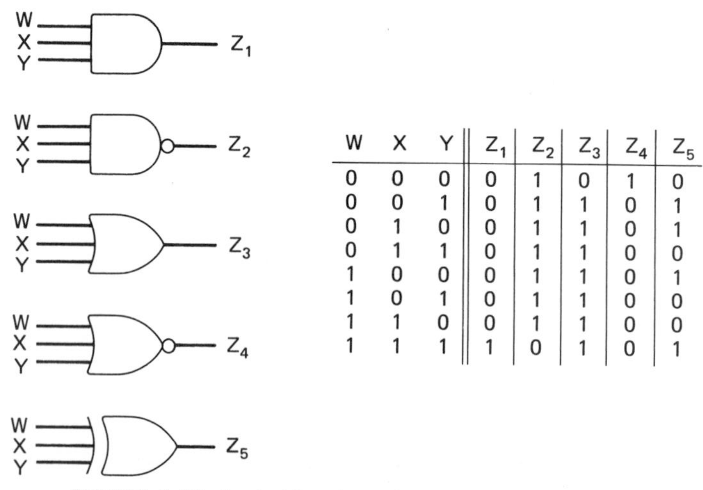

W	X	Y	Z_1	Z_2	Z_3	Z_4	Z_5
0	0	0	0	1	0	1	0
0	0	1	0	1	1	0	1
0	1	0	0	1	1	0	1
0	1	1	0	1	1	0	0
1	0	0	0	1	1	0	1
1	0	1	0	1	1	0	0
1	1	0	0	1	1	0	0
1	1	1	1	0	1	0	1

FIGURE 1-30 *Logical functions of typical three-input gates.*

1.3.4 Digital Signals Applied to Logic Gates

Thus far, fixed logic levels have been applied to gate inputs to obtain a fixed logic level at the output. In this section, we consider the application of a periodic waveform to gate inputs and observe the resulting output waveform. A waveform, in general, conveys the way in which an electrical signal varies with respect to time. The electrical parameter indicative of logic levels in most digital circuitry is voltage, so the waveforms of interest to us will be voltage waveforms. A ***digital signal*** is a waveform whose voltage at any point in time will be precisely one of a group of select discrete levels. In particular, our chief concern is with digital signals that have two states. Such signals are called ***binary digital signals*** and have a voltage at any point in time that is precisely one of two possible values. One value will correspond to logic '1', while the other value will correspond to logic '0'. Electronic logic in which a logic '1' is represented by the higher of two voltages is called ***positive logic***, while electronic logic in which a logic '1' is represented by the lower of two voltages is referred to as ***negative logic***. Most modern systems have adopted the convention of positive logic.

Actual electronic components can be obtained to within only a specified percentage tolerance of their ideal characteristics. For example, standard resistors are assured to be within 10% of their labeled value by the manufacturer. This inexactness is the result of slight variations in materials or processing when components are manufactured. If the material properties of a component vary, so also will its electrical properties. For this reason, it is unreasonable to expect a manufactured electronic logic gate to be capable of producing exact voltages at its output terminal. Instead of producing exact voltages, the circuits have been designed so that a *tolerance range* is specified for logic value voltages. This is illustrated in Figure 1-31 for the Transistor-Transistor-Logic (TTL) family of electronic logic. In this way, as long as a voltage falls within a tolerance range, it is interpreted by a logic gate as precisely a '1' or a '0', allowing for component variations in the electronics. Gate input voltages that fall between the logic '1' and logic '0' tolerance ranges are in the ***transition zone*** and may cause a gate output voltage to be in the transition zone (this is undesirable).

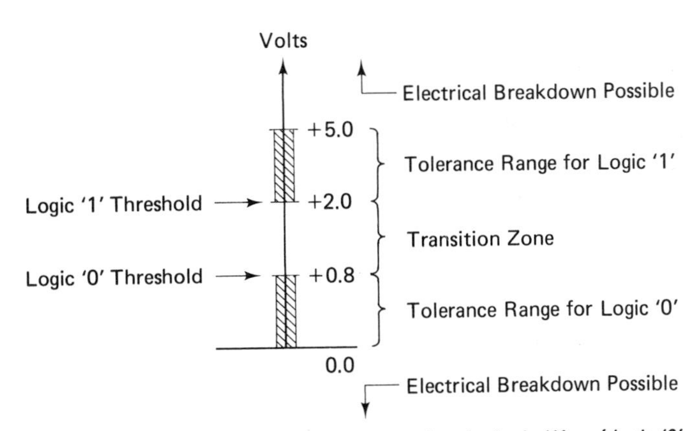

FIGURE 1-31 *Acceptable voltage ranges for the logic '1' and logic '0' states in the TTL family of electronic logic.*

Voltages above the logic '1' range or below the logic '0' range may induce electrical breakdown in the logic gate circuitry.

The borderline between the logic '0' range and the transition zone is referred to as the logic '0' ***threshold.*** Similarly, the borderline between the logic '1' range and the transition zone is referred to as the logic '1' threshold. These ideas are also illustrated in Figure 1-31.

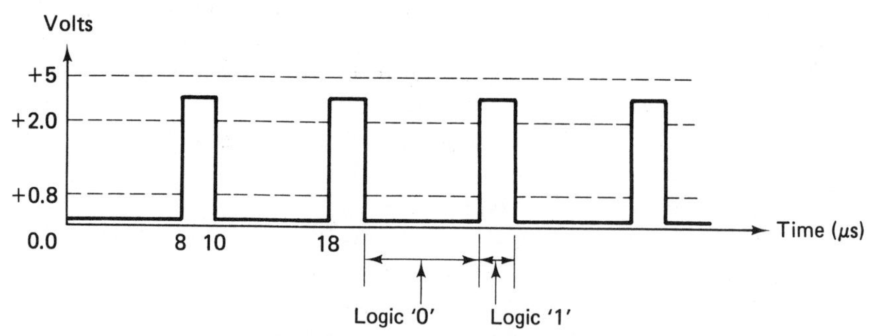

FIGURE 1-32 *Periodic digital signal.*

As an example of a binary digital signal, consider the waveform shown in Figure 1-32. Note first that the voltage alternately switches from a low level to a high level, and vice versa. The stable low- and high-level voltages fall within the normal range for TTL circuits, so for such circuits, this signal would be alternately changing between the '0' and '1' states. Because the pulses in this signal are identical in shape and equally separated in time, the signal is ***periodic.*** By studying the waveform, we note that the distance between two similar points on the waveform, say the rising edges, is 10 μs (1 $\mu s = 10^{-6}$ second). This quantity is a measure of the waveform's *period.* The *frequency* of a waveform is reciprocally related to its period as stated by Eq. 1-1.

$$F = \frac{1}{P} \tag{1-1}$$

where F is the frequency in hertz and P the period in seconds.

We additionally note that for each 10-μs period, the signal is in the logic '1' state for 2 μs. The ***duty cycle*** of a periodic digital signal is a measure of the percentage of time the signal is '1' compared to a full period, as stated by Eq. 1-2.

$$D = \frac{\tau_1}{P} \times 100 = \frac{P - \tau_0}{P} \times 100 \tag{1-2}$$

where D is the duty cycle, P the period, τ_1 the time spent in the '1' state over one period, and τ_0 the time spent in the '0' state over one period.

The duty cycle of the waveform depicted in Figure 1-32 is, therefore, 20%.

Of course, it is impossible for a voltage signal to change value instantly without going through a smooth transition from one value to another. Even though a digital signal may appear to change instantly on an oscilloscope display, the edges can always be magnified enough to see that there actually is a gradual change from one level to another. With reference to a binary digital signal, this means that a gradual change must be made between logic values, as illustrated in Figure 1-33. Often, it is important to know just how fast a waveform changes between levels. As we will see later, a digital signal whose change between levels is too slow may cause an erroneous operation. The ***rise time*** of a binary digital signal is a measure of the time needed to change from the logic '0' state to the logic '1' state. We consider the *difference* between the '0' and '1' voltage levels as the span of voltage that must be crossed when changing state. By convention, the rise time is defined to be the time needed to pass from 10% of the span to 90% of the span. This is also shown in Figure 1-33. The ***fall time*** is similar to the rise time except that it is related to the '1'-to-'0' change. It is necessary to consider the 10% and 90% points on the waveform instead of the actual full low and full high levels for ease and consistency in reading the waveform. In an actual signal, we will often find some overshoot when a transition occurs; the signal will not smoothly approach its final value, but will go beyond and then settle back. Considering the 10% and 90% points as our reference frame, we will not be confused by such distortions. As a final note, whenever rise time or fall time is measured, we examine the actual '1' and '0' voltage levels of the waveform to determine the voltage span. We do not consider exactly where within the tolerance range a '1' or '0' voltage lies.

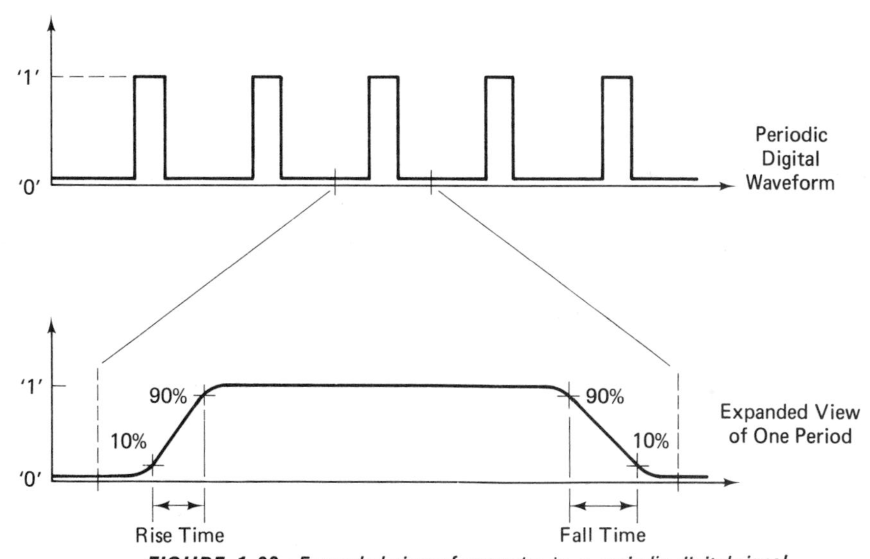

FIGURE 1-33 *Expanded view of one pulse in a periodic digital signal.*

The rise time and fall time of a digital signal should be as short as possible since it is undesirable to move through the transition zone too slowly. Should the signal hesitate unnecessarily long in the transition zone, the logic gate that is being driven by the signal may not respond decisively.

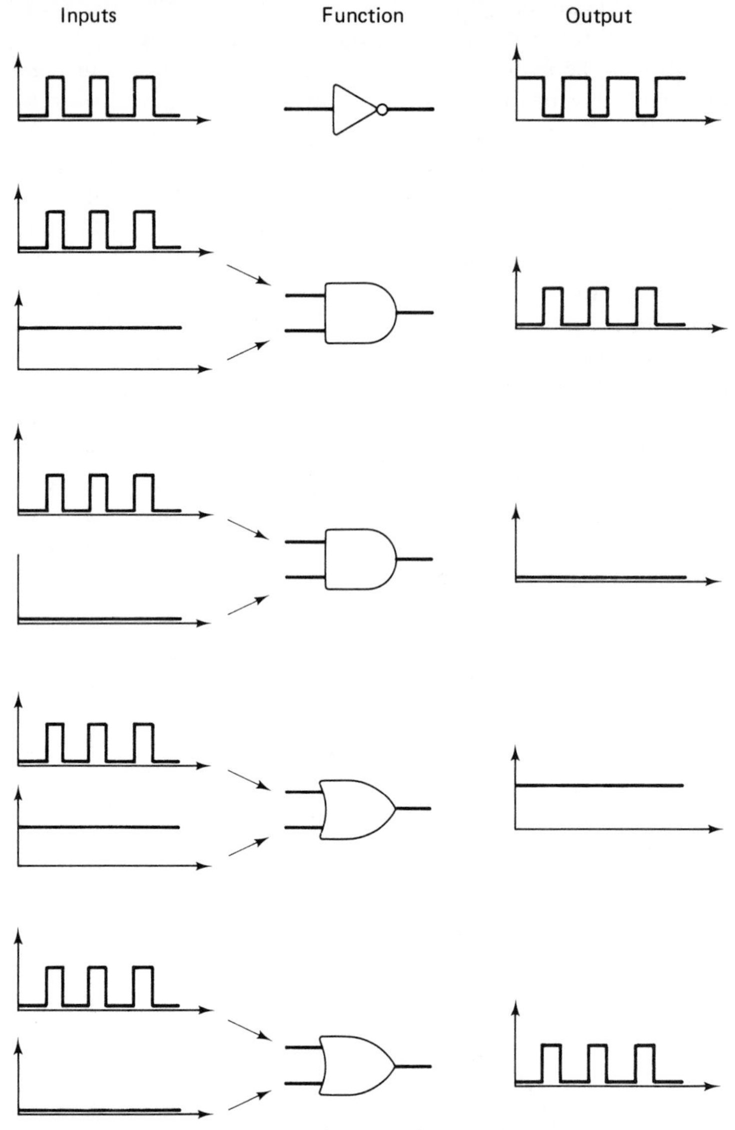

FIGURE 1-34 Application of periodic digital signals to logic gates.

A binary digital signal applied to a logic gate can simply be considered to be the application of a time sequence of '1's and '0's to the gate. We will now examine the response of various logic gates to periodic digital signals. Figure 1-34 illustrates the effect of applying a periodic digital signal to the three basic logic functions.

Digital signals can be combined in various ways using logic gates. It is possible, for example, to OR two waveforms using a two-input OR gate, as shown in Figure 1-35.

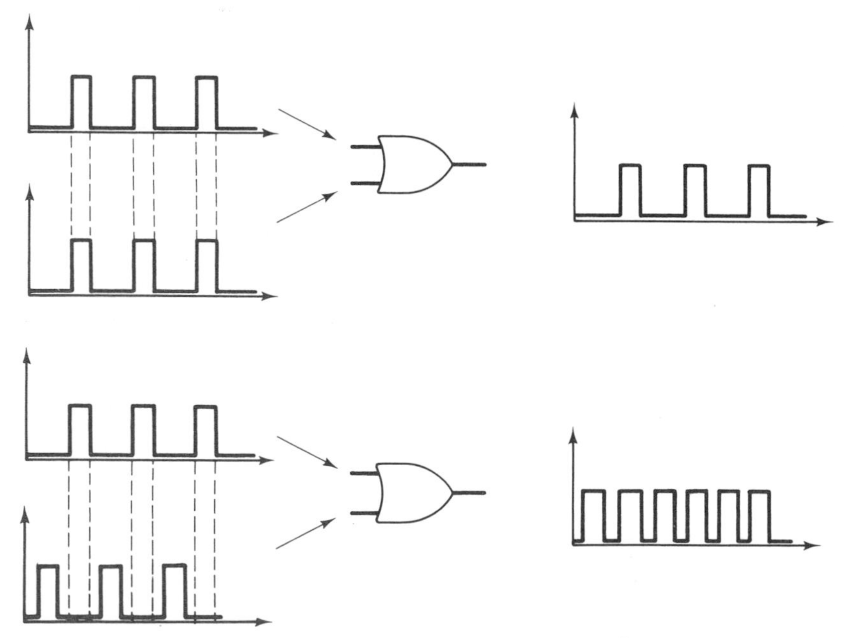

FIGURE 1-35 Combining two waveforms with an OR gate.

1.3.5 Electronic Considerations

When working with electronic digital circuits in the laboratory, it is important to realize some fundamental constraints and limitations imposed by electronic logic. For each electronic logic family, the constraints will be different because of different components and circuitry used to construct the basic logic functions. The important considerations can be presented, though, without making reference to any one logic family.

(1) Power supply voltage: In order to permit proper operation of logic gates, the power supply voltage must be maintained within specified limits. Failure to do so may result in improper circuit operation, or the destruction of components via voltage breakdown. When working in the laboratory with digital logic, always check the power supply voltage first, before operating the logic, and before searching for more complex problems if a malfunction occurs.

(2) Fanout: The maximum number of gate inputs that can be driven by the output of a single gate is limited, and is specified by the gate's ***fanout***. Exceeding the fanout restriction may cause improper gate operation or burnout of the driving gate. Fanout may be increased over the normal limit by using a special buffer gate, or connecting two gates in parallel, as shown in Figure 1-36. The increase in fanout for a buffer gate varies and is specified by the manufacturer. It is preferable to use a buffer gate when normal fanout must be increased.

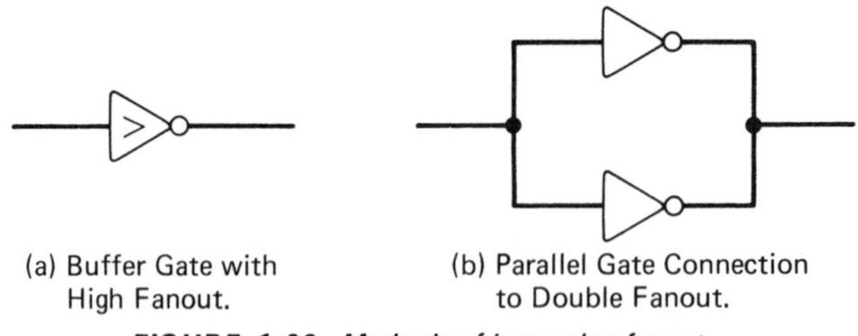
(a) Buffer Gate with High Fanout.
(b) Parallel Gate Connection to Double Fanout.

FIGURE 1-36 Methods of increasing fanout.

(3) Wire length: Connecting one gate to another with an excessively long wire is undesirable because the signal may become seriously degraded. Signal reflections, attenuation, and ringing are some of the effects that may result and will be discussed fully in a later chapter. For the TTL family, 18 in. or less is acceptable for a signal line. It is important to realize that the ground return is part of the signal circuit and must not be significantly longer than the signal line itself. This is illustrated in Figure 1-37.

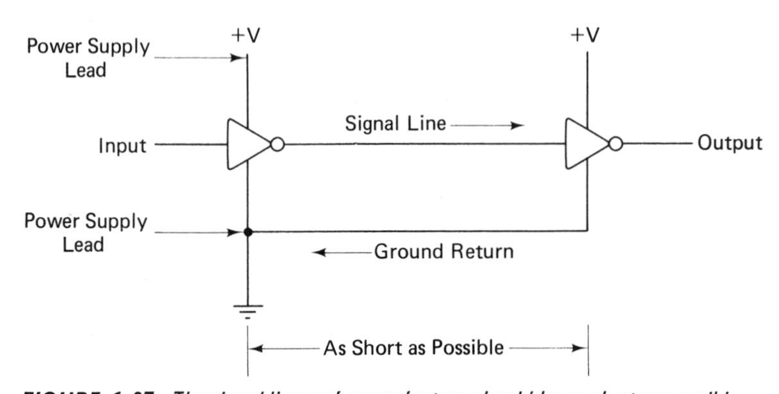

FIGURE 1-37 The signal line and ground return should be as short as possible.

(4) Parallel wires: When one signal line carries a digital waveform, other signal lines that are located close by and are parallel to the driving line may become ***electromagnetically coupled*** to the driving line. This will result in a fraction of the signal on the driving line appearing on the nearby parallel lines, even though no wired connection exists between them. To minimize coupling, avoid running lines in a parallel path. If it is necessary to do so, keep them as far apart as possible, or interleave ground lines between signal lines. Figure 1-38 illustrates these ideas.

(5) Propagation delay: No matter how fast an electronic logic gate can switch states, a certain amount of time is necessary for a change in the input to cause a change in the output. This is known as ***propagation delay***, and is depicted in Figure 1-39. We select a point midway between the logic '0' and logic '1' voltage levels as our reference frame when measuring propagation delay.

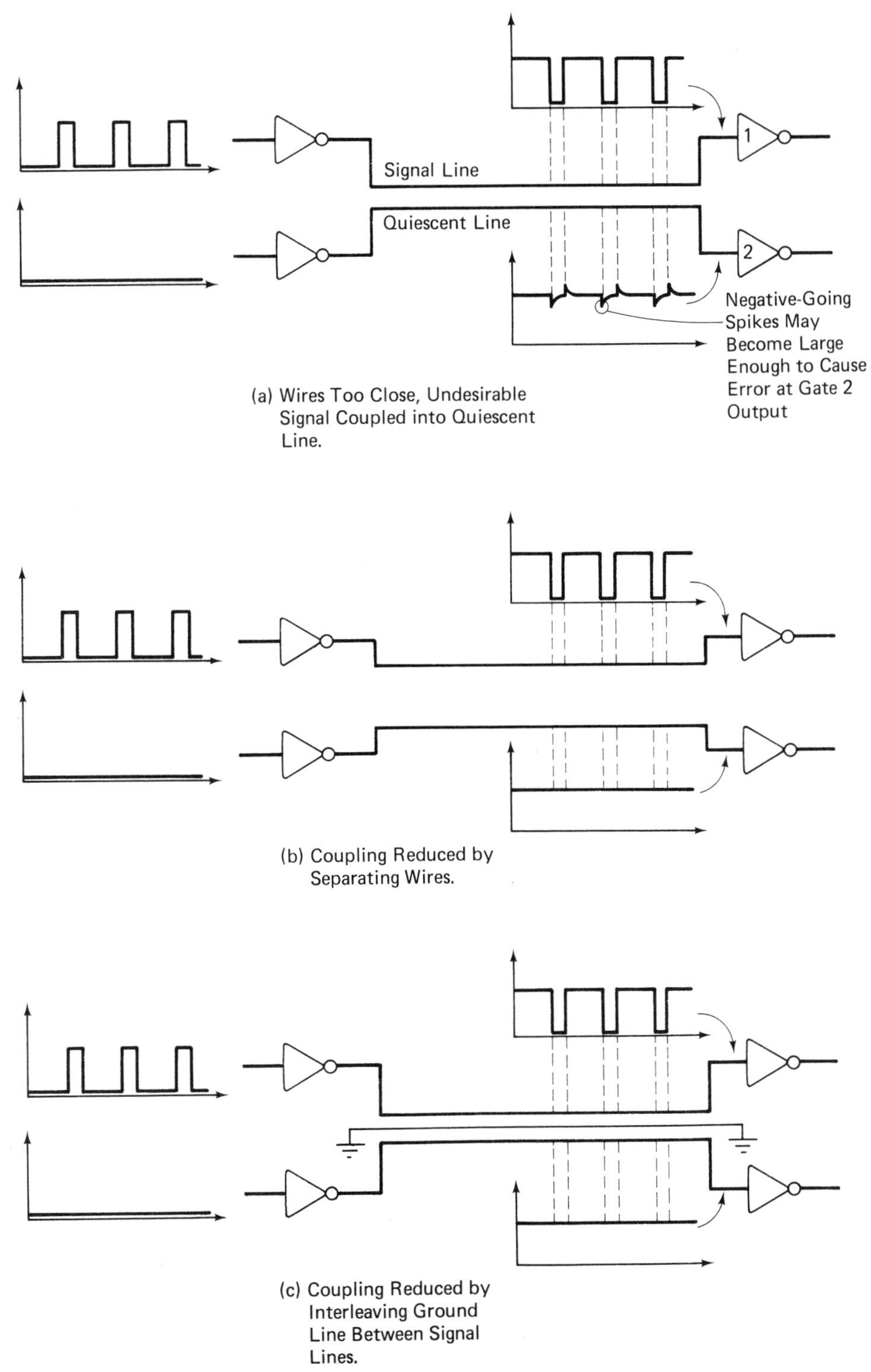

FIGURE 1-38 *Effect of electromagnetic coupling on closely spaced wires, and methods of reducing coupling.*

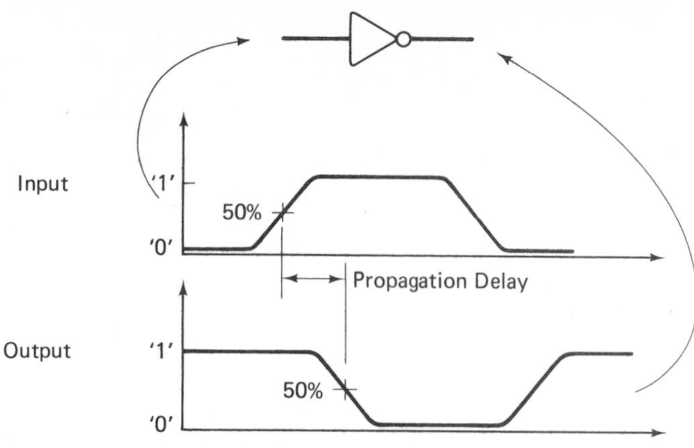

FIGURE 1-39 Propagation delay of a logic gate.

(6) Noise margin: A digital signal may have a noise signal added to it through coupling, power line transients, or other sources. Fortunately, small noise signals can be tolerated because of the width of the tolerance zones for digital signals. As long as an added noise signal does not cause the total signal voltage to pass a threshold level, no errors will occur. This is illustrated in Figure 1-40. ***Noise margin*** is defined to be the maximum noise voltage that can be added to (or subtracted from) a digital signal before passing a threshold.

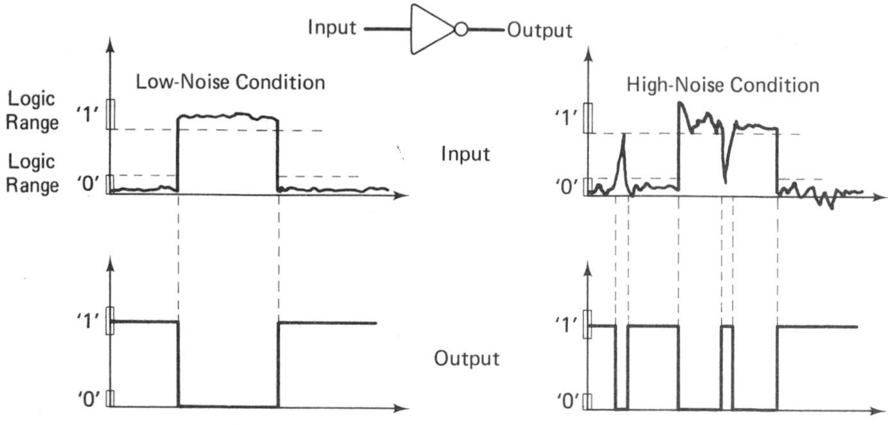

FIGURE 1-40 Effect of noise on digital signals.

SUMMARY

Analog information processing systems operate on information using an electrical parameter, such as voltage, that is proportionally related to a measured quantity, such as velocity. Analog signals may smoothly span a continuous range of voltages and are not restricted to occupy only fixed levels of voltage or current. In contrast, digital information processing systems operate on information using a numerically coded representation of a measured quantity. Digital signals are confined to fixed, discrete levels of voltage or current, each level of which represents a specific numerical value. In particular, binary digital signals may possess one of two possible values. Digital electronics concerns the generation, processing, and storage of digital signals, chiefly binary digital signals, and establishes a framework for the analysis, design, and construction of digital information processing systems.

The study of digital electronics involves two basic topics: the symbolic description of logic circuits and the electronic implementation of such circuits. The symbolic description of logic functions is highly independent of the means of electronic implementation and will survive unforeseen technological advances.

Elementary logic functions can be demonstrated with simple switch circuits, but are primarily constructed with transistor electronic circuits. Logic gates are electronic circuits that implement basic logic functions. Graphical symbols for logic gates convey a specific gate logic function, for instance AND, OR, or INVERT, but do not specify exact circuitry. A logic family, such as TTL, is a collection of logic gates and large-scale digital functions built around a common fundamental circuit structure. Technological advances in electronics will affect logic families by improving circuit characteristics but will have an insignificant effect on the gate symbology or basic logic functions.

Periodic digital signals provide a time sequence of numerical values, represented by voltage or current levels, that repeat after a certain quantity of time has elapsed. Periodic digital signals may be applied to a logic gate to generate a new signal. The new signal may be a combination of two or more input signals, or may be the result of the masking of one input signal by the others.

Relays can be used as memory elements when a feedback path is provided through a relay's own switch network. In this way, a relay may be energized by an external means, and then hold itself in the energized state regardless of the condition of the original external stimulus. Logic circuits that possess basic logic functions *and* feedback paths are referred to as sequential circuits, and may be constructed with relays or electronic logic. Circuits that are composed only of basic logic functions without feedback paths cannot have memory and are referred to as combinational logic circuits.

Switch networks are limited primarily by their slow speed, large size, and high power consumption. These characteristics are greatly improved by using the transistor as a switch element. Since a transistor can be made to appear conductive or nonconductive, depending on the electrical state of an input terminal, its use as an electrically controlled switch element is evident. Transistor logic also has limitations that must be respected. Logic gates require specific operating voltage ranges, offer limited fanout, cannot drive wires of unlimited length, contribute a propagation delay to the signal, and are susceptible to noise. Awareness of these factors is essential for effective logic circuit design.

NEW TERMS

Analog Information Processing
Digital Information Processing
Information Translation
Information Manipulation
Information Storage
Truth Table
Logic Circuit
Logic State
Logic Function
  AND
  OR
  INVERT
  NAND
  NOR
  XOR
Logic Gate
Single-Pole Single-Throw (SPST) Switch
Single-Pole Double-Throw (SPDT) Switch
Relay
Binary Number
Binary Variable
Intermediate Variable
Complement
Combinational Circuit
Sequential Circuit
Boolean Algebra
Voltage Mode
Current Mode
Inversion Circle
Equivalent Function
Universal Gate
Digital Signal
Binary Digital Signal
Transition Zone
Periodic Digital Signal
Rise Time
Fall Time
Propagation Delay
Fanout
TTL
Noise Margin
Electromagnetic Coupling
Threshold
10% Point
90% Point
Positive Logic
Negative Logic

PROBLEMS

1-1 Classify each of the following systems as an analog information processing system, a digital information processing system, or neither.

(a) FM radio

(b) Television receiver

(c) Slide rule

(d) Calculator

(e) Electric typewriter

(f) Electronic organ

(g) "Hot" warning light on automobile dash

(h) Megaphone

1-2 State which of the waveforms in Figure P1-2 can be classified as binary digital signals. Acceptable voltage ranges for logic '1' and logic '0' are given on each vertical axis. Assume that the horizontal scale has been adjusted so that only vertical passage through the transition zone is permissible for a valid digital signal.

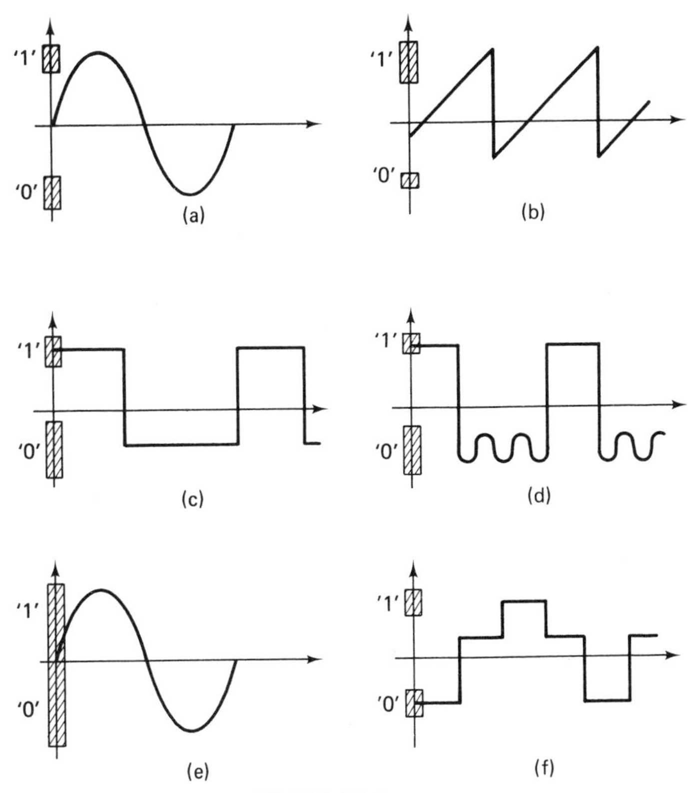

FIGURE P1-2

1-3 Determine the truth table that describes each of the switch circuits shown in Figure P1-3.

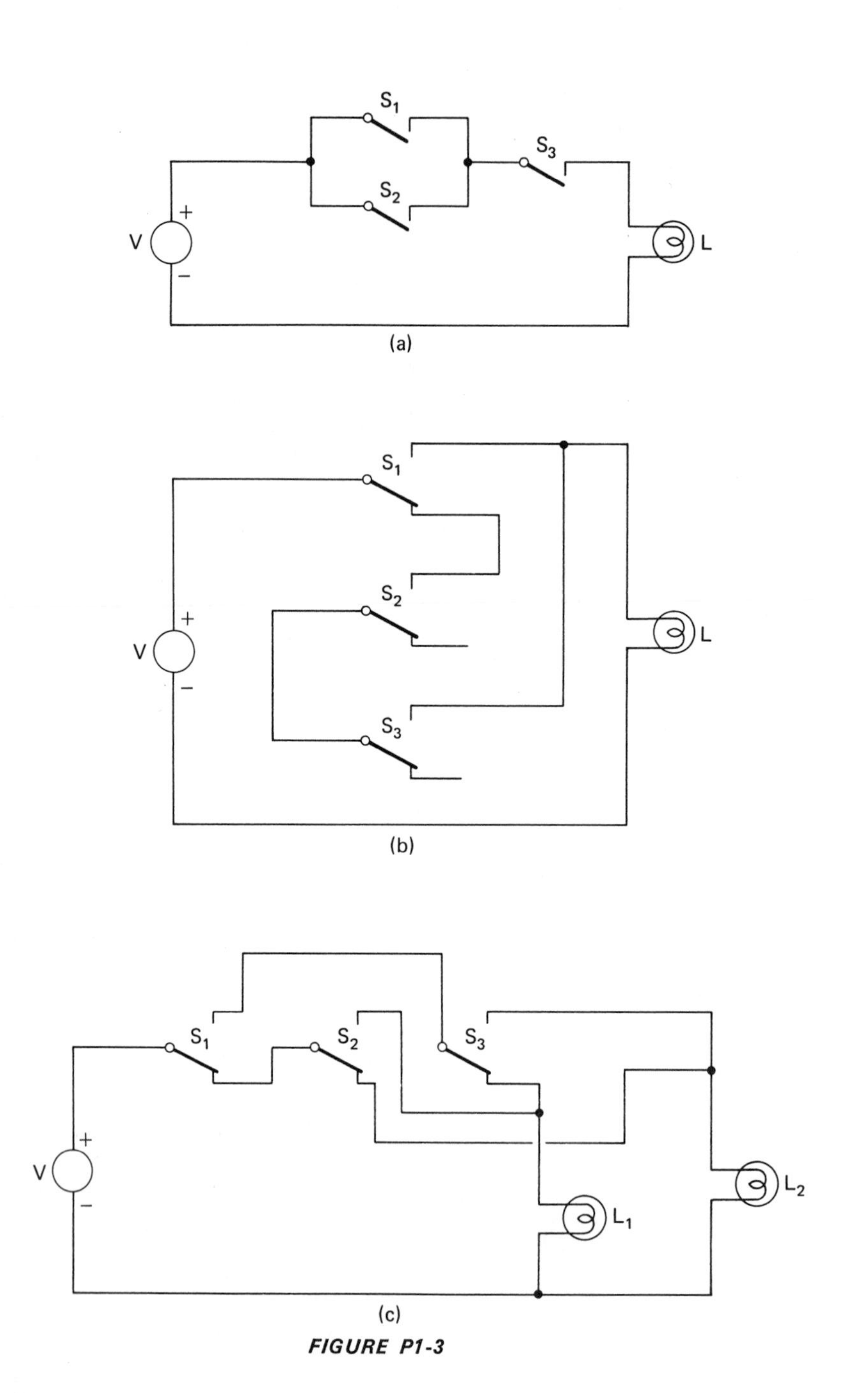

FIGURE P1-3

1-4 Determine a switch circuit that is described by each of the truth tables depicted in Figure P1-4.

(a)

S	L
0	1
1	0

(b)

S	L
0	1
1	1

(c)

S_1	S_2	S_3	L
0	0	0	0
0	0	1	0
0	1	0	0
0	1	1	0
1	0	0	1
1	0	1	0
1	1	0	0
1	1	1	1

(d)

S_1	S_2	S_3	S_4	L
0	0	0	0	0
0	0	0	1	0
0	0	1	0	0
0	0	1	1	0
0	1	0	0	0
0	1	0	1	1
0	1	1	0	1
0	1	1	1	1
1	0	0	0	0
1	0	0	1	1
1	0	1	0	1
1	0	1	1	1
1	1	0	0	0
1	1	0	1	1
1	1	1	0	1
1	1	1	1	1

FIGURE P1-4

1-5 Describe the behavior of each of the relay circuits illustrated in Figure P1-5. Write a truth table for each circuit; a binary variable may or may not be needed in each case.

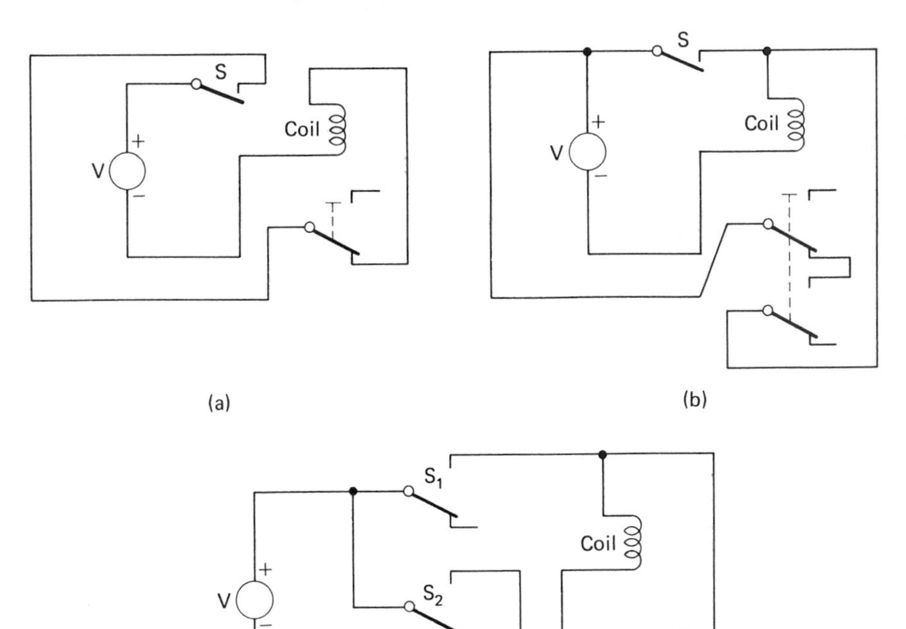

FIGURE P1-5

1-6 Determine the truth tables for each of the logic gate circuits illustrated in Figure P1-6. Indicate all intermediate variables in your truth tables.

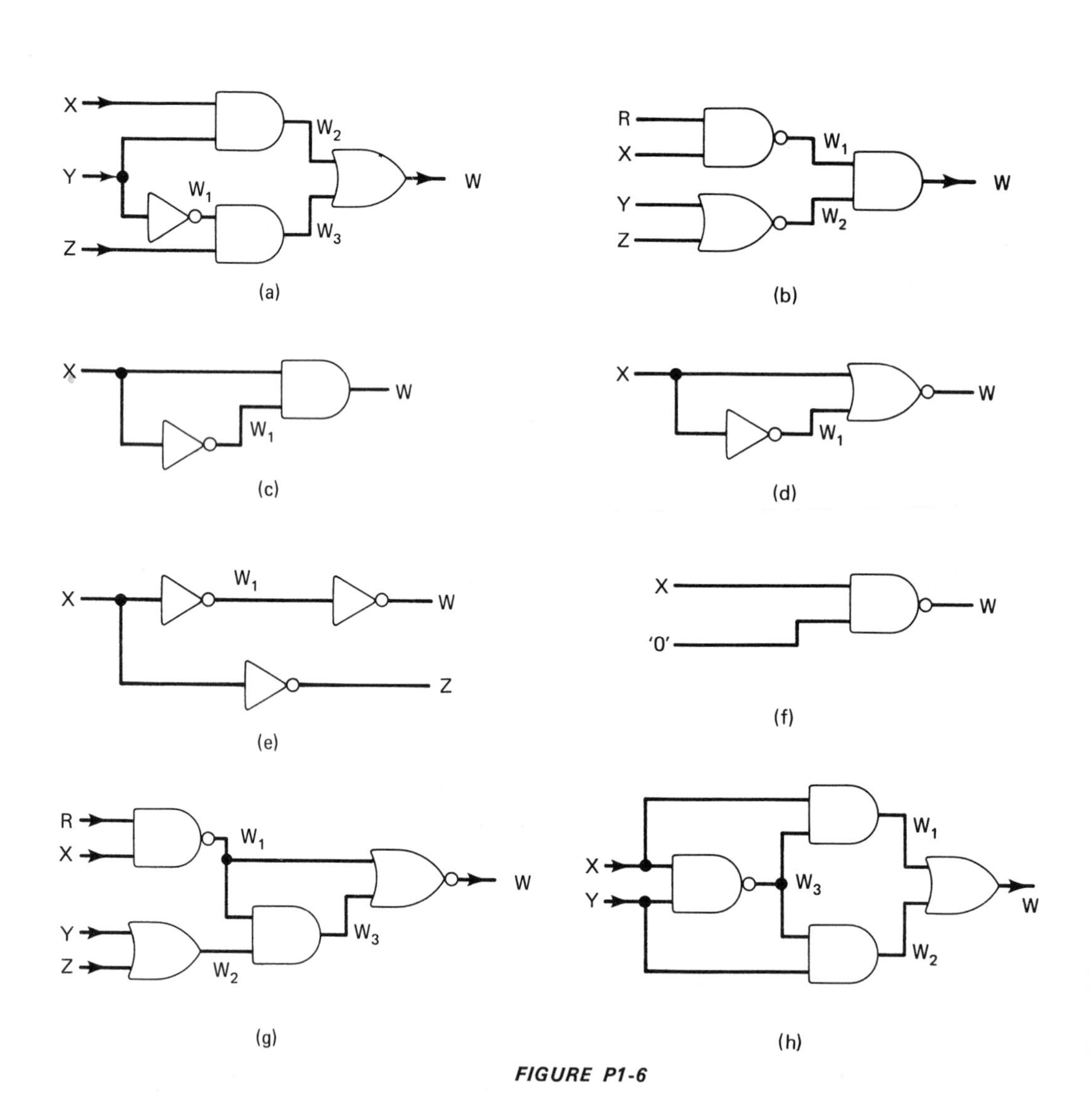

FIGURE P1-6

1-7 Can the logic circuits shown in Figure P1-7 be simplified so that in each case the same truth table is created using fewer gates? If so, determine the simplified circuit. Any kind of gate that has been described may be used in the simplified circuit.

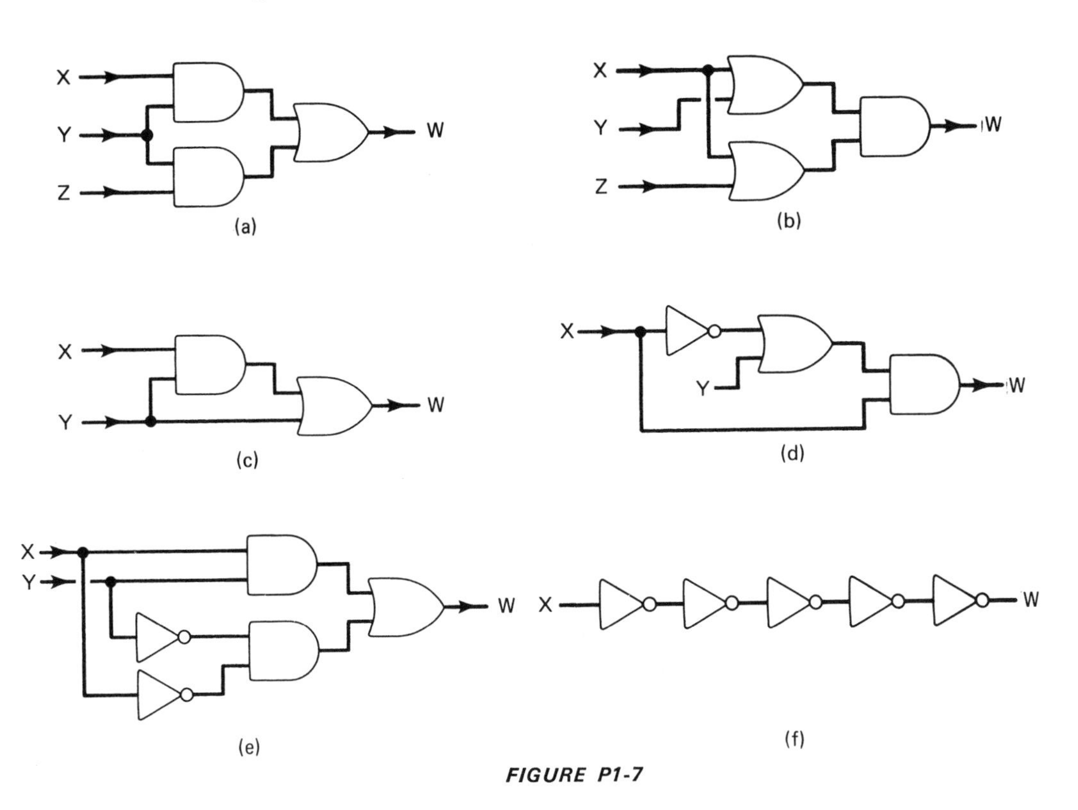

FIGURE P1-7

1-8 Implement each of the functions shown in Figure P1-8, using a minimum number of two-input NAND gates.

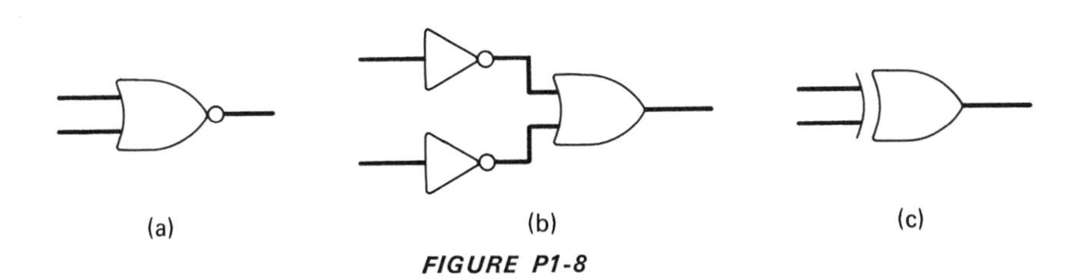

FIGURE P1-8

1-9 Demonstrate that a NOR gate is a universal gate.

1-10 Determine the operation of the following sequential circuit by tracing the sequence of changes in the output for the sequence of input changes shown in the table of Figure P1-10. Assume that the initial conditions are $X = 0$, $Y = 1$, and $W = 0$, and work down the table. In what ways is this sequence table different from a truth table?

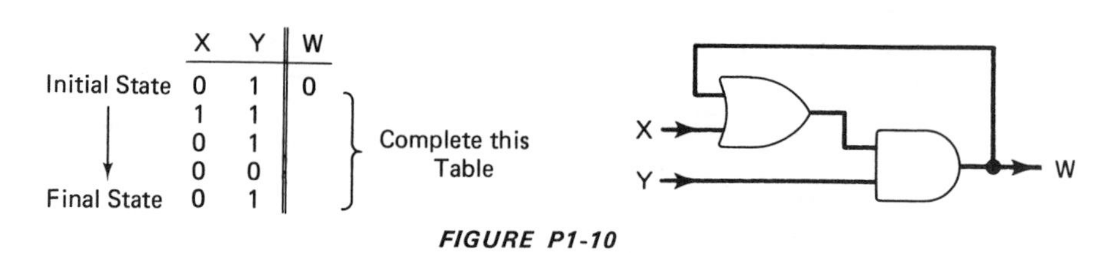

	X	Y	W
Initial State	0	1	0
	1	1	
	0	1	
	0	0	
Final State	0	1	

FIGURE P1-10

1-11 Using a minimum number of gates, detect the state '0 0 0' on the output lines of the digital circuit shown in Figure P1-11. Can it be done with one gate? Multiple-input gates may be used.

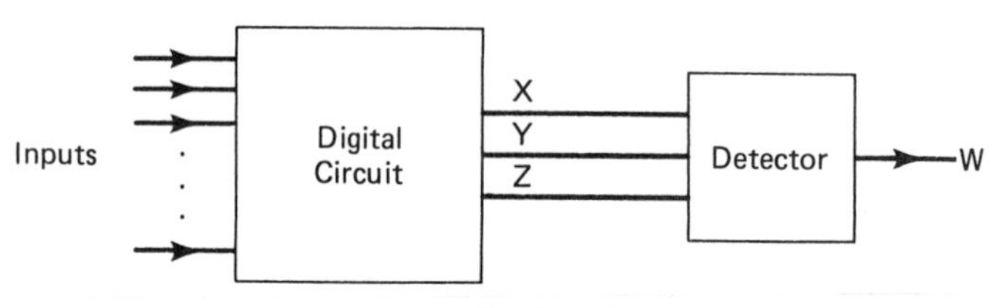

Detector Function: W = 1 if X, Y, and Z = 0
W = 0 otherwise

FIGURE P1-11

1-12 Find the rise time, fall time, and frequency of the binary digital signal illustrated in Figure P1-12.

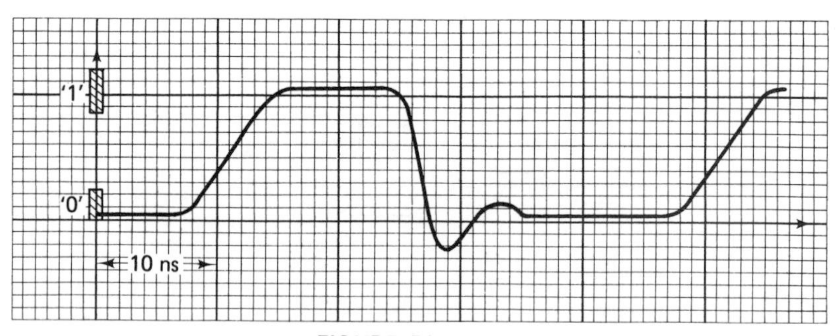

FIGURE P1-12

1-13 Digital signals are combined using various logic gates as shown in Figure P1-13. Sketch the results.

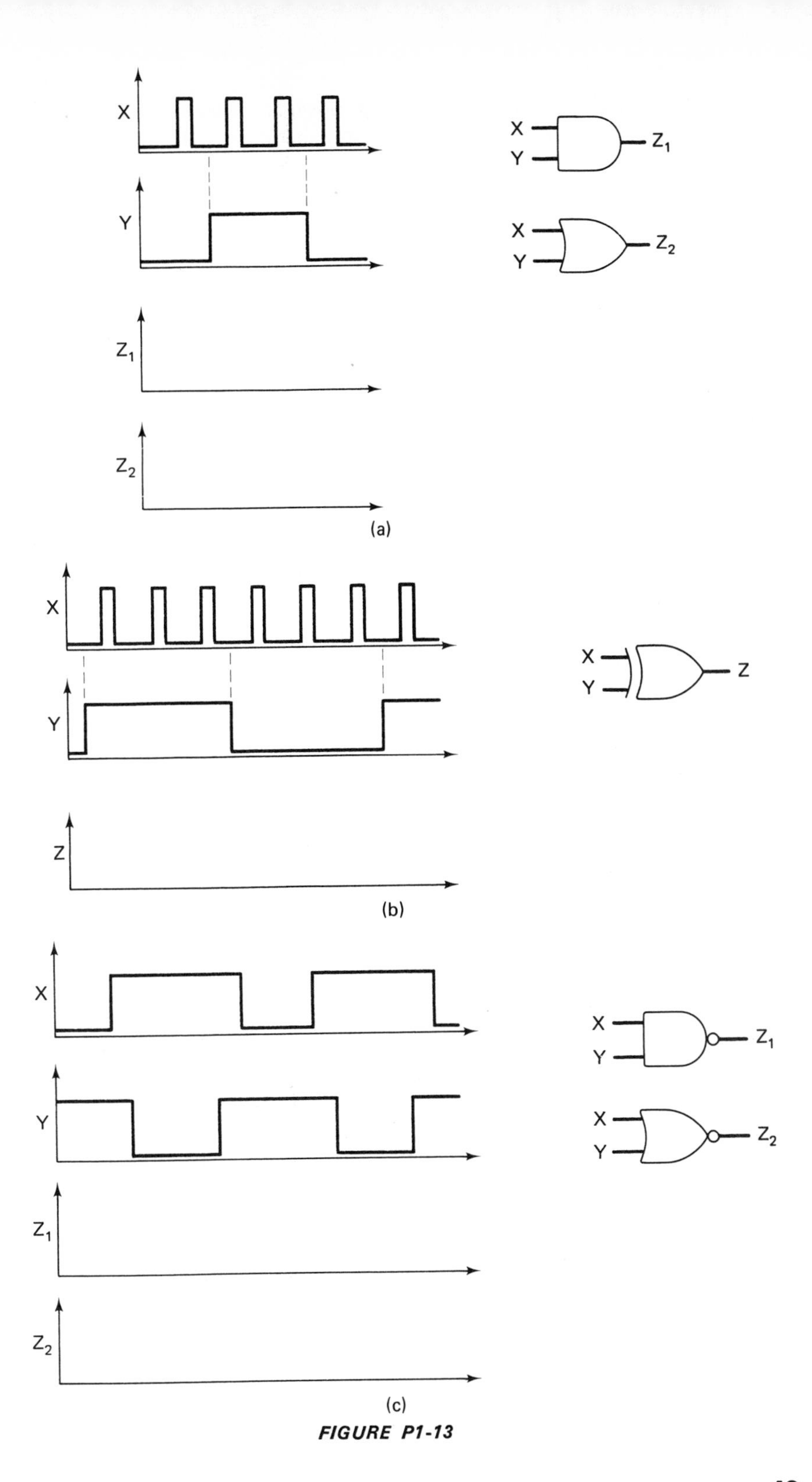

FIGURE P1-13

1-14 Sketch the output of the circuit shown in Figure P1-14. Assume that the propagation delay of each inverter is 10 ns, and that Z is initially '0'.

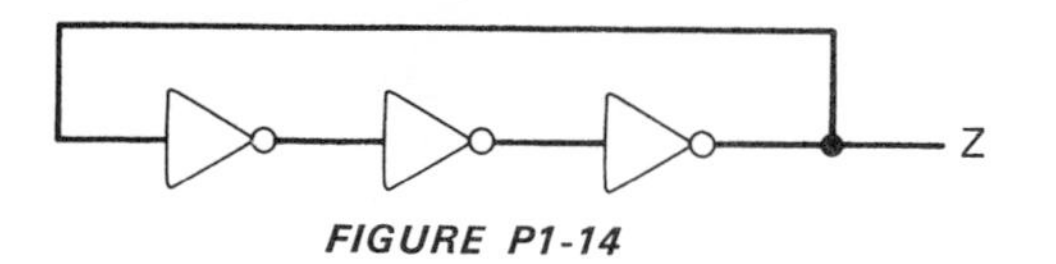

FIGURE P1-14

1-15 What problem might result if two outputs are tied together as shown in Figure P1-15? Would this problem be analyzed using algebraic or electronic techniques?

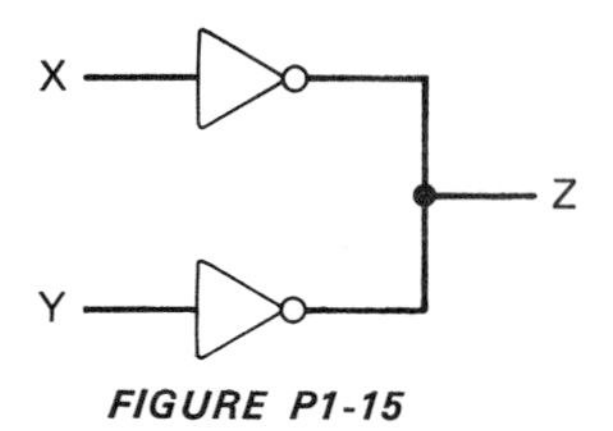

FIGURE P1-15

1-16 It is desired to control a light from three different locations. At each location is a two-position switch with a contact arrangement that is to be determined. Design a switch circuit that will control the light in this way. [*Hint:* A double-pole double-throw switch (Figure P1-16) will be needed.]

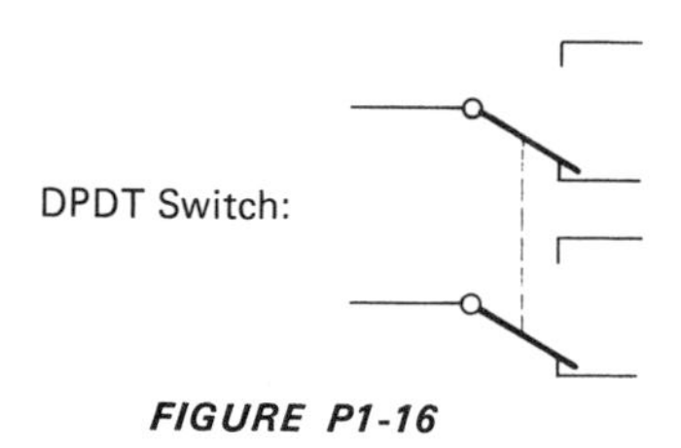

FIGURE P1-16

1-17 Using logic gates of any kind, design a logic circuit that has three inputs and one output (Figure P1-17), and operates in such a manner that when any one of the inputs changes logic state (from a '1' to a '0', or from a '0' to a '1') the output changes its logic state. This circuit is the logic gate equivalent of the three-switch light control circuit in Problem 1-16. (*Hint:* First consider solving the problem if there were two inputs, then generalize the solution to cases with more than two inputs.)

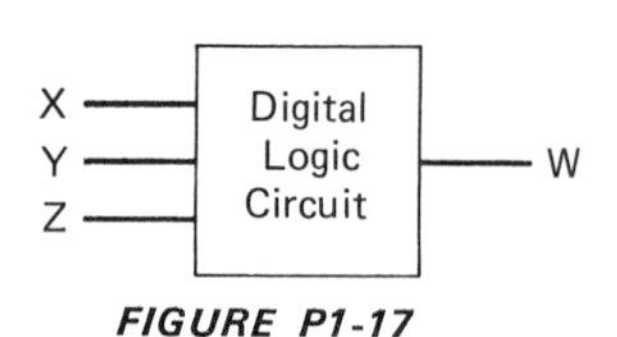

FIGURE P1-17

1-18 By using an electrically latched relay as a memory element, the control of a light from three remote locations can be achieved. At each location will be two pushbutton switches, one pushbutton pressed to turn the light on and the other pressed to turn the light off. Design a circuit that will accomplish this.

1-19 **(a)** The schematic for a four-position selector switch is shown in Figure P1-19(a). Design a digital logic circuit using any number and any type of logic gates so that the same selection function is performed on digital signals. Two control lines, C_1 and C_2, are provided. The logic levels on these lines determine which one of the four outputs in selected. Output lines that are not selected should produce a constant logic '0'.

(b) The switch selector can be reversed as shown in Figure P1-19b so that one of four inputs can be connected to a single output. Digital logic circuits, however, cannot be reversed in this manner but instead allow the flow of information only in one direction. Design a new selector circuit using any number and any type of logic gates so that the reverse selection function is achieved. C_1 and C_2 are digital control lines whose logic levels determine which of the four inputs is selected.

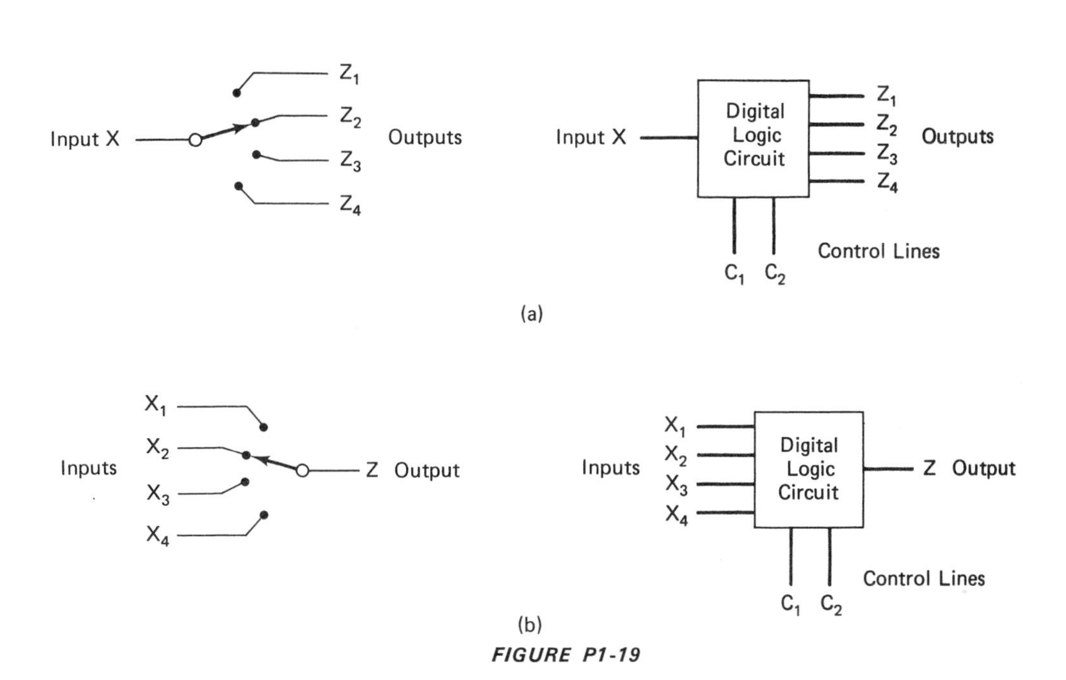

FIGURE P1-19

1-20 Two two-input NOR gates can be connected to obtain a simple two-state memory element. The circuit behaves according to the table given in Figure P1-20. Determine exactly how the NOR gates should be connected and where the inputs and outputs are located so that the operation specified in the table is realized.

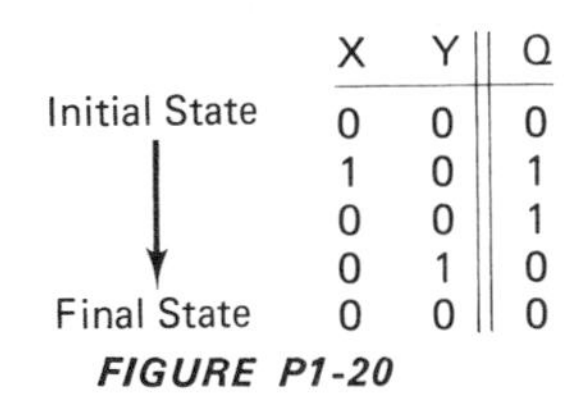

	X	Y	Q
Initial State	0	0	0
	1	0	1
	0	0	1
	0	1	0
Final State	0	0	0

FIGURE P1-20

2 BOOLEAN ALGEBRA

2.1 MOTIVATION FOR BOOLEAN ALGEBRA

Logic functions can range in complexity from simple relationships between one or two variables, to detailed conditional relationships among three, four, or more variables. When it becomes necessary to construct digital circuitry whose operation follows a prescribed logic function, certain factors arise that confine our choice of circuitry. In particular, it is not surprising to learn that complex logic functions involve proportionally more digital circuitry than do simple ones. Unfortunately, since there often is not a one-to-one correspondence between a logic function and a related gate structure, we must select one of many possible gate structures to obtain the desired logical operation.

Several different factors may be involved in our selection of the most desirable circuitry. For example, it may be necessary to

1. Minimize the number of logic gates in a circuit.
2. Minimize the cost of a circuit (which does not necessarily mean minimizing the number of logic gates).
3. Design a circuit using only one type of basic logic gate (such as a NAND gate).
4. Arrive at a circuit with minimum propagation delay (which might require parallel, redundant gates).

Other constraints may exist also. As a first step, it is usually very desirable to minimize the number of logic gates required in the construction of a digital circuit. It is valid to ask these questions: "How is it possible to tell whether a given gate structure is minimal?" and "How does one know that certain gates can be removed from a circuit without changing its logic function?" Normally, it is not obvious that a circuit is minimal, or that certain gates may be removed from the circuit without changing its operation.

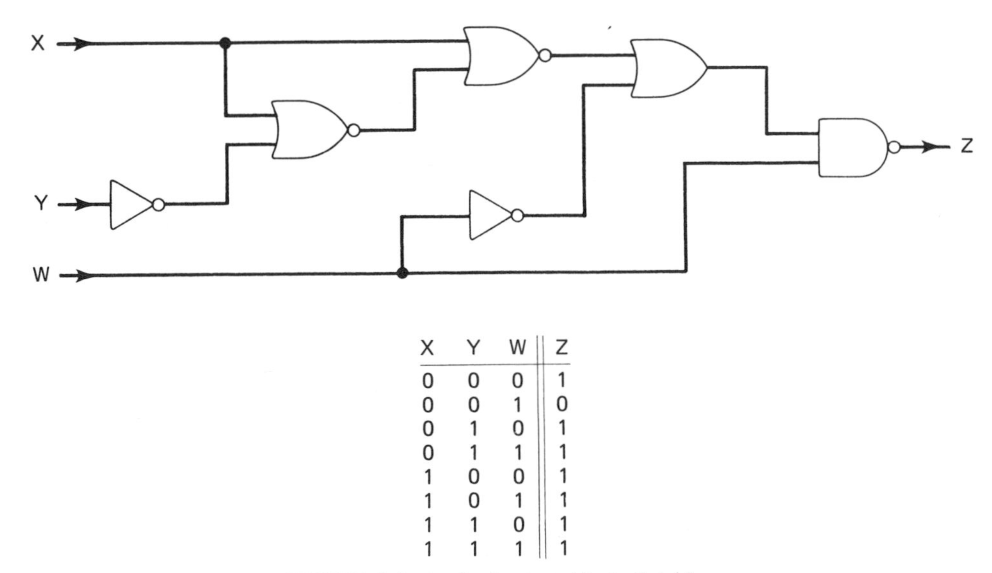

X	Y	W	Z
0	0	0	1
0	0	1	0
0	1	0	1
0	1	1	1
1	0	0	1
1	0	1	1
1	1	0	1
1	1	1	1

FIGURE 2-1 *Logic circuit and its truth table.*

It has been found that by representing symbolically the interconnection of logic elements, it is possible to engage mathematical rules that tell us precisely how reductions in circuitry can be brought about. Boolean algebra provides a means by which logic circuitry may be expressed symbolically, manipulated, and reduced. This chapter describes how logic functions can be so represented, and develops simple mathematical rules that define how we are permitted to manipulate the symbolic expressions.

To illustrate the problem of gate minimization, consider the circuit and corresponding truth table of Figure 2-1. Test several input combinations to convince yourself that the truth table is correct. This circuit is made up of six logic gates of four types. It would be difficult to determine by inspection whether this circuit is the simplest possible circuit (that is, uses the minimum number of gates), and if it is not, to determine the simplest circuit. By expressing the logic circuit as a Boolean equation and applying properties of Boolean algebra, as we shall learn to do in this chapter, the simplified circuit of Figure 2-2 is obtained. Note that in this case, only two gates of two types are required, offering a considerable savings in electronic hardware over the circuit of Figure 2-1.

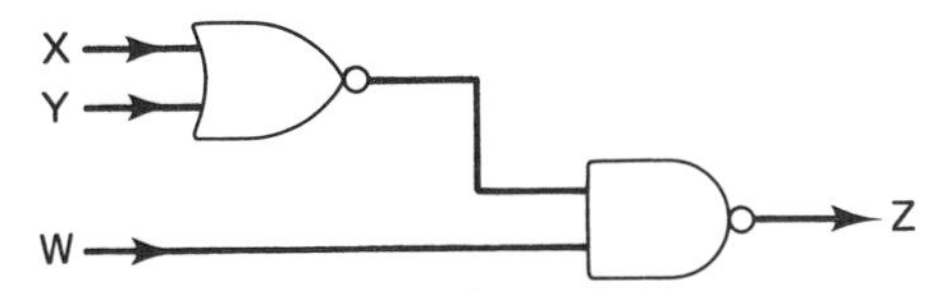

FIGURE 2-2 *Circuit utilizing fewer gates than that of Figure 2-1 but producing an identical logic function.*

Boolean algebra will be useful to us in the following ways:

1. Logic functions may be expressed as algebraic equations, offering more compactness and clarity than truth tables provide.
2. An algebraic equation, not necessarily minimal, can be obtained from a truth table by applying certain techniques of Boolean algebra.
3. A Boolean equation may be simplified, leading to a minimal gate structure.
4. A gate structure may be obtained from a Boolean equation, permitting us to transform a reduced equation into a desired logic circuit.

With the simple notation that Boolean algebra affords us, we will be able to analyze and design digital circuits on paper. In doing so, time and effort will be saved over purely experimenting to bring about the desired result. Once the elements of Boolean algebra are described, our primary emphasis will be on techniques of minimization. Algebraic and graphical methods of minimization will be investigated in this chapter and the next, with practical applications being demonstrated whenever possible.

2.2 BOOLEAN CONSTANTS, VARIABLES, AND FUNCTIONS

Binary digital systems operate on signals that can exist in one of two possible states. Boolean algebra relies on the numerical symbols '0' and '1' to represent these states. Although electrically, a particular state may be defined by a small range of voltage or current, mathematically it is important to know only whether a signal is within that range or not. If it is, the numerical symbol for that state is asserted; if it is not, the complementary value is asserted. As long as a valid signal may exist in only one of two possible states, it is correct to assume that if the signal is not in one state, it must be in the other.

A signal that does not change its state in time is considered a ***constant*** signal. The Boolean value of a constant signal will always be the same: either always '0' or always '1'. On the other hand, a signal that does change its state in time must be considered a ***variable*** signal, since it is seen to vary between one state and the other. The Boolean value of a variable signal is changeable and may be '0' at some times and '1' at other times. A constant signal will generally be referred to as just a constant, while a variable signal will generally be referred to as just a variable. Variables are represented in Boolean algebra by alphabetic letters, such as 'X', 'Y', 'Z', 'Q', and so forth. Figure 2-3 summarizes the relationship between constants and variables.

Boolean Constants:	'0', '1'	Numerical symbols representing fixed values that never change.
Boolean Variables:	'X', 'Y', 'Z', 'Q', etc.	Alphabetic symbols representing variable values. Each symbol may assume either a value of '0' or '1' at prescribed times.

FIGURE 2-3 *Relationship between constants and variables.*

We learned in Chapter 1 that a logic function could be uniquely specified by its truth table. The AND function, for example, would be defined by writing a truth table with two input columns and one output column. All possible combinations of the two binary inputs would be listed in the input columns, and the resulting values that correspond to the AND function would appear in the output column. This procedure exhaustively defines the AND function using the constants '0' and '1'. The AND function may be expressed algebraically in a more brief form, however. This is done using Boolean variables and the special function symbol '·', representing the AND function. Figure 2-4 illustrates the relationship between the AND truth table, its algebraic expression, and its graphical expression. Note that in the algebraic expression, it is necessary that we understand the meaning of the function symbol before a meaning for the entire expression is understood. To do so, we must ultimately refer back to the truth table. The algebraic expression, then, is not a free-standing definition, but just an abbreviated way of stating a logic function.

X	Y	Z
0	0	0
0	1	0
1	0	0
1	1	1

(a) Exhaustive Definition Using Constants.

$$X \cdot Y = Z$$

(b) Algebraic Expression Using Variables and the Function Symbol '·'.

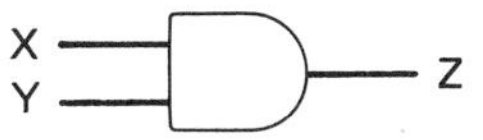

(c) Graphical Expression Using the AND Gate.

FIGURE 2-4 *The AND function expressed with a truth table, algebraically, and graphically.*

X	Y	Z
0	0	0
0	1	1
1	0	1
1	1	1

(a) Exhaustive Definition Using Constants.

$$X + Y = Z$$

(b) Algebraic Expression Using Variables and the Function Symbol '+'.

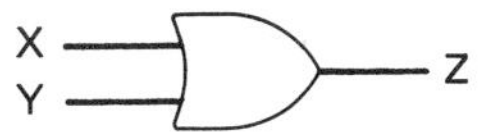

(c) Graphical Expression Using the OR Gate.

FIGURE 2-5 *The OR function expressed with a truth table, algebraically, and graphically.*

Figure 2-5 depicts the relationship between the OR truth table, its algebraic expression, and its graphical expression. In this case, the function symbol '+' is used to represent the OR function.*

* The '+' used in Boolean algebra to represent the OR function means something quite different from the '+' used in normal algebra to represent addition; be careful not to get the meanings confused.

X	Z
0	1
1	0

$\overline{X} = Z$

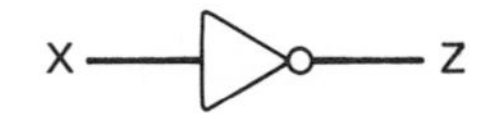

(a) Exhaustive Definition Using Constants.

(b) Algebraic Expression Using Variables and the Function Symbol '‾'.

(c) Graphical Expression Using the INVERT Gate.

FIGURE 2-6 *The INVERT function expressed with a truth table, algebraically and graphically.*

Figure 2-6 shows the INVERT function in the form of a truth table, an algebraic expression, and a graphical expression. The overbar, '‾', is used to indicate inversion algebraically. In contrast to the AND and OR functions, the INVERT function operates on *only one* binary variable (X in Figure 2-6).

We have seen that the AND and OR functions have been given special function symbols. This enables them to be expressed simply in a Boolean equation. Recall from Chapter 1, however, that other possible functions may exist also, for example the NAND and NOR functions. This addresses the general question: How many functions are possible with two binary variables? The question is answered with Table 2-1. First, note that four different combinations of two binary inputs are permitted. These input values are the ***arguments*** of each function, and are in the leftmost column of the table. A ***functional result*** must specify a '0' or '1' for each pair of arguments. We know that in the AND function, for instance, all functional results are '0' except when both arguments are '1', in which case the functional result is '1'. By stating successively each possible combination of four binary output digits, as shown in Table 2-1, we see that there are sixteen possible functions. Not all of these functions are familiar. Of greatest importance is the AND function (Z_2) and the OR function (Z_8). This is so because of their usefulness in logical machines and in reasoning. Furthermore, the AND, OR, and INVERT functions can be combined in various ways to produce any of the remaining functions in Table 2-1. It is because of the importance of the AND and OR functions that special symbols have been used to express them conveniently in algebraic form, namely '$\cdot$' for AND and '+' for OR. Using these function symbols, and the overbar, '‾', for INVERT, an algebraic system can be developed. Compound functions utilizing many logic gates can then be algebraically expressed, simplified, and translated into an optimal gate structure.

TABLE 2-1

Large truth table, showing all sixteen possible functions of two variables.

Arguments		Possible Functions															
		Null	*AND*					*XOR*	*OR*	*NOR*	*EQU*					*NAND*	*Identity*
X	Y	Z_1	Z_2	Z_3	Z_4	Z_5	Z_6	Z_7	Z_8	Z_9	Z_{10}	Z_{11}	Z_{12}	Z_{13}	Z_{14}	Z_{15}	Z_{16}
0	0	0	0	0	0	0	0	0	0	1	1	1	1	1	1	1	1
0	1	0	0	0	0	1	1	1	1	0	0	0	0	1	1	1	1
1	0	0	0	1	1	0	0	1	1	0	0	1	1	0	0	1	1
1	1	0	1	0	1	0	1	0	1	0	1	0	1	0	1	0	1

Note that a special symbol could be adopted for each one of the two argument functions in Table 2-1. This would permit each function to be stated briefly in algebraic form. For example,

$$X \Delta Y = Z$$

(↑ Special function symbol)

Boolean algebra, however, has been developed using special symbols only for AND, OR, and INVERT. Since all functions in Table 2-1 that are different from AND and OR can be expressed in terms of AND, OR and INVERT, nothing is excluded because of this self-imposed limitation on the number of special function symbols we use.

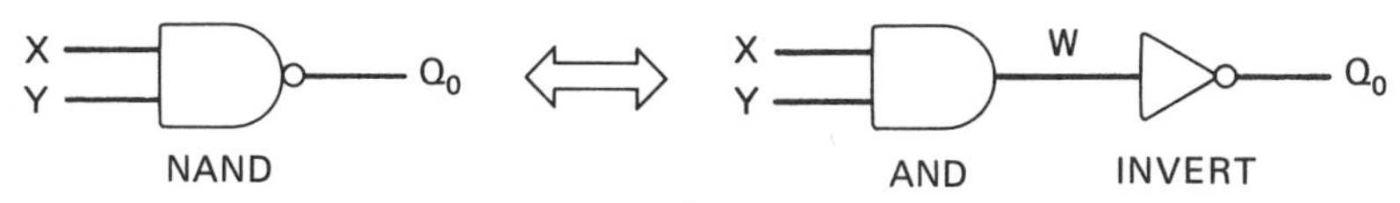

FIGURE 2-7 The NAND gate and its equivalent.

Recall that the NAND function can be considered to be a combination of AND followed by INVERT, as shown in Figure 2-7. To illustrate how a Boolean equation for the NAND function can be written using only the AND and INVERT function symbols, we consider the intermediate variable 'W' in Figure 2-7. We can express W as in Eq. 2-1, since W results from the AND of X and Y.

$$W = X \cdot Y \tag{2-1}$$

We also know that $Q_0 = \bar{W}$, and therefore can express Z as

$$Q_0 = \overline{X \cdot Y} \tag{2-2}$$

by substituting $X \cdot Y$ for W. Consequently, the NAND function is expressed by Eq. 2-2 using only the algebraic symbols for AND and INVERT. In a similar fashion, we can express the NOR function.

$$Q_1 = \overline{X + Y} \tag{2-3}$$

More complex functions can be written that are made up of many different combinations of basic functions. One such possibility, for example, is algebraically expressed as shown in Eq. 2-4.

$$Q_2 = (\overline{(X + Y) \cdot W}) + ((X \cdot Y) \cdot W) \tag{2-4}$$

If X and Y in Eq. 2-2 were specified to be certain values, $X = 1$ and $Y = 1$ for example, Q_0 could be evaluated exactly. We note that since the entire quantity '$X \cdot Y$' is covered by the INVERT overbar, $X \cdot Y$ must be computed first, and the result complemented to obtain Q_0. Thus, for $X = 1$ and $Y = 1$, $Q_0 = \overline{1 \cdot 1} = \bar{1} = 0$.

Recall that to obtain the result of $1 \cdot 1$, the AND of two 1's, we need only look at the truth table, or simply remember the operation of the AND function. In a similar way, we can compute Q_1 in Eq. 2-3. If $X = 1$ and $Y = 0$, then $Q_1 = \overline{1 + 0} = \overline{1} = 0$. We see here that when X and Y are exactly specified, an exact numerical result can be found. This is done by referring back to the truth table for the specific function that relates X and Y. In general, whenever all of the variables in an equation are specified to be exact numerical values, the entire equation can be evaluated to either a '1' or a '0'. This is true no matter how complex an equation may be. The action of reducing a Boolean equation from $Q =$ (complex expression) to $Q = 1$ or $Q = 0$, when the variables are exactly specified, is an act of ***simplification***. Unfortunately, it is an uncommon circumstance when the variables of an equation are exactly specified as '0's or '1's. This does not mean, however, that simplification of the expression is impossible. Actually, the power of Boolean algebra rests in its ability to permit general equations to be simplified *without* having to know exact values for the variables. In Section 2.3, we will see how a set of properties can be developed that may be applied to *any* general Boolean equation. In some cases, an equation can be greatly simplified by applying these properties, while in other cases simplification will not be possible. In any event, the result of simplifying a Boolean expression will be another Boolean expression, not necessarily a '0' or '1', that contains fewer variables and fewer function symbols than the original expression.

2.3 USEFUL PROPERTIES OF BOOLEAN ALGEBRA

In this section, a number of algebraic properties will be developed that describe those means by which a Boolean expression may be rearranged validly. Such rearrangement permits the reordering or compression of algebraic statements. Rather than relying only on mathematical proofs to demonstrate that a property is accurate, switch networks will be used as well to verify intuitively the result. In this way, a close relationship is seen between actual logic circuits and the symbolic means used to express them.

2.3.1 Generalizing a Switch Circuit

The logic AND function was first demonstrated with a series electrical circuit, as shown in Figure 2-8. This circuit behaved as an AND function only because of the series connection of the switches, not depending on the voltage source, wire, or lamp.

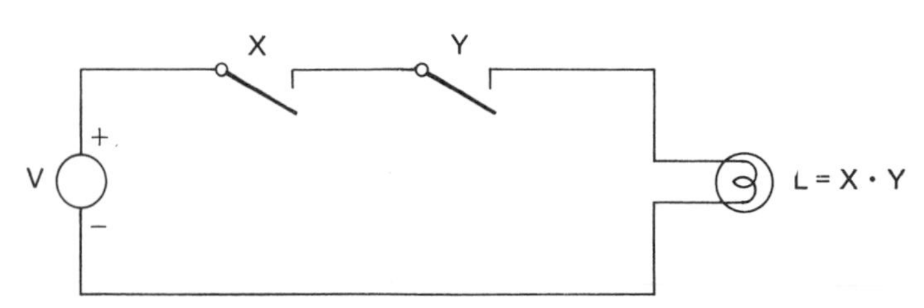

FIGURE 2-8 *The logic AND function demonstrated with a switch circuit.*

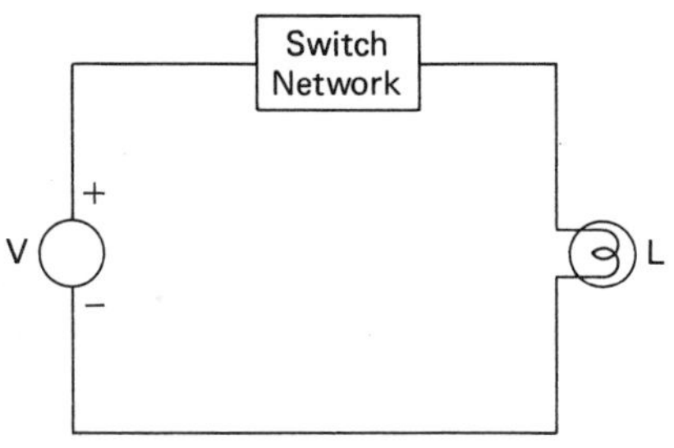

FIGURE 2-9 Logic circuit in which the switch connections are not specified.

Other logic functions were demonstrated by simply connecting the switches differently. Figure 2-9 shows how the switch connections can be represented in a general way using only a "black box" in the schematic in place of the connections. If the ***transmittance*** of this box is '1', the lamp will glow; whereas if the transmittance of this box is '0', the lamp will be dark. Since the switch network is the only part of the circuit in Figure 2-9 that determines the circuit's logical function, we can simplify our discussion in the future by concerning ourselves only with the switch network. Figure 2-10 illustrates a generalized switch network.

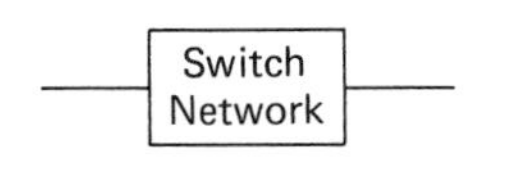

(1) Transmittance = '1' Means that Current Can Flow Through the Box.
(2) Transmittance = '0' Means that Current Cannot Flow Through the Box.

FIGURE 2-10 Generalized switch network.

2.3.2 The Symbols of Boolean Algebra and Their Relation to Switch Elements

Boolean algebra requires the use of a set of special symbols. They fall into four basic catagories:

1. Symbols for constants: '0' and '1'.
2. Symbols for variables: 'X', 'Y', 'Z', 'W', 'Q', etc.
3. Symbols for functions: '$\cdot$', '$+$', '$\bar{\ }$'.
4. Symbol for equality: '$=$'.

Figure 2-11 illustrates how we will relate these symbols to corresponding electrical circuits.

It is important to note several things about the electrical relationships of Figure 2-11. First, a complemented variable, as in item (4) of Figure 2-11, is drawn as a switch, but opening upward instead of downward. This is consistent with our convention adopted in Chapter 1 in which up (toward the top of the page) represents logic '1'. In this way, an uncomplemented variable in the '1' state corresponds to a switch with a transmittance of '1'. However, a complemented variable in the '1' state corresponds to a switch with a transmittance of '0'.

Description	Algebraic Symbol(s)	Electrical Representation
(1) The Constant '0'	'0'	Open Circuit Transmittance = 0
(2) The Constant '1'	'1'	Short Circuit  Transmittance = 1
(3) A Variable	'X'	Switch 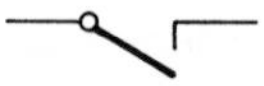 Transmittance Depends on Position of Switch
(4) A Complemented Variable	'$\bar{X}$'	Switch  Transmittance Depends on Position of Switch
(5) A Function	'X · Y'	Series Connection of Switches 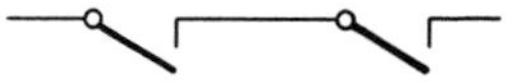 Transmittance Depends on Position of Switches
(6) An Equation	Z = X · Y	Switch Network [Z] Equals Particular Connection of Switches 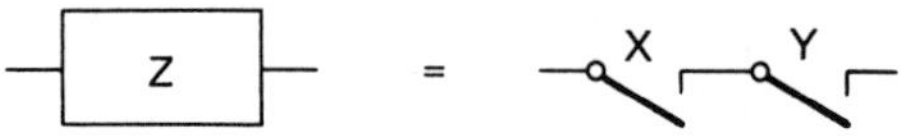 [X, Y]

FIGURE 2-11 *Relationship between algebraic symbols and electrical representations.*

As a second point, it should be understood that a box labeled "switch network" as in item 6 of Figure 2-11, may contain as one of its components *another* box labeled "switch network." That is, a single variable symbol such as 'X' or 'Y' need not necessarily correspond to a single switch. For example, the equation

$$Z = X \cdot Y \tag{2-5}$$

may correspond electrically to

where 'X' itself is another switch network. Remember that when all variables are specified to be particular values, any switch network, no matter how complex, can be evaluated to a transmittance of either '0' or '1', and nothing else.

2.3.3 Boolean Postulates

The only requirement that must be complied with in developing a mathematical system is that the system be *self-consistent*. That is, it should be impossible to disprove one rule of the system by using another rule of the system. It is *not* a requirement, however, that the mathematical system necessarily bear any relationship to the physical world. Fortunately, Boolean algebra happens to be closely related to the physical world, since it was originally developed in an effort to model reasoning. Our refinement of the original algebra of George Boole is clearly related to switch networks. To make more believable the properties of modern Boolean algebra, we will verify each property that is developed by relying on a corresponding switch network. In this way, a property that allows simplification of an algebraic expression is proven by showing that one switch network may be rewired to obtain an equivalent network. The equivalent network may itself contain fewer switches, or lead to a network that contains fewer switches.

In the development of a system of Boolean algebra, it is necessary to start with some initial assumptions, known as ***postulates***. The postulates that we assert originate from the three basic logic functions of AND, OR and INVERT, whose truth tables are shown for reference in Figure 2-12. Figure 2-13 states ten postulates which provide the foundation upon which our system of Boolean algebra is based. In addition, the corresponding electrical connections which relate to each postulate are shown. Note that if the transmittance of one switch network is the same as another network under the same switch settings, the networks are equivalent. Each postulate is a specific instance of one of the truth table entries for AND, OR, or INVERT. The postulates, however, are expressed as algebraic equations, using the appropriate function symbols.

AND

X	Y	Z
0	0	0
0	1	0
1	0	0
1	1	1

OR

X	Y	Z
0	0	0
0	1	1
1	0	1
1	1	1

INVERT

X	Z
0	1
1	0

FIGURE 2-12 *AND, OR, and INVERT truth tables shown for reference.*

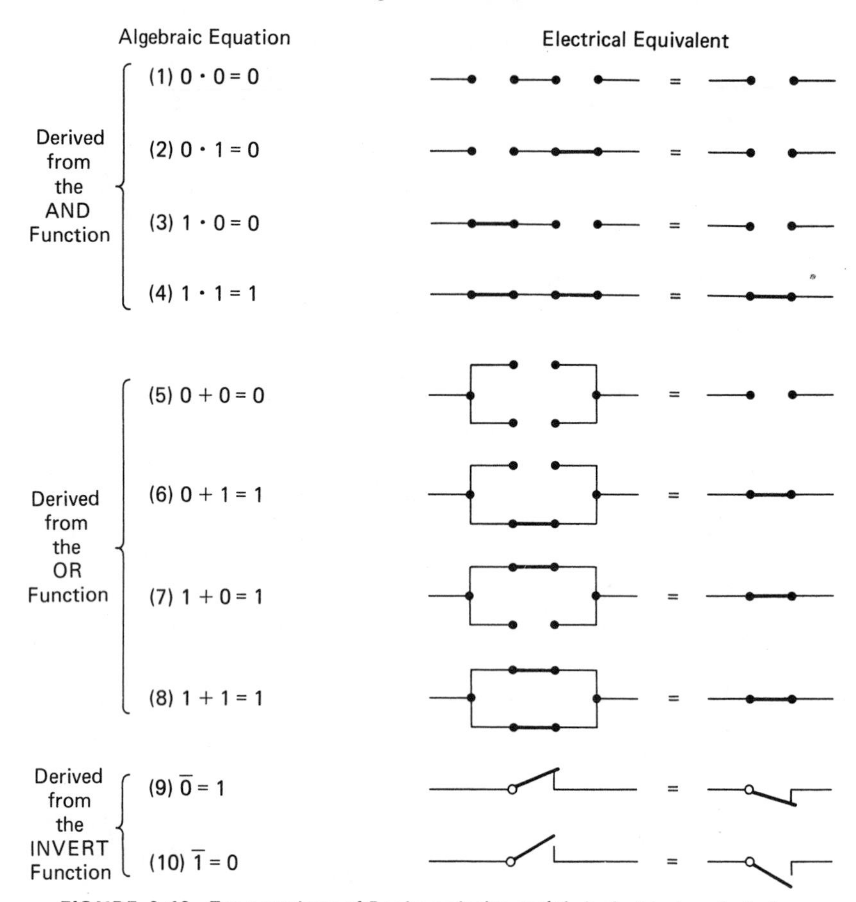

FIGURE 2-13 *Ten postulates of Boolean algebra and their electrical equivalents.*

2.3.4 Boolean Properties Derived from the Postulates

We now develop properties involving Boolean variables. A ***property*** is written as an equation whose left side contains a variable in combination with a constant or other variables, and whose right side is logically equivalent, but utilizing a different arrangement of variables or constants. Often, the right side of the equation contains fewer variables or constants. Boolean properties validate the rewriting of an algebraic expression into a simpler form. Each property is founded on one or more of the postulates. ***Proof*** that the property is accurate is offered by

1. Accounting for each binary value that a variable in the property may have.
2. Evaluating the equation for all combinations of variable values to show that, in each case, equality results.

The property is then verified by considering an equivalent switch circuit. Realize that as soon as a property is proven, it becomes a legitimate part of the mathematical system and may be used in any subsequent proofs. A summary of twenty-five properties is given in Table 2-3 (page 71). To gain some sense of direction, it may be helpful to preview this summary before studying the proofs.

PROPERTY 1

$$\boxed{X \cdot 0 = 0}$$

Proof: Case I: If $X = 0$, then $X \cdot 0 = 0 \cdot 0$,
$0 \cdot 0 = 0$ by Postulate 1.

Case II: If $X = 1$, then $X \cdot 0 = 1 \cdot 0$,
$1 \cdot 0 = 0$ by Postulate 3.

Therefore, $X \cdot 0 = 0$ for all X.

Verification:

Comment: Anything in series with an open circuit results in a transmittance of '0'.

PROPERTY 2

$$\boxed{0 \cdot X = 0}$$

Proof: (left as an exercise for the student)

Comment: Anything in series with an open circuit results in a transmittance of '0'. Comparing this result to Property 1, we also see that the position in which the open circuit is located does not matter.

PROPERTY 3

$$\boxed{X \cdot 1 = X}$$

Proof: Case I: If $X = 0$, then $X \cdot 1 = 0 \cdot 1$,
$0 \cdot 1 = 0$ by Postulate 2.

Case II: If $X = 1$, then $X \cdot 1 = 1 \cdot 1$,
$1 \cdot 1 = 1$ by Postulate 4.

Therefore, $X \cdot 1 = X$ for all X.

Verification:

Comment: The transmittance of anything in series with a short circuit is not affected.

PROPERTY 4

$$\boxed{1 \cdot X = X}$$

Proof: (left as an exercise for the student)

Comment: The transmittance of anything in series with a short circuit is not affected. Comparing this result with Property 3, we also see that the position in which the short circuit is located does not matter.

PROPERTY 5

$$\boxed{X + 0 = X}$$

Proof: Case I: If $X = 0$, then $X + 0 = 0 + 0$,
$0 + 0 = 0$ by Postulate 5.

Case II: If $X = 1$, then $X + 0 = 1 + 0$,
$1 + 0 = 1$ by Postulate 7.

Therefore, $X + 0 = X$ for all X.

Verification:

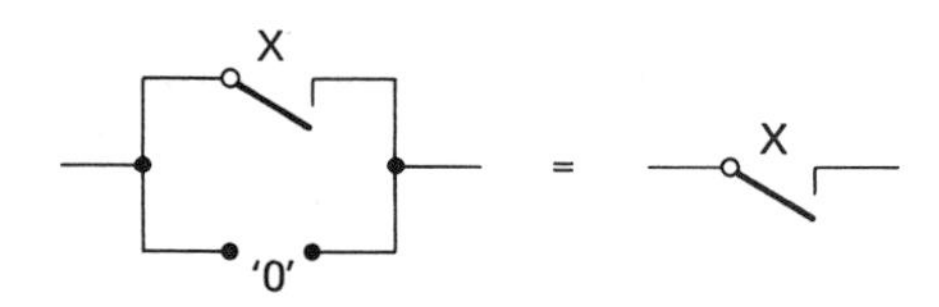

Comment: The transmittance of anything in parallel with an open circuit is not affected.

PROPERTY 6

$$\boxed{0 + X = X}$$

Proof: (left as an exercise for the student)

Comment: The transmittance of anything in parallel with an open circuit is not affected. Comparing this result with Property 5, we also see that the position in which the open circuit is located does not matter.

PROPERTY 7

$$\boxed{X + 1 = 1}$$

Proof: Case I: If $X = 0$, then $X + 1 = 0 + 1$,
$0 + 1 = 1$ by Postulate 6.

Case II: If $X = 1$, then $X + 1 = 1 + 1$,
$1 + 1 = 1$ by Postulate 8.

Therefore, $X + 1 = 1$ for all X.

Verification:

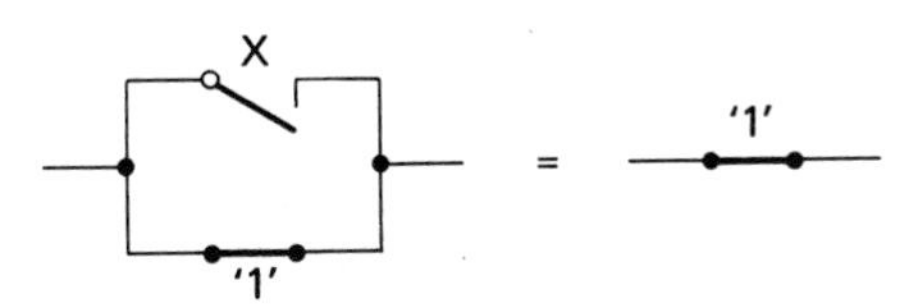

Comment: Anything in parallel with a short circuit results in a transmittance of '1'.

PROPERTY 8

$$\boxed{1 + X = 1}$$

Proof: (left as an exercise for the student)

Comment: Anything in parallel with a short circuit results in a transmittance of '1'. Comparing this result with Property 7, we see that the position in which the short circuit is located does not matter.

PROPERTY 9

$$\boxed{X \cdot X = X}$$

Proof: Case I: If $X = 0$, then $X \cdot X = 0 \cdot 0$,
$0 \cdot 0 = 0$ by Postulate 1.

Case II: If $X = 1$, then $X \cdot X = 1 \cdot 1$,
$1 \cdot 1 = 1$ by Postulate 4.

Therefore, $X \cdot X = X$ for all X.

Verification:

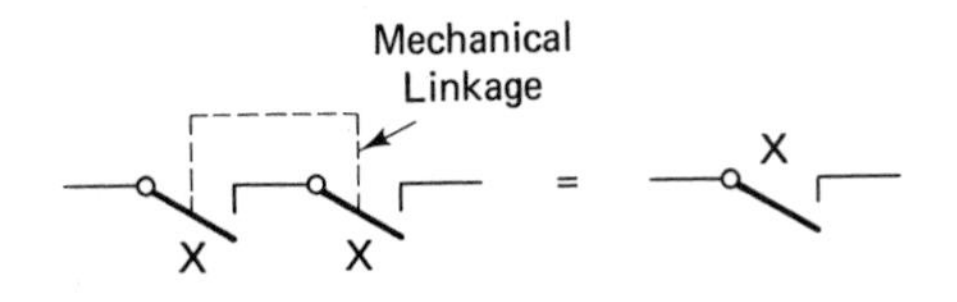

Comment: Two switches in series that are identically controlled can be replaced by a single switch.

PROPERTY 10

$$X \cdot \bar{X} = 0$$

Proof: (left as an exercise for the student)

Comment: Two switches in series that are controlled in a complementary manner result in an overall transmittance of 0; when one switch is closed, the other is open, and vice versa, and they will never both be closed at the same time.

PROPERTY 11

$$X + X = X$$

Proof: Case I: If $X = 0$, then $X + X = 0 + 0$,
$0 + 0 = 0$ by Postulate 5.

Case II: If $X = 1$, then $X + X = 1 + 1$,
$1 + 1 = 1$ by Postulate 8.

Therefore, $X + X = X$ for all X.

Verification:

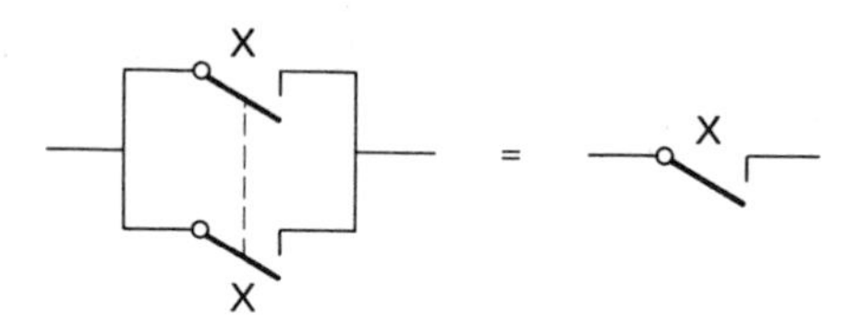

Comment: Two switches in parallel that are identically controlled can be replaced by a single switch.

PROPERTY 12

$$X + \bar{X} = 1$$

Proof: (left as an exercise for the student)

Comment: Two switches in parallel that are controlled in a complementary manner result in an overall transmittance of 1; one switch will always provide a path for current flow.

PROPERTY 13

$$\boxed{\bar{\bar{X}} = X}$$

Proof: Case I: If $X = 0$, then $\bar{\bar{X}} = \bar{\bar{0}}$,
$\bar{\bar{0}} = \bar{1}$ by Postulate 9,
$\bar{1} = 0$ by Postulate 10.

Case II: If $X = 1$, then $\bar{\bar{X}} = \bar{\bar{1}}$,
$\bar{\bar{1}} = \bar{0}$ by Postulate 10,
$\bar{0} = 1$ by Postulate 9.

Therefore, $\bar{\bar{X}} = X$ for all X.

Verification:

$\bar{\bar{X}}$ = X

Comment: Complementing a Boolean variable an even number of times (two, four, six, etc.) does not change the value of the variable.

These first thirteen properties involve only the single variable X. Note that we could have represented the variable with any letter, such as Y, Q, or A, and all the properties will remain valid. Also note that the variable does not necessarily have to represent a single switch, but could represent any desired switch network. Consider the property $X \cdot 1 = X$, for example. If X represented the parallel combination of two switches, Y and Z, algebraically written $X = Y + Z$, the property $X \cdot 1 = X$ could be rewritten $(Y + Z) \cdot 1 = Y + Z$. Verifying this with switches, we have

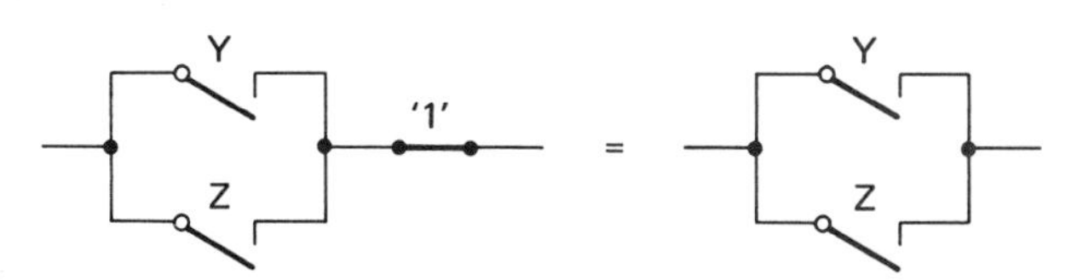

It is very important to realize that the properties apply equally well to combinations of variables as they do to single variables. We shall find this an important feature when reducing compound algebraic expressions.

By considering the AND and OR functions to be complementary functions, an interesting feature can be observed in any of the Boolean postulates. Namely, the symbol-by-symbol complement of a postulate equation yields one of the other postulates. For instance, Postulate 3 states that $1 \cdot 0 = 0$. Complementing every symbol, we have $\bar{1} \,\bar{\cdot}\, \bar{0} = \bar{0}$. Since AND and OR are considered complementary functions, $\bar{\cdot} = +$, and $\bar{+} = \cdot$. Thus, the postulate equation is rewritten as $0 + 1 = 1$, which is Postulate 6. Because of this feature, Postulates 3 and 6 are referred to as ***dual*** postulates. Duality also applies in a similar sense to the properties.

Properties 14 through 19 involve two or three variables and are identical to several properties of ordinary algebra. Their validity in Boolean algebra will again be verified with switch networks as was done in Properties 1 through 13. Also, having proven the validity of the first thirteen properties, it is acceptable to use these properties, in addition to the postulates, to prove any new properties. In general, every time a property is proven, it becomes a valid part of the mathematical structure and may be used to prove any new properties.

PROPERTY 14

$$\boxed{X \cdot Y = Y \cdot X}$$

Proof: Case I: If $Y = 0$, then $X \cdot Y = X \cdot 0$,

$Y \cdot X = 0 \cdot X$,

$X \cdot 0 = 0$ by Property 1,

$0 \cdot X = 0$ by Property 2.

Thus, $X \cdot 0 = 0 \cdot X$.

Case II: If $Y = 1$, then $X \cdot Y = X \cdot 1$,

$Y \cdot X = 1 \cdot X$,

$X \cdot 1 = X$ by Property 3,

$1 \cdot X = X$ by Property 4.

Thus, $X \cdot 1 = 1 \cdot X$.

Therefore, $X \cdot Y = Y \cdot X$ for all X and Y.

Verification:

Comment: The order in which a series combination of switches is connected does not matter.

PROPERTY 15

$$\boxed{X + Y = Y + X}$$

Proof. Case I: If $Y = 0$, then $X + Y = X + 0$,

$Y + X = 0 + X$,

$X + 0 = X$ by Property 5,

$0 + X = X$ by Property 6.

Thus, $X + 0 = 0 + X$.

Case II: If $Y = 1$, then $X + Y = X + 1$,

$$Y + X = 1 + X,$$
$$X + 1 = 1 \text{ by Property 7,}$$
$$1 + X = 1 \text{ by Property 8.}$$

Thus, $X + 1 = 1 + X$.

Therefore, $X + Y = Y + X$ for all X and Y.

Verification:

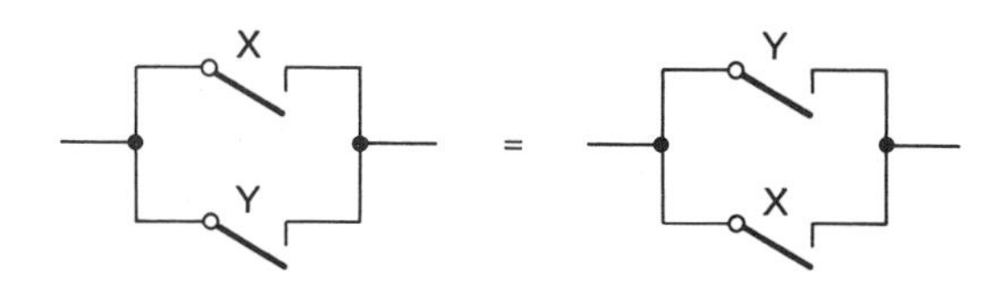

Comment: The order in which a parallel combination of switches is connected does not matter. Properties 14 and 15 together constitute what is known as the ***commutative law.***

We have seen that there is a close relationship between simple switch networks and algebraic expressions. Specifically, we note that a parallel switch connection is expressed with the OR function symbol, '+', and that a series switch connection is expressed with the AND function symbol, '·'. More complex switch networks in which a multitude of series and parallel connections are present can also be expressed algebraically. The algebraic result is a ***compound expression.*** A compound expression will include more than one function symbol. To indicate the order in which operations are to be performed in a compound expression, parentheses are often used. For example, the expression $X \cdot (Y + Z)$ indicates that the OR of Y and Z is to occur first, and the result combined with X using an AND function. With the same variables and function symbols, we can write another compound expression $(X \cdot Y) + Z$. In this case, the parentheses indicate that X and Y are to be combined using an AND function, and the result combined with Z using an OR function. Figure 2-14 illustrates

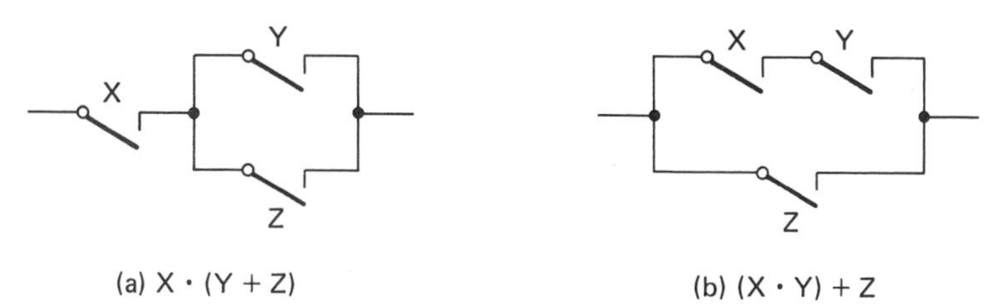

(a) $X \cdot (Y + Z)$ (b) $(X \cdot Y) + Z$

FIGURE 2-14 *The use of parentheses in a compound expression; the circuit in (a) is* not *equivalent to the circuit in (b).*

how these two expressions are different. Special operations relating to the use of parentheses are proven in the remaining properties. With these new properties, compound expressions can be expanded easily, or simplified.

PROPERTY 16

$$X \cdot (Y + Z) = (X \cdot Y) + (X \cdot Z)$$

Proof: Case I: If $X = 0$, then $X \cdot (Y + Z) = 0 \cdot (Y + Z)$,
$0 \cdot (Y + Z) = 0$ by Property 2.
$0 = 0 + 0$ by Postulate 5, and
$0 + 0$ can be written as
$(0 \cdot Y) + (0 \cdot Z)$ by Property 2.

Thus, $0 \cdot (Y + Z) = (0 \cdot Y) + (0 \cdot Z)$.

Case II: If $X = 1$, then $X \cdot (Y + Z) = 1 \cdot (Y + Z)$,
$1 \cdot (Y + Z) = Y + Z$ by Property 4.
$Y + Z$ can be written as
$(1 \cdot Y) + (1 \cdot Z)$ by Property 4.

Thus, $1 \cdot (Y + Z) = (1 \cdot Y) + (1 \cdot Z)$.

Therefore, $X \cdot (Y + Z) = (X \cdot Y) + (X \cdot Z)$ for all X, Y, and Z.

Verification:

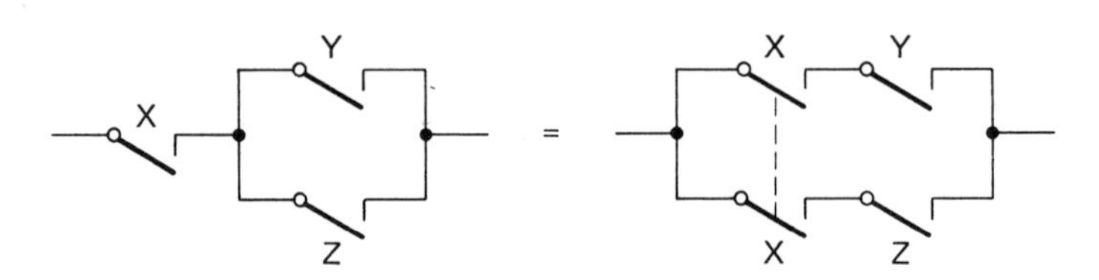

Comment: Boolean variables *distribute* with the AND function in the same sense as normal algebra variables distribute with multiplication.

PROPERTY 17

$$X + (Y \cdot Z) = (X + Y) \cdot (X + Z)$$

Proof: (left as an exercise for the student)

Properties 16 and 17 together constitute what is known as the ***distributive law.***

PROPERTY 18

$$X \cdot (Y \cdot Z) = (X \cdot Y) \cdot Z$$

Proof: Case I: If $X = 0$, then $X \cdot (Y \cdot Z) = 0 \cdot (Y \cdot Z)$,
$0 \cdot (Y \cdot Z) = 0$ by Property 2.
Since $0 = 0 \cdot Y$ and $0 = 0 \cdot Z$ by Property 2,
then $0 = 0 \cdot Z = (0 \cdot Y) \cdot Z$.

Thus, $0 \cdot (Y \cdot Z) = (0 \cdot Y) \cdot Z$.

Case II: If $X = 1$, then $X \cdot (Y \cdot Z) = 1 \cdot (Y \cdot Z)$,
$1 \cdot (Y \cdot Z) = Y \cdot Z$ by Property 4.
Since $Y = 1 \cdot Y$ by Property 4,
then $Y \cdot Z = (1 \cdot Y) \cdot Z$.

Thus, $1 \cdot (Y \cdot Z) = (1 \cdot Y) \cdot Z$.

Therefore, $X \cdot (Y \cdot Z) = (X \cdot Y) \cdot Z$ for all X, Y, and Z.

Verification:

X Y Z = X Y Z

Comment: The order in which switches are connected in series does not matter.

PROPERTY 19

$$X + (Y + Z) = (X + Y) + Z$$

Proof: (left as an exercise for the student)

Properties 18 and 19 together constitute what is known as the ***associative law***.

PROPERTY 20

$$X + (X \cdot Y) = X$$

Proof: Case I: If $X = 0$, then $X + (X \cdot Y) = 0 + (0 \cdot Y)$,
$0 + (0 \cdot Y) = 0 + 0$ by Property 2,
$0 + 0 = 0$ by Postulate 5.

Thus, $0 + (0 \cdot Y) = 0$.

Case II: If $X = 1$, then $X + (X \cdot Y) = 1 + (1 \cdot Y)$,
$1 + (1 \cdot Y) = 1 + Y$ by Property 4,
$1 + Y = 1$ by Property 8.

Thus, $1 + (1 \cdot Y) = 1$.

Therefore, $X + (X \cdot Y) = X$ for all X and Y.

Verification:

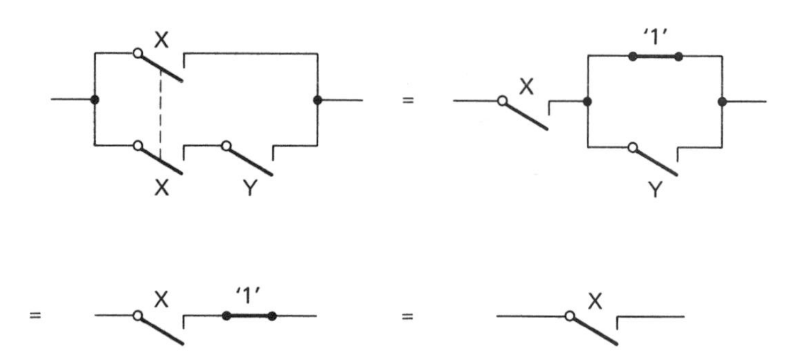

Comment: This property is known as a property of *absorption.*

PROPERTY 21

$$X \cdot (X + Y) = X$$

Proof: (left as an exercise for the student)

PROPERTY 22

$$X + (\bar{X} \cdot Y) = X + Y$$

Proof: Case I: If $X = 0$, then $X + (\bar{X} \cdot Y) = 0 + (\bar{0} \cdot Y)$,
$0 + (\bar{0} \cdot Y) = 0 + (1 \cdot Y)$ by Postulate 9,
$(1 \cdot Y) = Y$ by Property 4.

Thus, $0 + (\bar{0} \cdot Y) = 0 + Y$.

Case II: If $X = 1$, then $X + (\bar{X} \cdot Y) = 1 + (\bar{1} \cdot Y)$,
$1 + (\bar{1} \cdot Y) = 1 + (0 \cdot Y)$ by Postulate 10.
Since $0 \cdot Y = 0$ by Property 2,
then $1 + (0 \cdot Y) = 1 + 0$.
Now, since $1 + 0 = 1$ by Postulate 7, and
since $1 = 1 + Y$ by Property 8,
we can write $1 + 0 = 1 = 1 + Y$.

$$\text{Thus, } 1 + (\bar{1} \cdot Y) = 1 + Y.$$

$$\text{Therefore, } X + (\bar{X} \cdot Y) = X + Y \text{ for all } X \text{ and } Y.$$

Verification:

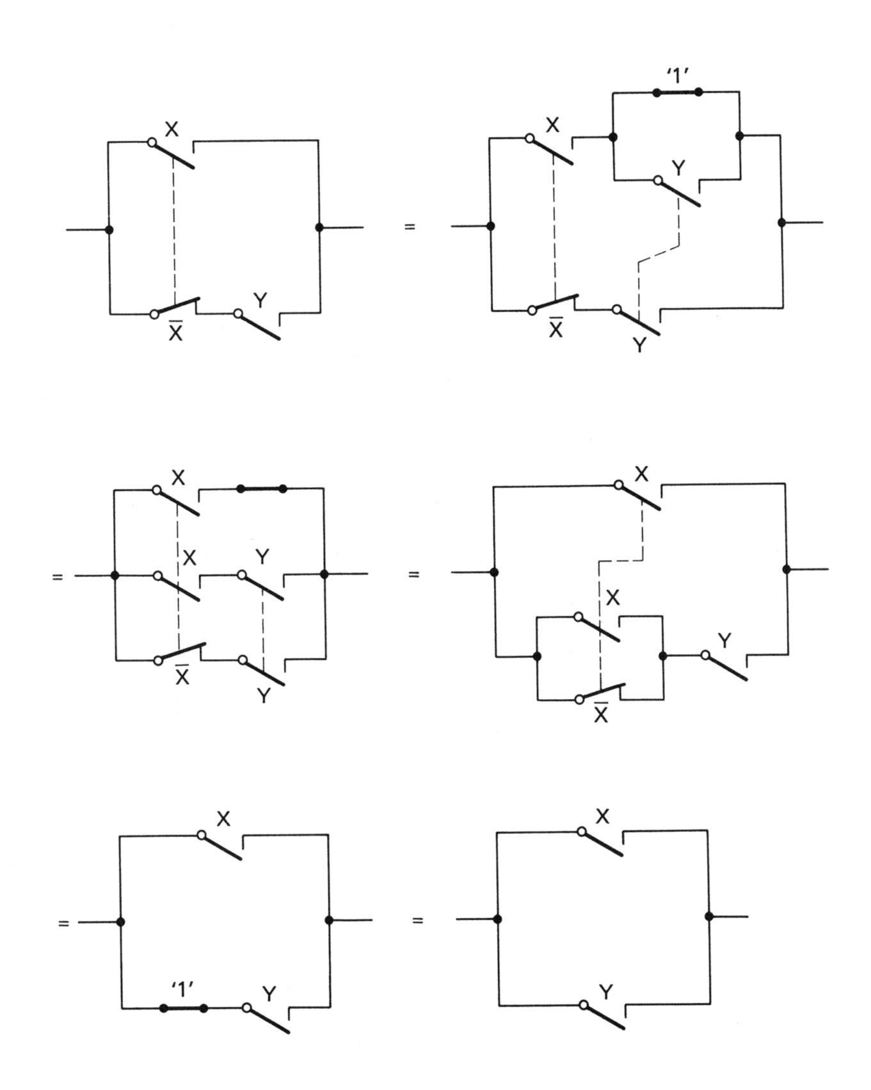

PROPERTY 23

$$\boxed{X \cdot (\bar{X} + Y) = X \cdot Y}$$

Proof: (left as an exercise for the student)

PROPERTY 24

$$\boxed{\overline{X + Y} = \bar{X} \cdot \bar{Y}}$$

Proof: Case I: If $X = 0$, then $\overline{X + Y} = \overline{0 + Y}$.
Since $0 + Y = Y$ by Property 6,
then $\overline{0 + Y} = \bar{Y}$.
Now, since $\bar{Y} = 1 \cdot \bar{Y}$ by Property 4, and
since $1 \cdot \bar{Y} = \bar{0} \cdot \bar{Y}$ by Postulate 9,
we can write $\bar{Y} = 1 \cdot \bar{Y} = \bar{0} \cdot \bar{Y}$.
Thus, $\overline{0 + Y} = \bar{0} \cdot \bar{Y}$.

Case II: If $X = 1$, then $\overline{X + Y} = \overline{1 + Y}$.
Since $1 + Y = 1$ by Property 8,
then $\overline{1 + Y} = \bar{1}$.
Now, since $\bar{1} = 0$ by Postulate 10, and
since $0 = 0 \cdot \bar{Y}$ by Property 2,
we can write $\bar{1} = 0 = 0 \cdot \bar{Y}$.
Finally, by Postulate 10, $0 \cdot \bar{Y} = \bar{1} \cdot \bar{Y}$.
Thus, $\overline{1 + Y} = \bar{1} \cdot \bar{Y}$

Therefore, $\overline{X + Y} = \bar{X} \cdot \bar{Y}$ for all X and Y.

Verification: (It is difficult to show inversion of a switch network function, as is required by $\overline{X + Y}$. Thus, this one is verified using truth tables.)

X	Y	$X + Y$	$\overline{X + Y}$	X	Y	$\bar{X}$	$\bar{Y}$	$\bar{X} \cdot \bar{Y}$
0	0	0	1	0	0	1	1	1
0	1	1	0	0	1	1	0	0
1	0	1	0	1	0	0	1	0
1	1	1	0	1	1	0	0	0

Thus, $\overline{X + Y} = \bar{X} \cdot \bar{Y}$.

PROPERTY 25

$$\boxed{\overline{X \cdot Y} = \bar{X} + \bar{Y}}$$

Proof: (left as an exercise for the student)

Properties 24 and 25 together constitute what is known as ***DeMorgan's law***. These relationships are very important and will be used frequently.

The twenty-five properties that have been discussed supply us with some general rules that can be applied to simplify Boolean expressions. Several examples of simplification are now considered.

Algebraic statement	*Justification*
$Z = (X \cdot Y) + [\bar{X} \cdot (Y + X)]$	Given. Brackets may be used instead of parentheses for clarity.
$= (X \cdot Y) + [(\bar{X} \cdot Y) + (\bar{X} \cdot X)]$	Distributive law (Property 16)
$= (X \cdot Y) + [(\bar{X} \cdot Y) + 0]$	Property 10
$= (X \cdot Y) + (\bar{X} \cdot Y)$	Property 5
$= (Y \cdot X) + (Y \cdot \bar{X})$	Commutative law (Property 14)
$= Y \cdot (X + \bar{X})$	Distributive law (Property 16)
$= Y \cdot 1$	Property 12
$= Y$	Property 3

Therefore, $Z = Y$ is the simplest Boolean equation.

Algebraic statement	*Justification*
Method 1:	
$Z = (X \cdot \bar{Y}) \cdot [(X \cdot Y) + \bar{Y}]$	Given
$= [(X \cdot \bar{Y}) \cdot (X \cdot Y)] + [(X \cdot \bar{Y}) \cdot \bar{Y}]$	Distributive law (Property 16)
$= [(X \cdot X) \cdot (Y \cdot \bar{Y})] + [X \cdot (\bar{Y} \cdot \bar{Y})]$	Associative law (Property 18)
$= [(X \cdot X) \cdot 0] + [X \cdot (\bar{Y} \cdot \bar{Y})]$	Property 10
$= 0 + [X \cdot \bar{Y}]$	Property 1, Property 9
$= X \cdot \bar{Y}$	Property 6

Method 2:

$Z = (X \cdot \bar{Y}) \cdot [(X \cdot Y) + \bar{Y}]$	Given
$= (X \cdot \bar{Y}) \cdot [\bar{Y} + (X \cdot Y)]$	Commutative law (Property 15)
$= (X \cdot \bar{Y}) \cdot [\bar{Y} + (Y \cdot X)]$	Commutative law (Property 14)
$= (X \cdot \bar{Y}) \cdot [\bar{Y} + X]$	Property 22

The last step was important. Note that although Property 22 is written as $X + (\bar{X} \cdot Y) = X + Y$, it is also true if we complement X: $\bar{X} + (X \cdot Y) = \bar{X} + Y$. Finally, it is still true if X and Y are interchanged: $\bar{Y} + (Y \cdot X) = \bar{Y} + X$.

$= (X \cdot \bar{Y}) \cdot (X + \bar{Y})$	Commutative low (Property 15)
$= [(X \cdot \bar{Y}) \cdot X] + [(X \cdot \bar{Y}) \cdot \bar{Y}]$	Distributive law (Property 16)
$= [(X \cdot X) \cdot \bar{Y}] + [X \cdot (\bar{Y} \cdot \bar{Y})]$	Commutative law (Property 15) Associative law (Property 18)
$= [X \cdot \bar{Y}] + [X \cdot \bar{Y}]$	Property 9
$= X \cdot \bar{Y}$	Property 11

As demonstrated by these examples, the specific rules that are applied, and the sequence in which they are applied, will vary with different Boolean expressions. To give a fixed, mechanical procedure that always results in a minimal algebraic expression is not possible. Rather, it is necessary to develop experience by working out many problems. In combination with several more worked examples, a number of problems are provided at the end of the chapter. The more of these that are worked out, the easier it will be to know when and how to reduce a Boolean expression.

2.3.5 Conventions that Simplify the Writing of Boolean Expressions

The examples that have been discussed thus far illustrate how parentheses and brackets are used to specify which functional operations are to be performed first when reducing an algebraic expression. To simplify the writing of compound expressions, a convention has been adopted that allows us to use fewer parentheses than we might otherwise need. This is done by letting the AND function have greater *precedence* than the OR function. That is, whenever the elimination of parentheses leaves uncertainty about which operation, AND or OR, should be performed first, we automatically do the AND first. For example, the expression

$$(X \cdot Y) + Z$$

could be written without parentheses as

$$X \cdot Y + Z$$

and have the same meaning. On the other hand, the expression

$$X \cdot (Y + Z)$$

cannot be rewritten without parentheses, since doing so would lead to an erroneous result under the ***AND precedence*** convention.

Another convention that simplifies the writing of Boolean expressions involves the ***implied AND*** function. This means that the AND function symbol, '$\cdot$', need not be written every time we wish to express the AND function. Thus, '$X \cdot Y$' can be rewritten simply as 'XY'. As long as only one letter is used to represent a Boolean variable, there will be no ambiguity.

TABLE 2-2

Use of accepted conventions to abbreviate the writing of Boolean expressions.

Normal Form	*Abbreviated Form*
$[(X \cdot Y) + Z] \cdot [(X + Y) \cdot Z]$	$(XY + Z)(X + Y)Z$
$[(A + B) + (C \cdot D)] \cdot \bar{B}$	$(A + B + CD)\bar{B}$
$[(A \cdot B) + (A \cdot C)] + (B \cdot C)$	$AB + BC + AC$
$[(X + Y) \cdot Z] + [(X + \bar{Z}) + \bar{Y}]$	$(X + Y)Z + X + \bar{Z} + \bar{Y}$

The AND precedence convention, and the implied AND convention, are similar to the symbolic abbreviations of normal algebra. Table 2-2 gives several instances of how Boolean expressions can be stated more simply. A summary of the postulates and properties of Boolean algebra is given in Table 2-3. Abbreviated forms are used when necessary, to add clarity.

TABLE 2-3

Summary of Boolean algebra postulates and properties.

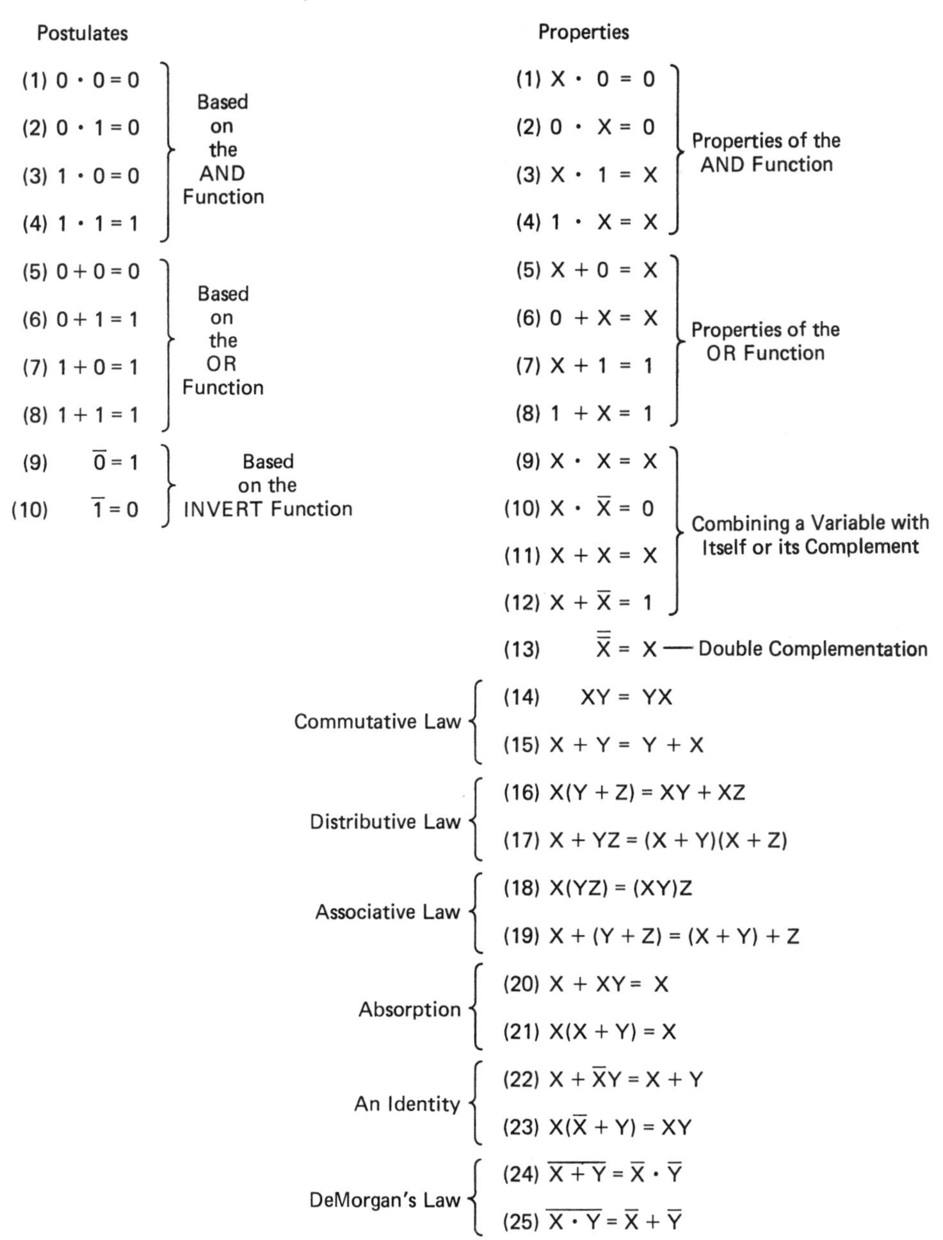

Postulates

Postulate	Basis
(1) $0 \cdot 0 = 0$	Based on the AND Function
(2) $0 \cdot 1 = 0$	
(3) $1 \cdot 0 = 0$	
(4) $1 \cdot 1 = 1$	
(5) $0 + 0 = 0$	Based on the OR Function
(6) $0 + 1 = 1$	
(7) $1 + 0 = 1$	
(8) $1 + 1 = 1$	
(9) $\overline{0} = 1$	Based on the INVERT Function
(10) $\overline{1} = 0$	

Properties

Properties of the AND Function	(1) $X \cdot 0 = 0$
	(2) $0 \cdot X = 0$
	(3) $X \cdot 1 = X$
	(4) $1 \cdot X = X$
Properties of the OR Function	(5) $X + 0 = X$
	(6) $0 + X = X$
	(7) $X + 1 = 1$
	(8) $1 + X = 1$
Combining a Variable with Itself or its Complement	(9) $X \cdot X = X$
	(10) $X \cdot \overline{X} = 0$
	(11) $X + X = X$
	(12) $X + \overline{X} = 1$
Double Complementation	(13) $\overline{\overline{X}} = X$
Commutative Law	(14) $XY = YX$
	(15) $X + Y = Y + X$
Distributive Law	(16) $X(Y + Z) = XY + XZ$
	(17) $X + YZ = (X + Y)(X + Z)$
Associative Law	(18) $X(YZ) = (XY)Z$
	(19) $X + (Y + Z) = (X + Y) + Z$
Absorption	(20) $X + XY = X$
	(21) $X(X + Y) = X$
An Identity	(22) $X + \overline{X}Y = X + Y$
	(23) $X(\overline{X} + Y) = XY$
DeMorgan's Law	(24) $\overline{X + Y} = \overline{X} \cdot \overline{Y}$
	(25) $\overline{X \cdot Y} = \overline{X} + \overline{Y}$

It is valid to ask: How does one know when to stop trying to reduce an expression? In general, the object of algebraic reduction is to minimize the number of variables, or occurrences of a variable, in an expression. This corresponds to minimizing the number of function symbols in an expression, and thereby minimizing the number of switches, or logic gates, in a circuit realization. Occasionally, it is possible to obtain more than one simplified form of an expression, each being equivalent in the number of variables or the number of function symbols used. In such cases, the final result may depend on other constraints in the problem, or may not matter. One form that is frequently used is the ***minimum sum of products*** (MSP) form. With the MSP form, the final result of a simplification is written without parentheses. This is accomplished by expanding the final result, if necessary, using the distributive law. All parentheses, then, are eliminated. Table 2-4 depicts how several reduced Boolean expressions can be rewritten in MSP form. Note that MSP form is not necessarily the simplest possible way in which an expression can be written. "Simple," in this case, means "the minimum number of function symbols" in an expression. In Table 2-4, for instance, the reduced expression $X(\bar{Y} + \bar{Z})$ requires two function operations: an OR and an AND. Its equivalent MSP expression $X\bar{Y} + X\bar{Z}$, however, requires three function operations: an OR and two AND's. Note also that an expression without parentheses is *not necessarily* in MSP form. The expression must be (1) reduced as much as possible, *and* (2) written without parentheses. Expressions that are just written without parentheses but not reduced are in the ***sum of products*** (SP) form. We will see that SP and MSP forms offer many advantages over other forms in relating logical ideas to Boolean equations.

TABLE 2-4

Reduced expressions rewritten in MSP form.

Reduced Expressions	*MSP Form*
$X(\bar{Y} + \bar{Z})$	$X\bar{Y} + X\bar{Z}$
$A(BD + \bar{C}(\bar{B} + D))$	$ABD + A\bar{B}\bar{C} + A\bar{C}D$
$(X + Y)(Z + W)$	$XZ + XW + YZ + YW$

2.4 SIMPLIFICATION OF COMPOUND EXPRESSIONS

In this section, we devote out attention to applications of the properties discussed in Section 2.3. This involves straight algebraic reduction and the translation of logic circuitry into algebraic equations. Specific examples are solved and discussed, and correspond to the types of problems found at the end of the chapter. In all algebraic reductions, specific properties will be referenced to justify each step. The properties are numbered according to the summary of properties in Table 2-3.

2.4.1 Algebraic Reduction

The first five examples in this section involve straight algebraic reduction. All conventions allowed in the abbreviation of Boolean expressions are in force, and occasionally several properties will be invoked in one step of a reduction.

Algebraic equation	*Justification*
$Z = X + \bar{X}Y + \bar{Y}\bar{X}$	Given
$= X + \bar{X}Y + \bar{X}\bar{Y}$	14
$= X + \bar{X}(Y + \bar{Y})$	16
$= X + \bar{X}(1)$	12
$= X + \bar{X}$	3
$= 1$	12

We conclude that the original equation evaluates to logic '1' regardless of the state of the variables. Said another way, the switch network corresponding to the original equation has a transmittance of '1' regardless of how the switches are set. Thus, if the equation was derived from a switch network, we could replace the switch network with a single short circuit without changing its effect.

Algebraic equation	*Justification*
$Z = X(\bar{X} + Y)(\bar{Y} + \bar{X})$	Given
$= (X\bar{X} + XY)(\bar{Y} + X)$	16
$= (0 + XY)(\bar{Y} + \bar{X})$	10
$= XY(\bar{Y} + \bar{X})$	6
$= XY\bar{Y} + XY\bar{X}$	16
$= 0 + YX\bar{X}$	10, 14
$= 0 + 0$	10
$= 0$	Postulate 5

In this case, the original equation evaluates to logic '0' regardless of the state of the variables. If the equation was derived from a switch network, we could replace the switch network with a single open circuit without changing its effect.

Algebraic equation	*Justification*
$Z = W(Y + W)(X + Y + W)$	Given
$= (WY + WW)(X + Y + W)$	16
$= (WY + W)(X + Y + W)$	9
$= (W + WY)(X + Y + W)$	15
$= W(X + Y + W)$	20
$= WX + WY + WW$	16
$= WX + (WY + W)$	9
$= WX + (W + WY)$	15
$= WX + W$	20
$= W + WX$	15
$= W$	20

The original equation, which contained six occurrences of variables and five occurrences of logic function operators, is reduced to a single-variable equation with no logic function operators.

	Algebraic equation	*Justification*
Method 1:	$Z = (\bar{X} + \bar{Y})\bar{W} + \overline{XY}$	Given
	$= (\bar{X} + \bar{Y})\bar{W} + (\bar{X} + \bar{Y})$	25
	$= (\bar{X} + \bar{Y})\bar{W} + (\bar{X} + \bar{Y}) \cdot 1$	3
	$= (\bar{X} + \bar{Y})(\bar{W} + 1)$	16
	$= (\bar{X} + \bar{Y})(1)$	7
	$= \bar{X} + \bar{Y}$	3
Method 2:	$Z = (\bar{X} + \bar{Y})\bar{W} + \overline{XY}$	Given
	$= (\bar{X} + \bar{Y})\bar{W} + (\bar{X} + \bar{Y})$	25
	$= (\bar{X} + \bar{Y}) + (\bar{X} + \bar{Y})\bar{W}$	15
	$= \bar{X} + \bar{Y}$	20
Method 3:	$Z = (\bar{X} + \bar{Y})\bar{W} + \overline{XY}$	Given
	$= (\bar{X} + \bar{Y})\bar{W} + (\bar{X} + \bar{Y})$	25
	$= \bar{X}\bar{W} + \bar{Y}\bar{W} + \bar{X} + \bar{Y}$	16
	$= (\bar{X} + \bar{X}\bar{W}) + (\bar{Y} + \bar{Y}\bar{W})$	15
	$= (\bar{X}) + (\bar{Y})$	20
	$= \bar{X} + \bar{Y}$	MSP form

This example illustrates how any one of several procedures may be used to obtain the reduced form of an equation. Since the same result is obtained regardless of the method used, it does not matter which one you choose. Some methods, however, are shorter than others. It is often a good idea to reduce the same equation by several methods, thereby verifying your results; always arriving at the same answer makes it very likely that the answer is correct.

Algebraic equation	*Justification*
$Z = (\overline{XW} + Y)(\overline{\bar{X} + \bar{W}})$	Given
$= (\overline{XW} + Y)(\bar{\bar{X}}\bar{\bar{W}})$	24
$= (\overline{XW} + Y)(XW)$	13
$= (\overline{XW})(XW) + Y(XW)$	16
$= (XW)(\overline{XW}) + Y(XW)$	14
$= 0 + YXW$	10
$= YXW$	6

It is important to note how Property 10 was used in this example. Rather than applying DeMorgan's law to $\overline{XW}$, we note that XW and $\overline{XW}$ are complements of each other, regardless of what the logic values of X and W actually are. Since each occurrence of a single variable in the table of properties can be replaced by a compound expression, we see that Property 10 can be written equivalently as $(XW) \cdot (\overline{XW}) = 0$. Thus, the procedure in this example is justified. As a more convincing argument, one may reduce the equation by expanding $\overline{XW}$ with DeMorgan's law and, following the proper steps of reduction, arrive at an identical answer. In the next example, we will complement a Boolean equation, and put the result in MSP form.

Algebraic statement	*Justification*
$Z = A\bar{B}(C + D) + B\bar{C}$	Given
$\bar{Z} = \overline{A\bar{B}(C + D) + B\bar{C}}$	Equation complemented
$= [\overline{A\bar{B}(C + D)}] \cdot [\overline{B\bar{C}}]$	24
$= [\overline{A\bar{B}} + (\overline{C + D})] \cdot [\bar{B} + \bar{\bar{C}}]$	25
$= [\bar{A} + \bar{\bar{B}} + \bar{C}\bar{D}] \cdot [\bar{B} + C]$	25, 24, 13
$= [\bar{A} + B + \bar{C}\bar{D}] \cdot [\bar{B} + C]$	13
$= \bar{A}\bar{B} + \bar{A}C + B\bar{B} + BC + \bar{C}\bar{D}\bar{B} + \bar{C}\bar{D}C$	16
$= \bar{A}\bar{B} + \bar{A}C + 0 + BC + \bar{C}\bar{D}\bar{B} + C\bar{C}\bar{D}$	10, 14
$= \bar{A}\bar{B} + \bar{A}C + BC + \bar{B}\bar{C}\bar{D} + 0$	10, 14, 5
$= \bar{A}\bar{B} + \bar{A}C + BC + \bar{B}\bar{C}\bar{D}$	5

Although this equation seems to be in MSP form, it really is not.

$\bar{Z} = \bar{A}\bar{B} + \bar{A}C + BC + \bar{B}\bar{C}\bar{D}$	From above
$= \bar{A}\bar{B} + \bar{A}C \cdot (1) + BC + \bar{B}\bar{C}\bar{D}$	3
$= \bar{A}\bar{B} + \bar{A}C(B + \bar{B}) + BC + \bar{B}\bar{C}\bar{D}$	12*
$= \bar{A}\bar{B} + \bar{A}CB + \bar{A}C\bar{B} + BC + \bar{B}\bar{C}\bar{D}$	16
$= \bar{A}\bar{B}(1 + C) + BC(\bar{A} + 1) + \bar{B}\bar{C}\bar{D}$	16, 14
$= \bar{A}\bar{B} + BC + \bar{B}\bar{C}\bar{D}$	8, 7 (MSP)

* It is not obvious that this is the step that should have been taken. We will see in the next chapter, however, that a special graphic method for reducing Boolean equations makes such tricks unnecessary.

Now, we will complement the result of part (a) to verify that the original equation is obtained.

$$\bar{Z} = \bar{A}\bar{B} + BC + \bar{B}\bar{C}\bar{D} \qquad \text{From part (a)}$$

$$\bar{\bar{Z}} = \overline{\bar{A}\bar{B} + BC + \bar{B}\bar{C}\bar{D}} \qquad \text{Equation complemented}$$

$$= (\overline{\bar{A}\bar{B}}) \cdot (\overline{BC}) \cdot (\overline{\bar{B}\bar{C}\bar{D}}) \qquad 24$$

$$= (A + B)(\bar{B} + \bar{C})(B + C + D) \qquad 25$$

$$= (A + B)[\bar{B}B + \bar{B}C + \bar{B}D + \bar{C}B + \bar{C}C + \bar{C}D] \qquad 16$$

$$= (A + B)[\bar{B}C + \bar{B}D + \bar{C}B + \bar{C}D] \qquad 10, 6$$

$$= A\bar{B}C + A\bar{B}D + A\bar{C}B + A\bar{C}D + B\bar{B}C + B\bar{B}D + B\bar{C}B + B\bar{C}D \qquad 16$$

$$= A\bar{B}(C + D) + A\bar{C}B + A\bar{C}D + B\bar{C} + B\bar{C}D \qquad 16, 10, 6, 14, 9$$

$$= A\bar{B}(C + D) + A\bar{C}D + B\bar{C}(A + B + 1) \qquad 15, 16$$

$$= A\bar{B}(C + D) + A\bar{C}D + B\bar{C} \qquad 7, 3$$

This is almost what is needed. To eliminate the extra term $A\bar{C}D$, we expand it as follows.

$$\bar{\bar{Z}} = A\bar{B}(C + D) + B\bar{C} + A\bar{C}D \cdot (1) \qquad 3$$

$$= A\bar{B}(C + D) + B\bar{C} + A\bar{C}D(B + \bar{B}) \qquad \text{12 [see footnote in part (a)]}$$

$$= \underset{①}{A\bar{B}C} + \underset{②}{A\bar{B}D} + \underset{③}{B\bar{C}} + \underset{④}{A\bar{C}DB} + \underset{⑤}{A\bar{C}D\bar{B}} \qquad 16$$

Factoring products (1) and (5), and products (3) and (4), the following is obtained.

$$\bar{\bar{Z}} = A\bar{B}(C + \bar{C}D) + B\bar{C}(1 + AD) + A\bar{B}D \qquad 16$$

$$= A\bar{B}(C + D) + B\bar{C} + A\bar{B}D \qquad 22, 8$$

$$= A\bar{B}(C + D + D) + B\bar{C} \qquad 16$$

$$= A\bar{B}(C + D) + B\bar{C} \qquad 11$$

Thus, $\bar{\bar{Z}} = Z$, not a surprising result.

2.4.2 Switch Networks

To simplify the drawing of switch circuit schematics, a new symbol for switches is adopted, illustrated in Figure 2-15. Using this new symbol, switch networks can be represented more compactly and are easier to read. This is demonstrated in Figure 2-16, in which a switch circuit is drawn using both the functional symbol and the new, abstract symbol.

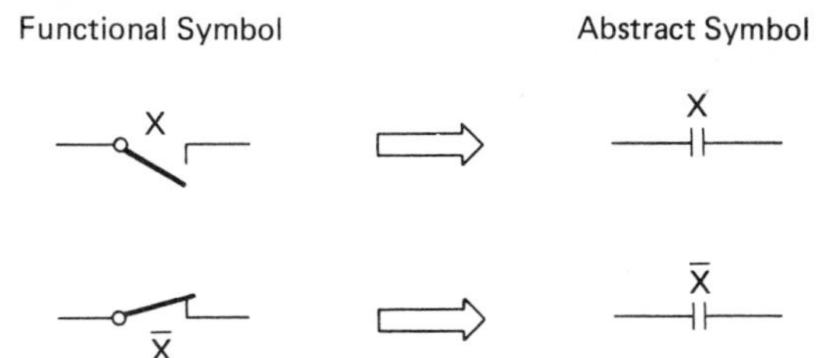

FIGURE 2-15 Simplified graphical symbol for a switch contact.

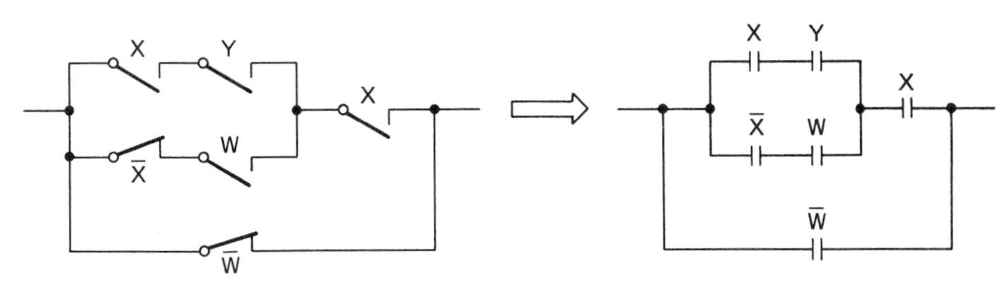

FIGURE 2-16 Rewriting a switch network using the symbol '⊣⊢' for switch contacts.

In the next two examples, we will determine the MSP form Boolean equation for the given switch networks. Remember that MSP form requires that the equation be reduced.

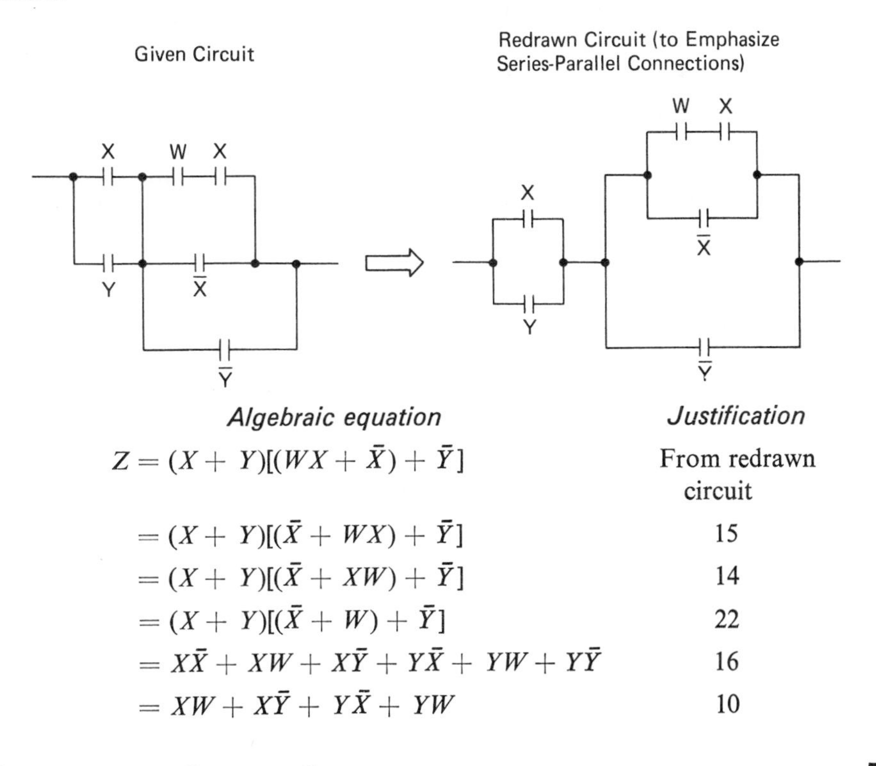

Algebraic equation	*Justification*
$Z = (X + Y)[(WX + \bar{X}) + \bar{Y}]$	From redrawn circuit
$= (X + Y)[(\bar{X} + WX) + \bar{Y}]$	15
$= (X + Y)[(\bar{X} + XW) + \bar{Y}]$	14
$= (X + Y)[(\bar{X} + W) + \bar{Y}]$	22
$= X\bar{X} + XW + X\bar{Y} + Y\bar{X} + YW + Y\bar{Y}$	16
$= XW + X\bar{Y} + Y\bar{X} + YW$	10

Given circuit:

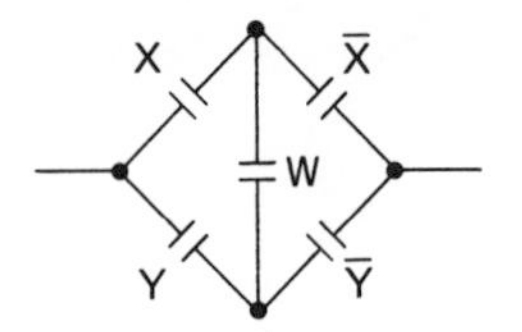

Since this network is not in series-parallel form, we must arrive at the circuit's Boolean equation by considering every possible path that would result in a transmittance of '1'. First, label the switch locations:

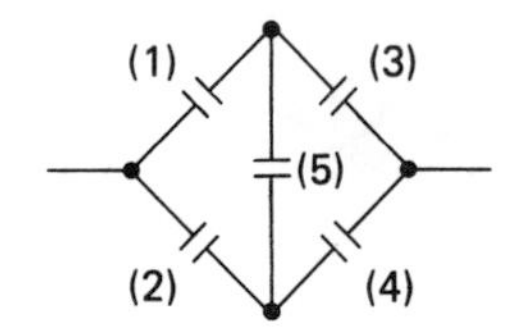

The possible paths are

(a) 1, 3 (c) 2, 4
(b) 1, 5, 4 (d) 2, 5, 3

The corresponding Boolean expressions for the possible paths are

(a) $X\bar{X}$ (c) $Y\bar{Y}$
(b) $XW\bar{Y}$ (d) $YW\bar{X}$

Since any one, or any simultaneous combination of these paths will result in a network transmittance of '1', we obtain the Boolean equation for the network by combining the possible paths with an OR function.

$$Z = X\bar{X} + XW\bar{Y} + Y\bar{Y} + YW\bar{X}$$

Reducing the equation results in

$$Z = W(X\bar{Y} + Y\bar{X})$$

Thus, an equivalent network can be drawn in series-parallel form.

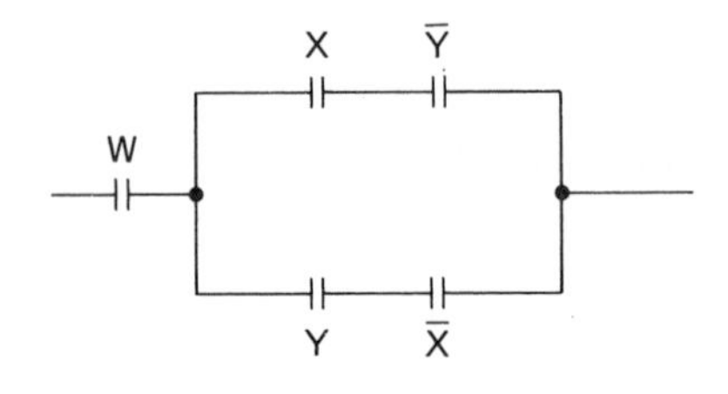

In MSP form, we have

$$Z = WX\bar{Y} + WY\bar{X}$$

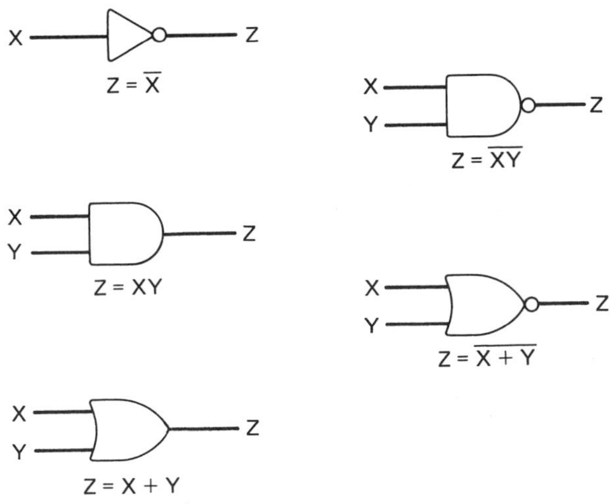

FIGURE 2-17 *Logic gate symbols and their corresponding algebraic representations.*

2.4.3 Logic Gate Networks

The techniques of Boolean algebra may be applied to electronic logic as easily as they have been applied to switch networks. It is really easier when dealing with electronic logic because of a clear correspondence between logic gate symbols and their algebraic function symbols. Figure 2-17 relates the basic logic gate symbols to their corresponding algebraic representation. Since any gate input can be driven by another gate's output, the ability to form gate networks becomes evident. Generally, a logic gate network is drawn so that the flow of information is from left to right. Thus, inputs to a logic gate network will be found on the left of a schematic drawing, and outputs will be found on the right. This convention is followed whenever possible to facilitate "reading" the schematic. Furthermore, the left-to-right rule makes it easier for us to find the algebraic equation of the total network. Consider the next example, in which we will find a Boolean equation that corresponds to a logic gate network.

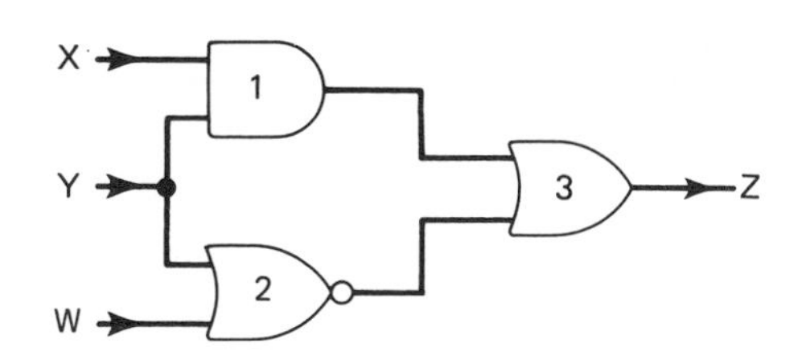

We wish to obtain an equation that relates the output variable, Z, to the input variables, X, Y, and W. Note that Z is the output of OR gate (3), and that to arrive at Z, two quantities must be combined with an OR function symbol: $Z = Q_1 + Q_2$. The unknown quantities, however, are themselves the outputs of logic gates and must be determined.

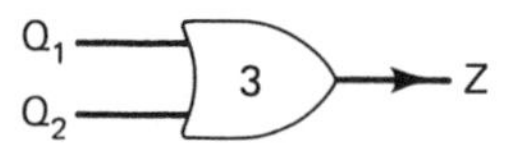

Redrawing the original network and labeling the intermediate lines Q_1 and Q_2, we have

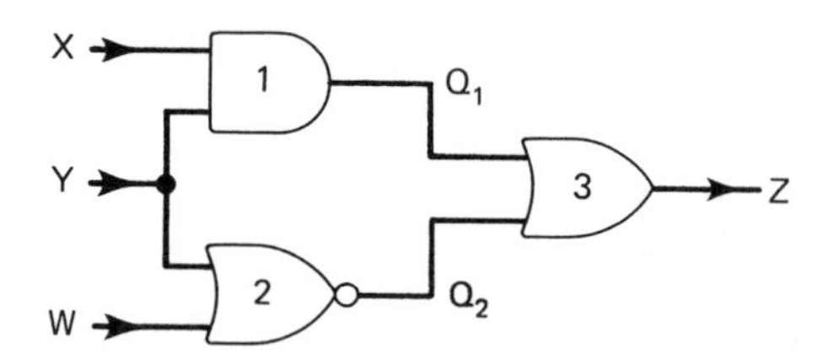

From this, we see that $Q_1 = XY$ and $Q_2 = \overline{Y + W}$. Having deduced the intermediate variables Q_1 and Q_2, a Boolean equation for this circuit can be written:

$$Z = XY + \overline{Y + W}$$

Expanding the equation with DeMorgan's law, we see that it is not possible to make any algebraic reduction:

$$Z = XY + \bar{Y}\bar{W}$$

Since no parentheses are shown, the equation is also in MSP form.

Here, we will determine whether the following logic gate network can be simplified so that it may be realized with fewer logic gates. If it can be simplified in this way, we will find the minimal logic circuit.

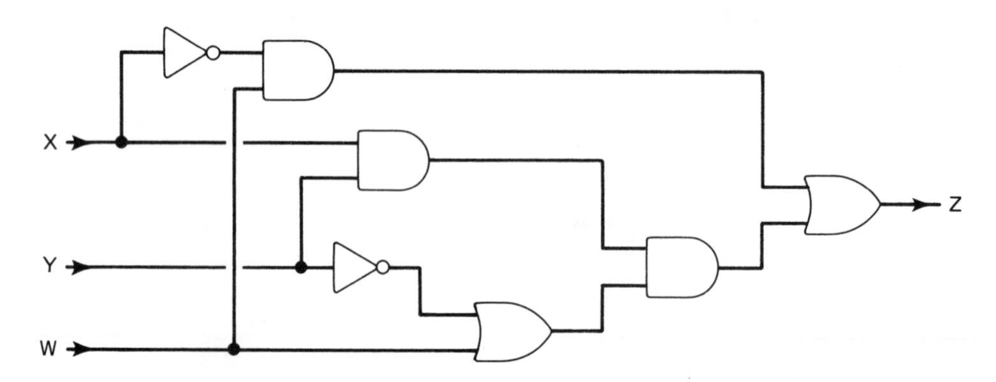

First, we must find a Boolean equation that describes the network. This requires that some intermediate variables be determined. Adding additional symbols to the schematic to identify the intermediate variables, we have

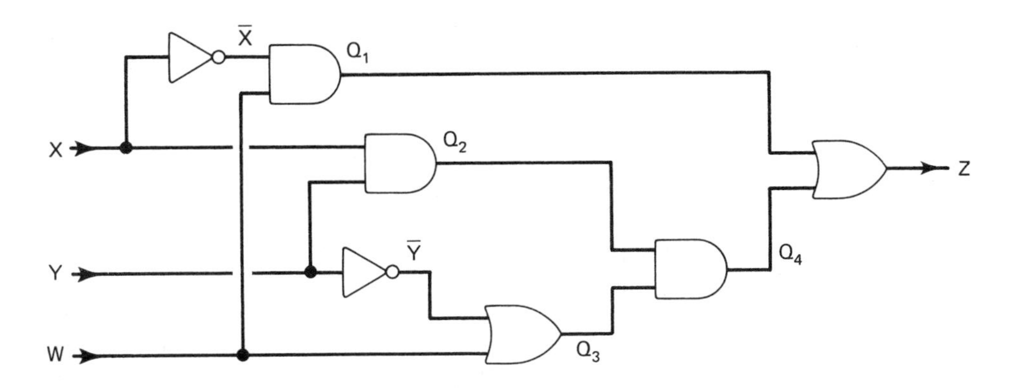

Working from left to right (input to output),

$$Q_1 = \bar{X}W$$
$$Q_2 = XY$$
$$Q_3 = \bar{Y} + W$$
$$Q_4 = Q_2 \cdot Q_3 = XY(\bar{Y} + W)$$
$$Z = Q_1 + Q_4 = \bar{X}W + XY(\bar{Y} + W)$$

The last equation expresses the output variable, Z, in terms of the input variables, X, Y, and W. This equation, however, is not minimal. By applying the properties of Boolean algebra, a reduced equation is obtained.

$$Z = W(\bar{X} + Y) = W \cdot (\bar{X} + Y)$$

Reconstructing the logic circuit from the reduced equation requires a two-input OR gate, a two-input AND gate, and an INVERT gate. This is determined by noting that the equation contains the function symbols '·', '+', and '¯', each occurring once. To arrive at the reduced circuit, X is complemented to obtain $\bar{X}$, $\bar{X}$ and Y are combined with an OR gate, then the result and W are combined with an AND gate to produce Z. The reduced function requires three logic gates instead of seven, a considerable savings.

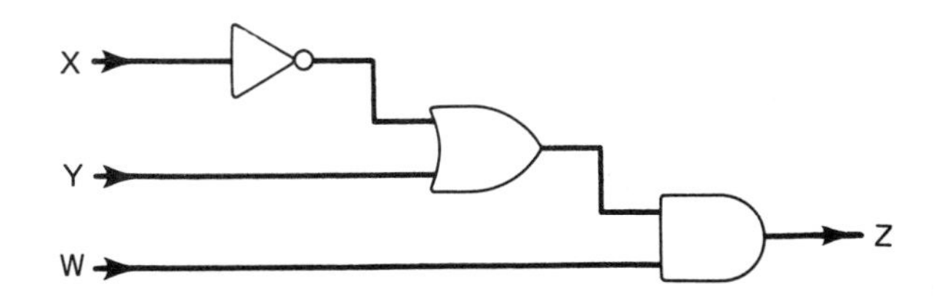

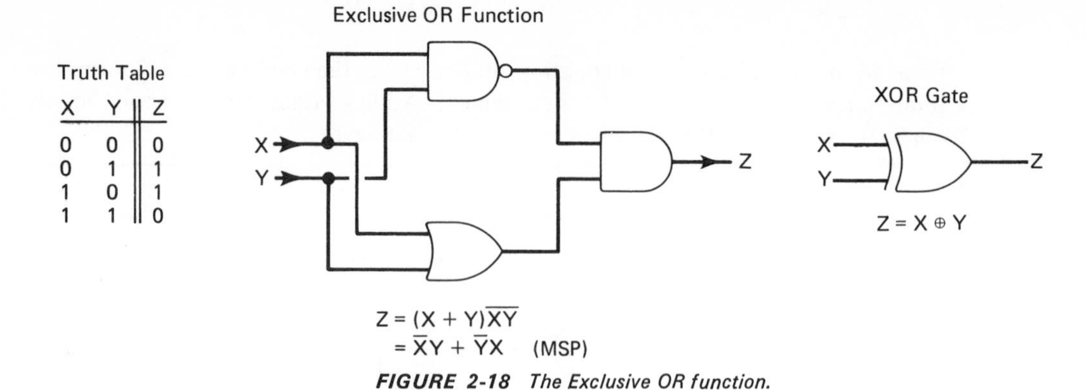

X	Y	Z
0	0	0
0	1	1
1	0	1
1	1	0

FIGURE 2-18 *The Exclusive OR function.*

We saw in Chapter 1 how an Exclusive OR function is different from a simple OR function. Recall that the XOR function produces a logic '1' whenever either one of two inputs is '1', *but not* when both are '1'. This last feature is what makes it different from the simple OR (also known as the Inclusive OR). Although the XOR function can be realized by an AND, OR, INVERT network, it is used so frequently that a special gate symbol has been adopted for it. Similarly, in Boolean algebra, a special function symbol is used to express an XOR function. Figure 2-18 relates the XOR truth table, its fundamental construction, its special gate symbol, and the corresponding algebraic equations.

The XOR function possesses several useful algebraic properties. They are summarized in Table 2-5.

TABLE 2-5

Algebraic properties of the XOR function.

$$X \oplus 0 = X$$
$$X \oplus 1 = \bar{X}$$
$$X \oplus Y = Y \oplus X$$
$$X \oplus X = 0$$
$$X \oplus \bar{X} = 1$$
$$(X \oplus Y) \oplus Z = X \oplus (Y \oplus Z)$$

Studying the first two properties in Table 2-5, we note that it is very easy, using an XOR gate, to cause a logic variable to become complemented, or to allow it passage through the gate unchanged. This is done by using one XOR input as a control input and the other as the logic variable input. Figure 2-19 illustrates the technique.

FIGURE 2-19 *Use of the XOR function to cause uncomplemented or complemented transmission of a logic variable.*

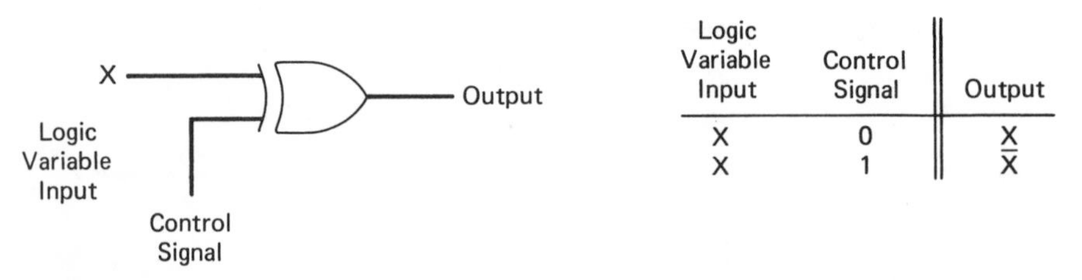

Logic Variable Input	Control Signal	Output
X	0	X
X	1	$\bar{X}$

2.5 RELATING TRUTH TABLES TO BOOLEAN EQUATIONS

Initially, we described logical functions with truth tables, showing for each possible combination of input conditions a specific output condition. Specifying an input/output relationship in the form of a truth table is very explicit in that every allowable combination of inputs and output is clearly indicated. More recently, we have seen how a Boolean equation can also describe a logic function. Using Boolean algebra, binary variables are represented by alphabetic characters, and the elementary functions of AND, OR, and INVERT are represented by special function symbols. The Boolean algebra postulates and properties specify the rules by which these variable and function symbols may be combined and manipulated. A particular logic function, then, calls for a certain arrangement of such symbols in an equation. Since a logic function can be expressed either with a truth table or an equation, it is justifiable to ask: Is there any advantage of one form over the other?

One evident advantage of Boolean equations over truth tables results from the compact nature of an equation; much less writing is necessary to express a given logic function with a Boolean equation than with a truth table. Furthermore, there is no direct relationship between a truth table and a logic gate network, whereas there is in a Boolean equation. The kinds of problems that are encountered in digital electronics often require fluency in relating truth tables to equations. For this reason, we address the following questions:

1. How can a related truth table be obtained from a given Boolean equation?
2. How can a related Boolean equation be obtained from a given truth table?

2.5.1 Obtaining a Truth Table from a Boolean Equation

A Boolean equation is one of the most general ways that a functional relationship, relating to logic, can be expressed. Because the value of a Boolean variable can be either '0' *or* '1', functional relations can be compressed into more brief forms than would be required if only Boolean constants were used. Having available just a Boolean equation, it may sometimes be necessary to obtain a specific output value for given, specific input values. A specific output value is found in a very simple way, calling for two basic steps:

1. Replace every instance of an input variable in an equation with a constant. The constants that are used are the given, specific input values for which the output is sought.
2. Using the Boolean postulates, reduce the equation of constants, resulting from step 1, to a single value. This single value is the corresponding output.

For example, if we were given the equation

$$Z = X + Y\bar{W}$$

and we needed to know the value of Z when $X = 0$, $Y = 1$, and $W = 0$, we first replace every instance of the variables with their assigned constant values, and then reduce the result.

$$\begin{aligned} Z &= X + Y\bar{W} \\ &= 0 + (1 \cdot \bar{0}) \\ &= 0 + (1 \cdot 1) \\ &= 1 \cdot 1 \\ &= 1 \end{aligned}$$

Thus, for $X = 0$, $Y = 1$, and $W = 0$, we see that $Z = 1$.

To obtain a truth table, the same procedure is used many times, once for each allowable input combination. In each case of a new input combination, a specific output will be found. By repeating the substitution-reduction procedure for all input combinations, a truth table will be compiled. As an example of this, consider the equation

$$Z = X(Y + \bar{W})$$

Entry Number	X	Y	W	Z
(1)	0	0	0	
(2)	0	0	1	
(3)	0	1	0	
(4)	0	1	1	
(5)	1	0	0	
(6)	1	0	1	
(7)	1	1	0	
(8)	1	1	1	

FIGURE 2-20 *The input sequence that will be applied to the equation $Z = X(Y + \bar{W})$.*

We first note that there are three input variables, and therefore that a corresponding truth table must contain eight entries to account for all possible inputs. This is illustrated is Figure 2-20. To determine Z for entry (1), substitute the input constants $X = 0$, $Y = 0$, and $W = 0$ into the given equation. This results in

$$\begin{aligned} Z &= X(Y + \bar{W}) \\ &= 0 \cdot (0 + \bar{0}) \\ &= 0 \end{aligned}$$

Thus, for entry (1), $Z = 0$. Proceeding in a similar way for all of the remaining entries, we obtain the completed truth table, as shown in Figure 2-21.

An alternative procedure that may be used to determine the truth table corresponding to a Boolean equation involves simply reasoning when the output should be '1'. Using the equation $Z = X(Y + \bar{W})$ again, we note that Z *must* be '0' as long as X is '0', since $0 \cdot (\text{anything}) = 0$. Thus, the first four entries could not possibly be '1'.

Entry Number	X	Y	W	Z
(1)	0	0	0	0
(2)	0	0	1	0
(3)	0	1	0	0
(4)	0	1	1	0
(5)	1	0	0	1
(6)	1	0	1	0
(7)	1	1	0	1
(8)	1	1	1	1

FIGURE 2-21 *The truth table that is obtained by applying the input sequence of Figure 2-19 to the equation $Z = X(Y + \bar{W})$.*

Next, we note that when X is '1', Z is '1' only if Y is '1' *or* W is '0'. This is true for entries (5), (7), and (8). Thus, $Z = 1$ for entries (5), (7), and (8), and $Z = 0$ for all other entries. The recommended way of obtaining a truth table from an equation uses the reasoning approach, and calls for two steps.

1. Place the Boolean equation in MSP form.
2. For each product in the MSP equation, reason which truth table entries should be '1'.

Recall that *each product* in an MSP equation contributes a logic '1' for certain input combinations. Bringing all the products together with an OR operation "assembles" the desired function from individual parts. Using the equation $Z = X(Y + \bar{W})$ once more, we first put the equation in MSP form:

$$Z = XY + X\bar{W}$$

For the first product, we reason that $Z = 1$ whenever X *and* Y are '1'. Referring to Figure 2-20, we find this to be true for entries (7) and (8). Remember that since W is not involved in this product, its value cannot affect where we locate the logic '1' values for Z. Said another way,

$$\begin{aligned} XY &= XY(W + \bar{W}) \\ &= XYW + XY\bar{W} \end{aligned}$$

meaning that W can be either '0' or '1' for the XY product to contribute a '1' to the function. For the second product, $X\bar{W}$, we reason that $Z = 1$ whenever X is '1' *and* W is '0'. Again referring to Figure 2-20, we find this to be true for entries (5) and (7). This time, Y is not involved in the product and can be ignored. Note that overlap between the two products occurs in entry (7). That is, entry (7) receives a '1' from both products. This often happens with MSP equations and in no way causes entry (7) to be different from entries (5) and (8) where we also have $Z = 1$.

As a second example, consider the equation

$$Z = \underset{(1)}{X} + \underset{(2)}{WY} + \underset{(3)}{\bar{W}\bar{Y}} \qquad \text{Product terms identified}$$

X	Y	W	Z	Product That Contributes to This Entry
0	0	0	1	(3)
0	0	1	0	
0	1	0	0	
0	1	1	1	(2)
1	0	0	1	(1) (3)
1	0	1	1	(1)
1	1	0	1	(1)
1	1	1	1	(1) (2)

FIGURE 2-22 *Truth table derived from the equation* $Z = X + WY + \overline{WY}$

Since the equation is already in MSP form, we need only reason where $Z = 1$ for each product. First, note that $Z = 1$ whenever $X = 1$, regardless of Y or W. Second, $Z = 1$ if W *and* Y are '1'. Third, $Z = 1$ if W *and* Y are '0'. Figure 2-22 depicts the truth table corresponding to $Z = X + WY + \bar{W}\bar{Y}$, and shows which products contribute '1's to the overall function.

2.5.2 Obtaining a Boolean Equation from a Truth Table

Practical problems in digital electronics often require that a verbally described situation be transformed into a Boolean equation. A frequent and useful intermediary in this process is a truth table. Thus, the verbal description is translated into a truth table, which is then expressed as a Boolean equation. Once the problem is stated as an equation, all the advantages of Boolean algebra are at our disposal to manipulate and reduce the equation. A minimal logic gate structure may then be found, if indeed that is what is needed to solve the problem. Section 2.5.1 addressed the problem of obtaining a truth table from an equation; this section addresses the converse problem, that of obtaining an equation from a truth table.

Consider the truth table of Figure 2-23. Note that three entries, (1), (3), and (6), contribute a logic '1' to the function. All other entries contribute a logic '0'. To obtain an equation, we need only write a product term for each entry that contributes a logic '1', and then assemble the function by connecting the products with a logic OR. For entry (1), $Z = 1$ if X *and* Y *and* W are '0'. The corresponding product term is, therefore, $\bar{X}\bar{Y}\bar{W}$. For entry (3), $Z = 1$ if W *and* X are '0' *and* Y is '1'. The corresponding product term is $\bar{W}\bar{X}Y$. For entry (6), $Z = 1$ if W *and* X are '1' and Y is

Entry Number	X	Y	W	Z
(1)	0	0	0	1
(2)	0	0	1	0
(3)	0	1	0	1
(4)	0	1	1	0
(5)	1	0	0	0
(6)	1	0	1	1
(7)	1	1	0	0
(8)	1	1	1	0

FIGURE 2-23 *Truth table to be analyzed.*

'0'. The corresponding product term is $WX\bar{Y}$. In general, if an entry contributes a logic '1' to the function, a product term is called for. That product is made by

1. Combining with an AND all the input variables.
2. Selecting for each variable in the product an overbar, or no overbar, so that when the *input values* of that entry are substituted, the product reduces to a value of logic '1'.

Finally, the products are brought together with the OR operation. For the truth table of Figure 2-23, we have

$$Z = \bar{X}\bar{Y}\bar{W} + \bar{X}Y\bar{W} + X\bar{Y}W$$

This equation is in the sum-of-products (SP) form. To arrive at the MSP form, the equation is reduced.

$$\begin{aligned} Z &= \bar{X}\bar{Y}\bar{W} + \bar{X}Y\bar{W} + X\bar{Y}W \\ &= \bar{X}\bar{W}(\bar{Y} + Y) + X\bar{Y}W \\ &= \bar{X}\bar{W} + X\bar{Y}W \qquad \text{(MSP)} \end{aligned}$$

In many cases, the sum-of-products equation that is obtained *will not* be minimal. Thus, the properties of Boolean algebra may have to be applied to place the equation in MSP form.

As a second example, consider the truth table of Figure 2-24. Four entries result in $Z = 1$. Thus, we have

$\bar{X}\bar{Y}\bar{W}$ for entry (1)

$\bar{X}YW$ for entry (4)

$X\bar{Y}\bar{W}$ for entry (5)

$X\bar{Y}W$ for entry (6)

The SP equation is

$$Z = \bar{X}\bar{Y}\bar{W} + \bar{X}YW + X\bar{Y}\bar{W} + X\bar{Y}W$$

Reducing this equation, we obtain

$$Z = \bar{Y}\bar{W} + \bar{Y}X + \bar{X}YW$$

FIGURE 2-24 *Truth table to be analyzed.*

Entry Number	X	Y	W	Z
(1)	0	0	0	1
(2)	0	0	1	0
(3)	0	1	0	0
(4)	0	1	1	1
(5)	1	0	0	1
(6)	1	0	1	1
(7)	1	1	0	0
(8)	1	1	1	0

SUMMARY

Boolean algebra provides a means by which logic circuitry may be expressed symbolically, manipulated, and reduced. A binary digital signal that does not change in time is considered a constant signal, and is associated with an unchanging numerical value of '0' or '1'. A binary digital signal that does change in time is considered a variable signal, and is represented by an alphabetic letter such as 'X', 'Y', or 'Q'. Such alphabetic letters are referred to as Boolean variables and may assume numerical values of either '0' or '1'. Boolean variables and constants are operated on by Boolean functions. Ultimately, the operation of a function must be specified by a truth table. However, special function symbols have been adopted for the AND, OR, and INVERT functions, owing to their usefulness in expressing logic relationships. These special function symbols are not free-standing definitions as truth tables are, but simply offer an abbreviated means of engaging particular functions in a logic relationship. A simple Boolean expression specifies a distinct logic relationship between two variables, and utilizes a single function symbol. A compound Boolean expression specifies a logic relationship between multiple variables, and utilizes at least two function symbols.

A system of Boolean algebra consists of a set of initial assumptions, known as postulates, and a series of rules that state permissible acts of rearrangement in a Boolean expression. These rules are known as "properties," or "theorems," and are founded exclusively on the postulates. A valid system of Boolean algebra must be entirely self-consistent in that it should not be possible to demonstrate a contradiction between any two properties.

A switch network can be considered generally as an unspecified interconnection of switches possessing an input line and an output line. The logic value associated with the overall network relates to the state of conduction that exists between the input line and output line. A network capable of sustaining current flow between input and output has a transmittance of '1', while a network incapable of allowing current flow between input and output has a transmittance of '0'. Two switch networks that have equal values of transmittance under all possible combinations of similar switch settings are equivalent networks. Switch networks have been used to verify intuitively the equivalence of Boolean expressions.

Boolean expressions are understood with greater ease by adopting the AND Precedence and Implied AND conventions. AND Precedence permits the elimination of parentheses enclosing AND operations; any ambiguity that may result from the elimination of parentheses is resolved by performing all AND operations first, before performing OR operations. The Implied AND convention makes it unnecessary to connect AND arguments with the symbol '·'; eliminating the function symbol and requiring that all variables be represented by a single letter simplifies the writing of Boolean expressions.

The minimum sum of products (MSP) form of an expression conveys a minimized algebraic statement without parentheses. This standardizes the writing of reduced expressions, and in addition eases the act of reasoning the equation. "Reasoning the equation" relates to obtaining a corresponding truth table, and understanding the interdependencies between various products in the MSP expression.

NEW TERMS

Minimization
Boolean Constant
Boolean Variable
Boolean Function
Expression
Simple Expression
Compound Expression
Equation
Argument
Functional Result
Property
Simplification
Postulate
Proof
Transmittance
Commutative Law
Associative Law
Distributive Law
DeMorgan's Law
AND Precedence
Implied AND
Sum of Products (SP)
Minimum Sum of Products (MSP)
Series-Parallel Network
Dual Postulate

PROBLEMS

2-1 Algebraically reduce each of the following Boolean equations. Justify each step in your reduction with one or more of the properties in Table 2-3, and show the final result in MSP form.

(a) $Z = XW + (XY + \bar{Y})\bar{X}$

(b) $Z = W + XY + XW + \bar{W} + X$

(c) $Z = XW(XY + \bar{X}\bar{Y} + \bar{X}Y + X\bar{Y})$

(d) $Z = (X\bar{W})(W\bar{X})$

(e) $Z = (\bar{X} + W)(\bar{W} + X)$

(f) $Z = (W + Y)(W + Q) + W$

(g) $Z = X\bar{Y} + \bar{X}\bar{Y}Q$

(h) $Z = \bar{\bar{X}}(\overline{W + X})\bar{W} + X$

(i) $Z = Y + WX + WY + \bar{X}$

(j) $Z = \bar{X}(Y + \bar{W})[(\overline{\bar{W} + X}) + X]$

(k) $Z = A(\bar{A} + B)(\bar{A} + \bar{B} + C)(\bar{A} + \bar{B} + \bar{C} + D) \ldots$

(l) $Z = \overline{A + B + C + D \ldots}$

(m) $Z = \overline{\bar{A}B\bar{C}D\bar{E}F \ldots}$

(n) $Z = \overline{\overline{\overline{\bar{A}B}C}D}$

(o) $Z = \bar{W}X(W + \bar{X} + \bar{W}X)(W + \bar{X})$

(p) $Z = \bar{Y} + XYQW + Q\bar{W}$

2-2 Find $\bar{Z}$ for each of the following equations. Verify that your result is correct by complementing $\bar{Z}$ to obtain $\bar{\bar{Z}}$. Although $\bar{\bar{Z}}$ is equivalent to Z, some rearranging of $\bar{\bar{Z}}$ may be necessary to make it look identical, term for term, to Z. Note that the original equation for Z may not be in reduced form.

(a) $Z = X\bar{Y} + \bar{X}Y$

(b) $Z = XY + X\bar{Y} + \bar{X}Y$

(c) $Z = (X + \bar{Y})(\bar{Q} + W)(X + W)$

(d) $Z = X\bar{Y}W + \bar{X}Y\bar{W} + QXW$

2-3 Determine which of the following expressions are in MSP form. Reduce any expressions that are not minimal and write the result in MSP form.

(a) $X(\bar{Z} + XYZ)$

(b) $XY\bar{Z} + X\bar{Y}Z + \bar{X}YZ$

(c) $QW + \bar{W}XQ + \bar{X}\bar{Q}$

(d) $A + \bar{A}B + \bar{A}\bar{B}C + \bar{A}\bar{B}\bar{C}D$

(e) $XZ + Y\bar{Z} + \overline{YW}$

(f) $XY + \bar{Y}X + X\bar{Y} + \bar{Y}\bar{X}$

2-4 Determine an algebraic expression that describes each of the switch networks of Figure P2-4. If the expression can be algebraically reduced, do so and draw the new switch network that corresponds to the reduced equation.

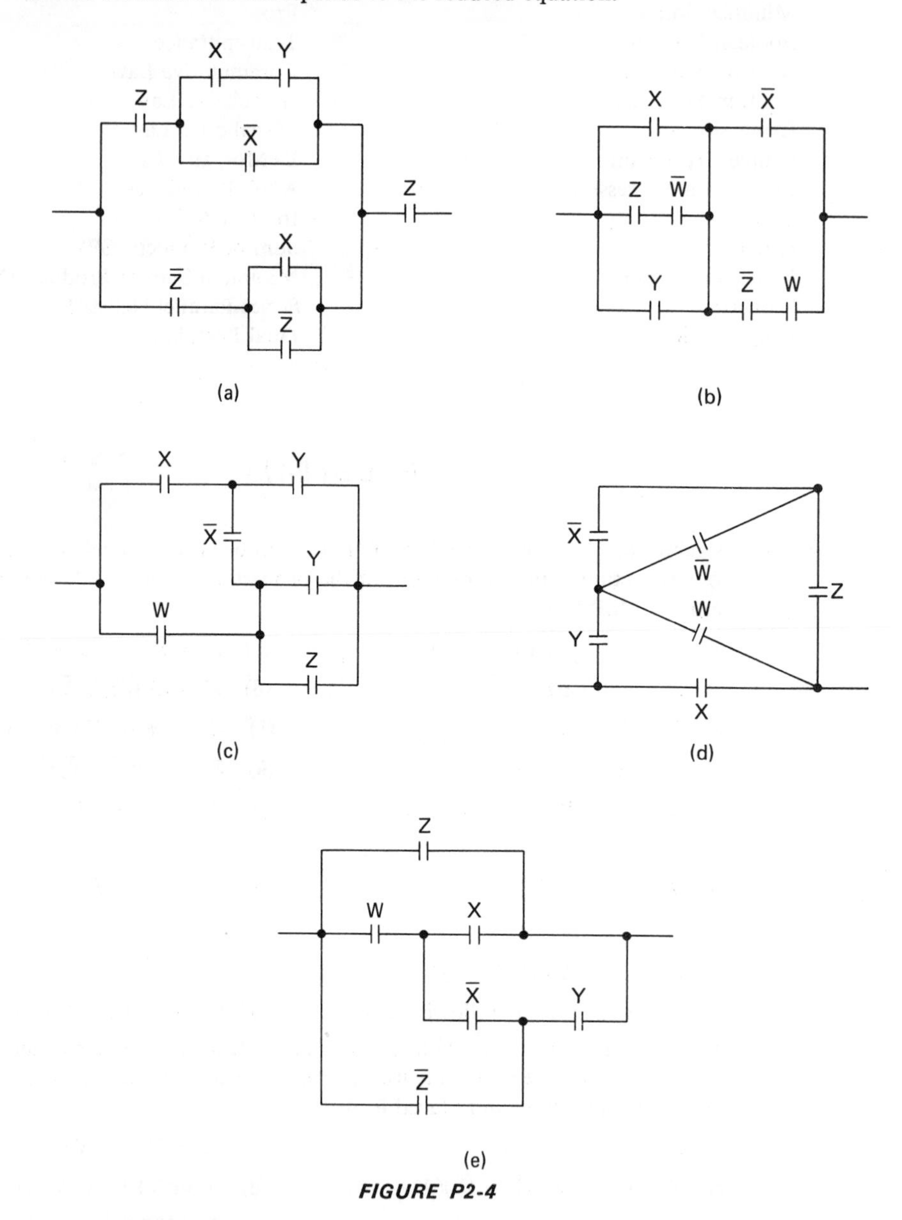

FIGURE P2-4

2-5 **(a)** Given a non-series-parallel switch network, is it ever possible for the current flow in a closed switch to reverse direction when other switches change state? If not, explain why. If so, give an example.

(b) Answer the same question as in part (a), except for a series-parallel switch network.

2-6 Is it possible to have a non-series-parallel logic gate network?

2-7 Determine an algebraic expression that describes each of the logic gate networks of Figure P2-7. If the expression can be algebraically reduced, do so and draw the new gate network that corresponds to the reduced equation. Remember that MSP form *does not* guarantee a minimal gate structure.

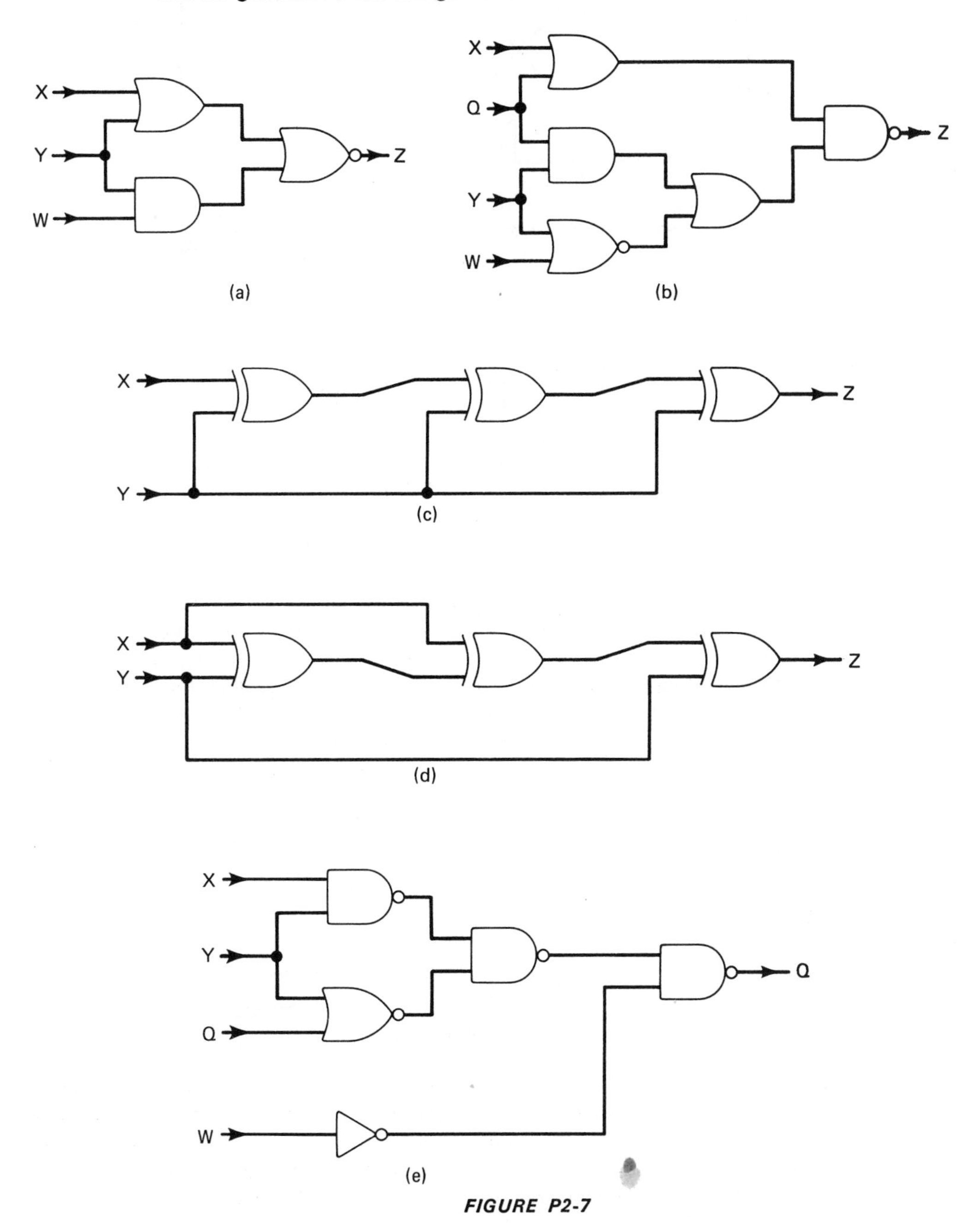

FIGURE P2-7

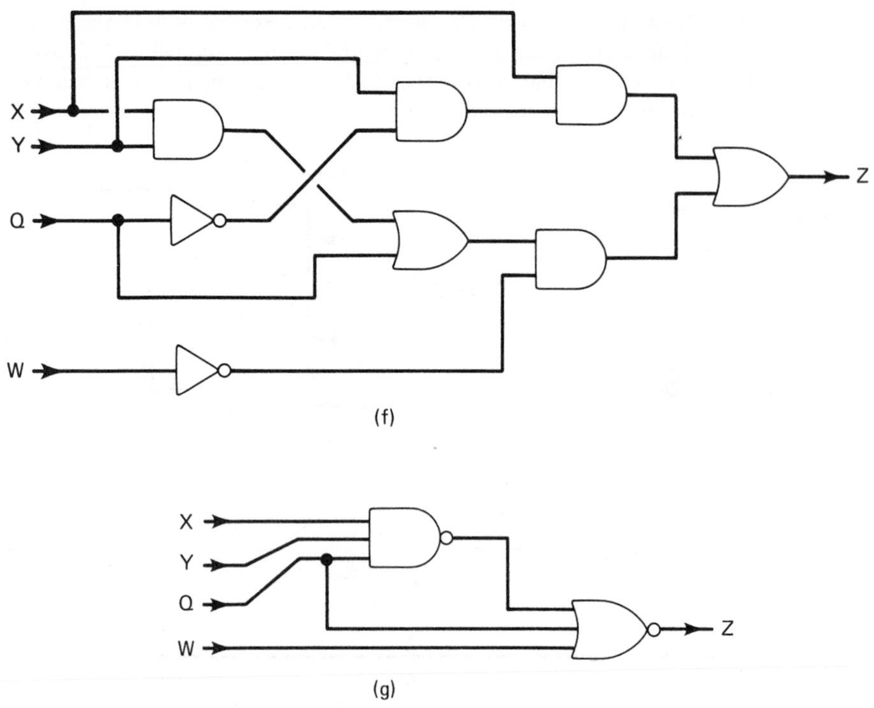

FIGURE P2-7 Continued

2-8 According to DeMorgan's law, the Boolean equation $Z = \overline{AB}$ is equivalent to $Z = \bar{A} + \bar{B}$. Using logic gate symbols, this equivalence can also be expressed as shown in Figure P2-8. Use DeMorgan's law to expand each of the following equations, thereby obtaining an equivalent form. Then, write the logic gate symbols for the original from and the equivalent form. Can you generalize a rule about transforming logic gate symbols to equivalent forms using DeMorgan's law?

(a) $Z = \overline{\bar{A}B}$ **(b)** $Z = \overline{\bar{A} + B}$

(c) $Z = \overline{A\bar{B}}$ **(d)** $Z = \overline{\bar{A} + \bar{B}}$

(e) $Z = \overline{A + B + C}$ **(f)** $Z = \overline{\bar{A}B\bar{C}}$

(g) $Z = \overline{A \oplus B}$ **(h)** $Z = \overline{\bar{A} \odot \bar{B}}$

FIGURE P2-8

2-9 It is possible to develop a form of Boolean algebra based only on the NAND function. Usuing the postulates of Table P2-9, and no others, prove properties (1) through (5) given in Table P2-9. Now, using your knowledge of the AND and OR functions, and any associated properties, prove properties (6) through (9) of Table P2-9. The symbol '∘' is used to represent algebraically the NAND function.

TABLE P2-9

NAND Postulates and properties.

NAND Postulates	*NAND Properties*
(1) $0 \circ 0 = 1$	(1) $X \circ Y = Y \circ X$
(2) $0 \circ 1 = 1$	(2) $X \circ X = \bar{X}$
(3) $1 \circ 0 = 1$	(3) $X \circ \bar{X} = 1$
(4) $1 \circ 1 = 0$	(4) $X \circ 1 = \bar{X}$
(5) $\bar{0} = 1$	(5) $X \circ 0 = 1$
(6) $\bar{1} = 0$	(6) $\overline{X \circ Y} = X \cdot Y$
	(7) $\bar{X} \circ \bar{Y} = X + Y$
	(8) $(X \circ Y) \circ \bar{Z} = XY + Z$
	(9) $(X \circ Y) \circ (Z \circ W) = XY + ZW$

2-10 The EQUIVALENCE function is useful whenever it is necessary to determine if two binary variables are equal. The truth table and logic gate symbol corresponding to this function are shown in Table P2-10. Using the symbol '$\odot$' to represent algebraically the EQUIVALENCE function, verify the properties listed in Table P2-10.

TABLE P2-10

The EQUIVALENCE function and properties.

EQUIVALENCE Function

X	Y	Z
0	0	1
0	1	0
1	0	0
1	1	1

EQUIVALENCE Gate

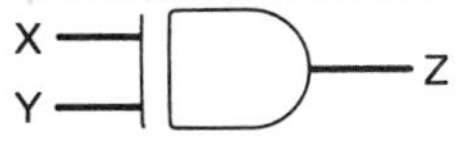

EQUIVALENCE Properties

(1) $X \odot 0 = \overline{X}$
(2) $X \odot 1 = X$
(3) $X \odot X = 1$
(4) $X \odot \overline{X} = 0$
(5) $X \odot Y = Y \odot X$
(6) $\overline{X \odot Y} = X \oplus Y$
(7) $\overline{X \oplus Y} = X \odot Y$

2-11 Simplify the networks in Figure P2-11. (*Hint:* Apply the properties of Table 2-5.)

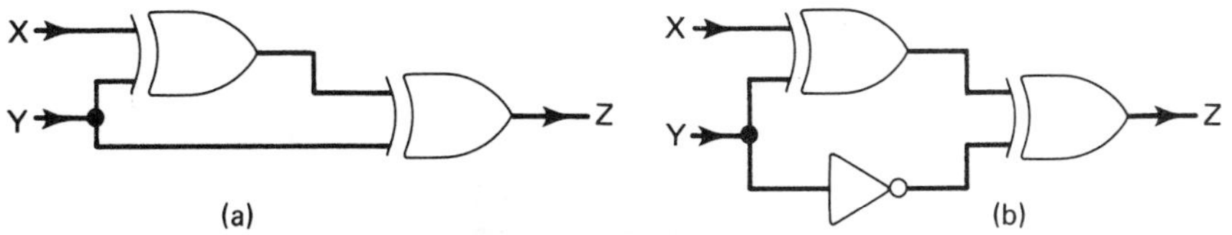

FIGURE P2-11

2-12 How will the logic circuit of Figure P2-12 behave if

(a) $X = 0$? **(b)** $X = 1$?

Assume that the propagation delay of the XOR gate is 20 ns.

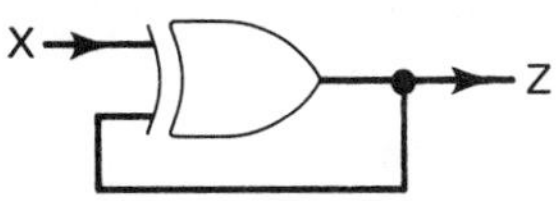

FIGURE P2-12

2-13 Draw a minimal gate network that implements each of the following equations. Any type of logic gate may be used.

(a) $Z = \bar{Y} + \bar{X}W$ (b) $Z = \bar{W} + XYW$

(c) $Z = \bar{X}Y + W\bar{X}$ (d) $Z = \bar{Y}\bar{X}W + \bar{X}Y\bar{W} + X\bar{Y}\bar{W} + XYW$

(e) $Z = \overline{XY} + \overline{YW} + \overline{XW} + XYW$ (f) $Z = X\bar{W}(X + Y)(W + \bar{X})$

2-14 Implement each of the following equations using a minimum number of two-input NAND gates. No other type of gate may be used.

(a) $Z = AB + CD$ (b) $Z = \bar{A} + B$

(c) $Z = (A + BC)A\bar{D}$ (d) $Z = X \oplus Y$

(e) $Z = \overline{XZ}$ (How would this circuit behave?)

2-15 Produce a corresponding, minimal logic gate network for each switch network shown in Figure P2-15. Use only AND, OR, and INVERT gates.

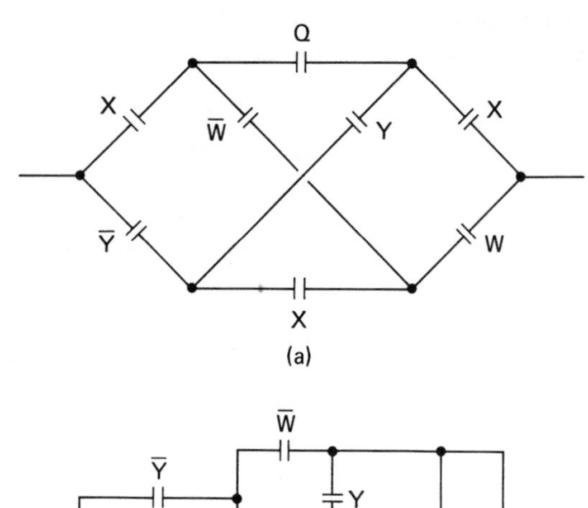

(a)

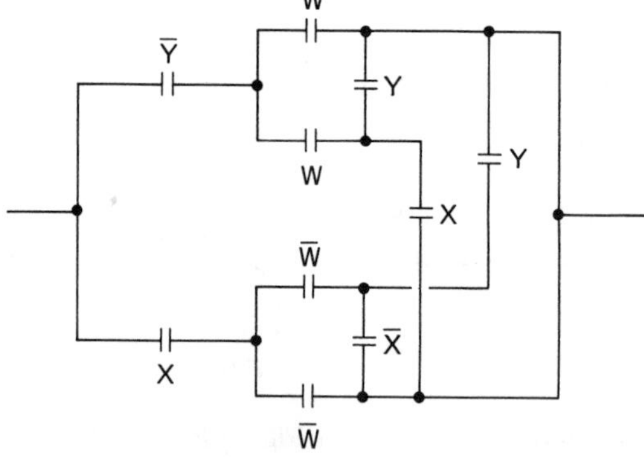

(b)

FIGURE P2-15

2-16 Determine the truth table that corresponds to each of the following Boolean equations.

(a) $Z = XYW + X\bar{Y}\bar{W} + \bar{X}Y\bar{W}$ (b) $Z = X\bar{W}(Y + X)$

(c) $Z = \bar{X}(Y + \bar{Y}\bar{W})$ (d) $Z = XW(XQ + \bar{X}(Y + \bar{Q})) + \bar{W}Q$

(e) $Z = A + \bar{A}BC + \bar{A}\bar{B}\bar{C}$

2-17 Find the MSP Boolean equation that corresponds to each of the following truth tables.

(a)

X	Y	W	Z
0	0	0	0
0	0	1	1
0	1	0	1
0	1	1	1
1	0	0	0
1	0	1	0
1	1	0	1
1	1	1	0

(b)

X	Y	Z
0	0	0
0	1	1
1	0	1
1	1	0

(c)

A	B	C	Q
0	0	0	0
0	0	1	1
0	1	0	1
0	1	1	0
1	0	0	0
1	0	1	1
1	1	0	1
1	1	1	0

(d)

Q	X	Y	W	Z
0	0	0	0	0
0	0	0	1	0
0	0	1	0	0
0	0	1	1	0
0	1	0	0	1
0	1	0	1	1
0	1	1	0	1
0	1	1	1	1
1	0	0	0	0
1	0	0	1	0
1	0	1	0	0
1	0	1	1	0
1	1	0	0	1
1	1	0	1	0
1	1	1	0	1
1	1	1	1	0

3 COMBINATIONAL LOGIC

3.1 DEFINITION OF COMBINATIONAL LOGIC CIRCUITS

Our primary concern in Chapters 1 and 2 has been to establish a framework through which fundamental logic concepts can be understood, and then expressed in accordance with a defined mathematical system. Throughout this discussion, the basic vehicle of explanation has been the combinational logic circuit, although only simple circuits have been involved. We are now in a position to examine more complex combinational networks. First, our definition of a combinational logic is formalized. Then, a very useful graphical technique of minimization is presented. Finally, and most important, we discuss some practical and realistic applications. It will be a satisfying experience, indeed, to see how the techniques that we have been studying are essential for effective solutions to these applications problems.

A ***combinational logic*** circuit is considered generally to possess a set of inputs, a *memoryless* logic network to operate on the inputs, and a set of outputs. The binary value of any output in a combinational logic circuit is determined solely by the present "combination" of binary input values, and has no dependence on prior input values. Furthermore, as long as the input values are maintained, the output values will be maintained. This general concept of a combinational logic circuit is illustrated schematically in Figure 3-1. Barring a circuit malfunction, combinational logic will always respond in the same way to a given input condition, and, therefore, may be fully described by a truth table. Note that the general model allows for a multitude of logic outputs, even though the actual circuitry that we have considered thus far has had

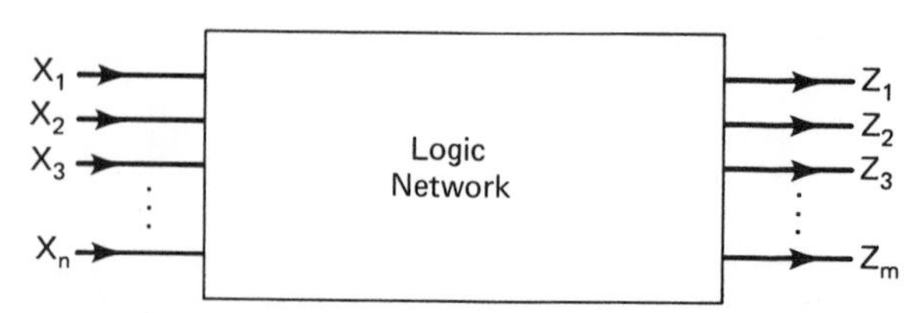

FIGURE 3-1 *General model of a combinational logic circuit.*

only one output. Also, the block labeled "Logic network" in Figure 3-1 places no restriction on the logic circuit's electrical components; they may be switches, transistors, logic gates, or any combination of these. Finally, note that the number of inputs and outputs is unspecified in the general model, denoted by the letter subscript 'n' on the last input variable, and the letter subscript 'm' on the last output variable. Although the exact number of inputs and outputs is unspecified, there must be a finite number of them.

A combinational logic circuit is defined mathematically by stating, for *each* possible output variable, a functional relationship to the input variables. The general definition is

$$Z_i = f_i(X_1, X_2, X_3, \ldots, X_n), \qquad 1 \leq i \leq m \tag{3-1}$$

As a specific example, consider the combinational logic circuit of Figure 3-2. Relating this circuit to the general model, we see that $n = 3$, and $m = 2$. Formulating the input/output relationship, we have

$$Z_1 = f_1(X_1, X_2, X_3) = X_1 X_2 \tag{3-2}$$

$$Z_2 = f_2(X_1, X_2, X_3) = X_1 X_2 + \bar{X}_2 \bar{X}_3 \tag{3-3}$$

The variables within the parentheses of the functional statement "$f_i(X_1, X_2, X_3)$" indicate simply all the input variables that are available for use in a Boolean equation that describes the function. It does not require, however, that all the variables actually be used in the equation. This is the case in Eq. 3-2, where only X_1 and X_2 are used.

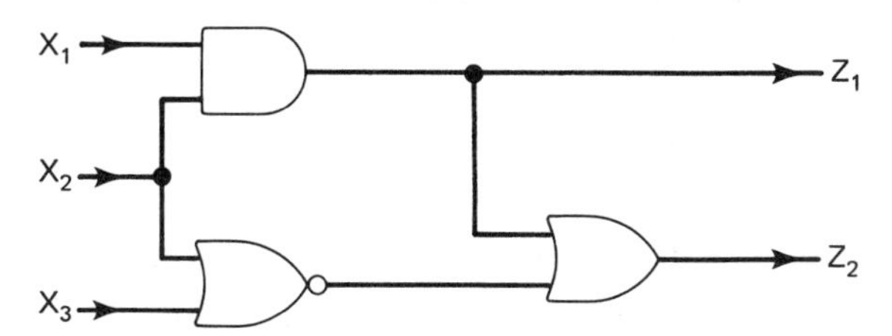

FIGURE 3-2 Combinational logic circuit with two outputs.

It is possible to treat one combinational logic circuit as a component within another larger combinational network. Such interconnections will always result in a valid combinational network, as described by the definition, as long as no internal feedback paths are present. When large systems are analyzed or designed, it is often desirable to consider the system as being made up of smaller subnetworks to ease understanding of the overall function.

3.2 KARNAUGH MAP TECHNIQUES

3.2.1 Motivation for Karnaugh Maps

The combinational logic circuit of Figure 3-3 is employed to obtain the output variable 'Z'. Our immediate goal is to investigate this circuit to determine whether

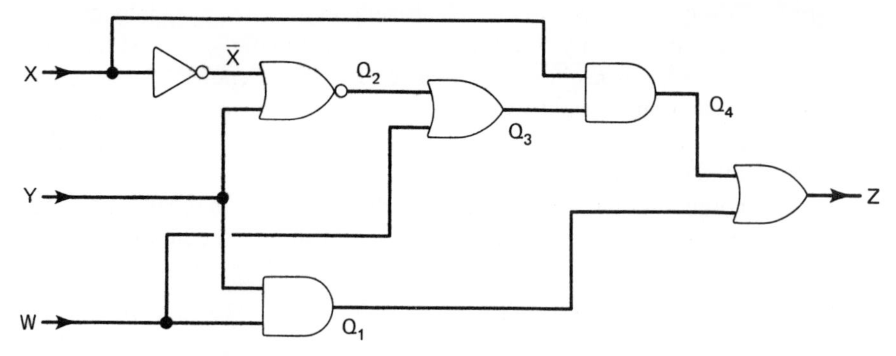

FIGURE 3-3 Combinational logic circuit to be analyzed.

'Z' can be obtained using fewer gates. Relying only on the methods presented in Chapter 2, it is necessary to

1. Obtain a Boolean equation corresponding to the given gate structure.
2. Reduce the equation as much as possible using the properties of Boolean algebra.
3. Re-create a gate structure corresponding to the reduced equation.

The intermediate variables indicated in Figure 3-3 are useful in determining the output variable, Z, in terms of X, Y, and W. From these intermediate variables, we develop Z in the following order:

$$Q_1 = YW$$
$$Q_2 = \overline{\bar{X} + Y}$$
$$Q_3 = Q_2 + W = (\overline{\bar{X} + Y}) + W$$
$$Q_4 = XQ_3 = X[(\overline{\bar{X} + Y}) + W]$$
$$Z = Q_4 + Q_1 = X[(\overline{\bar{X} + Y}) + W] + YW \tag{3-4}$$

It is important to realize that we have done no simplification as yet, we have just expressed Z in terms of X, Y, and W. Although it was possible to apply DeMorgan's law immediately to Q_2, we intentionally did not do so. This way, it is much easier to recheck results should an error become evident at a later time.

The equation for Z is now reduced, abiding the properties of Boolean algebra.

$$\begin{aligned} Z &= X[(\overline{\bar{X} + Y}) + W] + YW \\ &= X[\bar{\bar{X}}\bar{Y} + W] + YW \\ &= XX\bar{Y} + XW + YW \\ &= X\bar{Y} + XW + YW \qquad \text{(MSP, apparently)} \\ &= X(\bar{Y} + W) + YW \end{aligned} \tag{3-5}$$

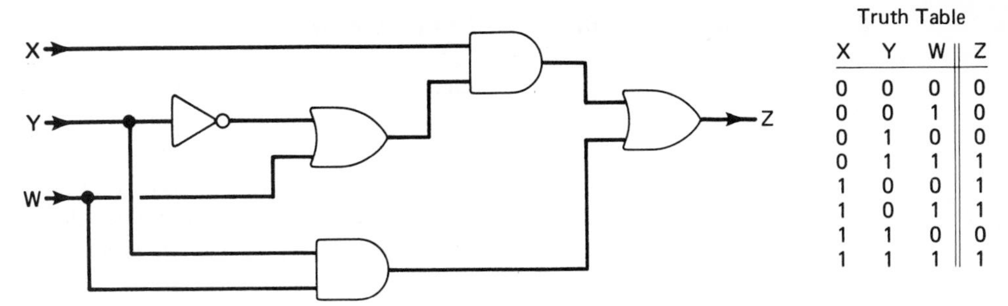

Truth Table

X	Y	W	Z
0	0	0	0
0	0	1	0
0	1	0	0
0	1	1	1
1	0	0	1
1	0	1	1
1	1	0	0
1	1	1	1

FIGURE 3-4 *A new circuit realizing the function of Figure 3-3, but using one less gate.*

With Z reduced to what seems to be MSP form, and factored, a corresponding gate structure is determined. This is shown in Figure 3-4. Following this reasonable procedure, we seem to have arrived at a minimal gate structure. Unfortunately, however, this circuit is not minimal. Consider the gate structure of Figure 3-5, and compare the truth table of this simpler circuit to the one of Figure 3-4. Since the truth tables for these two circuits are identical, their logic functions are equivalent, yet the circuit of Figure 3-5 has one less gate. The basic procedure used to arrive at the circuit of Figure 3-4 was correct, except that what looked like an MSP equation really was not minimal.

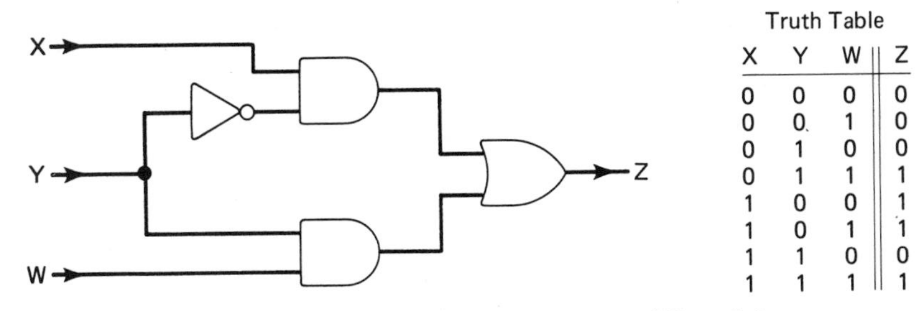

Truth Table

X	Y	W	Z
0	0	0	0
0	0	1	0
0	1	0	0
0	1	1	1
1	0	0	1
1	0	1	1
1	1	0	0
1	1	1	1

FIGURE 3-5 *Minimal circuit for the function of Figure 3-3.*

In order to see this, the following expansion is used.

$$\begin{aligned}
Z &= X\bar{Y} + XW + YW && \text{(apparent MSP)} \\
&= X\bar{Y} + XW(Y + \bar{Y}) + YW \\
&= X\bar{Y} + XWY + XW\bar{Y} + YW \\
&= X\bar{Y}(1 + W) + YW(X + 1) \\
&= X\bar{Y} + YW && \text{(true MSP)} \qquad \textbf{(3-6)}
\end{aligned}$$

Recreating a gate structure for the true MSP equation yields the circuit of Figure 3-5. The purpose of this example is to show that the steps needed for complete algebraic reduction may not always be obvious. As a result, the possibility for error is increased. It would be very desirable to have a method of reduction where the result could be believed with more confidence. It happens that there is a more reliable method than straight algebraic reduction. This improved method relies on a graphical description of Boolean functions, and is the subject of the next section.

3.2.2 *Graphical Description of Boolean Functions*

As we have learned, a Boolean function can be defined either by a truth table, or by an algebraic equation. Each form has its advantages. A truth table is quite explicit, stating for each possible input combination, a specific output value. A truth table, however, is not closely related to a corresponding gate structure, and can become unmanageable due to its size. An algebraic equation is more compact than a truth table, and corresponds more closely to a related gate structure. An equation, though, requires knowledge of the AND, OR, and INVERT functions, since it expresses the desired overall Boolean function in terms of interconnected elementary functions.

It is interesting to ask: Is there an intermediate form between a truth table and a Boolean equation that has the advantages of both? Unfortunately, such a hybrid is not known. However, it *is* possible to rearrange a truth table so that it bears a much closer relationship to its minimal Boolean equation. This is illustrated by an example.

Consider the truth table and corresponding MSP equation of Figure 3-6. The layout of this truth table can be rearranged as shown in Figure 3-7. Notice several features of the rearranged truth table. First, the new form is two-dimensional in nature.

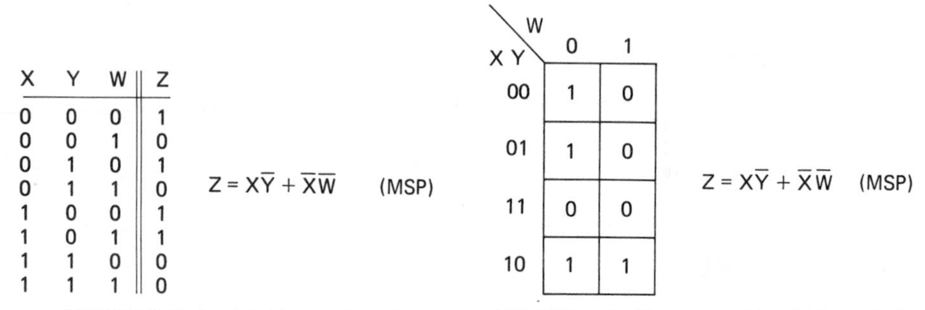

X	Y	W	Z
0	0	0	1
0	0	1	0
0	1	0	1
0	1	1	0
1	0	0	1
1	0	1	1
1	1	0	0
1	1	1	0

$Z = X\overline{Y} + \overline{X}\,\overline{W}$ (MSP)

FIGURE 3-6 *Truth table and its related MSP equation.*

XY \ W	0	1
00	1	0
01	1	0
11	0	0
10	1	1

$Z = X\overline{Y} + \overline{X}\,\overline{W}$ (MSP)

FIGURE 3-7 *The truth table of Figure 3-6 is rearranged.*

Rather than having a single output column as we do in a normal truth table, two columns are used, one for the case of $W = 0$, and the other for the case of $W = 1$. Because the new form is two-dimensional, we refer to it as a *map*. Each entry in the map is considered to have an *address* relating to the row and column in which it resides. The upper right corner of this map, for example, has an address '001', corresponding to $X = 0$, $Y = 0$, and $W = 1$. For each truth table entry, there is one corresponding address on the map. A single binary digit is recorded at each map address, determined by the value of the Boolean function under the input conditions of the address. Finally, it is important to note that the X, Y address column of the map *does not* progress in a natural binary order. Rather, the last two members of the sequence are different from a natural binary sequence. The result is that between any two adjacent members of the X, Y address column, no more than one digit will be different. Carrying this observation one step further, note that there is no more than a one digit difference between *any two successive addresses* of the entire map. As we shall see, this feature is crucial in relating a map to its corresponding Boolean equation.

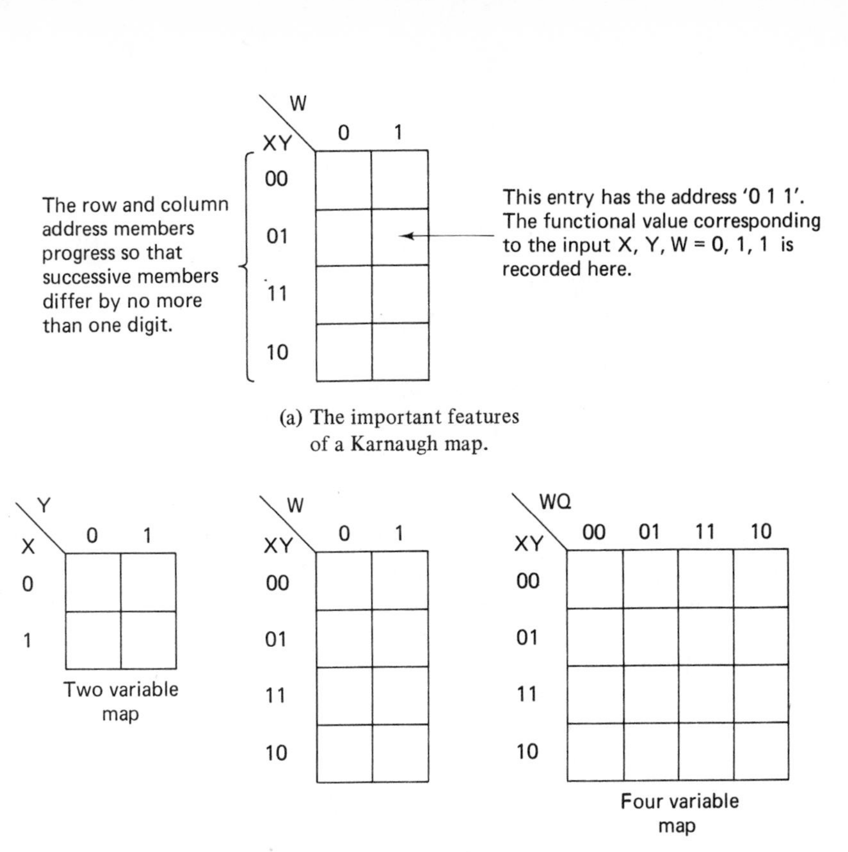

(a) The important features of a Karnaugh map.

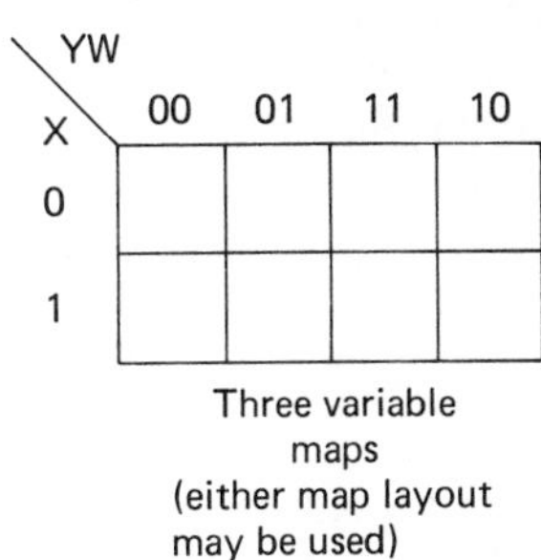

(b) Map structures for two, three, and four variable functions.

FIGURE 3-8

The map of Figure 3-7 is formally referred to as a *Karnaugh map*. Figure 3-8 illustrates the important parts of a Karnaugh map, and illustrates map structures for two-, three-, and four-variable functions.

Returning now to the function of Figure 3-7, we look for a relationship between the Karnaugh map and the Boolean equation. Labeling the product terms of the equation '(1)' and '(2)' respectively, we have

$$Z = \underset{(1)}{X\bar{Y}} + \underset{(2)}{\bar{X}\bar{W}} \tag{3-7}$$

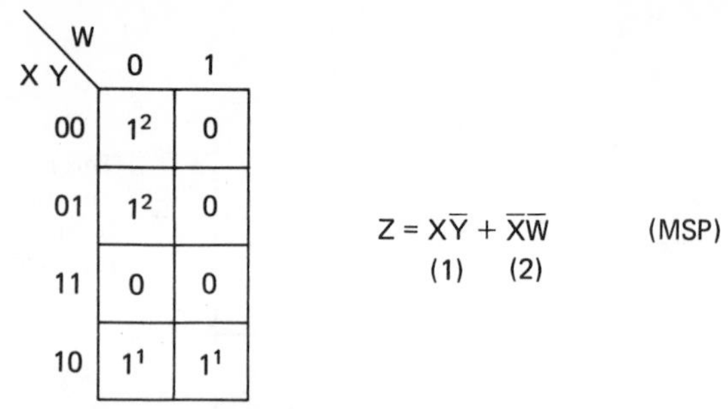

FIGURE 3-9 Each '1' entry in the Karnaugh map is labeled to show which product term in the Boolean equation is its source.

The first product, $X\bar{Y}$, evaluates to '1' whenever X is '1' *and* Y is '0', regardless of W. Thus, $X\bar{Y}$ results in '1' entries on the map for addresses $X, Y, W = 1\ 0\ 0$ and $1\ 0\ 1$. The second product, $\bar{X}\bar{W}$, evaluates to '1' whenever X is '0' *and* W is '0', regardless of Y. Thus, $\bar{X}\bar{W}$ results in '1' entries for addresses $X, Y, W = 0\ 0\ 0$ and $0\ 1\ 0$. Figure 3-9 labels each '1' entry in the Karnaugh map in accordance with the product term that contributes it. Note from Figure 3-9 that the '1' entries contributed by each product appear adjacent in the Karnaugh map. This is an important observation. As we shall see shortly, clusters of adjacent '1's on the map will guide us in selecting a minimal equation.

Recall from Section 2.5.2 that to obtain a Boolean equation from a truth table, a product term is found *for each* functional value of '1'. All product terms are assembled with an OR, and the equation is simplified. If that procedure were applied to the map of Figure 3-9, four product terms would be found. We would then algebraically simplify the result to reach MSP form. Now, is it possible to tell directly from the map what the MSP form will be? If it is, we will spare ourselves the task of algebraic reduction! The adjacency of the entries is the key to answering this question. Let us answer the broad question by addressing a series of simpler questions.

QUESTION 1

What is the special feature of a Karnaugh map that relates to adjacent entries?

ANSWER

The addresses of adjacent entries differ by only one digit.

QUESTION 2

It is known that an algebraic product term can be written for each single map entry of '1'. How are the algebraic product terms of two adjacent map entries related?

ANSWER

Consider the bottom row of the map in Figure 3-9. Here, two adjacent '1's are present. For the lower left entry, we have the product '$X\bar{Y}\bar{W}$'. For the lower right entry, we have the product '$X\bar{Y}W$'. When the overall equation is assembled, and then reduced, the following is obtained:

$$Z = X\bar{Y}\bar{W} + X\bar{Y}W + \text{(other terms)}$$
$$= X\bar{Y}(\bar{W} + W) + \ldots$$
$$= X\bar{Y} + \ldots \qquad \textbf{(3-8)}$$

Now, note that the variable that is algebraically eliminated between these two products corresponds to the digit in the entry address *that changes* as we move between adjacent '1' entries. The fact that an address digit changes as we move from one adjacent '1' entry to another indicates that the value of the variable corresponding to that digit *does not matter* in the reduced product term. Therefore, the presence of that variable in the reduced product term is unnecessary.

QUESTION 3

The answer to Question 2 provides some insight into a relationship between two adjacent map entries of '1' and the corresponding minimal algebraic product. Now, we must inquire into the occurrence of more than two adjacent entries of '1'. Specifically, how are the algebraic product terms of more than two adjacent map entries related?

X \ YW	00	01	11	10
0	0	1^1	1^2	0
1	0	1^3	1^4	0

FIGURE 3-10 *Karnaugh map to be analyzed. The superscripts correspond with products in Eq. 3-9.*

ANSWER

Consider the map of Figure 3-10. As has been done previously, we will write a product term for each entry of '1', combine the products with an OR function, and algebraically reduce the result. Finally, a relationship will be sought between the reduced equation and the map. Assuming the output variable is 'Z', the initial equation is

$$Z = \underset{(1)}{\bar{X}\bar{Y}W} + \underset{(2)}{\bar{X}YW} + \underset{(3)}{X\bar{Y}W} + \underset{(4)}{XYW} \qquad \textbf{(3-9)}$$

Reducing this equation, we have

$$Z = \bar{X}W(\bar{Y} + Y) + XW(\bar{Y} + Y)$$
$$= \bar{X}W + XW$$
$$= (\bar{X} + X)W$$
$$= W \qquad \text{(MSP)} \qquad \textbf{(3-10)}$$

In this case, notice that two variables have been eliminated in the reduced equation.

As we move between all possible combinations of adjacent '1' entries in Figure 3-10, we see that any address digit that *changes* during this movement corresponds to a variable that is *eliminated* in the reduced equation.

Generalizing on our observations thus far, it can be said that

1. Large clusters of adjacent '1's in a map mean the elimination of more variables.
2. The digits that change while moving between entries in a cluster of '1's relate to the variables that are eliminated in the reduced equation.

As useful as these generalizations may seem, there are some limitations that must be realized before a Karnaugh map procedure is established.

QUESTION 4

How is the MSP equation that corresponds to the map of Figure 3-11 different from that of Figure 3-10?

X \ YW	00	01	11	10
0	0	1^1	1^2	0
1	0	1^3	0	0

FIGURE 3-11 *Karnaugh map to be analyzed. The superscripts correspond with products in Eq. 3-11.*

ANSWER

This map is identical to the map of Figure 3-10 except for the '1 1 1' entry, which in this case is '0'. With an output variable 'Z', the initial equation is

$$Z = \underset{(1)}{\bar{X}\bar{Y}W} + \underset{(2)}{\bar{X}YW} + \underset{(3)}{X\bar{Y}W} \tag{3-11}$$

Reducing this equation, we have

$$\begin{aligned} Z &= \bar{X}W(\bar{Y} + Y) + X\bar{Y}W \\ &= \bar{X}W + X\bar{Y}W \\ &= W(\bar{X} + X\bar{Y}) \\ &= W(\bar{X} + \bar{Y}) \\ &= W\bar{X} + W\bar{Y} \qquad \text{(MSP)} \end{aligned} \tag{3-12}$$

The result of the reduction this time is quite different from that relating to the map of Figure 3-10, even though only one entry is different between these two maps. Also, it seems that our initial generalizations should be questioned, since moving among the cluster of three '1's in Figure 3-11 shows the same address digit changes as in

Figure 3-10, even though the MSP equations are different. Really, the generalizations need only be made more exact. In order to eliminate two variables from product terms of adjacent entries, as in Eq. 3-9, it must be proven that neither of the variables are needed to express algebraically the '1' entries on the map. This is shown in Eq. 3-9 when it is rewritten as

$$Z = W(\bar{X}\bar{Y} + \bar{X}Y + X\bar{Y} + XY) \tag{3-13}$$

Equation 3-11 rewritten in a similar way, however, is

$$Z = W(\bar{X}\bar{Y} + \bar{X}Y + X\bar{Y}) \tag{3-14}$$

revealing that the 'XY' product is missing, and therefore that one needed combination is absent. This prevents the products within the parentheses in Eq. 3-14 from evaluating to '1', and two variables cannot be eliminated. The actual MSP result that is obtained in Eq. 3-12 can be considered to be related to two overlapping clusters, each of two '1's. This is shown in Figure 3-12.

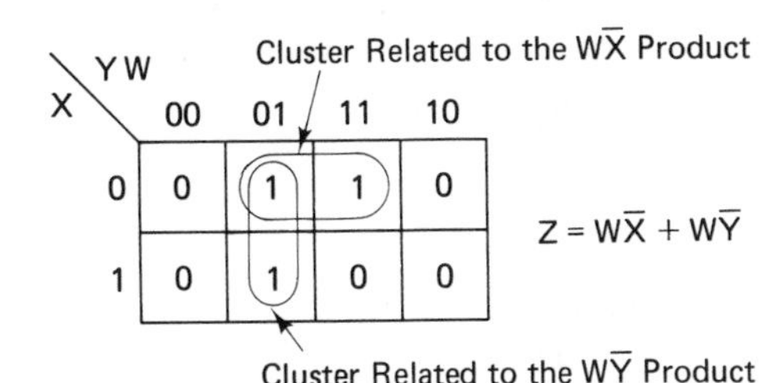

FIGURE 3-12 *The Karnaugh map of Figure 3-11 is considered as two overlapping clusters of two '1's.*

3.2.3 Determination of Minimal Equations

With the intuitive basis offered by the previous discussion, we will now formalize our results, and decide on a procedure to extract MSP equations from Karnaugh maps.

DEFINITIONS

1. A *Karnaugh map* is a graphical form of truth table consisting of a square or rectangular array of adjacent cells. A Boolean function utilizing 'n' variables requires a Karnaugh map made up of 2^n cells. Each cell possesses a unique address, specified by the row and column in which the cell resides. Rows and columns are labeled using a binary representation, and in a nonrepeating order such that succeeding row or column entries differ by no more than one digit. As a result, the full address of any cell differs by no more than one digit from the address of any adjacent cell. For a Boolean function expressed by 'n' variables, 'm' of those variables are associated with row labels, and the

remaining '$n - m$' variables are associated with the column labels. A single numerical value of '0' or '1' is entered in each cell. The exact value of the entry results from evaluating the Boolean function under the conditions of input specified by the address of the cell in which the entry is made. Valid Karnaugh map structures for two-, three-, and four-variable functions are illustrated in Figure 3-8(b). No more than two variables may be represented in any row or column of a Karnaugh map in order to guarantee that any two addresses different by one digit correspond to adjacent cells. This does not mean, however, that Karnaugh maps for functions of five or more variables are not possible. A five-variable map requires that we consider two four-variable maps, each being thought of as a different plane of a three-dimensional figure. Refer to Problem 3-5 for a further discussion of five-variable maps.

2. Two cells of a Karnaugh map are ***adjacent*** if their respective addresses differ by no more than one digit. Therefore,

   (a) Diagonal cells *are not* adjacent even though they have a corner point in common.

   (b) Cells that are not adjacent graphically, but exist at opposite extremities of a row or column, *are* adjacent numerically.

   These ideas are illustrated in Figure 3-13.

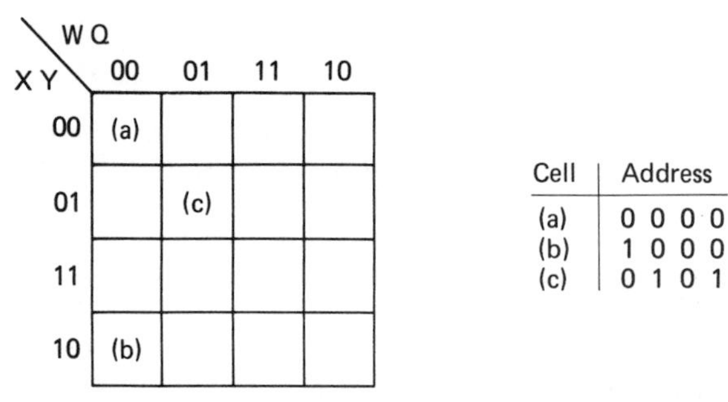

Cell	Address
(a)	0 0 0 0
(b)	1 0 0 0
(c)	0 1 0 1

FIGURE 3-13 *Cells (a) and (b) are adjacent, while cells (a) and (c) are not.*

3. A Boolean equation may be expressed in the sum of products (SP) form by applying the Distributive law until all parentheses are eliminated. Each product term in the SP expression is referred to as an ***implicant***. Each implicant relates to a single entry or a cluster of adjacent entries on a Karnaugh map. If an implicant relates to a cluster of adjacent entries, that cluster will be square or rectangular in shape, and consist of a quantity of cells that is an integer power of 2 (1, 2, 4, 8, . . .). An implicant is denoted on a Karnaugh map by an ***enclosure***. Figure 3-14 depicts several implicants. An implicant may overlap another implicant, as in Figure 3-12, or may be completely enclosed by another implicant, as in Figure 3-14.

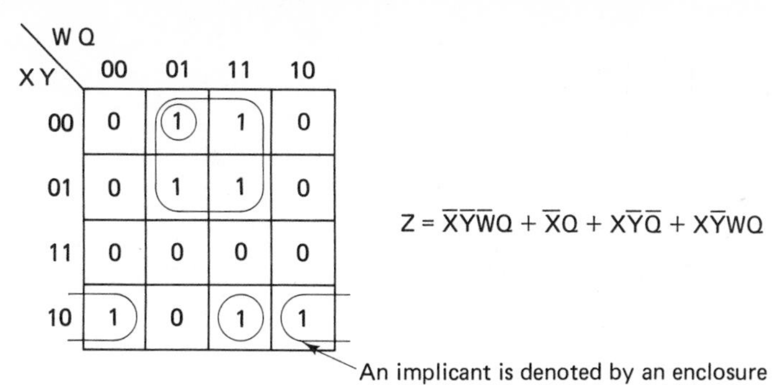

FIGURE 3-14 Boolean equation, and its corresponding Karnaugh map and implicants.

4. A ***prime implicant*** is an implicant that cannot be fully enclosed by a larger implicant on a Karnaugh map. A prime implicant encloses a maximum number of '1's, and, as with a nonprime implicant, is square or rectangular in shape enclosing a number of '1's that is an integer power of 2. Figure 3-15 compares prime implicants to nonprime implicants.

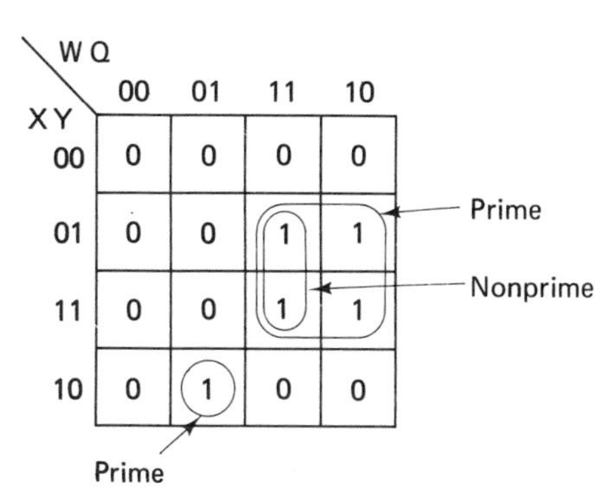

FIGURE 3-15 Comparison of prime implicants and nonprime implicants.

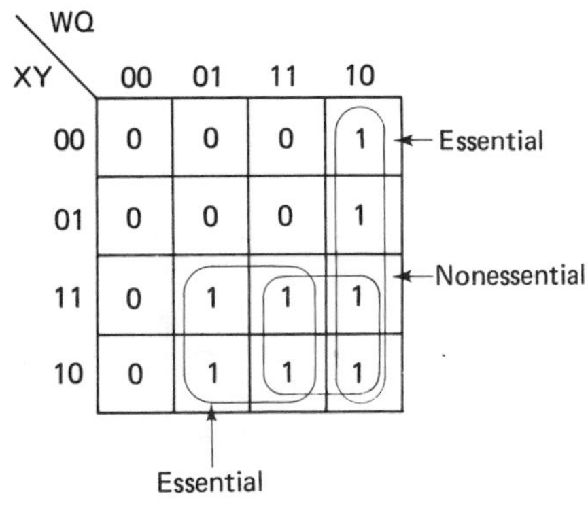

FIGURE 3-16 Comparison of essential prime implicants and a nonessential prime implicant.

5. In order to obtain a minimal algebraic expression from a Karnaugh map, it is required that every '1'-cell be enclosed by at least one prime implicant. If a particular '1' can be enclosed by only a single prime implicant, that enclosure denotes an ***essential prime implicant***. Figure 3-16 compares essential prime implicants to a nonessential prime implicant. Nonessential prime implicants can be removed without leaving any '1'-cells unenclosed. Nonessential prime implicants are, therefore, not needed.

Obtaining a minimal algebraic expression from a Karnaugh map involves two basic steps:

1. All '1'-cells are included in at least one essential prime implicant. On a Karnaugh map, this is denoted by every '1'-cell being within at least one

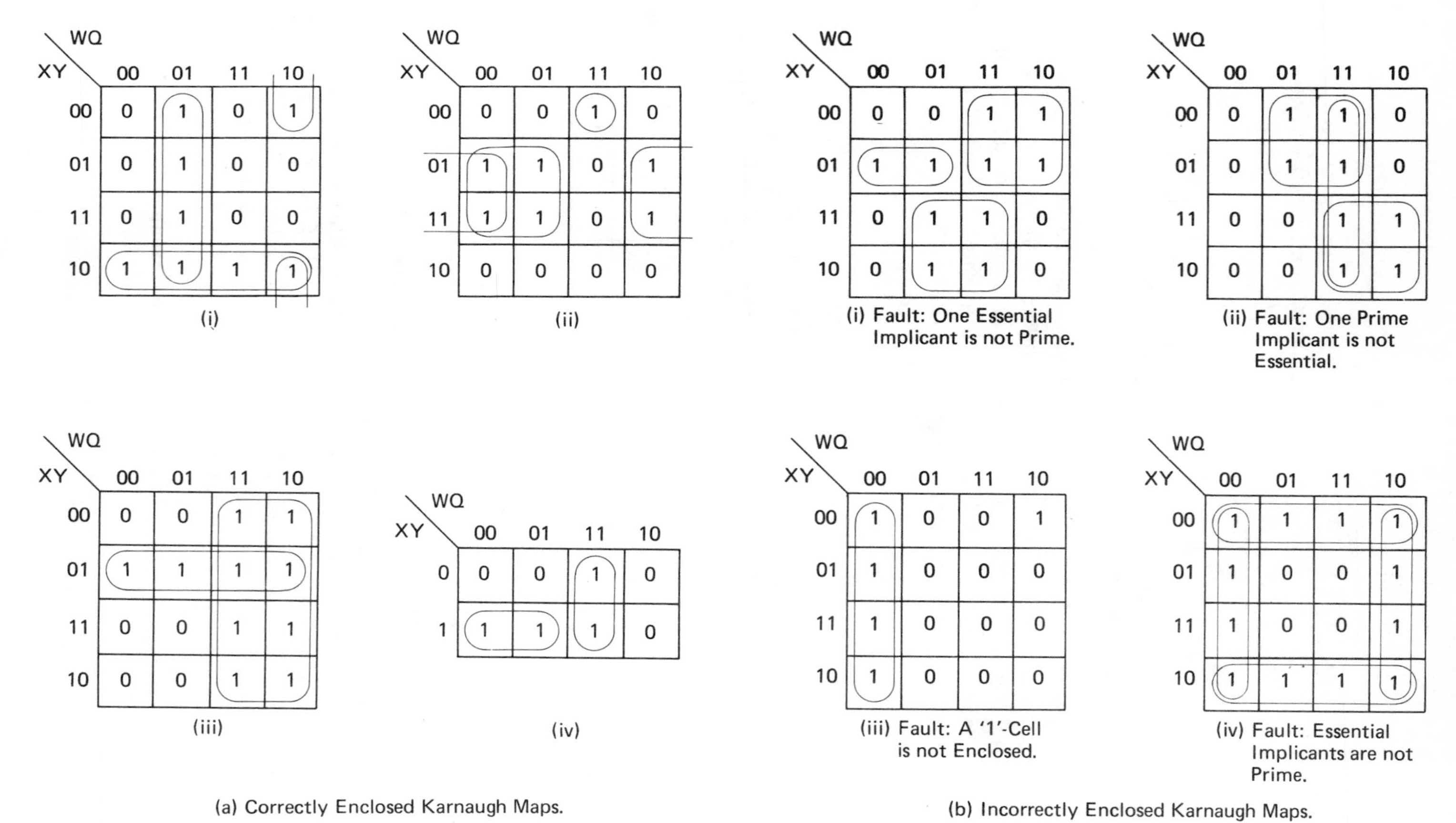

FIGURE 3-17 *Examples showing correctly and incorrectly enclosed Karnaugh maps; the correctly enclosed maps depict essential prime implicants.*

enclosure. It is permissible to enclose a '1'-cell more than once if it facilitates enlarging another enclosure.

2. Each essential prime implicant is translated into a particular product term. When a product term has been found for each essential prime implicant, the products are assembled using an OR function. The result is an MSP-form Boolean expression corresponding to the given logic function.

Relating to the first step, we are required to look over a given Karnaugh map and decide, as best we can, where the essential prime implicants are located. Several correctly and incorrectly enclosed Karnaugh maps are illustrated in Figure 3-17. Study this illustration carefully to be certain that you understand how essential prime implicants are determined.

Once the essential prime implicants have been decided, a product term must be written for each one. Bringing the products together with an OR function then yields the desired MSP form for the given Boolean function. We must ask now: How is an algebraic product term obtained from an essential prime implicant? Consider the Karnaugh map and essential prime implicant shown in Figure 3-18. Notice that if

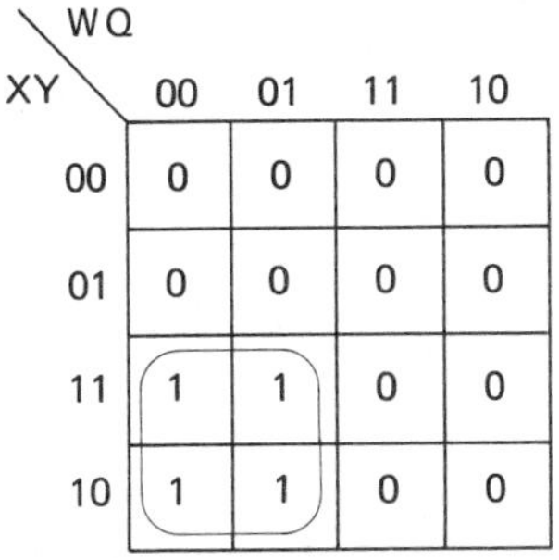

XY \ WQ	00	01	11	10
00	0	0	0	0
01	0	0	0	0
11	1	1	0	0
10	1	1	0	0

FIGURE 3-18 *Karnaugh map with one essential prime implicant.*

we move vertically from cell to cell *within the enclosure*, the address digit corresponding to the variable 'Y' changes between '1' and '0'. This means that in an unreduced SP equation, found by writing a product for each individual '1' entry, we would eventually be able to factor the equation to obtain

$$Z = (Y + \bar{Y})(\ldots) \tag{3-15}$$

Similarly, if we move horizontally from cell to cell within the enclosure, the address digit corresponding to the variable 'Q' changes between '1' and '0'. This means that in an unreduced SP equation, we would be able to further factor the equation to obtain

$$Z = (Q + \bar{Q})(Y + \bar{Y})(\ldots) \tag{3-16}$$

Since any variable connected in an OR with its complement is '1', the ability to factor Y and Q as in Eqs. 3-15 and 3-16 means that we completely eliminate these variables from the product term. Generalizing, then, any address digits that change as we move from cell to cell within an enclosure *are eliminated* in the product term for that enclosure. On the other hand, address digits that do not change while moving within the confines of an enclosure are needed in the product term for that enclosure. If an address digit is unchanging, two cases are possible:

1. If an address digit is '1' and does not change as we move within an enclosure, the corresponding variable is used in its *uncomplemented* form as a term in the minimal product.
2. If an address digit is '0' and does not change as we move within an enclosure, the corresponding variable is used in its *complemented* form as a term in the minimal product.

Applying these ideas to Figure 3-18, the minimal product corresponding to the essential prime implicant is $X\bar{W}$. The MSP equation, therefore, is

$$Z = X\bar{W} \tag{3-17}$$

Table 3-1 lists the correct MSP equations corresponding to the maps of Figure 3-17. In the case of Figure 3-17(b), where the implicants are not chosen properly, the equations of Table 3-1 relate to the *corrected* implicants.

TABLE 3-1

MSP equations corresponding to the maps of Figure 3-17. The equations given for Figure 3-17(b) relate to the corrected *implicants.*

(a)	(i)	$Z = \bar{W}Q + X\bar{Y} + W\bar{Q}\bar{Y}$
	(ii)	$Z = \bar{X}\bar{Y}WQ + Y\bar{W} + Y\bar{Q}$
	(iii)	$Z = \bar{X}Y + W$
	(iv)	$Z = X\bar{Y} + YW$
(b)	(i)	$Z = \bar{X}Y + \bar{X}W + XQ$
	(ii)	$Z = XW + \bar{X}Q$
	(iii)	$Z = \bar{Q}\bar{W} + \bar{X}\bar{Y}\bar{Q}$
	(iv)	$Z = \bar{Y} + \bar{Q}$

The rules involved in the use of Karnaugh maps for simplification are summarized.

SUMMARY OF DEFINITIONS AND RULES CONCERNING KARNAUGH MAPS

A Karnaugh map defines a Boolean function when each of its cells contains a specific binary digit, either '0' or '1'. To obtain the MSP expression related to the function,

(1) determine all essential prime implicants on the map, denoting them by enclosures, and (2) find a product term for each essential prime implicant, and connect all resulting products with an OR function. Equating the MSP expression to an output variable yields an equation.

I. Essential prime implicants are denoted by enclosures.
   1. Enclosures may be square or rectangular only, and must enclose a quantity of '1's that is an integer power of 2 (for example: 1, 2, 4, 8, 16).
   2. The enclosures must be as large as possible.
   3. Every '1'-cell on the map should be included in at least one enclosure. A '1'-cell may be a part of more than one implicant if it facilitates enlarging another enclosure.
   4. There must be as few enclosures as possible, but enough to cover all '1'-cells at least once.
   5. An enclosure may wrap around between borders of the Karnaugh map, since the borders have adjacent addresses.

II. Each essential prime implicant generates one product term in the MSP expression.
   1. The address digits that change as one moves from cell to cell within an enclosure correspond to variables that are ignored in the related product term.
   2. The address digits that do not change as one moves from cell to cell within an enclosure correspond to variables that are needed in the related product term. Variables that remain '1' within the enclosure are used in their *uncomplemented* form as a term in the product. Variables that remain '0' within the enclosure are used in their *complemented* form as a term in the product.

EXAMPLE 3-1: Reduce the following equation to MSP form using Karnaugh map techniques.

$$Z = \bar{X}[\bar{Y}Q + Y(\bar{W}Q + W)] + XYW\bar{Q}$$

To simplify the process of entering the function on a Karnaugh map, the equation is put in SP form and has the product terms labeled.

$$Z = \underset{(1)}{\bar{X}\bar{Y}Q} + \underset{(2)}{\bar{X}Y\bar{W}Q} + \underset{(3)}{\bar{X}YW} + \underset{(4)}{XYW\bar{Q}}$$

For each product, we reason when it should be '1', and then enter these values on a four-variable map.

(1) $\bar{X}\bar{Y}Q = 1$	whenever X and Y are '0' and Q is '1'
(2) $\bar{X}Y\bar{W}Q = 1$	whenever X and W are '0' and Y and Q are '1'
(3) $\bar{X}YW = 1$	whenever X is '0' and Y and W are '1'
(4) $XYW\bar{Q} = 1$	whenever X, Y, and W are '1' and Q is '0'

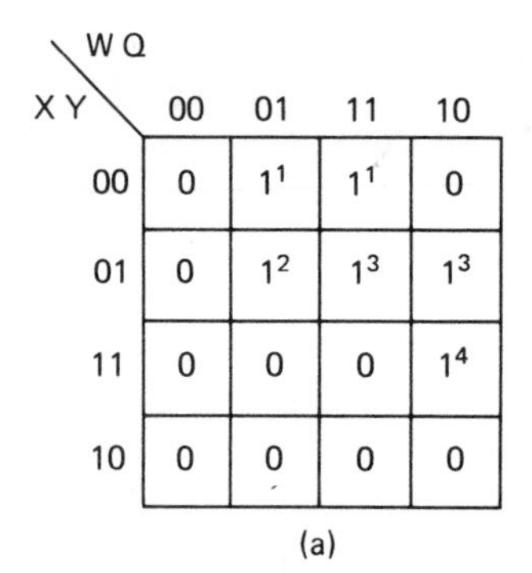

(a)

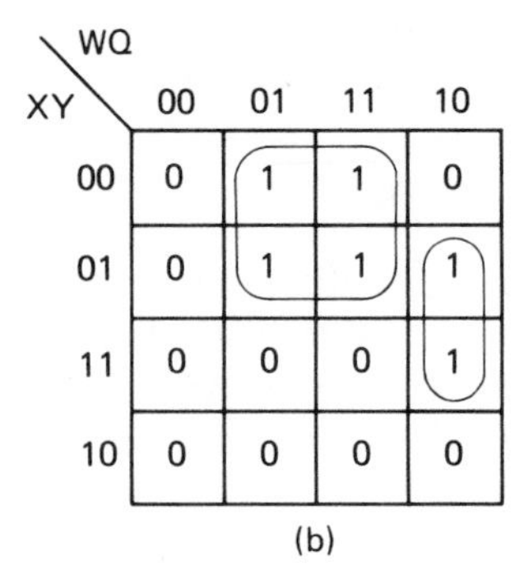

(b)

The essential prime implicants are now identified. Finally, a product is found for each enclosure, and an MSP equation assembled.

$$Z = \bar{X}Q + YW\bar{Q}$$

Note that no Boolean algebra was needed in this example, except in the initial expansion of the given equation, where the distributive law was used.

EXAMPLE 3-2: A Boolean function is defined by the truth table of Figure 3-19.

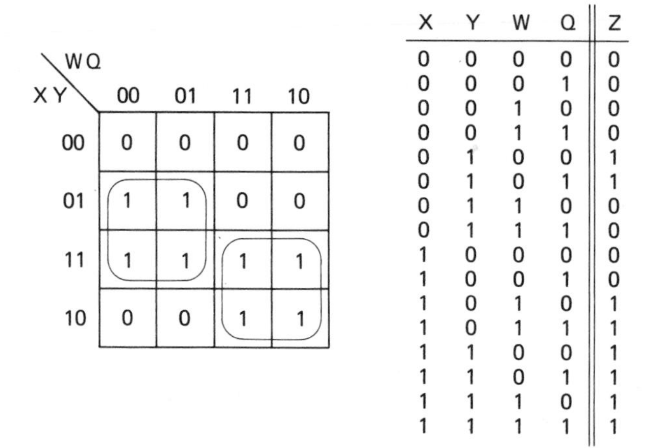

X	Y	W	Q	Z
0	0	0	0	0
0	0	0	1	0
0	0	1	0	0
0	0	1	1	0
0	1	0	0	1
0	1	0	1	1
0	1	1	0	0
0	1	1	1	0
1	0	0	0	0
1	0	0	1	0
1	0	1	0	1
1	0	1	1	1
1	1	0	0	1
1	1	0	1	1
1	1	1	0	1
1	1	1	1	1

FIGURE 3-19 *Truth table and corresponding Karnaugh map.*

(a) Using Karnaugh map techniques, determine the MSP equation that describes this function.

A Karnaugh map is drawn and the functional values are entered in accordance with the truth table. Two essential prime implicants are found and enclosed. From the enclosures, we determine the minimal products. Since there are two enclosures, two products are needed for an MSP equation. For that equation, we have

$$Z = Y\bar{W} + XW$$

(b) Find the complementary function, $\bar{Z}$, by complementing each cell of the Karnaugh map, finding new enclosures, and determining new products. Verify that this result is consistent with the result that would have been obtained by complementing algebraically the MSP equation of part (a), and applying DeMorgan's law.

First, a new Karnaugh map is obtained, and its implicants are indicated.

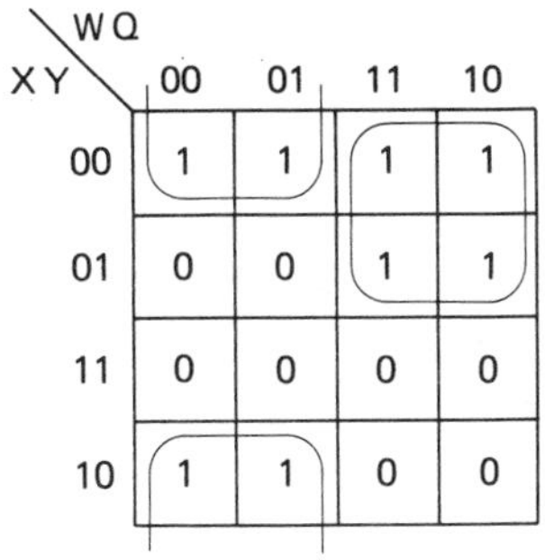

XY \ WQ	00	01	11	10
00	1	1	1	1
01	0	0	1	1
11	0	0	0	0
10	1	1	0	0

For two enclosures, two product terms are found. The equation is

$$\bar{Z} = \bar{Y}\bar{W} + \bar{X}W \qquad \text{(MSP)}$$

Algebraically, we start with the MSP equation found in part (a).

$$Z = Y\bar{W} + XW$$

Complementing the equation and applying DeMorgan's law, we have the following.

$$\begin{aligned} \bar{Z} &= \overline{Y\bar{W} + XW} \\ &= (\overline{Y\bar{W}})(\overline{XW}) \\ &= (\bar{Y} + \bar{\bar{W}})(\bar{X} + \bar{W}) \\ &= (\bar{Y} + W)(\bar{X} + \bar{W}) \end{aligned}$$

Expanding this result and reducing it to MSP form, the complementary equation is found.

$$\begin{aligned} \bar{Z} &= \bar{Y}\bar{X} + \bar{Y}\bar{W} + W\bar{X} + W\bar{W} \\ &= \bar{Y}\bar{X} + \bar{Y}\bar{W} + W\bar{X} \\ &= \bar{Y}\bar{X}(1) + \bar{Y}\bar{W} + W\bar{X} \\ &= \bar{Y}\bar{X}(W + \bar{W}) + \bar{Y}\bar{W} + W\bar{X} \\ &= \bar{Y}\bar{X}W + \bar{Y}\bar{X}\bar{W} + \bar{Y}\bar{W} + W\bar{X} \\ &= \bar{X}W(\bar{Y} + 1) + \bar{Y}\bar{W}(\bar{X} + 1) \\ &= \bar{Y}\bar{W} + \bar{X}W \end{aligned}$$

This algebraic result is equivalent to the graphical result obtained earlier.

3.2.4 Uniqueness

Occasionally, a Boolean equation is found for which *more than one* set of essential prime implicants are possible. Such a function is illustrated in Figure 3-20. When this happens, more than one correct MSP equation can be found, each with the same number of products and the same number of variables per product. These equivalent equations describe the same function, but differ in which variables are utilized. Boolean functions that cannot be described by a single MSP equation, such as the one of Figure 3-20, are termed ***nonunique***. On the other hand, those functions that can be described by a single MSP equation are termed ***unique***.

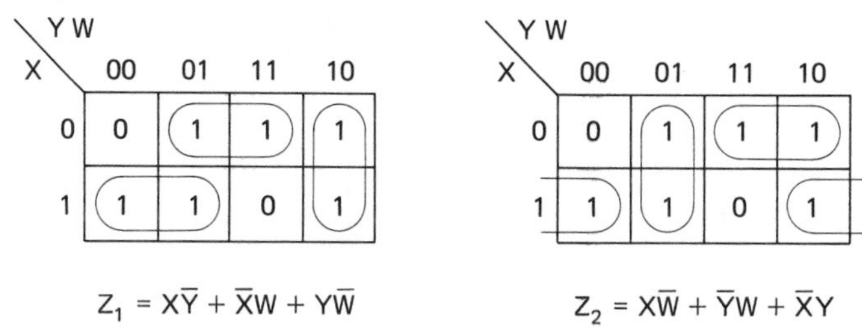

FIGURE 3-20 *Function for which two different but equally minimal sets of essential prime implicants are possible.*

If a particular nonunique Boolean function must be implemented with logic gates, it generally does not matter which of the possible MSP equations is selected. Sometimes, though, certain complemented variables or partial products required by one MSP form may already be available from existing circuitry. Thus, there may be an advantage under some circumstances in choosing one of the MSP forms over any others.

EXAMPLE 3-3: The function depicted by a Karnaugh map in Figure 3-21 is to be realized with a minimum number of two-input NAND gates. No other gates are available. Find the minimal network.

XY \ WQ	00	01	11	10
00	1	0	0	1
01	1	1	1	0
11	1	1	1	0
10	1	0	0	1

FIGURE 3-21 *Function to be realized with two-input NAND gates.*

First, we note that the map can be enclosed by two different sets of essential prime implicants:

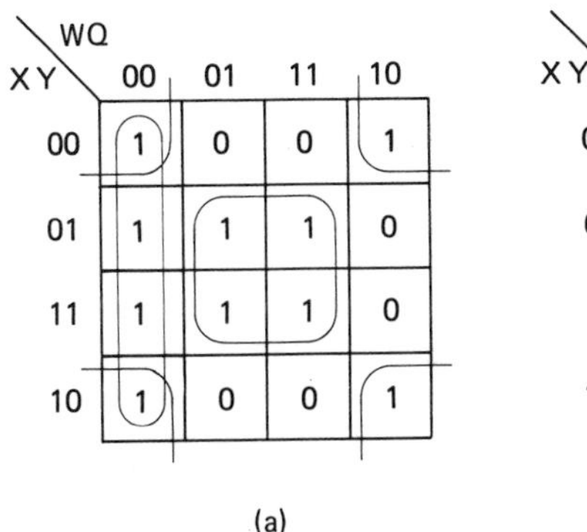

(a)

XY \ WQ	00	01	11	10
00	1	0	0	1
01	1	1	1	0
11	1	1	1	0
10	1	0	0	1

(b)

For map (a), we have

$$Z_a = \bar{Y}\bar{Q} + YQ + \bar{W}\bar{Q}$$

For map (b),

$$Z_b = \bar{Y}\bar{Q} + YQ + Y\bar{W}$$

Note that the equations are identical except for the third product. In order to find a minimal NAND network for the function, it is necessary to find the minimal NAND network for each of the two MSP equations, and select the one with fewest gates. The equations are transformed to NAND form by the reverse application of DeMorgan's law.

$$\begin{aligned} Z_a &= \bar{Y}\bar{Q} + YQ + \bar{W}\bar{Q} \\ &= \bar{Q}(\bar{Y} + \bar{W}) + YQ \\ &= \bar{Q}(\overline{YW}) + YQ \\ &= \overline{\overline{\bar{Q}(\overline{YW}) + YQ}} \\ &= \overline{\overline{\bar{Q}(\overline{YW})}(\overline{YQ})} \qquad \text{(NAND form)} \end{aligned}$$

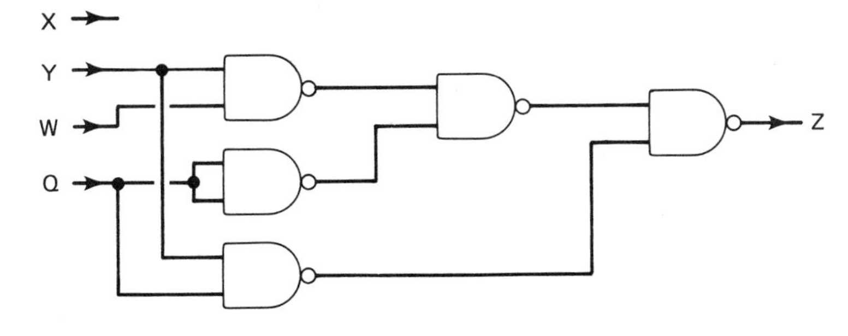

Five NAND's are required for Z_a.

$$\begin{aligned} Z_b &= \bar{Y}\bar{Q} + YQ + Y\bar{W} \\ &= \bar{Y}\bar{Q} + Y(Q + \bar{W}) \\ &= \bar{Y}\bar{Q} + Y(\overline{\bar{Q}W}) \\ &= \overline{\overline{\bar{Y}\bar{Q} + Y(\overline{\bar{Q}W})}} \\ &= \overline{(\overline{\bar{Y}\bar{Q}}) \cdot (\overline{Y(\overline{\bar{Q}W})})} \end{aligned}$$

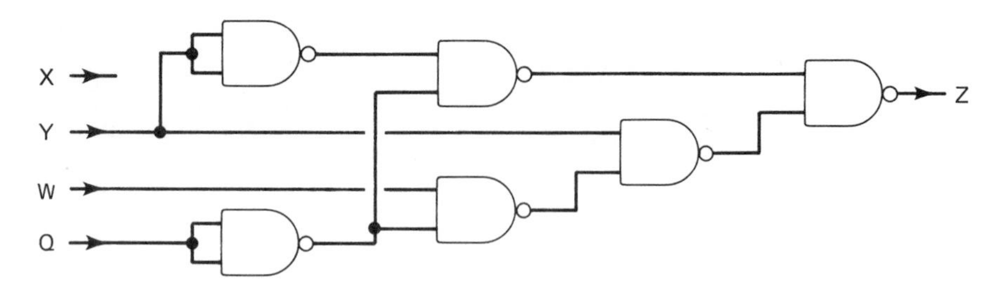

Six NAND's are required for Z_b. Therefore, we select Z_a, in the NAND form, as the minimal NAND realization.

3.2.5 "Don't Care" States

Many basic decision problems that are encountered in the design of digital circuitry may be solved with combinational logic. Often, a decision problem is described verbally, and must be transformed into a truth table before any circuitry can be developed. This involves selecting input and output variables, and assigning meaning to the numerical values of '1' and '0' that these variables may assume. Defining the problem in such a precise way forces us to account for all possible input conditions, since we can leave no truth table entries vacant. This, however, can be surprising because it may reveal certain assumptions that we made automatically in our mental image of the problem. In particular, while most input conditions require a certain specific output condition, there may be some input conditions where the exact output condition is unimportant. Whenever an output value may be '0' *or* '1' without affecting the desired logic operation, the output value is termed a ***don't care*** state, and is denoted by the symbol '$\emptyset$'. There is an important advantage to identifying "don't care" states with '$\emptyset$', rather than randomly assigning a fixed value to the condition. The advantage is realized when we attempt to minimize the function using Karnaugh map techniques. At that time '1' or '0' values may be assigned to each '$\emptyset$' state insofar as it facilitates enlarging the map enclosures. That is,

1. if a '$\emptyset$' occurs adjacent to a '1' on a Karnaugh map, and
2. if the '$\emptyset$' entry being '1' would permit enlarging an essential prime implicant,

we assign $\emptyset = 1$ and use the larger enclosure. On the other hand, if assigning $\emptyset = 1$ does not allow the enlargement of an essential prime implicant, we assign $\emptyset = 0$.

EXAMPLE 3-4: To illustrate the meaning of "don't care" states, consider the following situation. Paul, Joanne, and Carol live in the same house. Unfortunately, there are only two keys available that unlock the front door, one of which is kept by Paul, and the other by Joanne. The door is always locked when no one is home. However, when Carol is out and someone else is home, the door is left unlocked so that Carol can enter when she returns. Finally, if Carol is home, the door may be either locked or unlocked, since anyone else who may be out has a key. If Carol returns to an empty house, she must wait until Paul or Joanne come home to be let in. Paul, Joanne, and Carol are busy people leading independent lives, and cannot easily account for who is at home and who is not. Thus, it is decided to place a switch box inside the house near the front door. The switch box contains three switches, one for each member of the household. If a person leaves the house, he (or she) turns their switch on (logic '1'), and when they return home, their switch is turned off (logic '0'). It is necessary to design a combinational logic circuit that decides whether the front door should be locked or not. The doorknob on the front door is electrically lockable on the command of our logic circuit.

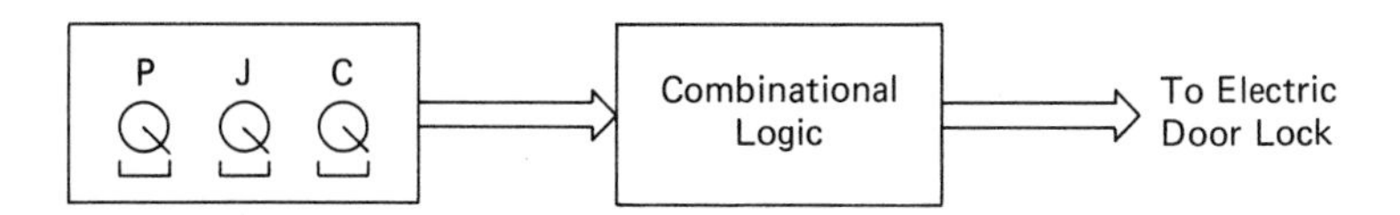

First, we must express the logical ideas in terms of a truth table. This is done in Figure 3-22(a). In order to obtain a minimal logic circuit for the desired operation, we transfer the truth table to a Karnaugh map, find the essential prime implicants, determine an MSP equation, and construct a logic circuit. This is shown in Figure 3-22(b). The solution requires a simple two-input AND. Note that Carol's switch is not even necessary.

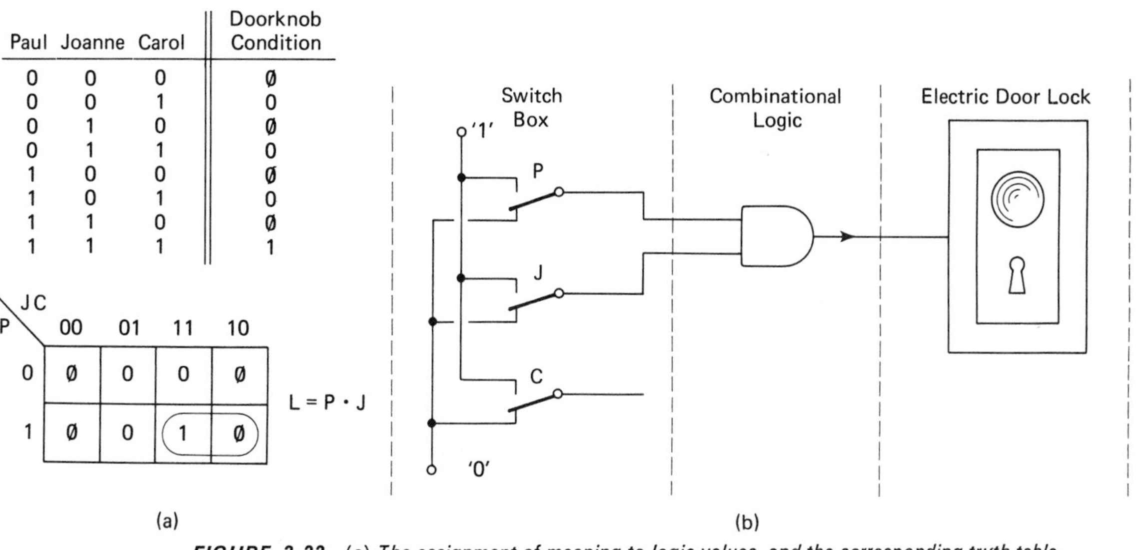

Paul	Joanne	Carol	Doorknob Condition
0	0	0	Ø
0	0	1	0
0	1	0	Ø
0	1	1	0
1	0	0	Ø
1	0	1	0
1	1	0	Ø
1	1	1	1

P \ JC	00	01	11	10
0	Ø	0	0	Ø
1	Ø	0	1	Ø

FIGURE 3-22 *(a) The assignment of meaning to logic values, and the corresponding truth table that fully expresses the relationships described verbally in the problem statement. (b) Solving the automatic lock problem.*

Table of Meanings

Inputs: '1' means that a person is out
'0' means that a person is at home.

Outputs: '1' means that the door should be locked
'0' means that the door should be unlocked
'Ø' means that it does not matter whether the door is locked or unlocked.

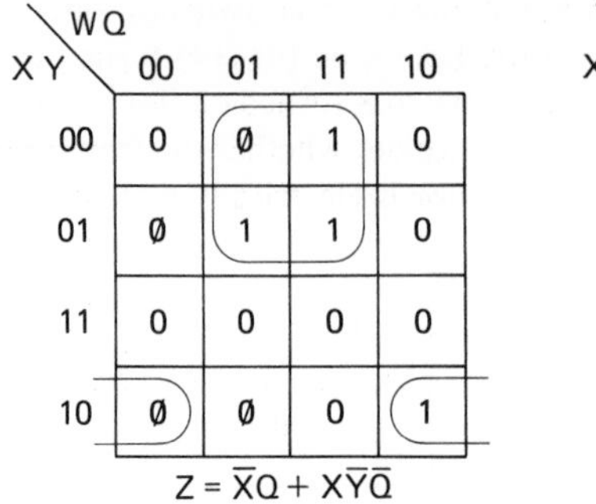

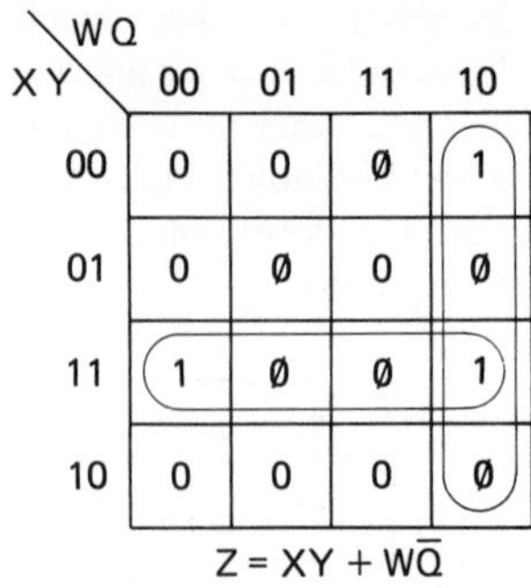

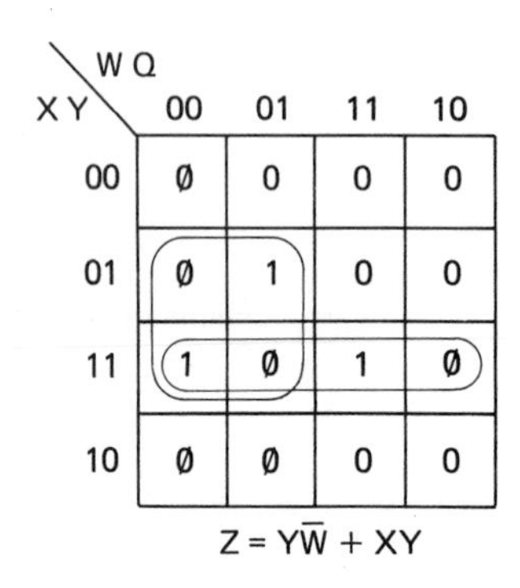

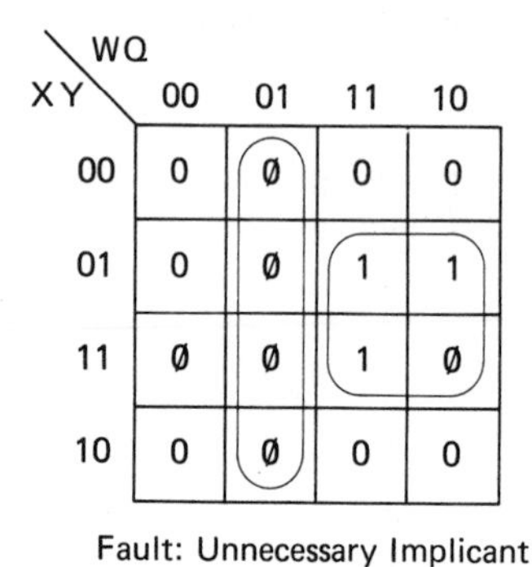

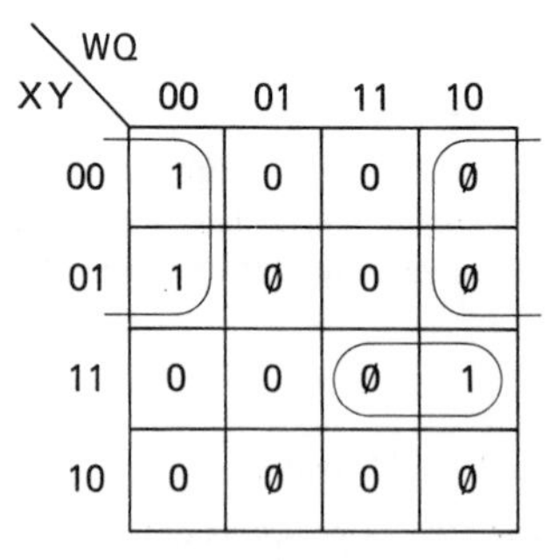

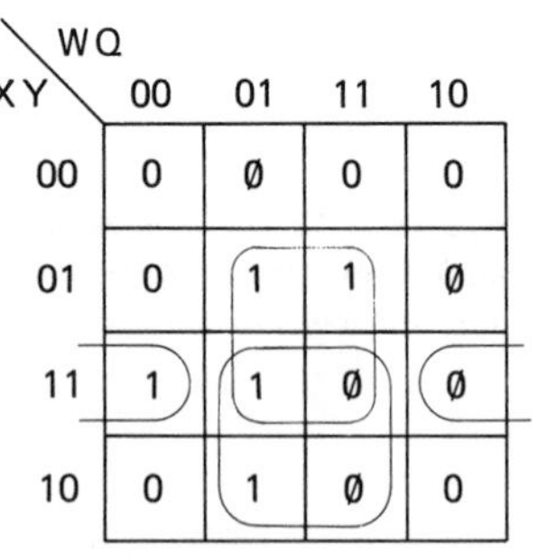

FIGURE 3-23 *(a) Correctly enclosed Karnaugh maps, and their corresponding MSP equations. (b) Incorrectly enclosed Karnaugh maps.*

Figure 3-23(a) illustrates how several Karnaugh maps are enclosed correctly when "don't care" states exist. Figure 3-23(b) shows serveral erroneously enclosed Karnaugh maps.

3.3 COMBINATIONAL LOGIC DESIGN

3.3.1 Evaluating the Problem and Planning Its Solution

At this point, all of the important analytical tools needed for the design of combinational logic networks have been presented. These analytical tools enable us to

1. Describe logic relationships with truth tables.
2. Express logic relationships with Boolean equations.
3. Manipulate and reduce equations using Boolean algebra.
4. Minimize logic relationships using Karnaugh map techniques.
5. Relate gate structures to Boolean equations.

In order to apply these tools effectively in designing combinational logic, it is necessary to develop an understanding of the design problem and plan a general course of action that will bring us to the solution. When such careful preparation is made, we need only execute our plan by calling to use the appropriate analytical tools. The major factors involved in evaluating a problem and planning its solution are described.

(1) Understanding the problem description: The description of a combinational logic problem may be verbal or mathematical. With a verbal description, some situation is presented in which people or events on the outside world are to cause a logic device to respond in a particular way. Often, verbal descriptions do not completely account for all possible events that may occur. Rather, the important input events are clearly described, while trivial or unlikely input events are not considered. It is the designer's responsibility to realize *all* possible input conditions, whether or not they relate to likely events. Having done so, those responses that are unspecified must either be inferred from common sense, or assumed to be unimportant and assigned "don't care" status.

If a problem is described mathematically, the task of realizing the problem is much easier. This is so because of the explicit nature of a truth table or Boolean equation. With a mathematical description, we are given a truth table or equation and must ultimately find a corresponding logic network. While the solution may not be straightfoward, the problem statement is.

(2) Analyzing the problem: Once the problem is understood, we must decide what actions are called for to bring about a solution. First, it must be determined that the problem really does involve combinational logic, and not sequential logic. If the operation of the desired circuit requires any kind of memory, a solution cannot be found using just combinational logic. Only when a fixed relationship exists between inputs and output(s) is a combinational logic solution applicable. When it has been decided that a combinational logic solution is called for, a plan of action must be found. Whether to make a truth table, or to minimize an equation with Boolean algebra or Karnaugh maps, or to use only NAND gates in the final circuit are factors that are considered. Deciding what analytical tools to apply and when to apply them comprises the plan of action.

(3) Formulating the problem: The first concrete steps taken to obtain a solution occur at this point. To begin, the number of input and output variables that are needed must be decided. If we are given an equation or truth table as the problem description, this decision is trivial. However, a verbal problem description may require some thought before the number of variables involved becomes clear.

In the case of a verbally described problem, meaning must be assigned to a variable's two values. For instance, in our door lock example given eariler, the output variable 'L' represented the condition of the door lock. The assigned meanings were $L = 0$ for an unlocked condition and $L = 1$ for a locked condition. Only when meaning is associated with the variable values can we transform a written description into a truth table.

Perhaps the most challenging step in formulating a verbally described problem is transforming the written problem statement into a truth table. The safest way to start this procedure is to write a truth table "skeleton," showing the input variables on the upper row and each possible binary combination below. For a three-variable problem, a skeleton table would appear as follows.

Input Variables			Output Variable
A	B	C	Z
0	0	0	
0	0	1	
0	1	0	
0	1	1	
1	0	0	
1	0	1	
1	1	0	
1	1	1	

By starting in this way, we are sure not to forget about any input conditions. Now, for each input condition, an output condition is decided and entered in the table. Usually, some degree of thought will be called for to unravel the verbal description. Before proceeding to minimization, all input conditions in the table should have an assigned output value of '0', '1', or '∅'.

Occasionally, several output variables are needed. In this case, all input conditions in the truth table should have an assigned value for *each* output variable. A truth table must be fully defined before proceeding to minimization. An example of a multiple output circuit is discussed in Section 3.3.5.

Certain problems seem simple enough so that an equation can be written directly from the verbal descritption. It is advised, however, that a truth table be formed anyhow to verify that all possible input conditions are involved correctly.

(4) Minimization: When we have a truth table, the remainder of the problem's solution is almost mechanical. By following the procedures established earlier in this chapter for Karnaugh maps, or in Chapter 2 for Boolean algebra, a reduced equation can be found that represents the function described by the truth table. Then, from the equation, a logic circuit is realized.

(5) Logic gate realization: It should be straightfoward at this point to obtain a combinational logic circuit from an equation. Several points should be noted, though. First, we may not always have free access to any kind of logic gate. Thus, it may be necessary to transform a reduced equation that would be directly implemented with AND, OR, or INVERT logic into a form suitable for other kinds of gates. This usually involves using DeMorgan's law. For example, the MSP equation

$$Z = X\bar{Y} + WY$$

is easily constructed with AND, OR and INVERT gates. However, if only NAND gates were available, the following transformation would be required.

$$\begin{aligned} Z &= X\bar{Y} + WY && \text{(given)} \\ &= \overline{\overline{X\bar{Y} + WY}} && \text{(recall that } Z = \bar{\bar{Z}}\text{)} \\ &= \overline{(\overline{X\bar{Y}}) \cdot (\overline{WY})} && \text{(DeMorgan's law)} \end{aligned}$$

The resulting equation is now suitable for NAND realization. Figure 3-24 compares the AND, OR, INVERT form to the NAND form.

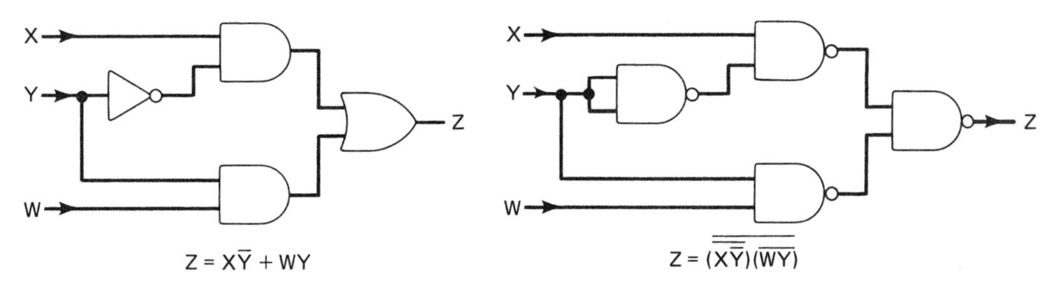

FIGURE 3-24 *Reduced equation implemented using different kinds of gates.*

A second point to be noted is that an MSP equation is not necessarily the form of the equation that should be implemented with logic gates. If possible, an MSP equation should be factored as much as possible before a gate structure is decided. For example, the MSP equation

$$Z = XY + \bar{W}Y$$

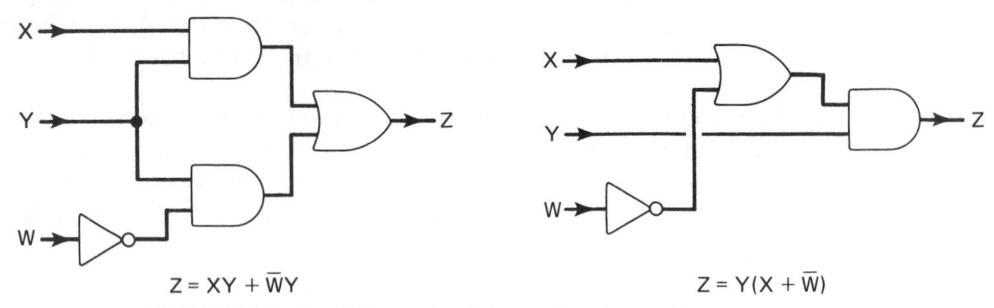

FIGURE 3-25 An MSP equation is factored to obtain a simpler gate structure.

would require three two-input gates, while its factored form,

$$Z = Y(X + \bar{W})$$

requires only two two-input gates. This is illustrated in Figure 3-25.

(6) Documentation: The end results should be carefully documented in writing and schematic drawings so that your solution can be clearly understood by other people. This is a crucial step, the failure of which may render a design useless. Unless others can understand your reasoning and results, the circuitry may be misunderstood or improperly applied.

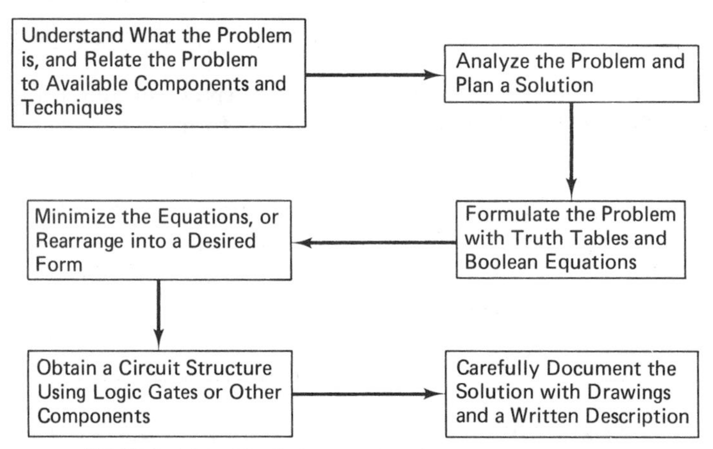

FIGURE 3-26 Block diagram describing the design process.

The steps that make up the design process are summarized in Figure 3-26. To demonstrate how these ideas are applied, a design example is presented.

EXAMPLE 3-5:

(1) Problem description: It is necessary to design a logic circuit that tests the operation of a traffic light. If the control circuitry for the traffic light malfunctions, it is possible that an invalid combination of signal lamps will appear. The sole purpose of

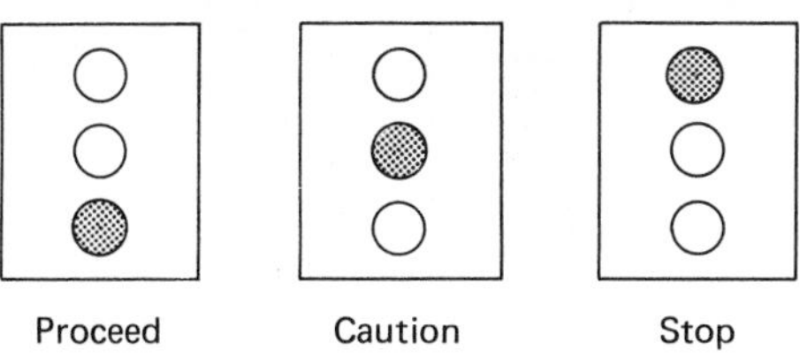

FIGURE 3-27 Valid combinations of signal lamps on a traffic light.

the test circuit is to detect any invalid combinations, and then generate an error signal that can be conveyed to the city maintenance crew. The valid combinations are shown in Figure 3-27.

(2) Analyzing the problem: First, note that memory is not required in the basic fault detection circuit. We need only detect an error condition and immediately produce an error signal. Practically, it might be desirable to have a simple latch memory involved so that a momentary error would result in a constant error signal. At this time, however, let us be concerned only with the detection of an error. For this, a combinational logic circuit is applicable. The plan of our solution will be to

(a) Deduce the possible error conditions.

(b) Write a truth table that depicts all possible signal lamp combinations, indicating in the output column whether a combination is valid or erroneous.

(c) Transfer the truth table to a Karnaugh map, decide on the essential prime implicants, and write an equation for the function.

(d) Realize the equation with a logic gate network.

We will assume that two-input gates of any type are available. Also, we will not be concerned with the electrical coupling of the 110-V ac lamp voltage to the low-voltage DC logic gate inputs until the final logic circuit is designed.

(3) Formulating the problem: The error detection circuit will sense the on-off condition of the three signal lamps in the traffic light, and generate an error signal if an invalid combination exists. We can conclude, then, that there must be three inputs and a single output. The input variables are chosen to be 'R' (red), 'A' (amber), and 'G' (green). The single output variable is labeled 'E' since it is asserted for an error condition. The natural assignment of meaning to these variables is '0' for an "off" lamp, '1' for an "on" lamp, and at the output, '1' for an error condition.

Since there are three lamps, we know that there are eight possible combinations of "on" and "off" that may exist. The valid combination were given in the problem description, Figure 3-27. Because only three of the eight possible combinations are valid, we must conclude that the remaining five combinations are invalid and should cause an error signal to be generated. A truth table may now be compiled, as is done in Figure 3-28.

(4) Minimization: A Karnaugh map is found, based on the truth table of Figure 3-28, and the essential prime implicants denoted. This is shown in Figure 3-29. An MSP equation is now derived from the map.

$$E = \bar{R}\bar{A}\bar{G} + RG + AG + RA \qquad \text{(MSP)} \qquad \textbf{(3-18)}$$

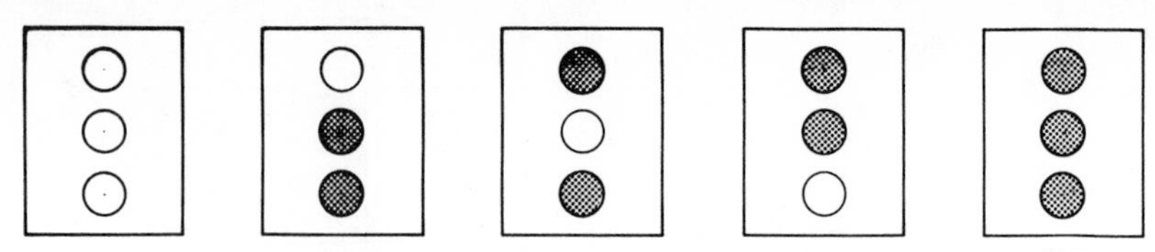

(a) Traffic Signal Conditions that are Erroneous.

R	A	G	E
0	0	0	1
0	0	1	0
0	1	0	0
0	1	1	1
1	0	0	0
1	0	1	1
1	1	0	1
1	1	1	1

(b) A Truth Table which Defines those Input Conditions that are to Result in the Generation of an Error Signal. For the Inputs, '1' Indicates that a Lamp is On. For the Output, '1' Indicates that an Error Signal Should be Produced.

FIGURE 3-28

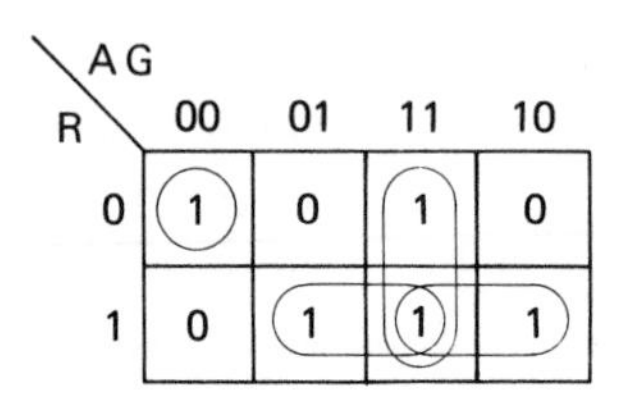

FIGURE 3-29 *The Karnaugh map related to the truth table of Figure 3-31.*

(5) Logic gate realization: Using only two-input gates, we see that three OR gates, five AND gates, and three INVERT gates (a total of eleven gates) would be required to realize the MSP equation. Assuming that two-input NOR gates are available, the MSP equation is factored and rearranged as follows:

$$E = \bar{R}\bar{A}\bar{G} + RG + AG + RA \qquad \text{(MSP, given)}$$
$$= \bar{R}\bar{A}\bar{G} + R(A + G) + AG$$
$$= \overline{R + A + G} + R(A + G) + AG$$

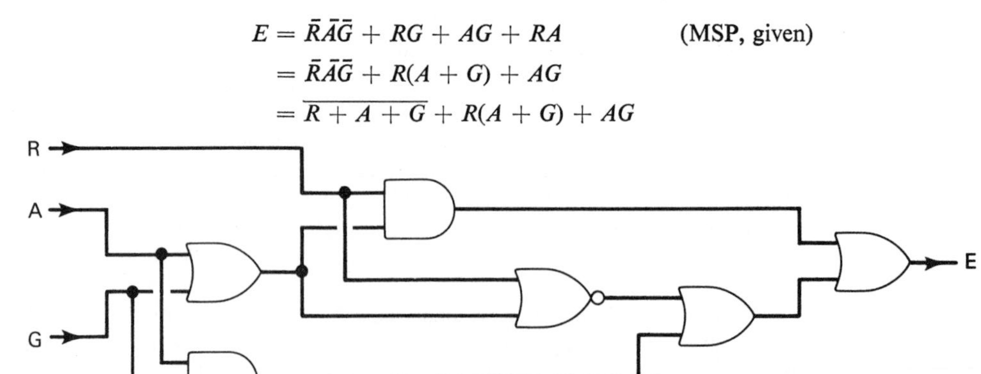

FIGURE 3-30 *Minimal gate structure for the traffic light error detection circuit.*

With this equation, a minimal gate structure, requiring only six gates, is obtained, as shown in Figure 3-30. The complete circuit, including coupling from the AC lamp power, is shown in Figure 3-31.

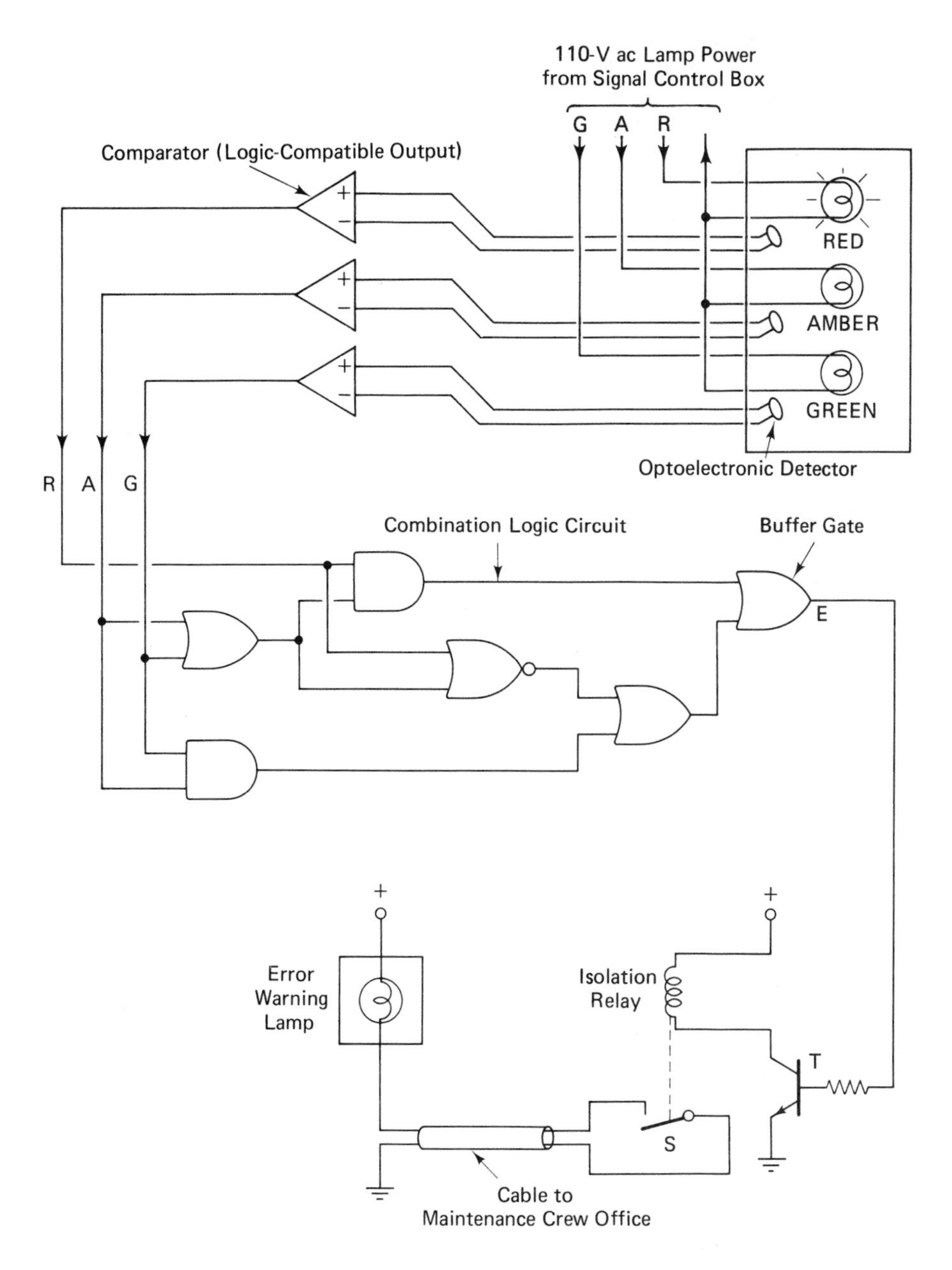

FIGURE 3-31 *Complete circuit for stoplight fault detector, shown in the valid red light state.*

(6) Documentation: A three-lamp traffic signal is represented in the upper right corner of Figure 3-31. The lamps must be supplied with 110-V AC power to glow, and will remain dark otherwise. A control box, not shown in the figure, causes power to be selectively applied to the lamps at periodic intervals so that proper operation of the traffic signal is achieved.

When a lamp is energized, a small percentage of the emitted light is captured by an optoelectronic detector, which responds to light by producing a small voltage. This voltage is also developed across the comparator inputs. The comparator contains an amplifier to boost the voltage to a valid logic '1' level, and also a Schmitt trigger circuit to ensure a sharp transition between logic states. In the absence of light, the detector produces no voltage, and the comparator's output is logic '0'.

The identification of valid and invalid lamp conditions is shown in the truth table of Figure 3-28. Rearranging the truth table into the form of a Karnaugh map, it is possible to select the essential prime implicants and arrive at an MSP equation, stated by Eq. 3-18. Using standard two-input logic gates, a minimal combinational circuit is developed and shown in Figure 3-30. The design of the combinational logic circuit provides that an error signal will be generated whenever an invalid signal lamp combination exists. The 'E' output of the combinational logic asserts a '1' only if an error condition occurs, and being '1', provides drive current to the base of transistor 'T'. The OR gate that has been selected to produce 'E' is a special gate electrically able to supply sufficient drive current to the transistor. This "buffer" gate provides greater output drive current than a normal unbuffered gate could provide. Only when drive current is applied to 'T' will current flow through the coil of the isolation relay, which in turn causes switch 'S' to close. An error warning lamp, installed in the office of the city maintenance crew, glows whenever the transmission line circuit to which it is connected becomes closed. This condition exists when the coil of the isolation relay is energized. When the traffic signal cycles through its normal sequence, it is possible that momentary errors may occur. This would be the case if the light emitted by an off-going lamp did not fade as fast as the light of an on-going lamp built up. Thus, the combinational logic would temporarily react as though two lamps were on—an undisputed error. However, the duration of this temporary state would not be sufficient to fully energize the relay coil, and this normal, momentary error *would not* be indicated on the error warning lamp. This circuit is capable of detecting both dead signal lamps, and defective control circuitry.

3.3.2 Alternative Forms of Logic Gate Symbols

One of the most important aspects of logic design is the drawing of a clear, readable logic diagram when the final design is completed. At first, this may seem to have nothing to do with logic design. Really, it is a very important factor in that if your design cannot be effectively communicated to other people, its value is greatly limited. There are certain obvious steps that can be taken to make a drawing clear. They include the use of a template for logic symbols, and line drawings made with a straightedge. One feature of a logic diagram that can make the difference between clear and obscure communication is the *symbolic form* in which a particular logic gate is expressed. DeMorgan's law allows us to express a given logic function in one of two equivalent forms. It is the choice of the most readable form with which we are concerned.

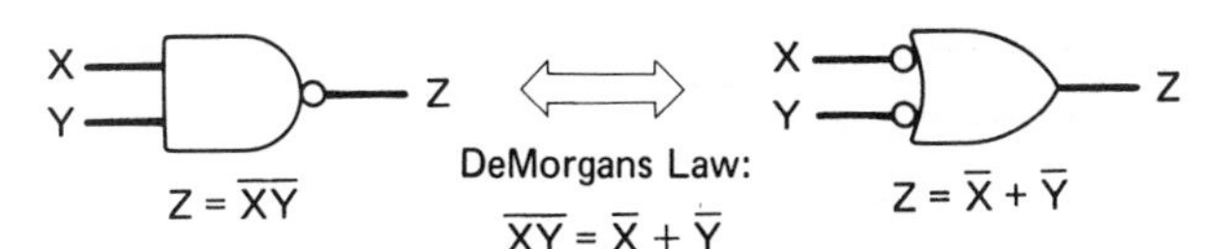

FIGURE 3-32 *Two different graphical symbols may be used to represent the NAND function. Each symbol is considered to be a NAND gate.*

Consider the NAND function, $Z = \overline{XY}$. We know that the correponding NAND gate is represented symbolically as an AND gate *with* an inversion circle at its output. By DeMorgan's law, however, $Z = \overline{XY}$ can also be written as $Z = \bar{X} + \bar{Y}$. Using the same inversion circle convention, this equivalent equation can be graphically written as an OR gate with an inversion circle at each input. This is illustrated in Figure 3-32. Why should one gate symbol be chosen over the other? The function

$$Z = XY + WQ$$

may be directly constructed with two two-input AND gates and a two-input OR gate. However, if only NAND gates are available, the circuit's construction is not quite as straightfoward. Figure 3-33 depicts both the AND–OR construction, and an equivalent NAND construction. Now, by drawing NAND gate 3 of Figure 3-33 with its equivalent symbol of Figure 3-32, we obtain the new schematic of Figure 3-34. Because the inversion circles at the inputs of gate 3 can be viewed as canceling the inversion circles at the outputs of gates 1 and 2, it is easy to comprehend the function of the NAND construction in terms of the AND–OR construction in Figure 3-33(a); whereas the NAND schematic of Figure 3-33(b) is much harder to follow.

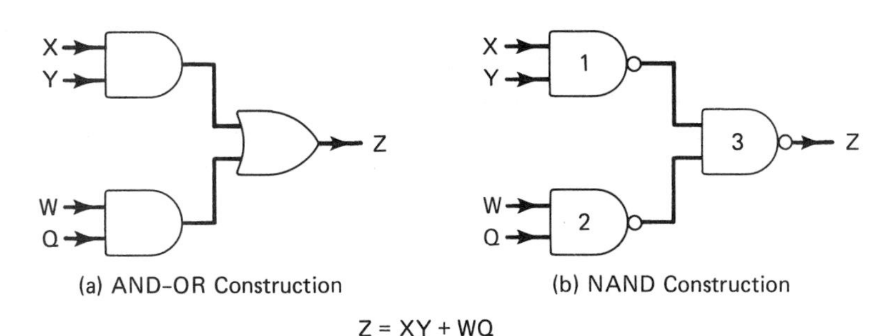

FIGURE 3-33 *A given function is realized using different kinds of gates.*

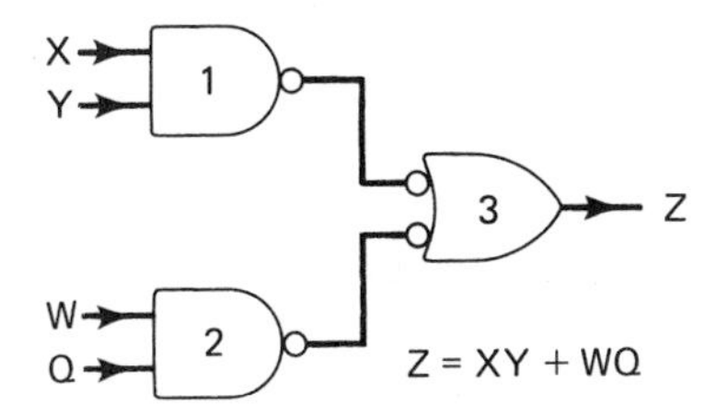

FIGURE 3-34 *A more readable drawing of the circuit in Figure 3-36(b).*

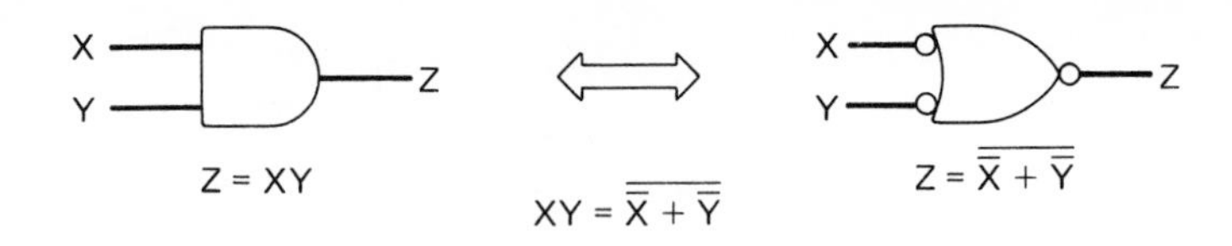

(a) The AND Function.

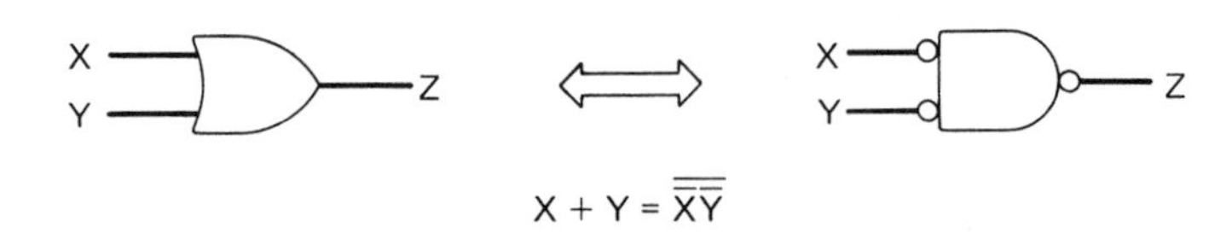

(b) The OR Function.

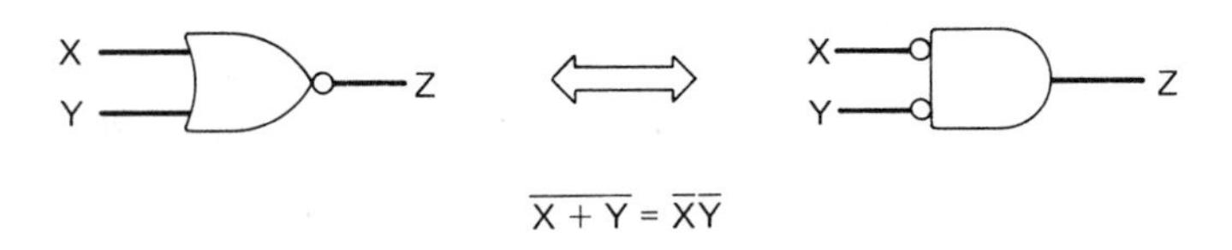

(c) The NOR Function

FIGURE 3-35 *Alternative graphical symbols for the AND, OR, and NOR functions.*

There are several variations of this technique that are also useful. In each case, DeMorgan's law is applied to express a given function in an alternative form. Figure 3-35 summarizes these variations. The alternative form of a logic gate symbol is selected whenever doing so allows us to comprehend the logic operation of a circuit more readily. A simple rule can be applied to arrive at the alternative form. If we consider the symbols "complements" of each other, then

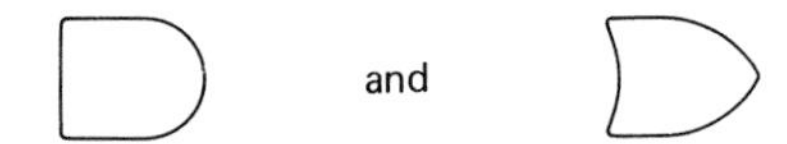

(a) Complement the symbol.

(b) Complement every input and output with an inversion circle.

The result is the DeMorgan equivalent form. Remember that two adjacent inversion circles cancel.

3.3.3 Multiple-Output Circuits

The general model of a combinational logic network portrays an unspecified interconnection of logic elements having an arbitrary number of inputs and outputs. The only restriction that was placed on the interconnected elements is that there be no

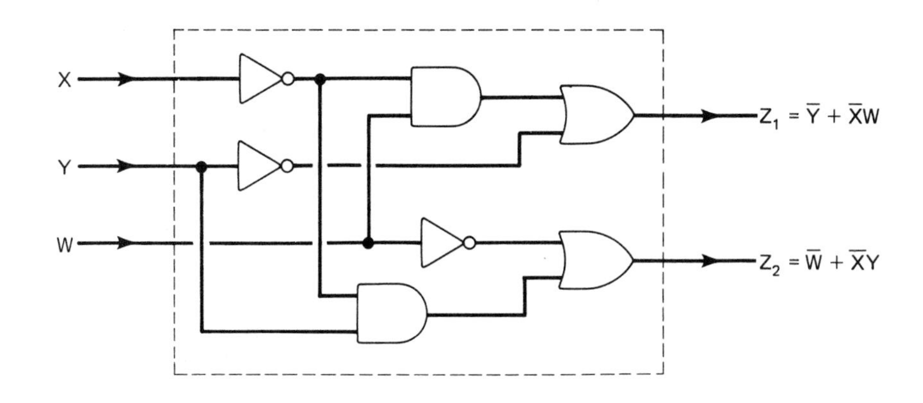

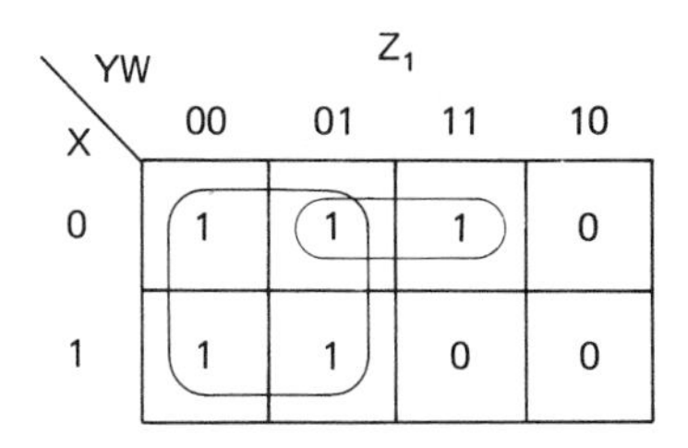

Z_2

X \ YW	00	01	11	10
0	1	0	1	1
1	1	0	0	1

(a)

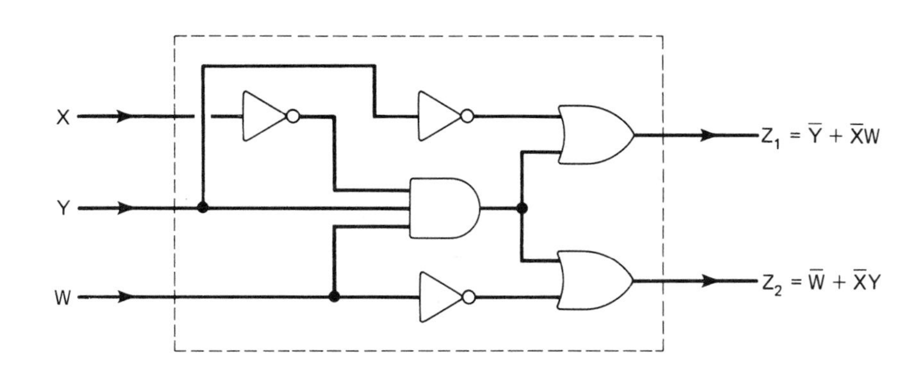

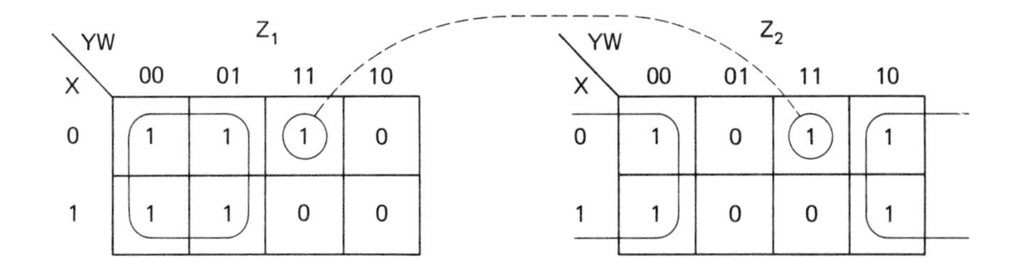

(b)

FIGURE 3-36 Multiple output circuit where (a) the functions are realized independently, and (b) where a common product is shared.

feedback paths. The combinational logic circuits that we have discussed thus far examined have had only a single output, however. Furthermore, all of our analysis and minimization techniques have assumed single output logic functions. It is frequently the case, though, that several output variables are derived from the same inputs. When this occurs, it may be possible to share certain intermediate product terms between several functions, and thereby eliminate some redundancy in the overall logic network. This is illustrated in Figure 3-36. There are several points that should be recognized in this example. First, it is necessary to specifically define what we mean by "minimal." Although the circuit of Figure 3-36(b) uses fewer gates and has fewer total inputs than that of Figure 3-36(a), it requires a three-input AND gate. Practically, it may be advantageous from an economic standpoint to use as many similar kinds of gates as possible. Thus, while the circuit of Figure 3-36(b) may be mathematically simpler, it is probably more expensive to build. Another practical consideration is related to reliability. That is, if product terms are shared between several functions, the failure of one commonly shared logic gate will cause failure in all these functions. In this case, the effect of circuit failure must somehow be accounted for in our definition of circuit cost, and thus in our criteria for minimality. In general, many factors can influence our understanding of "minimal." In addition to considering the number and type of logic gates, and reliability, other important factors include gate propagation delay, fanout, and the effect of a gate type on printed circuit layout. Finally, if it is decided that true algebraic minimization between multiple output logic is desired, regardless of gate type or reliability considerations, we must ask: How is it done? The reduction obtained in Figure 3-36(b) is not an obvious one, since the reduced form is obtained by first making two independent implicants smaller (more complex), and then recognizing that the result is common to both functions and can be shared. Thus, while we have made two implicants more complex, they can now be shared as *one* implicant, making the overall logic simpler.

3.3.4 Applications

(1) Multiplexer–Demultiplexer: When transmitting digital signals over distances greater than 3 ft, a number of new factors must be considered. First, steps must be taken to prevent signal distortion and interference effects. These steps include special interface circuits at the transmitting and receiving ends, and possibly the use of special cable. Second, the need to send several signal lines in parallel multiplies the cost of the interface circuitry and cable. Therefore, this need for providing parallel signal lines must be carefully weighed against the cost of doing so.

A very useful scheme has been developed that enables one transmission line to carry several digital signals. This scheme is referred to as ***time-division multiplexing*** and allows a transmission line to be shared by several signal lines. The "sharing" does not occur simultaneously in time-division multiplexing, but rather, the transmission line is used in short intervals at prescribed times by *each* signal source. Thus, a switching effect is observed as the multiplex circuitry alternately switches from one signal source to another. The technique of sharing one line by several lines is relatively uncomplicated. In its simplest form, two identically controlled selector switches are involved, one at each end of the transmission line. This is illustrated in Figure 3-37.

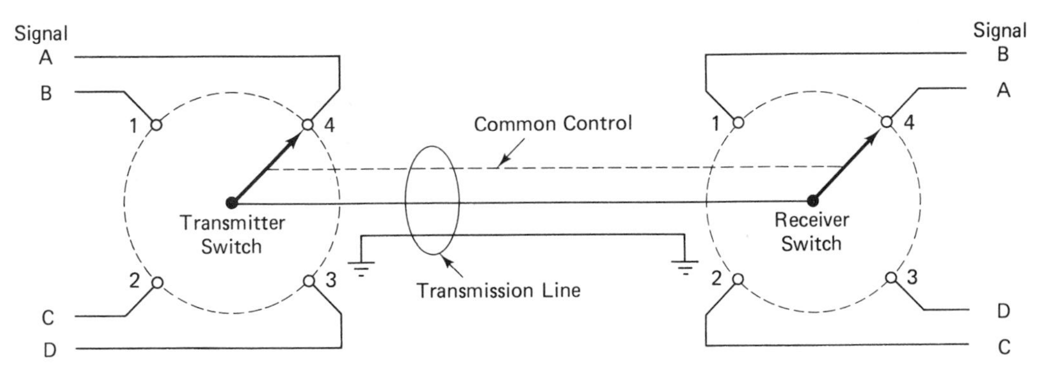

FIGURE 3-37 Sharing of one transmission line by four signal lines.

It is evident from Figure 3-37 that no two signal sources can be sent at exactly the same time; this is the sacrifice that is made when one transmission line is shared. Often, however, the signal lines are not constantly in use, but send information during some times and remain idle at other times. When this happens, the idle lines are ignored and the transmission line is shared only by the active lines. Thus, maximum use is made of the sole connection between transmitter and receiver. In sophisticated systems, it is possible to sample the digital source signals very rapidly, transmit the samples individually, and reassemble the signals at the receiver end using only the samples. The resulting effect is that of parallel transmission. Our primary concern in this section, though, is to investigate how the selection process is realized using logic circuitry instead of selector switches. The timing and synchronization problems of idle line elimination and sampling are deferred to a later chapter.

The circuitry that selects one of several signal source lines at the transmitting end is referred to as a ***multiplexer***. At the receiving end, a ***demultiplexer*** places the received signal on one of several output lines. In order that correct transmission occur between the selected source line and the corresponding destination line, the multiplexer and demultiplexer must be coordinated to change in the same way. In an actual multiplexed system, this means that one or more control lines must be provided in addition to the transmission line. When a small number of lines are to be multiplexed, the cost savings of a transmission line plus control lines may not offset the cost of parallel transmission lines with no control. If many lines are to be multiplexed, however, the cost savings can be substantial.

We are interested in the design of an electronic multiplexer/demultiplexer system using digital logic. With such a design, we will have the ability to select one of several binary digital signals for transmission over a single line. Because signal flow through digital logic components is in one direction only, it will not be possible to simply reverse inputs and outputs of the multiplexer design to obtain a demultiplexer, as we could do if selector switches were used. Instead, separate designs will be needed for the multiplexer and demultiplexer.

As a first step, consider the design of a two line–to–one line multiplexer. It can be represented generally as shown in Figure 3-38. The operation is to allow a signal present on the selected line to pass to the output, while the unselected line is ignored.

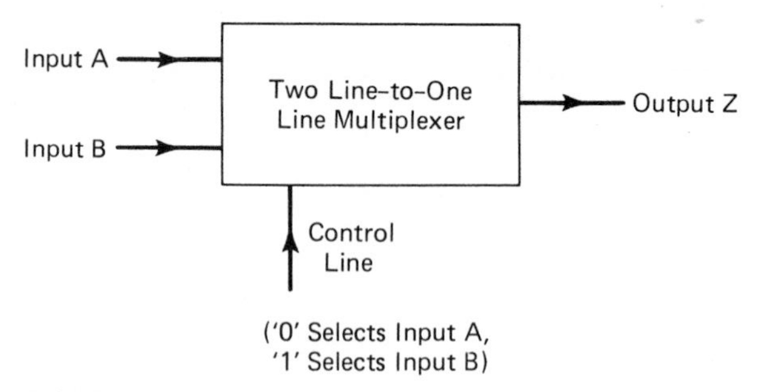

FIGURE 3-38 *General representation of a two line–to–one line multiplexer.*

If 'C' represents the control line, 'A' and 'B' the inputs, and 'Z' the output, an algebraic equation can be written to describe the desired operation.

$$Z = \bar{C}A + CB$$

If $C = 0$, then $CB = 0$ and $\bar{C}A = A$. Consequently, for $C = 0$, input A is selected. If $C = 1$, then $CB = B$, and $\bar{C}A = 0$. Thus, for $C = 1$, input B is selected. At no time will *both* or *neither* inputs be selected simultaneously. The minimal logic circuit for this equation is shown in Figure 3-39.

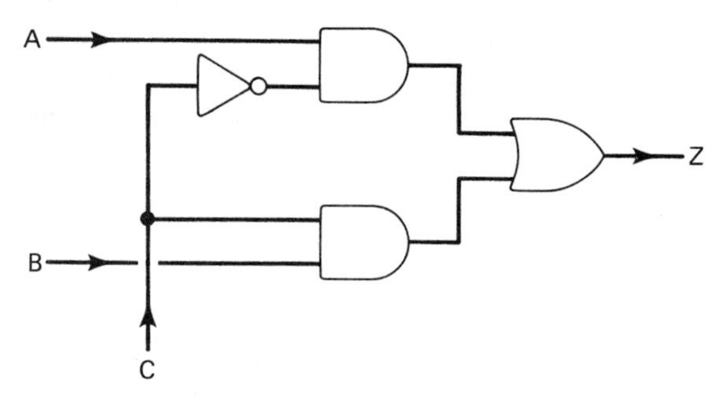

FIGURE 3-39 *Minimal logic circuit for a two line–to–one line multiplexer.*

More complex multiplexers can be designed by following the same basic pattern as with the two-input multiplexer. Namely, we start with the basic block representation as in Figure 3-38, showing inputs, output, and control lines, write an equation that describes the circuit's operation, and determine a minimal gate structure. Note that several control lines are needed when more than two inputs are present. The design of a four line–to–one line multiplexer is shown in Figure 3-40.

The demultiplexer design presents a slightly different problem. In this case, a single input is gated to one of several possible outputs. The general representation of a one line–to–two line demultiplexer is shown in Figure 3-41. The operation of the demultiplexer is to allow the input signal to be placed on only one of the output lines. The

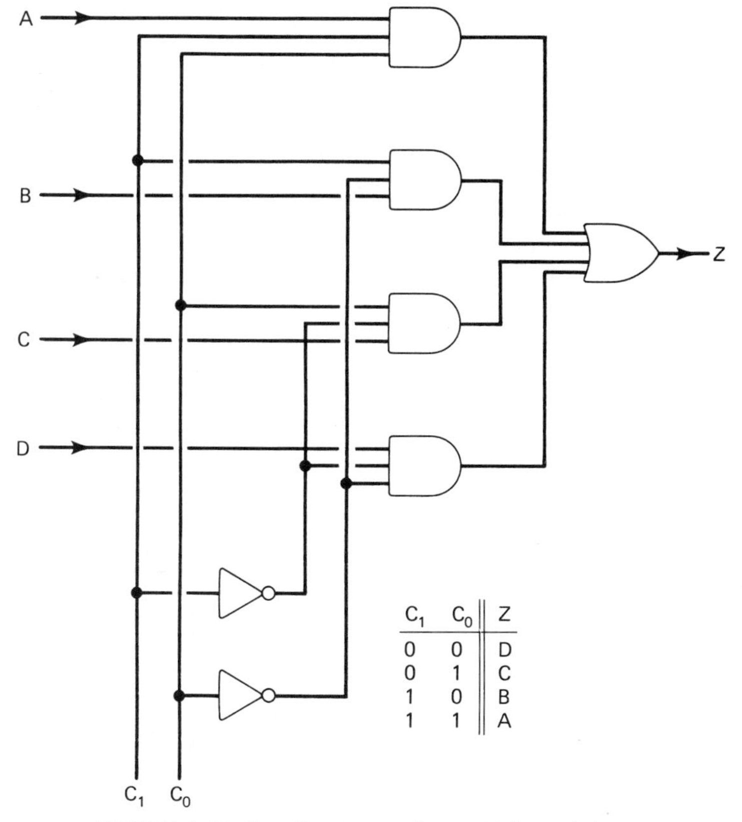

C_1	C_0	Z
0	0	D
0	1	C
1	0	B
1	1	A

FIGURE 3-40 *Four line–to–one line multiplexer design; unselected inputs are ignored.*

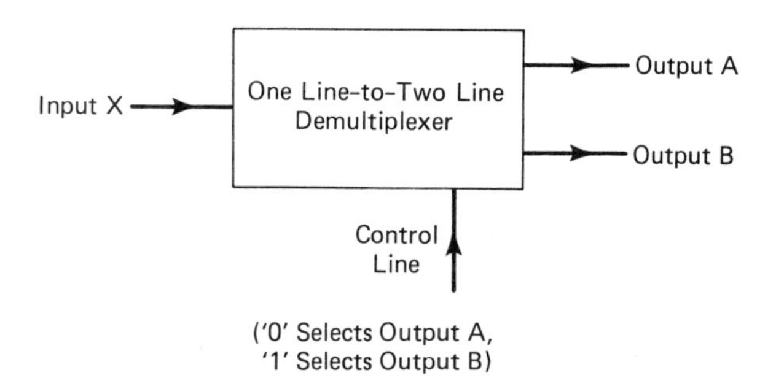

FIGURE 3-41 *General representation of a one line–to–two line demultiplexer.*

unselected outputs are to remain at logic '0'. If 'C' represents the control line, 'X' the input, and 'A and 'B' the outputs, two equations can be written to describe the desired operation.

$$A = \bar{C}X$$

$$B = CX$$

If $C = 0$, then $\bar{C}X = X$ and, therefore, output A receives input X. Under this condition, output B is a ways '0'. If $C = 1$, $CX = X$ and, therefore, output B receives X while output A is always to '0'. At no time will *both* or *neither* outputs be selected simultaneously. The minimal logic circuit for these equations is depicted in Figure 3-42.

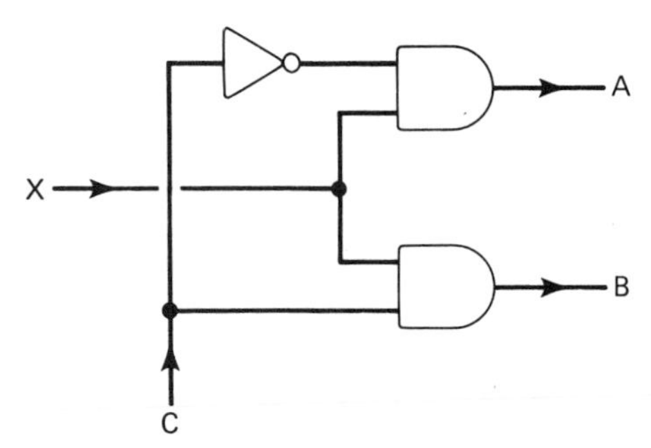

FIGURE 3-42 *Minimized logic circuit for a one line–to–two line demultiplexer.*

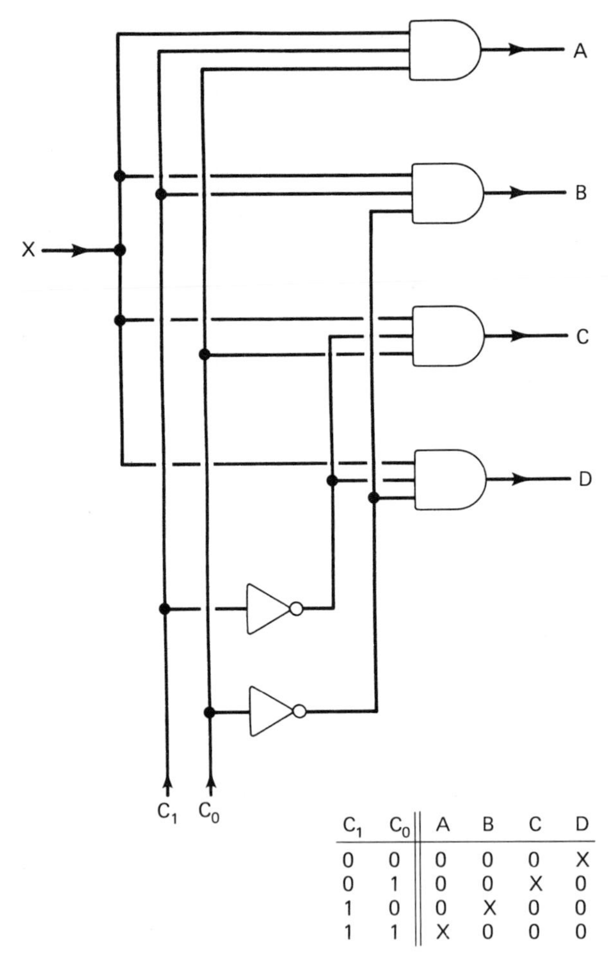

C_1	C_0	A	B	C	D
0	0	0	0	0	X
0	1	0	0	X	0
1	0	0	X	0	0
1	1	X	0	0	0

FIGURE 3-43 *One line–to–four line demultiplexer design; unselected outputs remain at logic '0'.*

Figure 3-43 shows the design for a one line–to–four line demultiplexer. Note that several control lines are needed to provide for the selection of more than two outputs.

(2) Seven-segment decoder: Many modern digital devices produce numeric results as their output. Typically, the numeric results appear on a visual display as decimal numbers. An electronic calculator and a digital voltmeter are examples of such devices. As we know, the actual digital circuitry within one of these devices can produce output only in terms of '1's and '0's. Thus, we must ask: How can the appearance of a decimal number be generated when only binary digits are produced at the output of a digital circuit?

It is possible to create the impression of a decimal digit by a particular combination of vertical and horizontal bars. Although the resulting image may not be as pleasing to the eye as a printed decimal digit, it is suitable for an electronic construction in which each bar is independently illuminated. Thus, by "turning on" the proper combination of vertical and horizontal bars, the image of a decimal digit is seen. Figure 3-44(a) depicts a seven-bar configuration, usually referred to as a ***seven-segment display***. Figure 3-44(b) illustrates how varying combinations of "on" segments give the appearance of decimal digits.

Since each segment of the numeric display can be independently illuminated, the problem of generating a decimal number has been reduced to one of generating a seven-digit binary number. The *position* of a digit in this binary number corresponds to one of the segments, and the binary value at that position corresponds to the segment's state of illumination. Thus, if the segment letters are related to digit position as

$$(a, b, c, d, e, f, g) \Longrightarrow (B_7, B_6, B_5, B_4, B_3, B_2, B_1),$$

then Table 3-2 indicates the appropriate seven-digit binary numbers. We assume that '0' represents an "off" segment and '1' represents an "on" segment.

TABLE 3-2

The relationship between decimal digits, "on" segments, and the corresponding seven digit binary numbers.

Decimal Digit	*Segments On*	*Seven Digit Binary Number*
0	a b c d e f –	1 1 1 1 1 1 0
1	– b c – – – –	0 1 1 0 0 0 0
2	a b – d e – g	1 1 0 1 1 0 1
3	a b c d – – g	1 1 1 1 0 0 1
4	– b c – – f g	0 1 1 0 0 1 1
5	a – c d – f g	1 0 1 1 0 1 1
6	a – c d e f g	1 0 1 1 1 1 1
7	a b c – – – –	1 1 1 0 0 0 0
8	a b c d e f g	1 1 1 1 1 1 1
9	a b c d – f g	1 1 1 1 0 1 1

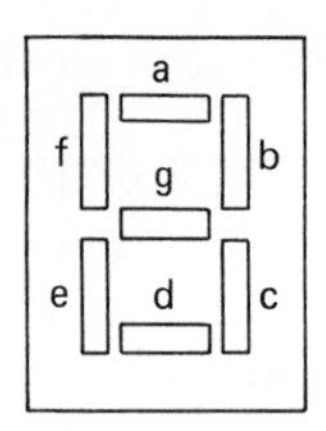

(a) Arrangement of Vertical and Horizontal Bars Needed to Produce the Decimal Digits Zero Through Nine. The Bars are Referred to as "Segments," and are Labelled 'a' Through 'g'.

Decimal Digit	Appearance	Segments On
0		a,b,c,d,e,f
1		b,c
2		a,b,d,e,g
3		a,b,c,d,g
4		b,c,f,g
5		a,c,d,f,g
6		a,c,d,e,f,g
7		a,b,c
8		a,b,c,d,e,f,g
9		a,b,c,d,f,g

(b) Various Combinations of "On" Segments Give the Appearance of Decimal Digits.

FIGURE 3-44

Although seven binary digits are required to represent decimal numbers in terms of seven segments, fewer binary digits are needed to represent simply ten different numeric symbols. In particular, with four binary digits, sixteen combinations of '1's and '0's are possible, and that is more than adequate to encode ten decimal digits. Now, a decision must be made by the designer of numeric output digital devices:

(a) Should the design treat each decimal digit as a seven-digit binary number?

(b) Should decimal numbers be encoded with fewer digits to save circuitry in the digital processor, and then on output, be passed through a seven-segment decoder for display purposes?

In most cases, it happens to be simpler to process small binary numbers than large ones. Consequently, most designers rely on seven-segment decoders when it becomes necessary to produce a decimal-digit output. Figure 3-45(a) relates the standard

Decimal Digit	Four-Digit Binary Code	Seven-Segment Code
0	0 0 0 0	1 1 1 1 1 1 0
1	0 0 0 1	0 1 1 0 0 0 0
2	0 0 1 0	1 1 0 1 1 0 1
3	0 0 1 1	1 1 1 1 0 0 1
4	0 1 0 0	0 1 1 0 0 1 1
5	0 1 0 1	1 0 1 1 0 1 1
6	0 1 1 0	1 0 1 1 1 1 1
7	0 1 1 1	1 1 1 0 0 0 0
8	1 0 0 0	1 1 1 1 1 1 1
9	1 0 0 1	1 1 1 1 0 1 1
–	1 0 1 0	Ø
–	1 0 1 1	Ø
–	1 1 0 0	Ø
–	1 1 0 1	Ø
–	1 1 1 0	Ø
–	1 1 1 1	Ø

(a) Table Relating the Minimal Decimal Digit Encoding to Seven-Segment Display Code.

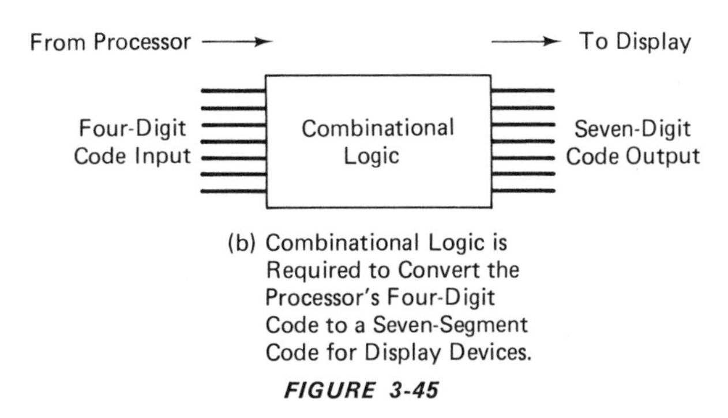

(b) Combinational Logic is Required to Convert the Processor's Four-Digit Code to a Seven-Segment Code for Display Devices.

FIGURE 3-45

four-digit code for decimal digits to the seven-digit binary numbers required for a display device. In order to realize this translation from four digits to seven digits, a multiple-output combinational logic circuit is needed, as shown in Figure 3-45(b). We now undertake to design this combinational logic circuit.

Several general observations are made before proceeding further. First, for each output variable, we must determine a minimal equation that relates it to the input variables. Karnaugh map techniques are most useful when there are three, four, or five input variables, so a Karnaugh map will be applied to determine the minimal equation for each output. Since there are seven output variables, we must find seven equations. Thus, the same basic procedure is applied seven times. Remember that each output corresponds to one segment of the display module, and each segment can glow only for certain input combinations. The Karnaugh maps are shown in Figure 3-46.

Because a multiple output combinational logic circuit is being designed, the opportunity for sharing certain product terms between output equations should be considered. If product sharing is possible, a reduction in the total number of logic gates will result.

Finally, note that six of the sixteen binary input codes are unassigned. Since no restriction is stated on the output values for these six codes, "don't care" status may be assumed. In some cases, it may be desirable to design a seven-segment decoder in which all segments are dark if an unassigned input code occurs. In our design, however, no such requirement is made. Therefore, the "don't care" entries will allow the circuitry to be simpler than it would be otherwise. Figure 3-46 develops independent MSP equations for each segment.

To determine what kind of product sharing is possible in this multiple output circuit, a table is made showing which product terms are associated with each equation, illustrated in Table 3-3. If a product term is used by more than one equation, it can be shared. The final design is shown in Figure 3-47.

TABLE 3-3

The use of certain product terms in more than one equation is investigated. The use of a product more than once means that it can be shared by several equations.

	Segment Equations Using This Product							
Product	*a*	*b*	*c*	*d*	*e*	*f*	*g*	*Number of Equations Using This Product*
$\bar{Q}\bar{Y}$	✓			✓	✓			3
QY	✓							1
$\bar{W}\bar{Q}$		✓				✓		2
WQ		✓						1
$\bar{Y}W$				✓			✓	2
$W\bar{Q}$				✓	✓		✓	3
$Y\bar{W}Q$				✓				1
$Y\bar{W}$						✓	✓	1
$Y\bar{Q}$						✓	✓	2

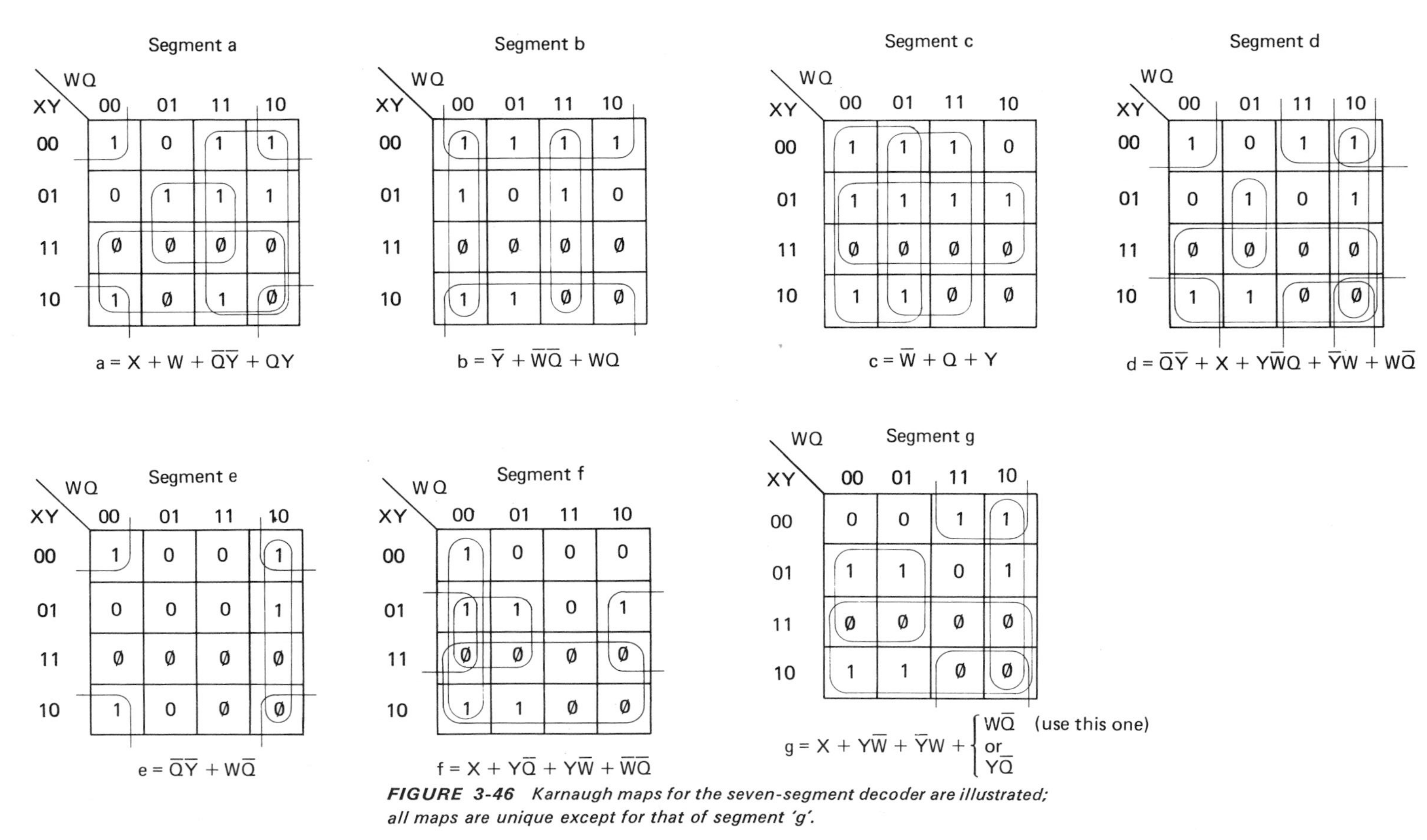

FIGURE 3-46 *Karnaugh maps for the seven-segment decoder are illustrated; all maps are unique except for that of segment 'g'.*

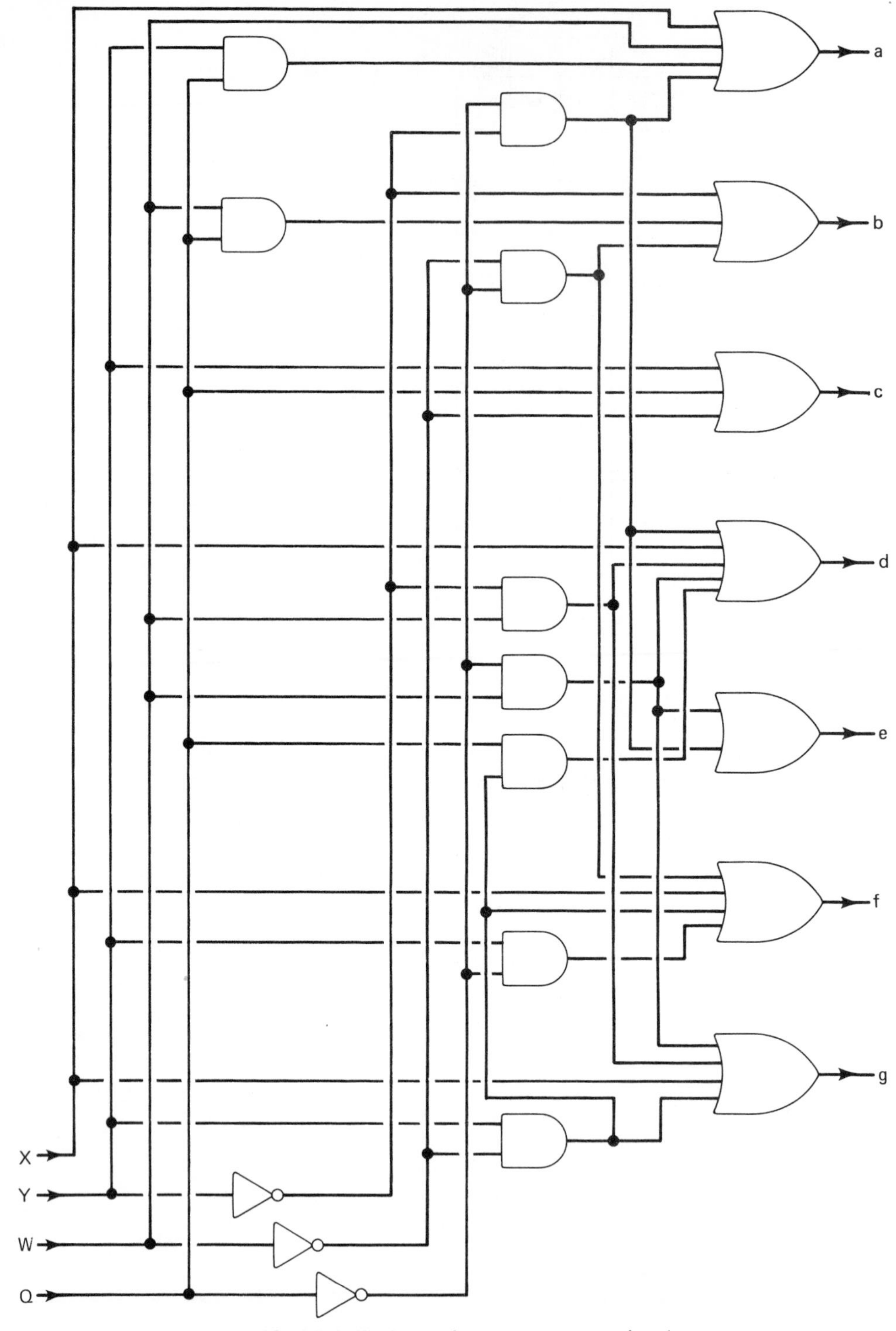

FIGURE 3-47 *Design for a seven-segment decoder.*

SUMMARY

A combinational logic circuit can be thought of generally as being a memoryless network of logic elements for which there are a finite number of inputs and outputs. Typically, logic circuits with two, three, or four inputs, and one output, have been discussed. However, logic circuits with more inputs and outputs are often encountered. A combinational logic circuit is characterized by a fixed relationship between inputs and outputs. This relationship is independent of any prior sequence of inputs applied to the circuit, and barring circuit malfunction, will not change. A large combinational network can be considered as being made up of a number of smaller interconnected subnetworks. Thinking of a large network as the interaction of many smaller networks often simplifies understanding the overall operation of a logic circuit.

To achieve greater confidence in our ability to obtain minimal algebraic expressions, Karnaugh maps have been developed. By expressing a Boolean function in a two-dimensional graphical form, as with a Karnaugh map, it is possible to select more reliably the minimum number of terms needed in an algebraic statement for the function. A Karnaugh map is a special rearrangement of a truth table in which the Boolean output values are entered in a matrix structure. Each cell of the matrix is identified by an address, where any two adjacent cells differ in address by only one digit. The address of a cell corresponds to the Boolean inputs, and the numerical value recorded in the cell corresponds to the related Boolean output. Clusters of adjacent '1's on a Karnaugh map lead to a single algebraic product term with fewer variables than would be obtained if each '1' entry of the cluster were expressed individually. Consequently, the largest possible clusters are identified on the map, following certain restrictions, to obtain an MSP algebraic equation.

In some cases, more than one MSP equation can be found for a Boolean function. Such functions are said to have MSP equations that are nonunique. Similarly, a Boolean function for which a single MSP equation is possible is said to have a unique MSP equation.

Verbally described logic problems often do not account for every possible input condition. When this happens, the logic values of unspecified output conditions must either be inferred from common sense, or assumed to be unimportant in affecting the desired logic operation. Unspecified output conditions whose exact values are unimportant are termed "don't care" states and are denoted by the symbol '∅'. When "don't care" states occur in Karnaugh maps, they may be assumed to be either '0' or '1' only insofar as it facilitates enlarging an enclosure. Once a '∅' has been replaced by a '0' or '1' in the determination of essential prime implicants, the Boolean function behaves according to this replacement. A '∅' entry in a Karnaugh map never results in a random output value in the final circuit design.

Combinational logic design is the procedure by which logic ideas and concepts are translated into workable circuitry. The techniques of Boolean algebra and Karnaugh maps are employed individually, or in combination, to minimize symbolically those circuit connections that are needed to obtain the desired result. Any logic design must be accompanied by a clear circuit description and set of schematics if other people are expected to understand the designer's intention. One factor involved in the drawing of logic diagrams is the form in which particular gate types are represented. DeMorgan's law can be applied to a gate symbol to produce an equivalent symbol in which the inputs, output, and symbol shape are complemented. An equivalent form is chosen whenever doing so simplifies the understanding of a logic gate network. Simplified understanding is possible by realizing that inversion circles at opposite ends of a wire can be viewed as canceling each other.

When multiple output logic circuits are encountered, it is sometimes possible to simplify the total gate requirement by sharing common subnetworks among several functions. Those logic elements suitable for sharing are revealed in each function's Karnaugh map by the presence of similarly located clusters of '1's. Many factors influence the meaning of "minimal" in our understanding of a minimal circuit. Thus, it is not possbile to provide a universal procedure to arrive at a minimal circuit. Generally, using the smallest number of gates with the fewest inputs is desirable.

NEW TERMS

Combinational Logic
Karnaugh Map
Minimal Circuit
Minimal Equation
Map Address
Adjacent Cells
Graphically Adjacent
Numerically Adjacent
Implicant
Enclosure
Prime Implicant
Essential Prime Implicant
Unique Function
"Don't Care" State
Multiple-Output Circuit
Multiplexer
Demultiplexer
Time-Division Multiplexing
Seven-Segment Display

PROBLEMS

3-1 Enter each of the following Boolean equations on an appropriate Karnaugh map. Do not attempt to minimize these equations; simply determine the binary value that must be recorded in each cell of the Karnaugh map. The given equations are not necessarily minimal.

(a) $Z = XY\bar{W} + \bar{X}Y\bar{W} + W\bar{Y}$

(b) $Z = XY + \bar{X}\bar{Y} + Q$

(c) $Z = (\overline{X + Y})(\overline{\bar{X} + \bar{W}})\overline{QW}$

(d) $Q = \bar{A}BC + A\bar{B}C + AB\bar{C}$

(e) $Z = (\bar{A} + \bar{B})(\bar{B} + \bar{C})(\bar{C} + \bar{D})(\bar{A} + \bar{D})$

(f) $Q = A + B + C + D$

3-2 Each different '0', '1' pattern that can exist on a Karnaugh map specifies a Boolean function. How many different '0', '1' patterns are possible on a four-variable Karnaugh map?

3-3 Figure P3-3 depicts four Karnaugh maps and eight Boolean equations. For each map, select all correct MSP equations that are present in the list.

(a)

XY \ WQ	00	01	11	10
00	1	0	0	1
01	0	1	1	0
11	0	1	1	0
10	1	0	0	1

(b)

XY \ WQ	00	01	11	10
00	1	1	1	1
01	1	0	0	1
11	0	1	1	0
10	0	1	1	0

(c)

XY \ WQ	00	01	11	10
00	1	0	1	1
01	0	1	1	0
11	0	1	1	0
10	0	0	1	1

(d)

XY \ WQ	00	01	11	10
00	1	0	0	1
01	1	Ø	1	Ø
11	Ø	1	Ø	1
10	1	0	0	Ø

$Z_1 = \overline{Y}W + YQ + \overline{X}\,\overline{Y}\,\overline{Q}$

$Z_2 = YQ + \overline{X}\,\overline{Y}\,\overline{Q} + \overline{Q}X\overline{Y}$

$Z_3 = Q\overline{Y} + \overline{X}\,\overline{Q} + XQ$

$Z_4 = Y + \overline{Q}$

$Z_5 = YQ + \overline{Y}\,\overline{Q}$

$Z_6 = WQ + YQ + \overline{Y}W + \overline{X}\,\overline{Y}\,\overline{Q}$

$Z_7 = XQ + \overline{X}\,\overline{Q} + \overline{X}\,\overline{Y}$

$Z_8 = (X \odot W \odot Q) + \overline{Y}\,\overline{Q}$

(See Problem 2-10 for a Definition of '⊙')

FIGURE P3-3

3-4 For each Boolean function shown in Figure P3-4, determine an MSP equation and indicate whether the equation is unique. If any nonunique equations are found, list all possible MSP equations.

FIGURE P3-4

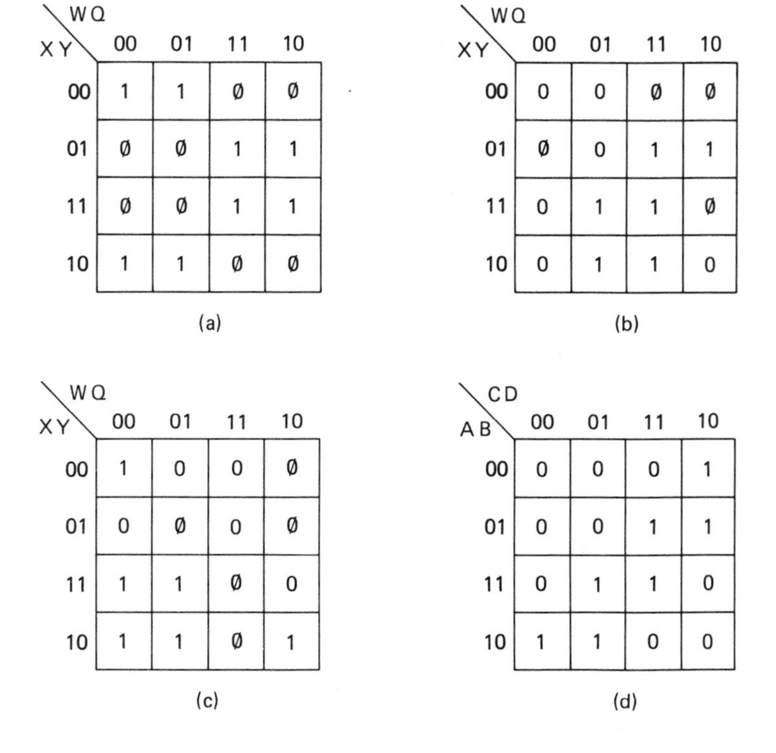

(a)

XY \ WQ	00	01	11	10
00	1	1	Ø	Ø
01	Ø	Ø	1	1
11	Ø	Ø	1	1
10	1	1	Ø	Ø

(b)

XY \ WQ	00	01	11	10
00	0	0	Ø	Ø
01	Ø	0	1	1
11	0	1	1	Ø
10	0	1	1	0

(c)

XY \ WQ	00	01	11	10
00	1	0	0	Ø
01	0	Ø	0	Ø
11	1	1	Ø	0
10	1	1	Ø	1

(d)

AB \ CD	00	01	11	10
00	0	0	0	1
01	0	0	1	1
11	0	1	1	0
10	1	1	0	0

3-5 The two four-variable maps drawn in Figure P3-5 are considered to represent one five-variable Boolean function. Find an MSP equation for this function, and indicate whether it is unique. Recall that our definition of cell adjacency requires only that adjacent cells differ by no more than one address digit. If you consider the two four-variable maps to be different levels of a three-dimensional figure, then the numerically adjacent cells will be geometrically adjacent also (except, of course, for the boundries).

E = 0

AB \ CD	00	01	11	10
00	0	0	0	1
01	1	1	0	0
11	1	1	0	1
10	0	0	0	1

E = 1

AB \ CD	00	01	11	10
00	0	0	0	1
01	1	0	0	1
11	1	0	0	1
10	0	0	0	1

FIGURE P3-5

3-6 Given a four-variable Karnaugh map, what is the relationship between the number of '1'-cells enclosed by an implicant, and the number of variables that are required in the product related to that implicant?

3-7 Is it necessary to know any properties of Boolean algebra when using Karnaugh map techniques to simplify an expression? Justify your answer.

3-8 Under what conditions, if any, might Karnaugh maps no longer be useful in simplifying algebraic expressions?

3-9 Can the map of Figure P3-9 be used as a Karnaugh map for a four-variable Boolean function? Why?

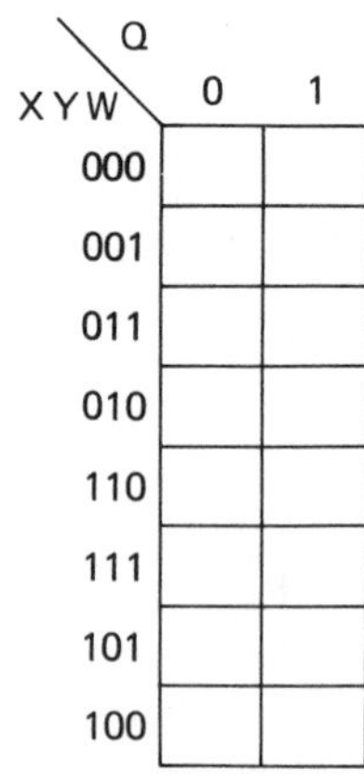

FIGURE P3-9

3-10 Is the function shown in Figure P3-10 a random waveform generator? Explain.

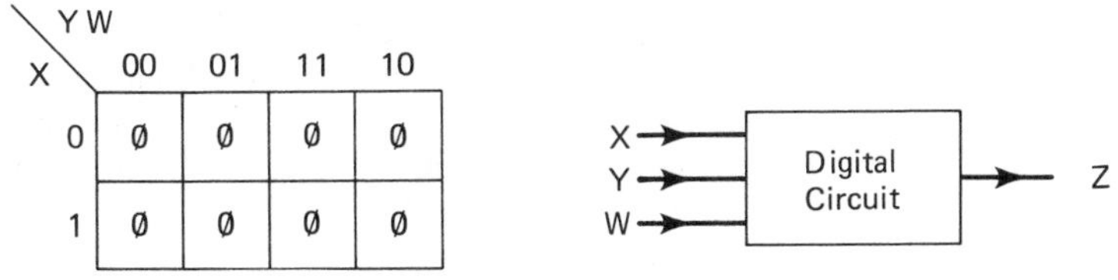

FIGURE P3-10

3-11 Describe the '1'-cell pattern that would appear on a three-variable Karnaugh map for the following functions.

(a) $Q = X \oplus Y \oplus Z$

(b) $Q = X \odot Y \odot Z$ (see Problem 2-10 for a definition of '$\odot$')

3-12 Design a two-variable Exclusive OR function with a minimum number of two-input NAND gates. An easily obtained network requires five gates; however, it can be done with four gates. When the four-gate version is found, describe the function that would result if every two-input NAND in this XOR design were replaced with a two-input NOR (assume the gate connections are identical).

3-13 Implement the traffic light truth table of Example 3-5 differently, assuming that EQUIVALENCE and Exclusive OR gates are available with any number of inputs.

3-14 It is necessary to design an overheat alarm for an oil-fired steam boiler. Three sensors are available. One sensor monitors the water temperature in the boiler, one monitors the chimney temperature, and one follows the on-off state of the burner. Figure P3-14 describes the logic operation of these sensors. An alarm signal should be generated whenever the burner is on and either the chimney or water temperature is too hot. Following the design process illustrated in Figure 3-26,

(a) Obtain a truth table that completely describes the operation of the alarm circuit.

(b) Transfer the truth table to a Karnaugh map and obtain a minimal equation.

(c) Construct the minimal circuit using any type of two-input gate.

Boiler Water

'0' Means Water is Within Normal Temperature Range
'1' Means Water is Too Hot

Chimney

'0' Means Chimney is Within Normal Temperature Range
'1' Means Chimney is too Hot

Oil Burner

'0' Means Burner is Off
'1' Means Burner is On

FIGURE P3-14

3-15 An airplane is to be equipped with a warning system that alerts the pilot under certain conditions of danger. The warning system monitors three instruments on the control panel. These instruments indicate the altitude, the airspeed, and the state of the landing wheels. Figure P3-15 indicates specifically the relationship of these inputs to logic states. Two conditions of danger must be detected. First, a "Landing Warning" lamp should glow if the airspeed is less than cruising speed and the landing wheels are not down, or if the wheels are down and the pilot is at the proper landing altitude but not at the proper landing speed. Second, an "Airframe Warning" lamp should glow whenever the airspeed is too fast, or when the wheels are down and the airspeed is above landing speed. Also, when either warning lamp glows, an alarm horn should sound to get the pilot's attention. Assume that only logic level voltages need be produced by the warning system, and that coupling the logic outputs to lamps and horn has already been done.

Altitude

'0' Means Landing Altitude
'1' Means Cruising Altitude

Wheels

'0' Means Wheels Up
'1' Means Wheels Down

Airspeed

'00' Means Too Slow
'01' Means Landing Speed
'10' Means Cruising Speed
'11' Means Too Fast

Increasing Speed ↓

FIGURE P3-15

3-16 An elevator system in a large building consists of six elevators. Cars 1 through 4 make up the primary system and are activated at all times. To save energy, cars 5 and 6 are completely shut down and are used only when the traffic is heavy. It is necessary to design a logic circuit that monitors the four primary cars and generates an "activate" signal when all four cars of the primary system are in use. The "activate" signal will cause the remaining two cars to become available for use. In addition, a "standby" signal should be generated for the cars when any three of the primary elevators are in use. A "standby" signal causes the motors of cars 5 and 6 to start, and an "activate" signal allows cars 5 and 6 to be used by the building occupants. Assume that four input lines are available, labeled X, Y, W, and Q, indicating the condition of the first four cars. A '1' on one of these input lines indicates that a car is in use, while a '0' indicates that a car is idle. Design a minimal logic circuit that provides the appropriate control signals to the standby cars. Any type of logic gates may be used. Assume that only logic-level voltages need be produced by the logic circuit, and that coupling to the elevator control system has been provided.

3-17 A seven-segment decoder design was discussed in Section 3.3.5. This design gave "don't care" output status to any four-digit input that did not correspond to a decimal digit. Determine the display pattern that would appear for each of the six unassigned four-digit inputs.

4 BINARY NUMBER OPERATIONS

4.1 BINARY REPRESENTATIONS

At the foundation of modern technology is man's ability to count. With this ability, it became possible to say such things as *how much* of something is available, *when* an event would occur, and *how far* two objects are separated. The physical world could then be thought of in a precise, well-defined manner, with mass, time, and space being measured in terms of basic units. Our ability to plan, to create, and to handle the most elementary aspects of our lives rests with our fundamental ability to count. We so often apply simple numerical concepts in the course of a typical day that the act becomes almost unconscious. As technological skills have advanced, so also have the numerical concepts with which the physical world is described. Modern numerical techniques have become extremely powerful in their ability to describe and predict properties of the physical environment.

Advances in technology generally bring about an *amplification* in the capacity of a human being. In particular, the development of digital electronics, and then the digital computer, have amplified man's ability to think, especially in thinking that pertains to arithmetic and numerical operations. Indeed, the degree to which a digital computer amplifies a human being's ability to perform numerical operations exceeds in amount almost every other enhancement of ability brought about by technological advance. Supersonic jet travel at 3000 miles per hour amplifies by a factor of 1000 the 3-mile/hour leisurely walking speed of a human being. The fastest computer, on the other hand, can perform about 70,000,000 additions *per second*, while a typical human mind would be hard-pressed to perform one addition per second.

The execution of numerical operations by electronic machines is a subject that requires an understanding of number systems and an understanding of digital electronics. This chapter examines many commonly understood properties of decimal numbers, and relates them to the binary number system that is used by digital circuitry. Through an understanding of this relationship, it will be possible to analyze and design digital circuits for numerical processing.

4.1.1 Number Systems

The simplest way in which a quantity of items can be represented on paper is to write a slash, a one, or some other symbol for each item of the group. For example, six items might be represented as "111111". While this scheme is useful for small numbers of items, say less than ten, it becomes extremely difficult to represent large quantities. Using a standard typewriter with single spacing between symbols and lines, almost two full $8\frac{1}{2}$- by 11-in. pages would be required to represent 10,000 items. Certainly, 10,000 items is not an unusually large quantity by modern standards. Faced with this problem of representing compactly large quantities of items, the earliest mathematicians devised a scheme of notation in which

1. Many different symbols are used, other than just a slash or a one.
2. The position of a symbol with respect to a fixed reference (e.g., the decimal point) weights the value of the symbol.

Such a scheme is referred to as a ***number system***. We are most familiar with the ***decimal number system***, which utilizes ten different elementary symbols: 0, 1, 2, 3, 4, 5, 6, 7, 8, and 9. These symbols may be combined according to certain rules to represent compactly any quantity of items. When considering whole items with no fractional part, a decimal point is unnecessary, and the resulting number is referred to as an *integer*. Integers are written along a line, with the location of the rightmost digit serving as a reference point for other digits. Now, if a counting sequence is imagined, we note that starting at zero, we can represent up to nine items before exhausting all of the single decimal symbols ('0' represents no items, and '9' represents nine items). To represent ten items, a second digit is needed, reusing one of the elementary symbols. Note, however, that the weight of the second digit is ten times the weight of the first digit. This is illustrated in Figure 4-1. Thus, a numeric symbol such as '1' has no absolute value of itself, but takes on a numerical value only when located in a particular digit position of a number. The weight of a decimal symbol *increases* by a

	Number	Quantity Represented
	0	None
	1	One
	2	Two
	3	Three
	4	Four
	5	Five
	6	Six
	7	Seven
	8	Eight
	9	Nine
The Symbol '1' in the Second-Digit Position Represents the Quantity "Ten," Not "One"	10	Ten plus None
	11	Ten plus One
	12	Ten plus Two
	13	Ten plus Three
	:	:

FIGURE 4-1 *Natural count sequence, showing how the position of a symbol determines its numerical value.*

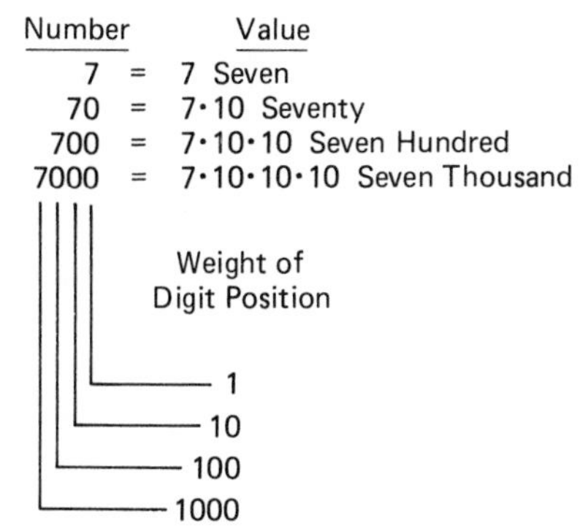

FIGURE 4-2 *The digit position of a symbol in a decimal number determines the numerical value of the symbol.*

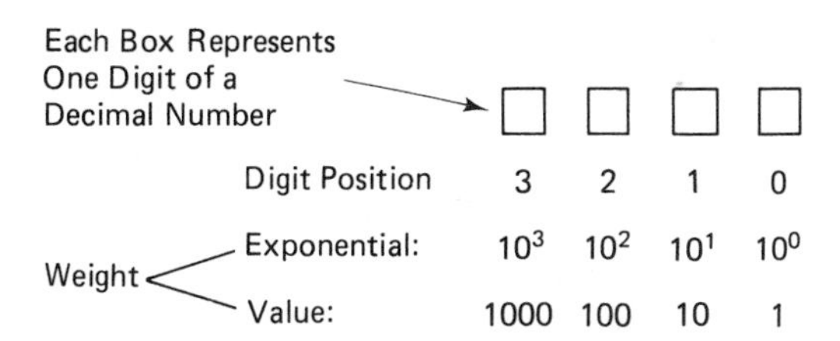

(a) The Digit Positions and their Weights in a Four-Digit Decimal Number

$$6992 = (6 \cdot 10^3) + (9 \cdot 10^2) + (9 \cdot 10^1) + (2 \cdot 10^0)$$
$$= 6000 + 900 + 90 + 2$$

$$2871 = (2 \cdot 10^3) + (8 \cdot 10^2) + (7 \cdot 10^1) + (1 \cdot 10^0)$$
$$= 2000 + 800 + 70 + 1$$

(b) Examples Showing How a Decimal Integer can be Considered to be the Sum of its Weighted Numerical Symbols.

FIGURE 4-3

factor of ten for each digit position it is moved to the left. Figure 4-2 illustrates this idea. To more clearly indicate the relationship between the position of symbol and its weight, the weight of the digit positions, as in Figure 4-2, is rewritten using an exponential form of notation. This is shown in Figure 4-3(a). An integer, then, can be considered to be the sum of weighted numeric symbols, with the weight of a given digit determined by its position in the number. Figure 4-3(b) illustrates this by example. Be careful not to confuse a multiplication dot with a decimal point!

When it is necessary to represent a fractional part of an item, integers are no longer satisfactory, since integers can represent only whole numbers. By adding a decimal point to the rightmost digit of an integer, and supplying additional digits to the right of the decimal point, fractional parts of a measured quantity can be expressed. In the

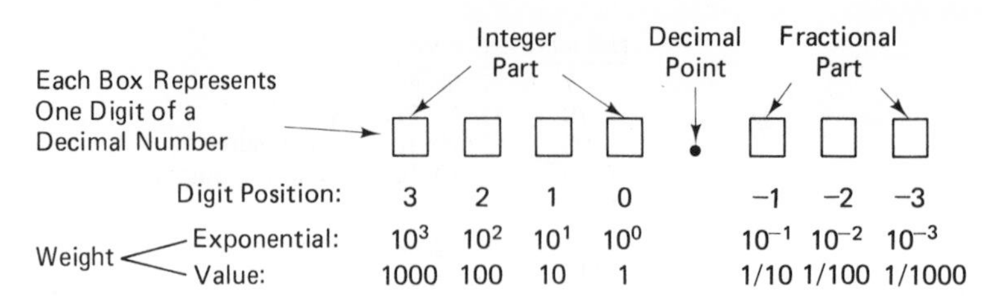

(a) The Digit Positions and their Weights in a Complete Decimal Number.

$$1298.5 = (1 \cdot 10^3) + (2 \cdot 10^2) + (9 \cdot 10^1) + (8 \cdot 10^0) + (5 \cdot 10^{-1})$$
$$= 1000 + 200 + 90 + 8 + 0.5$$

$$34.118 = (3 \cdot 10^1) + (4 \cdot 10^0) + (1 \cdot 10^{-1}) + (1 \cdot 10^{-2}) + (8 \cdot 10^{-3})$$
$$= 30 + 4 + 0.1 + 0.01 + 0.008$$

(b) Examples Showing How a Complete Decimal Number can be Considered to be the Sum of its Weighted Numeric Symbols.

FIGURE 4-4

case of fractions, the weight of a symbol *decreases* by a factor of ten for each digit position to the right that it is moved. Figure 4-4(a) illustrates the complete representation of a decimal number, and Figure 4-4(b) gives some examples. Although a seven-digit number is implied by Figure 4-4(a), a number can extend to any amount of digits to the left, and to the right, of the decimal point. Leading and trailing zeros tell us nothing new about the value of a number and are often eliminated.

It is possible to generalize our understanding of a decimal number by thinking of it as a sum of unspecified decimal coefficients weighted by position. Thus, any measurable quantity can be represented in decimal notation by 'S', as given in Eq. 4-1.

$$S = a_n 10^n + a_{n-1} 10^{n-1} + \ldots + a_2 10^2 + a_1 10^1 + a_0 10^0 + a_{-1} 10^{-1} + a_{-2} 10^{-2} + \ldots + a_{-m} 10^{-m} \quad \textbf{(4-1)}$$

In this equation, each coefficient is represented abstractly by the letter 'a', and the position of the coefficient within the number is represented by the subscript of each 'a'. The weighting factor that is to multiply each coefficient is given by '10' to some exponent. The value of the exponent is similarly determined by the position of the coefficient within the number. 'n' represents the most significant position needed to represent the integer part of the number, while '$-m$' represents the least significant position needed to represent the fractional part of the number. Each coefficient is a single digit that can be selected from the set of symbols {0, 1, 2, 3, 4, 5, 6, 7, 8, 9}. The fact that there are ten symbols in this set is exactly why the number '10' appears in the weighting factor. This number, '10', is referred to as the ***radix***, or alternatively as the ***base***, of the decimal number system. The radix always tells us the number of unique symbols that may be used in a number system.

It is possible, and sometimes desirable, to utilize number systems that are other than radix-ten. Indeed, if human beings were not born with ten fingers, we might very well be using a non-decimal-number system today. The following parts of this chapter examine properties of the radix-ten number system, and relate these properties to similar ones in the decimal system.

4.1.2 Decimal/Binary Relationships

If we hope to devise a means of processing numbers with electrical circuitry, it becomes necessary to represent numerical quantities by electrical signals. One direct way of doing so would be to have a span of voltage divided into ten discrete levels, where each level represents one symbol of the ten decimal symbols. Then, it would be possible to represent a five-digit decimal number with appropriate voltages on five different wires. Although this sounds reasonable, many difficulties would soon be discovered when an attempt is made to build electronic circuitry that responds to ten discrete voltage levels. First, complex and expensive circuitry is needed to correctly distinguish between ten different voltage levels. Second, only slight noise voltages need to be added to a ten-level signal to cause severe errors. Properties of semiconductor material limit the voltage range over which a high-speed switching transistor can operate effectively. For a given span of voltage, such as 5 V, divided into ten equally spaced intervals, a small noise voltage can easily superimpose on the true signal voltage, moving the total signal voltage through a threshold to another level. If the given span of voltage were divided into a smaller number of intervals, for instance two, much greater noise voltages could be tolerated before passing a threshold. Thus, with its complexity and susceptibility to noise, ten-level circuitry becomes impractical.

As we have seen in earlier chapters, binary digital systems operate using *two* discrete levels of voltage, represented by the symbols '0' and '1'. This is so because of the close relationship between true/false logic operations and bilevel signals. Also, digital circuitry that responds to two-level electrical signals is not extremely complex, and is relatively insensitive to electrical noise. For these reasons, it would be very desirable to represent numerical quantities with two-level signals. This means that a number system having only two symbols is needed. The ***binary number system*** is such a system, utilizing the symbol set {0, 1}, and is the basis of almost all modern digital equipment. A single binary digit is usually referred to as a ***bit***, which is a contraction of BInary digiT. To be able to understand binary digital systems that process numerical quantities, it is necessary that we become fluent with the binary number system. This involves binary counting, binary/decimal conversion, and binary arithmetic. The only difference between the binary system and the decimal system is the radix; the fundamental counting and arithmetic operations of each system are very similar.

Consider the problem of having to count a quantity of items using only the two symbols '0' and '1'. If we use a positional notation similar to the decimal system, we will apply the available symbols one at a time as the quantity of items we count increases. When all the symbols of the symbol set have been used once, a *second digit* is added to the number. This second digit reuses one of the symbols, but the position of the second digit has greater weight than the first digit. Consequently, a larger quantity can be represented. In this same sense, any number of digits can be added to form a simple numerical representation of any quantity of items. Figure 4-5 illustrates how fifteen items are counted using both a decimal notation and a binary notation.

Quantity of Items		Symbol Set: {0,1,2,3,4,5,6,7,8,9} Decimal Notation	{0,1} Binary Notation
Zero		0	0
One	/	1	1
Two	//	2	10
Three	///	3	11
Four	////	4	100
Five	/////	5	101
Six	//////	6	110
Seven	///////	7	111
Eight	////////	8	1000
Nine	/////////	9	1001
Ten	//////////	10	1010
Eleven	///////////	11	1011
Twelve	////////////	12	1100
Thirteen	/////////////	13	1101
Fourteen	//////////////	14	1110
Fifteen	///////////////	15	1111

FIGURE 4-5 *Comparison between decimal and binary counting.*

Since only two symbols are available in the binary number system, the system's radix is *two*. Thus, the weight that is assigned to a binary symbol in positional notation exactly *doubles* for every position to the left it is moved. Recall that in the decimal system, the radix is ten, and the weight that is assigned to a decimal symbol in positional notation increases by a factor of ten for every position to the left it is moved. Figure 4-6 illustrates how the weight of a binary symbol changes with respect to its position in a number. Notice that since fewer symbols are available in the binary system than in the decimal system, more digits are required to represent a given quantity with a binary number.

The conversion of a binary number to a decimal number simply involves multiplying each digit of the binary number by its appropriate weighting factor, and adding the results. The weight of digit positions in a binary integer is shown in Figure 4-7(a), and several examples demonstrating binary-to-decimal conversion are shown in Figure 4-7(b).

Fractional parts of an item may also be expressed in binary notation. In the binary system, however, a *binary point* is added to the right of the zero-position digit, and additional digits to the right of the binary point specify fractional parts of a whole quantity. In the case of the binary system, each successive digit to the right of the binary point has a value that is *one-half* the weight of the previous digit. Thus, the more fractional digits that a binary number has, the finer it can resolve a fractional amount. Figure 4-8(a) depicts the complete representation of a binary number with integer and fractional components, and Figure 4-8(b) gives some examples of binary-to-decimal conversion.

It is possible to generalize our understanding of a binary number by thinking of it as a sum of unspecified binary coefficients weighted by position. Thus, any measurable quantity can be represented in binary notation by 'R', as given in Eq. 4-2.

$$R = a_n 2^n + a_{n-1} 2^{n-1} + \ldots + a_2 2^2 + a_1 2^1 + a_0 2^0 + a_{-1} 2^{-1} + a_{-2} 2^{-2} + \ldots + a_{-m} 2^{-m} \quad \textbf{(4-2)}$$

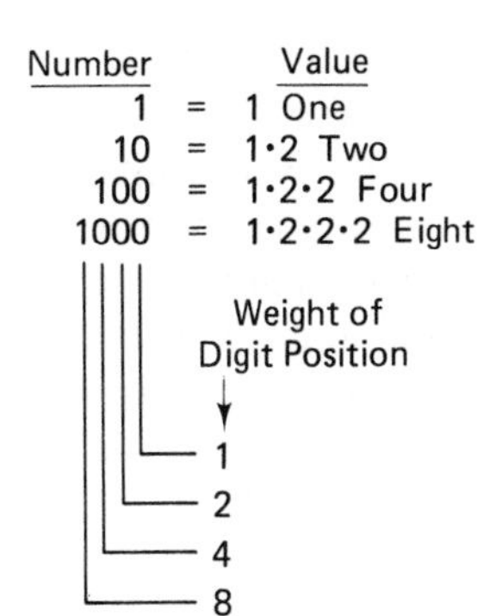

FIGURE 4-6 The digit position of a symbol in a binary number determines the numerical value of the symbol.

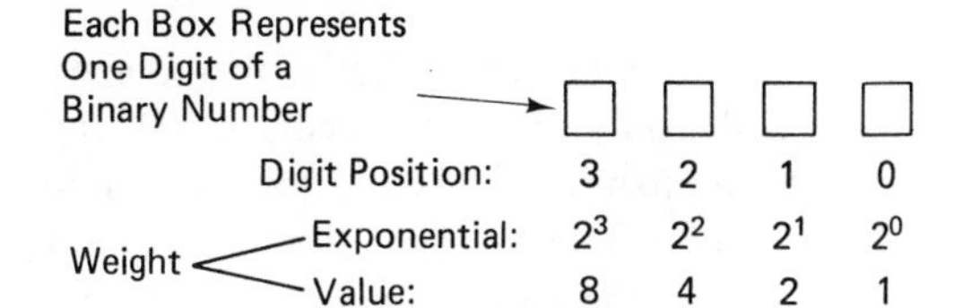

(a) The Digit Positions and their Weights in a Four-Digit Binary Number.

$$\begin{aligned}1100 &= (1 \cdot 2^3) + (1 \cdot 2^2) + (0 \cdot 2^1) + (0 \cdot 2^0)\\ &= (1 \cdot 8) + (1 \cdot 4) + (0 \cdot 2) + (0 \cdot 1)\\ &= 8 + 4 + 0 + 0\\ &= 12\end{aligned}$$

$$\begin{aligned}1001 &= (1 \cdot 2^3) + (0 \cdot 2^2) + (0 \cdot 2^1) + (1 \cdot 2^0)\\ &= (1 \cdot 8) + (0 \cdot 4) + (0 \cdot 2) + (1 \cdot 1)\\ &= 8 + 0 + 0 + 1\\ &= 9\end{aligned}$$

(b) Examples Showing the Conversion of a Binary Number to a Decimal Number.

FIGURE 4-7

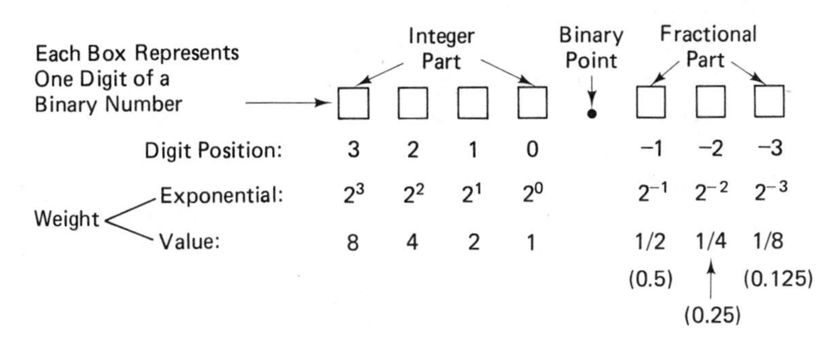

(a) Complete Representation of a Binary Number.

$$\begin{aligned}1101.011 &= (1 \cdot 2^3) + (1 \cdot 2^2) + (0 \cdot 2^1) + (1 \cdot 2^0) + (0 \cdot 2^{-1}) + (1 \cdot 2^{-2}) + (1 \cdot 2^{-3})\\ &= (1 \cdot 8) + (1 \cdot 4) + (0 \cdot 2) + (1 \cdot 1) + (0 \cdot 1/2) + (1 \cdot 1/4) + (1 \cdot 1/8)\\ &= (8 + 4 + 0 + 1 + 0 + 0.25 + 0.125\\ &= 13.375\end{aligned}$$

$$\begin{aligned}0111.101 &= (0 \cdot 2^3) + (1 \cdot 2^2) + (1 \cdot 2^1) + (1 \cdot 2^0) + (1 \cdot 2^{-1}) + (0 \cdot 2^{-2}) + (1 \cdot 2^{-3})\\ &= (0 \cdot 8) + (1 \cdot 4) + (1 \cdot 2) + (1 \cdot 1) + (1 \cdot 1/2) + (0 \cdot 1/4) + (1 \cdot 1/8)\\ &= 0 + 4 + 2 + 1 + 0.5 + 0 + 0.125\\ &= 7.625\end{aligned}$$

(b) Examples Showing the Conversion of a Binary Number to an Equivalent Decimal Number.

FIGURE 4-8

This equation is similar to the generalization of a decimal number given by Eq. 4-1, the only difference being in the radix, which in this case is 2. The digit of a binary number that has the lowest weight is referred to as the ***least significant bit***, often abbreviated "LSB." Similarly, the highest-weight digit in a binary number is referred to as the ***most significant bit***, abbreviated "MSB."

Because the binary symbols '0' and '1' are exactly identical to the first two decimal symbols, it is possible that confusion may result in the writing of a number. For example, when the number "10" is written, are we expressing the quantity two or ten? Of course, that depends on the number system that is being used. Whenever different number systems are used in the same equation, or in close proximity of each other on a written page, it is important to indicate the radix of the number so that confusion between different number systems is avoided. This is usually done by enclosing a number in parentheses, and writing a subscript to indicate the radix. The subscript is always interpreted as a *decimal* number. Thus,

$$(10)_2 = \text{two} \qquad \text{and} \qquad (10)_{10} = \text{ten}$$

In an equation, we can write

$$(101101)_2 = (45)_{10} \qquad \text{or} \qquad (15)_{10} = (1111)_2$$

Table 4-1 indicates a useful range of positive and negative powers of two.

We have seen that it is a relatively simple matter to convert a binary number into an equivalent decimal representation. All that is required is the multiplication of each binary digit by an appropriate power of two, as a weighting factor, and adding the results. On the other hand, converting a decimal number into an equivalent binary representation is somewhat more complex. The intuitive approach to this problem is to take a given decimal number and first, try to find the largest power of two that does not exceed the value of the decimal number. When such a power of two is found, we know that this power of two is one component of the sum of powers of two that will be used to represent the decimal number. Then, we would ask if the *next smallest* power of two fits into the *remainder* of the decimal number. If so, we will have found another component of the sum. Finally, when no decimal remainder is left, we will

TABLE 4-1

Positive and negative powers of two.

$2^0 = 1$	$2^{-1} = 0.5$
$2^1 = 2$	$2^{-2} = 0.25$
$2^2 = 4$	$2^{-3} = 0.125$
$2^3 = 8$	$2^{-4} = 0.0625$
$2^4 = 16$	$2^{-5} = 0.03125$
$2^5 = 32$	$2^{-6} = 0.015625$
$2^6 = 64$	$2^{-7} = 0.0078125$
$2^7 = 128$	$2^{-8} = 0.00390625$
$2^8 = 256$	
$2^9 = 512$	
$2^{10} = 1024$	
$2^{11} = 2048$	
$2^{12} = 4096$	

have obtained the complete sum of powers of two that is needed to represent the decimal number. Although this method produces valid answers, it is cumbersome. A simpler technique involves repeated division, and produces binary digits with the LSB coming first. To demonstrate this technique, an example is given showing the conversion of $(211)_{10}$ to an equivalent binary number.

Quotient Remainder	
$2 \overline{)211}$ = 105 + 1 (LSB)	(1) The given number is divided by 2, producing a quotient and remainder. The remainder of this first division is the LSB of the desired binary number.
$2 \overline{)105}$ = 52 + 1	(2) The quotient of the previous division is used as the dividend, and another division is carried out. Again, a quotient and remainder are produced. The remainder is the next-most-significant digit of the desired binary number.
$2 \overline{)52}$ = 26 + 0	(3) Successive divisions are made, with the remainder of each division contributing one more digit to the desired binary number.
$2 \overline{)26}$ = 13 + 0	
$2 \overline{)13}$ = 6 + 1	
$2 \overline{)6}$ = 3 + 0	
$2 \overline{)3}$ = 1 + 1	
$2 \overline{)1}$ = 0 + 1 (MSB)	(4) The division process is stopped when the quotient becomes zero. The remainder at this stage is the MSB of the binary number.

Thus, $(211)_{10} = (11010011)_2$.

To understand why this procedure yields correct results, consider applying it to the small decimal number '13'. Before doing anything, we can easily determine the binary number representing 'thirteen' by referring to Figure 4-5; $(13)_{10} = (1101)_2$. Now, using the repeated division method, but writing it in a different form, we have

$$\begin{aligned}(13)_{10} &= 2 \cdot 6 + 1 \\ &= 2 \cdot (2 \cdot 3 + 0) + 1 \\ &= 2 \cdot (2 \cdot (2 \cdot 1 + 1) + 0) + 1 \\ &= 2 \cdot (2 \cdot (2 \cdot (2 \cdot 0 + \underset{\downarrow}{1}) + \underset{\downarrow}{1}) + \underset{\downarrow}{0}) + \underset{\downarrow}{1} \\ &= \qquad\qquad (1 \quad 1 \quad 0 \quad 1)_2\end{aligned}$$

If a decimal number contains a fractional part, that fractional part must be converted to binary notation *independently* of the integer part, using a repeated multiplication technique. In this case, carry digits in the product which move across the decimal point into the integer part of the number indicate the bits of the desired fractional binary number. For example, to find the binary representation of the fractional decimal number '0.5625', the following procedure is used.

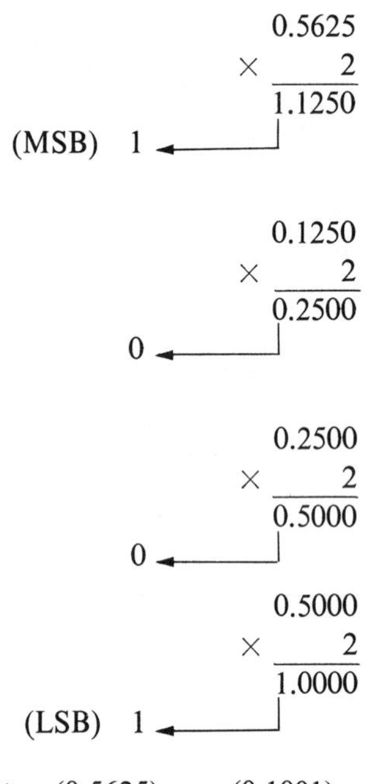

(1) The given number is multiplied by 2, producing a fractional product and integer carry. The integer carry of this first multiplication is the MSB of the desired fractional binary number.

(2) The fractional product of the previous multiplication is multiplied by two, producing a new fractional product and carry. The new carry is the next-least-significant bit of the desired binary number.

(3) Successive multiplications are made, with the carry of each multiplication contributing one more digit to the desired number.

(4) When the fractional product is zero, an exact binary representation for the decimal fraction has been achieved.

Thus, $(0.5625)_{10} = (0.1001)_2$.

Note that not all decimal fractions have an exact binary representation. In such cases, the fractional product will never become zero, and good judgment must be used to determine when sufficient binary digits have been found:

EXAMPLE 4-1: (a) Convert the decimal number '57.59375' into an equivalent binary number. We will treat the integer and fractional parts of the decimal number separately. First, for the integer part,

$$2\,\overline{)57} \quad 28 + 1 \quad \text{(LSB)}$$

$$2\,\overline{)28} \quad 14 + 0$$

$$2\,\overline{)14} \quad 7 + 0$$

$$2\,\overline{)7} \quad 3 + 1$$

$$2\,\overline{)3} \quad 1 + 1$$

$$2\,\overline{)\,1\,}\quad 0 + 1 \quad \text{(MSB)}$$

Thus, $(57)_{10} = (111001)_2$.

Second, for the fractional part,

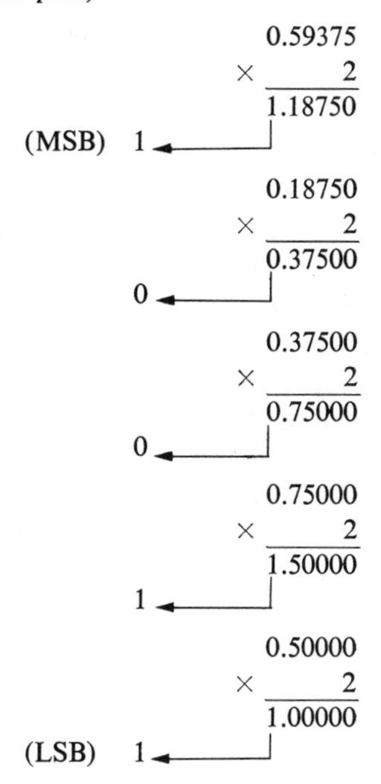

Thus, $(0.59375)_{10} = (0.10011)_2$.

Finally, $(57.59375)_{10} = (111001.10011)_2$.

(b) To verify that the decimal-to-binary conversion is correct, convert the resulting binary number back to a decimal number, and show that this decimal number is identical to the original one.

1 1 1 0 0 1 . 1 0 0 1 1

2^5	2^4	2^3	2^2	2^1	2^0	2^{-1}	2^{-2}	2^{-3}	2^{-4}	2^{-5}
32	16	8	4	2	1	0.5	0.25	0.125	0.0625	0.03125

$$32 + 16 + 8 + 0 + 0 + 1 + 0.5 + 0 + 0 + 0.0625 + 0.03125$$
$$= 57.59375$$

This result agrees with the original decimal number.

4.1.3 Negative Numbers

The concept of a "number line" is frequently used as a graphical description of a sequence of numbers. Zero is usually at the center of a number line, with positive numbers progressing to the right and negative numbers progressing to the left. This is shown in Figure 4-9. Notice that each number indicated on the number line can be considered to consist of a *magnitude* and a *sign*. The magnitude of a number can be

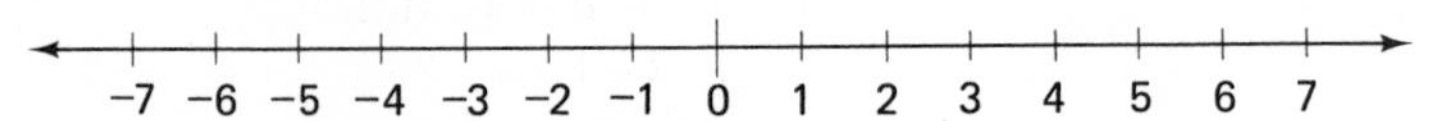

FIGURE 4-9 *Number line, labeled with decimal numbers.*

thought of as its geometric distance from zero, while the sign relates to its direction from zero on the number line. Identifying numbers in this manner is referred to as a ***sign-magnitude*** representation. Although we are most familiar with indicating the magnitude of a number in decimal notation, it may be indicated in any number system. In particular, the binary number system may be used in sign-magnitude form on a number line, as shown in Figure 4-10. Signed numbers that are to be processed by a digital electronic device must be represented in such a way that both the magnitude and sign of a number are shown. Since the sign of a number is either positive or negative, and nothing else (zero is usually considered to be a positive number), the simplest method of representing the sign of a binary number is to add an extra binary digit to the MSB side of the number. In the position of the ***sign bit***, '0' generally indicates a positive number, while '1' indicates a negative number. When a sign bit is added, leading zeros that would normally not be shown in the binary number are used to left-adjust the sign bit to a fixed digit position. Of course, with the sign bit fixed to a particular position, the magnitude of the number cannot grow beyond a maximum value. The sequence of four-bit signed binary numbers, expressed in sign-magnitude form, is shown in Figure 4-11.

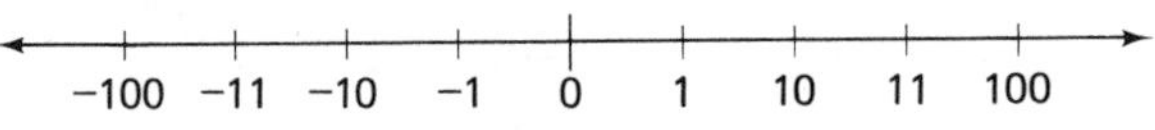

FIGURE 4-10 *Number line, labeled with binary numbers.*

Sign Bit: '0' Indicates a Positive Number
'1' Indicates a Negative Number

Magnitude: The Range is Zero Through Seven. The Magnitude Cannot Grow Above Seven with the Sign Bit Fixed to the Third Digit Position (Recall that the LSB is in the Zero Digit Position).

Magnitude ┐ Sign ┐

Signed Binary Number	Decimal Value
0111	+7
0110	+6
0101	+5
0100	+4
0011	+3
0010	+2
0001	+1
0000	0
1001	−1
1010	−2
1011	−3
1100	−4
1101	−5
1110	−6
1111	−7

FIGURE 4-11 *Sequence of four-digit signed binary numbers, using the sign-magnitude convention. The code '1000' is used as an alternative representation of zero.*

The sign-magnitude method is not unique in being able to express negative numbers. We shall see next how "complement" methods are possible. In our discussion of Boolean algebra, the *complement* of a binary value was indicated with an overbar, and defined as

$$\bar{0} = 1 \qquad \text{and} \qquad \bar{1} = 0$$

In the binary number system, a binary number can be considered to have a complement also. It is determined by simply taking the individual complement of *each bit* of the binary number. For example,

$$\overline{10100011} = 01011100 \qquad \text{and} \qquad \overline{11000} = 00111$$

If 'A' is a binary number, the bit-for-bit complement of 'A' is indicated as '$\bar{A}$', and is referred to as the ***one's complement*** of 'A'. An alternative scheme of representing signed binary numbers utilizes the one's-complement idea to express negative-number magnitude. This is shown in Figure 4-12 for the sequence of four-digit signed binary numbers.

Sign Bit: '0' Indicates a Positive Number
'1' Indicates a Negative Number

Magnitude: The Range is Zero Through Seven. Each Negative Number Magnitude is the One's Complement of the Corresponding Positive Number Magnitude.

Signed Binary Number (Sign, Magnitude)	Decimal Value
0111	+7
0110	+6
0101	+5
0100	+4
0011	+3
0010	+2
0001	+1
0000	0
1110	−1
1101	−2
1100	−3
1011	−4
1010	−5
1001	−6
1000	−7

FIGURE 4-12 *Sequence of four-digit signed binary numbers, using the one's-complement convention. The code '1111' results from certain one's complement arithmetic, and also represent zero.*

The ***two's complement*** of a binary number is defined to be the one's complement of that number *plus* one. Figure 4-13 illustrates the determination of two's-complement numbers. A third scheme of representing signed binary numbers makes use of the two's complement to express negative-number magnitude. This is shown in Figure 4-14.

Given Number	One's Complement		Add 1		Two's Complement
011	100	+	1	=	101
10100	01011	+	1	=	01100
111001	000110	+	1	=	000111

FIGURE 4-13 *Determination of the two's-complement of a binary number.*

Signed Binary Number	Decimal Value
0111	+7
0110	+6
0101	+5
0100	+4
0011	+3
0010	+2
0001	+1
0000	0
1111	−1
1110	−2
1101	−3
1100	−4
1011	−5
1010	−6
1001	−7
1000	−8

(Leftmost bit: Sign; remaining bits: Magnitude)

Sign Bit: '0' Indicates a Positive Number
'1' Indicates a Negative Number

Magnitude: The Range is Zero Through Seven for Positive Numbers, and Zero Through Eight for Negative Numbers. Each Negative Number Magnitude is the Two's Complement of the Corresponding Positive Number Magnitude.

FIGURE 4-14 *Sequence of four-digit signed binary numbers using the two's-complement convention.*

TABLE 4-2

Summary of common representations for signed binary numbers. Although four-bit numbers are shown in this table, the representations apply to signed binary numbers of any length. Note that the two's-complement scheme does not represent zero twice; '1000' is −8.

Signed Decimal	*Signed Binary*	*Sign-Magnitude*	*One's Complement*	*Two's Complement*
+7	+111	0111	0111	0111
+6	+110	0110	0110	0110
+5	+101	0101	0101	0101
+4	+100	0100	0100	0100
+3	+11	0011	0011	0011
+2	+10	0010	0010	0010
+1	+1	0001	0001	0001
0	0	0000	0000	0000
−1	−1	1001	1110	1111
−2	−10	1010	1101	1110
−3	−11	1011	1100	1101
−4	−100	1100	1011	1100
−5	−101	1101	1010	1011
−6	−110	1110	1001	1010
−7	−111	1111	1000	1001
−8	[a]	[a]	[a]	1000

[a] Not represented.

Although the sign-magnitude representation of signed binary numbers is easiest for us to read because of its similarity to signed decimal numbers, it is not well suited to the processing of binary numbers by electronic circuitry. On the other hand, one's-complement notation, and especially two's-complement notation, is extremely well suited to digital arithmetic circuitry. This fact will become clear in Section 4.2, where we discuss binary arithmetic. Table 4-2 summarizes the various representations of signed binary numbers that we have discussed.

4.1.4 Modulus Representation

In an abstract sense, the number line extends infinitely in both the positive and negative directions, and can be resolved into an infinitely fine grid of fraction points between any two whole-number points. However, any real, physical device used to manipulate numbers *does not* have infinite capacity to represent numbers, either in the extent of a number or in its resolution. Thus, the representation of a number on a physical device is limited by a maximum range and a maximum resolution. For example, the odometer on an automobile accumulates total mileage. Often, the maximum capacity of an odometer is 99,999 miles and its maximum resolution is 1 mile. When an attempt is made to represent a number that is beyond the capacity of the physical device, the most significant digits of the number are lost. In most odometers, it is not possible to represent 100,000 miles since only five integer digits have been manufactured into the device. Thus, only the five least significant digits of 100,000 will be displayed when this count is reached, and the mileage count will read 00,000. The maximum number of distinct integer quantities that a physical device (mechanical, electrical, or otherwise) can represent is referred to as the ***modulus*** of the device. The modulus of a five-digit odometer, for example, is 100,000. Note that although the capacity of the odometer is 99,999, its modulus is 100,000. Referring to the definition, "the maximum number of distinct integer quantities" that the odometer can represent is 100,000, namely 0 plus 1 to 99,999.

If we are to ever use the number-line concept in a discussion of actual counting and arithmetic devices, we must modify it somewhat to reflect the modulus nature of all such devices. One way to do this is to create a ***number circle*** to replace the number line. If we consider only positive integers for the moment, as would be the case for a counter, then a modulus-ten number circle would appear as in Figure 4-15. Moving clockwise around the number circle corresponds to incrementing a mod-ten counter.

FIGURE 4-15 *Modulus-ten number circle.*

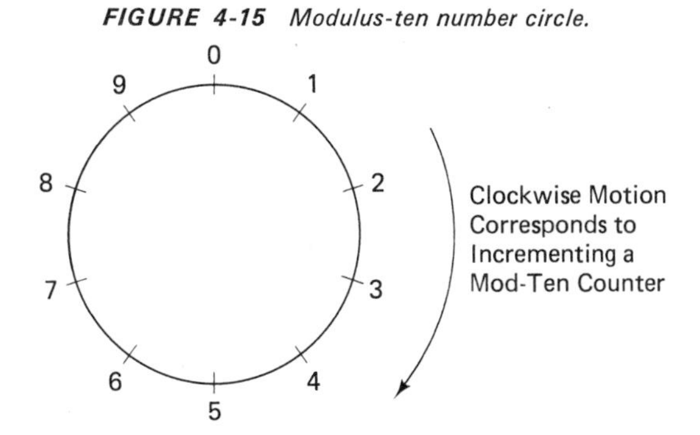

TABLE 4-3

Modulus counting in the decimal number system.

Number of Increments	READING ON A MODULUS COUNTER			
	Mod-Two	*Mod-Four*	*Mod-Ten*	*Mod-Sixteen*
0	0	0	0	0
1	1	1	1	1
2	0	2	2	2
3	1	3	3	3
4	0	0	4	4
5	1	1	5	5
6	0	2	6	6
7	1	3	7	7
8	0	0	8	8
9	1	1	9	9
10	0	2	0	10
11	1	3	1	11
12	0	0	2	12
13	1	1	3	13
14	0	2	4	14
15	1	3	5	15
16	0	0	6	0
17	1	1	7	1
18	0	2	8	2
19	1	3	9	3
20	0	0	0	4

TABLE 4-4

Modulus counting in the binary number system.

Number of Increments	READING ON A MODULUS COUNTER			
	Mod-Two	*Mod-Four*	*Mod-Ten*	*Mod-Sixteen*
0	0	0	0	0
1	1	1	1	1
2	0	10	10	10
3	1	11	11	11
4	0	0	100	100
5	1	1	101	101
6	0	10	110	110
7	1	11	111	111
8	0	0	1000	1000
9	1	1	1001	1001
10	0	10	0	1010
11	1	11	1	1011
12	0	0	10	1100
13	1	1	11	1101
14	0	10	100	1110
15	1	11	101	1111
16	0	0	110	0
17	1	1	111	1

When the tenth increment is received, our position on the number circle changes from '9' to '0'. Thus, '10' mod-ten is equal to '0', '11' mod-ten is equal to '1', '12' mod-ten is equal to '2', and so forth. Table 4-3 compares the mod-ten count sequence to several other possibilities.

Of course, the modulus concept can apply to any number system. Several examples of modulus count sequences in the binary system are shown in Table 4-4.

4.2 BINARY ARITHMETIC

The familiar, commonly used arithmetic operations of addition, subtraction, multiplication, and division take on a new appearance when viewed in terms of the binary number system. Although the fundamental arithmetic procedures are the same in both the decimal and binary systems, the actual manipulations differ in detail. With binary multiplication, for example, a new multiplication table must be considered. In this section we examine binary arithmetic and discuss some basic circuitry useful for binary addition and subtraction.

4.2.1 Addition

In any of the basic arithmetic operations, such as addition, subtraction, multiplication, and division, two initial quantities must exist to be arithmetically combined. In the arithmetic operation of addition, these quantities are the ***augend*** and the ***addend***. Combination by addition produces a ***sum***. This is symbolically represented as

$$\begin{array}{rl} A & \text{(augend)} \\ +\underline{B} & \text{(addend)} \\ C & \text{(sum)} \end{array}$$

where 'A' and 'B' are the numerical quantities to be added and 'C' is their sum. In terms of the number circle, addition corresponds to locating a point on the number circle that equals the value of one of the initial quantities, say 'A', and then *incrementing* beyond 'A' (clockwise) a number of steps equal to the value of the second initial quantity, 'B'. The sum is then read from the number circle. This is illustrated in Figure 4-16 with an example.

Recall that the number circle was introduced to reflect the limitation in capacity of any physical counting device. In arithmetic operations, the size of the number circle we choose to work with must be related to the maximum numerical quantities we expect to generate by the arithmetic operations. Such a number circle, however, would in general be extremely large when multidigit quantities are expressed. To avoid the complications of a large number circle, we consider the addition process as a sequence of one-digit additions, progressing from the least significant digit to the most significant digit. Of course, this is the way we treat addition when it is performed manually. In the decimal system, a mod-ten number circle is required to express one-digit addition. Figure 4-16 illustrated a mod-ten number circle. A number circle whose modulus is the same as the radix of the number system in which we are working is referred to as a ***radix circle***. Thus, the number circle of Figure 4-16 is a radix circle for the decimal system.

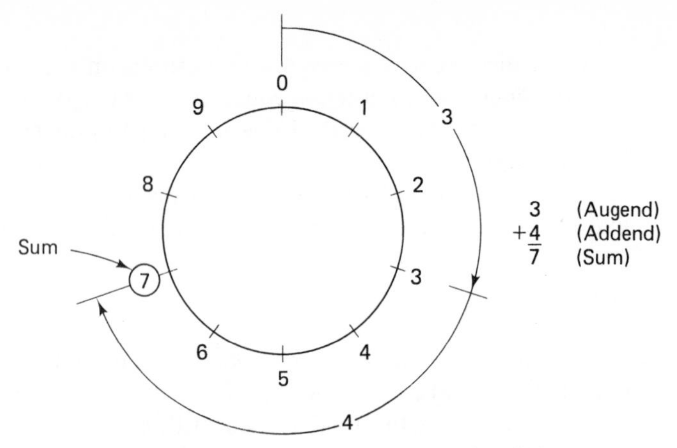

FIGURE 4-16 Addition on the number circle.

In the decimal system, there is exactly one notch on the radix circle for each of the ten unique decimal symbols. If two numbers are added, and their sum exceeds any one digit value on the radix circle, then only the least significant digit is indicated. In such cases, a ***carry*** digit is generated to represent the most significant digit of the sum. Figure 4-17 illustrates this idea. A table can be written that indicates all possible one-digit sums and carries for decimal addition. This information, shown in Table 4-5, was derived completely from one-digit addition on the decimal radix circle. As grade school students, we memorized this decimal addition table and now use it automatically whenever we perform an addition. If two multiple-digit numbers are added, we know how to apply the table digit by digit to obtain a complete multiple-digit sum. When one stage of addition receives a carry digit from a previous stage of addition, we know that the sum for the present stage is incremented by one. It is important to realize that for larger numbers, no new rules need to be learned or other tables referred to. Simply, the same rules are applied repeatedly, digit by digit, until a complete multidigit sum is achieved.

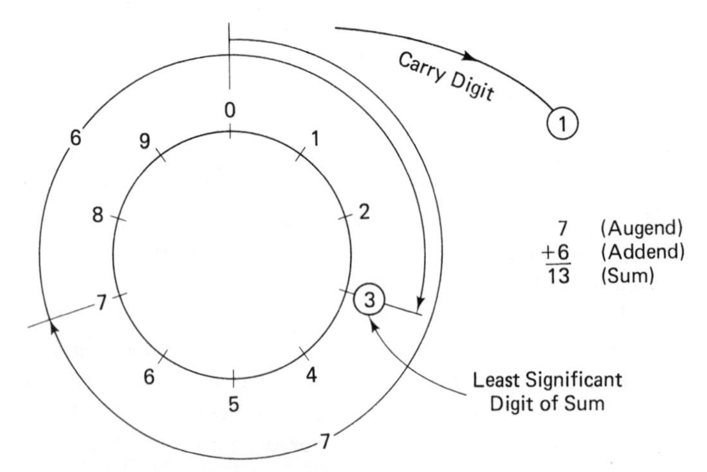

FIGURE 4-17 Addition on the decimal radix circle, with a carry generated.

TABLE 4-5

Decimal addition tables.

Sum Table (Augend across, Addend down)

Addend \ Augend	0	1	2	3	4	5	6	7	8	9
0	0	1	2	3	4	5	6	7	8	9
1	1	2	3	4	5	6	7	8	9	0
2	2	3	4	5	6	7	8	9	0	1
3	3	4	5	6	7	8	9	0	1	2
4	4	5	6	7	8	9	0	1	2	3
5	5	6	7	8	9	0	1	2	3	4
6	6	7	8	9	0	1	2	3	4	5
7	7	8	9	0	1	2	3	4	5	6
8	8	9	0	1	2	3	4	5	6	7
9	9	0	1	2	3	4	5	6	7	8

Carry Table (Augend across, Addend down)

Addend \ Augend	0	1	2	3	4	5	6	7	8	9
0	0	0	0	0	0	0	0	0	0	0
1	0	0	0	0	0	0	0	0	0	1
2	0	0	0	0	0	0	0	0	1	1
3	0	0	0	0	0	0	0	1	1	1
4	0	0	0	0	0	0	1	1	1	1
5	0	0	0	0	0	1	1	1	1	1
6	0	0	0	0	1	1	1	1	1	1
7	0	0	0	1	1	1	1	1	1	1
8	0	0	1	1	1	1	1	1	1	1
9	0	1	1	1	1	1	1	1	1	1

It is reasonable to believe that if we were using a number system other than the decimal system, the addition tables would change. Since we can form a radix circle for *any* number system, it is possible to form addition tables for any number system. Thus, although the specific addition tables will change from one system to another, the fundamental procedure for addition will not.

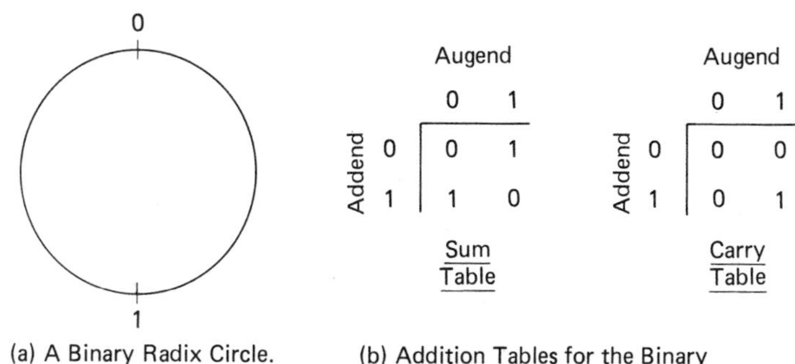

(a) A Binary Radix Circle.

(b) Addition Tables for the Binary Number System, Based on the Binary Radix Circle.

FIGURE 4-18

In the binary number system, two symbols are available, and thus a mod-two radix circle is needed to determine the tables for binary addition. A binary radix circle, and the corresponding tables of addition derived from it, are shown in Figure 4-18. When a multiple-digit binary sum is computed, it is necessary for any particular stage of addition to be able to account for a carry digit from the previous stage. We know that under the condition of a '1' carry from the previous stage, the sum is incremented. However, the tables in Figure 4-18(b) do not account for this. A complete addition table, drawn in the form of a truth table, is shown in Figure 4-19, and accounts for augend 'Y_n', addend 'X_n', and carry from a previous stage 'C_{n-1}' and depicts a sum 'S_n' and a new carry 'C_n'. This complete addition table is fully derived

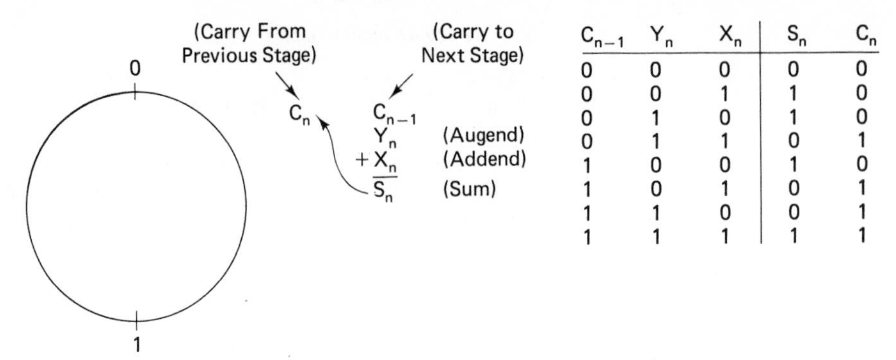

C_{n-1}	Y_n	X_n	S_n	C_n
0	0	0	0	0
0	0	1	1	0
0	1	0	1	0
0	1	1	0	1
1	0	0	1	0
1	0	1	0	1
1	1	0	0	1
1	1	1	1	1

FIGURE 4-19 *The complete binary addition table, as derived from the binary radix circle.*

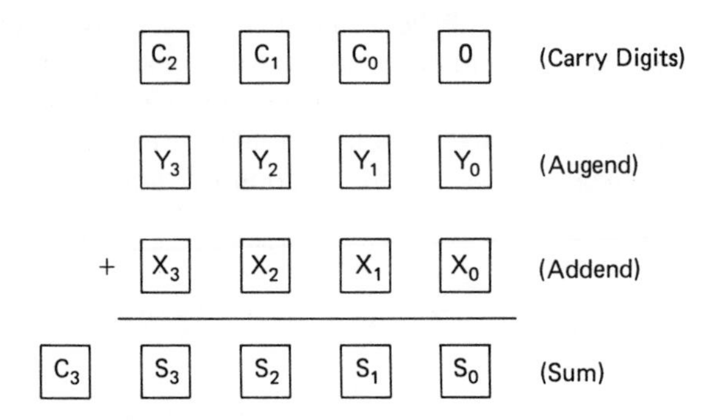

(a) Schematic Representation of Binary Addition. The Carry Digit into the First Addition Stage, C_0, is Always '0'.

Addition in Binary	Equivalent Addition in Decimal
1011 + 0110 = 10001	11 + 6 = 17
10100110 + 11110100 = 110011010	166 + 244 = 410
1001 + 111110 = 1000111	9 + 62 = 71

(b) Examples of Binary Addition, Verified by the Addition of Equivalent Decimal Numbers.

FIGURE 4-20

from the mod-two number circle by adding distances in a clockwise sense. The binary digit '1' represents a distance of one unit on the number circle, while the binary digit '0' represents a distance of zero units. For example, $0 + 1 + 1$ brings us back to the '0' notch on the radix circle and generates a carry digit. This corresponds to $C_{n-1} = 0$, $Y_n = 1$, and $X_n = 1$. Multiple-digit binary addition is schematically illustrated in Figure 4-20(a), with several examples given in Figure 4-20(b). Note in the examples that in order to determine S_n and C_n for any particular stage, we need only refer to the binary addition table in Figure 4-19 and look up these quantities.

The binary addition table of Figure 4-19 describes one representative stage of the binary addition process. Noting that this table can be viewed as a logic truth table, we see that a two-output combinational logic circuit can be designed to realize in electronic hardware the functional operation described in the table. Such a circuit design represents the needed hardware for one bit of binary addition. To produce a multiple-bit adder circuit, any number of identical one-bit adders can be cascaded. This is illustrated in Figure 4-21.

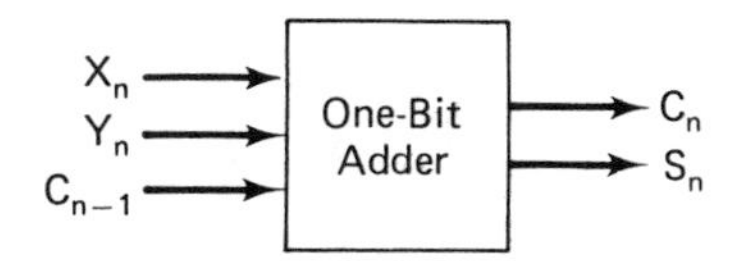

(a) The Block Representation of a One-Bit Adder. The Combinational Logic Needed to Implement this Block is Derived from the Truth Table of Figure 4-19.

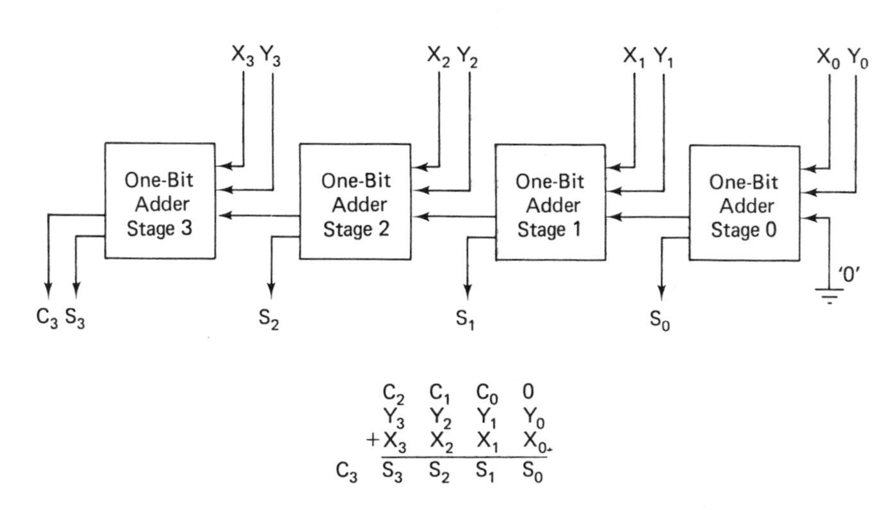

$$\begin{array}{rcccc} & C_2 & C_1 & C_0 & 0 \\ & Y_3 & Y_2 & Y_1 & Y_0 \\ + & X_3 & X_2 & X_1 & X_0 \\ \hline C_3 & S_3 & S_2 & S_1 & S_0 \end{array}$$

(b) The Cascading of Four One-Bit Adders to Produce a Four-Bit Adder. The Stage Zero Carry is Always '0' Since no Previous Stage Exists.

FIGURE 4-21

Our design for a binary adder must focus exclusively on the design of a one-bit adder block, since the basic circuit block will be duplicated many times in a multiple-bit binary adder. Following the procedures described in Chapter 3, we will express the truth table in the form of a Karnaugh map, determine a minimal equation, and implement it with a minimum number of logic gates. Referring to Figure 4-19, we have

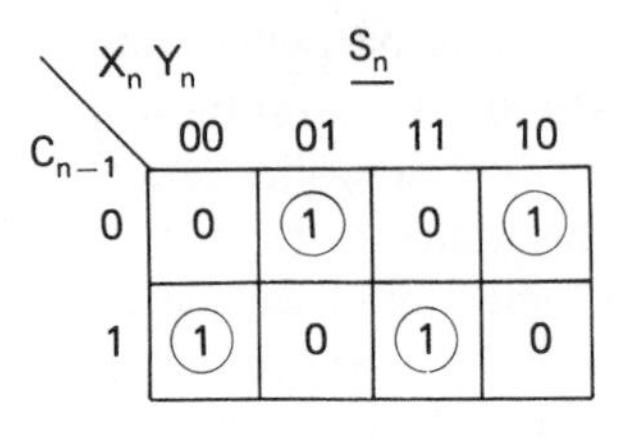

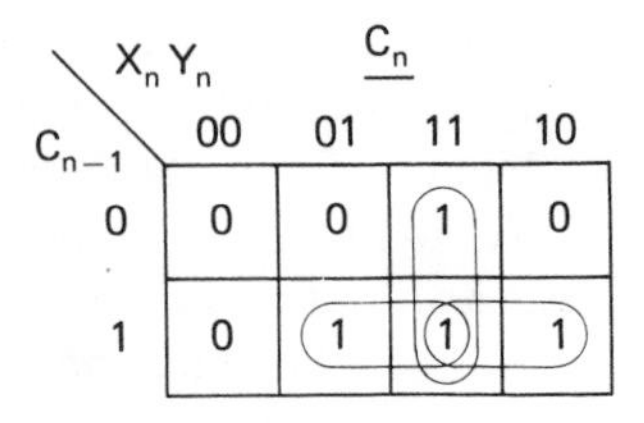

(a)

$$S_n = C_{n-1}\bar{X}_n\bar{Y}_n + \bar{C}_{n-1}\bar{X}_nY_n + C_{n-1}X_nY_n + \bar{C}_{n-1}X_n\bar{Y}_n \tag{4-3}$$

$$= C_{n-1} \oplus X_n \oplus Y_n \tag{4-4}$$

$$C_n = C_{n-1}Y_n + C_{n-1}X_n + X_nY_n \tag{4-5}$$

$$= C_{n-1}(X_n + Y_n) + X_nY_n \tag{4-6}$$

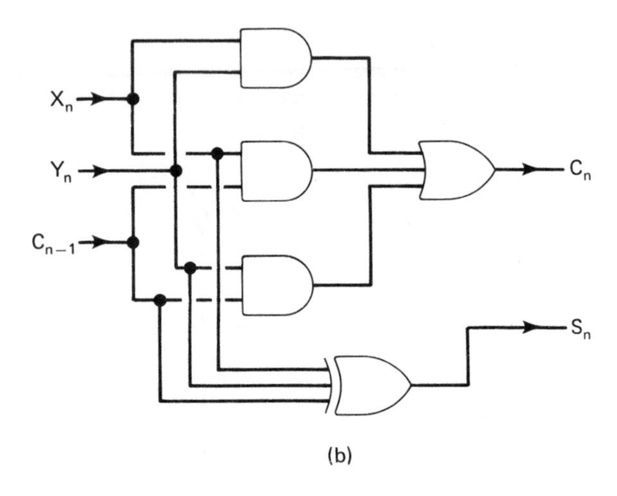

(b)

The exact gate structure we select to implement the equations for S_n and C_n depends on the types of logic gates that are available. Figure 4-22 demonstrates several other possibilities.

Derivation: Factoring Eq. 4-3, we can write

$$S_n = C_{n-1}(\bar{X}_n\bar{Y}_n + X_nY_n) + \bar{C}_{n-1}(\bar{X}_nY_n + X_n\bar{Y}_n)$$

Since $\bar{X}_n\bar{Y}_n + X_nY_n = \overline{(\bar{X}_nY_n + X_n\bar{Y}_n)}$, we can write

$$S_n = C_{n-1}\overline{(\bar{X}_nY_n + X_n\bar{Y}_n)} + \bar{C}_{n-1}(\bar{X}_nY_n + X_n\bar{Y}_n)$$

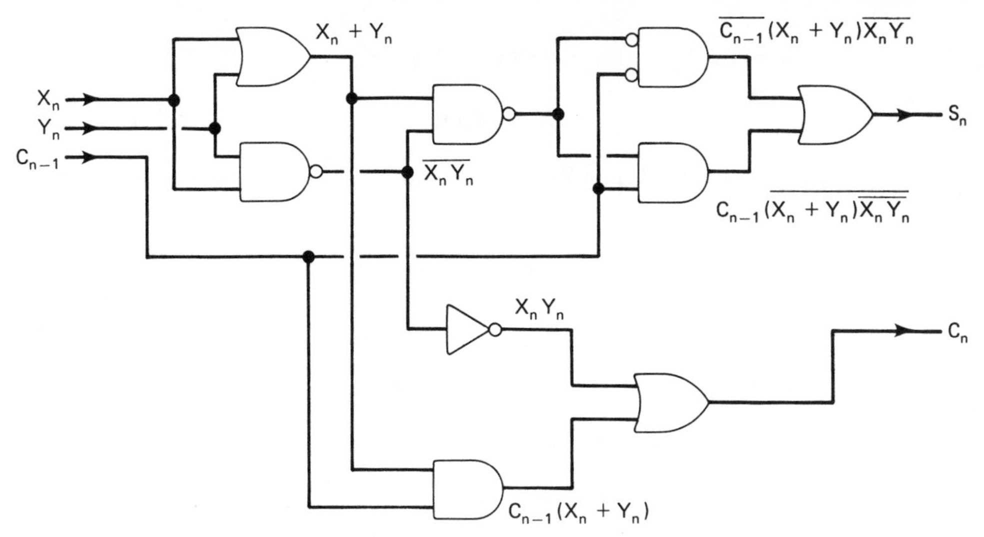

(a) *An adder circuit using only standard two-input gates (XOR gates are not available).*

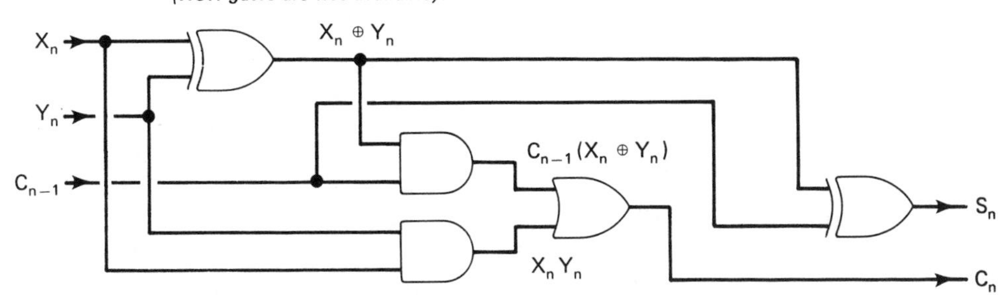

(b) *An adder circuit using standard two-input gates and two-input XOR gates.*

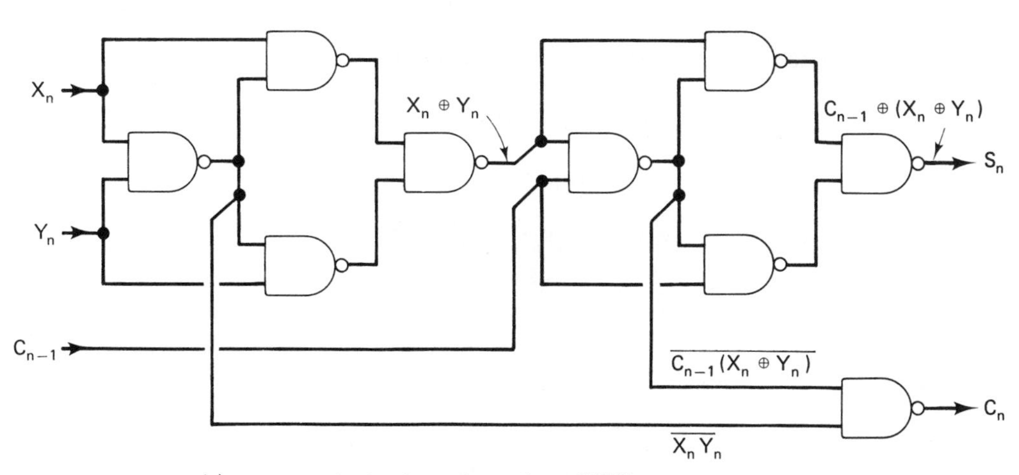

(c) *An adder circuit using only two-input NAND gates.*

FIGURE 4-22

Now, since $\bar{X}_nY_n + X_n\bar{Y}_n = (X_n + Y_n)\overline{X_nY_n}$,

$$S_n = C_{n-1}(\overline{(X_n + Y_n)\overline{X_nY_n}}) + \bar{C}_{n-1}((X_n + Y_n)\overline{X_nY_n}) \qquad \textbf{(4-7)}$$

Gate structure: Using Eq. 4-7 and Eq. 4-6 (derived earlier), the following gate structure is determined.

Derivation: Factoring Eq. 4-3, as in (a), we obtain

$$\text{(sum)} \quad S_n = C_{n-1}(\bar{X}_n\bar{Y}_n + X_nY_n) + \bar{C}_{n-1}(\bar{X}_nY_n + X_n\bar{Y}_n)$$

Since $\bar{X}_n\bar{Y}_n + X_nY_n = \overline{(\bar{X}_nY_n + X_n\bar{Y}_n)}$, we can write

$$S_n = C_{n-1}\overline{(\bar{X}_nY_n + X_n\bar{Y}_n)} + \bar{C}_{n-1}(\bar{X}_nY_n + X_n\bar{Y}_n)$$

If we let $Z = \bar{X}_nY_n + X_n\bar{Y}_n$, then

$$\begin{aligned} S_n &= C_{n-1}\bar{Z} + \bar{C}_{n-1}Z \\ &= C_{n-1} \oplus Z \\ &= C_{n-1} \oplus (\bar{X}_nY_n + X_n\bar{Y}_n) \\ &= C_{n-1} \oplus (X_n \oplus Y_n) \qquad \textbf{(4-8)} \end{aligned}$$

$$\text{(carry)} \quad C_n = C_{n-1}(X_n + Y_n) + X_nY_n \qquad \text{[restatement of Eq. (4-6)]}$$

However, since $X_n + Y_n = (X_n \oplus Y_n) + X_nY_n$,

$$\begin{aligned} C_n &= C_{n-1}((X_n \oplus Y_n) + X_nY_n) + X_nY_n \\ &= C_{n-1}(X_n \oplus Y_n) + C_{n-1}X_nY_n + X_nY_n \qquad \textbf{(4-9)} \\ &= C_{n-1}(X_n \oplus Y_n) + X_nY_n \end{aligned}$$

Gate structure: Using Eqs. 4-8 and 4-9, the following gate structure is determined.

Derivation: From Figure 4-22(b), we have

$$\text{(sum)} \quad S_n = C_{n-1} \oplus (X_n \oplus Y_n).$$

Given that $X_n \oplus Y_n = \bar{X}_nY_n + X_n\bar{Y}_n$, we can write

$$\begin{aligned} \bar{X}_nY_n + X_n\bar{Y}_n &= (\bar{X}_n + \bar{Y}_n)Y_n + X_n(\bar{X}_n + \bar{Y}_n) \\ &= \overline{X_nY_n}Y_n + X_n\overline{X_nY_n} \qquad \text{(by DeMorgan's law)} \\ &= \overline{\overline{(\overline{X_nY_n}Y_n)} \cdot \overline{(X_n\overline{X_nY_n})}} \qquad \text{(by DeMorgan's law)} \\ &= Z \end{aligned}$$

$$S_n = \overline{\overline{(\overline{C_{n-1}Z}Z)} \cdot \overline{(C_{n-1}\overline{C_{n-1}Z})}} \qquad \textbf{(4-10)}$$

Thus, the XOR function can be constructed with a simple NAND gate structure.

$$\text{(carry)} \quad C_n = C_{n-1}(X_n \oplus Y_n) + X_nY_n \qquad \text{[restatement of Eq. (4-9)]}$$

$$= \overline{\overline{C_{n-1}(X_n \oplus Y_n)} \cdot \overline{X_nY_n}} \qquad \textbf{(4-11)}$$

Gate structure: Using Eqs. 4-10 and 4-11, the following gate structure is determined.

In the first stage of a multiple bit adder, the carry input is always '0' since there are no preceding stages. Because of this, we can simplify the logic for the first stage as follows.

$$S_n = C_{n-1}\bar{X}_n\bar{Y}_n + \bar{C}_{n-1}\bar{X}_nY_n + C_{n-1}X_nY_n + \bar{C}_{n-1}X_n\bar{Y}_n$$
[restatement of Eq. (4-3)]

$$C_n = C_{n-1}(X_n + Y_n) + X_nY_n$$
[restatement of Eq. (4-6)]

Now, for the first stage, $n = 0$ and $C_{n-1} = C_{-1} = 0$. Thus,

$$S_0 = (0 \cdot \bar{X}_0\bar{Y}_0) + (\bar{0} \cdot \bar{X}_0Y_0) + (0 \cdot X_0Y_0) + (\bar{0} \cdot X_0\bar{Y}_0)$$

$$= \bar{X}_0Y_0 + X_0\bar{Y}_0 \qquad \textbf{(4-12)}$$

$$C_0 = 0 \cdot (X_0 + Y_0) + X_0Y_0$$

$$= X_0Y_0 \qquad \textbf{(4-13)}$$

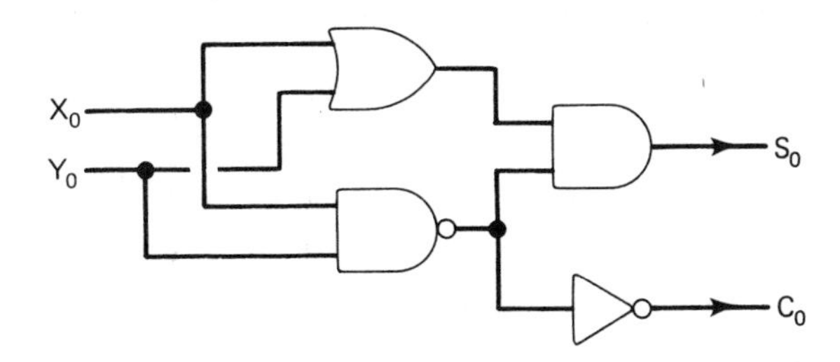

The circuitry for the first stage is now reduced to four two-input gates. This reduced circuit is commonly referred to as a ***half adder***, while the normal unreduced one-bit adder of Figure 4-22 is referred to as a ***full adder***. A full adder can be viewed as the interconnection of two half adders, as shown in Figure 4-23.

FIGURE 4-23 Full adder viewed as the interconnection of two half adders.

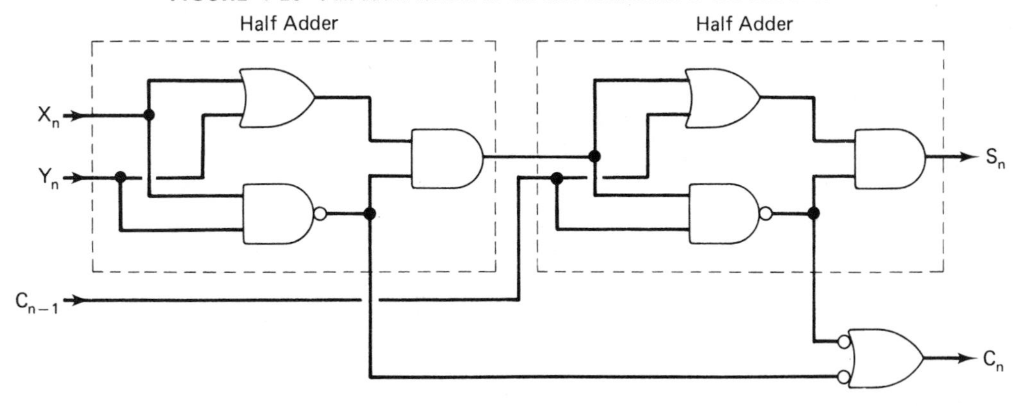

4.2.2 Subtraction

Following addition, the next most frequently used arithmetic operation is subtraction. The initial numerical quantities that are to be combined by subtraction are the ***minuend*** and the ***subtrahend***. Combination by subtraction produces a *difference*. This is symbolically represented as

$$\begin{array}{rl} A & \text{(minuend)} \\ -\underline{B} & \text{(subtrahend)} \\ C & \text{(difference)} \end{array}$$

where 'A' and 'B' are the numerical quantities to be subtracted and 'C' is their difference. Subtraction can be described with the number circle in a way similar to the description of addition. For subtraction, the minuend is located on the number circle and decremented (counterclockwise) by a number of steps equal to the value of the subtrahend. The difference is then read from the number circle. This is illustrated in Figure 4-24.

Subtraction as well as addition can be performed on a radix circle. In subtraction, however, we *decrement* by a number of steps equal to the subtrahend. Thus, rather than running the risk of exceeding the maximum value indicated on the radix circle, in subtraction it is possible for the difference to fall below the *minimum* value indicated on the number circle, namely '0'. Whenever counterclockwise movement on the radix

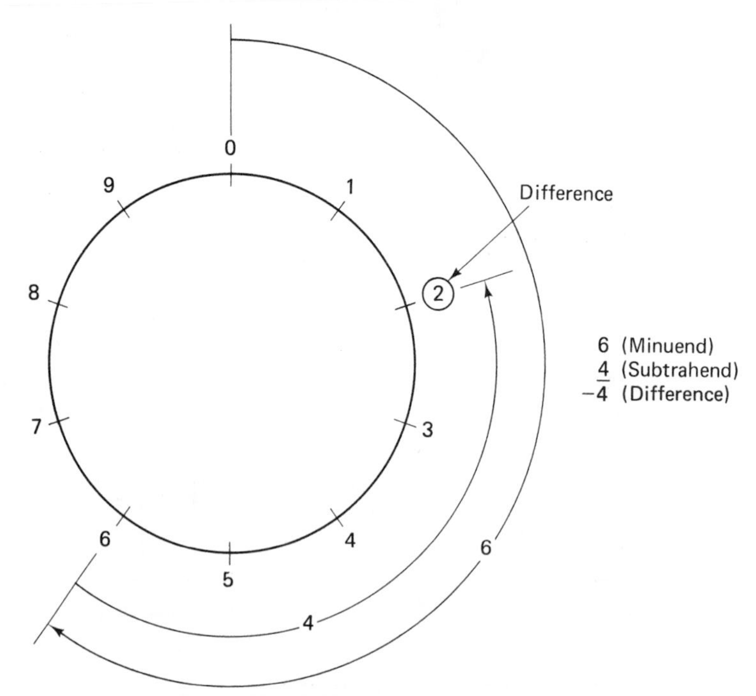

FIGURE 4-24 *Subtraction on the number circle.*

TABLE 4-6

Decimal subtraction tables.

Difference Table (Minuend across, Subtrahend down)

Subtrahend \ Minuend	0	1	2	3	4	5	6	7	8	9
0	0	1	2	3	4	5	6	7	8	9
1	9	0	1	2	3	4	5	6	7	8
2	8	9	0	1	2	3	4	5	6	7
3	7	8	9	0	1	2	3	4	5	6
4	6	7	8	9	0	1	2	3	4	5
5	5	6	7	8	9	0	1	2	3	4
6	4	5	6	7	8	9	0	1	2	3
7	3	4	5	6	7	8	9	0	1	2
8	2	3	4	5	6	7	8	9	0	1
9	1	2	3	4	5	6	7	8	9	0

Borrow Table (Minuend across, Subtrahend down)

Subtrahend \ Minuend	0	1	2	3	4	5	6	7	8	9
0	0	0	0	0	0	0	0	0	0	0
1	1	0	0	0	0	0	0	0	0	0
2	1	1	0	0	0	0	0	0	0	0
3	1	1	1	0	0	0	0	0	0	0
4	1	1	1	1	0	0	0	0	0	0
5	1	1	1	1	1	0	0	0	0	0
6	1	1	1	1	1	1	0	0	0	0
7	1	1	1	1	1	1	1	0	0	0
8	1	1	1	1	1	1	1	1	0	0
9	1	1	1	1	1	1	1	1	1	0

circle causes us to pass '0', a ***borrow*** digit is generated. This is similar to the generation of a carry digit in arithmetic addition when '0' is passed in a clockwise direction. Table 4-6 indicates all possible one-digit differences and borrows for decimal subtraction.

As with addition, subtraction can be performed on multidigit numbers by doing a sequence of one-digit subtractions, in each case "looking up" difference and borrow values from Table 4-6. Whenever a borrow is generated by one stage of subtraction, the difference computed for the next succeeding stage is decremented by one. Of course, the next stage may be caused to generate a borrow of its own when the difference for that stage and the borrow of the previous stage are accounted for. A problem is encountered in subtraction, however, if the subtrahend is greater than the minuend, and no succeeding stages exist from which to borrow. This problem can be viewed either as subtracting a larger number from a smaller one, or in terms of digit-by-digit subtraction as generating a borrow from the most significant stage of subtraction. We know that the result of such an operation would be a negative number, yet no values currently shown on any number circle represent negative numbers. Thus, a correct answer cannot be obtained. When subtraction is performed manually and we see that the subtrahend is greater than the minuend, we automatically interchange the order of subtraction so that normal rules work, but then assign a negative sign to the answer. Such a procedure is not easily done when an electronic device is to perform the subtraction. To investigate an alternative method, first consider a modified number circle on which negative numbers have been included, shown in Figure 4-25. This number circle was intentionally chosen to not be a radix circle for the decimal system so that the procedure for obtaining correct negative number results is emphasized. Note that the modified number circle depicts positive numbers on the outside of the circle and negative numbers on the inside. Now, the example of Figure 4-25 shows the subtraction of '9' from '4'. Moving to '4' on the number circle and decrementing '9' units, we arrive at a notch labeled '11' on the outside and '−5' on the inside. We see that the correct answer, '−5', is located on the inside of the circle. In general, it

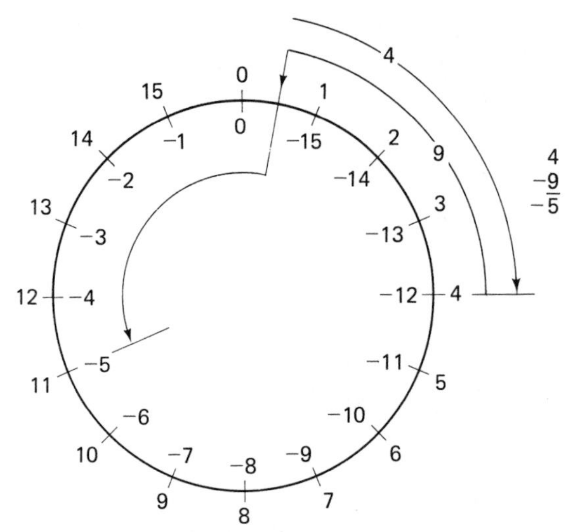

FIGURE 4-25 *Modified number circle that includes negative numbers.*

can be said that whenever '0' is crossed in a *counterclockwise* direction, the correct answer is to be read from the inside of the circle. It is useful to ask

> Can the negative numbers on the interior of the circle be determined from their corresponding values on the outside?

If this can be done, we need not label each notch with two numbers. The answer to the question is "yes," and the procedure is simply to subtract the modulus of the number circle from the positive (exterior) values to obtain the corresponding negative values. Specifically,

Exterior Reading		*Modulus*		*Interior Reading*
15	−	16	=	−1
14	−	16	=	−2
13	−	16	=	−3
12	−	16	=	−4
11	−	16	=	−5
		etc.		

To summarize, whenever '0' is crossed in a counterclockwise sense on a number circle, either (1) a borrow digit is generated for a following stage, or (2) if there is no following stage, the arithmetic difference will be a negative quantity. In the second case, a proper negative number value is obtained by subtracting the circle's modulus from the indicated positive number. The usefulness of the procedure might seem questionable at this time, since it is now necessary to subtract the modulus from a positive number to obtain the valid negative number result. Of course, subtraction is what we are trying to explain in the first place. As we shall see, however, it is surprisingly easy in the binary number system to obtain a valid negative number result, because subtraction of the modulus is not necessary.

Binary subtraction is similar to binary addition in several ways. First, a multidigit subtraction can be obtained by a sequence of one-digit subtractions, where each stage utilizes identical circuitry. Second, each subtraction stage requires three inputs and produces two outputs. These ideas are illustrated in Figure 4-26. The similarity ends here, however. Since we are performing a logical operation that is not the same as addition, the truth table for subtraction must be different than that for addition. More important, a borrow generated by the most-significant-digit stage of subtraction is not just attached to the left of the result, as a carry would be in the addition process. Rather, a borrow in this stage indicates that '0' has been passed in a counterclockwise direction on the number circle, and that the correct answer should be a negative number. In the decimal system, we found negative results from the inside of a modified number circle, and then proved that these negative numbers on the interior of the circle can be determined by subtracting the circle's modulus from each of the exterior positive numbers. In the binary system, the interior negative numbers are found by taking the *two's complement* of the exterior positive numbers. These important features of subtraction in the binary system are illustrated in Figure 4-27. Examples demonstrating binary subtraction are shown in Figure 4-28.

The subtraction of two numbers can be illustrated on a number circle as shown in Figure 4-29. In this figure, we have a mod-eight number circle, with the subtraction of '2' from '5' depicted. The correct answer, '3', is found on the number circle by moving clockwise five units, then moving counterclockwise two units, and finally reading the answer from the circle. Note, however, that we can also obtain the correct answer by moving forward (clockwise) six units instead of moving backward two units. On the mod-eight number circle, therefore, we can say that $5 - 2 = 5 + 6 \text{ (mod-eight)} = 3$. This is a very useful observation, since it means that we can convert a subtraction problem into an addition problem. Thus, by first converting a subtrahend into a suitable addend, the same circuitry can be used for subtraction as for addition. As well as simplifying the electronic hardware, the technique is very effectively applied to binary arithmetic in computers. A question that must be answered before we can proceed is

How is the subtrahend converted into a suitable addend?

This question is answered with the following derivation.

Four N-bit binary numbers are represented by the letters 'm', 's', 'd', and 'a'. These letters correspond to "minuend," "subtrahend," "difference," and "addend." Their modulus is equal to '$2N$', and will be referred to as 'k'. The subtraction problem is symbolically stated as

$$m - s = d \quad \textbf{(4-14)}$$

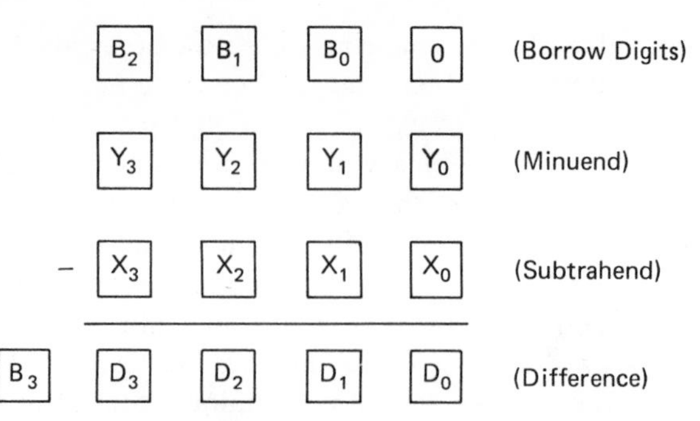

(a) Schematic Representation of Binary Subtraction. The Borrow Digit of the First Subtraction Stage is Always '0'. If $B_3 = 1$, the Answer as it Stands is Incorrect, since a Large Number Subtracted From a Smaller One Should Produce a Negative Result.

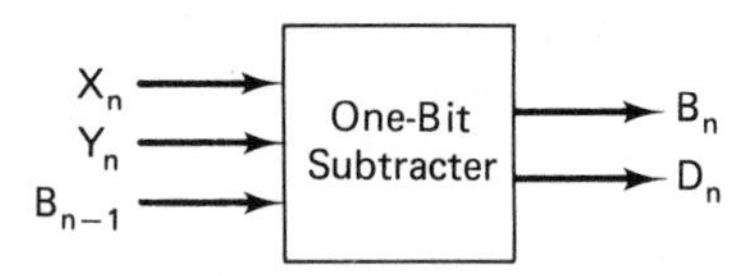

(b) Block Representation of a One-Bit Subtracter.

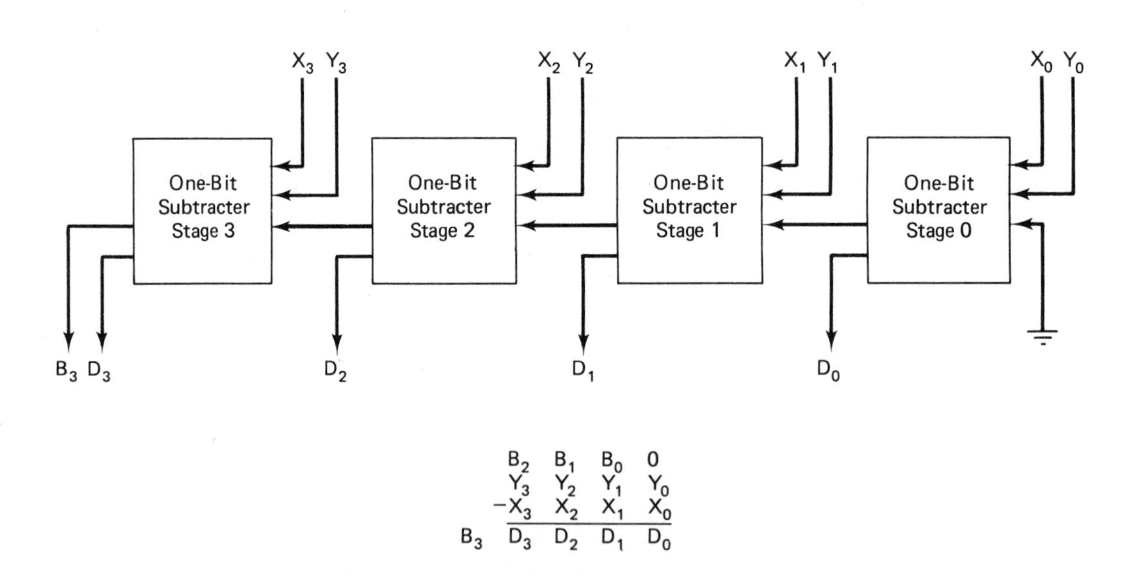

$$
\begin{array}{ccccc}
 & B_2 & B_1 & B_0 & 0 \\
 & Y_3 & Y_2 & Y_1 & Y_0 \\
 & -X_3 & X_2 & X_1 & X_0 \\
\hline
B_3 & D_3 & D_2 & D_1 & D_0
\end{array}
$$

(c) The Cascading of Four One-Bit Subtracter to Produce a Four-Bit Subtracter. The Stage Zero Borrow is Always '0' since no Previous Stage Exists.

FIGURE 4-26

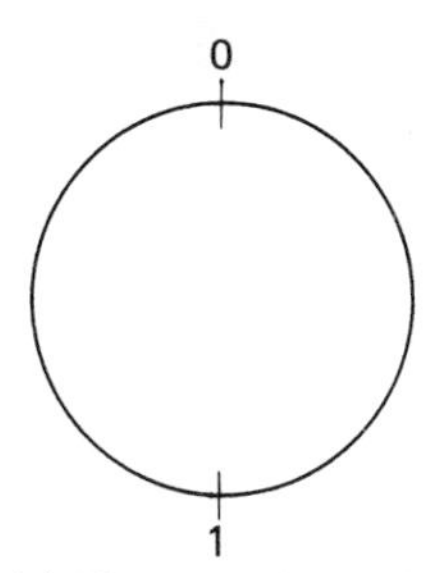

(a) Binary Radix Circle.

B_{n-1}	Y_n	X_n	D_n	B_n
0	0	0	0	0
0	0	1	1	1
0	1	0	1	0
0	1	1	0	0
1	0	0	1	1
1	0	1	0	1
1	1	0	0	0
1	1	1	1	1

(b) Complete Binary Subtraction Table, as Derived from the Binary Radix Circle.

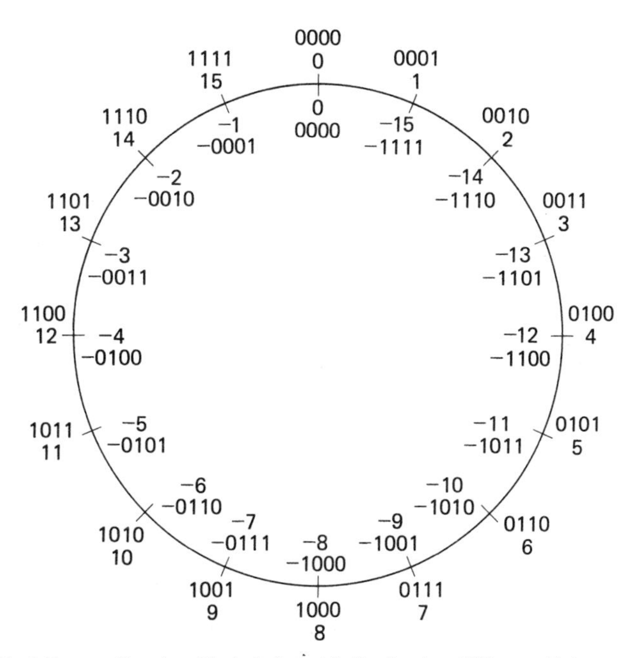

(c) Mod-Sixteen Number Circle Labeled in Decimal and Binary. Note that each Binary Number Inside the Circle is Found by Taking the Two's Complement of the Binary Number on the Outside of the Circle.

FIGURE 4-27

```
(Borrows)   1000

             1011     11
            -0111    - 7
            00100      4
```

```
(Borrows)   1110

             1100     12
            -0101    - 5
            00111      7
```

```
(Borrows)   1000

             1001      9
MSB         -1101    -13
Borrow -->  11100     -4

             0011  (One's Complement)
             0100  (Two's Complement)

            -0100  (Correct Answer)
```

```
(Borrows)   11000

             10010     18
            -01100    -12
            000110      6
```

```
(Borrows)   000000

             101101     45
            -  1100    -12
            0100001     33
```

```
(Borrows)   111110

              10000     16
MSB         -110101    -53
Borrow -->  1011011    -37

             100100  (One's Complement)
             100101  (Two's Complement)

            -100101  (Correct Answer)
```

FIGURE 4-28 *Examples of binary subtraction. A borrow from the most significant stage indicates that the two's complement of the result must be found, and a negative sign attached, for the correct answer.*

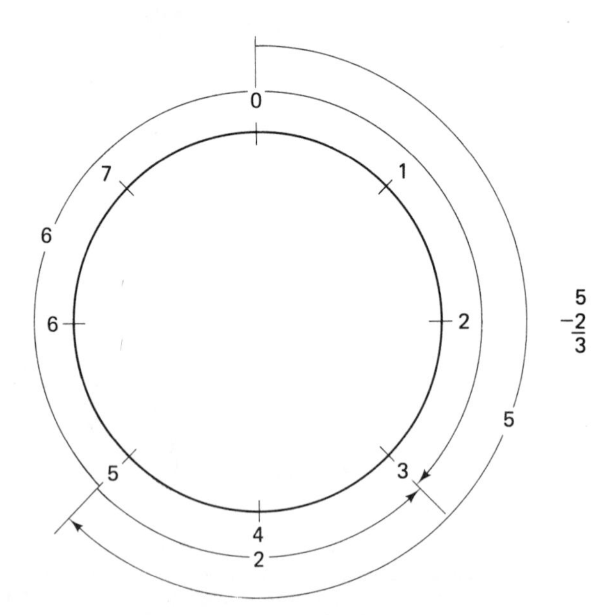

FIGURE 4-29 *Subtraction on a mod-eight number circle.*

It is desired to perform a modulus-k addition in which 'a' is added to 'm' to obtain the same result, 'd', as in the subtraction problem. Thus, we write

$$m + a = d \qquad \text{mod-}k \tag{4-15}$$

Therefore, we must investigate the relationship between 's' in Eq. 4-14, and 'a' in Eq. 4-15.

The ***modulus property*** states that if an N-bit binary number, say 'q', is added to its one's complement and the result incremented by 1, the modulus of that binary representation is produced:

$$q + \bar{q} + 1 = k \qquad \text{[the modulus property, (4-16)]}$$

where 'k' is the modulus of the binary representation.
Examples of the modulus property are:

(a) Four-bit representations:

$$\begin{array}{rl} 1\,0\,1\,1 + 0\,1\,0\,0 = & 1\,1\,1\,1 \\ & +1 \\ \hline & 1\,0\,0\,0\,0 = (16)_{10} \end{array}$$

thus, 1011 is a mod-sixteen number

$$\begin{array}{rl} 0\,0\,1\,0 + 1\,1\,0\,1 = & 1\,1\,1\,1 \\ & +1 \\ \hline & 1\,0\,0\,0\,0 = (16)_{10} \end{array}$$

thus, 0010 is a mod-sixteen number

(b) Seven-bit representation:

$$\begin{array}{rl} 1\,1\,0\,0\,0\,0\,1 + 0\,0\,1\,1\,1\,1\,0 = & 1\,1\,1\,1\,1\,1\,1 \\ & +1 \\ \hline & 1\,0\,0\,0\,0\,0\,0\,0 = (128)_{10}, \end{array}$$

thus, 1100001 is a mod-128 number

Applying the modulus property to 's', we have

$$s + \bar{s} + 1 = k \tag{4-17}$$

Solving for 's' produces

$$s = k - \bar{s} - 1 \tag{4-18}$$

Replacing the occurrence of 's' in Eq. 4-14 with its equivalent value in Eq. 4-18, we obtain

$$m - (k - \bar{s} - 1) = d \tag{4-19}$$

Rearranging this equation yields

$$m + (\bar{s} + 1) = d + k \tag{4-20}$$

Because 'd' is a mod-k number, we can say that $d + k = d$, mod-k. Thus,

$$m + (\bar{s} + 1) = d \qquad \text{mod-}k \tag{4-21}$$

Comparing this equation, term for term, with Eq. 4-13, we see that

$$a = \bar{s} + 1 \tag{4-22}$$

Thus, 'a' is the two's complement of 's', and this conclusion establishes the relationship between 'a' and 's' which was desired.

To summarize, subtraction may be performed by *adding* the two's complement of the subtrahend to the minuend. This procedure is referred to as ***two's-complement subtraction.*** The earlier examples of Figure 4-28 are recomputed in Figure 4-30 using two's-complement subtraction.

Subtraction		Two's-complement addition	Decimal
1011 −0111	⟹	1011 +1001 1|0100	11 − 7 4
10010 −01100	⟹	10010 +10100 1|00110	18 −12 6
1100 −0101	⟹	1100 +1011 1|0111	12 − 5 7
101101 − 1100	⟹	101101 +110100 1|100001	45 −12 33
1001 −1101	⟹	1001 +0011 0|1100 0011 (One's Complement) 0100 (Two's Complement) −0100 (Correct Answer)	9 −13 −4
10000 −110101	⟹	10000 +001011 0|011011 100100 100101 −100101 (Correct Answer)	16 −53 −37

FIGURE 4-30 *Examples of binary subtraction using the two's-complement method. The lack of a carry from the most significant digit indicates that the two's complement of the result must be found for the correct answer.*

It would be very valuable if a single method could apply to both addition and subtraction. If this were possible, the need to create one type of circuitry for addition and a separate type for subtraction would be eliminated. As was implied by two's-complement subtraction, such a method actually is possible, relying only on binary addition. Two important factors justify the method:

1. The subtraction of two numbers can always be viewed as the addition of a positive number to a negative number. More generally,

$$X + Y = (X) + (Y)$$
$$X - Y = (X) + (-Y)$$
$$-X + Y = (-X) + (Y)$$
$$-X - Y = (-X) + (-Y)$$

2. Because of the modulus nature of all physical counting devices, wraparound is possible on a number circle. Thus, negative numbers can be represented in two's-complement form as positive numbers, and combined with a second number according to the rules of addition.

If we represent the two's complement of a binary number 'X' as '$\hat{X}$', then signed binary arithmetic (addition and subtraction) can be considered in a general sense as shown in Figure 4-31. ***Overflow*** occurs whenever the reuslt of an arithmetic operation requires '$N + 1$' digits to express, while only 'N' digits are available in the physical hardware.

Arithmetic Notation		Addition of Signed Numbers		Addition of Two's-Complement Numbers	Possible Condition of Result
X + Y	=	(X) + (Y)	=	X + Y	Result Always Positive, Overflow Possible
X − Y	=	(X) + (−Y)	=	$X + \hat{Y}$	Result Either Positive or Negative, Overflow is not Possible
−X + Y	=	(−X) + (Y)	=	$\hat{X} + Y$	Result Either Positive or Negative, Overflow is not Possible
−X − Y	=	(−X) + (−Y)	=	$\hat{X} + \hat{Y}$	Result Always Negative, Overflow Possible

FIGURE 4-31 *Interpretation of signed binary arithmetic. $\hat{X}$ is the two's complement of $\hat{X}$.*

For the purpose of machine representation, the sign of a binary number will always be indicated as an additional bit following the most significant digit of the number. For example, a three-digit signed number is expressed as '$a_3a_2a_1a_0$', where 'a_3' is the sign bit and '$a_2a_1a_0$' represents the value of the number. For $a_3 = 0$ (a positive number), '$a_2a_1a_0$' is the natural binary representation of a numerical value; whereas, if $a_3 = 1$ (a negative number), then '$a_2a_1a_0$' is the two's complement of the numerical value. Recall that Table 4-2 described two's-complement signed numbers. Examples of signed binary arithmetic using the two's-complement technique are shown in Figure 4-32. In signed binary arithmetic, the sign bit is always treated as a normal digit in the binary number. Note that when the value of a sum cannot be fully expressed with only the three digits '$a_2a_1a_0$', overflow has occurred. Similarly, when the value of a difference cannot be fully represented by '$a_2a_1a_0$', overflow has also occurred. Overflow is possible only when adding two positive numbers or two negative numbers. Examining Figure 4-32, we see that overflow can be detected by the logic condition

$$V = X_nY_n\bar{S}_n + \bar{X}_n\bar{Y}_nS_n \tag{4-23}$$

where $n = 3$ for Figure 4-32. In general, 'n' refers to the most significant digit of a signed binary number, namely the sign bit.

Two's-Complement Representation	
+7	0111
+6	0110
+5	0101
+4	0100
+3	0011
+2	0010
+1	0001
0	0000
−1	1111
−2	1110
−3	1101
−4	1100
−5	1011
−6	1010
−7	1001
−8	1000

X + Y

```
  0010    2         0100     4
 +0011   +3        +0101    +5
  0101    5         1001    −7
                 (Overflow) (Error)
```

X − Y

```
  0111    7         0001     1
 +1111   −1        +1001    −7
  0110    6         1010    −6
```

−X + Y

```
  1010   −6         1010    −6
 +0111   +7        +0010    +2
  0001    1         1100    −4
```

−X − Y

```
  1110   −2      1010   −6      1001     −7
 +1101   −3     +1110   −2     +1000     −8
  1011   −5      1000   −8      0001     +1
                             (Overflow) (Error)
```

FIGURE 4-32 Examples of signed binary arithmetic.

4.2.3 Multiplication

Binary multiplication is more like decimal multiplication than any other binary arithmetic procedure is to its related decimal operation. This is so because the binary multiplication table is a subset of the decimal multiplication table. Figure 4-33 illustrates this idea. The procedure for multiplying two multidigit binary numbers is identical to that for decimal numbers, except that in binary multiplication, the partial products are added using binary addition. Several examples of binary multiplication are shown in Figure 4-34.

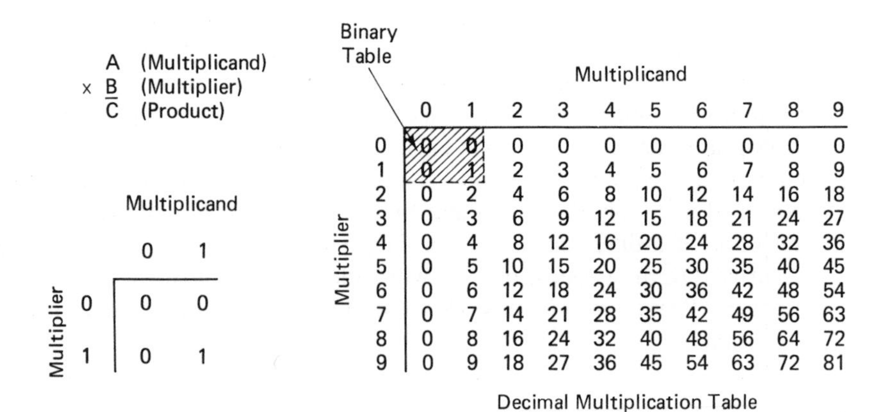

Binary Multiplication Table (Multiplier rows, Multiplicand columns)

	0	1
0	0	0
1	0	1

Decimal Multiplication Table (Multiplier rows, Multiplicand columns)

	0	1	2	3	4	5	6	7	8	9
0	0	0	0	0	0	0	0	0	0	0
1	0	1	2	3	4	5	6	7	8	9
2	0	2	4	6	8	10	12	14	16	18
3	0	3	6	9	12	15	18	21	24	27
4	0	4	8	12	16	20	24	28	32	36
5	0	5	10	15	20	25	30	35	40	45
6	0	6	12	18	24	30	36	42	48	54
7	0	7	14	21	28	35	42	49	56	63
8	0	8	16	24	32	40	48	56	64	72
9	0	9	18	27	36	45	54	63	72	81

Decimal Multiplication Table

Binary Multiplication Table

FIGURE 4-33 Comparison of the binary and decimal multiplication tables.

```
Multiplicand ---->     111      7              110      6
Multiplier   ---->   x 101     x5             1101    x13
                       111                     110
Partial            000                    000
Products          111                   110
Product ---->   100011     35          110
                                      1001110       78

                      1111     15          10001       17
                    x 1010    x10         x   11       x3
                      0000                 10001
                     1111                 10001
                    0000                  110011       51
                   1111
                 10010110     150
```

FIGURE 4-34 Examples of binary multiplication.

It is interesting to note that for every occurrence of a '1' digit in the multiplier, the multiplicand appears as a partial product. The final product, then, is made up of a sum of *shifted* versions of the multiplicand. This fact is the basis of the ***shift-and-add*** procedure for binary multiplication, a technique frequently used in computer programs that perform binary multiplication. However, the shift-and-add procedure requires a means of temporary storage for the partial products. Furthermore, since an '*n*'-digit multiplier calls for '*n*' additions, the result requires several cycles of computation. Thus, if we intend to construct a digital logic device to implement this technique, a sequential circuit will be needed. Although sequential circuits are not examined in detail until Chapter 7, we need not defer our discussion of a hardware binary multiplier until then. By thinking of the multiplier and multiplicand jointly as inputs to a combinational logic circuit, we can design a binary multiplier that does not require temporary storage of partial products, and produces a result after only several gate delays of time.

FIGURE 4-35 Truth table and block representation for a two-bit by two-bit multiplier circuit.

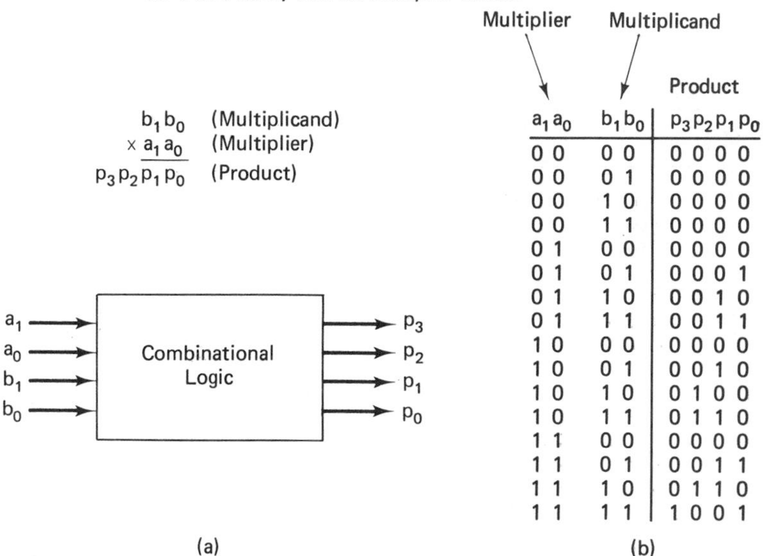

$a_1 a_0$	$b_1 b_0$	$p_3 p_2 p_1 p_0$
0 0	0 0	0 0 0 0
0 0	0 1	0 0 0 0
0 0	1 0	0 0 0 0
0 0	1 1	0 0 0 0
0 1	0 0	0 0 0 0
0 1	0 1	0 0 0 1
0 1	1 0	0 0 1 0
0 1	1 1	0 0 1 1
1 0	0 0	0 0 0 0
1 0	0 1	0 0 1 0
1 0	1 0	0 1 0 0
1 0	1 1	0 1 1 0
1 1	0 0	0 0 0 0
1 1	0 1	0 0 1 1
1 1	1 0	0 1 1 0
1 1	1 1	1 0 0 1

All possible combinations of two-bit multipliers and multiplicands are shown in the truth table of Figure 4-35. From this truth table, it is clear that a four-input, four-output combinational logic circuit can be designed to perform binary multiplication. The following equations are derived for the binary multiplier.

$$p_3: \quad p_3 = a_1 a_0 b_1 b_0 \qquad \text{no simplification is possible} \tag{4-24}$$

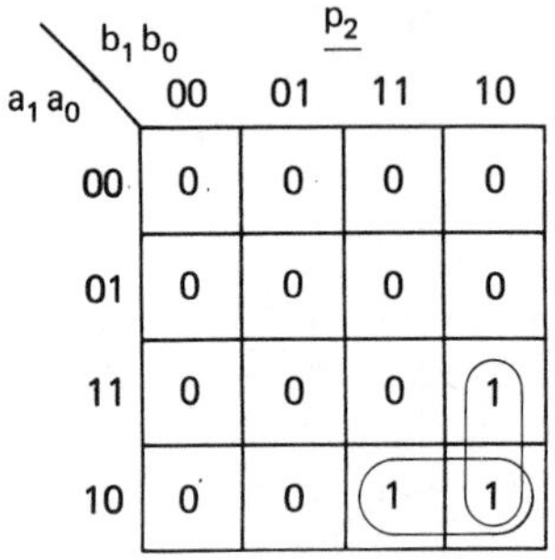

p_2

a_1a_0 \ b_1b_0	00	01	11	10
00	0	0	0	0
01	0	0	0	0
11	0	0	0	1
10	0	0	1	1

$$\begin{aligned} p_2 &= a_1\bar{a}_0 b_1 + a_1 b_1 \bar{b}_0 \\ &= a_1 b_1(\bar{a}_0 + \bar{b}_0) \\ &= a_1 b_1(\overline{a_0 b_0}) \end{aligned} \tag{4-25}$$

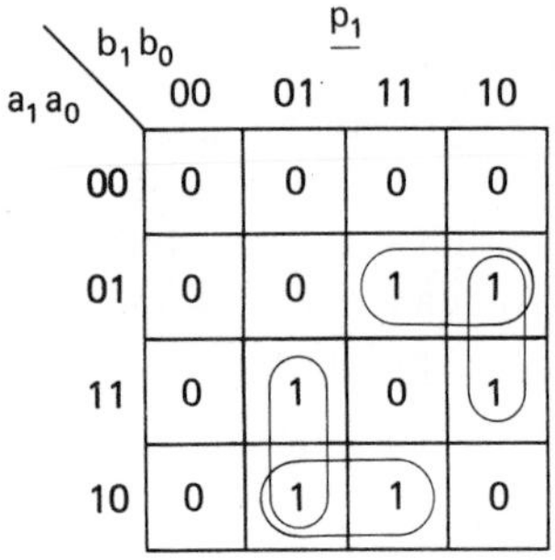

p_1

a_1a_0 \ b_1b_0	00	01	11	10
00	0	0	0	0
01	0	0	1	1
11	0	1	0	1
10	0	1	1	0

$$\begin{aligned} p_1 &= a_1\bar{b}_1 b_0 + a_1\bar{a}_0 b_0 + \bar{a}_1 a_0 b_1 + a_0 b_1 \bar{b}_0 \\ &= a_1 b_0(\bar{a}_0 + \bar{b}_1) + a_0 b_1(\bar{a}_1 + \bar{b}_0) \\ &= a_1 b_0(\overline{a_0 b_1}) + a_0 b_1(\overline{a_1 b_0}) \\ &= a_1 b_0 \oplus a_0 b_1 \end{aligned} \tag{4-26}$$

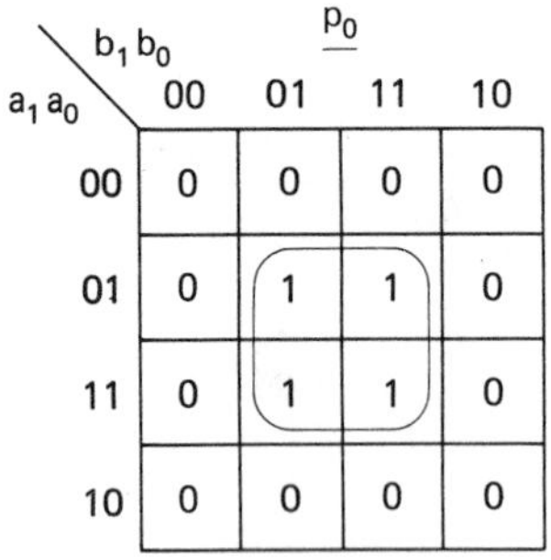

p_0

a_1a_0 \ b_1b_0	00	01	11	10
00	0	0	0	0
01	0	1	1	0
11	0	1	1	0
10	0	0	0	0

$$p_0 = a_0 b_0 \tag{4-27}$$

These equations are implemented with two input gates in Figure 4-36.

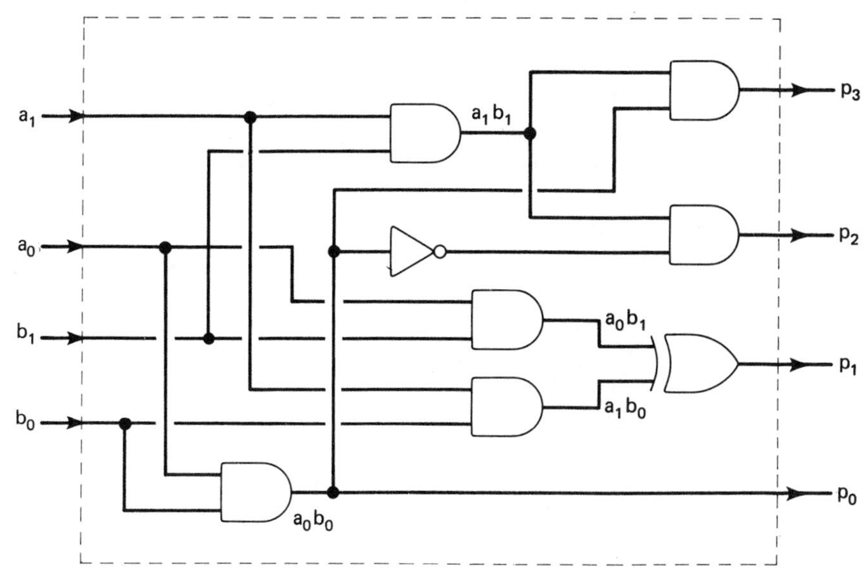

FIGURE 4-36 Combinational logic multiplier circuit.

An alternative way of deriving Eqs. 4-24 to 4-27 generalizes the process of multiplication to a series of bit-by-bit logic operations:

$b_1 \quad b_0$	(multiplicand)
$\times\ a_1 \quad a_0$	(multiplier)
$a_0b_1 \quad a_0b_0$	(first partial product)
$a_1b_1 \quad a_1b_0$	(second partial product)
a_0b_0	(p_0)
$a_0b_1 \oplus a_1b_0$	(p_1, the binary sum of b_1a_0 and a_1b_0, no carry from the previous stage)
$a_1b_1 \oplus (a_0b_1a_1b_0)$	(p_2, the binary sum of a_1b_1 and the carry from p_1)
$a_1b_1(a_0b_1a_1b_0)$	(p_3, only the carry from p_2)

$$
\begin{aligned}
p_3 &= a_1b_1(a_0b_1a_1b_0) \\
&= a_0b_1a_1b_0 \qquad \text{(same as Eq. 4-24)} \\
p_2 &= a_1b_1 \oplus (a_0b_1a_1b_0) \\
&= a_1b_1(\overline{a_0b_1a_1b_0}) + \overline{a_1b_1}(a_0b_1a_1b_0) \\
&= a_1b_1(\bar{a}_0 + \bar{b}_1 + \bar{a}_1 + \bar{b}_0) + (\overline{a_1b_1})(a_1b_1)(a_0b_0) \\
&= a_1b_1(\bar{a}_0 + \bar{b}_0) \qquad \text{(same as Eq. 4-25)} \\
p_1 &= a_0b_1 \oplus a_1b_0 \qquad \text{(same as Eq. 4-26)} \\
p_0 &= a_0b_0 \qquad \text{(same as Eq. 4-27)}
\end{aligned}
$$

4.2.4 Division

Division is the most complex of the four basic arithmetic operations. Binary division is somewhat simpler than decimal division because of the fewer number of symbols in the binary system. Yet, compared to other binary arithmetic operations, binary division still requires the most effort. For multiplication, addition, and subtraction, an exact result can always be found, given enough digits. For division, however, some results are impossible to express exactly with a finite number of digits. In the decimal number system, for example, $2 \div 3$ produces the repeating decimal result 0.6666 . . . . Thus, a minor degree of inaccuracy must be tolerated. Also, division requires much more time to be executed than any other basic arithmetic operation. Binary division performed in the arithmetic-logic unit of a typical computer requires at least 50% more time to be executed than a binary multiplication of the same number of bits.

We will examine two basic methods of binary division. The first is ***restoring division***, a shift-and-subtract technique very similar to the shift-and-add technique for binary multiplication. Restoring division in the decimal system is the familiar "long division" method learned in grade school. Several examples of binary restoring division are shown in Figure 4-37. Studying Figure 4-37, we see that since the divisor is three bits wide, we compare it to the dividend three bits at a time, starting at the most significant end of the dividend. Whenever the divisor can be subtracted from this three-bit part of the dividend without yielding a negative result, the subtraction is performed and a '1' is entered as a digit of the quotient. However, if the subtraction would yield a negative result, no subtraction is actually performed, and a '0' is entered as a digit of the quotient. This procedure is continued repeatedly on successive groups of three-bit partial dividends, until all integer dividend bits have been used, and a subtraction of divisor from partial dividend yields a zero result. It is possible that many fractional digits, possibly an infinite number, may need to be computed before the subtraction of divisor from partial dividend yields a zero result. In such cases, the division process is stopped when sufficient accuracy has been obtained, but before an exact result is found.

Note that in restoring division, it is necessary to perform a comparison *before* any subtraction of divisor from partial dividend is done. This is to ensure that the subtraction will not produce a negative result. In ***nonrestoring division***, no comparison is necessary, and the subtraction of divisor from partial dividend is performed regardless of the sign of the result. However, if a negative result is produced, then on the next stage of division, the divisor is *added* to the partial dividend. The quotient bits are determined by looking at the sign of the result of each subtraction (or addition). An example is shown in Figure 4-38. Note that since no comparison is necessary, the procedure will be executed more quickly.

The nonrestoring division method can be made more simple by using two's-complement subtraction. Thus, no actual subtraction is ever performed, only the addition of two's-complement numbers. An example of this is shown in Figure 4-39. Division is thus reduced to a sequence of shift and add operations.

```
             Quotient + Remainder
Divisor  |  Divided
```

(a) The Quantities of Division.

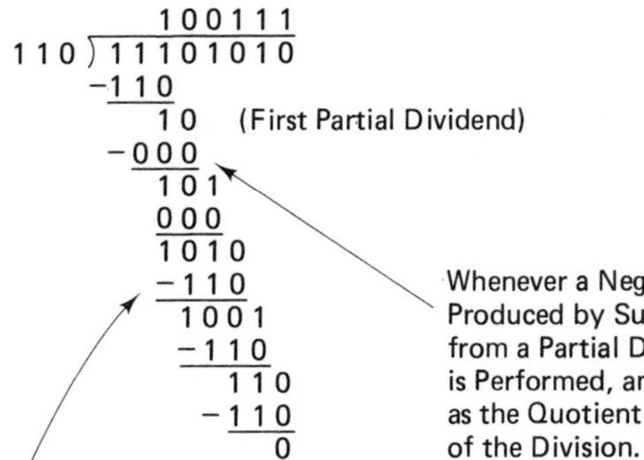


```
   39
6)234
 -18
   54
  -54
    0
```

Whenever a Negative Result would be Produced by Subtracting the Divisor from a Partial Dividend, no Subtraction is Performed, and a '0' is Entered as the Quotient Bit for that Stage of the Division.

Whenever a Positive Result would be Produced by Subtracting the Divisor from a Partial Dividend, the Subtraction is Performed, and a '1' is Entered as the Quotient Bit for that Stage of Division.

```
          11010.01
100 ) 1101001.00
     -100
       101
      -100
        10
       -000
        100
       -100
         01
        -000
          10
         -000
          100
         -100
            0
```

```
   26.25
4)105.00
 -8
  25
 -24
   10
   -8
    20
   -20
     0
```

(b) Examples of Restoring Division.

FIGURE 4-37

FIGURE 4-38 *Example of nonrestoring division.*

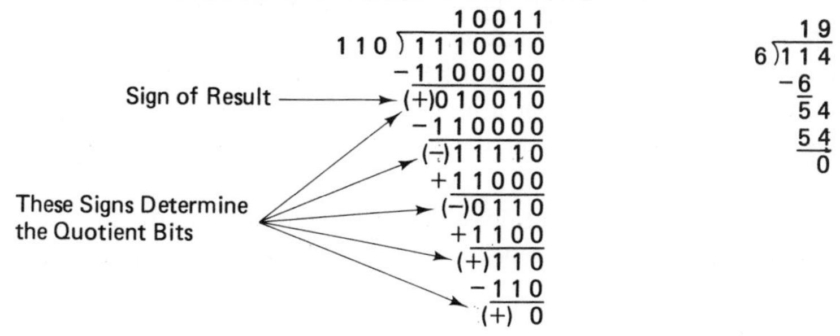


```
   19
6)114
 -6
  54
  54
   0
```

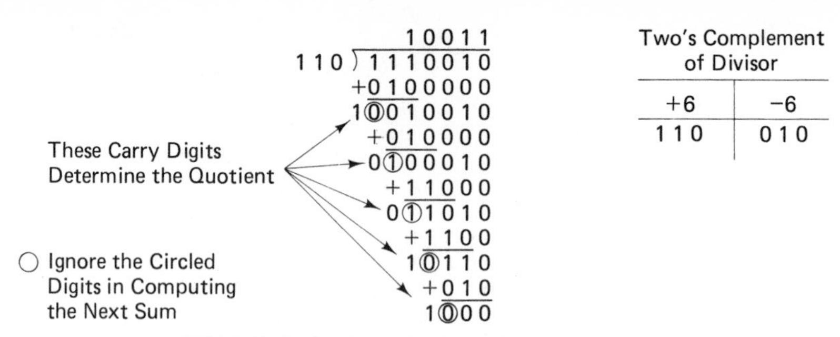

***FIGURE 4-39** Example of nonrestoring division, using two's-complement subtraction.*

4.3 BINARY CODES

We have seen how negative numbers can be represented in the binary system by several different methods. In particular, the sign-magnitude, one's-complement, and two's-complement methods have been examined. Each method has advantages that make it suitable for certain applications. The sign-magnitude convention, for example, is most desirable when binary numbers are to be converted into a decimal format and indicated on an electronic display device. The two's-complement convention, however, is the most efficient representation of binary numbers for arithmetic processing. Indeed, many other representations of binary numbers are possible besides these three. An important restriction that must be placed on any binary representation is that it *uniquely* represent numerical values. It would be unwise, for instance, to use one binary code to represent two different numerical values, since ambiguity would result. In this section, we are chiefly concerned with different binary representations of positive integers. We shall see how different binary codes are interpreted, and how varying applications make efficient use of the available codes.

4.3.1 Weighted Codes

From our previous work, we are most familier with the ***natural binary code.*** "Natural" in this case refers to the fact that the binary counting sequence 0, 1, 10, 11, 100, 101, . . . , exactly corresponds with the numerical values zero, one, two, three, four, five, ... Conversion of a natural binary number into an equivalent decimal number is easily performed by weighting each digit of the binary number by an appropriate integer power of two, and adding the results. Consequently, the natural binary code is classified as a ***weighted code.***

It is frequently necessary to represent the decimal digits by binary codes. Since ten decimal digits must be encoded, at least four binary digits will be required to uniquely represent them. A four-digit *natural* binary code commonly used to represent the decimal digits is referred to as the *8-4-2-1 code.* The numbers 8, 4, 2, and 1 correspond to the weight of the digit positions in the binary number. Several other weighted codes have been developed for the decimal digits and are illustrated in Figure 4-40. Several points should be made with regard to Figure 4-40. First, note that many different binary representations are possible for the ten decimal digits, but that in each case a unique binary code is associated with each decimal symbol. Second, at least four

Decimal Digit	8-4-2-1	4-2-2-1	2-4-2-1	Excess-3	Biquinary 5-0 4-3-2-1-0
0	0000	0000	0000	0011	01 00001
1	0001	0001	0001	0100	01 00010
2	0010	0010	0010	0101	01 00100
3	0011	0011	0011	0110	01 01000
4	0100	1000	0100	0111	01 10000
5	0101	0111	1011	1000	10 00001
6	0110	1100	1100	1001	10 00010
7	0111	1101	1101	1010	10 00100
8	1000	1110	1110	1011	10 01000
9	1001	1111	1111	1100	10 10000

FIGURE 4-40 *Codes for the decimal digits. Except for the Excess-3 code, all codes are weighted. The Excess-3 code is always three greater than the same 8-4-2-1 representation.*

binary digits are required to represent the ten decimal symbols. However, more than four binary digits may be used if there is an advantage in doing so. Finally, sixteen different binary codes are possible using four binary digits, yet only ten are needed for the decimal symbols. Thus, six binary codes will remain unassigned for any decimal encoding scheme using four binary digits.

Using one of the codes of Figure 4-40, a decimal number can be expressed by encoding each of its decimal digits individually, and combining the result into a long string of binary digits. For example, using the 8-4-2-1 code, the decimal number '375' can be represented as

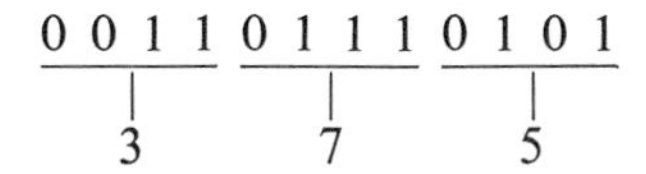

The number '375' expressed in this manner is said to be in ***binary-coded-decimal*** (BCD) form. The chief advantage of BCD numbers is the ease with which they are interconverted between binary and decimal representations. From the circuit designer's point of view, this has very favorable implications, since nearly all numerical processing circuitry must encode signals from a decimal system keyboard, and must present output information on a decimal display device (either an electronic display or a print mechanism). The circuitry needed to translate a four-digit binary code into the assertion of a single logic '1' on one of ten-output lines is not complex, and is illustrated by a block diagram in Figure 4-41. To decode multiple-digit BCD numbers, it is only necessary to duplicate the basic four line–to–ten line decoder circuitry once for every added digit of the BCD number. However, using a natural

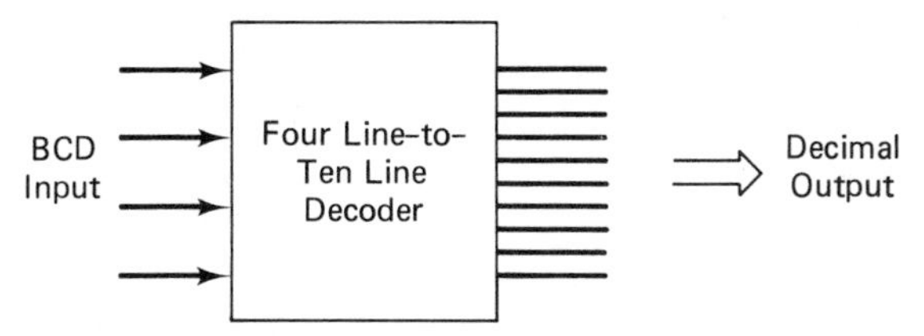

FIGURE 4-41 *Four line–to–ten line decoder used to convert a BCD input code into a decimal output code.*

binary encoding rather than BCD encoding, very complex circuitry is required. Note that natural binary encoding is a more efficient way of representing decimal quantities, in that no binary codes are wasted, as in BCD. Furthermore, numerical quantities must be in natural binary form for the most efficient binary arithmetic procedures to apply. Clearly, there is a trade-off between the BCD and natural binary systems. The ultimate selection of one scheme will depend on the desired speed of the circuitry, its interface requirements with input and output devices (keyboards, printers, etc.), and the ease and expense of adding capacity to the digital circuitry for more binary digits.

If BCD code is used in arithmetic circuitry, the rules of binary arithmetic that we have discussed can no longer be applied. Actually, BCD arithmetic is a curious mix between binary and decimal arithmetic, based on a mod-ten radix circle with binary encoding. In relation to the 4–2–2–1, 2–4–2–1, and Excess-3 codes, we briefly discuss one of the aspects of BCD arithmetic.

In the same sense that one's- and two's-complement numbers are useful in binary subtraction, ***nine's***- and ***ten's***-complement numbers are useful in decimal subtraction. The nine's complement of a decimal number is obtained by subtracting each digit of the number from nine. For example:

Decimal Digit	*Nine's Complement*
2	7
5	4
8	1

The ten's complement of a decimal number is obtained by incrementing by one the nine's complement of that number. The 4–2–2–1, 2–4–2–1, and Excess-3 codes shown in Figure 4-40 have the special property that the one's complement of the binary representation produces the corresponding nine's complement of the decimal representation. For example:

Decimal Digit	4–2–2–1	*One's Complement*	*Decimal Digit*
2	0 0 1 0	1 1 0 1	7
5	0 1 1 1	1 0 0 0	4
8	1 1 1 0	0 0 0 1	1

This feature makes these codes especially suitable for BCD arithmetic circuitry, much more so than the 8–4–2–1 BCD code. Because of the special complementing property of the 4–2–2–1, 2–4–2–1, and Excess-3 codes, they are often referred to as ***self-complementing*** codes.

4.3.2 Gray Code

Recall from our study of Karnaugh maps that the horizontal and vertical addresses of a sixteen-cell map progressed in a special sequence. This sequence is one in which

Two-Digit Gray	Three-Digit Gray	Four-Digit Gray	Natural Binary
00	000	0000	0000
01	001	0001	0001
11	011	0011	0010
10	010	0010	0011
	110	0110	0100
	111	0111	0101
	101	0101	0110
	100	0100	0111
		1100	1000
		1101	1001
		1111	1010
		1110	1011
		1010	1100
		1011	1101
		1001	1110
		1000	1111

FIGURE 4-42 *Two-, three-, and four-digit Gray code, compared with the four-digit natural binary code.*

no two adjacent addresses differ by more than one digit. Although only a two-digit code was involved in Karnaugh maps, it is possible to expand the sequence to any length. The result is referred to as the ***Gray code***, and is illustrated in Figure 4-42.

Any natural binary number can be converted to a corresponding Gray code number by combining adjacent binary digits with an Exclusive OR function, as shown in Figure 4-43(a). Conversion of a Gray code number into an associated natural binary number is somewhat more difficult, and is illustrated in Figure 4-43(b).

Example: Natural Binary Number: 1011

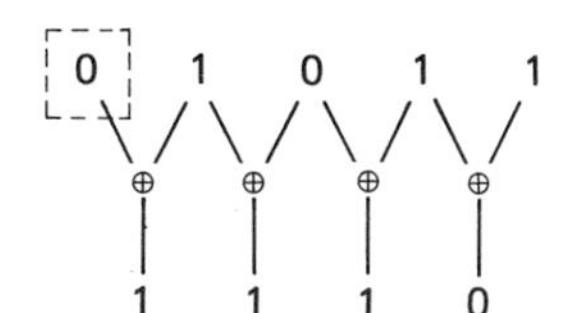

Corresponding Gray Code Number: 1110

(a) Conversion of a Natural Binary Number into a Gray Code Number.

Procedure: Sequential Addition Left to Right, without Carry, Starting with '0'.

Example: Gray Code Number: 1110

(i)	(ii)	(iii)	(iv)
1110	1110	1110	1110
+0	+01	+010	+0101
1	10	101	1011

Corresponding Natural Binary Number: 1011

(b) Conversion of a Gray Code Number into a Natural Binary Number.

FIGURE 4-43

4.3.3 Parity

The transmission of digital information from one point in space to another is always subject to distortion and electrical interference. Usually, the close proximity of components in a digital circuit the size of a typewriter, or less, precludes any real probability of transmission errors. However, if several digital circuits are separated by more than a few meters, an occasional transmission error can be expected. This is due to the enhanced effect of surrounding electrical noise, brought about because of the increased length of cable connecting the digital circuitry. Also, it is possible that digital information recorded on a semipermanent medium such as magnetic tape can become altered by irregularities in the properties of the recording surface. To avoid the undesirable effects of erroneous data, several methods have been developed to detect and correct transmission errors. The simplest methods merely detect the occurrence of an error, while more complex methods can actually impose corrections. Which method is chosen depends on the seriousness of the effect an error would have, the probability of error, the ability to retransmit, and other factors. We will examine a technique that allows the detection of a one-bit error in a multibit data word (often eight bits of data).

The ***parity*** convention requires that a single, additional bit be added to a data word. This bit is referred to as the ***parity bit***, and its value is determined by the number of '1' digits in the data word. In a system of ***even parity***, the parity bit is assigned a value of '1' to make the total number of '1' digits in the transmitted word even; whereas in the system of ***odd parity***, the parity bit is assigned a value of '1' to make the total number of '1' digits in the transmitted word odd. Figure 4-44 illustrates even and odd parity as applied to several arbitrary eight-bit data words. Only one system of parity (even or odd) is chosen and consistently used in a particular digital design.

Quantity of '1's in Data Word	Parity Bit	Eight-Bit Data Word
5	0	1 1 0 1 0 0 1 1
4	1	0 1 0 1 1 0 1 0
3	0	0 1 1 0 0 1 0 0
2	1	1 0 0 0 0 1 0 0
8	1	1 1 1 1 1 1 1 1
1	0	0 0 0 0 0 0 0 1

Odd-Parity System

Quantity of '1's in Data Word	Parity Bit	Eight-Bit Data Word
5	1	0 1 0 1 1 1 1 0
4	0	0 1 1 1 0 0 1 0
5	1	1 0 0 1 1 0 1 1
6	0	1 1 1 0 0 1 1 1
3	1	1 0 0 1 0 0 1 0
7	1	0 1 1 1 1 1 1 1

Even-Parity System

FIGURE 4-44 *Examples showing how an appropriate parity bit is added to an eight-bit data word.*

Now, if a one-bit error in any digit position occurs in the transmission or recording process, the actual parity of the received data word *will not* agree with the parity bit. Consequently, the occurrence of an error can be detected. However, since there is no way of knowing precisely which bit is erroneous, it is not possible to correct the error at the receiving end. Furthermore, if an even number of one-bit errors occur in a single data word, the parity of the transmitted word will not change and the error will go undetected. Fortunately, the probability of getting a single bit error is much greater than the probability of getting multiple bit errors. Thus, our chance of detecting any error at all is quite good using the parity bit technique. Figure 4-45 illustrates various possibilities for transmitted and received data, showing when errors can be detected. Figure 4-46 depicts a combinational logic circuit that can be used to generate a parity bit for an odd-parity system. The parity bit is usually added to the most-significant-digit side of the data word.

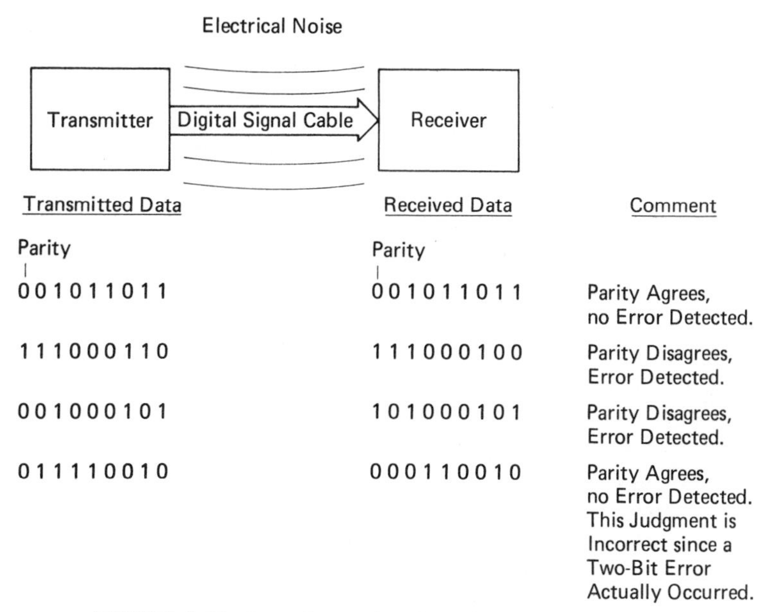

FIGURE 4-45 *Several examples of transmitted and received data in an odd-parity system.*

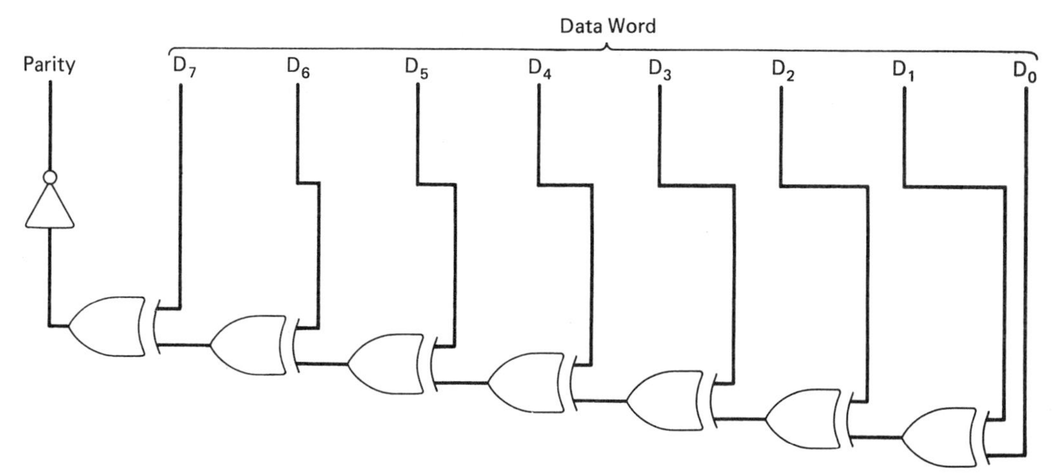

FIGURE 4-46 *Even-parity generator for an eight-bit data word.*

4.4 OCTAL AND HEXADECIMAL CODES

The use of digital systems and computers frequently requires the writing of binary numbers and sometimes the verbal communication of binary numbers to other people. Unfortunately, the binary number system is quite awkward for people to use, despite its suitability to electronic circuitry. Thus, it would be very desirable if an alternative system of coding could be found that represents binary numbers in a simple way, but also is easy for people to use. Several coding schemes have been adopted to meet this need. One of these schemes is the ***octal*** system of notation. The octal system is a radix-eight number system, consisting of the symbol set {0, 1, 2, 3, 4, 5, 6, 7}. The conversion from binary to octal notation is extremely simple, and requires only that we divide a given binary number into three-bit sections, beginning at the least-significant-digit side of the number. Each three-bit section of the binary number is then converted into an appropriate octal digit, using a 4–2–1 weighting. This is illustrated with several examples in Figure 4-47. Leading zeros are added to the most significant end of a binary number to complete a three-bit section, if necessary. It is clear from Figure 4-47 that octal notation is both simple to interpret in terms of binary digits, and easy to read from a human point of view.

Another scheme for compactly representing binary numbers is the ***hexadecimal*** system of notation. In this case, a binary number is divided into *four-bit* sections, instead of three-bit sections as in octal notation. With an 8–4–2–1 weighting, however, we use the last decimal symbol, '9', with the four-digit code '1001'. Since it is desirable to represent the remaining six four-bit codes with single character symbols, a new assignment of characters must be made. The full hexadecimal code is shown in Figure 4-48. Examples illustrating the conversion between binary and hexadecimal notation are shown in Figure 4-49. Again, as with octal notation, hexadecimal notation is both simple to interpret in terms of binary digits and easy to read. The hexadecimal system requires even fewer digits than the octal system to express binary numbers, although the code of Figure 4-48 must be remembered.

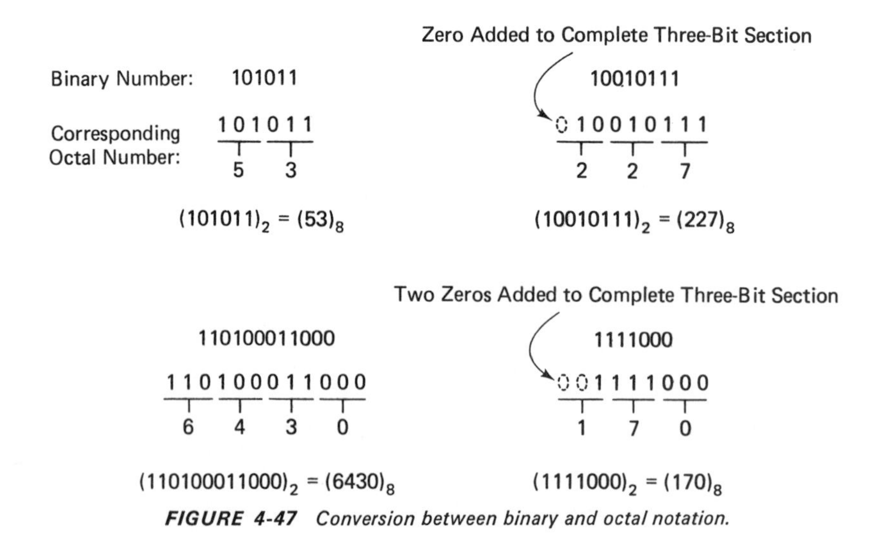

FIGURE 4-47 *Conversion between binary and octal notation.*

Four-Digit Binary Code	Hexadecimal Code
0 0 0 0	0
0 0 0 1	1
0 0 1 0	2
0 0 1 1	3
0 1 0 0	4
0 1 0 1	5
0 1 1 0	6
0 1 1 1	7
1 0 0 0	8
1 0 0 1	9
1 0 1 0	A
1 0 1 1	B
1 1 0 0	C
1 1 0 1	D
1 1 1 0	E
1 1 1 1	F

FIGURE 4-48 *Interpretation of four-bit binary code as single hexadecimal digits.*

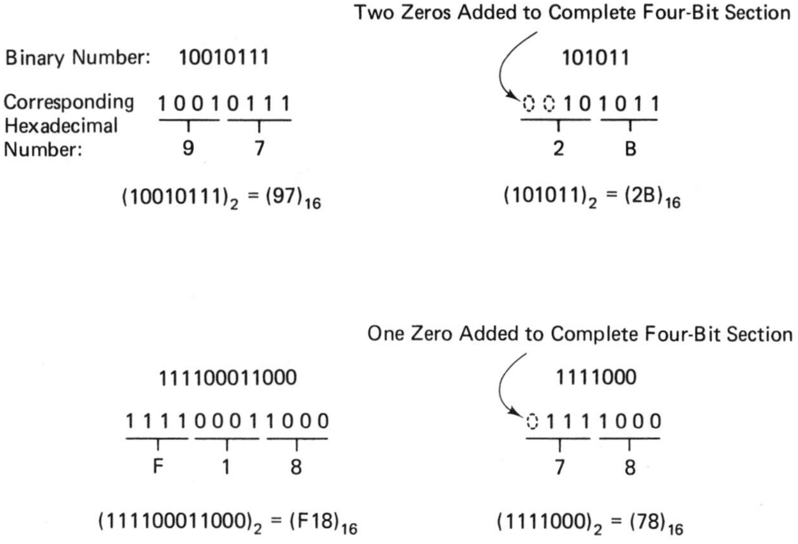

Two Zeros Added to Complete Four-Bit Section

Binary Number: 10010111 | 101011

Corresponding Hexadecimal Number: 1 0 0 1 0 1 1 1 → 9 7 | 0 0 1 0 1 0 1 1 → 2 B

$(10010111)_2 = (97)_{16}$ $(101011)_2 = (2B)_{16}$

One Zero Added to Complete Four-Bit Section

111100011000 | 1111000

1 1 1 1 0 0 0 1 1 0 0 0 → F 1 8 | 0 1 1 1 1 0 0 0 → 7 8

$(111100011000)_2 = (F18)_{16}$ $(1111000)_2 = (78)_{16}$

FIGURE 4-49 *Conversion between binary and hexadecimal numbers.*

SUMMARY

Binary numbers permit us to measure and count using only the symbols '0' and '1'. Although binary numbers are less compact and less readily comprehended than decimal numbers, they are ideally suited to binary digital systems. The binary number system applies integer powers of two as weighting factors to a sequence of '0', '1' symbols. The sum of such a weighted sequence reveals the value of the binary number.

The most significant bit of a binary number is that digit whose weight is greatest. Although fractional numbers require weighting factors with negative exponents, the same rules that apply to whole numbers are applied to fractional numbers in the determination of value. Signed binary numbers usually possess an additional bit fixed to the MSB digit location to indicate the sign of the number. Typically, a sign bit of '0' represents a positive number, while '1' represents a negative number. The one's-complement and two's-complement forms are often applied to the magnitude part of a binary number to facilitate arithmetic processing.

The limitation in capacity of any real counting device is defined by the device's modulus, which represents the total number of different numerical states that can be indicated. The number circle is an adaptation of the number line that accounts for the modulus nature of all arithmetic hardware. Addition and subtraction can be performed on the number circle by relating distance around the circle to numerical value. Exceeding the maximum capacity of the number circle causes a carry or borrow digit to be generated. Complete arithmetic tables for addition and subtraction in any number system can be determined from a radix circle for that system. A particular radix circle is divided into as many sectors as there are symbols in a given number system, and is labeled according to these symbols.

Combinational logic can be designed to realize one-bit binary addition. By cascading '*n*' one-bit adders, it is possible to form a combinational logic circuit for '*n*' bits of binary addition. A similar design is possible for binary subtraction.

Binary subtraction is greatly simplified by expressing all subtractions as the addition of negative numbers, and then placing these negative numbers in two's-complement form. The subtraction problem is thereby converted into one of addition. This has very favorable implications for the digital circuit designer since the hardware required in a two's-complement signed arithmetic system is less than would be required if separate addition and subtraction circuits were built.

The binary multiplication table is contained within the decimal multiplication table with which we all are familier. For this reason, the process of binary muitiplication is quite easy to remember, and is different from decimal multiplication only in that binary addition must be used to combine the partial products. A shift-and-add procedure can be applied when designing hardware or writing computer programs for binary multiplication, eliminating any reference to the binary multiplication table.

Binary division can be done using one of several methods. Restoring division subtracts the divisor from partial dividend only if a nonnegative answer will result. Whether or not a subtraction is performed for a given partial dividend is what determines the quotient bit for the stage of division. In nonrestoring division, the subtraction of divisor from first partial dividend is always performed, and following stages will involve either subtraction or addition of the divisor depending on the sign of the current partial dividend. Nonrestoring division can be achieved using two's-complement addition, and thus has the advantage of being implemented as a shift-and-add procedure, similar to binary multiplication.

The correspondence between binary numbers and numerical value can be assigned in many different ways. In terms of arithmetic processing, the natural binary code is often used since it is the logical extension from a decimal number system to a two-symbol (binary) number system. The natural binary code is a weighted code in that a binary number value is determined by the weighted sum of its digits. The weighting factors in a natural binary code are successive integer powers of two. Other weighted codes are also possible. Nonweighted binary codes, such as the Gray code, do not reveal a natural association with numerical value, but are associated with value only insofar as arbitrary assignments are made.

Each digit of a decimal number can be encoded as a four-bit binary number, with the resulting sequence of four-bit numbers considered as a binary representation for the decimal number. This BCD coding scheme is very useful when binary quantities are displayed or printed in decimal form on actual output devices.

A parity bit is often attached to the MSB side of a binary number prior to transmission through a noise-prone environment. The parity bit is determined by the quantity of '1' digits in a binary data word. At the receiving end, disagreement between the parity bit and the actual parity of the data word indicates that a transmission error has occurred. The nature of the error, however, cannot be determined.

Octal and hexadecimal codes are frequently employed to more compactly express binary numbers. Groups of three (for octal) or four (for hexadecimal) binary digits are converted into a single symbol that is more easily written, spoken, and understood. The conversion always progresses from right to left. Leading zeros are added when it is necessary to complete a group of three or four bits.

NEW TERMS

Number System
Decimal Number System
Binary Number System
Radix
Base
Bit
Least Significant Bit
Most Significant Bit
Number Line
Number Circle
Sign-Magnitude
One's Complement
Two's Complement
Sign Bit
Modulus
Radix Circle
Carry Digit
Half Adder
Full Adder
Modulus Property
Augend
Addend
Sum
Minuend
Subtrahend
Difference
Two's Complement Subtraction
Overflow
Shift-and-Add
Borrow Digit
Restoring Division
Nonrestoring Division
Weighted Code
Natural Binary Code
Binary-Coded-Decimal Code
8-4-2-1 Code
Nine's Complement
Ten's Complement
Self-Complementing Code
Gray Code
Parity Bit
Even Parity
Odd Parity
Octal Code
Hexadecimal Code

PROBLEMS

4-1 Express the following decimal numbers as a sum of weighted decimal symbols.

(a) 1998

(b) 57.361

(c) 1000.0001

4-2 Determine the decimal numbers corresponding to the following sums.

(a) $(2 \cdot 10^{-2}) + (8 \cdot 10^{-1}) + (5 \cdot 10^{0}) + (6 \cdot 10^{1}) + (9 \cdot 10^{2})$

(b) $(3 \cdot 10^{3}) + (5 \cdot 10^{1}) + (1 \cdot 10^{-3})$

(c) $(2.1 \cdot 10^{3}) + (3.8 \cdot 10^{2}) + (9.5 \cdot 10^{1}) + (3.15 \cdot 10^{0})$

4-3 Convert the following binary numbers into decimal numbers.

(a) 101101 **(b)** 100.111

(c) 111.0011 **(d)** 10111.1101

(e) 101100011 **(f)** 01011101

4-4 Convert the following decimal numbers into binary numbers (do not compute more than four fractional digits).

(a) 210 **(b)** 406.8

(c) 19.125 **(d)** 1023

(e) 86 **(f)** 1.9751

4-5 Is the most significant bit of the binary number '001101.111' a '0' or a '1'?

4-6 Determine the one's complement and two's complement of the following binary numbers.

(a) 11010111 **(b)** 10010000

(c) 10110001 **(d)** 10000000

(e) 0110 **(f)** 0

4-7 Add each of the following binary number pairs, and verify your answers using decimal addition.

(a)
```
   10001
 + 01001
```
(b)
```
   1101
 + 0101
```
(c)
```
   01011111
 + 10111101
```
(d)
```
   1010010
 +     111
```
(e)
```
       1000
 + 11011111
```
(f)
```
   111101
 + 010011
```
(g)
```
   10111001
 + 01000111
```
(h)
```
   10101010
 + 01010101
```
(i)
```
   1101011
 + 1110101
```

4-8 **(a)** Perform the following binary additions and verify your answers using decimal addition.

(i)
```
   1011
 + 1011
```
(ii)
```
   10101010
 + 10101010
```
(iii)
```
   11111
 + 11111
```

(b) In each of the cases above, by a factor of how much is the sum larger than the addend or augend?

(c) Compare the '1', '0' bit sequence of the sums to the addend and augend in part (a), and describe the effect of doubling a binary number.

4-9 Subtract each of the following binary number pairs using the binary subtraction table of Figure 4-27(b), and verify your answers using decimal subtraction.

(a) 10010 − 01101 **(b)** 1101 − 1011 **(c)** 110010 − 010100

(d) 10000 − 01111 **(e)** 101100111 − 1101 **(f)** 100101 − 111010

(g) 1100 − 110101 **(h)** 00000 − 10101 **(i)** 101001 − 101001

4-10 **(a)** Perform the following binary subtractions.

(i) 1111 − 1011 (ii) 1111111 − 0110111 (iii) 111111111 − 110101011

(b) In view of the results of part (a), suggest a method of determining the two's complement of a binary number.

4-11 Convert each of the following decimal arithmetic problems into a signed binary arithmetic problem (see Figure 4-32). Compute a result in each case, being certain that all results are shown in *sign-magnitude* form.

(a) 4 + 3 **(b)** 6 + 7 **(c)** 7 − 3 **(d)** 5 − 6

(e) −2 + 4 **(f)** 0 − 8 **(g)** −8 + 5 **(h)** −1 − 6

(i) −5 − 2 **(j)** −8 − 8 **(k)** −7 + 7 **(l)** −4 − 4

4-12 Using the truth table of Figure 4-27(b), design a combinational logic subtracter circuit using a minimum number of two-input gates of any type (Exclusive OR gates may be used).

4-13 Multiply each of the following binary number pairs and verify your results using decimal multiplication.

(a) 1001 × 1101 **(b)** 1101101 × 110

(c) 1110 × 1001111 **(d)** 1010 × 1010

(e) 10001 × 1110 **(f)** 010011 × 000101

4-14 Divide the following binary number pairs using restoring division, and verify your results using decimal division.

(a) $101\overline{)11001}$ **(b)** $11\overline{)100001}$

(c) $110\overline{)100011}$ **(d)** $1110\overline{)1111110}$

4-15 Repeat Problem 4-14 using nonrestoring division.

4-16 Convert the following binary numbers to (i) octal numbers, and (ii) hexadecimal numbers.

(a) 100111110 (b) 10111000
(c) 1010011 (d) 0100010111
(e) 101101000 (f) 11101001

4-17 Convert the following octal numbers to hexadecimal numbers.

(a) 172 (b) 3115
(c) 67 (d) 2045
(e) 1776 (f) 177777

4-18 Convert the following hexadecimal numbers to octal numbers.

(a) 3C (b) 6BA5
(c) F1FC (d) 10A1
(e) 4BD (f) 98FA3

4-19 Convert the following decimal numbers to BCD numbers.

(a) 1510 (b) 81
(c) 32677 (d) 8791
(e) 3 (f) 1685

4-20 What is the modulus of a digital stopwatch that can count in decimal to 59.99 sec.?

4-21 For which number system(s) (binary, octal, decimal, hexadecimal) is the following equation a true statement?

$$10(11 + 100) - 11(10 + 0) = 1000$$

4-22 How many fractional digits will the decimal number 2^{-48} have? (Refer to Table 4-1.)

4-23 If 'A' and 'B' are four-digit binary numbers and '$\bar{A}$' and '$\bar{B}$' are their one's complements, determine which, if any, of the following equations are true. In this problem, '+' represents *addition*, not logic OR.

(a) $A + \bar{A} = 1111$ (b) $\overline{A + B} = \bar{A} + \bar{B} + 1$
(c) $\overline{\bar{A} + \bar{B}} = A + B$ (d) $A + \bar{B} \times \bar{A} + B$
(e) $A - B = (A + \bar{B} + 1)$ mod-16

4-24 An electronic counter displaying decimal numbers in initialized to zero. Each occurrence of a +5-V pulse at the input advances the counter once. What will be displayed if $(81)_{10}$ pulses are applied when the counter is

(a) mod-100 (b) mod-10
(c) mod-9 (d) mod-2
(e) mod-6

4-25 (a) Determine the addition and subtraction tables for the *radix-three* number system (symbol set {0, 1, 2}). (*Hint:* Use a mod-three radix circle.)

(b) Using the tables found in part (a), perform the following additions and subtractions, and verify your answers using decimal arithmetic.

(i) $\begin{array}{r} 120 \\ +\ \underline{211} \end{array}$ (ii) $\begin{array}{r} 1211012 \\ +\ \underline{1012122} \end{array}$

(iii) $\begin{array}{r} 1011010 \\ +\ \underline{0111100} \end{array}$ (iv) $\begin{array}{r} 212 \\ -\ \underline{121} \end{array}$

(v) $\begin{array}{r} 2212011 \\ -\ \underline{2122120} \end{array}$ (vi) $\begin{array}{r} 12021 \\ -\ \underline{21221} \end{array}$

4-26 Using the block representation of a one-bit adder *as a component*, as in Figure 4-21, and any number of logic gates of any type, design a four-bit two's-complement circuit. (*Hint:* It can be done with four one-bit adders and four gates.)

4-27 Given two four-bit binary numbers, design a comparator circuit to determine which number is larger. Two-input Exclusive OR gates may be used.

4-28 A number system is created with the symbol set $\{0, \Delta, \nabla, \Gamma\}$, corresponding to the values zero, one, two, and three, respectively.

(a) What is the radix of this number system?

(b) If a three-digit counter were made using this number system, what would the counter's modulus be?

(c) Convert the following $\{0, \Delta, \nabla, \Gamma\}$ numbers into decimal numbers.

(i) $\Gamma\ 0\ \nabla\ 0$ (ii) $\nabla\ \Gamma\ \Gamma\ \Delta$

(iii) $\Delta\ \nabla\ \nabla\ \Gamma\ .\ 0\ 0\ \Delta$ (iv) $0\ 0\ \nabla\ 0\ .\ 0\ 0\ \Delta$

(d) Convert the following decimal numbers into $\{0, \Delta, \nabla, \Gamma\}$ numbers.

(i) 0 (ii) 18

(iii) 31 (iv) 10.25

4-29 Redesign the even-parity generator of Figure 4-46 to minimize the propagation delay. Use only two-input Exclusive OR gates in the improved design.

4-30 The circuit block of Figure P4-30 converts four-bit natural binary numbers into four-bit Gray code numbers. Using this block as a component, design a circuit that converts four-bit Gray code numbers into four-bit natural binary numbers. Assume that any amount of binary-to-Gray converter blocks are available.

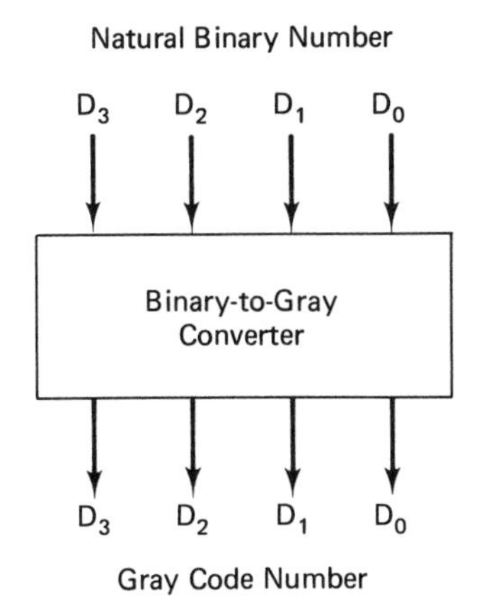

FIGURE P4-30

4-31 The circuit block of Figure P4-31 is a four-bit binary full adder. Using this block as a component, design a circuit that performs two's-complement subtraction. Assume that the minuend and subtrahend are available as unsigned positive integers (binary, of course), and that any type of logic gates are available. (*Hint:* It can be done using only four inverters.)

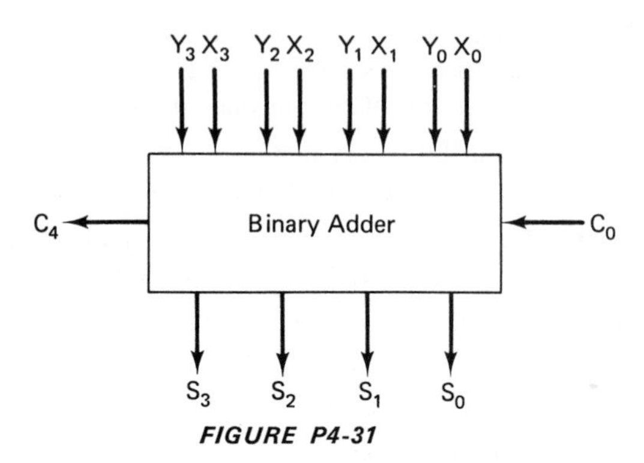

FIGURE P4-31

5 FLIP-FLOPS

5.1 OVERVIEW

Our primary concern in previous chapters has been with combinational logic circuits. For a given input condition, we have seen that the output of a combinational logic circuit will always be the same, regardless of when in time the input condition is applied. Consequently, a truth table can fully and explicitly describe the operation of such a circuit. The only timing considerations involved in combinational logic relate to the internal propagation delays of logic gates. Once the propagation delays have expired, the outputs will remain absolutely constant and predictable as long as the inputs are maintained. The nature of combinational logic prevents it from ever being able to retain a particular output condition after the input values that brought about that condition have been removed. To have such a memory feature, though, is extremely important and, indeed, is necessary in most practical digital circuit applications. ***Sequential circuits*** satisfy this need for memory, and constitute a second basic area of study in digital circuitry.

Sequential circuits are functionally different from combinational circuits in that the *prior* sequence of inputs, as well as the *present* input conditions, determine the *present* output conditions. Sequential circuits are structurally different from combinational circuits in that feedback loops can be observed in the circuit connections. Because a sequence of input conditions determine the output conditions that are produced, ***timing diagrams***, which show changes in the value of a binary variable with respect to time, will become very meaningful in describing the operation of sequential circuits. Timing diagrams will be useful in analyzing short-lived conditions of a logic circuit brought about by varying propagation delays in its circuit components, and also in observing the relationship between signals when clock pulses are periodically applied to a circuit to trigger a change of state.

The relay hold circuit, which was introduced in Section 1.2.3, begins our study of sequential circuits in Chapter 5, and leads to the development of the flip-flop. Chapter 6 is devoted to digital counting circuits, and describes a special type of sequential circuit (the binary counter) that is often employed in large digital systems.

Formal techniques for the analysis of more complex, and very interesting, sequential circuits are presented in Chapter 7, where minimum circuitry configurations are sought. These formal techniques expand from the intuitive framework that we will have developed in Chapters 5 and 6 and lend power to our ability to design sequential circuits, in the same sense that Boolean algebra lends power to our ability to design combinational logic.

5.2 THE LATCH

5.2.1 The Relay Hold Circuit

A simple memory circuit utilizing a relay, switches, and a voltage source was introduced in Chapter 1. This circuit is redrawn in Figure 5-1, and is often referred to as a relay hold circuit. With the switches set to $S_1 = 0$ and $S_2 = 1$, two unchanging

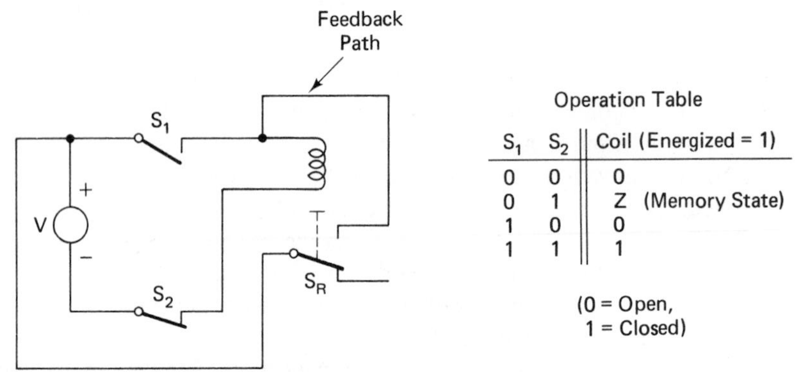

Operation Table

S_1	S_2	Coil (Energized = 1)
0	0	0
0	1	Z (Memory State)
1	0	0
1	1	1

(0 = Open,
1 = Closed)

FIGURE 5-1 Relay hold circuit.

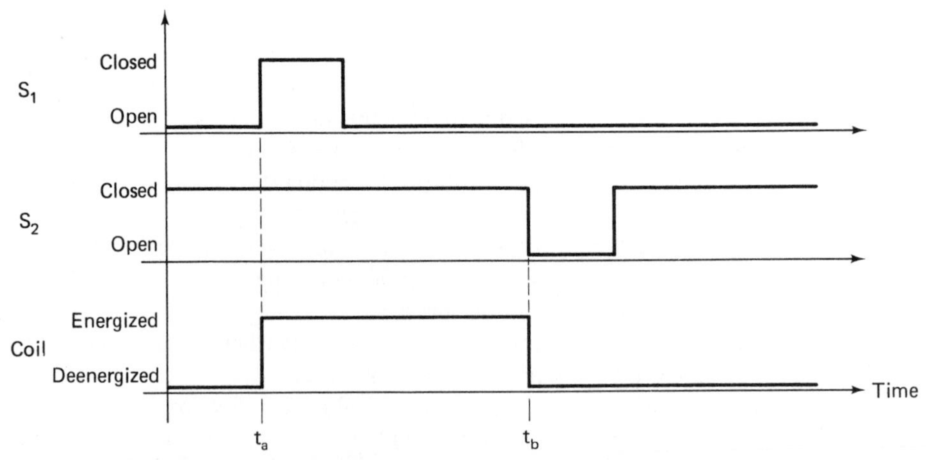

FIGURE 5-2 Timing diagram for the relay hold circuit of Figure 5-1. 't_a' and 't_b' are independent of each other. Also, propagation delays in the coils are not shown.

conditions for the coil are possible; it may be either energized or deenergized. Exactly which condition exists depends upon the switch setting *just prior* to entering the $S_1 = 0, S_2 = 1$ input state. Consequently, the relay hold circuit can retain a condition into which it was placed at an earlier time, and therefore exhibits the feature of memory. Thus, this circuit can be classified as a sequential circuit. Structurally, we note the feedback path in Figure 5-1 between 'S_R' and the relay coil. It is through this feedback path that the coil, once having been energized by the closing of 'S_1', will hold itself on, independently of 'S_1'. Figure 5-2 illustrates the timing relationships for one cycle of operation of the relay hold circuit.

5.2.2 Cross-Coupled Gates

The behavior of the relay hold circuit as a memory element can also be realized using logic gate components. Such a logic gate circuit would have all of the advantages over the relay hold circuit that logic gates generally have over relays, namely high speed, low power, ability to be microminiaturized, reliability, and insensitivity to vibration. A logic gate circuit whose function is identical to the relay hold circuit of Figure 5-1 is shown in Figure 5-3, where logic variables 'S' and 'R' in the gate circuit

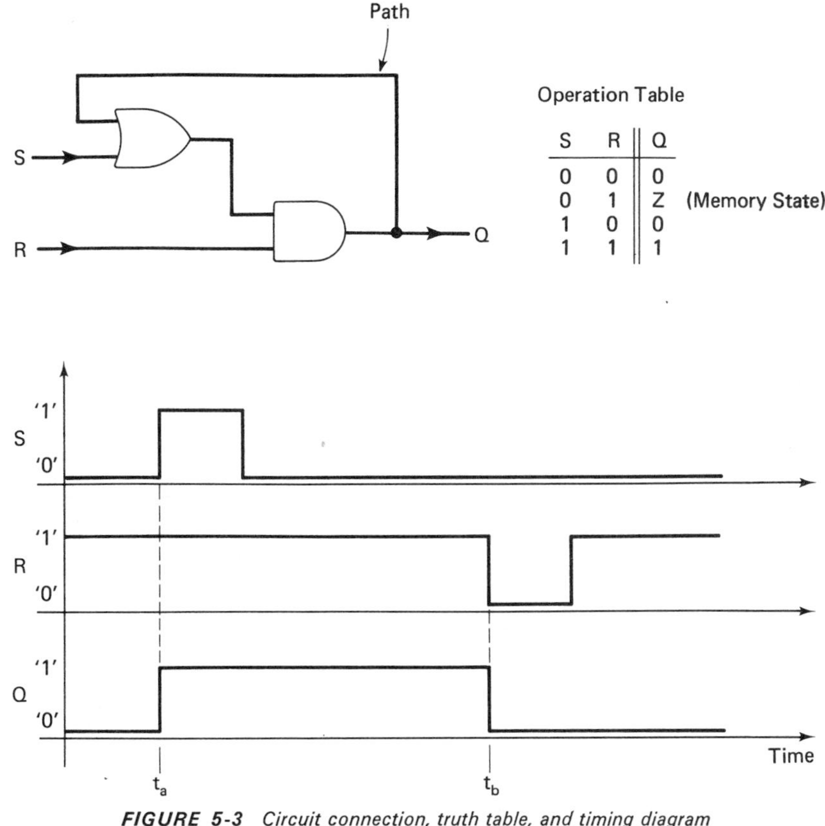

S	R	Q	
0	0	0	
0	1	Z	(Memory State)
1	0	0	
1	1	1	

FIGURE 5-3 Circuit connection, truth table, and timing diagram for a logic gate equivalent to the relay hold circuit.

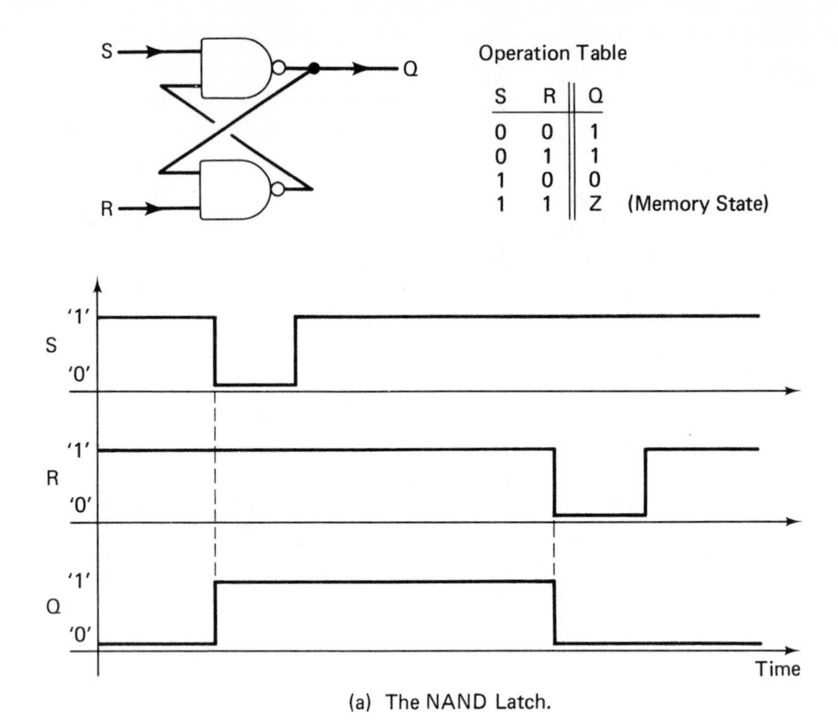

(a) The NAND Latch.

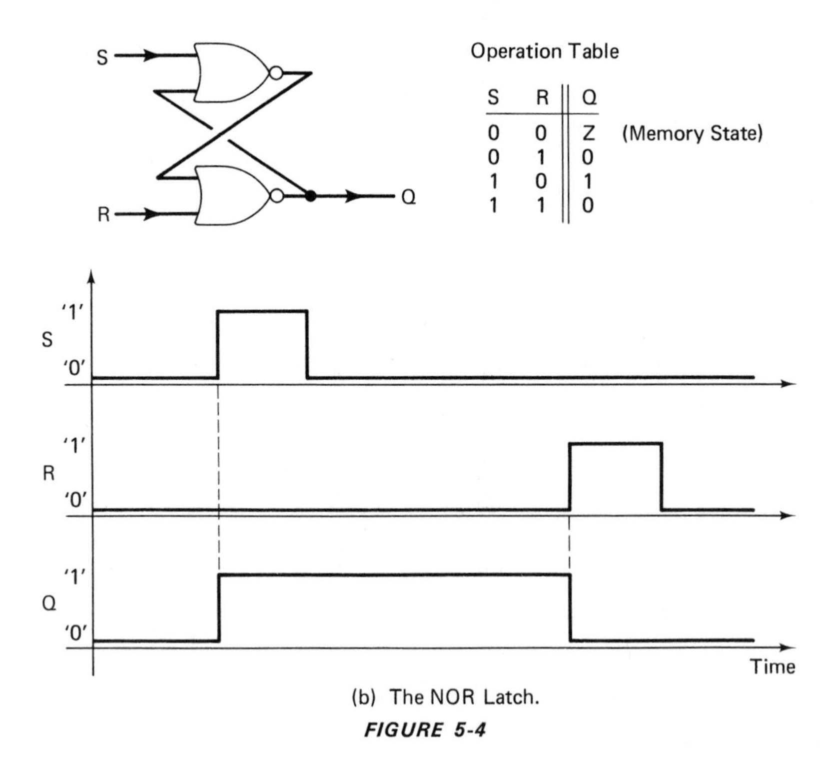

(b) The NOR Latch.

FIGURE 5-4

correspond, respectively, to switches 'S_1' and 'S_2' in the relay circuit. This circuit is commonly referred to as a *latch*, and is characterized by the fact that it has two stable states. The inputs are labeled 'S' and 'R' to correspond with the words "set" and "reset." By definition, the latch is in the "set" state when $Q = 1$, and enters this state when the 'S' input is asserted. For the AND–OR latch of Figure 5-3, the asserted condition for 'S' is $S = 1$, while the asserted condition for 'R' is $R = 0$. The latch circuit, which may also be implemented using NAND and NOR gates, is the most basic form of memory element, and is a fundamental component of all more complex digital memory devices. We shall use it often in our digital circuit designs, and will find it in one form or another as a component in many digital circuit applications.

By observing the truth table of Figure 5-3 and noting the feedback path in the gate connections, it is evident that the latch is a sequential circuit. Using different types of gates, different latch circuits are possible. Figure 5-4 illustrates two alternative circuit forms of the latch. In each case, however, only cosmetic changes have been made; the basic operation table, and the nature of the latch as a sequential circuit, remain the same. In comparing the latch circuits of Figures 5-3 and 5-4, it is important to note that three of the four input conditions force the 'Q' output to a particular value (either '0' or '1'), while only one input condition is passive and allows the circuit to hold its last forced value. The passive condition can be thought of as the memory state of the circuit, and is labeled as such in the truth tables of Figure 5-3 and 5-4. For each of the three latch circuits presented, the input values corresponding to the memory state are different. For instance, the NAND latch requires the condition $S = 1$ and $R = 1$ for the memory state, while the NOR latch requires $S = 0$ and $R = 0$ for the memory state. It is this fact that may make one form of the latch more desirable than other forms in certain applications.

5.2.3 Switching Dynamics

Timing diagrams can be drawn in different ways to emphasize different features of circuit performance. In one case, it might be necessary to show every detail of a logic waveform, depicting rise time, fall time, overshoot, ringing, and propagation delay. Such a detailed timing diagram is illustrated in Figure 5-5. In other cases, only the basic, overall features of a waveform need to be drawn, showing only the logic

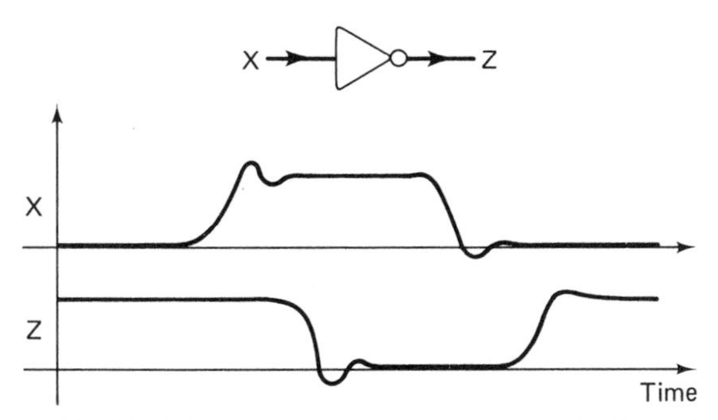

FIGURE 5-5 *Timing diagram, showing many details of the input and output waveforms.*

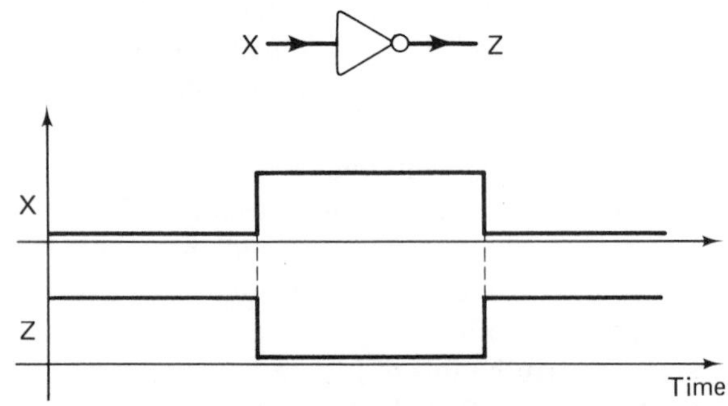

FIGURE 5-6 *Timing diagram, showing only the important logic relationships between waveforms.*

relationships between variables. This is illustrated in Figure 5-6, and is similar to the timing diagrams of Figure 5-3 and 5-4. Often, it will be necessary to analyze in great detail the timing relationships in certain circuits. As we shall see, this is particularly true when the output of a circuit is erroneous for only very brief moments. In such cases, the propagation delay of the logic gate components is an important feature, and should be expressed in any timing diagrams. Figure 5-7 illustrates how a timing diagram can be drawn to emphasize propagation delay. In another case, perhaps ringing should be emphasized. The detail that is used in a timing diagram depends on just what is being shown. It is very important to stress the pertinent features of a waveform, while not obscuring the drawing with excessive detail. In most cases, it is sufficient to supply a simple, functional timing diagram with propagation delay, while excluding from the waveform features that are more of an electrical nature such as rise time and fall time.

To enhance our understanding of latch circuits, the NAND latch of Figure 5-4(a) will be more carefully examined, paying special attention to the propagation delay of the NAND gates. Figure 5-8 depicts a NAND latch and the timing diagram corresponding to one set/reset cycle of operation. In this figure, an output labeled '$\bar{Q}$' is also available, and is normally the complement of 'Q'. Because of the NAND gate propagation delays, the latch outputs do not change instantly when a 'set' or 'reset' signal is applied. As we can see in Figure 5-8, one gate delay is required between the application of 'set' and the change of 'Q' from '0' to '1', while *two* gate delays are required between the application of 'reset' and the change of 'Q' from '1' to '0'. Furthermore, we can see that for the duration of one gate delay during each output change, 'Q' is *not* the complement of '$\bar{Q}$'.

The propagation delay of a gate is usually very small when compared to the rate of change of input signals in many logic circuits. However, the accumulation of even a few of these brief propagation delays over several cascaded stages in a circuit can result in a critical error. Thus, it can be important in digital circuit analysis to be aware of the exact sequence of changes, and to know when a consecutive sequence of changes due to propagation delay may have an undesirable effect on circuit performance. We will investigate these issues in more depth in Chapter 8 when we study digital circuit fault analysis.

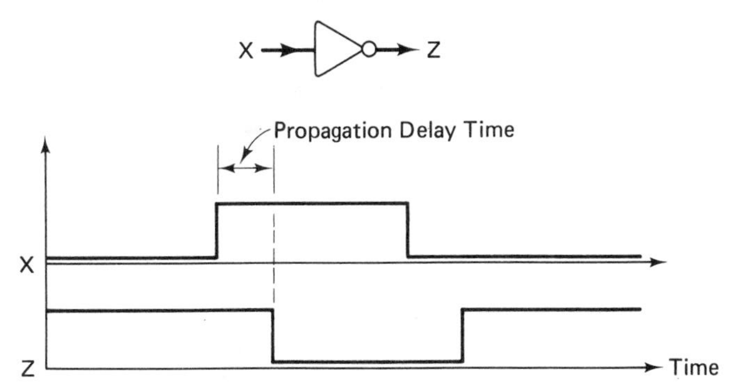

FIGURE 5-7 Timing diagram, showing the logic relationship between two waveforms, and emphasizing propagation delay.

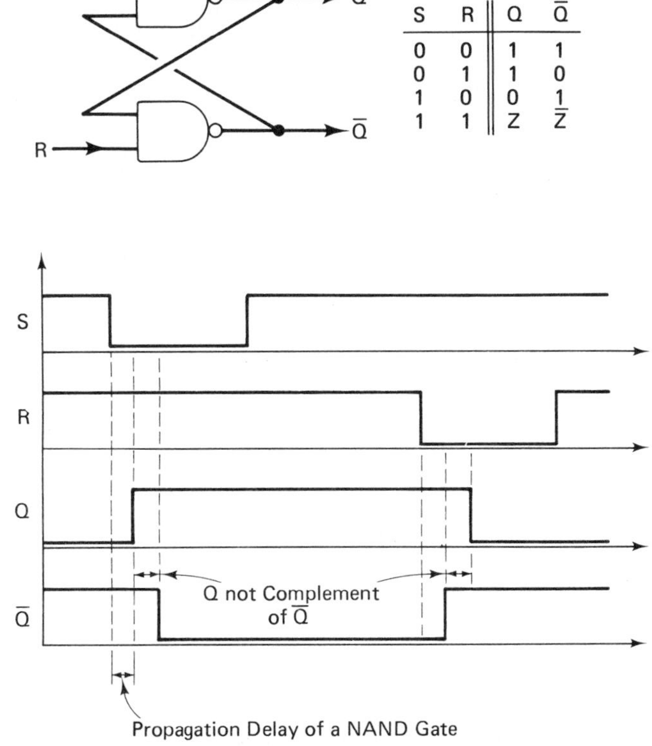

Operation Table

S	R	Q	$\overline{Q}$
0	0	1	1
0	1	1	0
1	0	0	1
1	1	Z	$\overline{Z}$

FIGURE 5-8 Detailed analysis of the NAND latch, emphasizing the propagation delay that is present in actual logic gates.

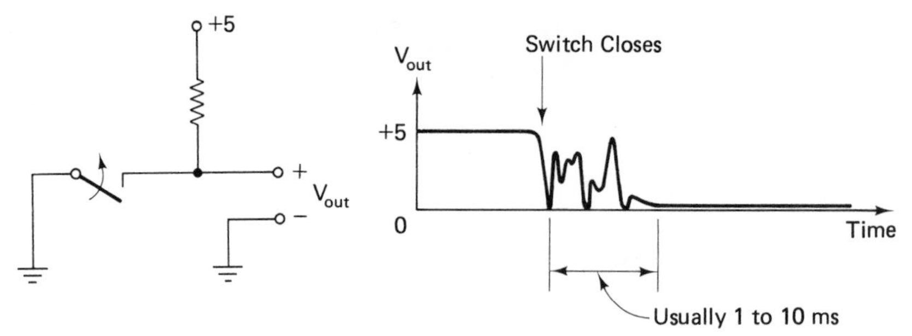

FIGURE 5-9 *The closure of a switch produces an intermittent electrical connection when the movable wiper first meets the contact.*

5.2.4 An Application for the Latch

Toggle and pushbutton switches are frequently used in digital circuits so that human operators can control some aspect of the circuit's function. A typical mechanical switch consists of a movable metal wiper and one or more stationary contacts. When the switch is caused to change position, the metal wiper does not instantly settle on a contact, but rather bounces on the contact several times before making a continuous electrical connection. The duration of the bounce is normally less than 10 ms. Figure 5-9 illustrates the ***switch bounce*** problem. If a logic gate input is driven directly from the switch output of Figure 5-9, many unwanted transitions between '0' and '1' will occur at the gate's output, as shown in Figure 5-10. This is so because the response time of a logic gate is much faster than the rate of switch bounce. Consequently, the gate's output will respond to the switch transient as if the transient were a valid signal. Switch bounce also occurs when the wiper and contact separate, although the bounce in this case is usually less.

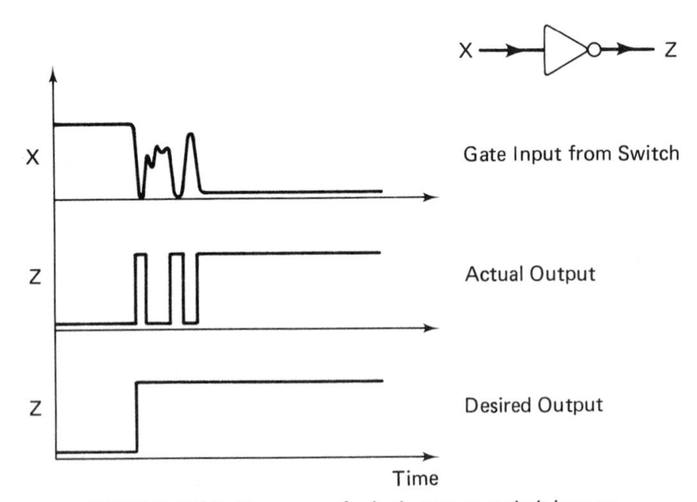

FIGURE 5-10 *Response of a logic gate to switch bounce.*

A NAND latch can be used as a means of correcting the switch bounce problem. The circuit of Figure 5-11 demonstrates the technique. Note, however, that a single-pole double-throw (SPDT) switch is required; the circuit cannot operate properly with a SPST switch. Studying the circuit of Figure 5-11, we see that the switch wiper may be in one of three different positions:

(a) Touching the upper contact only.

(b) Touching neither contact.

(c) Touching the lower contact only.

As the switch changes position, the wiper moves through the sequence '*a-b-c*', or '*c-b-a*'. Consider the wiper to be initially in the '*a*' position, with $Q = 0$. Noting that if a contact is not touched by the wiper, it assumes logic '1', we see that as the wiper moves from '*a*' to '*b*', the bouncing has no effect whatsoever on the 'Q' or '$\bar{Q}$' outputs.

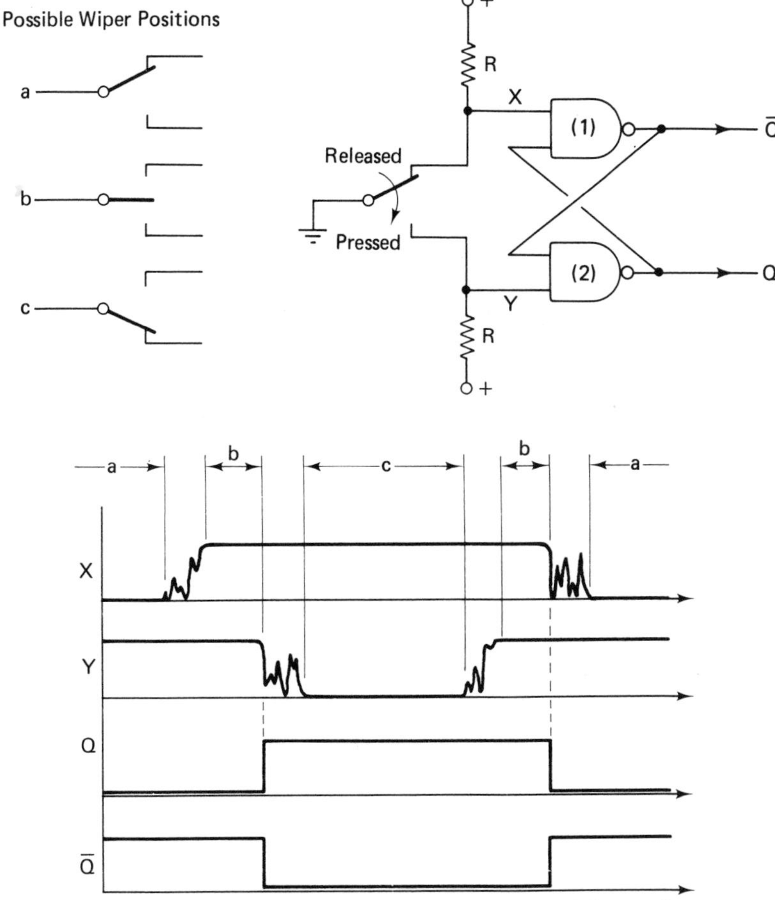

FIGURE 5-11 A NAND latch is used to prevent switch bounce from adversely affecting logic-level transitions.

This is so because the second input to gate (1) is held at '0' by the 'Q' output, thus maintaining the output of gate (1) in the '1' state. In the transition from 'b' to 'c', the wiper also bounces before making solid contact, but in this case, the logic state is changing from '1' to '0' at 'Y'. On the *first* '1' to '0' bounce, the '0' at 'Y' causes 'Q' to change to '1', and then '$\bar{Q}$' to change to '0'. The '0' on '$\bar{Q}$' now overrides any further transitions at 'Y' that may occur due to continued bouncing. Therefore, this first '1' to '0' change causes the latch to switch states and hold. Note that the gates can change state much faster than the bounce period. A similar argument is valid for the '*c-b-a*' transition. It is important to note that the debouncer latch circuit *will not* work properly if either

1. The wiper bounces between positions 'a' and 'c', or
2. The combined propagation delay of the NAND gates is *longer* than the period of the contact bounce.

The latch circuit is one member of a class of circuits called ***flip-flops***. A flip-flop can also be referred to as a ***bistable multivibrator***, where "bistable" indicates simply that two stable states are possible ($Q = 0$, and $Q = 1$). The latch circuit has many other important applications besides switch debouncing. For instance, it is the basis for clocked flip-flops (examined in Section 5.3), and digital counters (examined in Section 5.3 and Chapter 6). We shall see that the latch is as important to sequential circuits as the AND, OR, and INVERT gates are to combinational logic.

5.3 CLOCKED FLIP-FLOPS

5.3.1 The Clocked S-R Latch

In the previous section, we analyzed the basic latch circuit, and noted that it could change state almost immediately upon assertion of the 'S' or 'R' inputs, pending only the gate propagation delays. Many circumstances arise, however, when it is desirable to have flip-flops change at only certain times, and in many cases synchronously with respect to other circuit elements. This need is satisfied with *clocked flip-flops*, the simplest of which is the ***clocked S-R latch*** shown in Figure 5-12. The label 'CK' is commonly used to abbreviate "clock." In this circuit, the condition $CK = 1$ enables the

FIGURE 5-12 Clocked S-R latch.

Operation Table

CK	S	R	Q
0	Ø	Ø	No Change
1	0	0	No Change
1	0	1	0
1	1	0	1
1	1	1	$Q = \bar{Q} = 1$

(When CK → 0 from this State, the Circuit may Stabilize at Either Q = 0 or Q = 1)

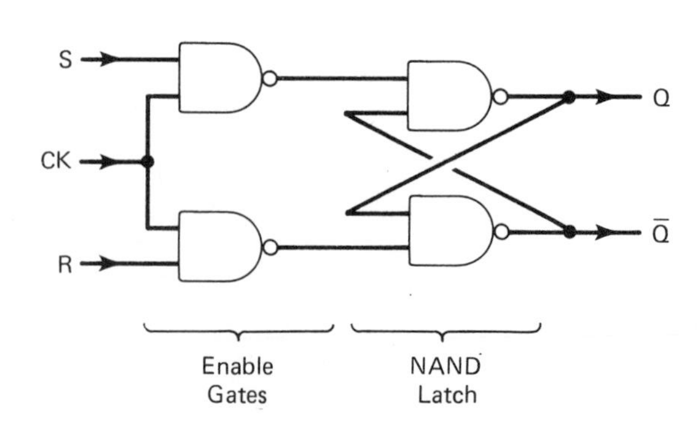

'S' and 'R' inputs, and under this condition, the circuit behaves like a normal NAND latch (in this case, however, 'S' and 'R' are asserted with '1', not '0' as in the unclocked NAND latch). If $CK = 0$, no changes can occur in the state of the flip-flop, regardless of any changes on the 'S' and 'R' inputs. This circuit is also referred to as an S-C flip-flop, where 'S' and 'C' represent "set" and "clear."

A variation of the clocked S-R latch is the ***clocked D latch***, shown in Figure 5-13. In this circuit, a single 'D' input, representing "data," replaces the 'S' and 'R' inputs of the S-R latch. Thus, for $CK = 1$, the logic value of 'Q' exactly equals that of 'D', and will follow changes in 'D'. For $CK = 0$, the value of 'Q' will be maintained at the last value prior to 'CK' becoming '0'. With a 'D' input instead of 'S' and 'R' inputs, it is not possible to obtain the undesirable output of $Q = \bar{Q} = 1$, which would result in the clocked S-R latch if $S = 1$ and $R = 1$ while the clock input is asserted.

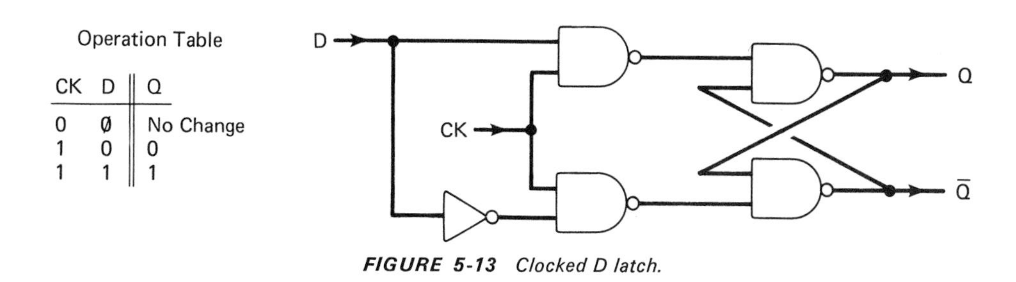

FIGURE 5-13 Clocked D latch.

5.3.2 Master-Slave Devices

The 'CK' input to the S-R latch of Figure 5-12 can be viewed as an "enable" input. This is true in that while $CK = 1$, the S-R latch behaves normally (the enabled condition), and while $CK = 0$, the latch outputs are held constant at the last stable value prior to 'CK' becoming '0'. We see also that in the clocked S-R latch, the inputs are asserted by logic '1', rather than by logic '0' as is the case for the NAND latch. This is due only to the inversion provided by the enable gates in Figure 5-12. As long as $CK = 1$, the 'Q' output will behave in accordance with the latch truth table, and will follow any changes in the inputs. In many flip-flop applications, it is desirable, and often necessary, to minimize the period of time that the device outputs can change. In the case of the clocked S-R latch, such operation could be obtained by applying very short duration clock pulses. As we shall see, however, applying clock pulses that are only several gate delays in duration, which would produce the desired effect, is a difficult and impractical solution.

One of the most important flip-flop functions that we will encounter is that of the toggle flip-flop, whose operation is illustrated in Figure 5-14. This case will motivate our desire to minimize the period of time that the device outputs can change. We see in Figure 5-14 that a block representation for the toggle flip-flop is given without specifying exact circuitry. At this time, we must ask the question

> What circuitry goes into the block of Figure 5-14(a) to produce toggle operation?

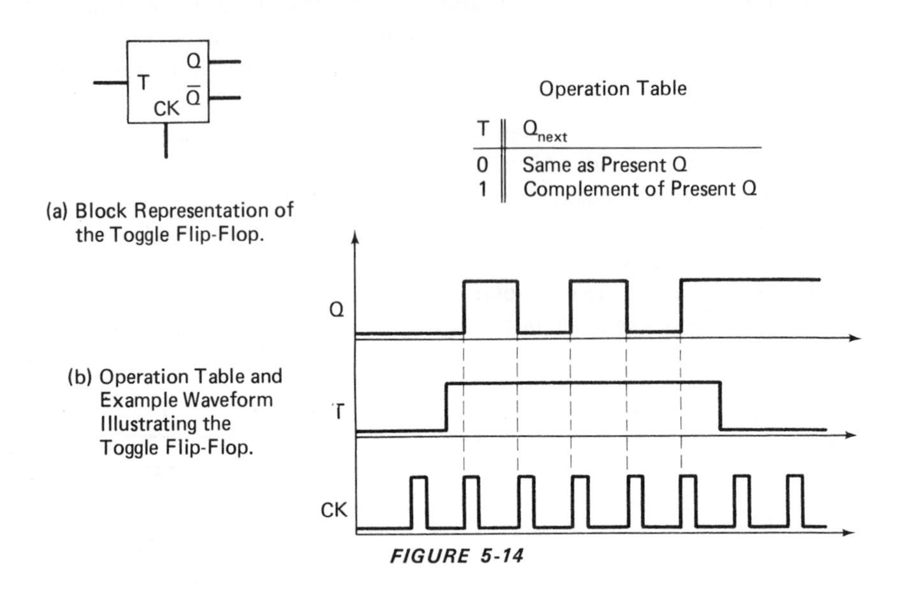

(a) Block Representation of the Toggle Flip-Flop.

(b) Operation Table and Example Waveform Illustrating the Toggle Flip-Flop.

FIGURE 5-14

The operation table of Figure 5-14(b) tells us that with $T = 1$, the 'Q' output must have its present state complemented whenever a clock pulse is received. Put another way, "complementing" the output means that if $Q = 1$, the device should be *reset* on the next clock pulse, while if $Q = 0$, the device should be *set* on the next clock pulse. Using a clocked S-R latch as a building block to which additional circuitry may be connected, we can work on a preliminary design. In tabular form, a sequence of changes can be written:

Present State		Proper Excitation for Next State	
Q	$\bar{Q}$	S	R
0	1	1	0
1	0	0	1
0	1	1	0
1	0	0	1
⋮		⋮	

This table implies that a feedback connection should be made between 'Q' and 'R', and '$\bar{Q}$' and 'S', as shown in Figure 5-15. Note, however, that the width of the clock pulse in this circuit is crucial. If $CK = 1$ for too long, the circuit will complement *constantly* (oscillate). On the other hand, if the period is too short, no change in the circuit's output will occur. These ideas are illustrated in Figure 5-16.

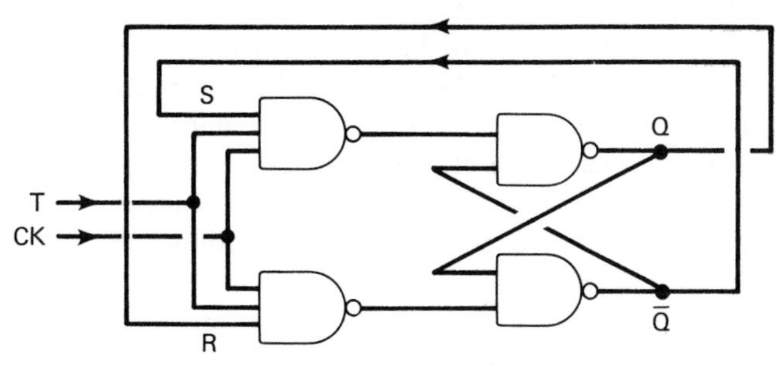

FIGURE 5-15 Feedback connection for an S-R latch to produce toggle flip-flop operation. The circuit works properly, however, for only extremely short clock pulses.

FIGURE 5-16 The operation of the circuit in Figure 5-15 is illustrated for varying durations of the clock input. The actual waveforms viewed on an oscilloscope would be smooth, with rounded corners, instead of the sharply edged waveforms drawn here, because of the capacitive effects in the circuitry.

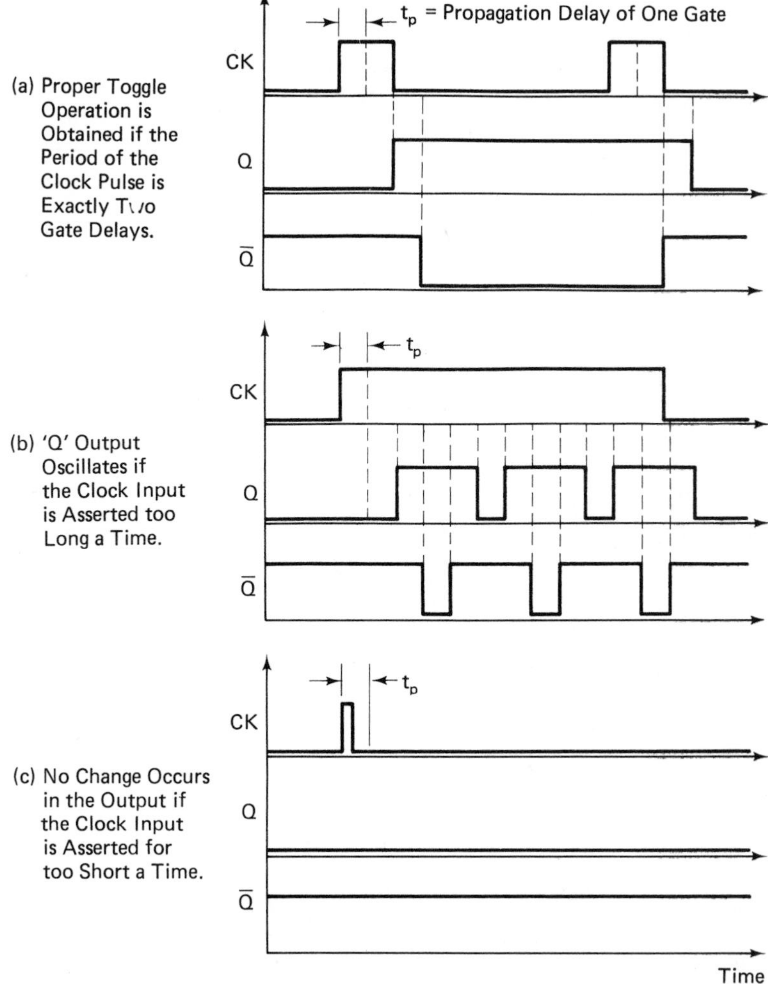

The strict timing requirement on the duration of a clock pulse is due entirely to the nature of the direct feedback from the 'Q', '$\bar{Q}$' outputs to the 'R', 'S' inputs. In words, the present state of the outputs determines the next state they will have, but this next state takes effect only two gate delays after the present state is first reached. When this next state takes effect, the cycle of change will begin again, unless the 'CK' (enable) signal is deasserted (forced to '0'). It would be very desirable, then, to have some controllable level of isolation between changes in the 'Q', '$\bar{Q}$' outputs and changes in the 'R', 'S' inputs. With such isolation, the direct feedback transmission from outputs to inputs would be broken, and the strict timing requirement on the duration of the clock pulse would no longer be necessary. A circuit that provides the desired isolation is shown in Figure 5-17, and is referred to as a ***master-slave flip-flop***. The master-slave flip-flop is made up of two clocked S-R latches. The *master* receives external inputs on the 'T' and 'CK' lines, while the *slave* is inserted in the feedback path to provide isolation between the outputs and the 'S' and 'R' inputs. The master and slave are never *both* enabled at the same time, because the inverter separates the slave clock from the master clock. This feature is exactly what is responsible for the isolation between master and slave. With $CK = 1$, the master is loaded with the input information derived from 'T', 'Q', and '$\bar{Q}$'; however, the slave remains inactive. On the transition of 'CK' from '1' to '0', the master is disabled and the slave is enabled. Thus, at this time the contents of the master are transferred to the slave, and the new 'Q', '$\bar{Q}$' outputs become valid.

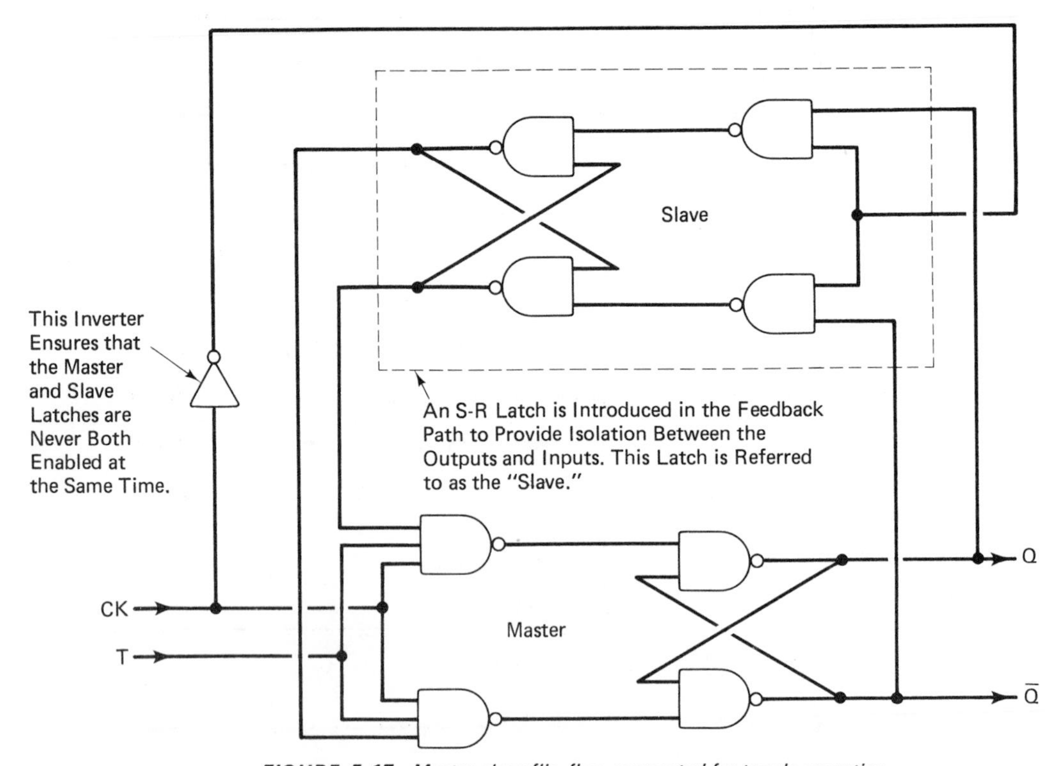

FIGURE 5-17 *Master-slave flip-flop, connected for toggle operation.*

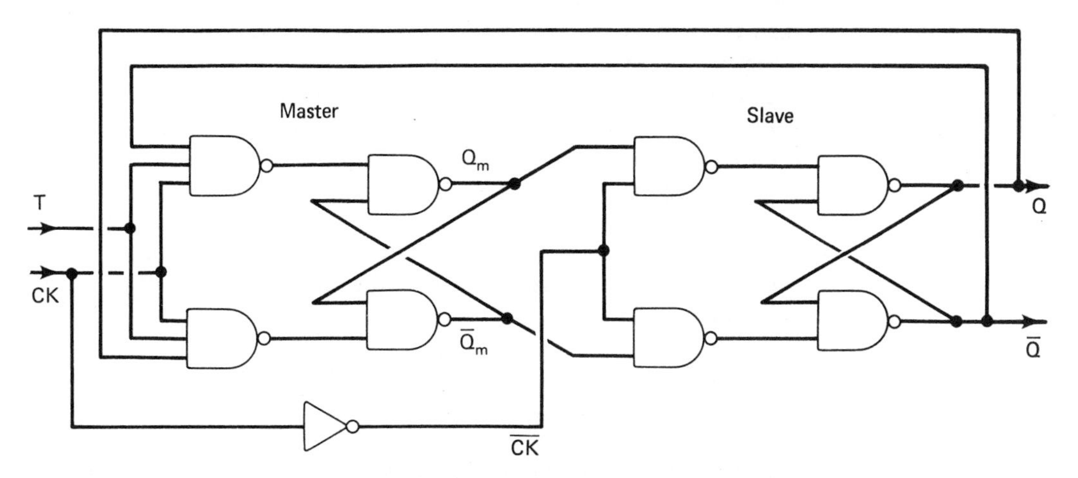

(a) Master-Slave Toggle Flip-Flop, Drawn for Left-to-Right Signal Flow Through the Logic Gates.

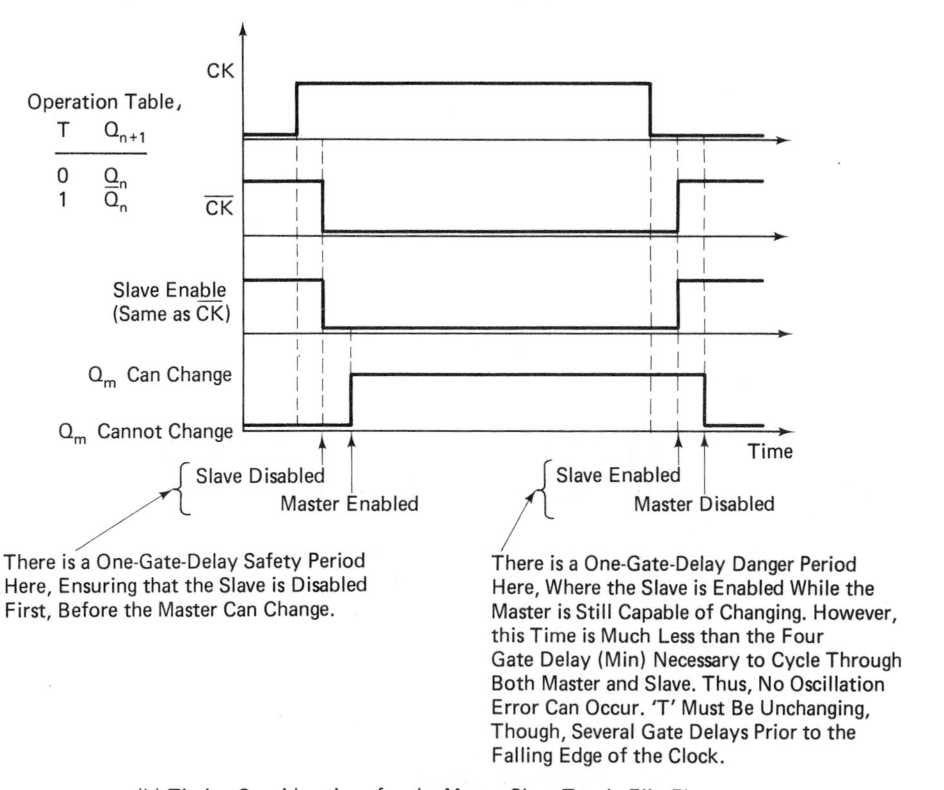

(b) Timing Considerations for the Master-Slave Toggle Flip-Flop.

FIGURE 5-18

A more conventional drawing for master-slave devices is illustrated in Figure 5-18(a); this circuit is identical to the one of Figure 5-17 except that the actual device outputs are taken from the slave rather than from the master. An *operation table* is also given in Figure 5-18(a) to indicate how the flip-flop will respond to '0' and '1' conditions on the 'T' input. Typically, 'Q_n' represents the present output, while 'Q_{n+1}' represents the output one clock period later. Detailed timing considerations for the circuit of Figure 5-18(a) are depicted in Figure 5-18(b).

In our study of master-slave devices, we must ask the following question:

> Can the input(s) of a master-slave flip-flop change at any time in the course of a clock cycle, or must they be held stable at certain key moments?

Figure 5-19 helps answer this question for the toggle master-slave flip-flop. For this device, we assume that a falling edge on 'CK' transfers data from master to slave. The four points that are labeled in this timing diagram indicate the four possible ways of changing the input at various times in the course of a clock cycle. Namely,

1. '0' to '1' on 'T' while $CK = 0$.
2. '1' to '0' on 'T' while $CK = 0$.
3. '0' to '1' on 'T' while $CK = 1$.
4. '1' to '0' on 'T' while $CK = 1$.

Operation Table

T	Q_{n+1}
0	Q_n
1	$\overline{Q}_n$

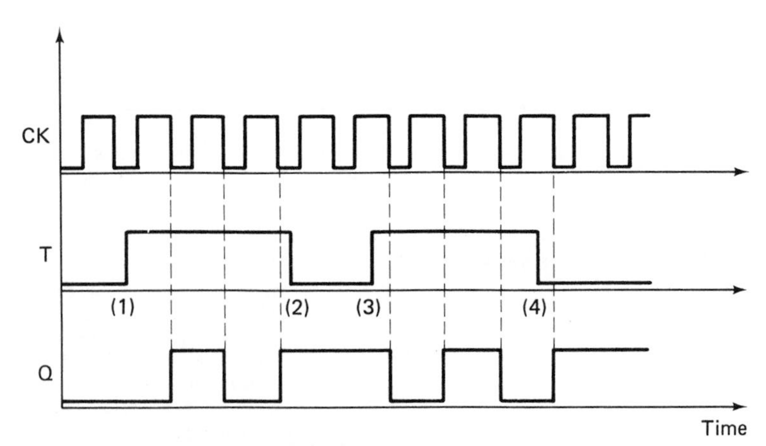

FIGURE 5-19 *Timing diagram for a toggle master-slave flip-flop, showing four important points in the waveform. At (4), the device does not behave according to its operation table.*

By studying the diagram of Figure 5-19, we see that the toggle master-slave flip-flop behaves properly, according to its operation table, at points (1), (2), and (3). However, at point (4), although the 'T' input is brought to '0' prior to the '1' to '0' transition on CK, the 'Q' output complements once more before being held constant. Thus, in this one instance, the operation table is invalid. The reason for this behavior is that the slave is loaded when $T = 1$ and $CK = 1$. The slave, though, *cannot* be returned to its prior state if 'T' is then brought to '0' while 'CK' is maintained at '1'. Consequently, the output complements once more. Although this point may seem trivial, it is responsible for the development of a second family of clocked flip-flops that do not have this characteristic. Devices in this second family are referred to as ***edge-triggered*** flip-flops, and always produce output values that are exactly consistent with the flip-flop operation table. An edge-triggered device can be triggered to change state on the rising edge of 'CK' or on the falling edge of 'CK', depending on its design. Circuits and operating characteristics for edge-triggered devices will be discussed in more detail in the next section.

The discrepancy between the operation table and the actual operation of the toggle master-slave flip-flop was illustrated at point (4) in the timing diagram of Figure 5-19. The chance of such erroneous operation can be minimized by reducing the duration of time that $CK = 1$ in a clock period. Changing the clock waveform in this way results in a sequence of very narrow pulses being applied to 'CK'. Thus, master-slave flip-flops are often referred to as ***pulse-triggered*** devices.

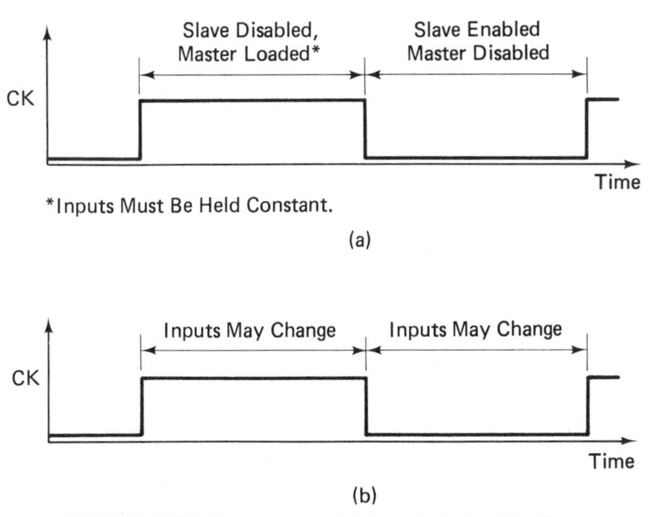

FIGURE 5-20 Summary description of clocked flip-flops.

It is now possible to classify the various types of clocked flip-flops that have been introduced. Figure 5-20 defines level-sensitive devices, master-slave devices, and edge-triggered devices, and illustrates the relationship between clock signals and the circuit's operation.

1. Level-sensitive devices—circuit outputs can change at any time while $CK = 1$, and will follow changes in the inputs.

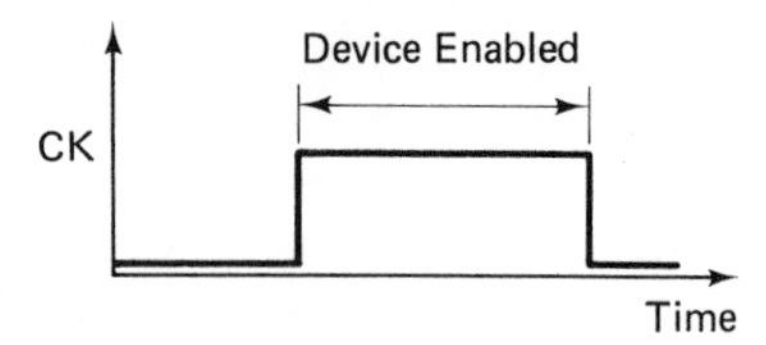

2. Synchronous devices—circuit outputs can change *only* on a rising edge or falling edge (either one but not both) of 'CK', and will not follow changes in the inputs between edges.

   (a) Master-slave devices—the master is loaded according to the input condition when $CK = 1$, then the master is disabled and the slave enabled when 'CK' changes to '0'. This describes a falling-edge master-slave device. A rising-edge master-slave device is obtained by complementing 'CK'. The inputs must remain constant while the master is enabled.

   (b) Edge-triggered devices—the circuit inputs are *sampled* on the rising edge or falling edge (either one but not both) of 'CK', and from this sampling the outputs are established. Thus inputs may change in any way before or after the triggering edge, but should not change simultaneously with the triggering edge.

The S-R and toggle flip-flops can be combined into a single, universal device referred to as the ***J-K flip-flop***. The circuit and operation table for a J-K master-slave flip-flop are shown in Figure 5-21. It should be noted that for the *J-K* input values of '00', '01', and '10', the device behaves like an S-R flip-flop. With the *J-K* input values of '11', however, the device behaves like a toggle flip-flop. Also, the J-K flip-flop can be designed as either a master-slave device, as shown in Figure 5-21, or as an edge-triggered device.

In drawing complex digital circuit schematics, it is important not to obscure the circuit's basic function with excessive detail. Toward this end, various complex circuit components are often represented as block elements. The flip-flop is a perfect example of this. Therefore, rather than drawing the detailed logic schematic for a flip-flop every time we wish to use one in a design, we will express the flip-flop in block form as illustrated in Figure 5-22. This block representation of the flip-flop is the first of many basic digital modules to be introduced so that excessive detail in schematics is avoided. Furthermore, flip-flops are usually available from manufacturers as modules, so it is almost never necessary to construct one from logic gates. Using the block form of Figure 5-22, we illustrate how the J-K flip-flop can be used to produce S-R-, T-, and D-type operation in Figure 5-23.

It is often necessary to initialize a flip-flop prior to its operating in a digital circuit. Such a need arises, for example, when system power is first applied. To accomplish the initialization, one or two additional inputs are provided, referred to as ***set-direct*** and ***clear-direct***, and are illustrated in Figure 5-24. It is important to note that the set-direct and clear-direct inputs *override* the clock input, and take effect immediately when asserted. If both 'SD' and 'CD' are applied at the same time, both 'Q' and '$\bar{Q}$' will become '1'. Set-direct may also be referred to as *preset*, while clear-direct is sometimes referred to as just *clear* or *reset*. Figure 5-25 illustrates how a master-slave flip-flop circuit can be modified to include 'SD' and 'CD' inputs.

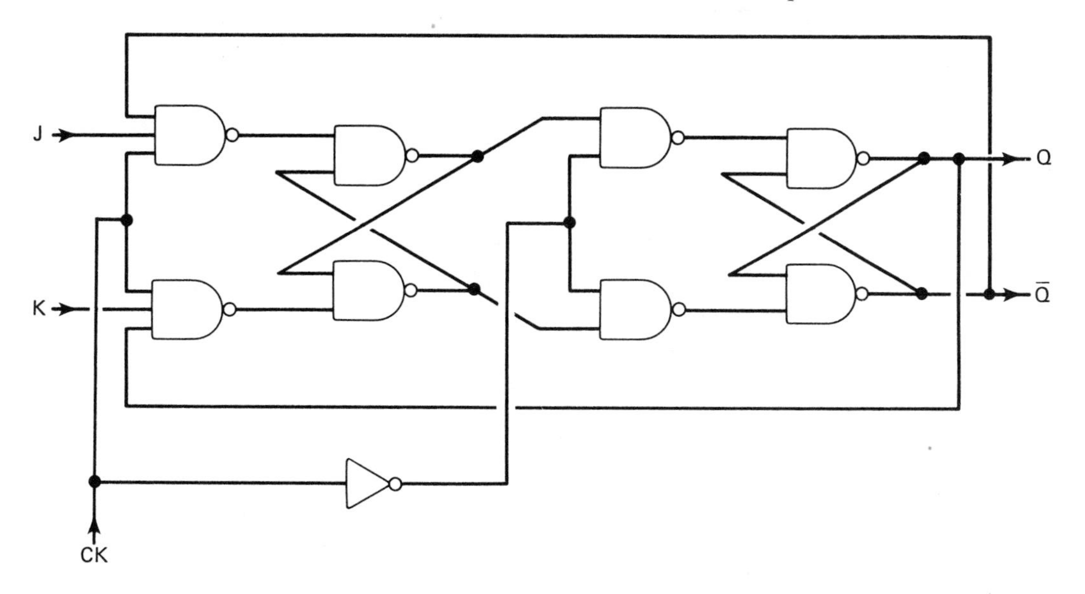

Operation Table

J	K	Q_{n+1}	
0	0	Q_n	(No Change)
0	1	0	
1	0	1	
1	1	$\bar{Q}_n$	(Complement)

FIGURE 5-21 *J-K master-slave flip-flop and its operation table.*

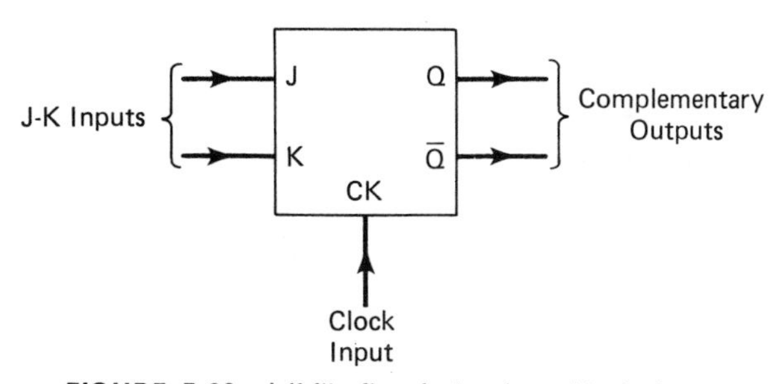

FIGURE 5-22 *J-K flip-flop depicted as a block element.*

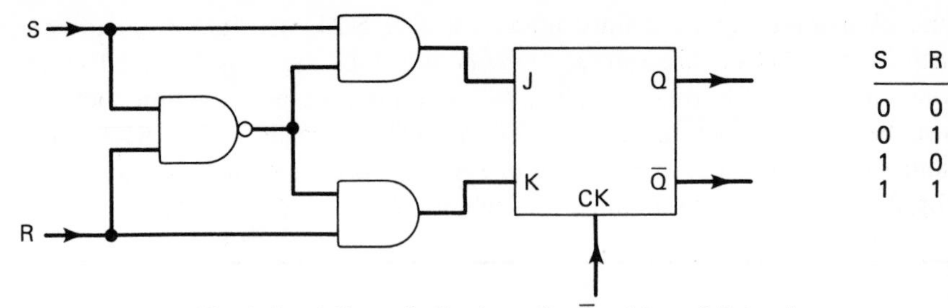

S	R	Q_{n+1}
0	0	Q_n
0	1	0
1	0	1
1	1	Q_n^*

*S = 1, R = 1 Normally Produces $Q = \overline{Q} = 1$ in an S-R Latch. In this Case, However, the State Remains Unchanged.

(a) S-R Flip-Flop.

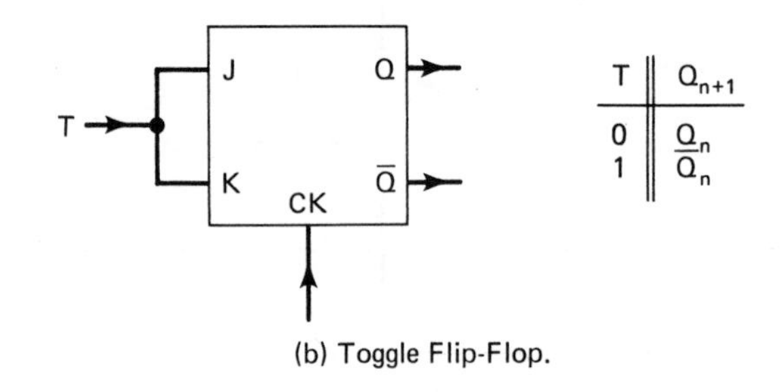

T	Q_{n+1}
0	Q_n
1	$\overline{Q}_n$

(b) Toggle Flip-Flop.

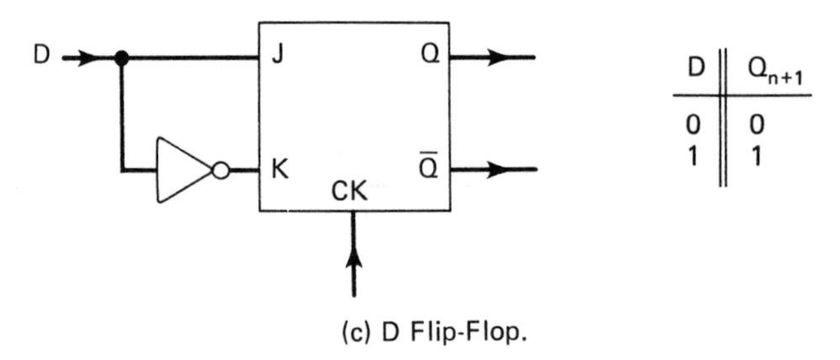

D	Q_{n+1}
0	0
1	1

(c) D Flip-Flop.

FIGURE 5-23 *S-R, T, and D operation is obtained from a J-K device.*

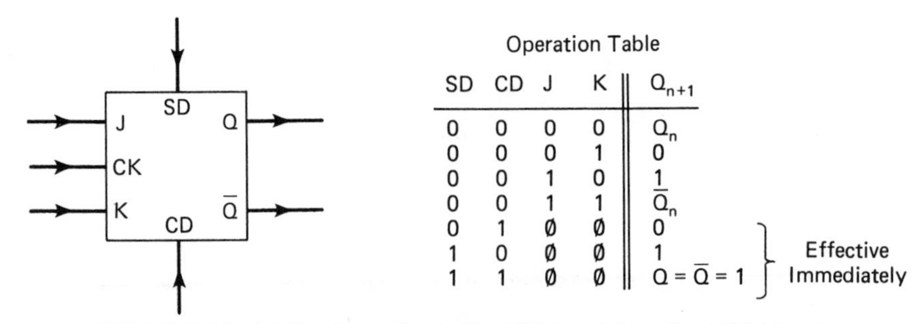

Operation Table

SD	CD	J	K	Q_{n+1}	
0	0	0	0	Q_n	
0	0	0	1	0	
0	0	1	0	1	
0	0	1	1	$\overline{Q}_n$	
0	1	Ø	Ø	0	Effective Immediately
1	0	Ø	Ø	1	Effective Immediately
1	1	Ø	Ø	$Q = \overline{Q} = 1$	Effective Immediately

FIGURE 5-24 *J-K flip-flop with set-direct (SD) and clear-direct (CD) inputs.*

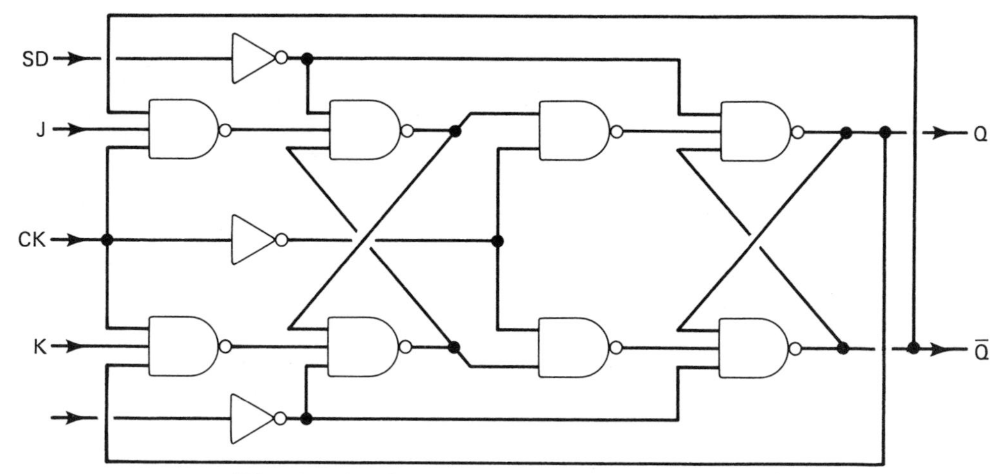

FIGURE 5-25 J-K master-slave circuit with 'SD' and 'CD' inputs.

5.3.3 Edge-Triggered Devices

One notable characteristic of the master slave flip-flop is that its inputs cannot reliably be changed while '*CK*' is asserted.* There are many applications where this characteristic is undesirable, and over restrictive on circuit operation. This is particularly true when a digital circuit has inputs that can change in an unsynchronized way with respect to internal timing signals. For instance, a pushbutton mounted on the control panel of a digital instrument can be depressed at any time. We cannot require, however, that the person operating this instrument push or release the button only when the internal clock signal is deasserted, which may occur 100,000 times per second (or more) and last for 5 μs (or less). Edge-triggered devices overcome this undesirable restriction of master-slave devices by making the inputs effective *only at the time* of the triggering edge. Thus, changes in the input values before and after the triggering edge are ignored.

Many different circuits have been developed for edge-triggered flip-flops, one of which is shown in Figure 5-26(a). Although this is a complex circuit with several feedback paths, the important features of the circuit can be easily described. Note the division of the circuit into three basic sections, as shown in Figure 5-26(a). The "output latch" retains the present state of the flip-flop, and produces 'Q' and '$\bar{Q}$' device outputs. We recognize the output latch as simply a pair of cross-coupled NAND gates, a familiar connection. The output latch can be operated only by the "control latch." At the moment of the rising edge on '*CK*', next-state information derived from the 'J-K control logic" is loaded into the control latch and *held*. Immediately thereafter (one gate delay), feedback lines 'F_1' and 'F_2' take effect to *block* any possible changes in the '*J-K*' inputs. This prevents any changes on the inputs from affecting the new state of the control latch after the rising edge of '*CK*'. The J-K control logic deter-

* Set-direct and clear-direct inputs are not included in this statement, since they may be reliably asserted at any time.

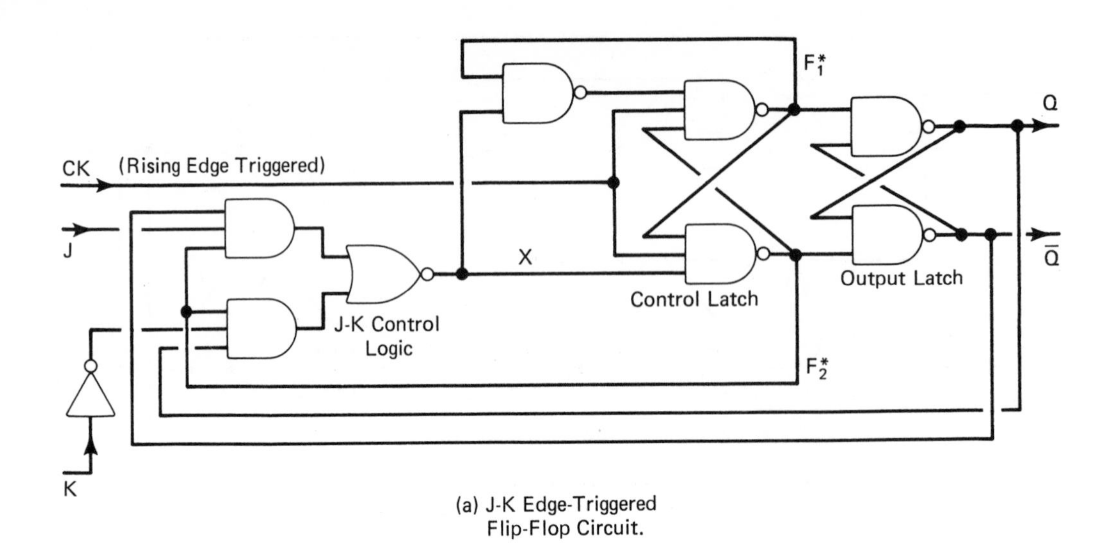

(a) J-K Edge-Triggered Flip-Flop Circuit.

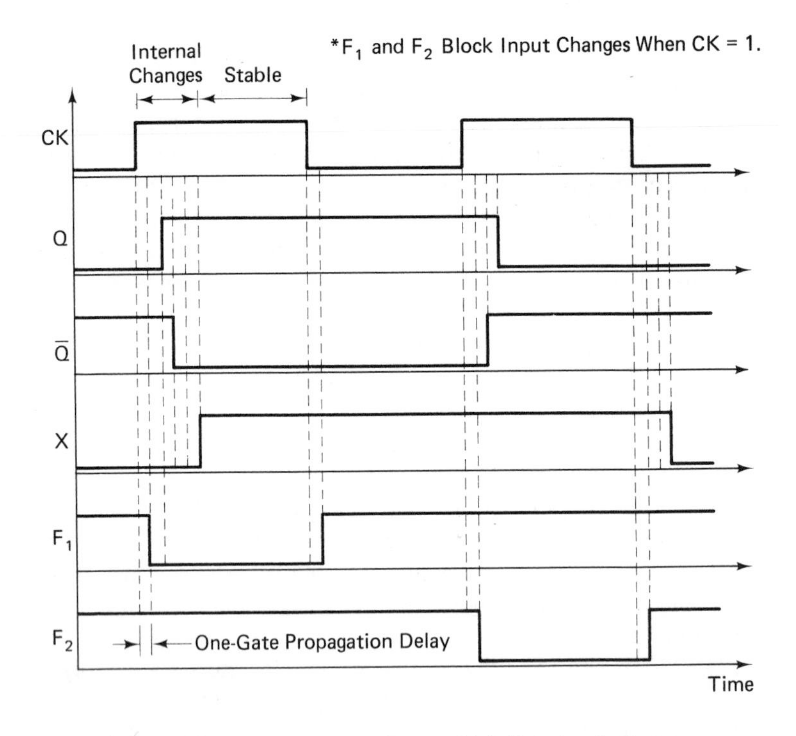

(b) Detailed Timing Diagram for the Circuit of (a), Showing the State Changes that would Occur in the Course of Two Clock Periods. It is Assumed that J = 1 and K = 1, thus Placing the Device in the Toggle Mode.

FIGURE 5-26

mines how the control latch should be loaded so that, based on the inputs and the present outputs, the device behaves according to the operation table for a J-K flip-flop. Figure 5-26(b) depicts the detailed timing considerations for two clock periods of change, assuming that $J = 1$ and $K = 1$.

5.3.4 Switching Dynamics

All logic gates that are constructed from actual physical components, such as transistors, have an inherent propagation delay associated with them. Advances in semiconductor technology are occurring constantly and provide new logic gates with smaller propagation delays than previous generations. No matter how sophisticated the technology becomes, however, propagation delay can never be eliminated completely, and will always constitute a measure of logic gate switching speed. Since flip-flop circuits are constructed from logic gate circuitry, it is reasonable to expect that flip-flops will also possess some basic propagation delay, and consequently, will have a fundamental limit on switching speed. The amount of the propagation delay in a flip-flop will be the sum of the delays through those various circuit stages which make up the device, and will vary depending on the particular flip-flop design. In the circuit of Figure 5-26, for example, we see that three gate delays must elapse after the triggering edge before both 'Q' and '$\bar{Q}$' become valid. It is important to note, however, that manufacturers usually provide flip-flops as whole circuit modules, ready to use. As a result of the flip-flop module being designed as a whole, rather than as an interconnection of individual logic gates, certain electronic advantages are possible that reduce the overall propagation delay to somewhat less than the sum of individual gate delays. The idea of propagation delay in a flip-flop circuit is illustrated in Figure 5-27.

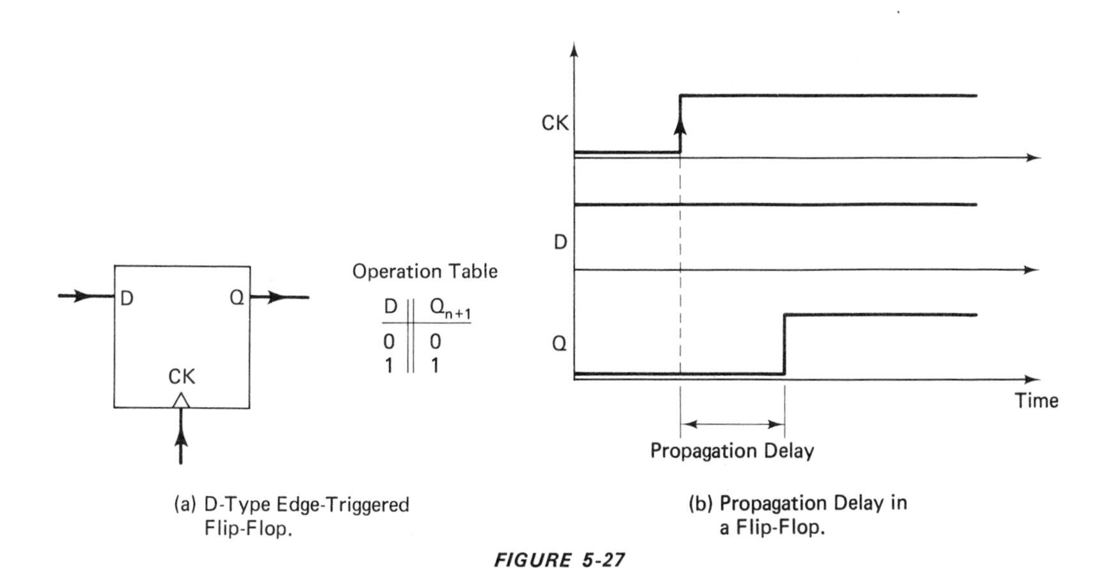

(a) D-Type Edge-Triggered Flip-Flop.

(b) Propagation Delay in a Flip-Flop.

FIGURE 5-27

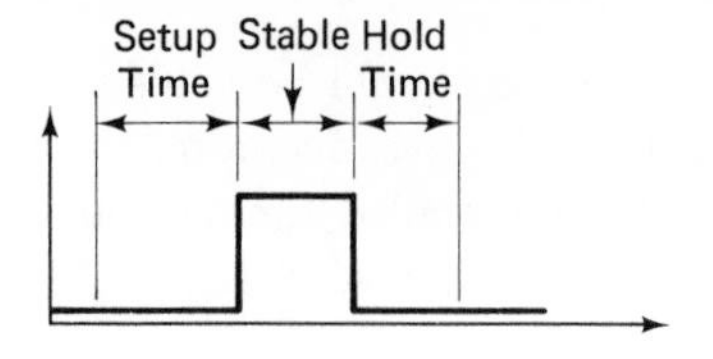

(a) For a Master-Slave Circuit.

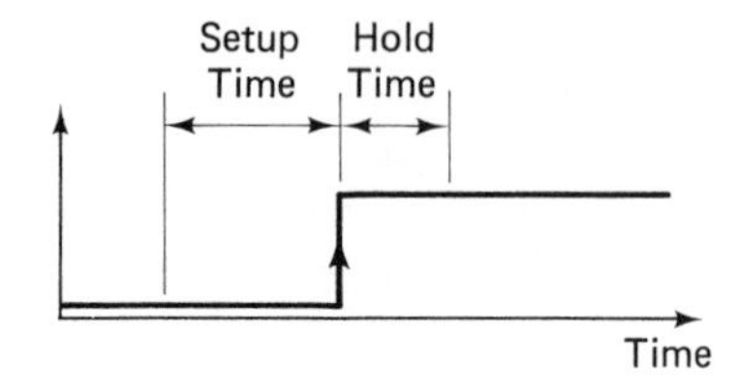

(b) For an Edge-Triggered Circuit.

FIGURE 5-28 *The meaning of "setup" and "hold" time.*

In addition to propagation delay, other timing considerations are necessary, and relate to how close in time the flip-flop inputs can change with respect to the triggering condition of the clock. The ***setup time*** is the minimum time that the device inputs must be stable (unchanging) *prior* to the triggering condition of the clock. The ***hold time*** is the minimum time that the device inputs must be stable *after* the triggering condition of the clock. These ideas are illustrated in Figure 5-28 for master-slave and edge-triggered devices. If setup and hold conditions are violated, the flip-flop may not behave according to its truth table. Some possible error conditions that may result are shown in Figure 5-29. Normally, the setup and hold times are a very small fraction of the total clock period.

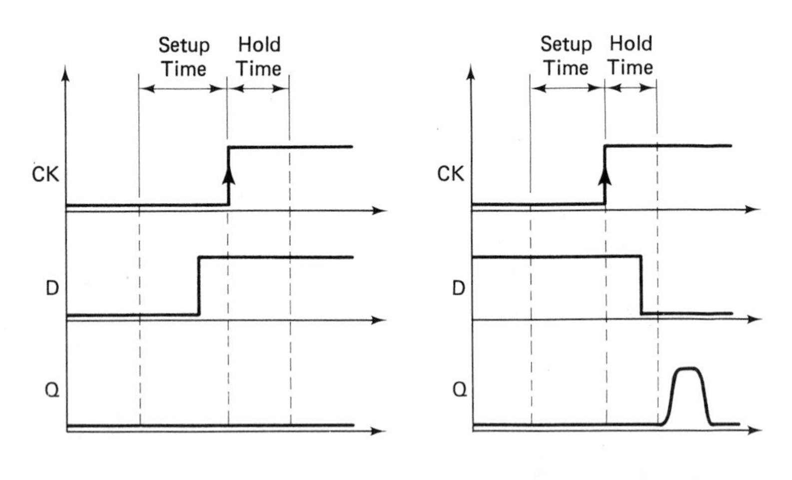

(a) Device Fails to Set, No Change in 'Q'.

(b) Device Fails to Set, Glitch Produced on 'Q'.

FIGURE 5-29 *Possible error conditions that can occur if setup and hold times are violated.*

5.3.5 Symbol Conventions

Flip-flops usually appear in logic diagrams and schematics as block elements, with little attention being paid to their inner workings. Not surprisingly, a number of conventions have come into use to describe the different features and characteristics of these block element flip-flops. In this section, we describe the symbolism accepted and frequently used in logic diagrams.

A flip-flop symbol must indicate clearly a number of different device characteristics. In particular, the symbol must show

1. The flip-flop function—'T', 'D', 'J-K', or 'S-R'.
2. Whether the device is level sensitive or synchronous.
3. Whether the circuit is master-slave or edge-triggered.
4. Whether set-direct or clear-direct inputs are available.
5. The active signal levels required for the inputs to become effective.

To fulfill these requirements, the following conventions have been generally accepted.

1. A flip-flop's basic structure is a square or a rectangle, with function inputs on the left and device outputs on the right. For example:

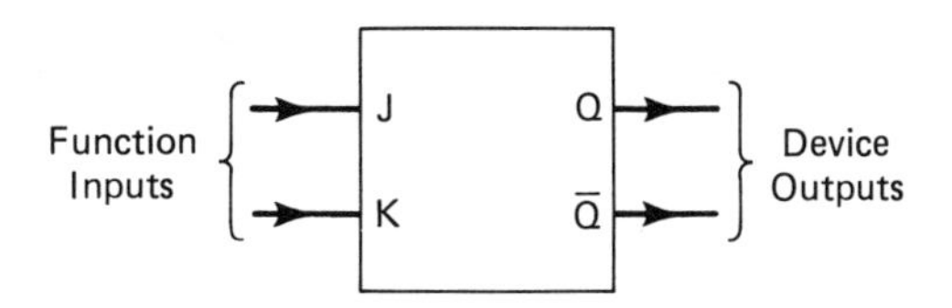

2. The clock input is shown on the left or on the bottom of the square, and is abbreviated '*CK*'. The clock input is *not* abbreviated '*CL*' to avoid confusing it with 'clear direct'. Synchronous flip-flops are denoted by a triangular notch shown on the clock input. Level-sensitive devices do not have a notch at their clock input. For example:

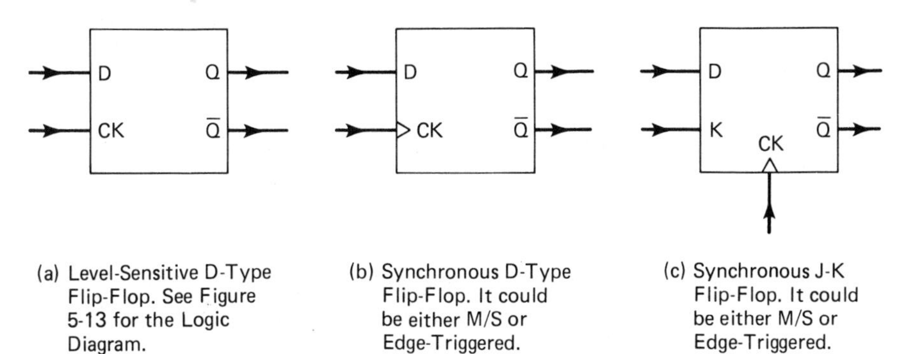

(a) Level-Sensitive D-Type Flip-Flop. See Figure 5-13 for the Logic Diagram.

(b) Synchronous D-Type Flip-Flop. It could be either M/S or Edge-Triggered.

(c) Synchronous J-K Flip-Flop. It could be either M/S or Edge-Triggered.

3. Master-slave flip-flops are explicitly stated as being such, using the words "master-slave," "pulse-triggered," or the abbreviation "M/S." All synchronous devices that are not explicitly denoted as master-slave circuits are assumed to be edge-triggered devices. For example:

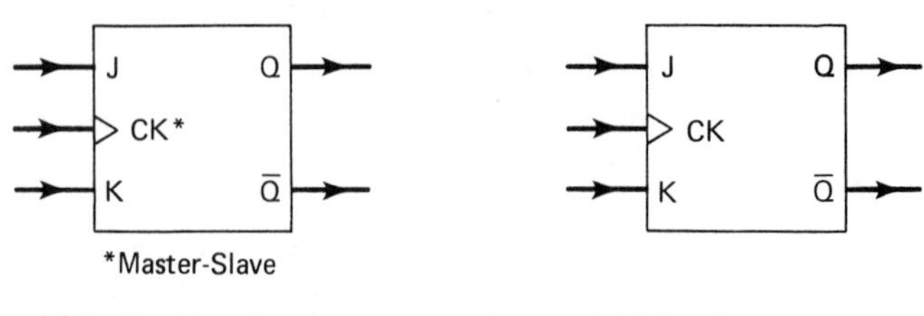

(a) Master-Slave Device.

(b) Edge-Triggered Device.

4. If a flip-flop has 'set-direct' or 'clear-direct' inputs, they are shown in the upper or lower part of the square, and are abbreviated '*SD*' (or '*PR*') for 'set-direct' and '*CD*' (or '*CLR*') for 'clear-direct'. For example:

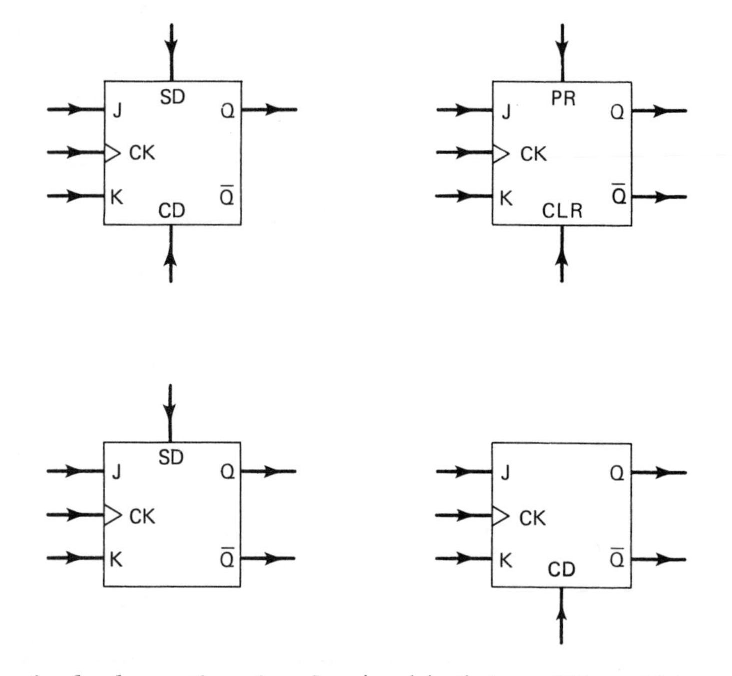

5. The ***active level*** or ***active edge*** of a signal is that condition which causes a change to occur in a circuit (not necessarily a flip-flop circuit). A circuit in which 'set-direct' is asserted with a logic '1' has an *active high* input for this signal. In general, an active high input requires that a logic '1' be applied to cause the input to become asserted, while a logic '0' will produce no effect. In the case of an *active low* level input, logic '0' causes the input to become asserted, while logic '1' has no effect. If a flip-flop is triggered on the rising

edge of 'CK', the rising edge of 'CK' is the *active edge* and the falling edge has no effect. Similarly, if a flip-flop is triggered on the falling edge of 'CK', the falling edge is active and the rising edge has no effect. *Inversion circles* are used to indicate active *low* levels, or active *falling* edges, while the absense of inversion circles indicates active *high* levels, or active *rising* edges. For example:

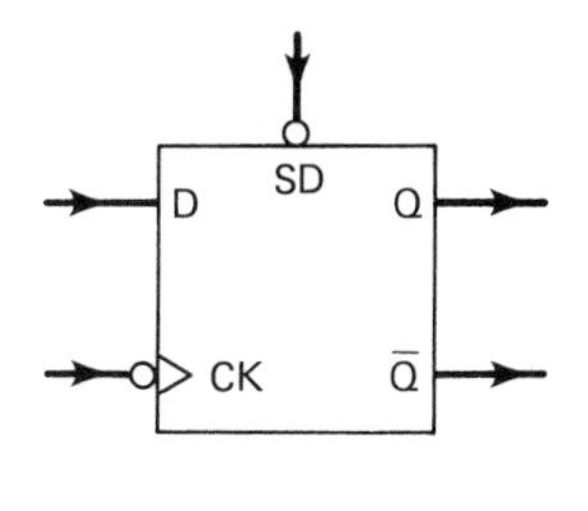

(a) Edge-Triggered D-Type Flip-Flop, Responsive on the Falling Edge of 'CK', and caused to Become Set to '1' (Q = 1) when 'SD' is Logic '0'.

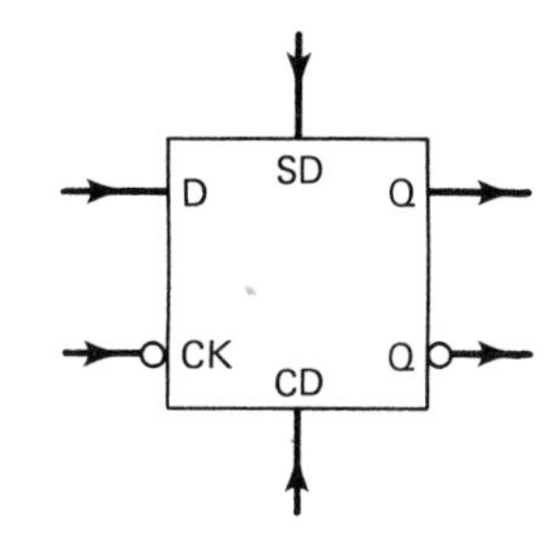

(b) Level-Sensitive D-Type Flip-Flop, Enabled when 'CK' is Logic '0', and caused to Become Set to '1' (Q = 1) when 'SD' is Logic '1', or Cleared to '0' (Q = 0) when 'CD' is Logic '1'.

An *overbar* may be used instead of an inversion circle to indicate the active condition of a signal, as is the case with the '$\bar{Q}$' output. For example:
An INVERT gate may be inserted in any signal line to complement the active condition of the signal. For example, the following flip-flop circuits are equivalent:

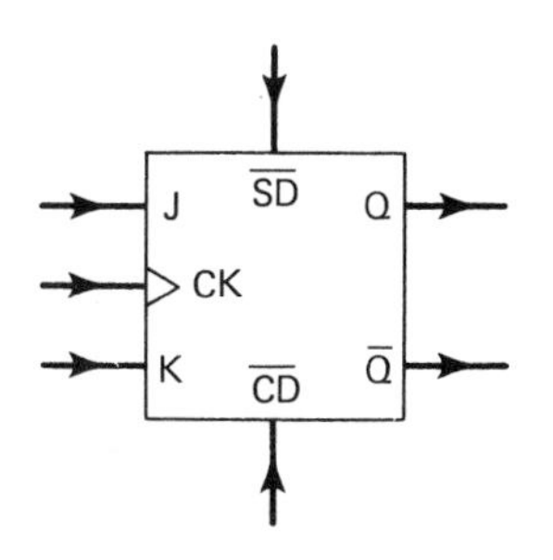

(a) Edge-Triggered J-K Flip-Flop, Responsive on the Rising Edge of 'CK', and caused to Become Set to '1' (Q = 1) when 'SD' is Logic '0', or Cleared to '0' (Q = 0) when 'CD' is Logic '0'.

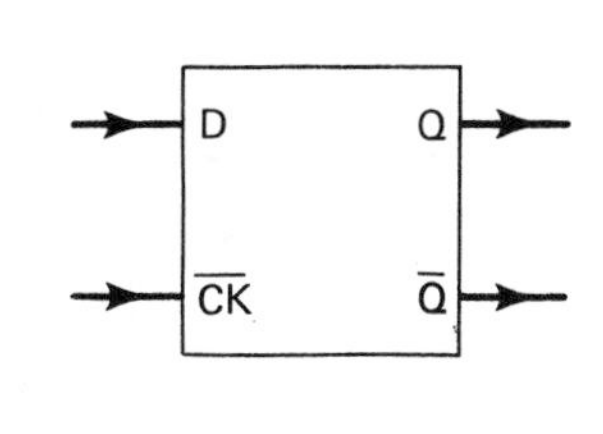

(b) Level-Sensitive D-Type Flip-Flop, Enabled when 'CK' is Logic '0'.

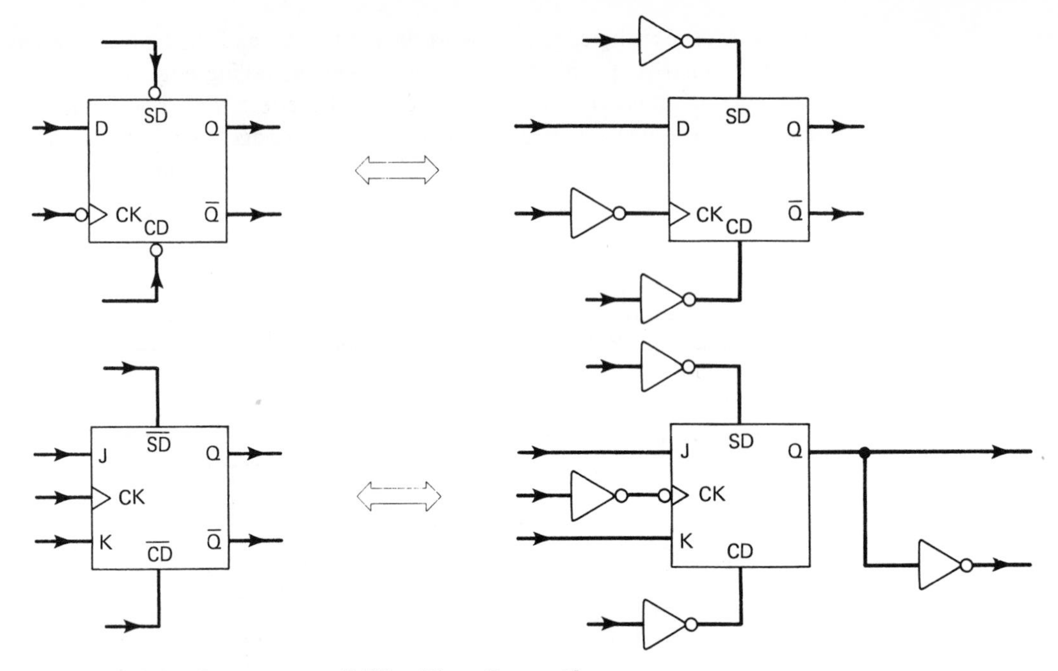

5.3.6 *Summary of Flip-Flop Operations*

To review the important characteristics of the flip-flop devices presented in Sections 5.2 and 5.3, a brief summary of operating features is presented. Propagation delays are not shown in timing diagrams to avoid obscuring the main issues.

(1) Level-sensitive devices: This is the most basic memory element, capable of retaining one of two possible states. Two complementary outputs are usually available; however, if only one is available, the other can be derived from it. The output value can change almost instantly when the inputs are changed, slowed only by gate propagation delay, and is changed independently of any clock signal. The basic latch is a component in almost all other flip-flops and memory devices. A typical timing diagram is shown that illustrates the operation of the NAND latch.

(a) The Latch

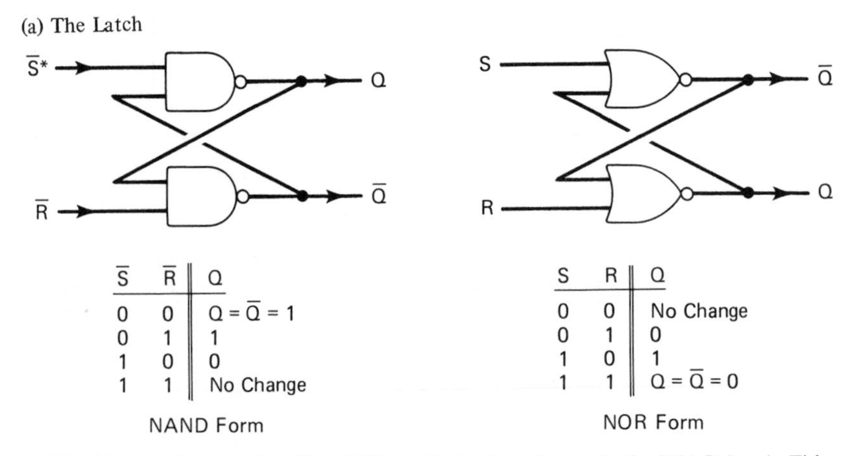

$\overline{S}$	$\overline{R}$	Q
0	0	$Q = \overline{Q} = 1$
0	1	1
1	0	0
1	1	No Change

NAND Form

S	R	Q
0	0	No Change
0	1	0
1	0	1
1	1	$Q = \overline{Q} = 0$

NOR Form

*The Overbar Denotes that 'S' and 'R' are Active Low Inputs in the NAND Latch. This is Consistent with our Convention of Section 5.3.5, and Indicates that 'Q' is Set to '1' when 'S' is Asserted '0'.

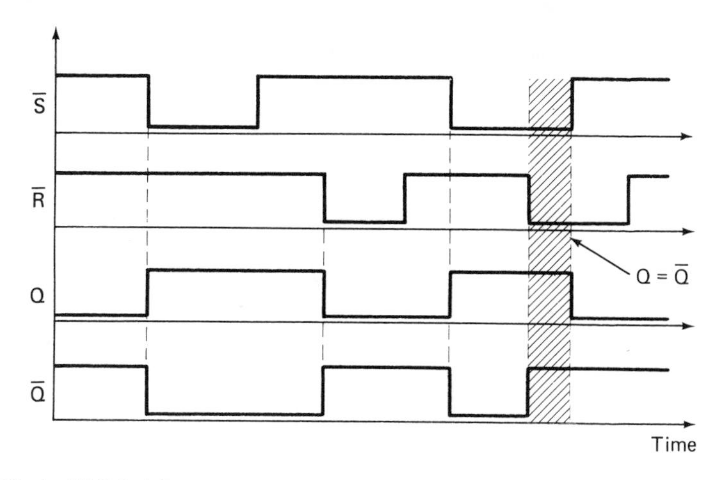

(b) The Clocked S-R Latch

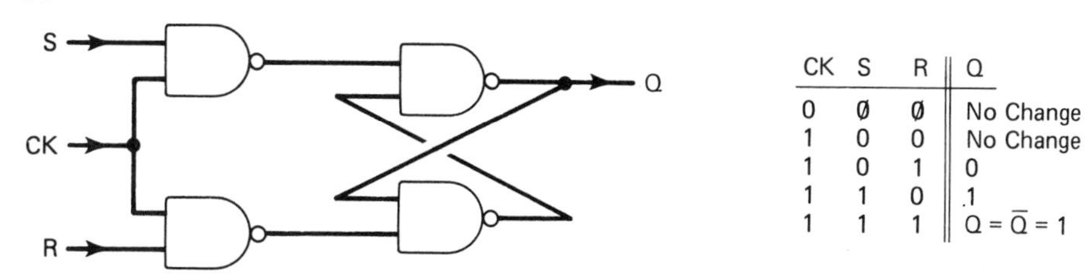

CK	S	R	Q
0	Ø	Ø	No Change
1	0	0	No Change
1	0	1	0
1	1	0	1
1	1	1	$Q = \bar{Q} = 1$

Two control gates are added to the basic latch to provide an "enable" signal, labeled '*CK*', for the device, and to provide inversion for the inputs. The inversion allows the '*S*' and '*R*' inputs to be *active high*, as opposed to the '$\bar{S}$' and '$\bar{R}$' inputs of the basic NAND latch. While $CK = 1$, the device responds as a normal latch, but with $CK = 0$, the device outputs are unchanging regardless of '*S*' and '*R*' input values. The timing diagram that follows shows how '*CK*' serves as an "enable" signal in the clocked S-R latch.

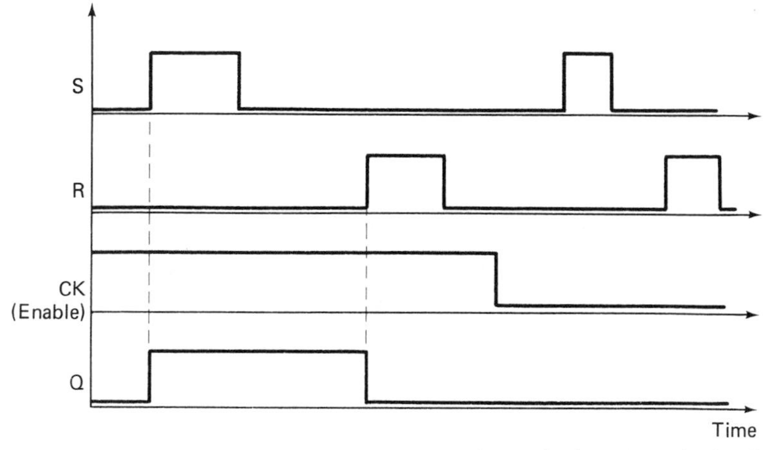

* The overbar denotes that '*S*' and '*R*' are *active low* inputs in the NAND latch. This is consistent with our convention of Section 5.3.5, and indicates that '*Q*' is set to '1' when '*S*' is asserted '0'.

(2) Synchronous devices: In all of the following cases, many different circuit designs are possible, using both master-slave and edge-triggered structures. Therefore, we show each device as a block element, and describe the general operating characteristics. In each case, 'set-direct' and 'clear-direct' inputs are possible, but are not shown.

(a) The D-Type Flip-Flop

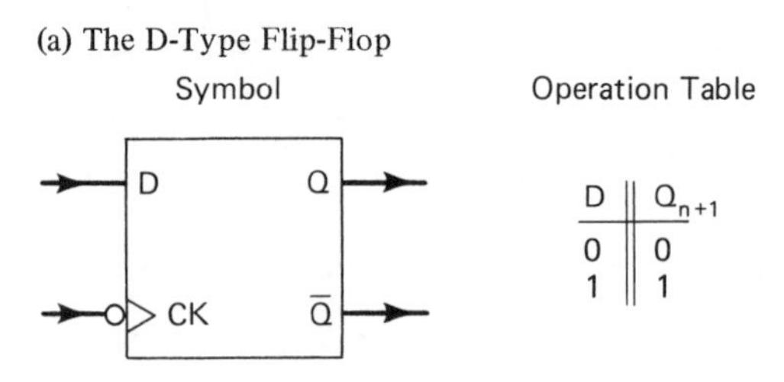

The 'Q' output of a D-type flip-flop always follows the 'D' input, but is changed only on the active edge of 'CK' (which in this case is the falling edge). The following timing diagram illustrates this idea for both edge-triggered and master-slave devices.

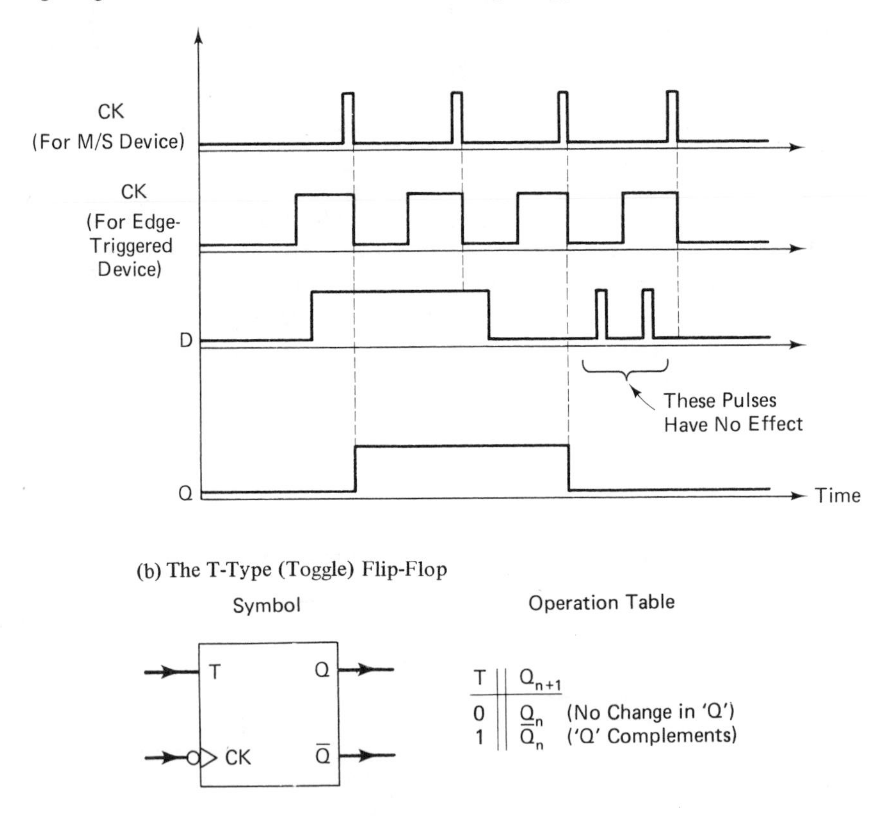

The 'Q' output of a toggle flip-flop remains constant if $T = 0$, and complements if $T = 1$. The output is capable of changing only on the active edge of 'CK' (which in this case is the falling edge). The following timing diagram illustrates this idea for both master-slave and edge-triggered devices.

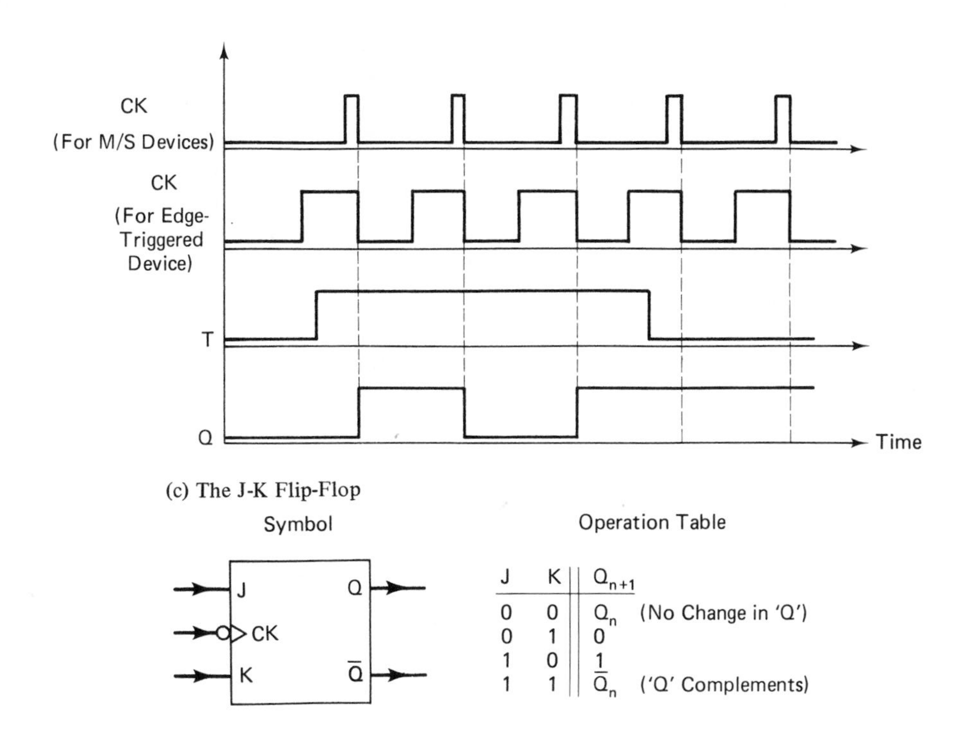

(c) The J-K Flip-Flop

Operation Table

J	K	Q_{n+1}
0	0	Q_n (No Change in 'Q')
0	1	0
1	0	1
1	1	$\overline{Q}_n$ ('Q' Complements)

The J-K flip-flop combines the advantages of the D and T flip-flops in one device. Its output can be set, reset, complemented, or left unchanged according to the operation table given above. These changes can occur only on the active edge of '*CK*' (which in this case is the falling edge). The following timing diagram illustrates these ideas for both master-slave and edge-triggered devices.

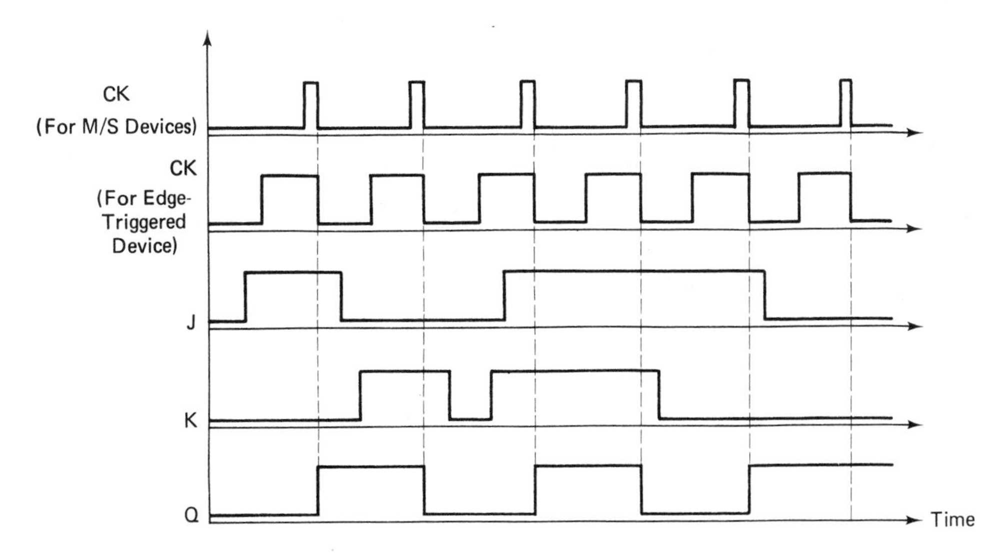

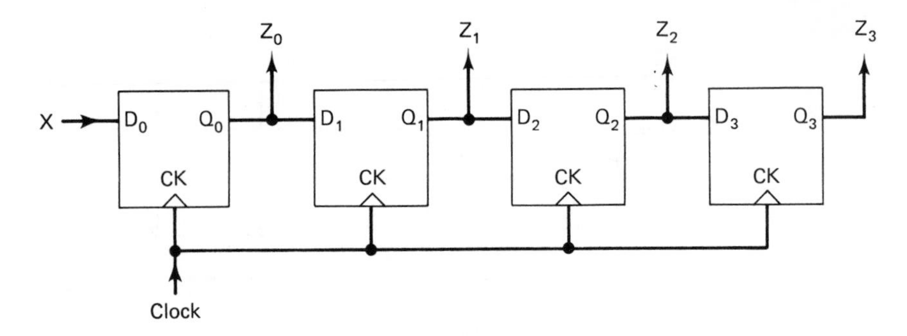

(a) Four-Stage
Shift Register.

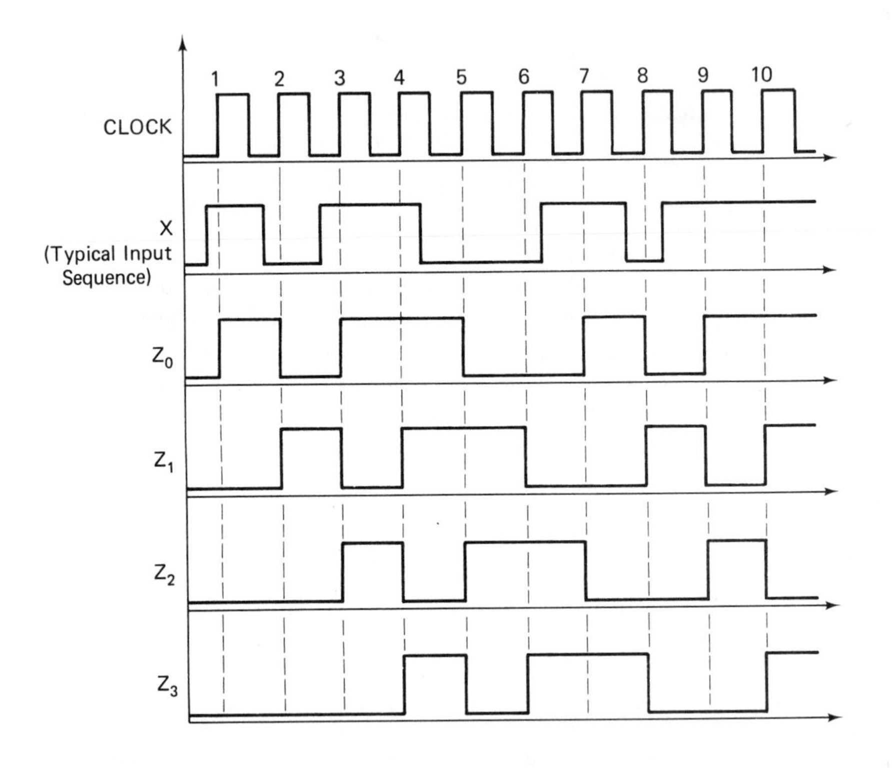

(b) Waveforms
Associated with
the Four-Stage
Shift Register.

FIGURE 5-30

5.3.7 Flip-Flop Applications

Shift register: A cascade of D flip-flops can be connected to form a ***shift register*** circuit. Figure 5-30(a) depicts a four-stage shift register, while Figure 5-30(b) shows how a typical input sequence propagates through successive stages of the circuit. At first glance, the circuit of Figure 5-30(a) might lead us to believe that the value of the '*X*' input will immediately propagate through all four stages when the first clock pulse occurs. In fact, this *does not* happen, as is shown by the timing diagram in Figure 5-30(b). To understand why this is true, several important features of the circuit should be noted. First, all of the flip-flops are edge-triggered devices and are sensitive to the rising edge of 'CLOCK'. Since all of the flip-flops are connected to the same clock line, they will all change state synchronously. Second, each flip-flop has a given amount of propagation delay associated with it, even if the value is small. Thus, we see that when a rising edge occurs on 'CLOCK', the value that is loaded into a particular stage of the shift register is the value of that stage's input *at the instant* of the rising edge, not after the propagation delay of the previous stage. In other words, a given stage is loaded *before* the previous stage can change state, even if it is just a few nanoseconds before. It is, therefore, not possible for an input on '*X*' to immediately propagate through all four stages on the first rising edge. Connecting the '*Z*' outputs in Figure 5-30(a) to indicator lamps, and applying the input shown in Figure 5-30(b), we would observe the pattern of changes shown in Figure 5-31.

Time ↓	Z_0	Z_1	Z_2	Z_3
Initial Condition	○	○	○	○
(1)	●	○	○	○
(2)	○	●	○	○
(3)	●	○	●	○
(4)	●	●	○	●
(5)	○	●	●	○
(6)	○	○	●	●
(7)	●	○	○	●
(8)	○	●	○	○

○ Means "Off" Lamp
● Means "On" Lamp

FIGURE 5-31 *The pattern of changes that would appear on indicator lamps connected to the shift register outputs of Figure 5-32(a).*

Shift registers are very useful as ***parallel-to-serial converters***, in which a multitude of binary values are loaded at one time into the register and then individually shifted out. A four-bit parallel-to-serial converter is illustrated in Figure 5-32. Note that in this circuit, 'preset' and 'clear' inputs are required on each flip-flop, in addition to a small amount of external logic. Since '*PR*' and '*CL*' override the clock, no shifting can occur during parallel load. Only when the parallel load input is deasserted can data be shifted out.

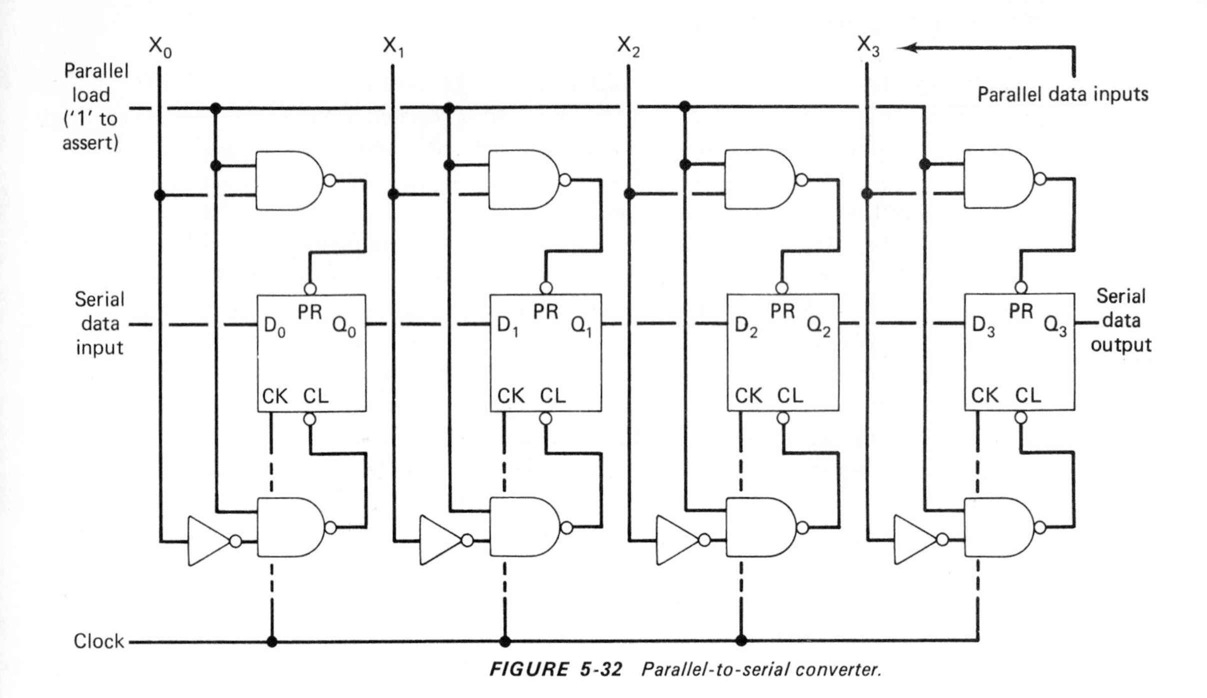

FIGURE 5-32 *Parallel-to-serial converter.*

A ***serial-to-parallel converter*** performs the reverse function of a parallel-to-serial converter, and that is to receive a multitude of binary values one at a time and compile them into a whole binary word for parallel output. To accomplish this, we need only a normal shift register with parallel outputs; 'preset' and 'clear' inputs are not needed. The circuit that was presented earlier in Figure 5-30(a) could be used as a serial-to-parallel converter. Figure 5-33 illustrates how parallel/serial conversions can reduce cable costs in the long-distance transmission of digital information. Note, however, that the net rate of data transfer from transmitter to receiver is much less when serial transmission is employed, given a fixed bit-rate per single channel.

Binary counter: Figure 5-34 illustrates how the output of a toggle flip-flop complements in value when successive rising edges are applied to 'CK'. We first note that this simple circuit is acting as a ***frequency divider***. Specifically, the output frequency is exactly one-half the input frequency. This is seen in the timing diagram of Figure 5-34 by observing that the period of the 'Q' output is twice as long as that of 'CK'. Now, we can reason that by connecting a second flip-flop in cascade with the first, but clocked from the 'Q' output of the first, we should be able to halve the frequency again. This is, indeed, the case, and is illustrated in Figure 5-35. Clearly, we can cascade any number of toggle flip-flops to obtain successive division of the frequency by two. For 'n' flip-flop stages, we have

$$f_n = f_{\text{in}}(\tfrac{1}{2})^n, \qquad n \geq 1 \tag{5-1}$$

where 'f_n' is the output frequency of the nth stage, and 'f_{in}' is the clock input frequency.

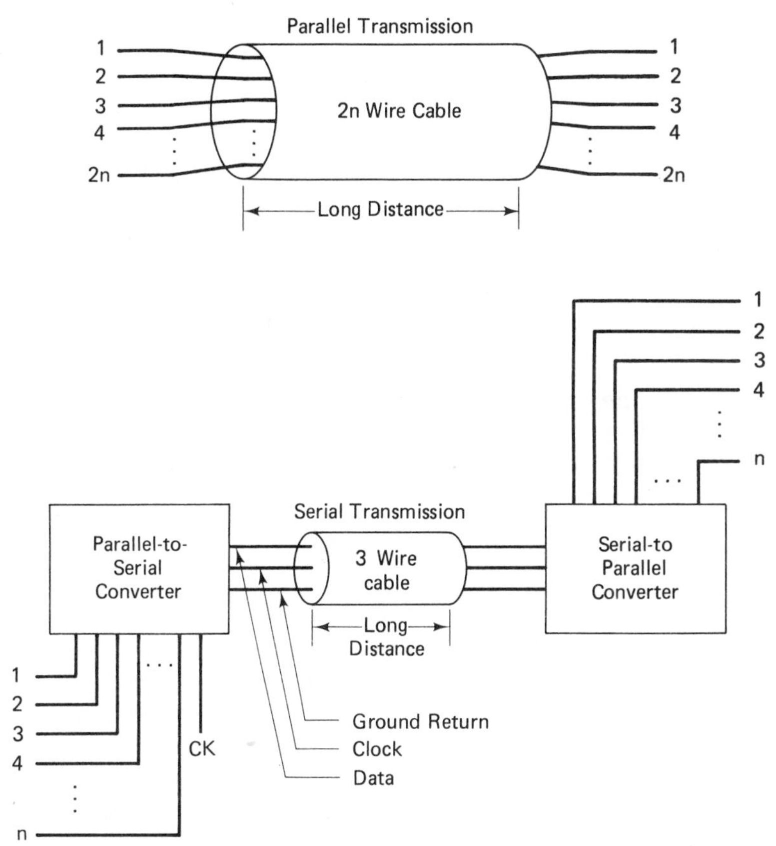

FIGURE 5-33 Parallel and serial data transmission schemes for 'n' digital channels. In the parallel transmission method, each channel usually requires its own ground return for satisfactory noise and crosstalk immunity. However, in some cases, a ground return may be shared, reducing the number of wires in parallel transmission from '2n' to 'n + 1'. In certain telephone cables, a value of n = *1000 is not unusual.*

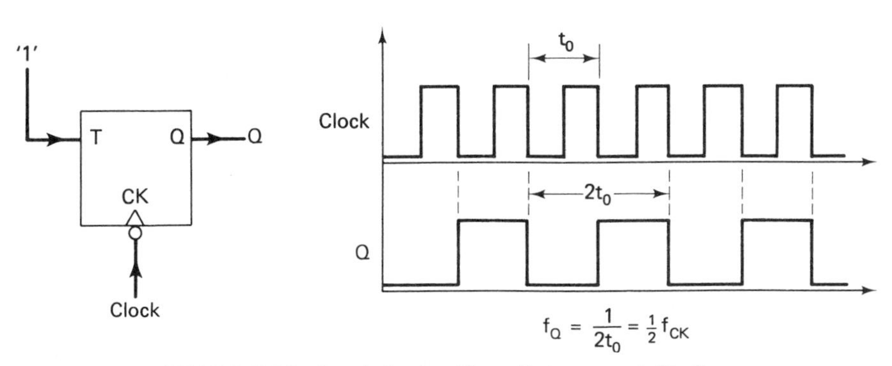

FIGURE 5-34 A periodic signal is applied to a toggle flip-flop.

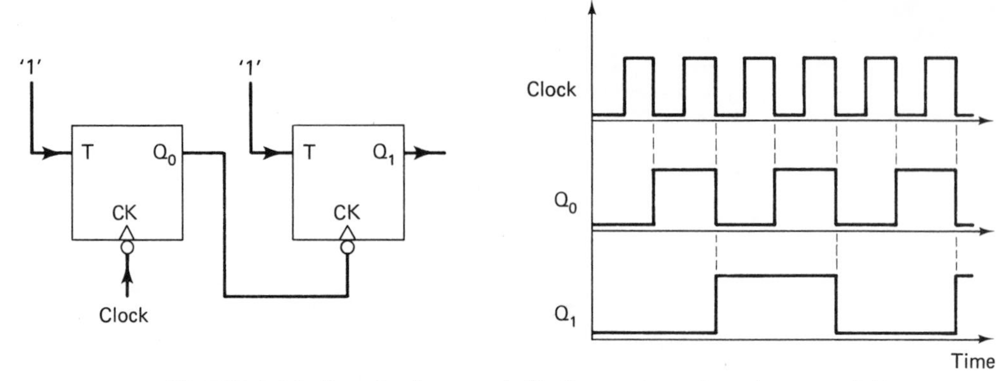

FIGURE 5-35 *Cascade of two toggle flip-flops, connected as a frequency divider.*

Consider a three-stage frequency divider, as shown in Figure 5-36(a). The signal flow in this circuit has been reversed so that signals propagate from right to left. Doing so makes the sequence table more meaningful. As we can see from Figure 5-36(b), the frequency divider circuit is acting as a mod-eight natural binary counter. Actually, the circuit of Figure 5-36(a) is usually though of as a ***ripple-carry counter***, rather than as a frequency divider, although it is both. This is the simplest kind of binary counter, yet its simple design has some disadvantages. In particular, the transition from one state to the next requires clock signal propagation through each of the successive flip-flops that make up the counter. For large, multistage counters, the accumulated delay can be quite long, which is very undesirable in many binary counter applications. Chapter 6 examines binary counters in much more detail, and presents an alternative design that eliminates the propagation problem.

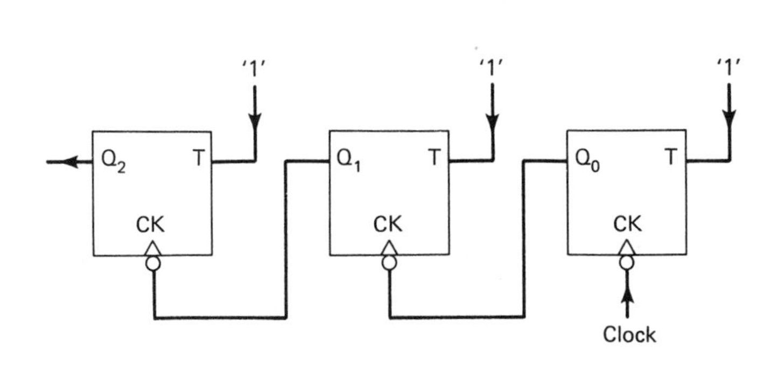

(a) Three-Stage Frequency Divider.

Number of Falling Edges on CK	FF Outputs		
	Q_2	Q_1	Q_0
0	0	0	0
1	0	0	1
2	0	1	0
3	0	1	1
4	1	0	0
5	1	0	1
6	1	1	0
7	1	1	1
8	0	0	0
9	0	0	1
10	0	1	0
11	0	1	1
12	1	0	0

(b) Sequence of Changes that Occur in the 'Q' Outputs of the Frequency Divider Circuit as Clock Pulses are Applied.

FIGURE 5-36

5.4 MONOSTABLE MULTIVIBRATORS

5.4.1 General Description

Flip-flops are sometimes referred to as *bistable multivibrators*, where "bistable" is derived from the fact that a flip-flop has *two* stable states. A ***monostable multivibrator*** is a device similar to a flip-flop, and has one stable state and one momentary state. In this case, "monostable" is derived from the fact that there is only *one* stable state. When a monostable multivibrator is triggered, the device's 'Q' output is set to logic '1', and remains set for a predetermined length of time. When this period of time has elapsed, the 'Q' output automatically returns to logic '0', and the device has then returned to its stable state. A resistor and capacitor are added to the monostable circuit to serve as timing components, and are usually wired externally to specified pins of the monostable package. These timing components precisely determine what the delay period will be. Monostable multivibrators are almost always *edge-triggered* devices, thereby making the width of the input pulse unimportant to the circuit's operation. Figure 5-37 demonstrates the operation of a positive edge-triggered monostable device. A monostable multivibrator is sometimes referred to as a "one-shot," or just a "monostable."

FIGURE 5-37 *Operation of a positive edge-triggered monostable multivibrator. An external resistor and capacitor are added to determine the period 't_0'.*

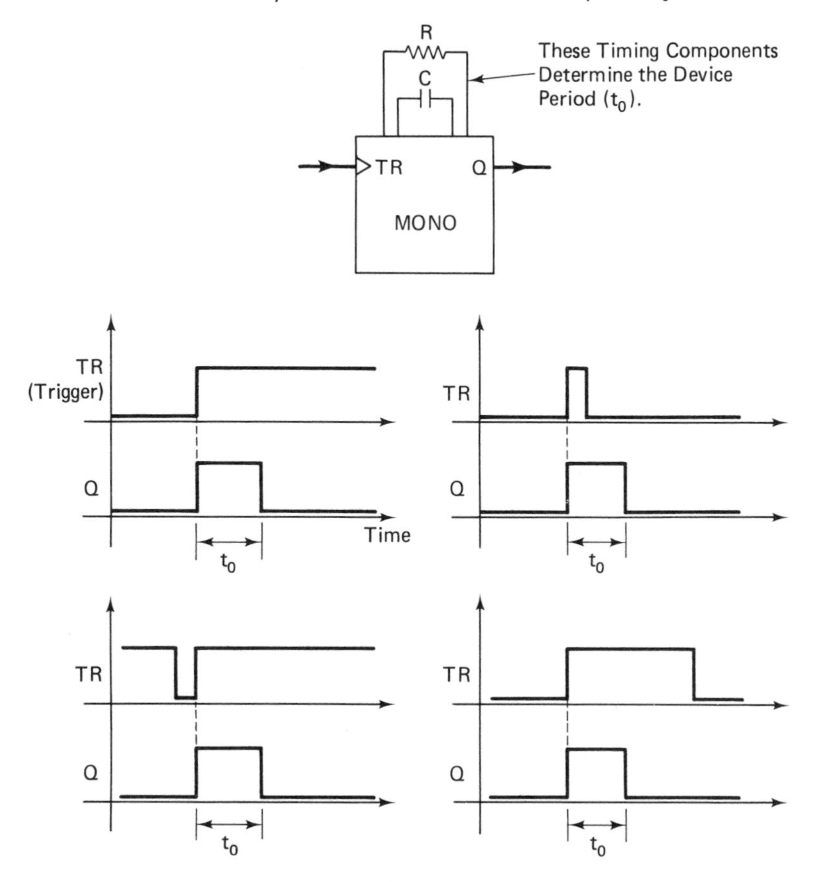

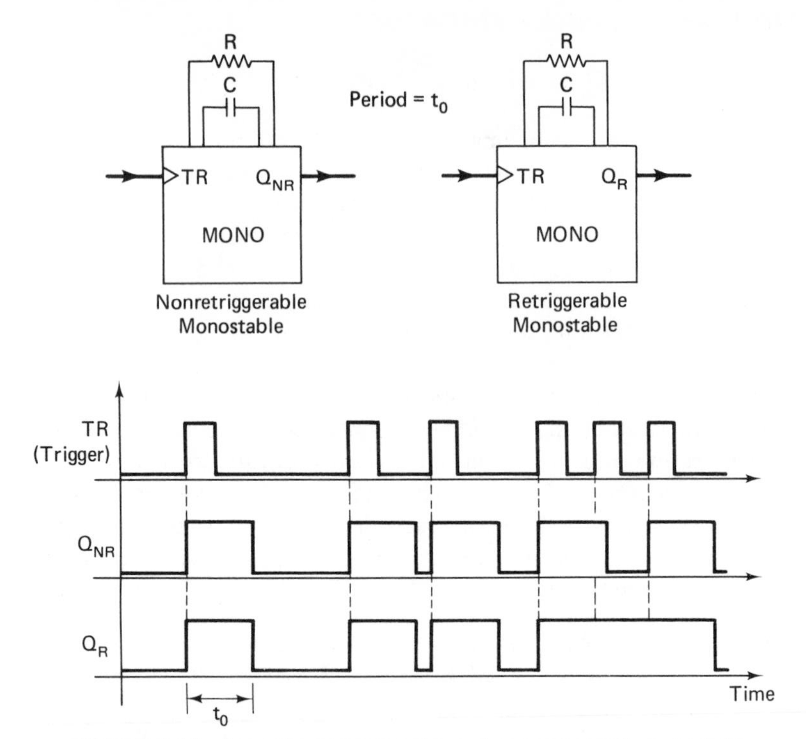

FIGURE 5-38 Operation of retriggerable and nonretriggerable monostable multivibrators under identical input conditions.

Two basic types of monostable circuits are usually available, and operate differently only under certain types of pulse train inputs. ***Retriggerable*** devices restart the timing sequence if another triggering edge is applied before the end of the present period. On the other hand, ***nonretriggerable*** devices ignore any further triggering edges until completion of the present period. Figure 5-38 illustrates how retriggerable and nonretriggerable monostable multivibrators behave under identical input conditions.

5.4.2 An Application

Heart rate monitor: Hospital patients with heart problems often require constant heartbeat monitoring using an electrocardiograph machine. The output of such a machine is shown in Figure 5-39 and depicts a large pulse for every major pumping contraction of the heart muscle. Normally, the period of this waveform is about 0.83 s, corresponding to 72 beats per minute. Although some variation is acceptable in the normal heart rate, a dangerous condition would exist for a patient if his or her heart rate were considerably faster or considerably slower than 72 beats per minute. Thus, it would be very useful to have an electronic heart rate monitor that would sound an alarm if the heart rate became too fast or too slow. In practice, such a monitor circuit would be quite complex, allowing for heart rate irregularities, false beats, and other factors. For the sake of example, however, we can design a simple digital circuit that satisfies the basic requirement of a heart rate monitor.

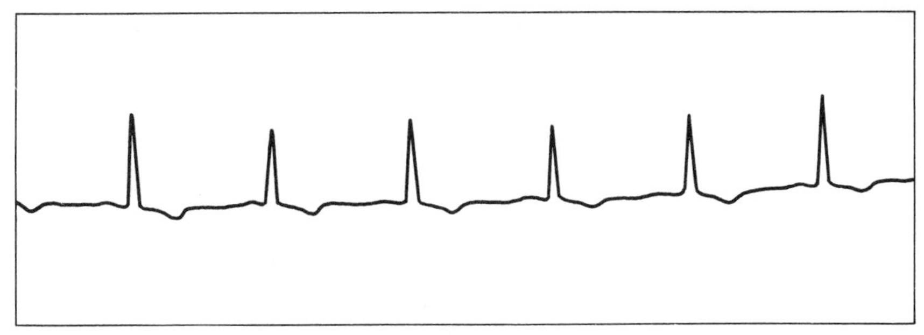

FIGURE 5-39 Electrocardiogram waveform of a normal heart.

Such a circuit is shown in Figure 5-40(a) and incorporates monostable multivibrators in its design. The related waveforms are depicted in Figure 5-40(b). A slow heart rate, medically termed "bradycardia," is detected by monostable (1) not being retriggered. The resulting switch of 'Q_R' from '1' to '0' sets the bradycardia alarm latch. For a normal heart rate, 'Q_R' is always '1'. A speeding heart rate, medically referred to as "tachycardia," is detected by the period of monostable (2) not having elapsed when another heart pulse is received. NAND gate (3) then sets the tachycardia alarm latch. For a normal heart rate, 'Q_{NR}' will always be '0' when a heart pulse arrives. Note that monostable (2) is triggered on the falling edge of 'T'. Once an alarm latch is set, it will remain set until the reset button is pressed.

FIGURE 5-40(a)

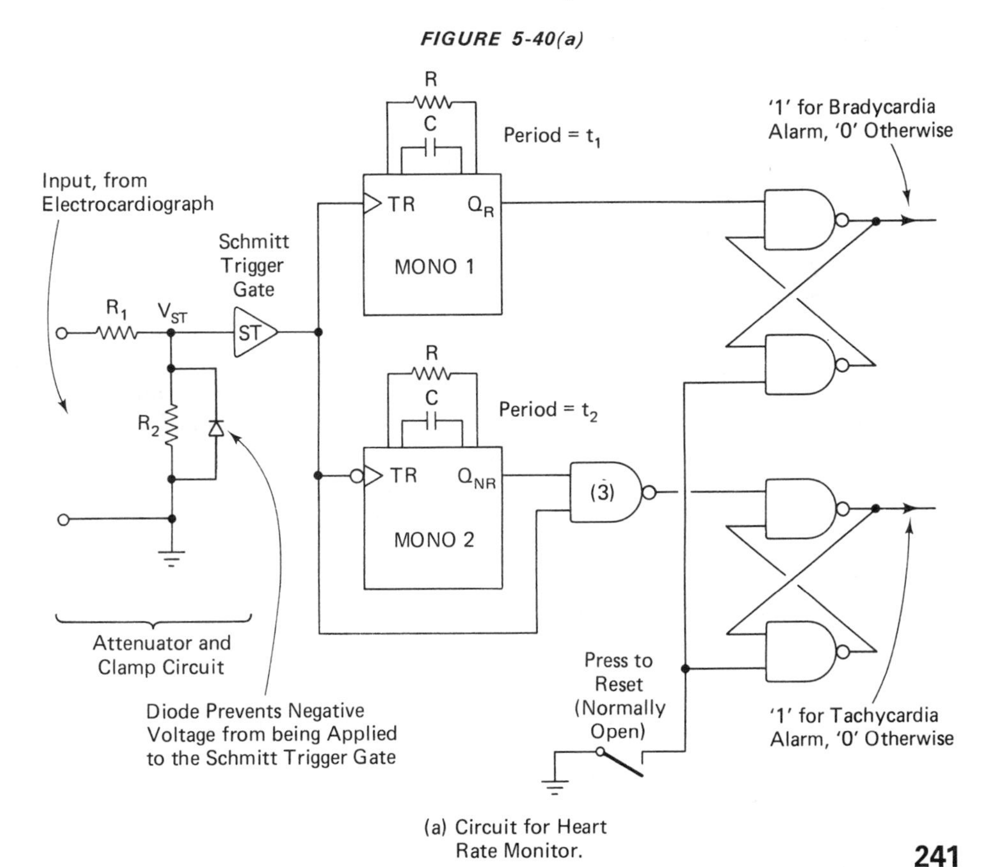

(a) Circuit for Heart Rate Monitor.

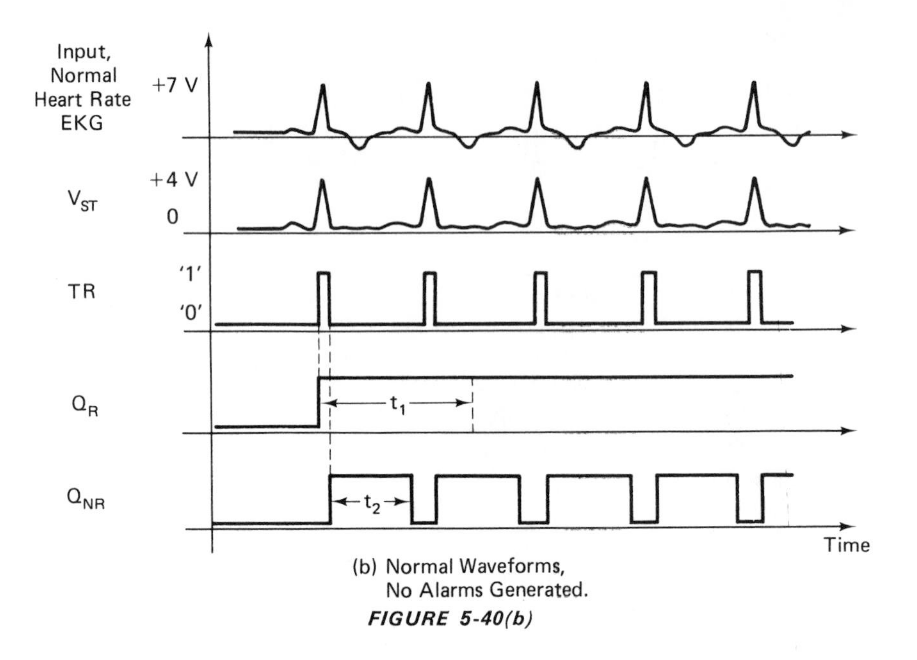

(b) Normal Waveforms,
No Alarms Generated.
FIGURE 5-40(b)

5.5 ASTABLE MULTIVIBRATORS

5.5.1 Periodic Waveforms

In several of this chapter's applications problems, and in earlier chapters, periodic waveforms such as a pulse train were assumed to be available to operate clocked flip-flops and other components. At this time, we will discuss how such waveforms are generated, and the important aspects of periodic waveforms that should be controlled in digital circuit applications. Our discussion will begin with a description of exactly what is meant by a periodic waveform, and will lead to an examination of several circuits that produce the desired signals.

In general, a periodic waveform is characterized by a repeating sequence of identical waveform elements. Each element portrays the variation of voltage* versus time over a duration of time referred to as the ***period***. In the case of a binary digital signal, the voltage value of the signal can be maintained at only one of two possible voltage values. These two voltage values are often referred to as *states*. In real signals, it is not possible to pass from one state to another instantly, and instead, a smooth transition must always occur. Even if a digital signal appears on an oscilloscope to move from one level to another instantly, the magnification (sweep speed) can always be increased to a point whereby the transition is visible. Recall from Chapter 1 that the transition time from a low voltage level to a higher one is referred to as the *rise time*, while the transition time from high voltage level to low is referred to as *fall time*.

* "Current" can be interchanged with "voltage" in the following description if current mode logic is being used.

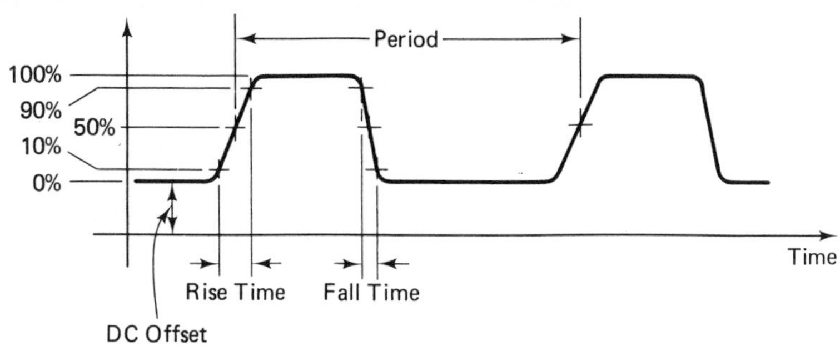

FIGURE 5-41 Detailed measures that define a periodic digital signal.

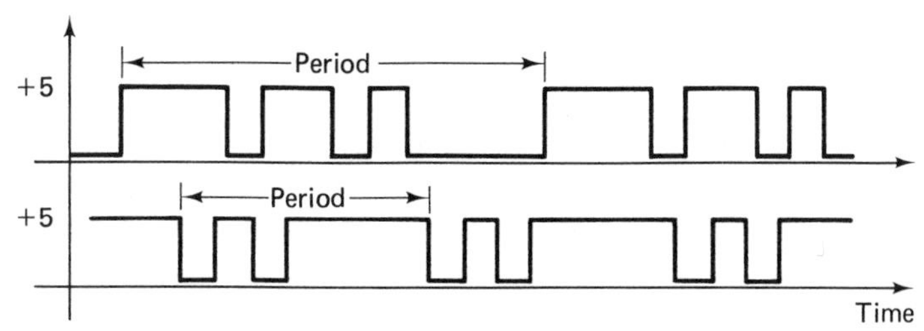

FIGURE 5-42 Periodic digital signals. In this case, the rise time and fall time are much less than each signal period.

Often, rise time and fall time are much less than the signal's period. Also involved in the character of a periodic binary digital signal are its amplitude, its midpoint value (related to the threshold of the logic circuitry), and its duty cycle. Figure 5-41 depicts these important measures. In many families of digital circuitry, the ***DC offset*** value is chosen to be zero volts, and the high-level value is some small positive voltage, such as 5 V. Using this convention, some other possiblities for periodic digital signals are shown in Figure 5-42.

5.5.2 Astable Circuits

Figure 5-43 illustrates two periodic waveforms often encountered in digital circuitry. The ***square wave*** is just a special case of a pulse train, with $T_{\text{high}} = T_{\text{low}}$. We would now like to examine several circuits that produce periodic waveforms of this

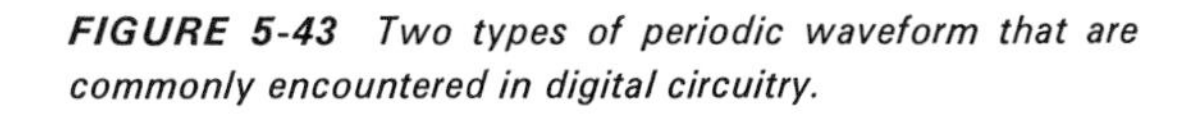

FIGURE 5-43 Two types of periodic waveform that are commonly encountered in digital circuitry.

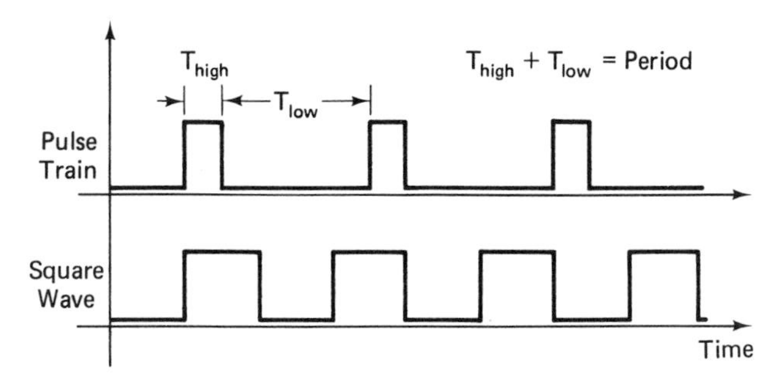

type. The basic objective of such a circuit is to transform its constant, DC input from the power supply into an oscillating signal. A circuit that possess this characteristic is referred to as an ***astable multivibrator***, or alternatively as a clock circuit, and is illustrated in Figure 5-44 in the form of a block element. "Astable" literally means "without stable states."

A simple logic circuit that produces a square wave output is shown in Figure 5-45(a), with its associated waveforms in Figure 5-45(b). This circuit is referred to as a ***ring oscillator***. Although this is a simple and reliable circuit, we have very little control over the output frequency, which is determined by the propagation delay of the logic gates and the number of gates in the loop. Since we often must carefully select the frequency of an astable circuit, a ring oscillator is usually unsatisfactory.

Figure 5-46(a) depicts a simple, reliable, and highly flexible astable circuit. In this case, the frequency of oscillation is determined primarily by the resistor and capacitor timing components. Figure 5-46(b) illustrates one charge/discharge cycle and the associated waveforms. An astable circuit that relies on resistors and capacitors for the oscillation frequency is generally referred to as ***RC oscillator***. The exact

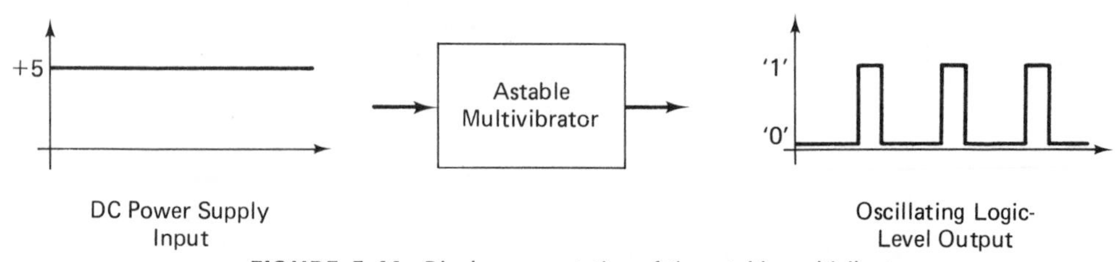

FIGURE 5-44 Block representation of the astable multivibrator.

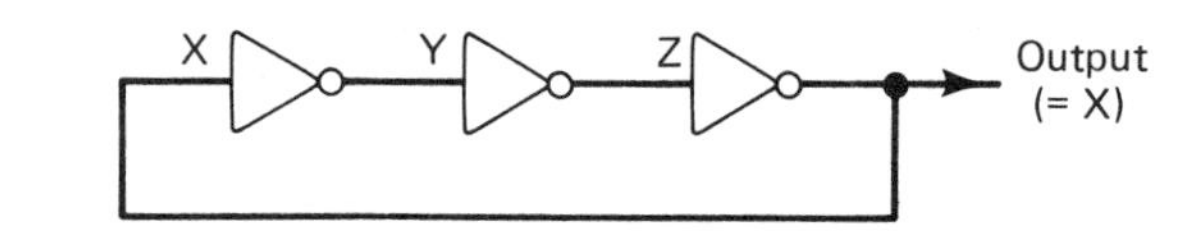

(a) Ring Oscillator Circuit. Each Gate has a Propagation Delay of 't_0'.

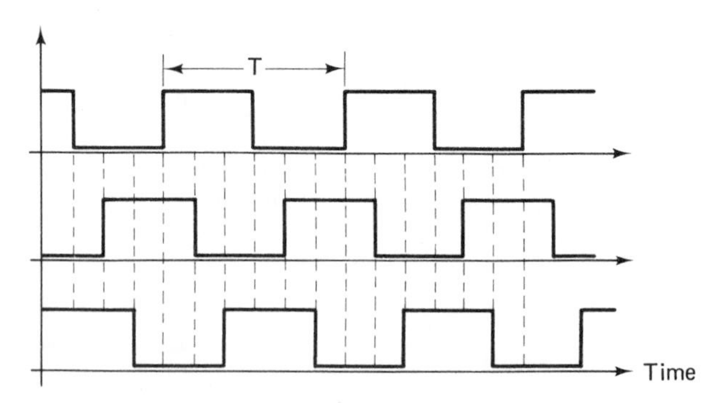

(b) Timing Diagram Associated with the Ring Oscillator. Note that the Period, 'T', of the Output Waveform is Exactly $6t_0$.

FIGURE 5-45

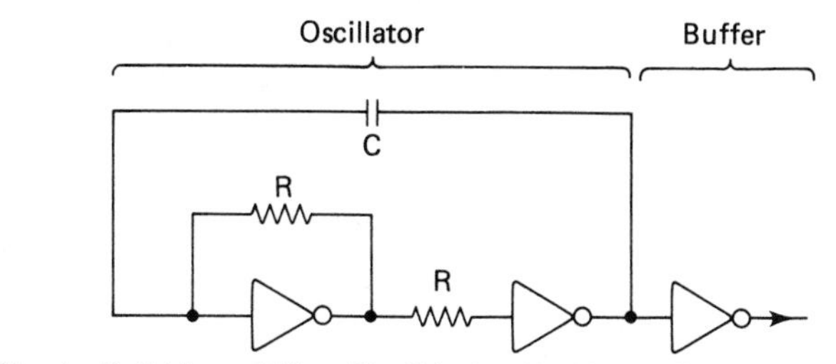

(a) Simple, Reliable, and Very Flexible Astable Circuit. For TTL Circuits, Each 'R' is About 400 Ω, and 'C' is Determined by Eq. 5-2.

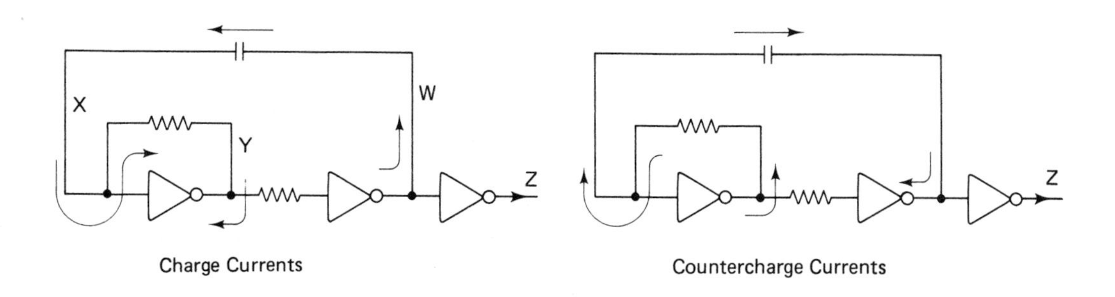

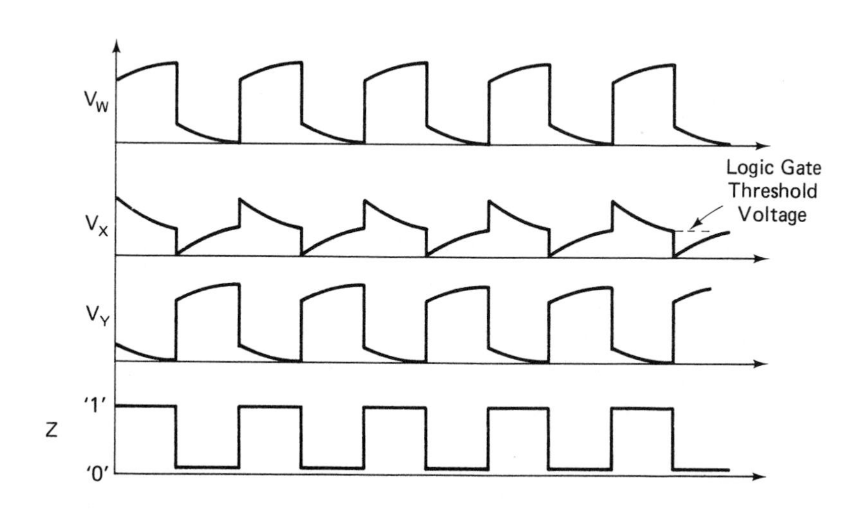

(b) Charge and Countercharge Currents are Illustrated with Associated Voltage Waveforms. In these Drawings, we Assume that the Logic Gate Input is of Relatively High Impedance when Compared to the Logic Gate Output.

FIGURE 5-46

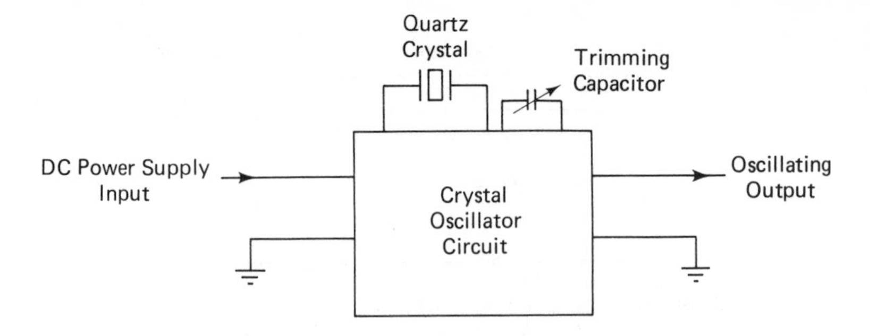

(a) Block Representation of a Crystal Oscillator.

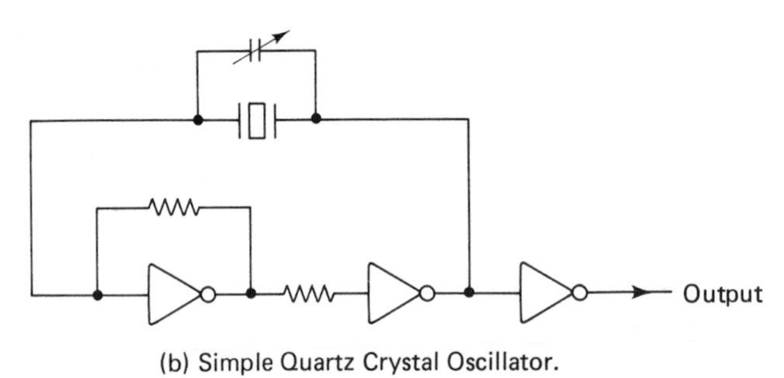

(b) Simple Quartz Crystal Oscillator.

FIGURE 5-47 *(a) Block representation of a crystal oscillator; (b) Simple quartz crystal oscillator.*

relationship between the oscillation frequency and the 'R' and 'C' components depends in part on the electrical characteristics of the logic gates. For the standard TTL family of components, a resistance of approximately 400 Ω produces the relationship

$$f = \frac{1}{0.001C} \tag{5-2}$$

where 'C' is measured in microfarads and 'f' is measured in hertz. There is virtually no upper limit on the value of 'C'; for $R = 390\ \Omega$ and $C = 28{,}000\ \mu F$, the period was found to be approximately 28 s. Thus, extremely low frequencies can be produced. The maximum frequency possible for this circuit is limited by the gate propagation delays, and by the 'R' and 'C' components not behaving properly at very high frequencies. For $R = 390\ \Omega$ and $C = 470$ pF, the period was found to be approximately 500 ns.

A ***crystal oscillator*** is very similar to an *RC* oscillator, except that the primary frequency determining component is a quartz crystal. Because of the physical nature of a quartz crystal, an extremely accurate and stable frequency can be produced by a crystal oscillator, making it more desirable than other astable circuits for certain critical timing applications. Good crystal oscillator circuits are generally more complex than *RC* circuits, and, therefore, are not usually constructed from basic components when needed. Rather, crystal oscillator circuits are typically available as whole circuit elements, requiring only the addition of a quartz crystal. A trimming capacitor can be added to allow very slight frequency adjustment. Figure 5-47(a) depicts a block representation of a quartz crystal oscillator circuit. Crystal oscillators are generally used to produce frequencies from 100 kHz to 150 MHz or greater. The output may be in the form of a sine wave or a digital pulse train, depending on the circuit design. Of course, the frequency of a digital pulse train can always be divided using a counter circuit, so that arbitrarily low frequencies can be generated. Figure 5-47(b) follows the RC circuit design of Figure 5-46, replacing the capacitor with a quartz crystal.

5.5.3 An Application

Pulse synchronizer: Clock signals often need to be applied to a circuit and then removed on the command of some external signal. A very simple way to accomplish this using a single AND gate is shown in Figure 5-48(a). In Figure 5-48(b), we see possible waveforms that might be produced. Since the control input can change at any time without being synchronized with the astable pulses, it is possible that the gated output may produce truncated (incomplete) initial and final pulses, as shown in Figure 5-48(b). This is very undesirable, since the truncation may be so severe that the affected pulses are reduced in width to the size of a glitch, as shown in

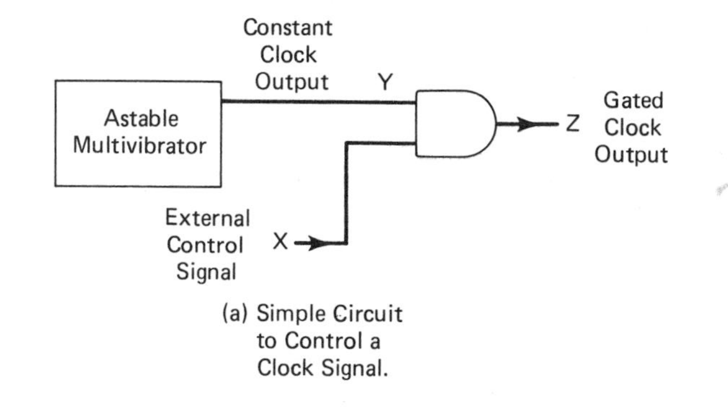

(a) Simple Circuit to Control a Clock Signal.

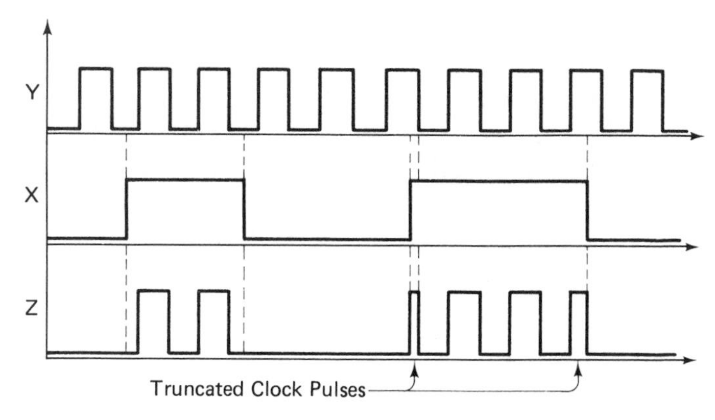

(b) Possible Waveforms for the Clock Control Circuit.

FIGURE 5-48

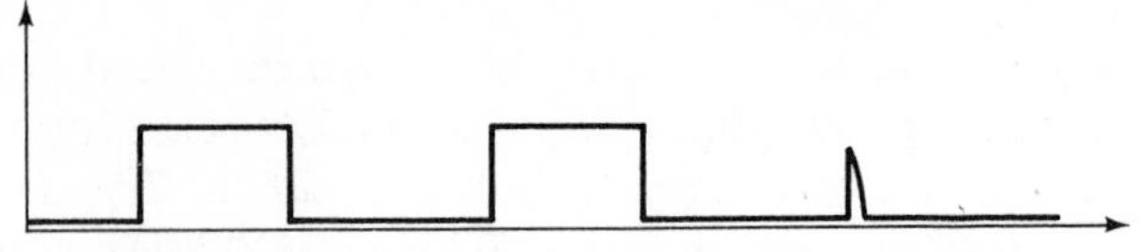

FIGURE 5-49 Extreme truncation turns a normal pulse into a glitch.

FIGURE 5-50

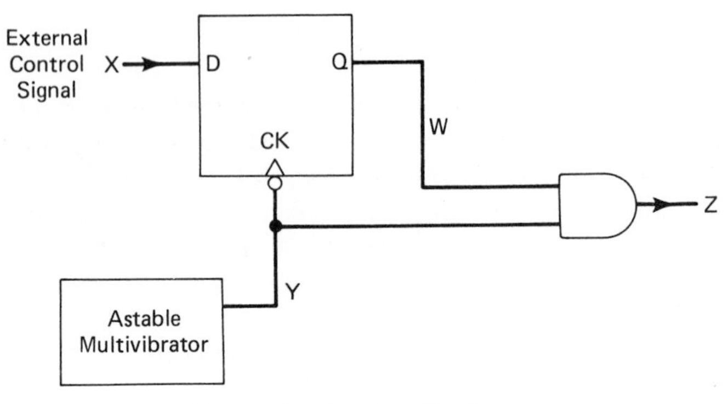

(a) Pulse Synchronizer Circuit.

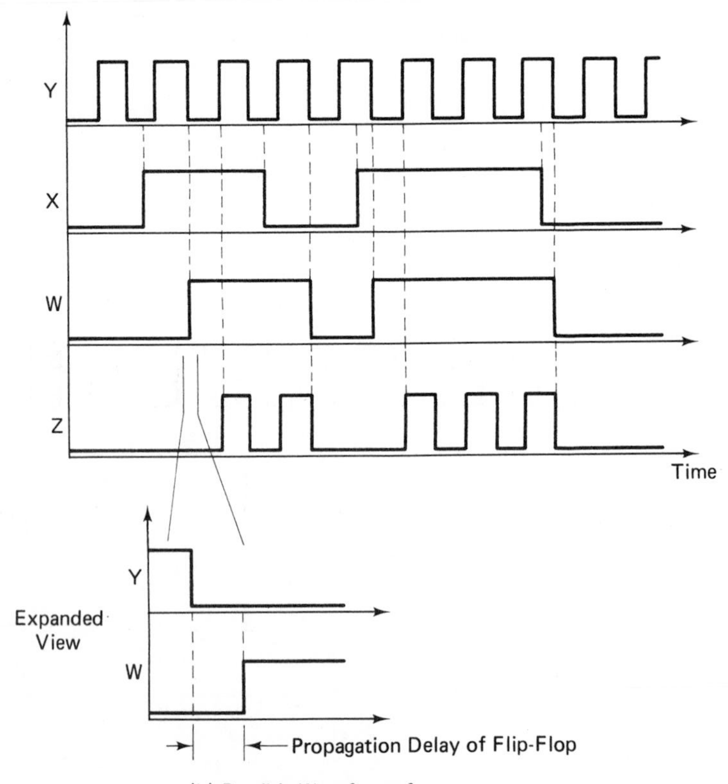

(b) Possible Waveforms for the Pulse Synchronizer.

Figure 5-49. A glitch may be of sufficient size to trigger some devices connected to the gated output, but not long enough to trigger other devices. Thus, erratic circuit operation may result. A ***pulse synchronizer*** circuit is used to eliminate the truncation problem, and is illustrated in Figure 5-50(a). Associated waveforms are depicted in Figure 5-50(b), and demonstrate how pulse synchronizer prevents truncation. In this circuit, the D flip-flop permits application of the control signal to the AND gate only on the falling edge of '*CK*'. This is one of those critical timing situations that always works in our favor, since when the propagation delay of the flip-flop has expired, the logic '1' value previously on '*CK*' will have changed to logic '0'. Thus, no overlap is possible and no truncation results. We note, however, that the output pulses are always delayed by one period of '*CK*' from when the control signal is applied.

SUMMARY

Flip-flops are one-bit memory elements that are used frequently in digital systems. All flip-flops consist of combinational logic components connected in a feedback configuration, and consequently are classified as sequential circuits. The simplest flip-flop is the latch, and is realized, in one form, by connecting two NAND gates in a cross-coupled manner. The clocked latch operates in a similar way to the unclocked latch, except that an enable signal is required prior to its responding to input changes. Synchronous flip-flops, such as T, D, and J-K devices, can change only when a clock pulse is applied to the flip-flop. The advantage of synchronous devices over nonsynchronous devices such as the latch lies in their ability to be coordinated with other system components by way of a clock line. Master-slave circuits require that no input changes occur while the clock input is asserted, whereas edge-triggered circuits are much more versatile and require only that the device inputs not be changed at the moment of the triggering edge. Synchronous flip-flops are drawn in schematics as block elements, and are usually available from manufacturers as whole modules. Thus, it is never really necessary to know the exact logic connections for a synchronous device, as long as it behaves according to manufacturer specifications. With any block elements available as circuit modules, it is valid to assume that electronic design simplifications have been made in the course of preparing a microminiature integrated circuit. Thus, a logic diagram describing a manufactured block element can be only an approximation.

All flip-flops are subject to propagation delay, as are logic gates. Propagation delay should always be accounted for in timing diagrams whenever a critical sequence of changes is to be described in a logic circuit design. Propagation delay should not be shown, however, if it does not meaningfully contribute to the circuit's description. Carefully drawn timing diagrams can often lead to the discovery of design errors and transient conditions that may cause circuit malfunction. Setup and hold times are additional parameters that describe synchronous devices, and demand that no input changes be made just before and just after the triggering condition of '*CK*'. Changing an input variable contrary to these rules may result in improper flip-flop operation.

Certain symbol conventions have been generally accepted by digital circuit designers, and should be followed, whenever possible, in the drawing of schematic diagrams. Particularly important is the indication of active levels and active edges by the use, or nonuse, or inversion circles. Active low inputs, or active falling edge inputs, are denoted by an inversion circle on the flip-flop block representation, or by an overbar in the input label.

Monostable multivibrators are a special type of flip-flop possessing one stable state and one momentary state. The momentary state is entered by the application of a triggering edge, and persists for an amount of time determined by external timing components added to the device. The momentary state usually can be adjusted to be as short as a few nanoseconds, and as long as several minutes. Retriggerable devices will restart the timing period should a second triggering edge be received before the end of the current period. Nonretriggerable devices, on the other hand, ignore all subsequent transitions on the trigger input until the current period is complete.

Astable multivibrators produce periodic digital signals for use in timing, counting, and synchronizing. The input of an astable circuit is nothing more than the power supply, while the output is a pulse train or square wave capable of driving a multitude of digital circuit components. The *RC* oscillator is often used since it is extremely simple and has an easily adjustable output frequency. A good crystal oscillator is somewhat more complex than an *RC* oscillator, yet it offers very good frequency stability. Crystal oscillators are much less sensitive to temperature and power supply variations, and are used most often in critical timing applications.

NEW TERMS

Timing Diagram
Latch
  NAND Latch
  NOR Latch
Switch Bounce
Flip-Flop
Bistable Multivibrator
Clocked S-R Latch
Clocked D Latch
Master-Slave Flip-Flop
Edge-Triggered Flip-Flop
Pulse-Triggered Flip-Flop
J-K Flip-Flop
T Flip-Flop
D Flip-Flop
Set Direct
Clear Direct
Setup Time
Hold Time
Active Level
Active Edge
Level-Sensitive Device
Shift Register
Parallel-to-Serial Converter
Serial-to-Parallel Converter
Frequency Divider
Ripple-Carry Counter
Monostable Multivibrator
  Retriggerable Device
  Nonretriggerable Device
Period
Dc Offset
Square Wave
Astable Multivibrator
Ring Oscillator
RC Oscillator
Crystal Oscillator
Pulse Synchronizer
Synchronous Device

PROBLEMS

5-1 Explain why a latch cannot be designed without any feedback paths.

5-2 Plot 'S', 'R', and 'Q' versus time for the latch circuit of Figure P5-2. Indicate what the asserted condition is ('1' or '0') for 'S' and for 'R'. Your timing diagram should account for the propagation delay times of the gates.

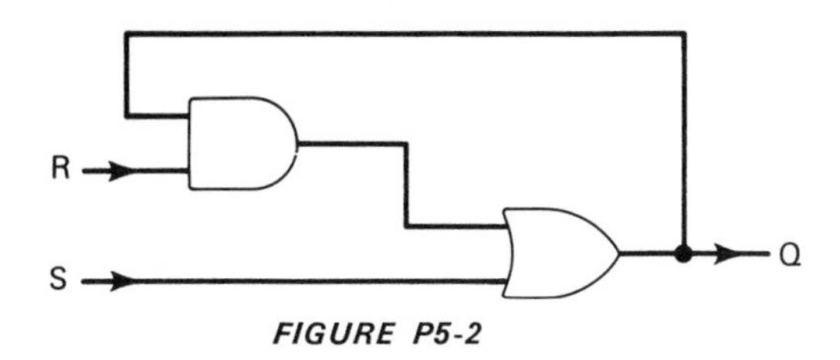

FIGURE P5-2

5-3 Is it possible to design a latch circuit using standard logic gates in which 'Q' and '$\bar{Q}$' are always exact complements of each other? Refer to Figure 5-8.

5-4 Can the switch debouncing circuit of Figure 5-11 be redesigned to use a SPST switch and only logic gates and resistors? Explain.

5-5 One of the simplest types of latch circuits is made by connecting two NAND gates as shown in Figure P5-5. Are there one or two feedback paths in this circuit? Redraw the circuit to prove your answer.

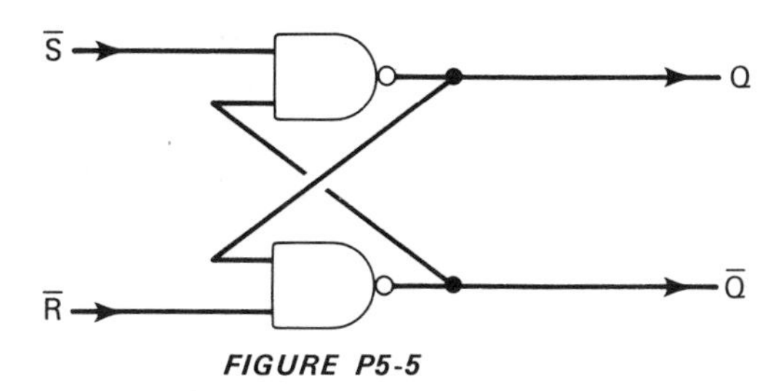

FIGURE P5-5

5-6 Describe the operation of the circuit shown in Figure P5-6. Explain how this circuit can be used to simulate the flipping of a coin. Assume that $Y = 0$ corresponds to "heads" and $Y = 1$ corresponds to "tails."

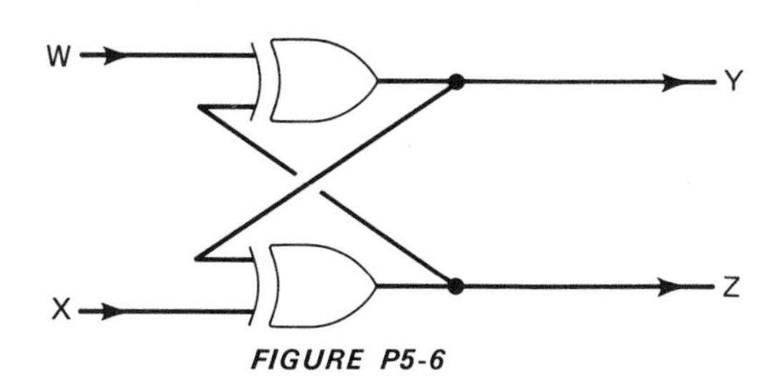

FIGURE P5-6

5-7 Design a switch debouncing circuit using a SPDT switch, two resistors, and two NOR gates. As part of the explanation of your design, draw a complete timing diagram, similar to that of Figure 5-11.

5-8 Design a clocked S-R latch using NOR gates only. The latch must obey the following operation table:

CK	S	R	Q
0	$\emptyset$	$\emptyset$	No change
1	0	0	No change
1	0	1	0
1	1	0	1
1	1	1	$Q = \bar{Q} = 0$

5-9 Figure 5-25 illustrates how the basic J-K master-slave flip-flop circuit can be altered to include set-direct and clear-direct inputs. These inputs override the clock and take effect immediately when asserted. Modify the circuit of Figure 5-25 so that the 'SD' and 'CD' inputs *do not* override the clock and take effect only after a clock pulse has been applied.

5-10 A priority circuit is depicted in Figure P5-10. In this drawing, 'S_1' and 'S_2' are normally '0'. When one of them becomes '1', the corresponding 'Q' output is set to '1', and any further changes on either of the 'S' inputs are blocked. Thus, once one 'Q' output is set, both 'Q' outputs are inhibited from changing. Such a circuit could be used, for example, in a swimming race to indicate which swimmer arrives at the finish line first.

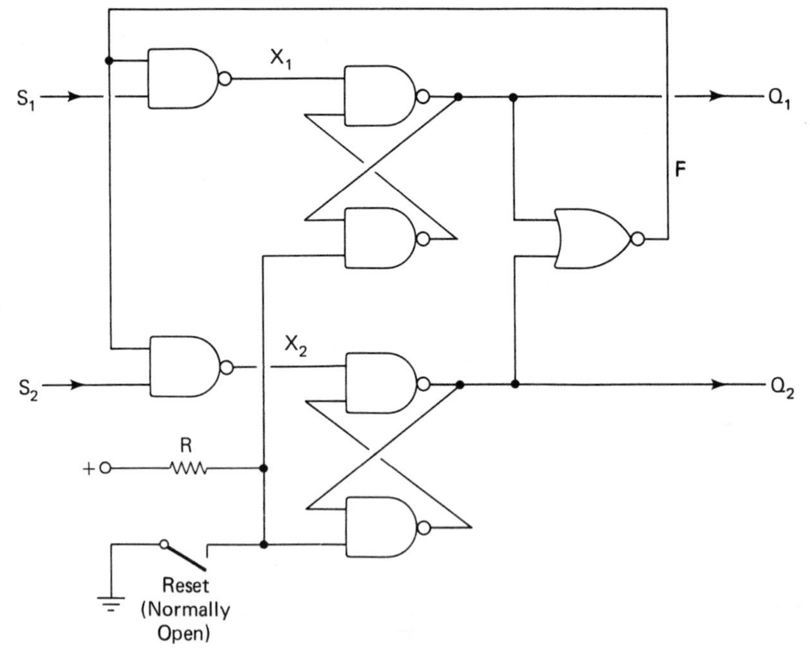

FIGURE P5-10

In that application, the input variables would be derived from photoelectric sensors, and the 'Q' outputs would cause corresponding lamps to glow when set to '1'. Assuming that all gate delays are 10 ns and that the circuit is initially reset, plot a timing diagram from variables S_1, S_2, X_1, X_2, F, Q_1, and Q_2, under the condition where 'S_1' changes from '0' to '1' 100 ns before 'S_2' changes. Under what conditions can 'Q_1' and 'Q_2' both be set at the same time?

5-11 In Figure 5-19, it was shown that the toggle master-slave flip-flop does not behave according to its operation table if the 'T' input is changed while $CK = 1$. Does this also apply to a D-type master-slave flip-flop? Explain.

5-12 Design a master-slave toggle flip-flop, similar to that of Figure 5-18, using two and three input NOR gates only. Be certain to indicate the proper sense for all inputs and outputs (for example, CK or $\overline{CK}$).

5-13 Using a D-type flip-flop and one logic gate, design a circuit that performs exactly as does the circuit of Figure P5-13(a).

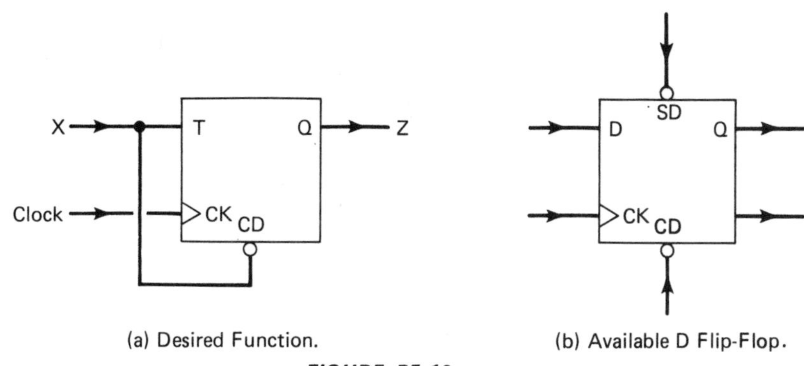

(a) Desired Function. (b) Available D Flip-Flop.

FIGURE P5-13

5-14 A pulse train is available in which the period is 10 μs and the duty cycle is 20%.

(a) For how many microseconds per period is the signal logic '1'?

(b) What is the frequency of this signal?

(c) Design a simple circuit that converts this signal into a square wave whose period is 20 μs.

(d) Design a simple circuit that converts this signal into a square wave whose period is 10 μs.

(e) Assume that your circuit of part (d) is in use and that the period of the input pulse train suddenly changes to 20 μs. Will the duty cycle of the output change? If so, what will it become?

5-15 A J-K edge-triggered flip-flop is available with a set-direct input only. We wish to obtain an identical device, but with a clear-direct input only. Using no logic gates and only the given device, show how the desired operation can be obtained. Refer to Figure P5-15. This problem demonstrates that the meaning of the signal labels on a flip-flop can be simply a matter of interpretation.

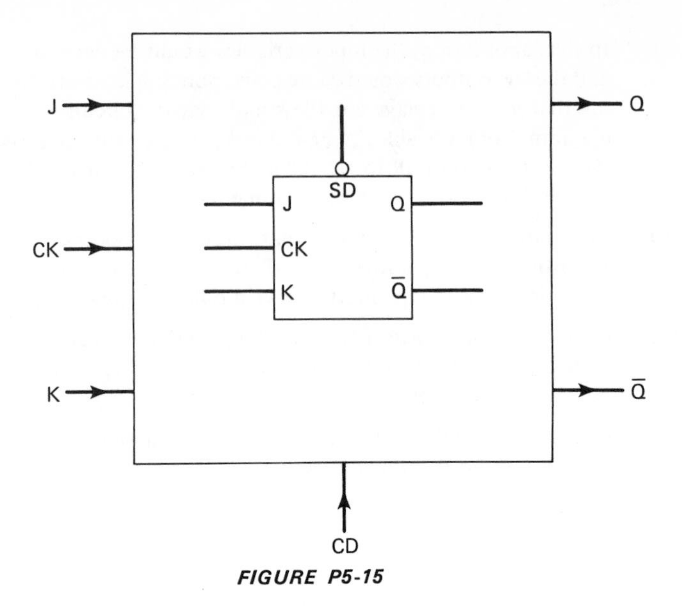

FIGURE P5-15

5-16 Using the device shown in Figure P5-16(a) and any needed logic gates, show how the operation of the devices in Figure 5-16(b) can be obtained.

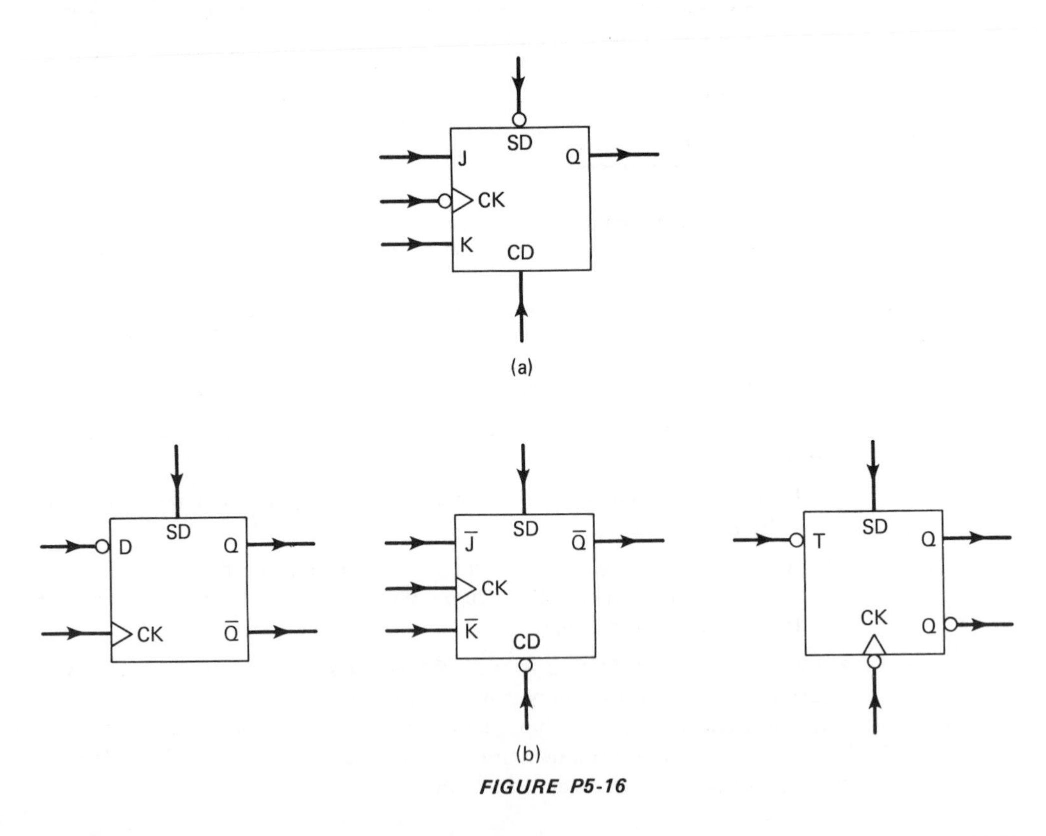

FIGURE P5-16

5-17 **(a)** Using a T-type edge-triggered flip-flop and any type of logic gates, design a circuit that behaves like a D-type edge-triggered flip-flop.

(b) Using a D-type edge-triggered flip-flop and any type of logic gates, design a circuit that behaves like a J-K edge-triggered flip-flop.

5-18 A *ring counter* is a shift register circuit in which the output of the last flip-flop stage is fed back to the input of the first stage. Such a circuit is shown with four stages in Figure P5-18(a).

(a) Describe how this circuit can be used as a frequency divider. Indicate all possible factors by which the '*CK*' input frequency can be divided, and how the initial state of the flip-flops should be set so that these divisions are obtained.

(b) Often, a single '1' is circulated in a ring counter to perform sequencing operations. When system power is turned on, however, it is not likely that only one flip-flop will come on set, and the others reset. Rather, a more random distribution of initial states can be expected. Design a circuit in which an initializing pulse can be applied to the ring counter so that it is preset to the condition $D_0 = 1$, $D_1 = D_2 = D_3 = 0$. Assume that '*SD*' and '*CD*' inputs are available.

(c) Describe the operation of the circuit shown in Figure P5-18(b).

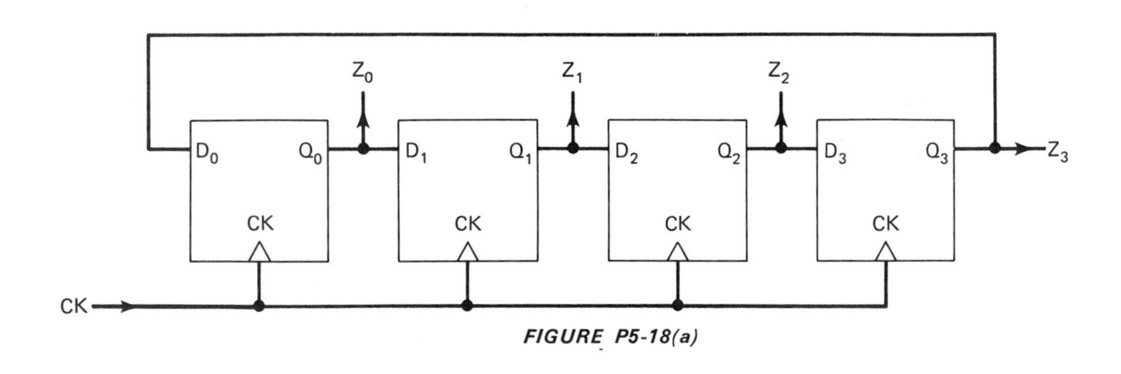

FIGURE P5-18(a)

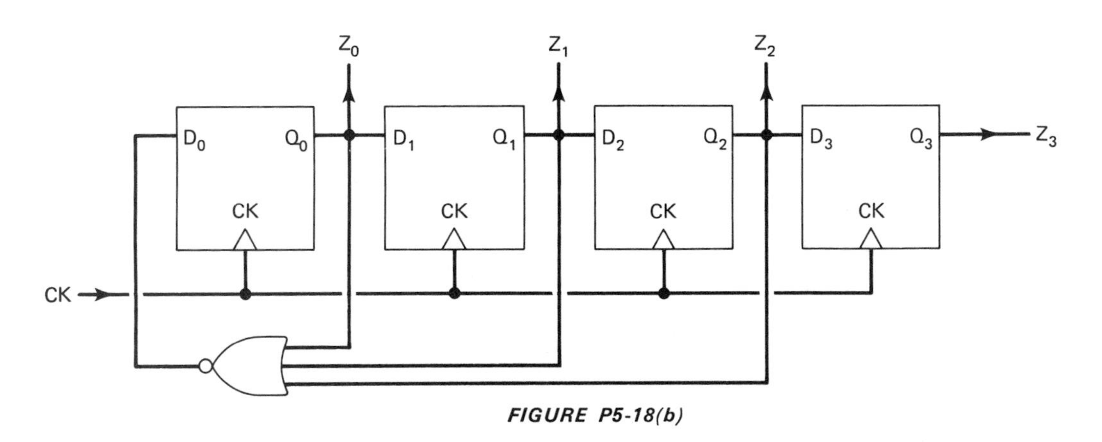

FIGURE P5-18(b)

5-19 Describe the operation of the circuit shown in Figure P5-19.

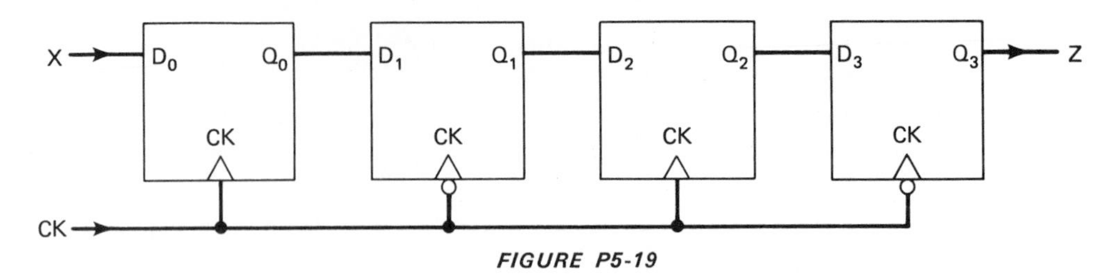

FIGURE P5-19

5-20 It was demonstrated in Figure 5-36 that a three-stage frequency divider can be viewed as a natural binary up counter.

(a) Complement each of the 'Q' outputs in the table of Figure 5-38(b) and note that a natural binary *down* count sequence results. Thus, by simply observing the '$\bar{Q}$' outputs, a natural binary down counter is obtained.

(b) Describe a second way in which the circuit of Figure 5-38(a) can be converted into a natural binary down counter.

5-21 Using two monostable multivibrators with 'Q' and '$\bar{Q}$' outputs, and with periods 'T_1' and 'T_2', respectively, design an astable circuit. Draw a timing diagram to describe the operation of your design. Be sure to denote the astable period in terms of 'T_1' and 'T_2'.

5-22 A Geiger counter is used to measure radioactive decay, and produces an audible "click" whenever radiation of high energy passes through a Geiger tube sensor. Although the rate at which clicks are heard is very random, the *total* number of clicks received over a period of time is a good measure of the surrounding radiation. In this problem, it is necessary to design a digital circuit that monitors the Geiger counter and generates an alarm when the level of radiation exceeds a safe limit. Our criterion for safety will allow for no more than fourteen clicks over a 1-sec. interval. The key components of this design are a four-stage ripple counter and a monostable multivibrator. Assume that the Geiger counter has a digital output and produces one pulse for every audible click. Also assume that radiation will not adversely affect the electronic logic gates.

5-23 Design an astable circuit using only *one* monostable multivibrator, resistor and capacitor components, and a Schmitt trigger gate.

5-24 Explain why the circuit of Figure P5-24 is not a good pulse synchronizer design.

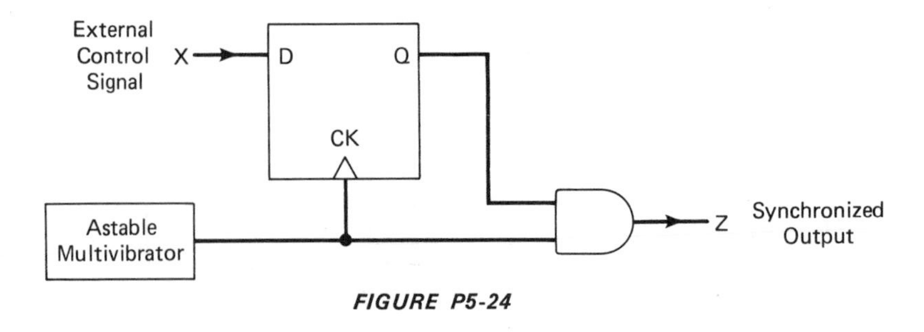

FIGURE P5-24

6 COUNTER ANALYSIS AND DESIGN

6.1 INTRODUCTION

6.1.1 Motivation

Timing, sequencing, and the counting of events are all functions for which a digital circuit may be designed. A timing problem might require that a high-frequency pulse train, such as the output of a 10-MHz crystal oscillator, be divided to produce a pulse train of a much lower frequency, say 1 Hz. This application is required in a precision digital clock, where it is not possible to build a crystal oscillator whose natural frequency is 1 Hz. A sequencing problem would arise if, for instance, it became necessary to apply power to various components of a large machine in a specific order. The starting of a rocket motor is an example where the energizing of fuel pumps, ignition, and possibly explosive bolts for staging must follow a critical order. Measuring the flow of auto traffic on a roadway is an application in which an event (the passage of a vehicle) must increment a tally. This can be done automatically with an electronic counter triggered by a photocell or road sensor. In this way, the total number of vehicles passing a certain point can be counted.

Timing, sequencing, and the counting of events are functions that can be realized using digital counters. Like flip-flops, counters can retain an output state after the input condition which brought about that state has been removed. Consequently, digital counters are classified as sequential circuits. While a flip-flop can occupy one of only two possible states, a counter can have many more than two states. In the case of a counter, the value of a state is expressed as a multidigit binary number, whose '1's and '0's are usually derived from the outputs of internal flip-flops that make up the counter. It is reasonable to believe, and indeed a fact, that the number of states a counter may have is limited only by the amount of electronic hardware that is available. As a result of the many possible states a counter may have, the complexity and variety of counter circuits can be very great.

6.1.2 Count Sequences

A *state* is thought of generally as a particular internal condition that a circuit may have. A state is usually expressed as a binary number, or equivalently as a string of '1's and '0's. The binary digits that describe a state normally correspond to the outputs of internal flip-flops. As clock pulses are applied to a digital counter, we can observe transitions from one state to another, the exact order of which defines a ***count sequence***. In most of the counter circuits that will concern us, there will be exactly one binary digit in the state value for every flip-flop in the circuit. A counter circuit with three flip-flops, for example, is described by three-digit states. Thus, given the number of flip-flops present in the circuit, we can decide the maximum possible number of states. In the case of a counter circuit that contains three flip-flops, for instance, we know that there can be at most eight unique states. They are

0 0 0   0 0 1   0 1 0   0 1 1

1 0 0   1 0 1   1 1 0   1 1 1

A ***flow graph*** describes a particular count sequence and is composed of states taken from the pool of all possible states. A typical count sequence is that of a three-bit natural binary down counter and is described by the following flow graph:

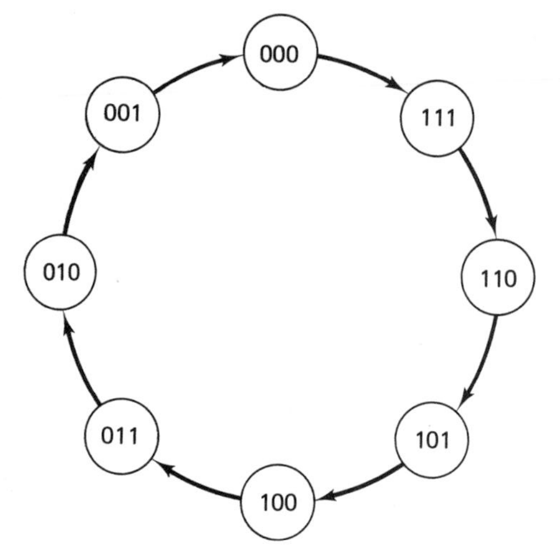

It is valid to have a count sequence that has fewer than the maximum number of states. One possibility is

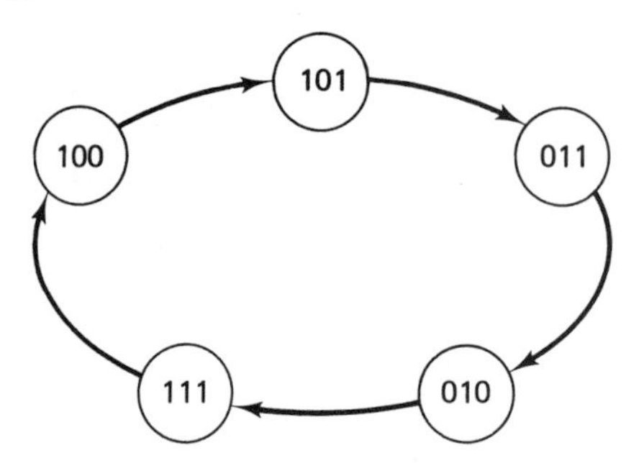

The outputs of a counter may also be applied to combinational logic of a specific design to translate one count sequence into another. Thus, it is possible to obtain the following variation of a basic three-digit, eight-state natural binary up counter:

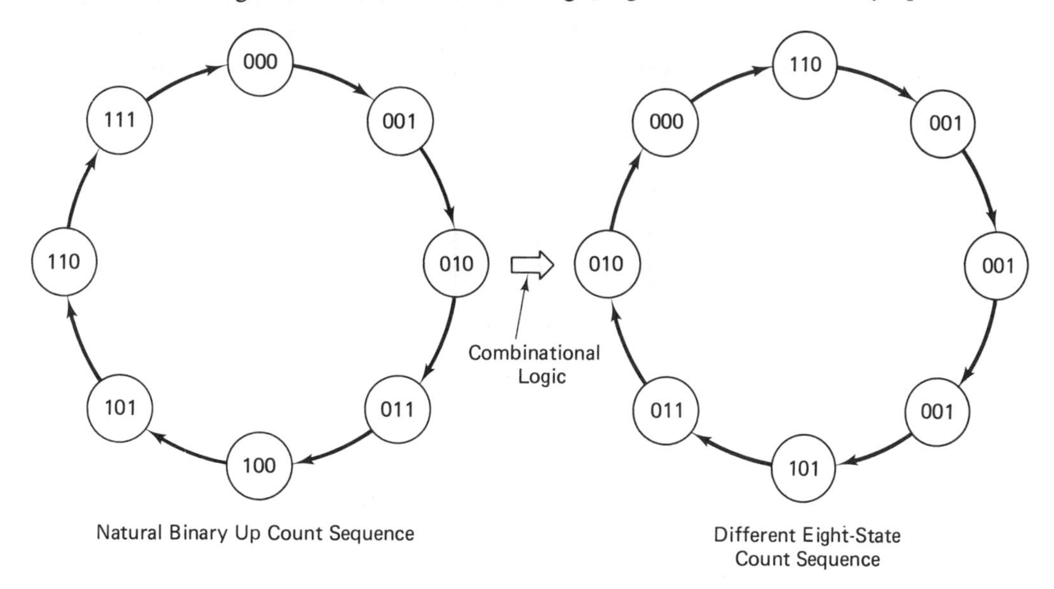

Whenever we *analyze* a counter circuit, one of the important objectives will be to determine the count sequence, if it is not already available. Also, if a counter is malfunctioning, important clues to the problem can often be found by comparing the desired count sequence to that actually observed. In the *design* of a counter circuit, a count sequence will be given, and we must determine a minimal circuit that realizes this sequence. For both analysis and design, we will use a flow graph to express simply the nature of a digital counter, even though the circuit details are not revealed by it.

6.1.3 General Description

Digital counters are usually made up of a cascaded interconnection of flip-flops, as we have seen in the ripple counter of Chapter 5. Any number of flip-flops may be involved in a counter, as demanded by the counter's application. In some cases, many counter stages are necessary to provide a long count sequence, while in other cases, only several stages are required. Any counter, whatever its function, is provided with a "clock" input to which a digital pulse train can be applied. When the counter receives a clock pulse, it is caused to advance from its present state to the next state in the count sequence. Applying clock pulses in rapid succession, as with a pulse train, will cause the counter to quickly move from state to state and to cycle through its count sequence. The basic digital counter can be represented as a block element, illustrated in Figure 6-1. Optional inputs that may be found on digital counters include:

1. ***Parallel load*** inputs—such inputs permit the counter to be preset to an initial state, prior to beginning the count sequence. These inputs usually include one

data line for each flip-flop, and a control line to place the counter into either the parallel load mode or the count mode.

2. An *enable* input—this permits the counter to either proceed with the count sequence or to freeze at the present state regardless of clock pulses.
3. A *clear* input—when asserted, the clear input forces the counter into the zero state (all flip-flops reset).

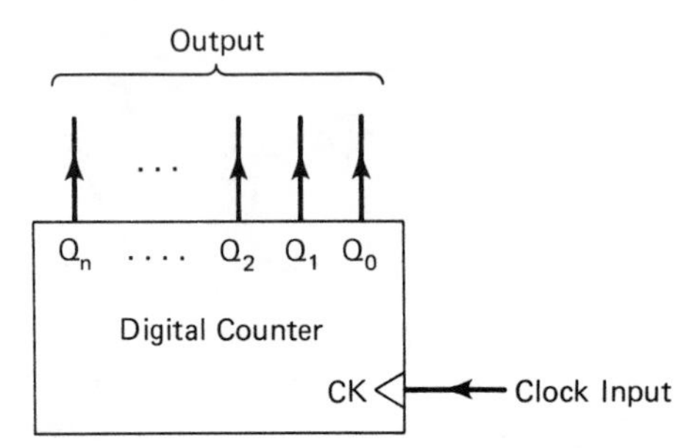

FIGURE 6-1 *Basic digital counter, shown as a block element. In this counter, there are a total of n + 1 stages.*

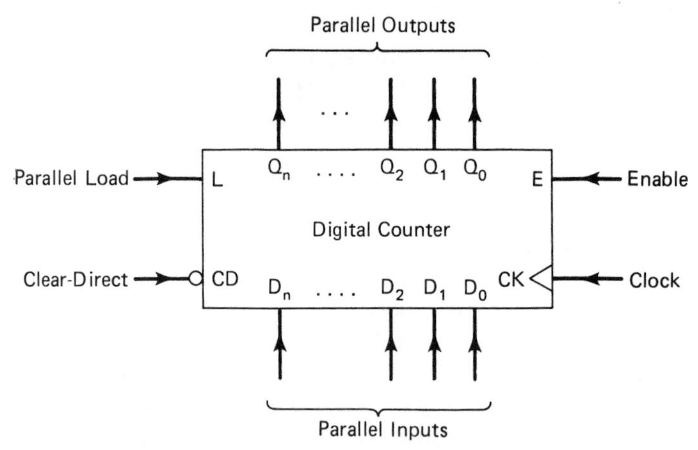

FIGURE 6-2 *Digital counter with additional features.*

A counter with these features is shown as a block element in Figure 6-2.

With this simple introduction, we have gained a basic understanding of counters, from a block element point of view. We will now delve into the circuitry that makes counters possible and study several interesting examples in which counters can be applied. Although counters can be purchased as whole modules from manufacturers, in accordance with the block element description of Figure 6-2, much is to be gained by examining the details of counter design.

6.2 RIPPLE COUNTERS

6.2.1 Analysis

Our earliest notion of a counter circuit came in Section 5.3, where we saw how a frequency divider can be viewed as a natural binary up counter. This very simple form of counter circuit is referred to as a ripple-carry counter, or a "ripple counter" for short. The name is derived from the fact that a triggering edge must pass from stage to stage when a clock pulse is applied. The resulting transition time necessary for the counter to stabilize at its new state can vary from one flip-flop propagation delay (example: 0000 ⟶ 0001) to several or more flip-flop propagation delays (example: 1111 ⟶ 0000). Figure 6-3 depicts a three-stage ripple counter and illustrates how propagation delay through the flip-flop stages affects the output.

Two problems arise as a result of the skew effect in ripple counters. First, the speed of a ripple counter is limited by *the sum* of all flip-flop propagation delays. That is, the amount of time between clock pulses must be long enough to guarantee that all flip-flops are in their proper state. This amount of time increases in direct proportion to the number of stages in the counter. Thus, ripple counters with many stages would necessarily be very slow when compared to the propagation delay of a single stage.

The second problem may occur whenever a particular state is detected. Note that the desired flow graph for a three-digit natural binary up counter consists of eight states following the natural binary count sequence. However, upon studying the detailed timing diagram of Figure 6-3, we see that several erroneous states are momentarily entered while the propagation delays are in progress. This is illustrated in Figure 6-4. Although the duration of these erroneous states is usually small when compared to the clock period, they may falsely trigger other circuit elements connected to the counter. The counter's application determines whether or not these transient states have an undesirable effect. If incandescent lamps were being driven by the counter outputs, then a transient state lasting several microseconds (or less) could not possibly cause the lamp filaments to glow; this normally takes several tens of milliseconds. However, if a three-input NOR gate is used to detect the '000' state, an erroneous output would result whenever each of the two momentary '000' states is entered. This is illustrated in Figure 6-5. In particular, whenever an output state is detected in a ripple counter, extreme caution must be used to ensure that

1. The decoded state is not also falsely detected as a transient state.
2. If such a transient state is detected, it does not adversely affect the operation of other circuit elements.

It may be that the NOR detector of Figure 6-5 drives only an incandescent lamp, in which case the transients would never be seen. However, if the detector was driving a clocked flip-flop, or another counter, the false outputs would surely cause false triggering. This problem can be avoided by synchronizing the detector with the clock input. Because of their restricted speed, and the detection problem, ripple counters

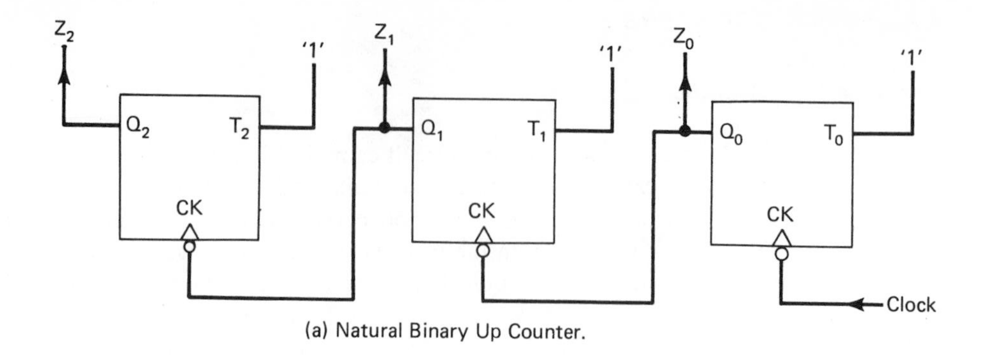

(a) Natural Binary Up Counter.

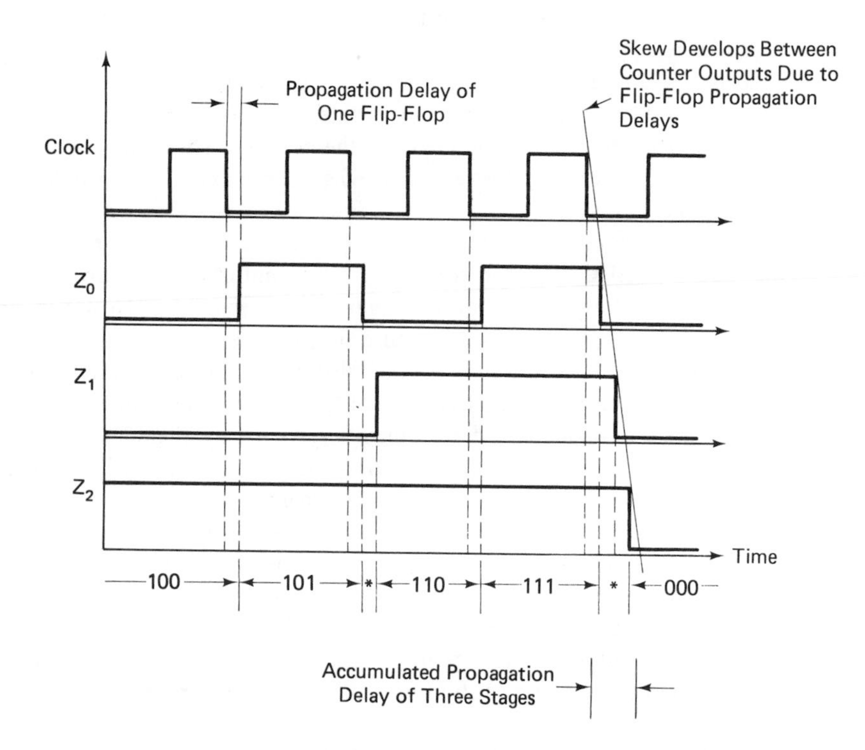

(b) Timing Diagram for the Binary Up Counter, Showing How Propagation Delay Accumulates Causing a Skew to Develop Between the Counter Outputs. The State of the Counter After Every Clock Pulse is Shown Below the Waveforms. For Example, for $Z_2 = 1$, $Z_1 = 0$, and $Z_0 = 0$, the State is '100'. Erroneous States are Indicated with an Asterisk (*).

FIGURE 6-3

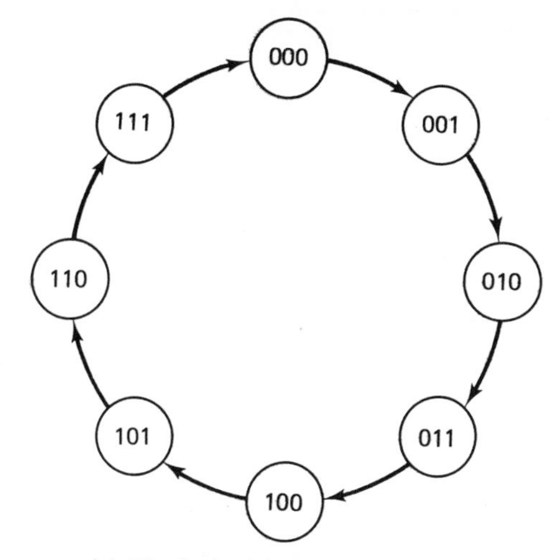

(a) The Desired Count Sequence for a Natural Binary Up Counter.

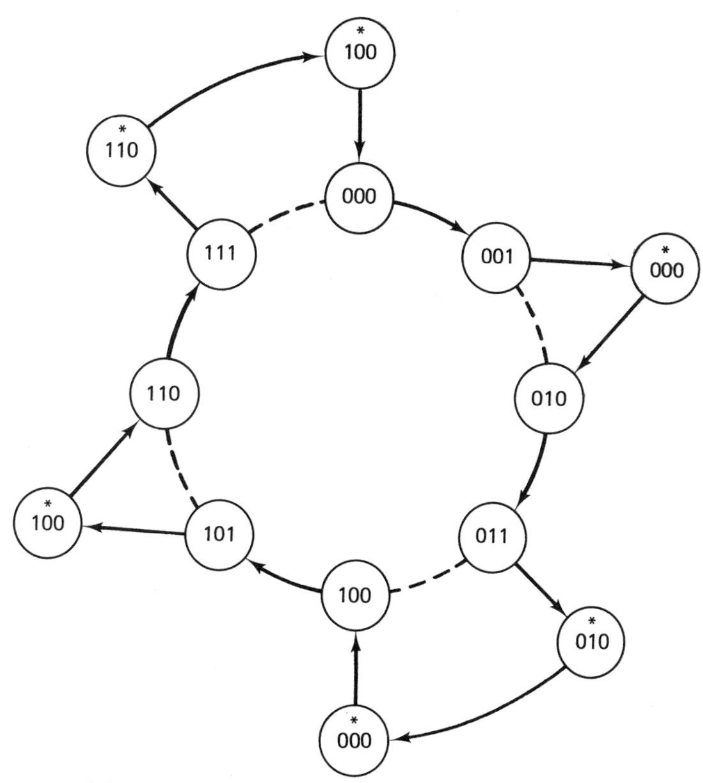

(b) The Actual Count Sequence for a Ripple Counter. Each State Marked with an Asterisk (*) is a Transient State Lasting No More than one Flip-Flop Propagation Delay.

FIGURE 6-4

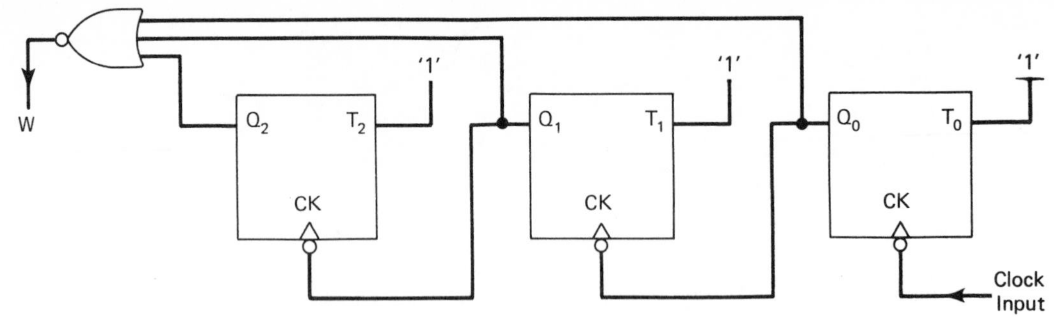

(a) Ripple Counter with '000' Detector.

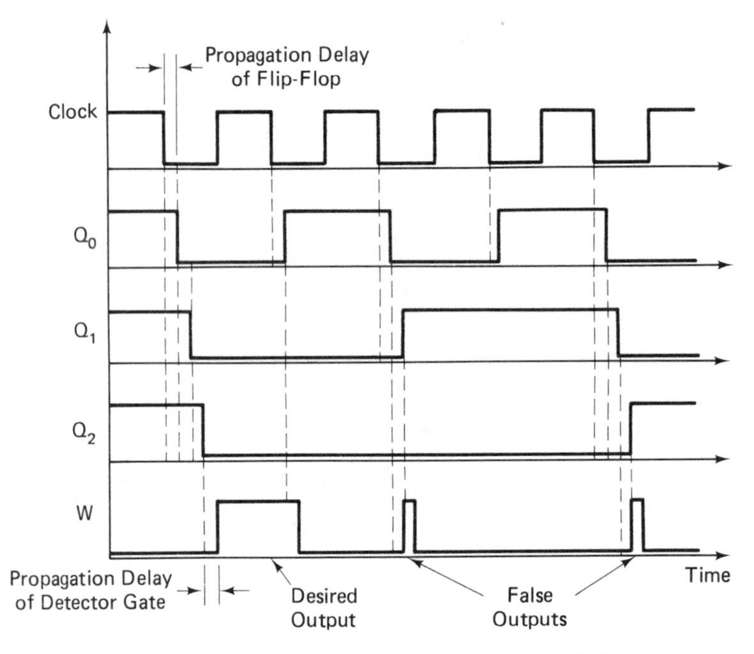

(b) Waveforms Showing False Output Signals from the '000' Detector.

FIGURE 6-5

FIGURE 6-6 *Circuit connections for a down counter.*

Z_2 Z_1 Z_0

'1' '1' '1'

Q_2 T_2 Q_1 T_1 Q_0 T_0

CK CK CK

Clock Input

Note the Absence of Inversion Circles

have very limited usefulness. One exception, though, is frequency division by integer powers of 2, which does not require state detection. We shall see in Section 6.3 how synchronous counters can be designed to avoid the speed and output skew limitations of ripple counters.

6.2.2 Design

Ripple counters normally take only one of two possible forms: either natural binary up counters, or natural binary down counters. Consequently, the design problem is greatly simplified. In each case, only toggle flip-flops are required, and no external combinational logic is needed. In the up-counter version, the 'Q' output of one stage must drive an *inverted* 'CK' input on the next stage, as we have previously shown in Figure 6-3. The down-counter version is different in that the 'Q' output of one stage drives a *noninverted* 'CK' input on the next stage. A down-counter circuit is illustrated in Figure 6-6.

6.2.3 An Application

An electronic musical instrument is required to produce a wide selection of musical tones using an oscillator bank or other means. In the case of an electronic organ, the keyboard may span as many as seven octaves, or more, of distinct pitches varying in frequency between about 30 Hz and 4000 Hz (these are the fundamental frequencies; overtones can go much higher). Illustrated in Figure 6-7 are two middle-range octaves of the musical scale, as viewed from the keyboard of an electronic organ. Observe in Figure 6-7 that exactly twelve notes are included in one octave, and that the frequencies of the higher octave are exactly twice the frequencies of corresponding lower octave notes (this would be true even if the frequencies were not rounded to whole numbers). Thus, we see that A_0 is 440 Hz and that A_1 is 880 Hz. This exact 1:2 relationships between one octave and the next highest octave can be very effectively applied

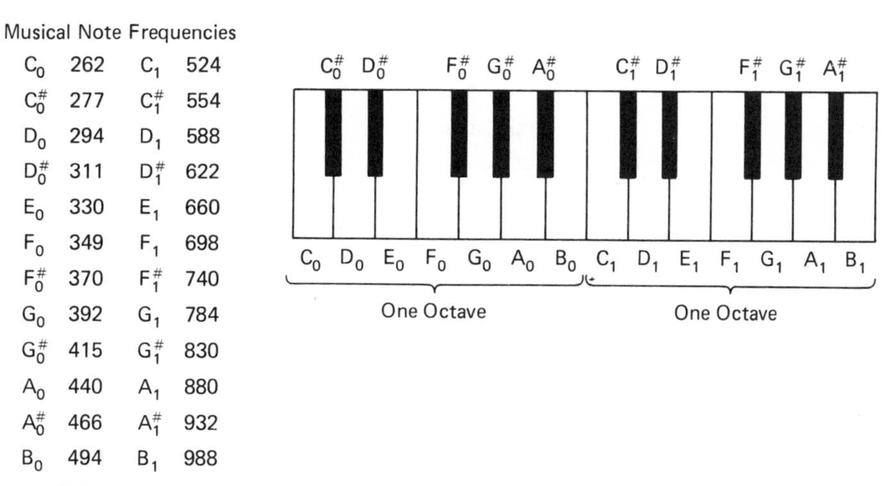

Musical Note Frequencies

C_0	262	C_1	524
$C_0^\#$	277	$C_1^\#$	554
D_0	294	D_1	588
$D_0^\#$	311	$D_1^\#$	622
E_0	330	E_1	660
F_0	349	F_1	698
$F_0^\#$	370	$F_1^\#$	740
G_0	392	G_1	784
$G_0^\#$	415	$G_1^\#$	830
A_0	440	A_1	880
$A_0^\#$	466	$A_1^\#$	932
B_0	494	B_1	988

FIGURE 6-7 Two middle-range octaves of the musical scale, and the corresponding frequency of each note. The frequencies are rounded to whole numbers for simplicity.

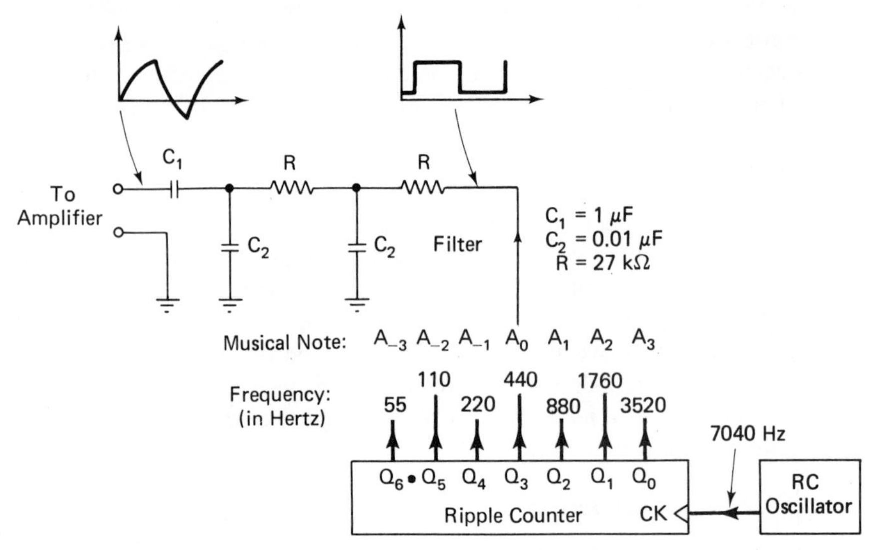

(a) Typical Oscillator-Frequency Divider Network Used to Produce all *'A'* Frequencies. For Each of the Other Eleven Notes of the Upper Octave, the Oscillator Frequency has a Different Value. A Typical Filter Network is Shown that Converts the Square Wave into a More Pleasant Sounding Signal.

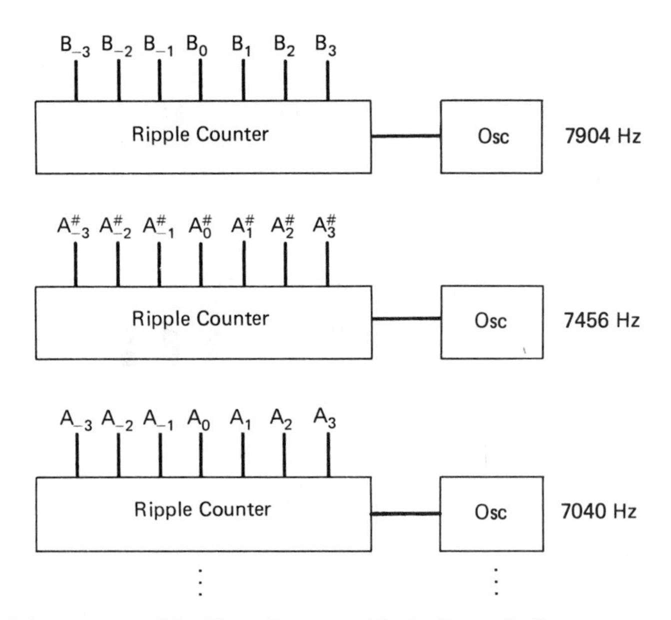

(b) Arrangement of Oscillator-Frequency Divider Networks Necessary to Produce Seven Octaves of the Musical Scale. Three of Twelve Networks are Shown.

FIGURE 6-8

in the design of an electronic musical instrument. In particular, the fundamental frequencies of the highest octave can be generated using twelve *RC* oscillators, and for each of the twelve high-octave frequencies, a ripple counter can be used as a frequency divider to generate the remaining notes of lower octaves. The square waves produced by the digital citcuitry can be filtered and amplified to produce pleasant-sounding musical tones. This scheme is illustrated in Figure 6-8. Other more sophisticated methods of digitally producing and modifying musical tones are possible, and involve digital-to-analog conversion. Some of these ideas will be presented in Chapter 9.

6.3 SYNCHRONOUS COUNTERS

6.3.1 Analysis

The output skew problem associated with ripple counters can be very serious if certain states need to be detected in the count sequence. ***Synchronous counters*** have been developed to overcome this problem. In synchronous counters, all component flip-flops are caused to change state at the *same time*, regardless of where in the cascade they are located. Thus, all flip-flops in in a synchronous counter *must* be triggered from the same clock line. Since all flip-flops are triggered on every clock pulse, added combinational logic is necessary to assert, or deassert, the flip-flop inputs (not necessarily T-type) at the appropriate times. These ideas are demonstrated in Figure 6-9, where a three-stage synchronous up counter is shown. The combinational logic is chosen so that for any stage, its toggle input is asserted only *when the 'Q' outputs of all previous stages are* **'1'**. This rule follows from study of the natural binary count sequence, for example the three-digit sequence of Figure 6-9(b). Here, we see that a given stage should *complement* its present value whenever all previous stages are in the '1' state. Based on this idea, a general equation can be written to describe the needed combinational logic for an '*m*'-stage synchronous counter designed with toggle flip-flops. For the '*n*'th stage,

$$T_n = Q_{n-1} \cdot Q_{n-2} \dots Q_2 \cdot Q_1 \cdot Q_0, \qquad 1 \leq n \leq m-1 \qquad \textbf{(6-1)}$$

For the '0'th stage, 'T_0' is always '1'. Using Eq. 6-1, the equations for a five-stage synchronous up counter are determined.

$$T_0 = 1$$
$$T_1 = Q_0$$
$$T_2 = Q_1 \cdot Q_0$$
$$T_3 = Q_2 \cdot Q_1 \cdot Q_0$$
$$T_4 = Q_3 \cdot Q_2 \cdot Q_1 \cdot Q_0$$

If the counter will not be used for extremely high speed operation,* these equations can be minimized as follows.

* See Problem 6-6.

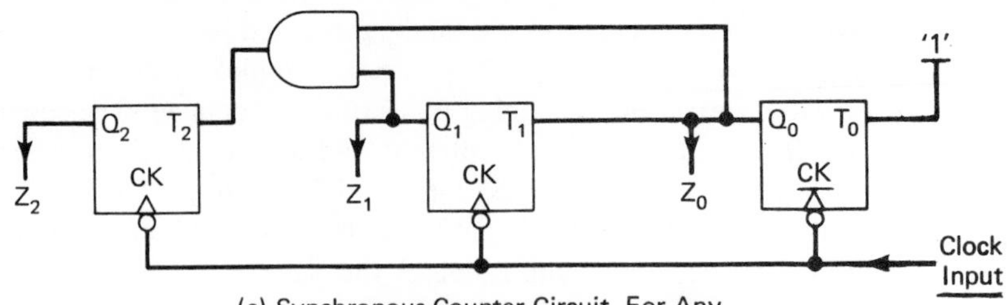

(a) Synchronous Counter Circuit, For Any Stage, a Toggle Input is Asserted Only When All Previous 'Q' Questions are '1'.

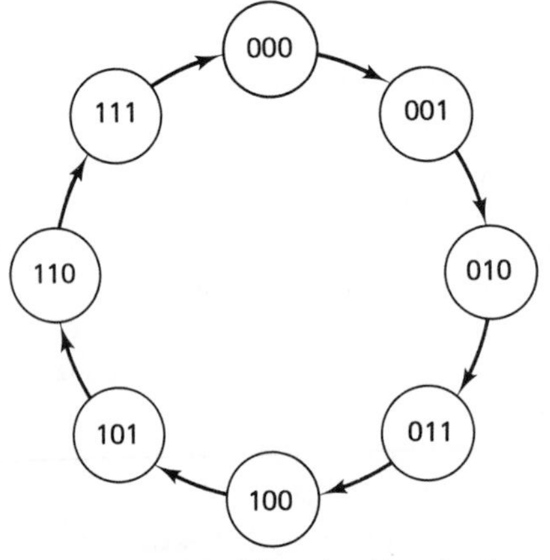

(b) Actual State Transitions that Describe the Synchronous Counter, Assuming that all Flip-Flop Delays are Equal. Note that no False States are Present.

Present State			Excitation for Next State		
Q_2	Q_1	Q_0	T_2	T_1	T_0
0	0	0	0	0	1
0	0	1	0	1	1
0	1	0	0	0	1
0	1	1	1	1	1
1	0	0	0	0	1
1	0	1	0	1	1
1	1	0	0	0	1
1	1	1	1	1	1

(c) Table Showing How Flip-Flops are Asserted to Reach the Next State.

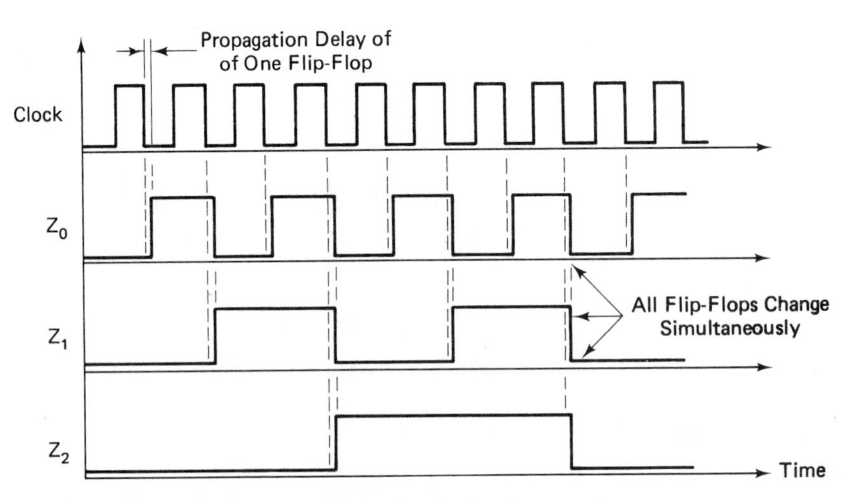

(d) Timing Diagram for the Synchronous Counter.

FIGURE 6-9

$$T_0 = 1$$
$$T_1 = Q_0$$
$$T_2 = Q_1 \cdot Q_0$$
$$T_3 = Q_2 \cdot T_2$$
$$T_4 = Q_3 \cdot T_3$$

This permits the use of two-input gates and allows the design of a ***universal synchronous counter stage***. Figure 6-10 shows the minimized five-stage design, and Figure 6-11 illustrates the idea of a universal stage. In the universal counter stage design, 'T_{n-1}' is often referred to as the ***carry input*** and 'T_n' as the ***carry output***.

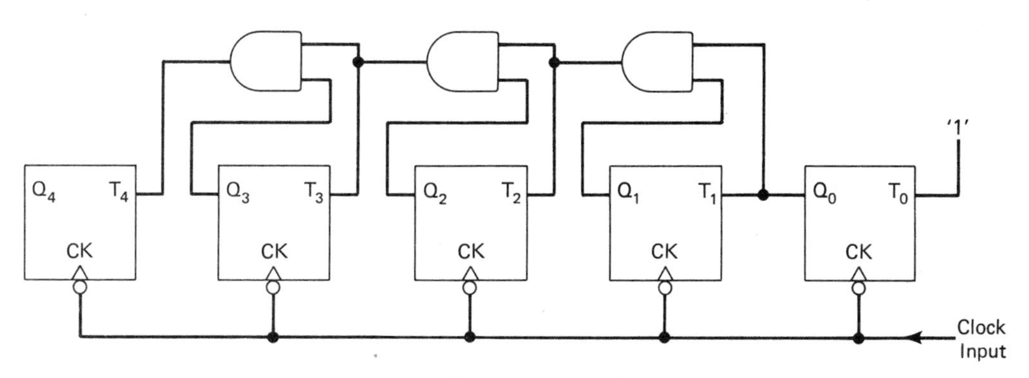

FIGURE 6-10 *Five-stage synchronous counter.*

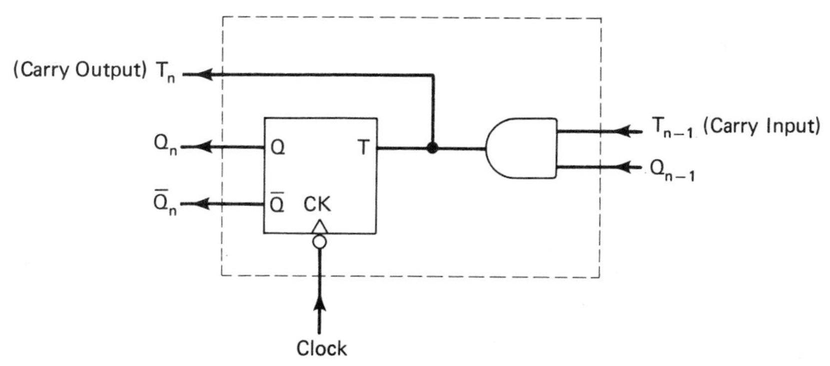

FIGURE 6-11 *Universal counter stage design.*

It is interesting and important to note that any number of counter stage blocks can be assembled into a complete counter, the whole of which itself can be considered as a block element. Actually, this is often the way manufacturers provide counter circuits, that is, as multistage counter modules that can be interconnected to form larger counters of any length. For example, synchronous counter modules of four or eight stages are usually available. Figure 6-12 depicts a four-stage counter module

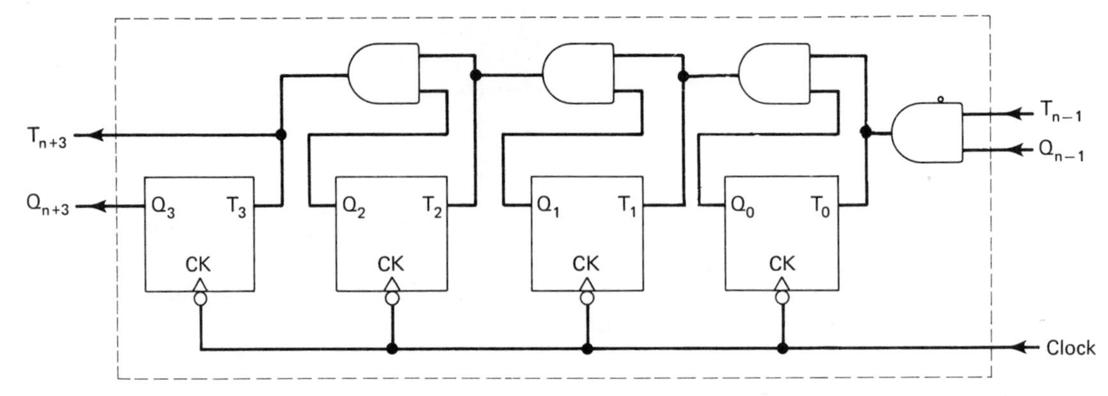

FIGURE 6-12 *Four-stage counter module design.*

design. As with the manufacture of flip-flop modules, counter modules are subject to electronic simplification when designed as a single semiconductor circuit. Thus, the logic schematics that are drawn for these modules can only be approximations. Whatever the actual circuit, however, the function is identical.

6.3.2 Design

Thus far, all of the counter circuits that we have discussed have realized ***maximal length*** natural binary sequences. Our three-stage synchronous counter of Figure 6-9, for instance, possessed eight states, the maximum number possible. Frequently, it is necessary to have counters with fewer than the maximum number of states, but more states than could be had with one less flip-flop. The five-state sequence of Figure 6-13 is an example. It is our object in this section to investigate methods of design that give us greater control over selecting the length of the count sequence.

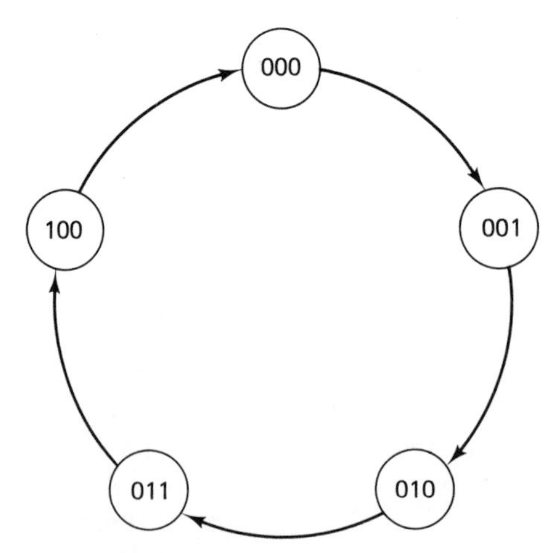

FIGURE 6-13 *Five-state count sequence.*

In the last section, we discovered a feature of synchronous up counters that can be readily applied to their design. In particular, we found that for a given counter stage, the proper toggle excitation for that stage results from the AND of all 'Q' outputs of previous stages. Using only the needed AND gates in a counter design, such as in Figure 6-10, the simplest form of synchronous counter results and produces a maximal length natural binary sequence. After the highest binary number in the count sequence is reached, the beginning state of the count sequence is reentered and the cycle begins over. To have a count sequence with fewer than the maximum number of states, it is necessary to

1. Detect the last *desired* state in the count sequence.
2. Intervene with the flip-flop excitations so that the proper next state is entered.

In effect, we are "short-circuiting" the count sequence, as illustrated in Figure 6-14.

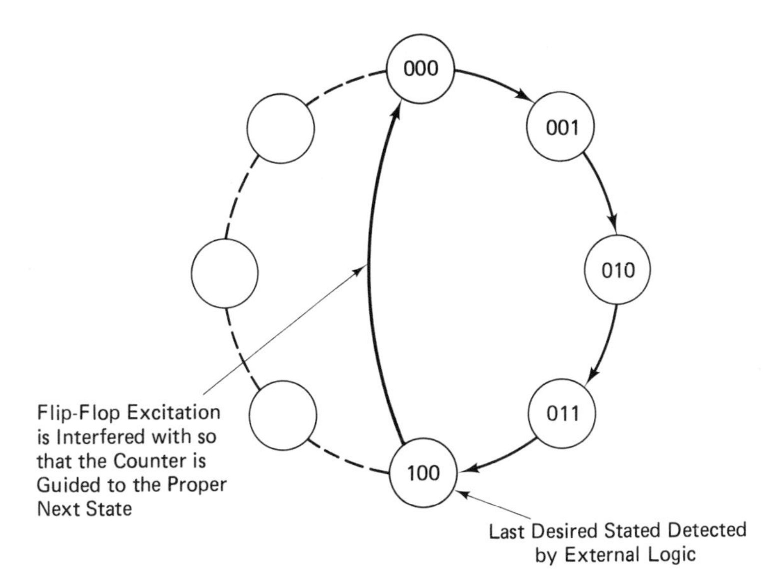

FIGURE 6-14 *The five-state sequence of Figure 6-13 is shown as a shortened maximal-length sequence.*

To actually implement the five-state sequence of Figure 6-14, we must detect the state '100' and then interfere with the natural excitation of the 'T' inputs so that the transition 100 ⟶ 000 is realized. The detection of '100' results by realizing the equation

$$Z = Q_2 \cdot \bar{Q}_1 \cdot \bar{Q}_0 = \overline{\bar{Q}_2 + Q_1 + Q_0}$$

which can be done with a NOR gate and one inverter (if '$\bar{Q}$' outputs are available on the flip-flops, no inverters are necessary). In guiding the flip-flops from the detected state to the next desired state, we must compare the natural transition that would occur if no logic were added, to the transition that is desired. Schematically,

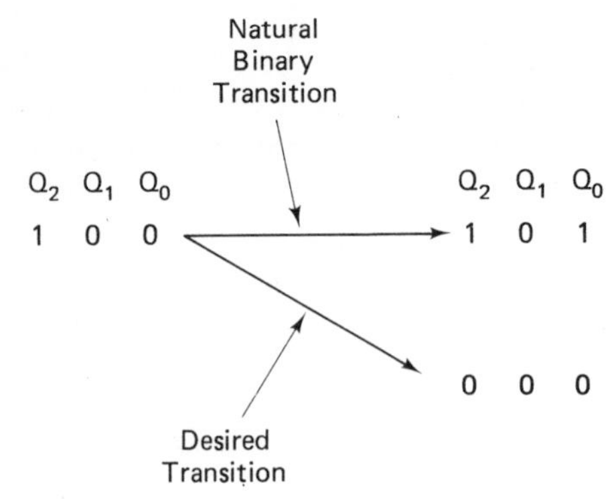

First, for 'Q_0' we argue:

If we do not interfere with 'T_0', then 'Q_0' will naturally complement, from $Q_0 = 0$ to $Q_0 = 1$. However, we want 'Q_0' to remain '0'. Therefore, when the last state in the sequence is detected, 'T_0' must be forced to '0' to prevent complementing on the next clock pulse. This is accomplished by applying an AND function to whatever normally feeds 'T_0', and logic '0'. Thus, the following circuit is obtained:

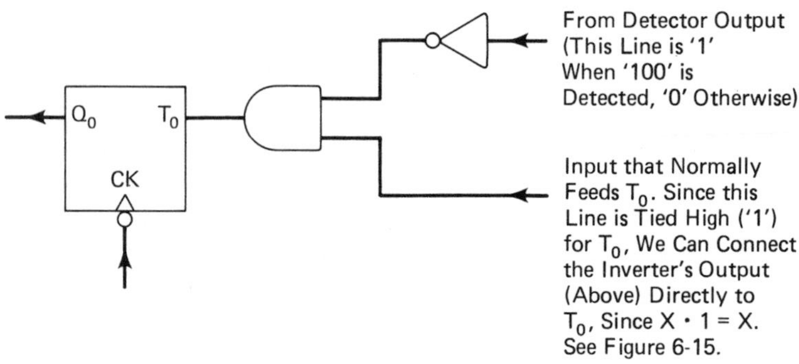

Second, for 'Q_1' we argue:

If we do not interfere with 'T_1', then 'Q_1' will remain '0'. Since 'Q_1' should remain '0' in the desired transition, no action need be taken to change the normal input of 'T_1'.

Finally, for 'Q_2' we argue:

If we do not interfere with 'T_2', then 'Q_2' will remain '1'. However, we want 'Q_2' to complement from $Q_2 = 1$ to $Q_2 = 0$. Therefore, when the last state in the sequence is detected, 'T_2' must be forced to '1' to cause the complementing of 'Q_2' on the next clock pulse. This is accomplished by applying an OR function to whatever normally feeds 'T_0', and logic '1'. Thus, the following circuit is obtained:

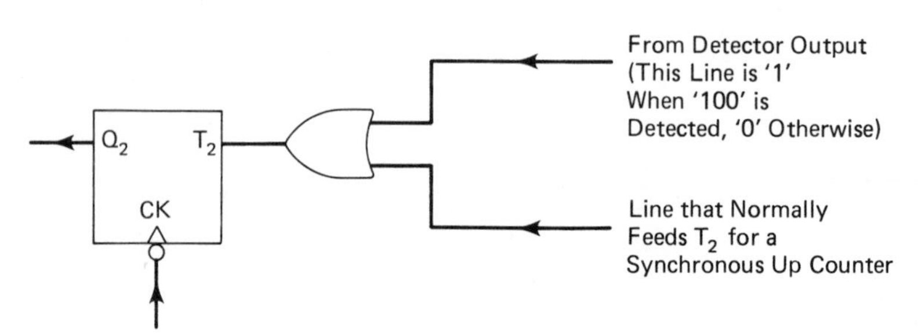

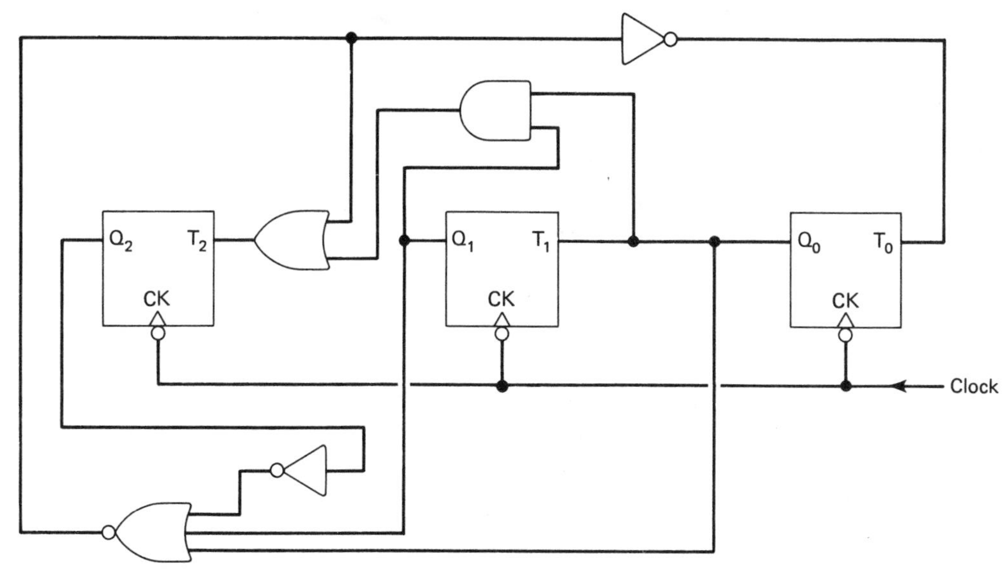

FIGURE 6-15 *Five-state synchronous counter design. It is assumed that the flip-flops do not have '$\bar{Q}$' outputs.*

Our final design for the five-state synchronous counter is shown in Figure 6-15.

The technique that was used in this design is referred to as the ***detect and steer*** method, and will be of great use to us in designing arbitrary length natural binary up *and* down counters. In summary, the detect and steer method requires that

1. We always start with a synchronous natural binary counter.
2. The last desired state of the count sequence be detected.
3. Each flip-flop be either
   (a) Forced to deviate from its otherwise natural transition, or
   (b) Left alone, because the natural transition happens to correspond to the desired transition.

Forcing a flip-flop to deviate from its otherwise natural transition is accomplished by either

1. Applying a '1' to an otherwise '0' value 'T' input, using an OR function, to induce the 'Q' output to complement on the next clock pulse, or
2. Applying a '0' to an otherwise '1' value 'T' input, using an AND function, to prevent the 'Q' output from complementing on the next clock pulse.

Whenever '$\bar{Q}$' outputs are available, they should be used as needed to minimize the amount of external logic.

As a second example, consider the count sequence shown in Figure 6-16. The design process is begun by first noting where the natural binary sequence is broken.

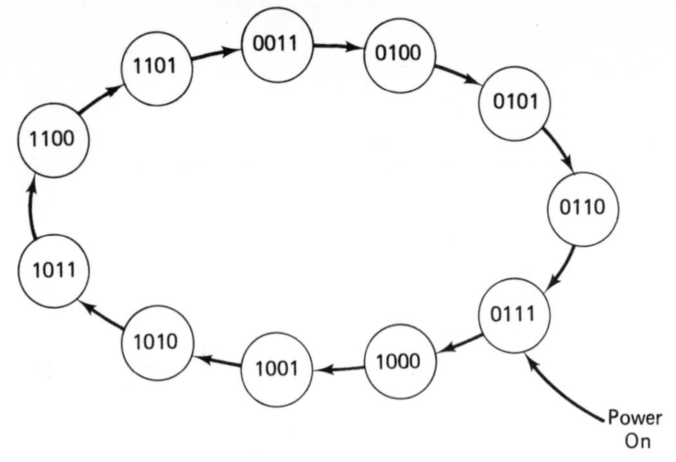

FIGURE 6-16 *Eleven-state count sequence.*

We see that this occurs in the transition 1101 ⟶ 0011. Thus, the state '1101' must be detected. A transition schematic is drawn to determine which flip-flops should be forced and which may be left alone.

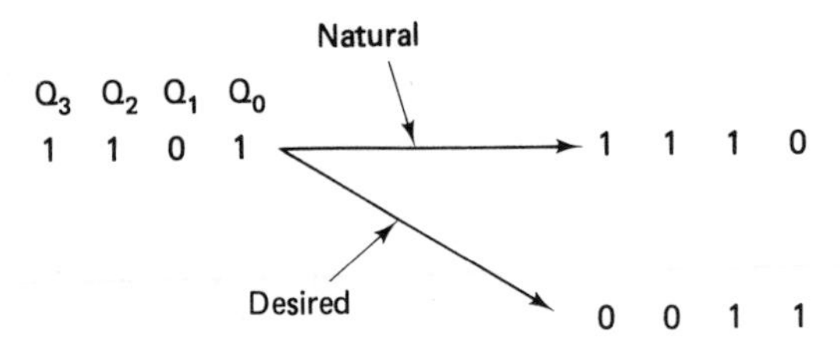

From this transition schematic, we determine that

1. 'Q_0' should be forced to *not* complement.
2. 'Q_1' need not be affected.
3. 'Q_2' should be forced to complement.
4. 'Q_3' should be forced to complement.

According to these decisions, the circuit of Figure 6-17 is designed.

6.3.3 Applications

Precision monostable: To review briefly, a monostable multivibrator is a two-state device in which one state is stable and the other is momentary. The momentary state is entered by the application of a trigger pulse (or edge), and its duration is controlled by timing components (a resistor and capacitor) that are added externally. Unfortunately, the resistance and capacitance of the timing components vary slightly with temperature and age, among other things. Thus, it would not be reasonable to expect a long-term stability of more than several percent in the monostable's period. For certain applications, such a variation in period would be unacceptable, and a better timing circuit must be found. In this application problem, we will design a digital circuit utilizing counters, flip-flops, and a crystal oscillator, that operates

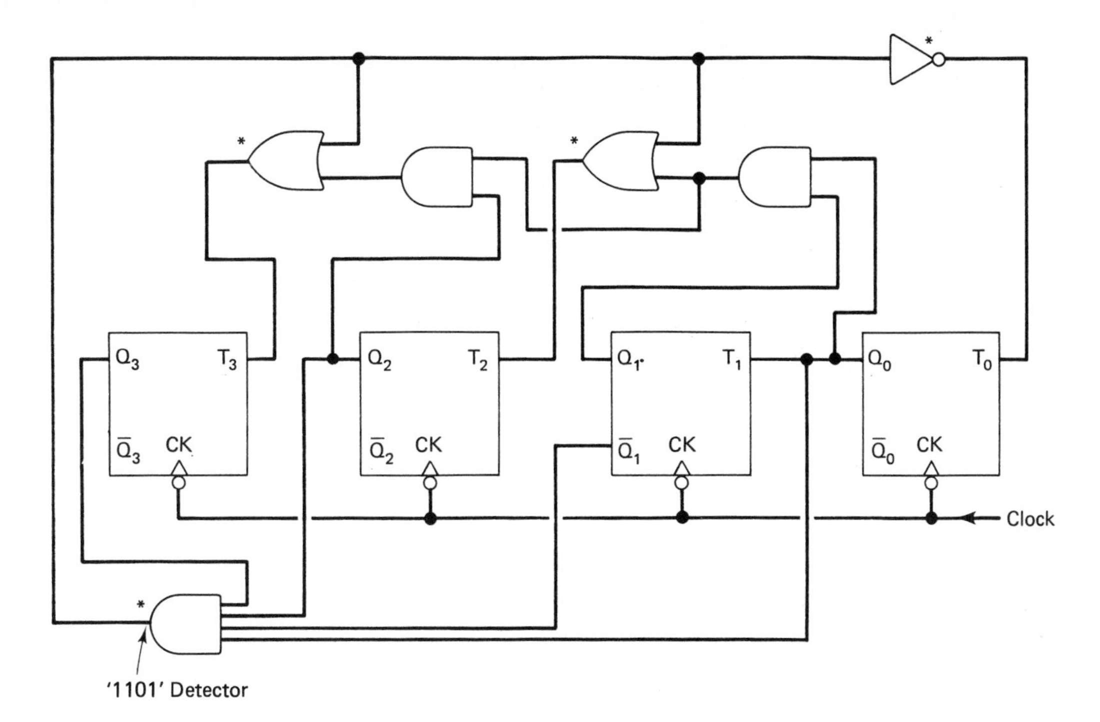

FIGURE 6-17 Circuit design for the eleven-state count sequence of Figure 6-16. Asterisks (*) mark logic gates that were added to a natural binary up counter to achieve the desired eleven-state sequence.

exactly like a nonretriggerable monostable, and whose period will be maintained within $\pm 0.005\%$ of the design value. This circuit will be referred to as a *precision monostable*.

A 2-MHz crystal oscillator is available as a module and produces a square wave that is electrically compatible with the digital circuitry that will be used. The long-term stability of the crystal oscillator is specified to be $\pm 0.005\%$ over a temperature range of $-30°C$ to $+60°C$. Flip-flops and logic gates of any type are available in any number; however, we wish to use a minimum of components. No counter modules are allowed in the design. The period of the precision monostable is to be *one second*.

The design effort consists of two basic problems:

1. Determining a counter circuit that accumulates a count of 2 million, so that one cycle through the full count sequence occurs for every 2 million input periods from the crystal oscillator.
2. Adding suitable control logic so that
   (a) The count sequence is begun on the rising edge of a "trigger" input (this will be the '*TR*' input of the precision monostable).
   (b) The counting process is not disturbed if additional rising edges are applied to the trigger input (this feature makes the precision monostable *nonretriggerable*).
   (c) The count sequence halts when a count of 2 million is reached.

If '2 000 000' were an integral power of '2', our counter design would be extremely simple, since it would reduce to a large ripple counter. Unfortunately, the closest we can come is '2^{21}', which is equal to '2 097 152'. Thus, some form of synchronous counter must be used to allow a count sequence length that is a nonintegral power of two. It is interesting to note, however, that we can divide '2 000 000' by '2' a number of times, using a ripple counter, *before* a more complex synchronous counter is required. Thus, by reducing the 2-MHz crystal frequency with a frequency divider (ripple counter), many fewer stages will be required in the more complex synchronous counter design. To determine the maximum permitted amount of frequency division, the following table is written:

Stage Number	*Output Frequency*
(input)	2 000 000 Hz
0	1 000 000
1	500 000
2	250 000
3	125 000
4	62 500
5	31 250
6	15 625 ←
7	7 812.5
8	3 906.25

We cannot divide the frequency below '15 625' since the synchronous counter can have only whole states (we could not build a synchronous counter with '7 812.5' states). Thus, a seven-stage ripple counter (Q_0 through Q_6) will be needed.

The synchronous counter that is now required must have '15 625' states, so that the cycle time is precisely 1 s. To determine the required number of stages, we convert '15 625' into a binary number (refer to Section 4.1.2) and find that

$$(15\ 625)_{10} = (11\ 1101\ 0000\ 1001)_2$$

Thus, fourteen stages are needed.

According to the discussion of synchronous counters in the last section, a detector circuit satisfying Eq. 6-2 is needed.

$$Z = Q_{13} \cdot Q_{12} \cdot Q_{11} \cdot Q_{10} \cdot \bar{Q}_9 \cdot Q_8 \cdot \bar{Q}_7 \cdot \bar{Q}_6 \cdot \bar{Q}_5 \cdot \bar{Q}_4 \cdot Q_3 \cdot \bar{Q}_2 \cdot \bar{Q}_1 \cdot Q_0 \quad \textbf{(6-2)}$$

Clearly, implementing this equation will require a great deal of combinational logic. The detector can be simplified considerably, though, if we recognize that many of the possible states for the fourteen-stage up counter will *not* be used (namely, any states above count '11 1101 0000 1001'). Thus, we can eliminate a large number of variables in Eq. 6-2, leaving just enough to uniquely identify the maximum count. Equation 6-3 is the result.

$$Z = Q_{13} \cdot Q_{12} \cdot Q_{11} \cdot Q_{10} \cdot Q_8 \cdot Q_3 \cdot Q_0 \quad \textbf{(6-3)}$$

Further simplification in the detector is possible if we offset the count sequence so that the length is the same, but in such a way that the last state in the sequence requires less combinational logic to detect. Doing so, of course, requires that the sequence start at a point other than the 'zero' state. Using the detect and steer method, the offset idea may or may not require more *steering logic*, and may possibly cancel any gain from the simplified detector. In our design, however, we will make use of 'clear-direct' and 'set-direct' inputs to the counter's flip-flops to *preset* the starting state, making steering logic unnecessary. Therefore, simplifying the detector by offsetting the count sequence will definitely result in a logic savings. Presently, the count sequence spans

(first state) 00 0000 0000 0000 (0_{10})
↓
(last state) 11 1101 0000 1001 $(15\ 625_{10})$

Detection of the last state requires a seven-input AND function. Now, we can offset the count sequence to span

00 0000 1111 0111 (247_{10})
↓
11 1110 0000 0000 $(15\ 872_{10})$

In this case, a five-input AND function is needed to detect the last state. If one more flip-flop were available, the offset could be chosen to result in an even simpler design. Refer to Problem 6-9. Our final design for the fourteen-stage presettable counter is shown in Figure 6-18 (the needed control logic for the precision monostable has not been discussed yet and is not shown here).

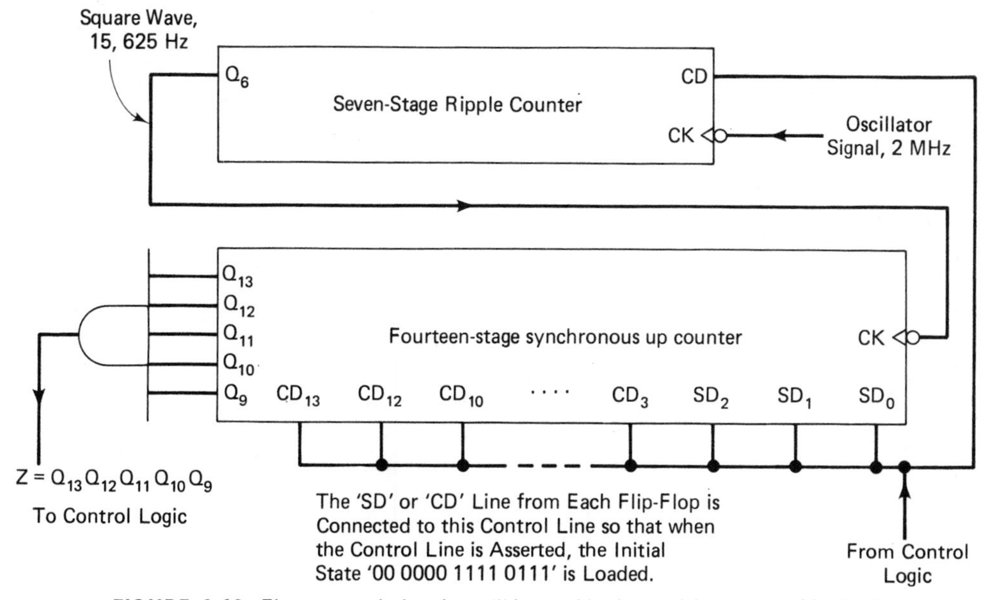

FIGURE 6-18 *The counter design that will be used in the precision monostable circuit.*

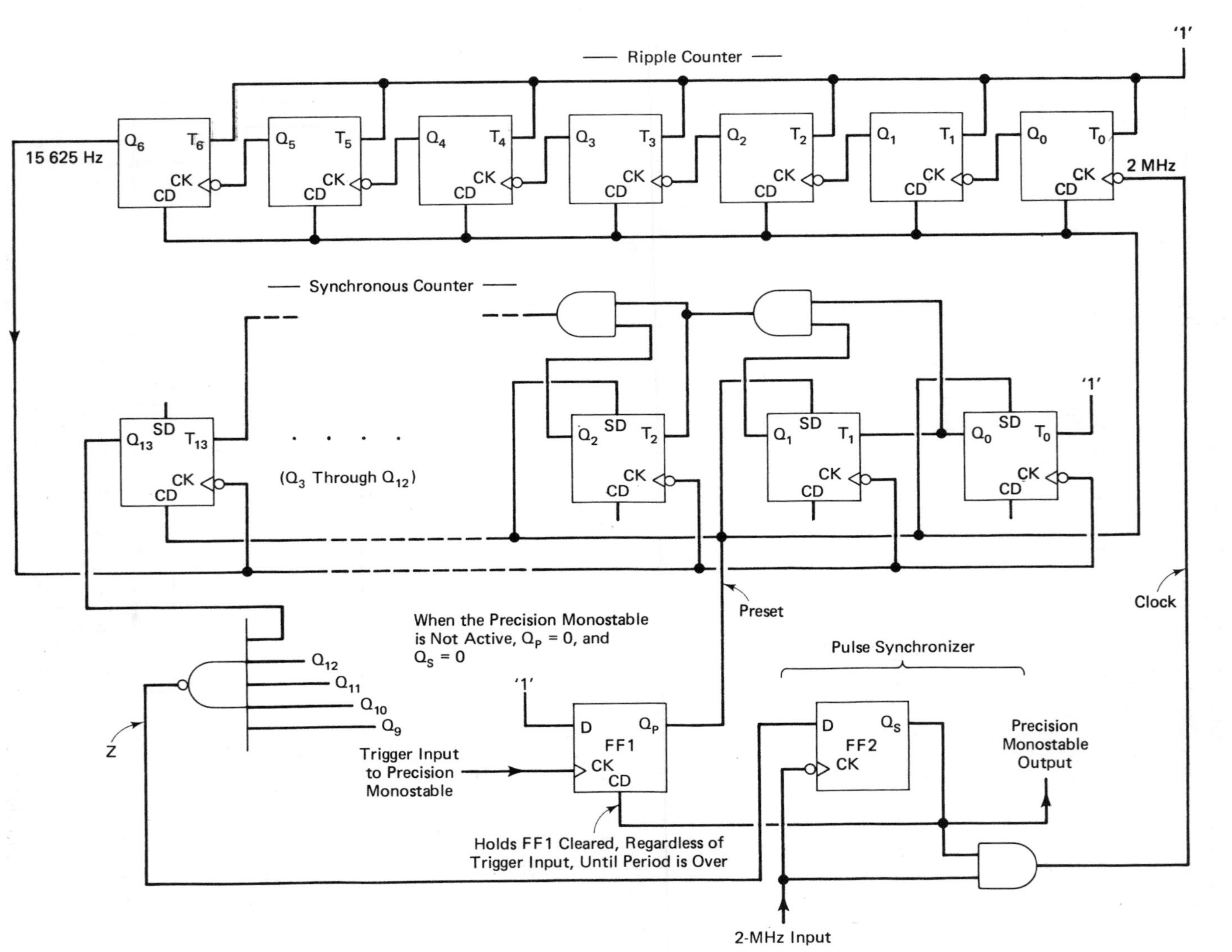

FIGURE 6-19 *Counters and control logic for the precision monostable.*

The necessary control logic must accept a *rising edge* input, which will correspond to the "trigger" of our precision monostable. On the rising edge, the counter should be preset and clock pulses enabled so that the counter begins progressing through its sequence. When the last state is reached, the detector should cause further clock pulses to be blocked. Note that if secondary rinsing edges are applied to the "trigger" input before the present count sequence has finished, they should be ignored.

Figure 6-19 depicts the counters and control logic. In the "stable" state of the precision monostable, the counters are frozen in the last state of the count sequence, and the detector output, 'Z', is asserted ($Z = 0$). Consequently, the pulse synchronizer output, 'Q_S', is '0', and clock pulses remain blocked. Since 'Q_S' is the precision monostable's output, the output is also '0'. When a rising edge is applied to the trigger input, 'Q_P' becomes '1', thereby causing the ripple counter to be cleared and the synchronous counter to be preset to the first state. Within 500 ns (one period of the 2-MHz clock), the pulse synchronizer output becomes '1' and the clock pulses to the ripple counter are enabled. The seventh-stage output of the ripple counter, 'Q_6', drives the clock input of the synchronous counter, causing the synchronous counter to advance on every '1' to '0' transition of 'Q_6'. Five hundred nanoseconds after the last state in the count sequence is reached, 'Q_S' and the precision monostable's output become '0', and the circuit is returned to its original state. The complete design for the precision monostable is shown in Figure 6-19.

Frequency meter: The precision monostable can be used as a component in a *digital frequency meter* circuit. Such a circuit will cause a number to be displayed that corresponds to the frequency of a periodic input signal (actually, the value displayed will correspond to the number of rising or falling edges per second). For this problem, we will assume that a digital pulse train is the desired input and that it is necessary to display frequencies between 1 and 99 999 Hz, with a resolution of ± 1 Hz. Our approach will be to produce an exact 1-s time "window" that will be used to gate the input pulse train. The gated result can then be fed to a counter, initially set to the 'zero' state, which accumulates the total number of pulses. Consequently, the value of the counter at the end of 1 s will correspond to the number of rising (or falling) edges in the input signal in 1 s. This value is, simply, the signal's frequency. Figure 6-20 illustrates the idea. Note that a pulse synchronizer would actually be used to gate the input signal; this ensures that no input pulses are truncated.

The precision monostable and the pulse synchronizer are familiar circuits which we have discussed previously. So also are the seven-segment decoders, which will be needed for the visual output of the frequency (refer to Chapter 3). Thus, our primary concern must focus on the counter circuit of Figure 6-20. Since it is necessary to produce a *decimal* number output (not binary), we must give serious thought to the type of counter circuit that we wish to employ. The decision does not regard whether the counter is ripple carry or synchronous, but rather, it concerns the nature of the count sequence. One choice would involve a natural binary up counter consisting of seventeen stages. It would be designed with a count sequence starting at '0' and ending at '1 1000 0110 1001 1111' ($= 99\ 999_{10}$). This would be our choice without further question *if* the visual output is to be displayed in *binary* notation. To make a useful measuring instrument, though, the visual output should appear in *decimal* notation,

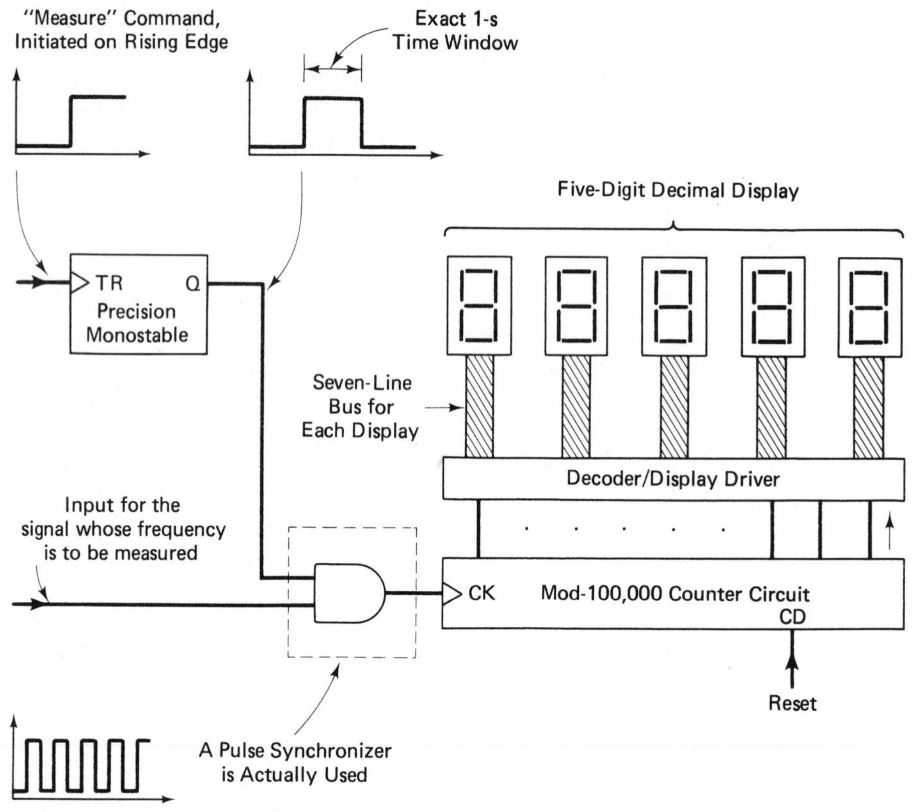

FIGURE 6-20 *Simple digital frequency meter.*

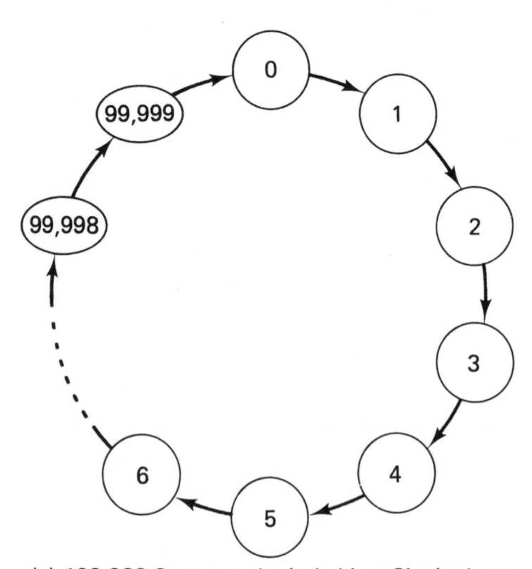

(a) 100,000 States are Included in a Single, Large Natural Binary Counter.

FIGURE 6-21(a)

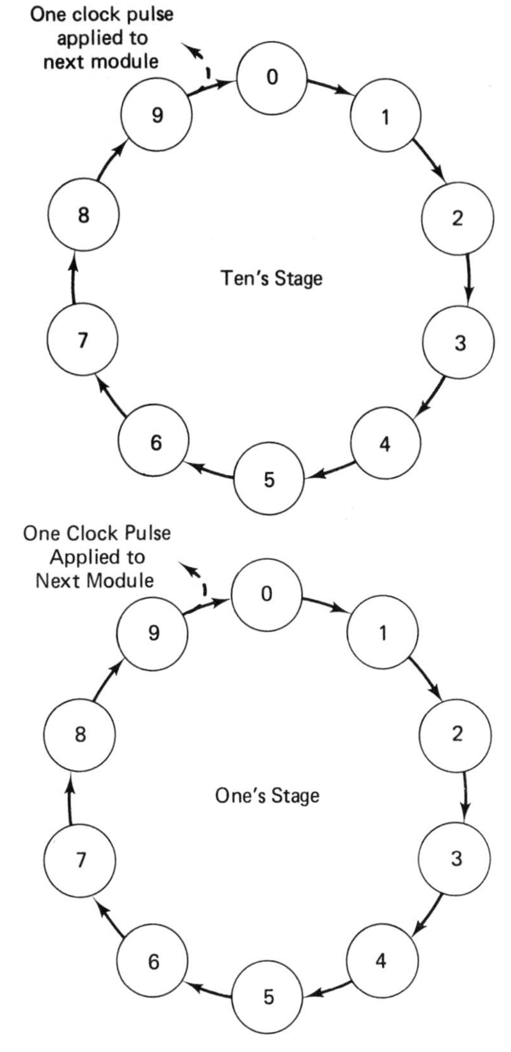

(b) Ten states are included in each of five BCD counters; only two ten-state flow graphs are shown. By interpreting the state value of each counter as a decimal digit, 100,000 combinations are possible.

FIGURE 6-21(b)

which is much easier for us to read and use. Now, if a natural binary counter is used, a special combinational logic translator circuit would be necessary to convert the natural binary output into a BCD (binary coded decimal) form suitable for seven-segment decoder circuits. Such a translator would be quite complex, requiring much additional circuitry. A better solution would be to design a counter module that *counts in BCD*; five modules of this type connected in cascade would then constitute the overall counter circuit. Figure 6-21 compares the natural binary flow graph to the BCD flow graph.

Using five decades of *identical* BCD counters, our design effort involves only one ten-state counter. The interconnection of five such circuits will make up the five-digit frequency counter. Figure 6-22 depicts the ten-state flow graph for a BCD counter.

Referring to the transition schematic in Figure 6-22, the detect and steer method can be used to find that

(a) 'Q_0' need not be affected by steering logic.

(b) 'Q_1' should be inhibited from complementing.

(c) 'Q_2' need not be affected by steering logic.

(d) 'Q_3' should be forced to complement.

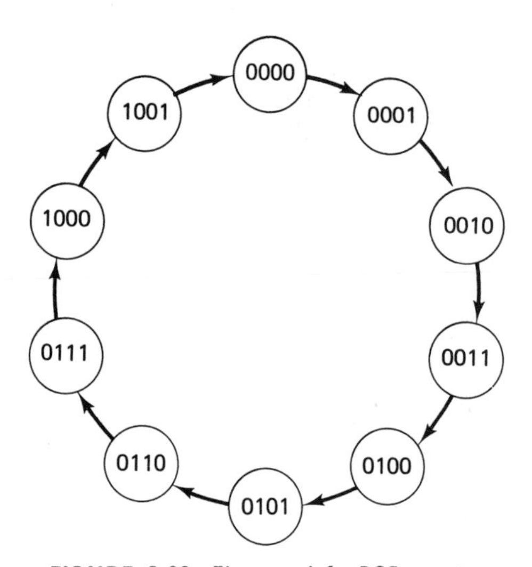

FIGURE 6-22 *Flow graph for BCD counter.*

Figure 6-23 shows a counter circuit that detects '1001' and is steered to '0000' based on the requirements of (a) through (d) above. Note that in the flow graph of Figure 6-21(b), the second decade is to be incremented *only* when the previous decade "overflows" from '1001' to '0000'. At all other times, the second decade should be inhibited from counting. Since the design will be synchronous, all '*CK*' inputs of all decades will be connected to a common clock. Thus, rather than blocking clock pulses, the counter is inhibited by forcing the '*J-K*' inputs to '0'. We accomplish this by inserting an AND function between the steering logic and each '*J-K*' input, as shown in Figure 6-24(a). The detector output serves as an indication of overflow, and constitutes the "carry" output. This is also shown in Figure 6-24(a). The BCD counter is often shown as a block element, and is illustrated in Figure 6-24(b).

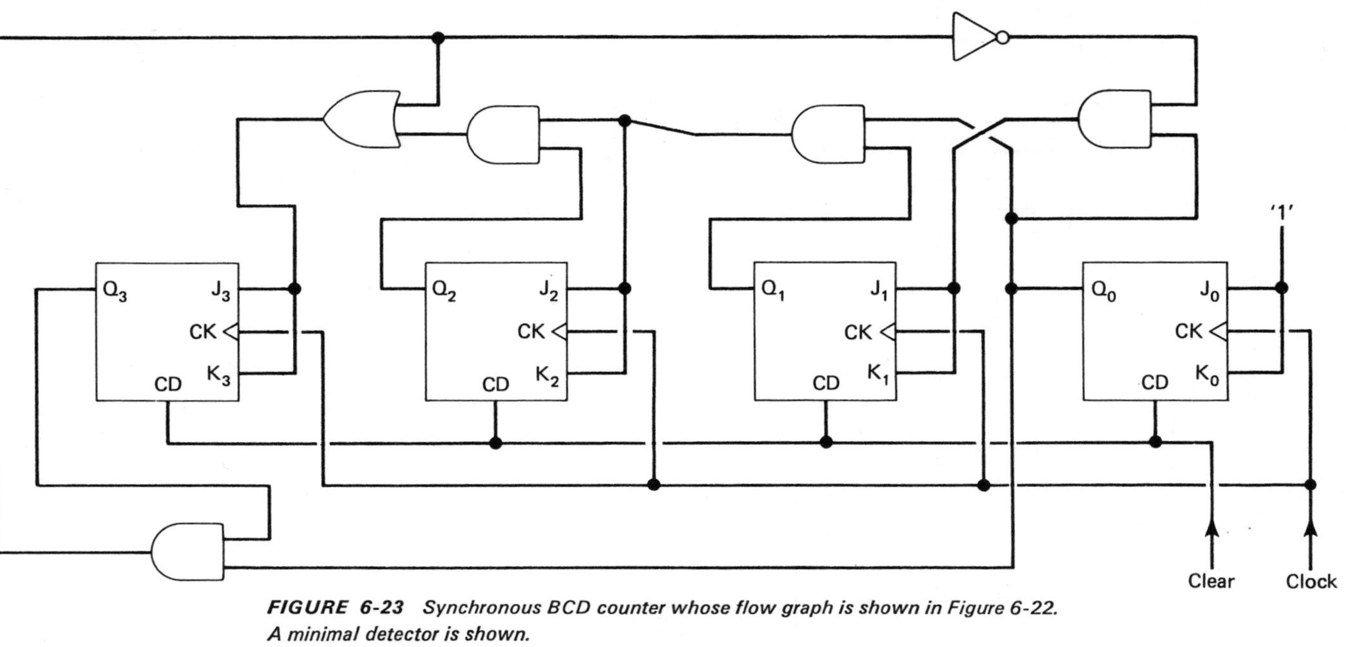

FIGURE 6-23 *Synchronous BCD counter whose flow graph is shown in Figure 6-22. A minimal detector is shown.*

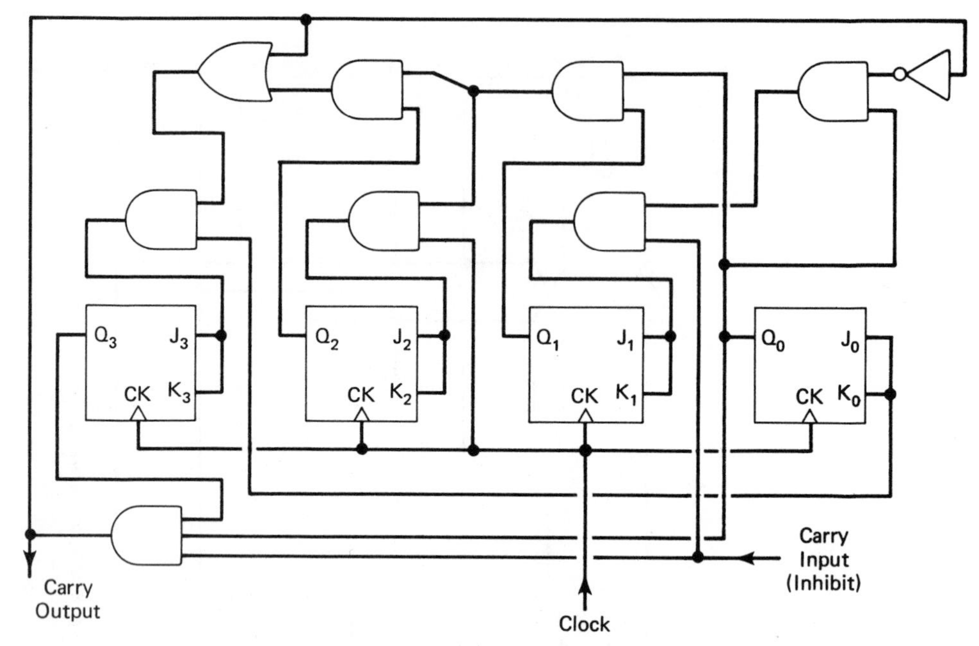

(a) Circuit for a BCD Counter Module.

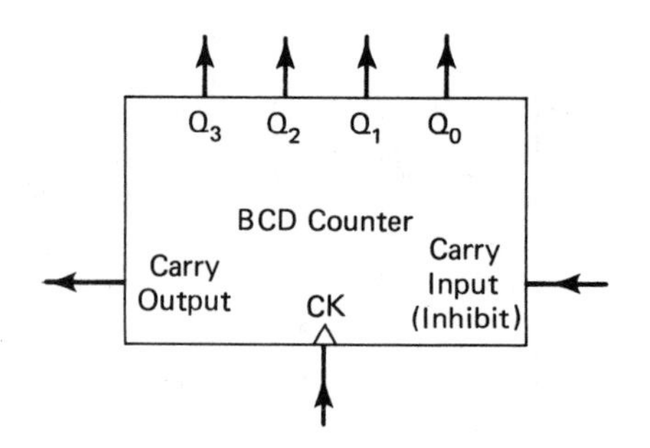

(b) BCD Counter Depicted as a Block Element.

FIGURE 6-24

Figure 6-25 illustrates how five BCD counters are interconnected to form the desired decade counter for our frequency meter design. Normally, a clear-direct input would also be available on each BCD counter to permit initialization. With the counter design decided, all module components for the frequency meter have been discussed, and we need only to assemble them according to the block diagram of Figure 6-20. As a guard against overflow from the fifth decade, it is advisable to include an additional flip-flop, as shown in Figure 6-25, that becomes set to '1' whenever the measured frequency exceeds 99 999 Hz. The 'Q' output of this flip-flop might cause an indicator lamp to glow on the display if overflow occurs.

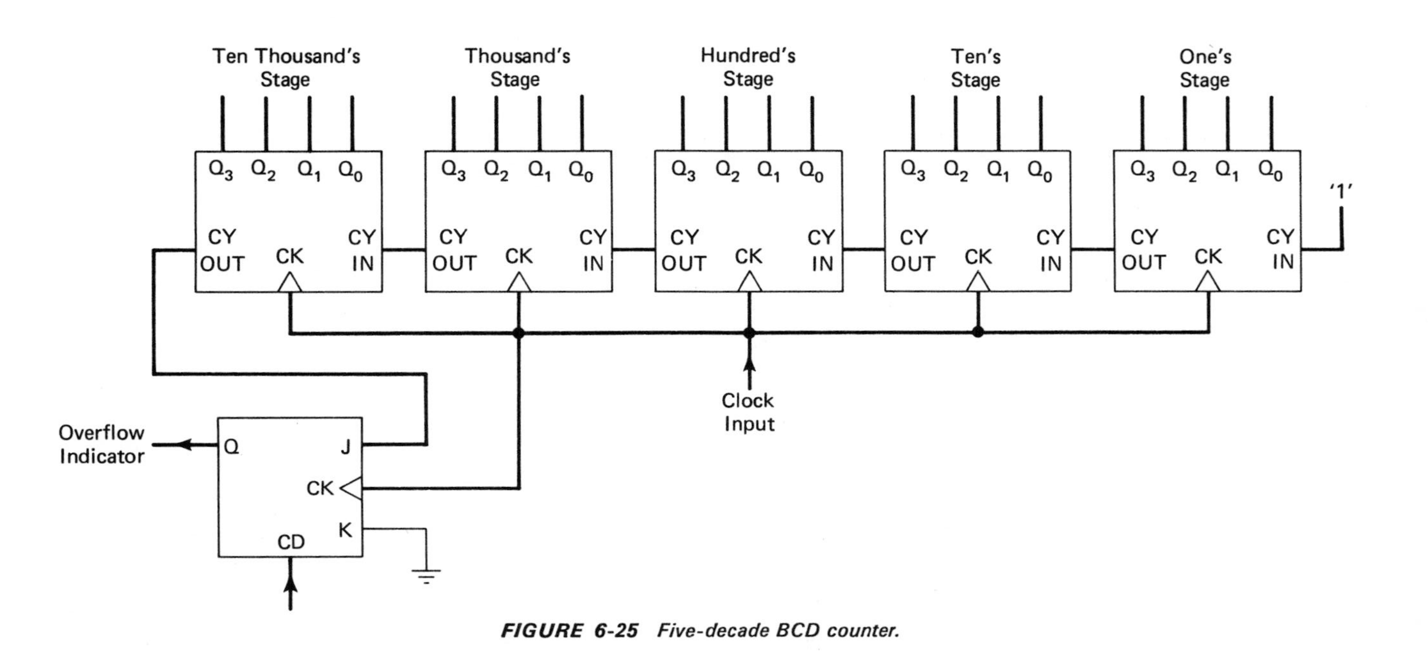

FIGURE 6-25 *Five-decade BCD counter.*

6.4 PROGRAMMED COUNTERS

6.4.1 Motivation

We have examined in some detail the principles of natural binary counters and have seen several important applications. The simplest natural binary counter circuits require toggle flip-flops only, not external combinational logic. Their count sequences are of maximal length in that the largest possible number of states is obtained for the number of flip-flops used. It was shown how external combinational logic could be added to transform the simple ripple counter into a synchronous counter, thereby increasing the maximum speed and permitting greater control over the length of the count sequence. In all cases, however, the resulting count sequence was natural binary in nature, either incrementing or decrementing in value from one state to the next. In this section, we will show how counter circuits can be designed in which the produced count sequences are *not necessarily* natural binary in nature. The only restriction that will be placed on these more general count sequences is that no state be used *more than once*. Having such increased power to select the count sequence not only adds to our ability to solve digital problems with counters, but also strengthens our understanding of natural binary counters. As we shall see, applying the general counter design technique that we learn in this section to a natural binary count sequence will result in the same design which we intuitively developed in Section 6.3. Moreover, the general counter design procedure is an ideal preview of the more formal techniques of sequential circuit design that are presented in Chapter 7.

6.4.2 Analysis

A ***programmed counter*** does not necessarily follow a natural binary count sequence, nor does its count sequence necessarily consist of the maximum possible number of states. Actually, the flow graph of a programmed counter may depict one large closed cycle (the maximal length case), several smaller unconnected closed cycles, or a closed cycle with one or more branches. The term "programmed counter" derives from the fact that we, as designers, "program" the count sequence by following a certain procedure. It is a well-chosen term in that we will have as much control as possible in selecting the count sequence. Of course, with the greatly increased number of possible count sequences, we can expect the circuit complexity to become greatly increased also. Indeed, the circuitry will often become so intricate that our intuition fails, and more powerful, organized procedures must be followed. To demonstrate this point, consider the circuit of Figure 6-26. It is our objective here to determine the count sequence produced by this circuit and to construct an appropriate flow graph. The analysis procedure that we use on this circuit depends basically on one concept—that is, that we view *each* flip-flop individually as being *excited* by a network of combinational logic. The inputs to this combinational logic can be derived, in a general sense, from any of the flip-flop outputs in the circuit, including the device whose network is being analyzed. Figure 6-27 illustrates this idea. Thus, we can understand a programmed counter to be simply an array of flip-flops, each of which is driven by a net-

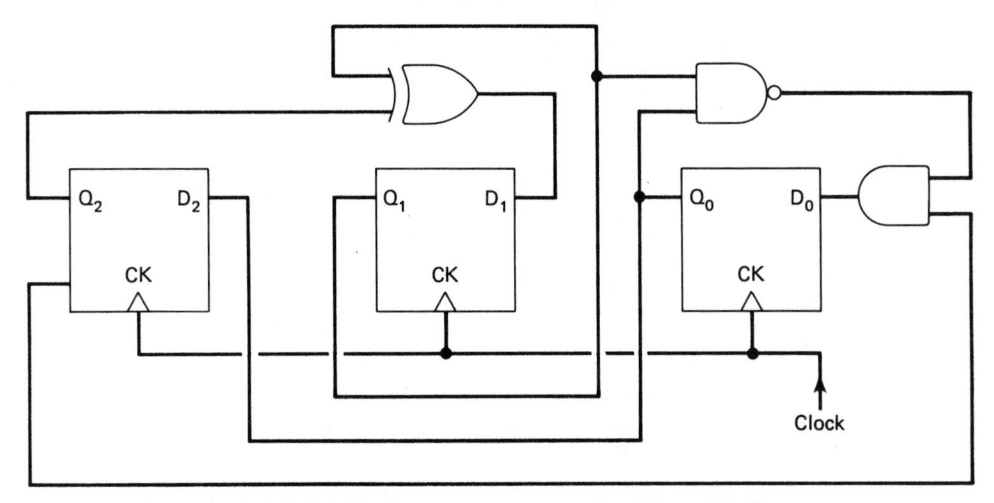

FIGURE 6-26 Programmed counter using 'D'-type flip-flops.

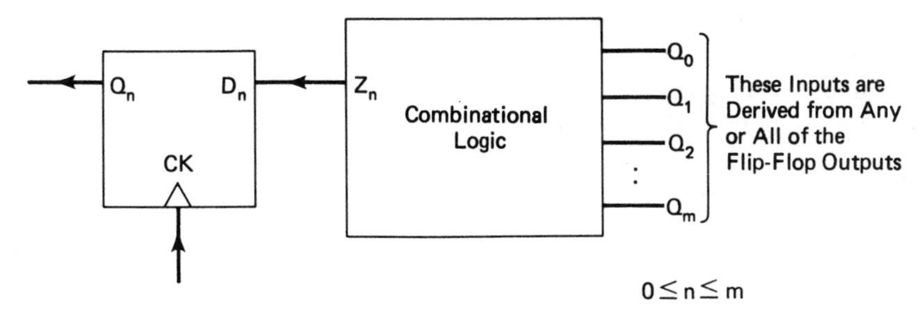

FIGURE 6-27 Generalized stage of a programmed counter. Each flip-flop in a programmed counter is individually viewed as though it is being excited by a network of combinational logic.

work of combinational logic. An *excitation equation* can be derived for each flip-flop and indicates how its combinational network stimulates the flip-flop input for each circuit state. Three excitation equations are necessary for the circuit of Figure 6-26. They are as follows:

$$\begin{aligned} D_0 &= \bar{Q}_2(\overline{Q_1 Q_0}) \\ &= \bar{Q}_2(\bar{Q}_1 + \bar{Q}_0) \\ &= \bar{Q}_2\bar{Q}_1 + \bar{Q}_2\bar{Q}_0 \end{aligned} \tag{6-4}$$

$$\begin{aligned} D_1 &= Q_1 \oplus Q_2 \\ &= Q_1\bar{Q}_2 + Q_2\bar{Q}_1 \end{aligned} \tag{6-5}$$

$$D_2 = Q_0 \tag{6-6}$$

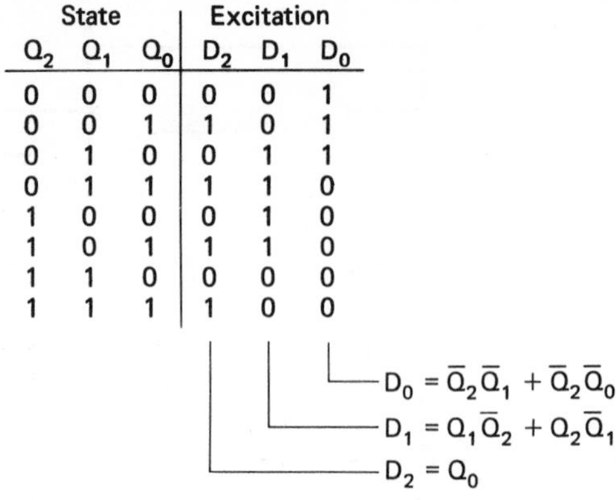

State			Excitation		
Q_2	Q_1	Q_0	D_2	D_1	D_0
0	0	0	0	0	1
0	0	1	1	0	1
0	1	0	0	1	1
0	1	1	1	1	0
1	0	0	0	1	0
1	0	1	1	1	0
1	1	0	0	0	0
1	1	1	1	0	0

FIGURE 6-28 *Excitation of the 'D' inputs for each state is derived from the excitation equations, Eqs. 6-4 to 6-6.*

An *excitation table* can now be constructed to show the binary values present at the '*D*' inputs for each possible state. This is shown in Figure 6-28. Now, since we are dealing with a synchronous circuit, no state changes will occur until a clock pulse is applied. Because 'D' flip-flops are used, the value of a '*Q*' output will become exactly the value of the corresponding '*D*' input when the clock pulse occurs. These new '*Q*' output values become the new state, and thereby cause new '*D*' values to be produced in preparation for the next clock pulse. Assuming that we start at '000', the following transitions will be observed, according to Figure 6-28.

	Present State	*Clock Pulse*	*Next State*	*Number of Clock Pulses Applied*
(initial condition)	0 0 0	⟶	0 0 1	1
	0 0 1	⟶	1 0 1	2
	1 0 1	⟶	1 1 0	3
	1 1 0	⟶	0 0 0	4
		Repeats ↓		

From these four transitions, we can begin to construct the flow graph:

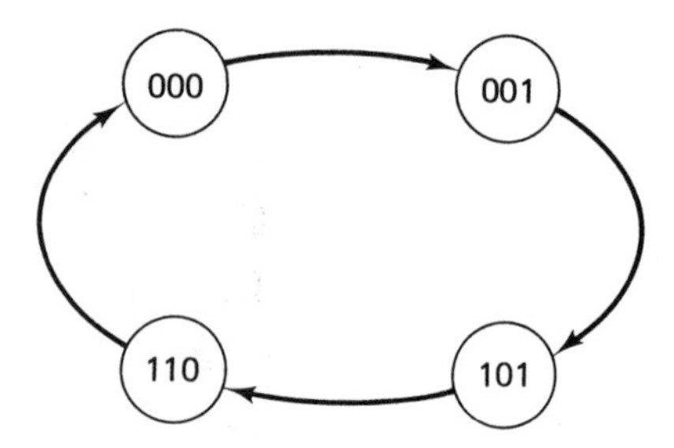

In this initial flow graph, we see that only four of the eight possible states are involved.

To have a complete flow graph, however, all eight states must be included. Referring to Figure 6-28, the first state we find that is *not* shown in the initial flow graph is '010'. We see that '010' leads to '011', which then leads to '110'. Since '110' is a member of the four-state cycle already determined, a branch has been discovered. Following this procedure for the remaining two states, we arrive at the complete flow graph shown in Figure 6-29. The interpretation of this flow graph reveals a four-state cycle and a four-state branch. No matter what the initial state may be, we can guarantee that in no more than four clock pulses, the closed cycle will have been entered. A branch state may be entered when power is first applied, by presetting, or from a previous branch state.

In the last example, a programmed counter utilizing D-type flip-flops was analyzed. Programmed counters may also be designed using T and J-K flip-flops. The analysis procedure is somewhat more complex, though, since there will no longer be a one-to-one relationship between flip-flop inputs and outputs, as there is when D-type devices are used. Excitation of a '*T*' input with '1', for example, must be viewed as causing the present state of the flip-flop to complement when a clock pulse occurs. If each D flip-flop shown in the circuit of Figure 6-26 is replaced with a T flip-flop *without* changing the excitation logic, a completely different count sequence is obtained. Determination of the flow graph for this new circuit is demonstrated in Figure 6-30. Since a count sequence is considered to be cyclic, we find only a single two-state sequence for the toggle flip-flop version. It is possible that a programmed counter will not have a count sequence at all, but instead a flow graph consisting of disjoint states and unconnected branches.

6.4.3 Design

Often, a particular count sequence is required of a programmed counter in the solution of a problem. We must ask, then:

> What steps are necessary to arrive at a counter design for the given sequence?

An initial step must be to decide how many flip-flops are necessary. This involves

(a) Noting how many states are in the sequence.

(b) Determining if any of the states are repeated.

Regarding (a), if there are '*m*' states in the sequence, we must select a number of flip-flops, '*n*', such that

$$2^{n-1} < m \leq 2^n \tag{6-7}$$

In other words, the next highest integer power of '2' above '*m*' is chosen, and the value of the exponent corresponds to the required number of flip-flops. Regarding (b), repeated states call for additional flip-flops, above the number specified by Eq. 6-7, or require that combinational logic be used to translate a nonrepeating state

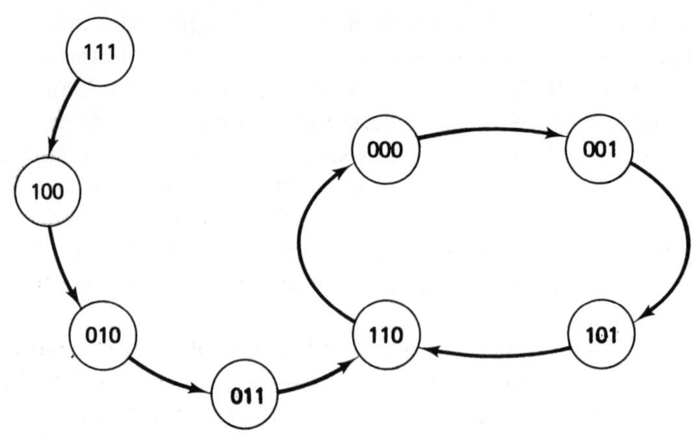

FIGURE 6-29 Flow graph that fully describes the programmed counter of Figure 6-26.

State			Excitation			Next State		
Q_2	Q_1	Q_0	T_2	T_1	T_0	Q_2	Q_1	Q_0
0	0	0	0	0	1	0	0	1
0	0	1	1	0	1	1	0	0
0	1	0	0	1	1	0	0	1
0	1	1	1	1	0	1	0	1
1	0	0	0	1	0	1	1	0
1	0	1	1	1	0	0	1	1
1	1	0	0	0	0	1	1	0
1	1	1	1	0	0	0	1	1

Note: When 'D' Flip-Flops are Used, the Next State is Exactly the Same as the Excitation, and this Column Need Not be Shown.

(a) Excitation Table for the 'T' Flip-Flop Version of Figure 6-26.

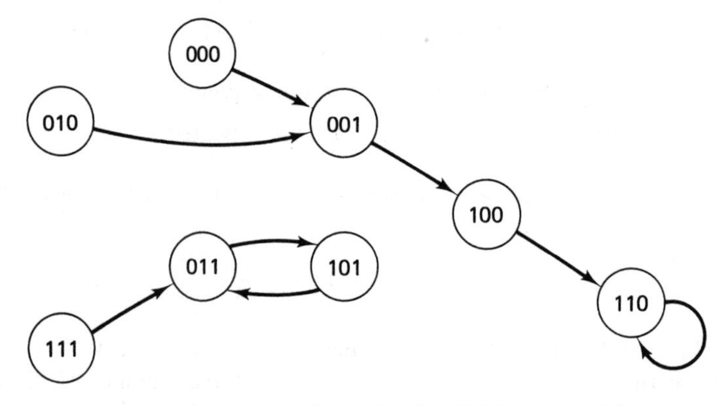

(b) Flow Graph Resulting from the State Transition of (a).

FIGURE 6-30

sequence into a repeating one. To simplify our discussion, only nonrepeating state sequences will be presented.

Beyond the initial step of selecting the number of flip-flops, the type of flip-flops must be chosen, and most important, the required combinational logic must be determined. In most cases, D-type flip-flops allow the simplest design procedure, and therefore, D-type devices will most often be used. Determining the needed combinational logic is not difficult and draws on our knowledge of Karnaugh maps and circuit minimization.

Recall the programmed counter of Figure 6-26. At that time, our goal was to deduce the count sequence from the circuit diagram. We proceeded to do this by first writing an excitation equation for each flip-flop and producing an excitation table corresponding to these equations. Since the excitation of 'D' flip-flops becomes their next state after a clock pulse is applied, we could easily find the transition sequence (count sequence) from the excitation table. Now, the design process for programmed counters works almost in the exact *reverse* order. From the given count sequence, we derive an excitation table, then excitation equations, and finally a circuit diagram. Figure 6-31 illustrates the basic procedure. As an example, we will design a

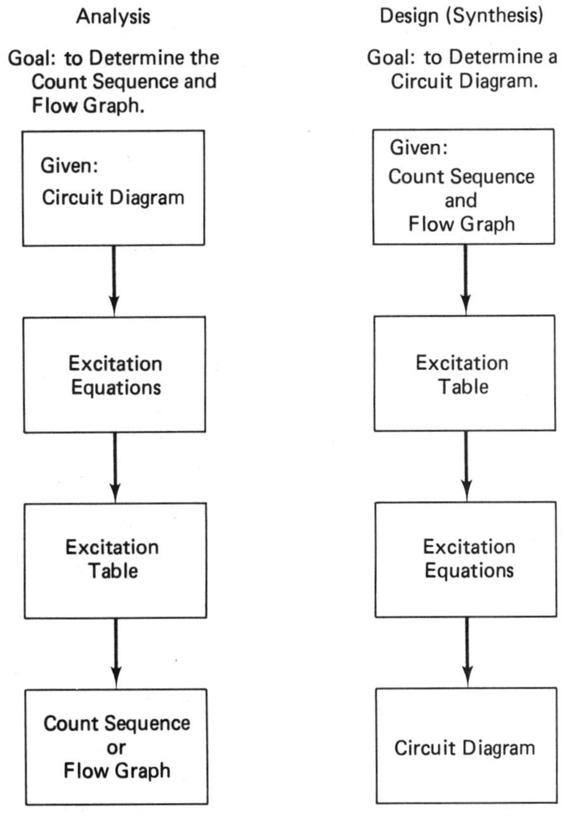

FIGURE 6-31 *Comparison of the needed steps for analysis and design.*

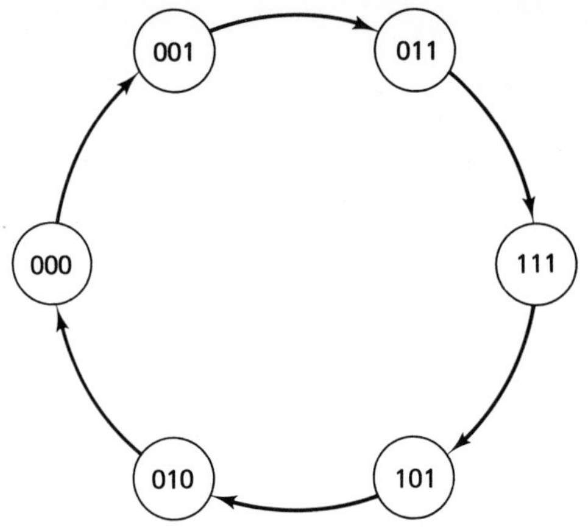

FIGURE 6-32 Given count sequence for which a counter circuit will be designed.

counter circuit using D-type flip-flops that produces the sequence shown in Figure 6-32. Since six states are involved, three flip-flops are necessary; by Eq. 6-7, we have

$$2^2 < 6 \leq 2^3 \quad \text{and thus} \quad n = 3$$

Figure 6-33 shows how the state transitions are used to form an excitation table. Excitation equations are derived from the excitation table using standard combinational logic design techniques. This is illustrated in Figure 6-34. Remember that a combinational circuit is required for each 'D' input, and that the excitation value for each input can be derived only from the 'Q_2', 'Q_1', and 'Q_0' outputs. The circuit diagram for this design is shown in Figure 6-35.

FIGURE 6-33 An excitation table is determined by listing all allowed transitions for the eight possible states.

State Transitions

000
001
011
111
101
010

↓

Repeats

Excitation Table

Present State			Next State			Required Excitation for the Next State*		
Q_2	Q_1	Q_0	Q_2	Q_1	Q_0	D_2	D_1	D_0
0	0	0	0	0	1	0	0	1
0	0	1	0	1	1	0	1	1
0	1	0	0	0	0	0	0	0
0	1	1	1	1	1	1	1	1
1	0	0**	Ø	Ø	Ø	Ø	Ø	Ø
1	0	1	0	1	0	0	1	0
1	1	0**	Ø	Ø	Ø	Ø	Ø	Ø
1	1	1	1	0	1	1	0	1

Notes: *Since 'D' Flip-Flops are Being Used, Values for the Required Excitation are Exactly the Same as the Next State.

**Because the Count Sequence Specifies Only Six of the Eight Possible States, the Two Unused States are Assigned "Don't Care" Status.

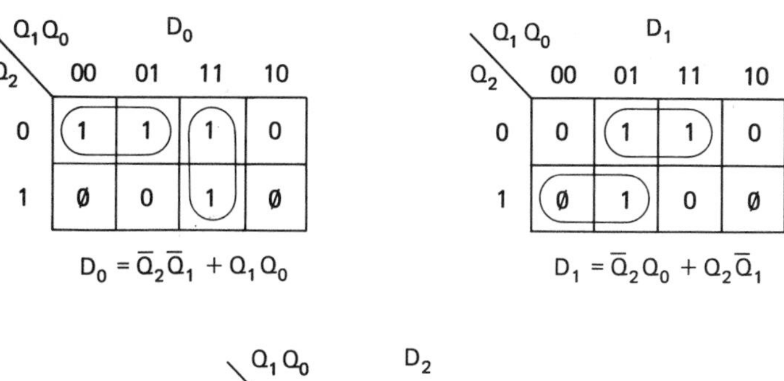

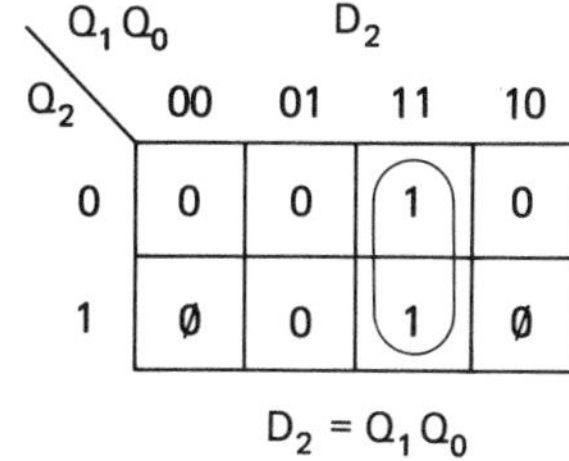

FIGURE 6-34 *Excitation equations are determined for the count sequence of Figure 6-33.*

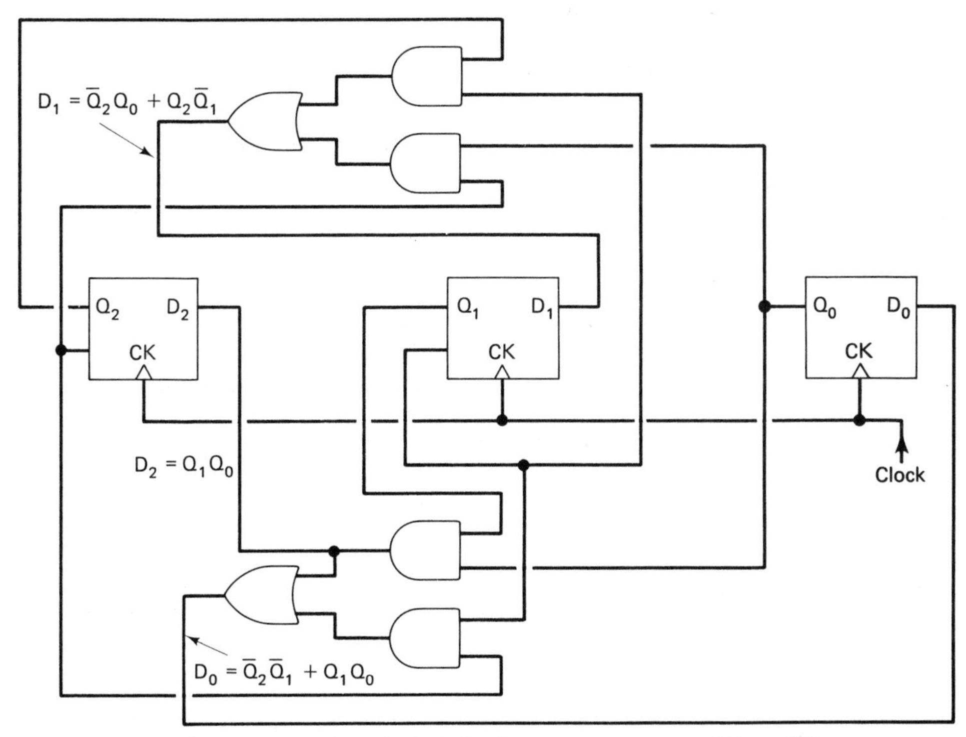

FIGURE 6-35 *Final circuit design for the count sequence of Figure 5-32.*

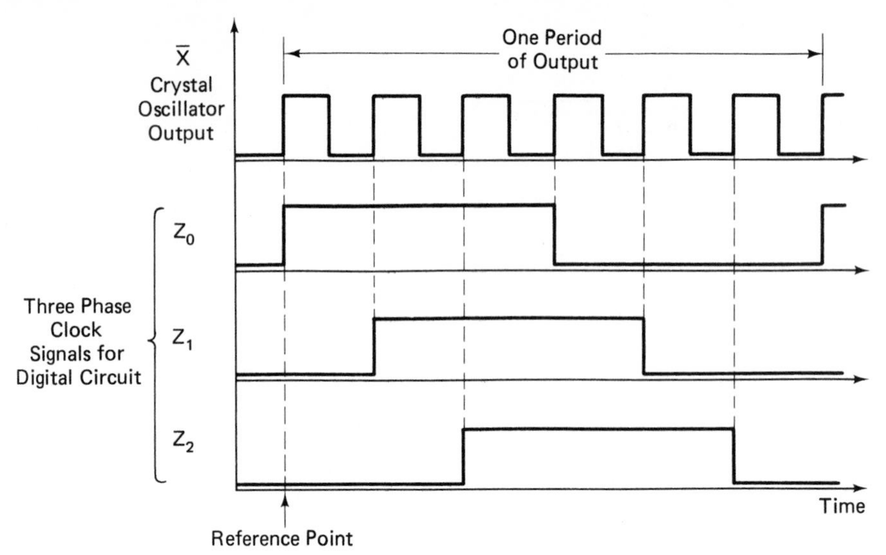

FIGURE 6-36 *Three square waves that are precisely, 0°, 60°, and 120° out of phase with respect to the indicated reference point.*

6.4.4 An Application

Three-phase clock generator: Some complex digital circuits require several clock inputs that are all square waves of the *same frequency*, but differ in phase (by a precise amount) with respect to one another. A set of three such square waves is illustrated in Figure 6-36. These three waveforms are to be the output of a circuit that we will design. This circuit will have as its sole input the crystal oscillator waveform, 'X', as shown in the figure.

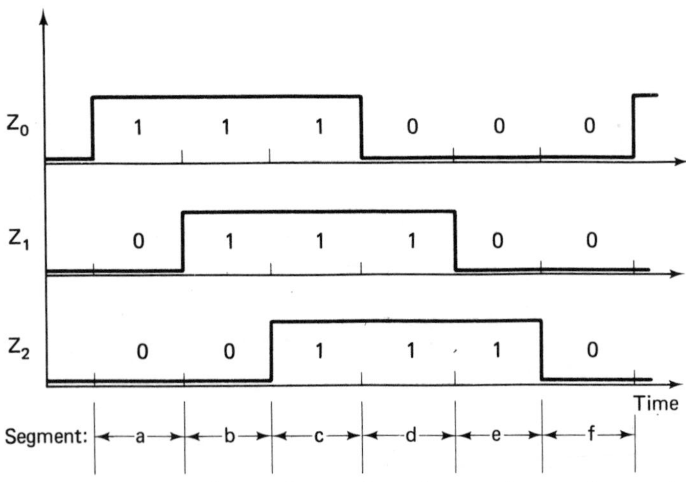

FIGURE 6-37 *Subdividing the waveforms into equally spaced segments, labeled 'a', 'b', 'c', etc.*

State Transitions
0 0 1
0 1 1
1 1 1
1 1 0
1 0 0
0 0 0
Repeats

(a) Six-State Count Sequence.

Present State			Next State			Required Excitation for the Next State		
Q_2	Q_1	Q_0	Q_2	Q_1	Q_0	D_2	D_1	D_0
0	0	0	0	0	1	0	0	1
0	0	1	0	1	1	0	1	1
0	1	0	Ø	Ø	Ø	Ø	Ø	Ø
0	1	1	1	1	1	1	1	1
1	0	0	0	0	0	0	0	0
1	0	1	Ø	Ø	Ø	Ø	Ø	Ø
1	1	0	1	0	0	1	0	0
1	1	1	1	1	0	1	1	0

(b) Excitation Table.

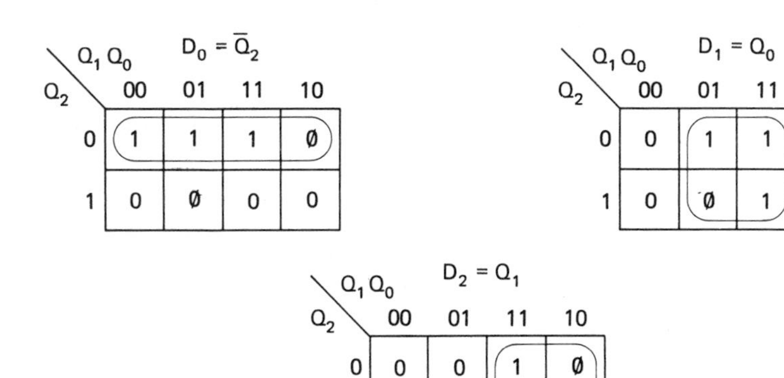

(c) Derivation of the Excitation Equations.

FIGURE 6-38

To begin, we can divide one period of 'Z' into six equally spaced segments. Refer to Figure 6-37. In each segment, the relationship between the waveforms is slightly different. Noting the binary values superimposed on the waveforms, it is immediately apparent that a binary sequence is present. One solution to this problem, then, is to design a programmed counter that follows this sequence, and use the counter's 'Q' outputs directly as outputs for the three-phase clock generator. The design procedure is illustrated in Figure 6-38. D flip-flops are used in the design, which is shown in its final form in Figure 6-39. Refer to Problem 6-19 for further study of this circuit.

FIGURE 6-39 *Three-phase clock generator circuit.*

SUMMARY

In its most basic form, a counter is a device that accepts a clock input (often a pulse train) and produces a multidigit binary output that is related to the total number of clock pulses that have been applied. A particular value for the multidigit binary output defines a state. The maximum possible number of states that a counter may have is limited only by the number of memory elements (flip-flops) that are involved in the counter's design. A count sequence specifies the exact order in which output states will appear and is conveniently expressed using a flow graph. A valid count sequence will be cyclic in nature and will not lead to any dead-end branch states. A count sequence may be entered, however, from a branch state.

In addition to a clock input, certain optional inputs may be found on a counter circuit. The "enable" input permits the count sequence to progress normally when asserted, and can inhibit further counting when deasserted. The "parallel load" input is used to preset the counter to an initial state, as defined by the values present on parallel input lines. The "clear-direct" input presets the counter to the zero state when asserted. Some natural binary counter modules provide an "up-down" control input, which will determine the direction of the natural binary count sequence.

Ripple counters are the simplest of all counters, requiring no gating logic. However, the propagation effect observed when changes of state occur make ripple counters unsuitable for many applications outside of frequency division. Synchronous counters overcome the propagation problem by requiring that all internal flip-flops be triggered synchronously. Gating logic is required, though, to ensure that the proper state transitions are realized. Synchronous counters are used whenever high-speed operation is necessary. Since false states will not occur in synchronous counters as long as the flip-flop propagation delays are equal, it is not possible for a detector gate to produce a false output.

The detect and steer method is used to shorten the length of a natural binary count sequence. This method requires that the last correct state be detected, and when this occurs, that the component flip-flops be excited by external logic in such a manner that the next desired state is reached on the next clock pulse.

Counters can be obtained as modules, with four or more stages internally connected. The design of a counter module is such that it can be cascaded with copies of itself to produce even larger counters. To facilitate the cascading of synchronous counter modules, carry input (enable) and carry output lines are made available. The carry output of one module is usually connected to the carry input of the next.

The design technique for programmed counters allows us to create counter circuits that do not necessarily follow the natural binary sequence. This can be extremely useful in certain sequencing applications where minimal logic circuits are required. By considering each flip-flop in a programmed counter as being excited by its own combinational network, the count sequence can be treated on a digit-by-digit basis. Thereby, an excitation equation is derived for each flip-flop, based only on the 'Q' output variables.

NEW TERMS

Count Sequence
Flow Graph
Parallel Load Input
Enable Input
Output Skew
State Detector
Synchronous Counter
Universal Counter Stage
Carry Input
Carry Output
Maximal Length Sequence
Detect and Steer
Natural Transition
Desired Transition
Steering Logic
BCD Counter
Programmed Counter
Excitation Equation
Excitation Table
Branch State

PROBLEMS

6-1 How many three-bit, eight-state nonrepeating count sequences are possible?

6-2 A three-stage ripple counter, such as the one shown in Figure 6-3, is used to produce a natural binary count sequence. Four detector gates are added to sense the states '010', '100', '110', and '111'. These gates are labeled 'A', 'B', 'C', and 'D', respectively. Assume that the propagation delay of the flip-flops is 20 ns and that '$\bar{Q}$' outputs are available.

(a) Which gates will produce erroneous outputs?

(b) For each detector that produces a false output, describe (1) where in the sequence each false output occurs, (2) how many false outputs there are in one cycle, and (3) the exact duration of each false output.

6-3 A '000' detector is added to a ripple counter that counts *down*. How many false outputs will be observed?

6-4 Design a sixteen-state synchronous natural binary *down* counter using T flip-flops. Based on your design, write a general equation for the excitation of the 'n'th stage in a synchronous down counter. Refer to Eq. 6-1.

6-5 Using three universal counter stages of the type shown in Figure 6-11 and a minimal amount of combinational logic, design a synchronous natural binary *up/down* counter. One additional input, 'U/D', should be present in your design. When $U/D = 1$, the counter should count "up" and when $U/D = 0$, it should count "down."

6-6 The synchronous counter of Figure 6-10 is used in a high-speed circuit where the clock period is 35 ns. Assume that each two-input AND gate has a propagation delay of 10 ns, that each flip-flop has a propagation delay of 20 ns, and that the setup and hold times are zero. Describe the output sequence that is observed on 'Q_0' through 'Q_4' if the counter starts in the zero state and the 35-ns clock is applied. Explain exactly why the counter does not work properly, and propose a change in the design to correct the problem. Higher-speed components may not be used.

6-7 Indicate how the precision monostable circuit of Figure 6-19 can be modified so that it is *retriggerable.*

6-8 Determine the flow graph corresponding to the three-digit programmed counter whose excitation equations are as follows:

$$D_2 = \bar{Q}_2$$
$$D_1 = \bar{Q}_1(Q_2 + Q_0)$$
$$D_0 = Q_2Q_0 + Q_1\bar{Q}_0$$

6-9 A ten-stage synchronous counter is used to count to $(1000)_{10}$. When the last state in the sequence is reached, a detector output of '1' causes control logic to inhibit further clock pulses so that the counter is halted. When an external signal is received by the control logic, the counter is cleared, and the count sequence is begun over, to be halted again at a count of $(1000)_{10}$. Refer to Figure 6-18 for a similar design.

(a) Determine the binary number corresponding to a count of $(1000)_{10}$. From this number, determine the minimal detector equation.

(b) Assume that each flip-flop in the counter is supplied with '*SD*' and '*CD*' inputs and that a 'carry out' is available on the counter. Show that by adding an appropriate offset to the sequence and connecting one additional flip-flop to the circuit, the detector logic can be eliminated.

6-10 Apply the detect and steer method to the design of a counter that follows the count sequence shown in Figure P6-10.

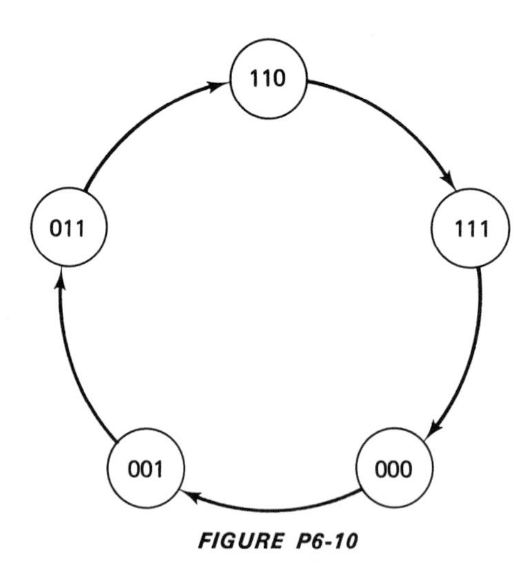

FIGURE P6-10

6-11 Use the detect and steer method to design a natural binary *down* counter that follows the sequence shown in Figure P6-11.

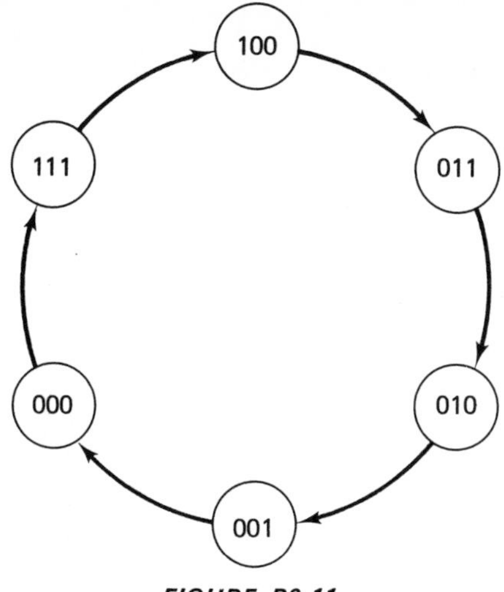

FIGURE P6-11

6-12 A synchronous natural binary up counter is built according to the circuit diagram of Figure 6-9(a). Two of the flip-flops used in the construction are made by manufacturer "A" and one is made by manufacturer "B". Since the flip-flops are made by different manufacturers, slightly different electrical characteristics can be expected. When the counter is tested, four undesired transient states are observed, each lasting 10 ns, as shown in Figure P6-12.

(a) What is the likely cause of these erroneous states?

(b) Determine which flip-flops are made by "B" and which by "A".

(c) What can be done to correct the problem with a minimum of changes?

FIGURE P6-12

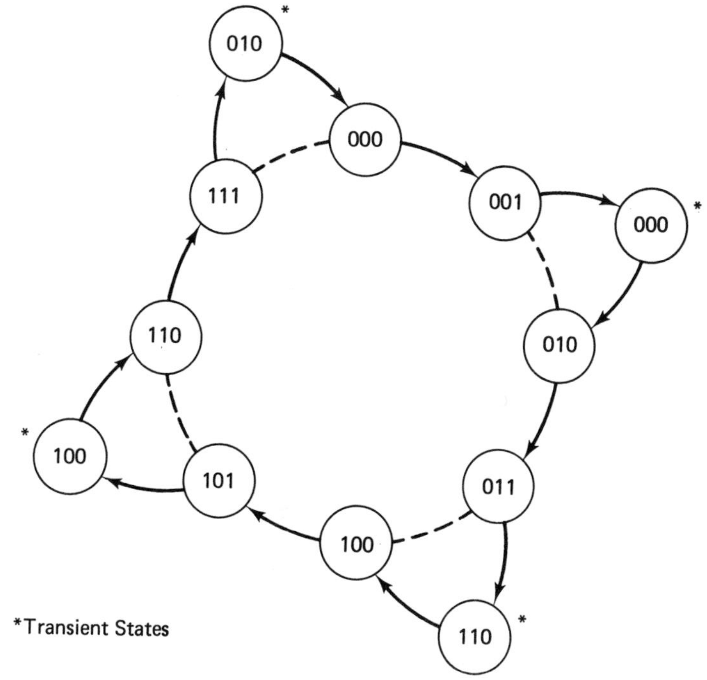

*Transient States

6-13 A radio receiver is tuned to detect an RF carrier of exactly 20.000 MHz. When such a carrier is present, an amplifier/shaper circuit in the receiver produces a 20.000-MHz square wave of sufficient power to drive a digital logic device. The resulting square wave is fed to a frequency divider (ripple counter), whose output is to drive a small speaker. Refer to Figure P6-13. How many stages must the frequency divider have in order to produce an audible square wave in the range 1000 to 2000 Hz?

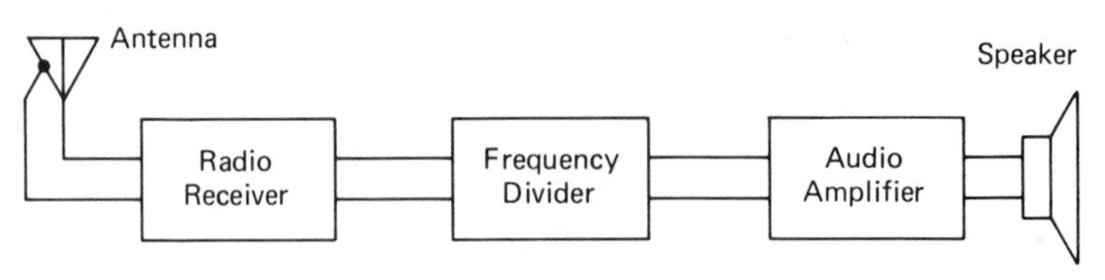

FIGURE P6-13

6-14 Figure P6-14 depicts a six-state counter design in which the clear-direct flip-flop inputs are used.

(a) With a flow graph, exactly describe the count sequence that is produced by this counter. Are there any transient states?

(b) There is a serious flaw in this design which may or may not result in counter malfunction. What is it?

(c) This design would be a particularly dangerous if the circuit were to be manufactured in the thousands from gate and flip-flop components whose electrical characteristics were not perfectly consistent. Redesign this counter using the detect and steer method to make the design less sensitive to component variations.

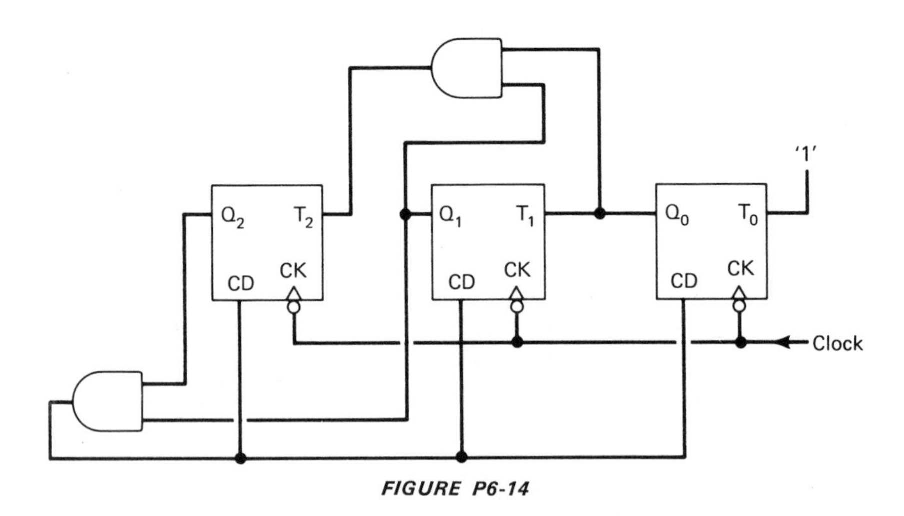

FIGURE P6-14

6-15 An electronic music box may be constructed using a counter, a decoder, and a tone generator, as shown in Figure P6-15(a). The binary counter advances through a sequence of states determined by its design, and as it does so, a melody is produced.

Figure P6-15(b) depicts how the sixteen codes of a four-stage counter are associated with musical tones.

(a) Design the programmed counter for a music box that is to play the tune of Figure P6-15(c). A silent note (no tone) separates the end of the piece from a repeat. Four 'D' flip-flops and a minimal amount of combinational logic should be used. Show flip-flop excitation equations only—a circuit diagram is not necessary.

(b) Why would it be impossible, using the design techniques we have studied, to construct a four-stage counter that plays the tune shown in Figure P6-15(d)?

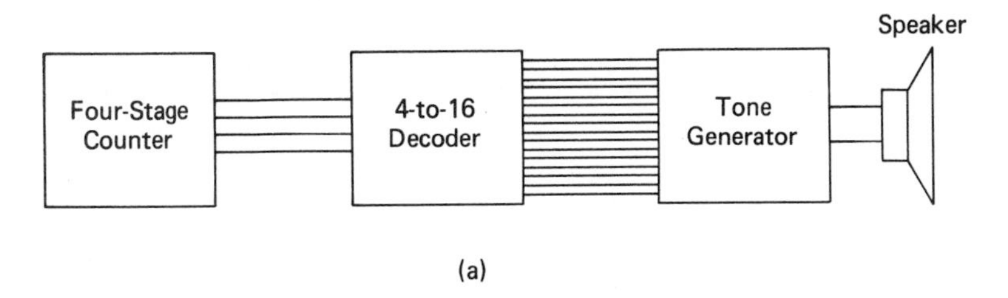

(a)

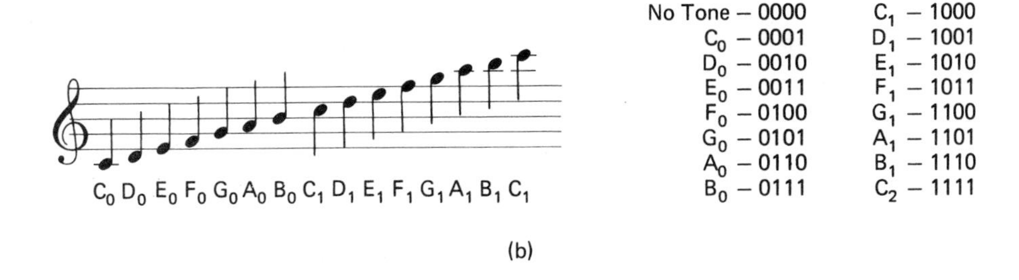

(b)

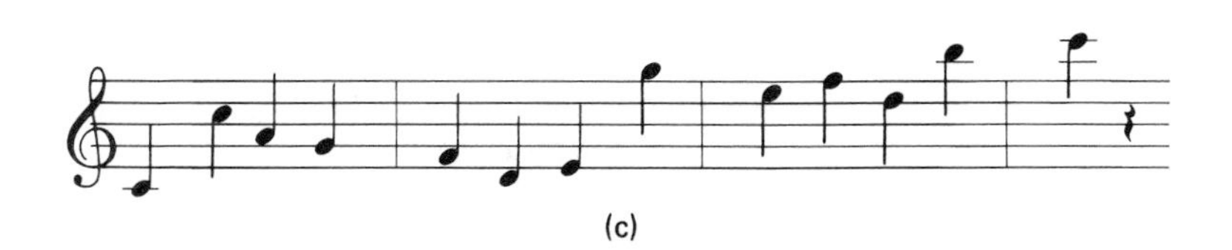

(c)

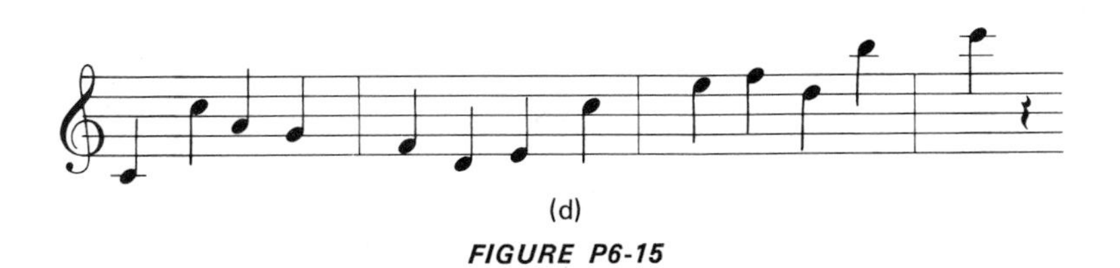

(d)

FIGURE P6-15

6-16 Determine the flow graph of the circuit shown in Figure P6-16. Note that this circuit produces an eight-state cyclic count sequence, even though 'Q_1' and '$\bar{Q}_1$' are not used.

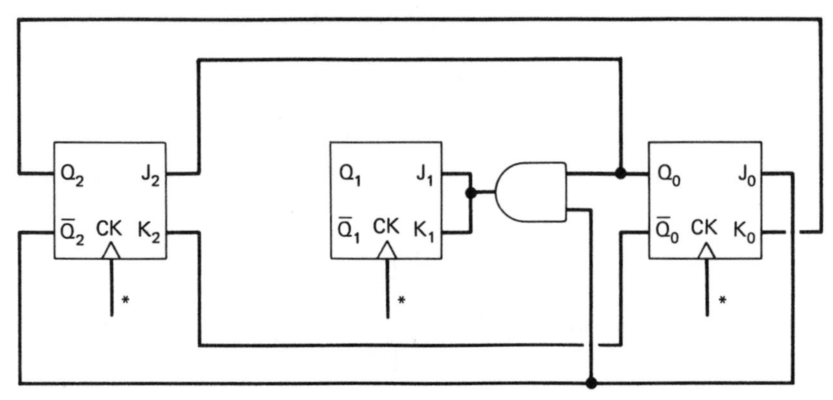

FIGURE P6-16

6-17 Design a three-bit Gray code counter using D flip-flops and combinational logic. Refer to Section 4.3.2 for a review of the Gray code.

6-18 Using the *programmed counter method*, design an eight-state natural binary up counter with T flip-flops. Verify that this circuit, which is minimal, follows exactly the same form as the excitation equations described by Eq. 6-1.

6-19 **(a)** Draw the complete flow graph for the three-phase clock generator of Figure 6-39, showing all eight states. Sketch the output waveforms appearing on Z_1, Z_2, and Z_3 if the circuit happens to enter the '010' or '101' state on power up.

(b) Modify the three-phase clock generator so that the intended six-state count sequence is always eventually reached, regardless of the initial state. The transitions that are chosen to lead '010' and '101' into the six-state count sequence should be such that a minimum amount of combinational logic is added. There are several possibilities.

6-20 **(a)** Determine the flow graph for the circuit of Figure P6-20.

(b) Assume that each T flip-flop in Figure P6-20 is replaced with a D flip-flop. Determine the flow graph for the resulting circuit.

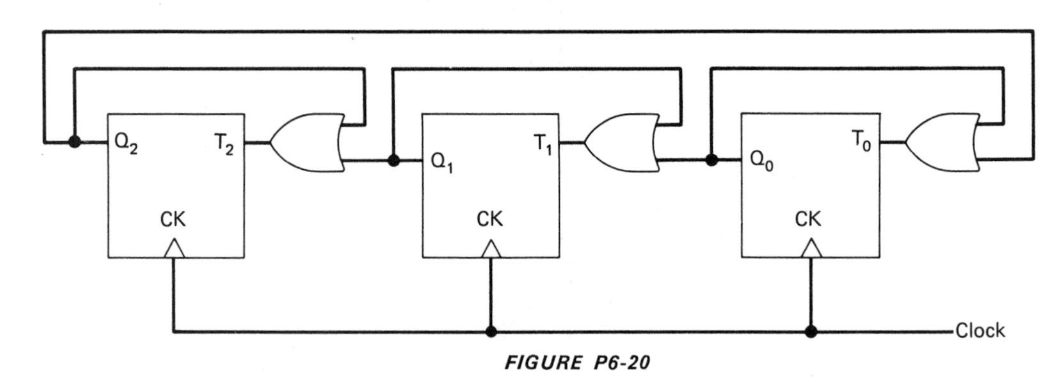

FIGURE P6-20

6-21 Design a two-phase clock generator using J-K flip-flops. The waveform of Figure P6-21 should be produced by your design.

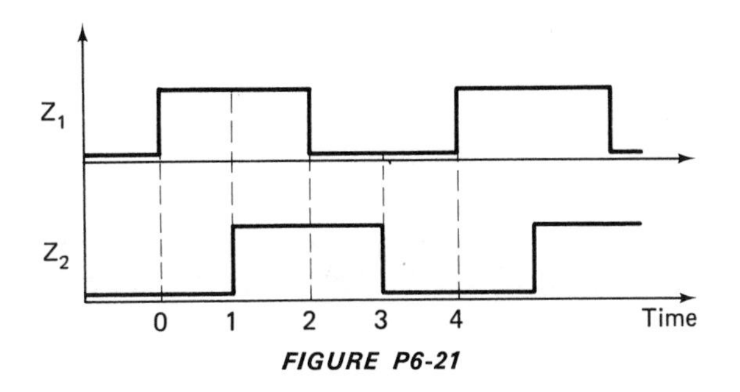

FIGURE P6-21

6-22 Using two J-K flip-flops and a minimal amount of combinational logic, design a circuit that produces the waveform shown in Figure P6-22.

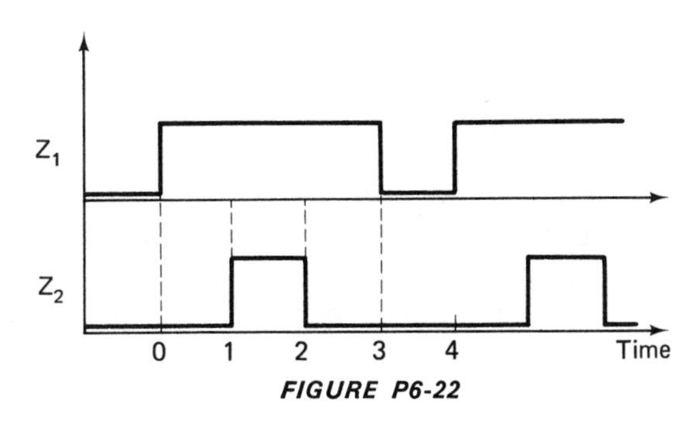

FIGURE P6-22

7 SEQUENTIAL CIRCUITS

7.1 OVERVIEW

Sequential circuits are a class of logic circuits in which the output conditions of the present time are determined by both present *and* past input conditions. In contrast, combinational logic circuits can respond only to inputs of the moment, and have no capacity to retain the effect of previous input conditions. This concept should not be new to us. In Chapter 5, we began an informal study of sequential circuits by analyzing flip-flops, and noted that flip-flops are the simplest of all memory devices in digital electronics. Flip-flop applications, such as shift registers and counters, are somewhat more complicated expressions of sequential circuits, but may nevertheless be handled by informal and intuitive techniques. In this chapter, we have reached a point in the study of sequential circuits where the circuit complexity can exceed our intuitive powers to understand. To meet this challenge, we will develop some formal techniques through which an understanding of complex sequential circuits can be achieved. While the circuits we analyze may be quite complex, the techniques we will learn to use for their analysis are surprisingly simple. It will be a most rewarding experience to begin with an intricate circuit design and watch its operation slowly unfold as succeeding stages of analysis progress. By reversing the order of the analysis procedure, sequential circuits of our own design can be realized. This adds an important and useful tool to our library of design techniques. Whether our concern is analysis or design, the primary intent of this chapter is to establish a connection between a sequential circuit's basic function, as it is shown in a flow graph, and the circuit's logic design.

7.2 DESCRIPTION OF SEQUENTIAL CIRCUITS

7.2.1 General Model

At the beginning of Chapter 3, in Figure 3-1, we saw how a combinational logic circuit could be represented in general terms as a block element. In such a general block element, inputs are shown at the left and outputs on the right. The flow of information is in one direction only—from inputs to outputs. No internal feedback paths and no internal memory are present. We recall also that the operation of a combinational logic circuit can be fully and completely described by a truth table. Now, in order to represent a sequential circuit in general terms, it is only necessary to make a slight alteration to the general model of a combinational circuit. Specifically, we must add external feedback paths to the combinational logic block, connecting certain outputs to inputs. This idea is illustrated in Figure 7-1. Take care to note in this figure how sequential circuit inputs and outputs are classified as either ***primary*** or ***secondary***, depending on whether they are involved in feedback paths. In general, primary variables serve as an interface to switches, indicator lamps, or other circuits.

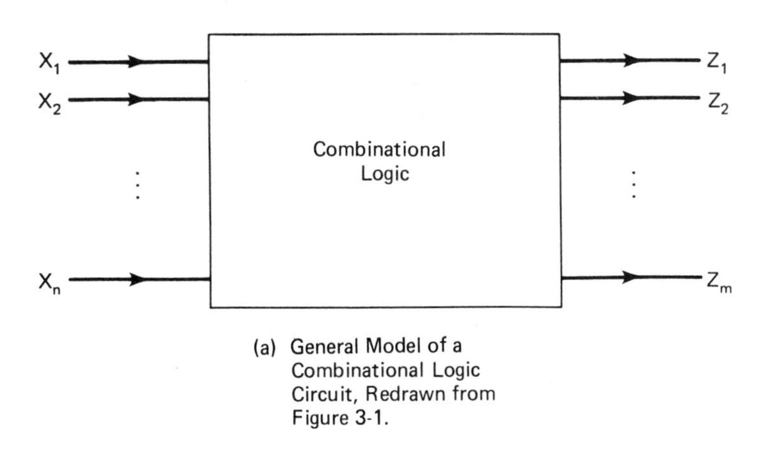

(a) General Model of a Combinational Logic Circuit, Redrawn from Figure 3-1.

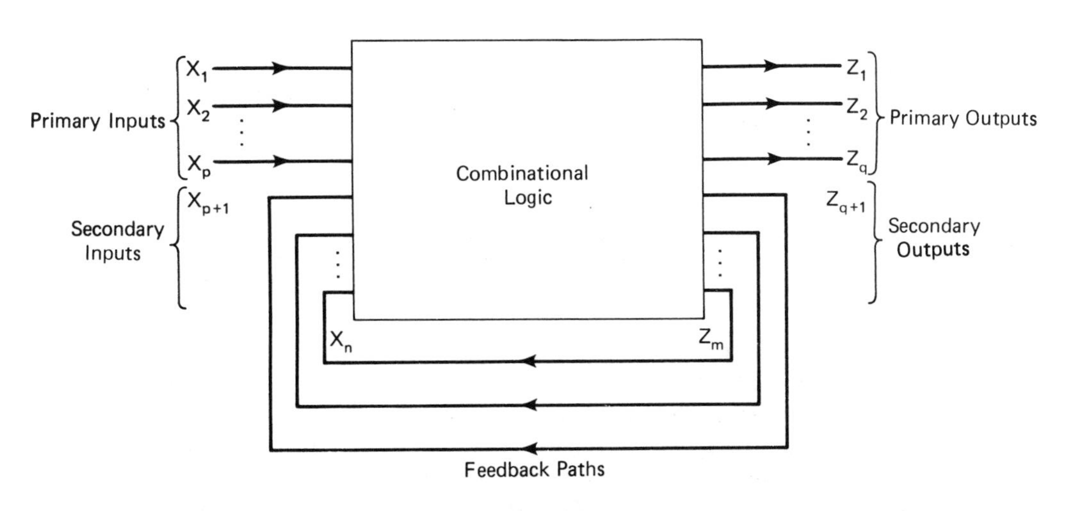

(b) General Model of a Sequential Circuit, Derived by Adding Feedback Paths to the Combinational Logic.

FIGURE 7-1

Secondary variables, on the other hand, are used only to establish feedback within the sequential circuit. The *state* of a sequential circuit is defined exclusively by the values of the secondary variables. A sequential circuit in which three secondary variables (feedback paths) are shown, for example, has exactly eight possible states. One goal of sequential circuit analysis is to determine all possible transitions that can occur between the various states. Another goal is to specify the primary output values that are generated during the state transitions and while a state is maintained.

Figure 7-2 shows how a simple NAND latch can be redrawn to conform to the general model of a sequential circuit. Note that the 'Q' output is *both* a primary and a secondary output.

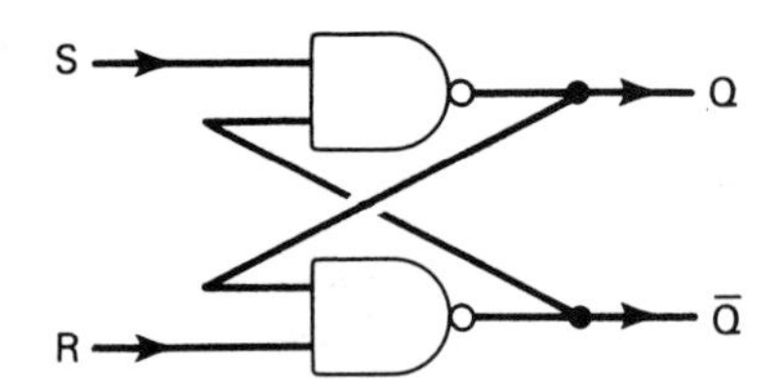

(a) Normal Schematic of a NAND Latch.

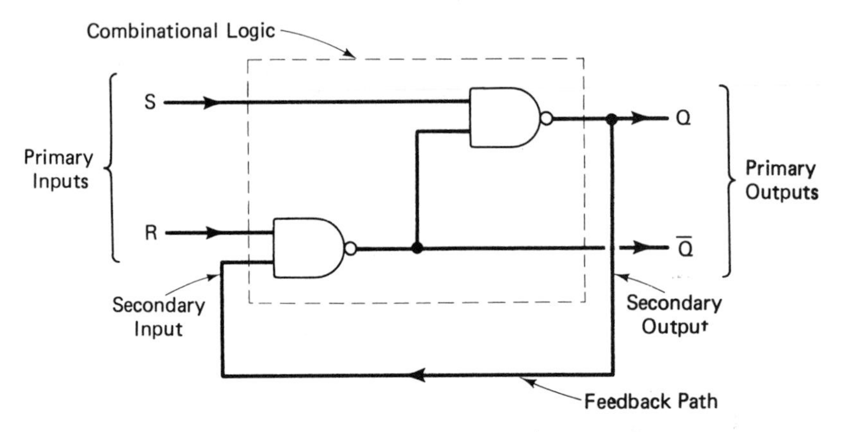

The Schematic of a NAND Latch is Redrawn to Conform to the General Model of a Sequential Circuit Given in Figure 7-1.

FIGURE 7-2

As we know, any combinational logic circuit constructed with real electronic components exhibits some fixed amount of propagation delay. Thus, when a sequential circuit is constructed, the propagation delay attributed to the combinational logic block introduces delay in the sequential circuit's performance. To greatly simplify the analysis procedure for sequential circuits, we will consider the delays present in the combinational logic to be *lumped together* into separate delay blocks. Therefore, we will now think of the combinational logic block as a connection of ideal (zero propagation delay) logic elements, whose output drives pure delay elements. These delay elements represent the combined effect of all gate delays that would normally be present within a realistic combinational logic block. Figure 7-3 illustrates these ideas.

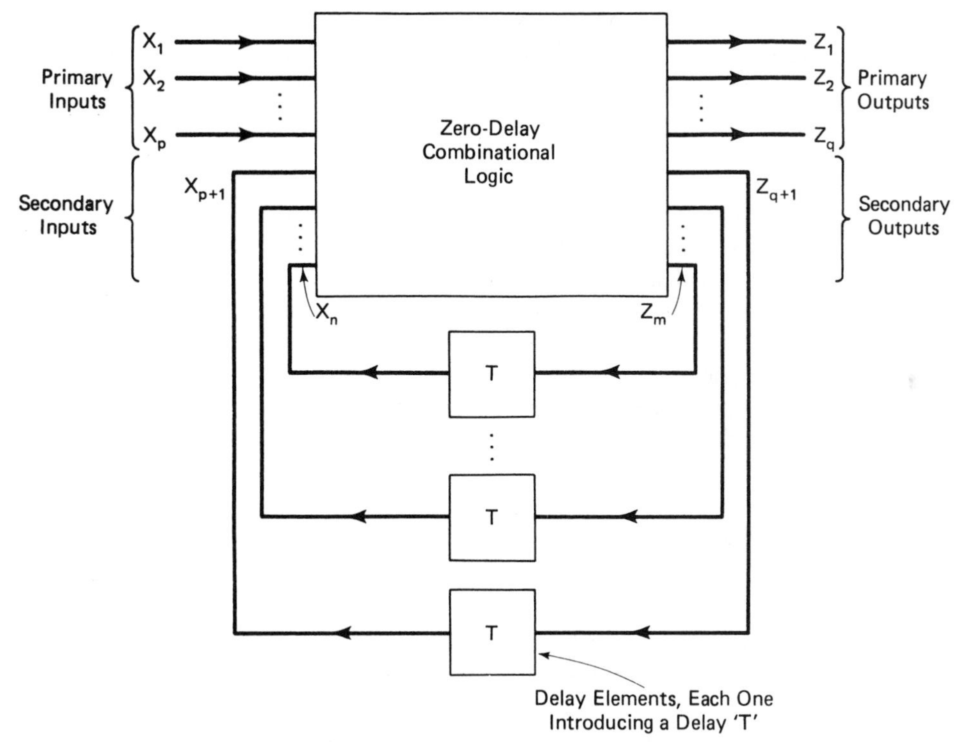

FIGURE 7-3 Propagation delay normally present in combinational logic is lumped into distinct delay elements, leaving a separate zero-delay combinational network.

We make the assumption here that *all* input-to-output paths possess equal propagation delay, even though this may not actually be the case. When our analysis procedure has been presented, we will reexamine this assumption to see how it may affect sequential circuit performance. To be consistent, delay elements should also be shown in the primary output paths. However, since the presence or absence of delay elements on the primary outputs has no effect on the analysis, they are omitted. In the lumped delay model of a sequential circuit, the *state* of the circuit is specified by the *outputs* of the delay elements. Of course, when no changes are occurring in the circuit, the output and input values of the delay elements are equal.

7.2.2 The Asynchronous and Synchronous Concepts

Consider the relationship between a simple NAND latch and an edge-triggered D flip-flop. Both are one-bit memory elements, and both are considered to be flip-flops. In the case of the NAND latch, however, changes in the S-R inputs can cause an almost instant change in the output, slowed only by the gate propagation delays. The edge-triggered D flip-flop, though, can change state only when a triggering edge is applied to its clock input. The NAND latch belongs to the class of ***asynchronous sequential circuits***. In such circuits, the only factor that determines the length of time between circuit states is gate propagation delay. The edge-triggered D flip-flops, on the other hand, belongs to the class of ***synchronous sequential circuits***. Any clocked device, such as a synchronous counter or shift register, is considered to be a synchronous sequential circuit. In this case, the length of time between circuit states is determined exclusively by the clock period. In synchronous circuits, the "clock" input is thought of as a control variable, *not* as a primary input.

The general model of a sequential circuit that was shown in Figure 7-3 is actually the representation of an asynchronous sequential circuit. This is so because the delay elements represent gate propagation delay only, and are in no way involved with an external clock signal. Figure 7-3 can be modified slightly to represent synchronous sequential circuits. To do this, we need to replace each delay element with a clocked flip-flop, as shown in Figure 7-4. A three-stage synchronous counter is redrawn in Figure 7-5 to conform with the general model of a synchronous sequential circuit. Note that one output is both a primary and a secondary output, and that there are no primary inputs. Also, T-type flip-flops are used in the synchronous counter instead of D-type devices.

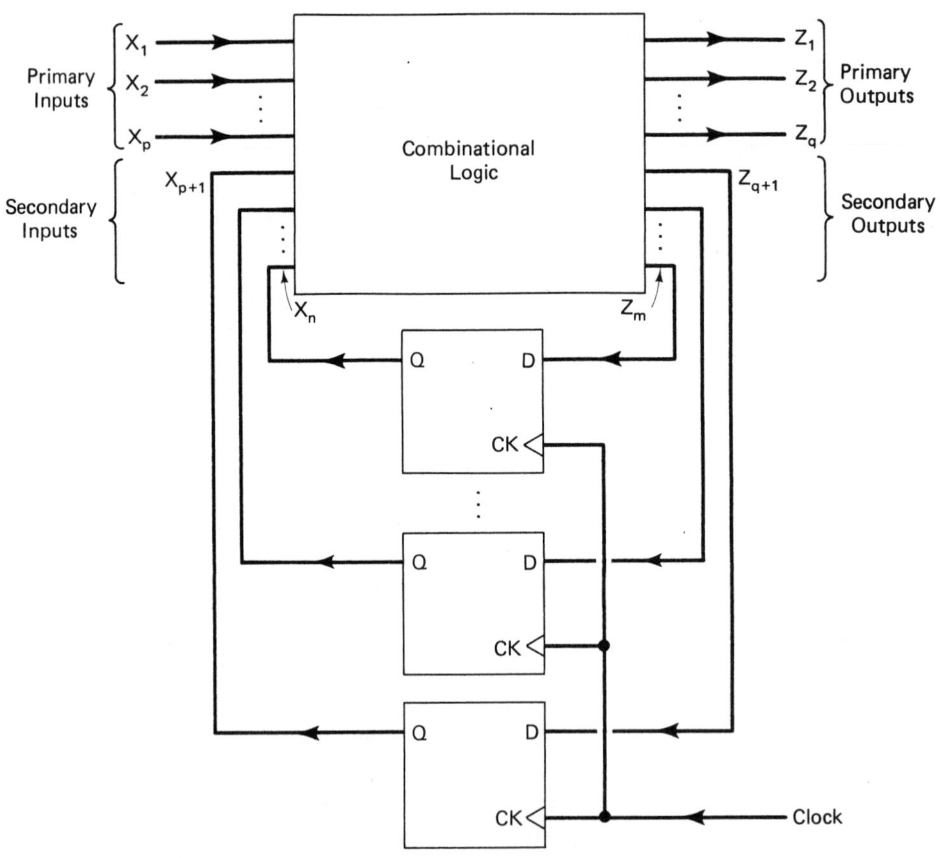

FIGURE 7-4 *General model for a synchronous sequential circuit, derived by replacing each delay element of Figure 7-3 with a clocked flip-flop.*

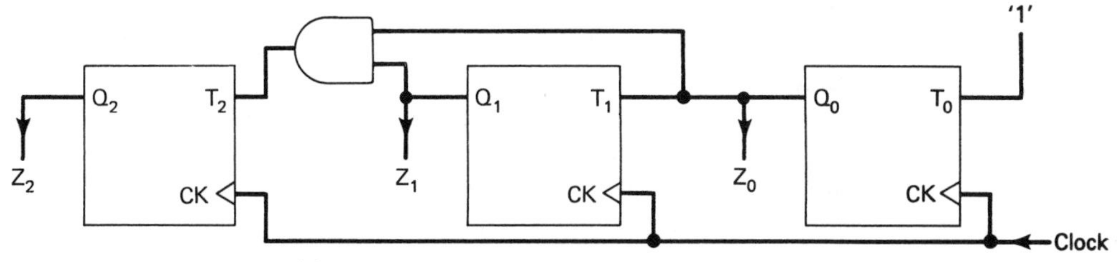

(a) Standard Three-Stage Synchronous Counter.

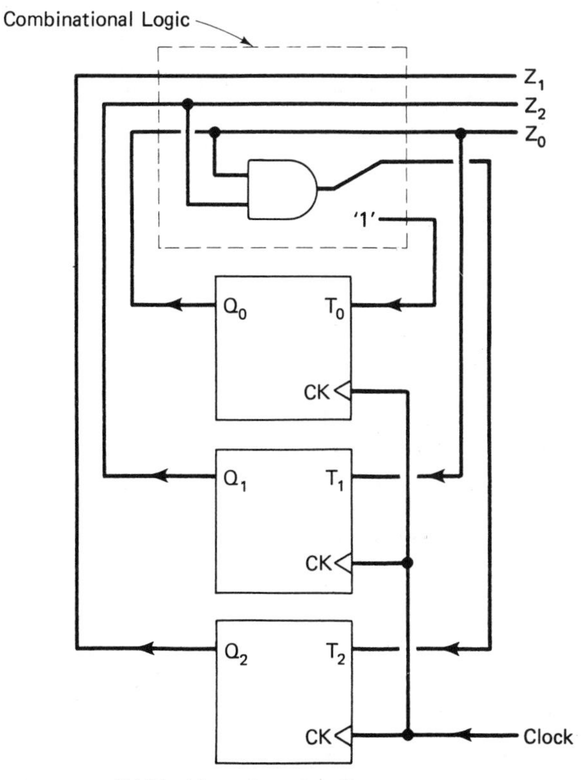

(b) The Three-Stage Synchronous Counter is Redrawn to Conform with the General Model of a Synchronous Sequential Circuit.

FIGURE 7-5

We will find that the analysis procedure for asynchronous and synchronous sequential circuits is virtually identical. However, certain practical considerations, which will be noted, make synchronous designs more attractive in terms of reliability and testing. We present the analysis technique using an asynchronous circuit as the vehicle, but keep in mind that the technique is also valid for synchronous circuits. This equivalence will be demonstrated by example in Section 7.4.

7.2.3 *The Relationship Between Counters and General Sequential Circuits*

Digital counters of all types must certainly be classified as sequential circuits. They are a special kind of sequential circuit, however, in that counters have no primary inputs. It is this feature which permits us to analyze and design counters in a simple, intuitive manner. From the beginning, we described a count sequence using a flow graph. We shall soon see that the operation of *all* sequential circuits can be described

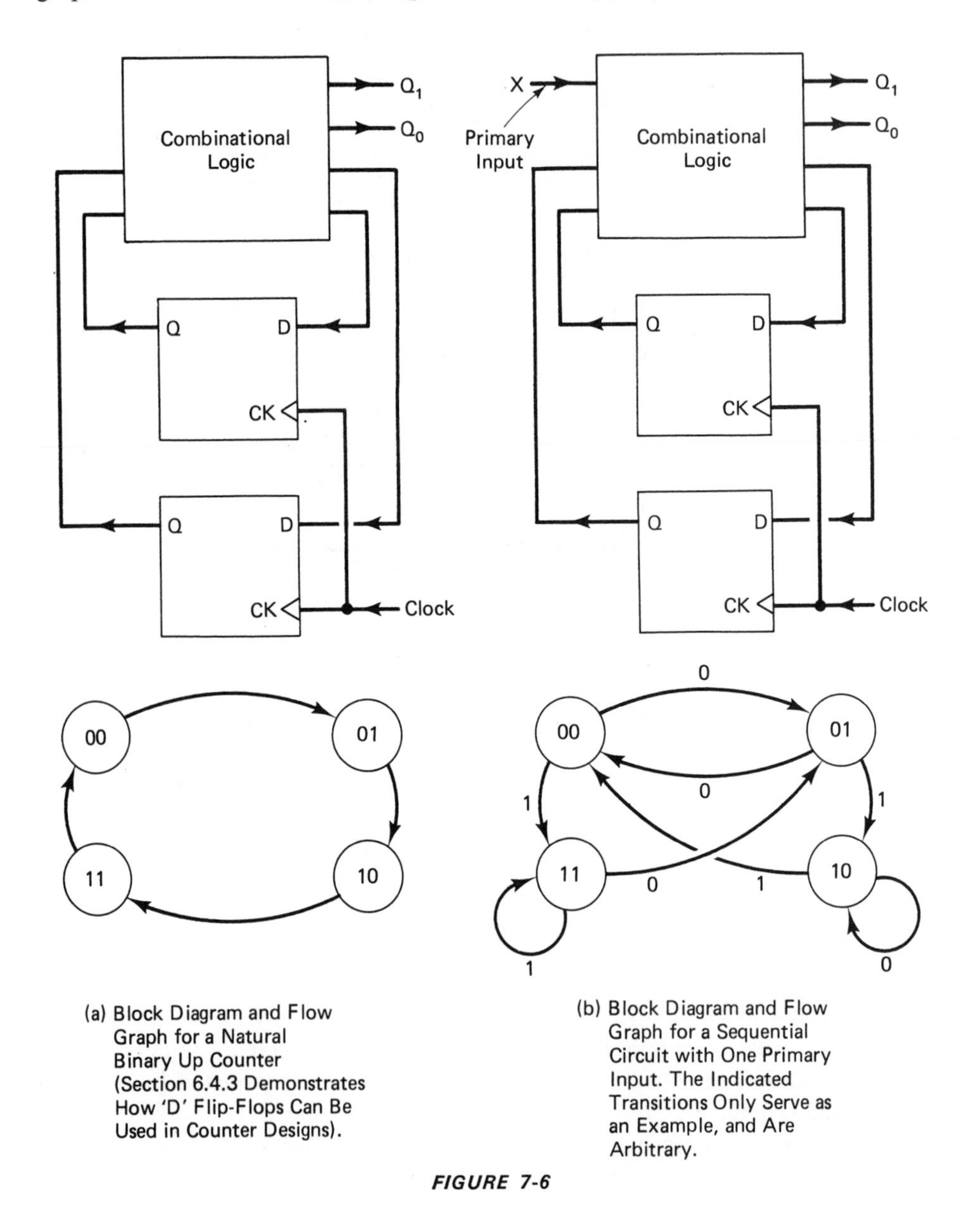

(a) Block Diagram and Flow Graph for a Natural Binary Up Counter (Section 6.4.3 Demonstrates How 'D' Flip-Flops Can Be Used in Counter Designs).

(b) Block Diagram and Flow Graph for a Sequential Circuit with One Primary Input. The Indicated Transitions Only Serve as an Example, and Are Arbitrary.

FIGURE 7-6

using a flow graph. Rather than being connected in just closed loops, the states of a sequential circuit with primary inputs will generally have many different interconnections. The exact path of transition from state to state will be dependent on the primary inputs. It is reasonable to believe, then, that a sequential circuit with no primary inputs, such as a counter, can have only one possible path of transitions. Figure 7-6 compares the block diagram and flow graph of a four-state counter to that of a typical four-state synchronous sequential circuit with one primary input. Note that the value of the primary input determines which one of the two possible exit paths from each state is to be followed.

7.3 ASYNCHRONOUS SEQUENTIAL CIRCUITS

7.3.1 Analysis

The ultimate goal of the analysis procedure is to arrive at a flow graph, as in Figure 7-6, for the sequential circuit under study. The only information available to us initially is the logic schematic, indicating how the combinational logic and feedback paths are connected. The desired goal is reached by executing the following five steps, in order: Given the circuit schematic,

1. Find the *excitation equations.*
2. Determine the *excitation table.*
3. Translate the excitation table into a *transition table.*
4. Derive the *state table* and *output table.*
5. Draw the flow graph.

Each of these steps will be explained in detail as we work through a simple example. Actually, some of this terminology should already be familiar from our study of programmed counters in Chapter 6. A slight amount of added detail is involved, though, so that sequential circuits in general, not just counters, can be accurately represented.

To illustrate the analysis procedure, one of the simplest possible asynchronous sequential circuits is chosen, namely the NAND latch. Thus, since we are completely familiar with this circuit, the results found at the end of each step will support our previous understanding of the NAND latch.

(1) Excitation equations: Figure 7-7 illustrates a NAND latch circuit, drawn to conform to the general model of a sequential circuit. Note that there are two ways in which the NAND latch can be drawn—one with only 'Q' feedback and the other with only '$\bar{Q}$' feedback. The analysis will work in either case and will show identical operation. Since no flip-flops are present in the feedback path, this sequential circuit is asynchronous. An ***excitation equation*** is a Boolean equation that relates a secondary output variable to primary and secondary input variables. One excitation equation is required for *each* secondary output. The excitation equations for an asynchronous sequential circuit are determined as follows:

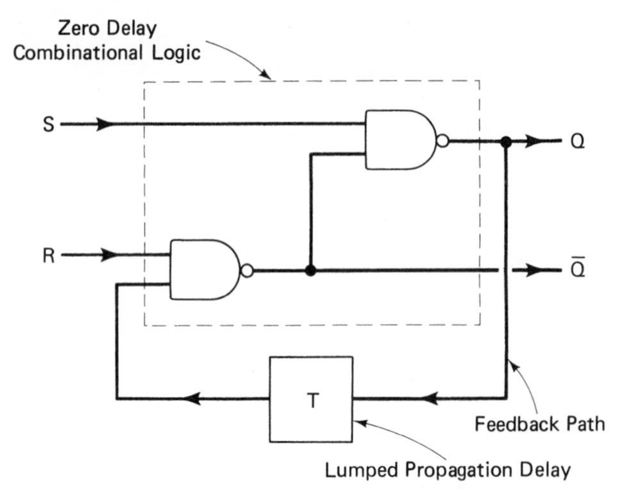

FIGURE 7-7 *The NAND latch, drawn to conform with the general model.*

(a) Break *each* feedback path in the sequential circuit around the lumped delay element.

(b) Label the secondary output side of each broken path with a *capital letter* variable, and the secondary input side with a corresponding *lowercase letter* variable. Of course, if more than one feedback path is present, different letters should be used so that variables may be distinguished.

(c) For each secondary output variable (capital letter), write a Boolean equation in terms of the primary and secondary input variables. These are the excita-

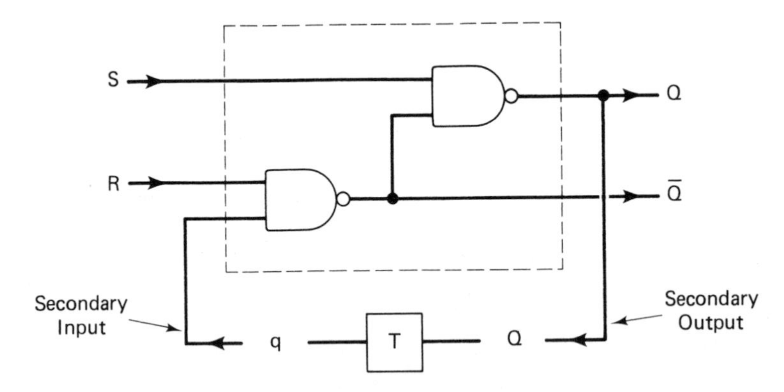

(a) Breaking the Feedback Path and Labeling the Secondary Output and Input Variables.

$$Q = (qR)S = qR + S$$

$$\boxed{Q = qR + S}$$

(b) The Excitation Equation for a NAND Latch

FIGURE 7-8

tion equations. Remember that the lumped delay element is only delay, and does not appear in the excitation equations.

Figure 7-8 illustrates how an excitation equation is determined for the NAND latch. Note that since only one feedback path is present in this circuit, only one excitation equation will be found. Also, it is important to realize that 'Q' and 'q' are always equal, *except* when a change in 'Q' occurs. At that time, 'Q' and 'q' remain unequal for 'T' nanoseconds ('T' is the lumped propagation delay), and the delay element is said to be *excited.* When 'T' expires, the delay element becomes *quiescent.* Figure 7-9 illustrates these ideas.

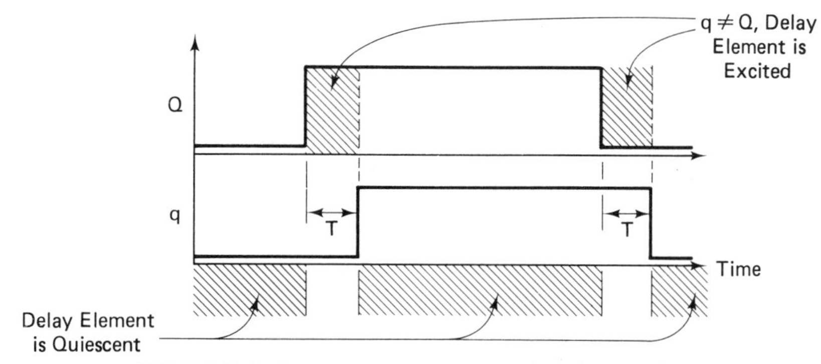

FIGURE 7-9 *Q = q at all times, except for 'T' units of time after a change in 'Q' has occurred.*

(2) Excitation table: The term "excitation table" was used in Chapter 6 in our discussion of programmed counters. In that context, an excitation table looked much like a truth table, and specified the binary values present on each of the flip-flop inputs prior to a clock pulse. From these values, the counter's next state could be predicted. The basic purpose of an excitation table is the same when analyzing a sequential circuit. In this case, however, *primary inputs* must also be accounted for. As a result, the excitation table becomes somewhat more complex. Note that in an asynchronous circuit, we speak of excitation of the *delay elements* since no flip-flops are present.

To ease the reading of a sequential circuit's excitation values, the excitation table is arranged as a two-dimensional structure, resembling a Karnaugh map. Row addresses correspond to secondary input values, while column addresses correspond to primary input values. The entry specified by row and column address digits defines the excitation observed on the delay elements. Each entry corresponds to one cell of a Karnaugh map and contains as many binary digits as there are delay elements. Recall that the input lines to the delay elements are the secondary outputs of the sequential circuit.

An excitation table for the NAND latch is shown in Figure 7-10, and was derived exclusively from the excitation equation. Whenever an excitation table is drawn, all primary inputs should be separated from secondary inputs. The usual convention is to place primary inputs on the top row and secondary inputs in the leftmost column, as we have described.

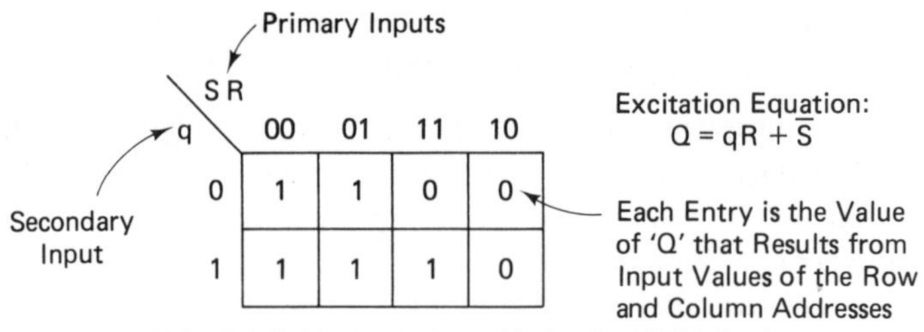

FIGURE 7-10 *Excitation table for the NAND latch.*

(3) Transition table: To understand the meaning of an excitation table, several important questions must be answered. First, we inquire:

> What causes changes of state to occur in a sequential circuit?

In a combinational circuit, we know that a change in the output conditions can be brought about by a change in the inputs. This is also true in a sequential circuit; however, the "inputs" are not limited to only the primary inputs, but include both primary *and* secondary inputs. Similarly, the "outputs" of a sequential circuit include both primary and secondary outputs. With this understanding, we see that a change in primary input values (which we control) can produce a change in secondary output values, which can then cause a change in secondary input values (which we do not control), resulting in yet another change in secondary output values, and so on. Thus, it is possible for a chain of events to occur in rapid succession, and perhaps for oscillator action to be set up. While it is correct to say that a change in input values can cause a change of state to occur, it is not the most fundamental answer to the original question.

Note that the excitation equations and excitation table express 'Q' (the secondary output) and 'q' (the secondary input) as though they are *different* variables, even though their values are different for only the very short delay period 'T'. This idea is crucial to the analysis and is the key to our understanding of sequential circuits. As long as 'Q' and 'q' are equal (either both '0' or both '1') and the primary inputs remain steady, the circuit is not inclined to change state, and is therefore *stable*. An initial change in the primary inputs, and any subsequent changes in the secondary inputs which cause 'Q' and 'q' to become unequal, results in the circuit being *unstable*. The instability of a circuit state is always resolved after the delay period 'T', at which time 'Q' and 'q' become equal. However, the new secondary input values may cause a new state of instability. The sequential circuit will change state repeatedly until a stable state is found and maintained. In some sequential circuits, only one state transition occurs before stability is restored. In other circuits, a chain of state transitions must occur before stability is restored. In yet other circuits, a closed loop of unstable states is entered, stability is never found, and oscillator action results. We will see examples of each possibility within the next few pages. Our original inquiry is answered now by saying that:

> A sequential circuit is caused to change state when its current state becomes unstable. The current state may become unstable as a result of a change in the

primary inputs, the secondary inputs, or both. Changes of state occur at a rate determined by the lumped delay period, 'T', and progress until stability is found and maintained. As long as the secondary outputs are exactly equal to the corresponding secondary inputs, the circuit is stable.

To support this statement by an example, let us now return to the NAND latch and its excitation table in Figure 7-10.

In order to determine the sequence of changes that can occur in the NAND latch as the inputs are changed, we must apply the stability criterion to the excitation table. In particular, we must investigate the relationship between 'Q' and 'q' for every entry in the excitation table, and deduce whether the entry represents a stable or an unstable condition. Figure 7-11(a) depicts the comparison of two typical entries and shows how one entry is stable and the other is unstable. A ***transition table*** is shown in Figure 7-11(b), and results by replacing each numeric entry in the excitation table with an open circle for stability or a solid circle for instability. Finally, the transition table is completed in Figure 7-11(c) by adding pointers to show how unstable states change

'q' and 'Q' are Not Equal, and Hence the Entry at q = 0, S = 0, and R = 1 is an Unstable State. After the Delay Period 'T', 'q' Changes from '0' to '1'.

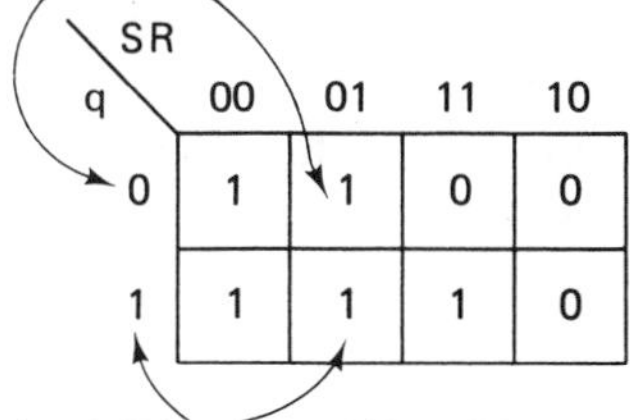

'q' and 'Q' are Both '1', and Hence the Entry at q = 1, S = 0, and R = 1 is a Stable State. No Further Transitions Will Occur Until a Primary Input is Changed.

(a) Excitation Table for a NAND Latch.

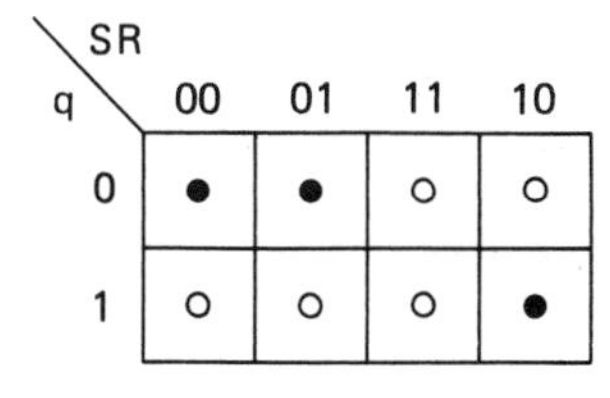

(b) In a Transition Table, Open Circles Represent Stable States, While Solid Circles Represent Unstable States.

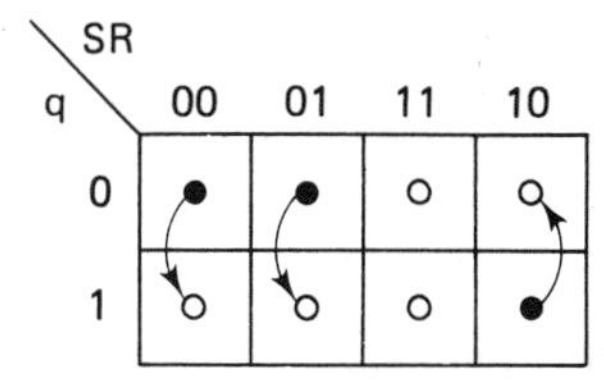

(c) The Transition Table is Completed by Entering Pointers to Show the Direction of Change for Unstable States.

FIGURE 7-11

in search of stability. Remember that when 'Q' and 'q' are unequal in an unstable state, 'q' will *become equal* to 'Q' after the delay period 'T'. Thus, we always know how to write the pointer in the transition table.

Let us pause for a moment and see if the transition table of Figure 7-11(c) agrees with our previous understanding of the NAND latch. First, we know that if both 'S' and 'R' inputs are '1', the latch is in the "memory" state—that is, either $Q = 0$ or $Q = 1$ can be maintained. In the transition table, under the column where 'S' and 'R' both equal '1', we find that the circuit is stable for either $Q = 0$ or $Q = 1$. This agrees perfectly with our earlier understanding. Now, assume that $Q = 0$ and 'S' is asserted to '0', with 'R' remaining '1'. We know that this should "set" the latch (cause 'Q' to become '1'). When (S,R) is changed from (1, 1) to (0, 1), the transition table indicates that the circuit first enters an unstable state $[(S,R) = (0,1),\ Q = 0]$, and then after the delay period, 'Q' changes to '1' and becomes stable. This also agrees with our earlier understanding of the NAND latch. If (S,R) now changes from (0,1) back to (1,1), a stable state is entered immediately, as we would expect, and no change in 'Q' occurs. In a similar manner, we can test the other possible transitions and find that in every case, the transition table supports our earlier understanding of the NAND latch.

(4) State table and output table: Having determined the transition table, it is a simple matter to form a ***state table***, which expresses the same information in a more abstract form. In a state table, each circuit state is labeled with an alphabetic letter, rather than with binary digits. Stability and instability are determined by comparing the letter in an entry with the row letter. If they are the same, the state is stable. If they are different, the state is unstable and a change of state will be observed. Figure 7-12 illustrates this idea.

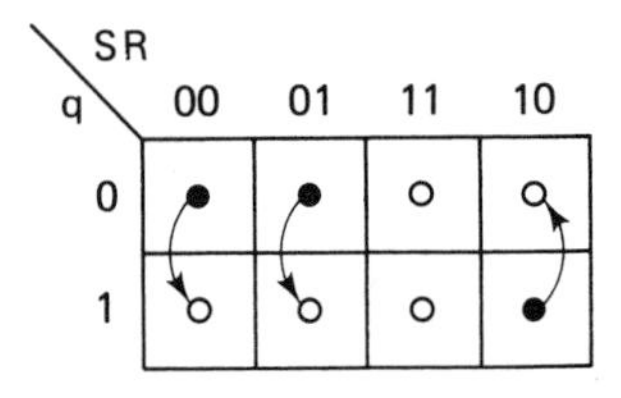

(a) Transition Table.

q \ SR	00	01	11	10
A	B	B	A	A
B	B	B	B	A

(b) State Table.

FIGURE 7-12

The ***output table*** provides information about the primary outputs, and is determined by ***output equations***. By following the same procedure as we did for the excitation equations, the output equations can be found. Referring back to Figure 7-8, the output equations are derived, and are given by Eqs. 7-1 and 7-2.

$$Q = qR + \bar{S} \tag{7-1}$$

$$\bar{Q} = \bar{q} + \bar{R} \tag{7-2}$$

Note that Eq. 7-1 is the same as the single excitation equation found for the NAND latch. This is so because the primary and secondary outputs for 'Q' are identical. '$\bar{Q}$' is only a primary output, and thus an excitation equation for it is not found. From Eqs. 7-1 and 7-2, an output table is determined and is shown in Figure 7-13.

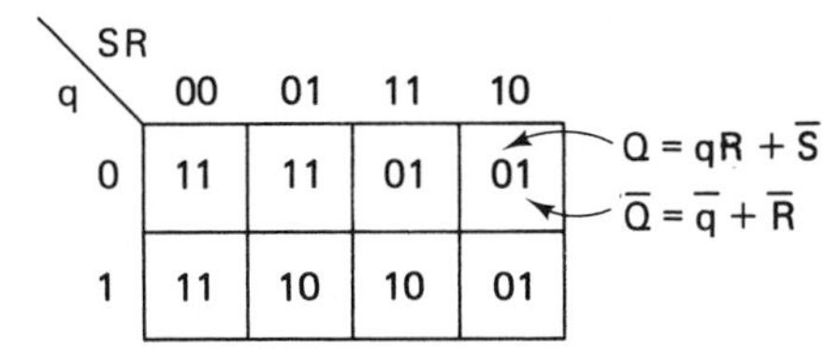

FIGURE 7-13 Output table for the NAND latch.

(5) Flow graph: As our final step, a flow graph is drawn to convey the sequential circuit's operation in a simple, graphic form. The type of flow graph that is used for general sequential circuits is very similar to that used for counters, except that now we must indicate primary input values that exist when a state transition occurs. Also, primary output values must be shown. The flow graph for our NAND latch is derived exclusively from the state table and output table, and is shown in Figure 7-14. Note how a slash (/) separates the primary inputs from the primary outputs. It is important to realize that the primary outputs take effect immediately when a change in input conditions occur, owing to the fact that in our lumped delay model, primary outputs are tapped *before* the delay elements. Thus, when a state is unstable, the primary outputs are valid before the delay period 'T' has expired.

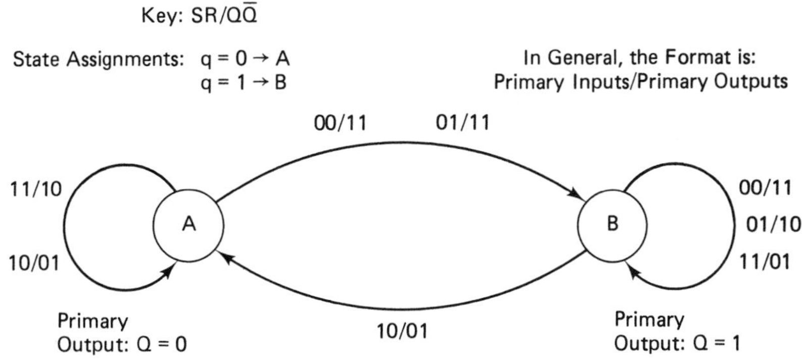

FIGURE 7-14 Flow graph for the NAND latch. Note in this case that 'Q' is both a primary and secondary output.

By studying Figure 7-14 in detail, we can verify that the NAND latch is completely represented. First, note that when the latch is in state 'A' (the "reset" state), it will remain in that state

1. If the primary inputs are both '1', or
2. If we attempt to reset the latch while it is already in the reset state.

Similarly, with the latch in state 'B' (the "set" state), it will remain in that state

1. If the primary inputs are both '1', or
2. If we attempt to set the latch while it is already in the set state.

A transition from 'A' to 'B' occurs when we apply $S = 0$, $R = 1$, thereby causing the latch to become "set" from the "reset" state. Note that on the primary outputs, $Q = \bar{Q}$ for the delay period 'T', and stabilizes to the correct value when state 'B' is entered. A transition from 'B' to 'A' occurs when we apply $S = 1$, $R = 0$, thereby causing the latch to become "reset" from the "set" state. Finally, whenever both inputs are '0', the condition $Q = \bar{Q} = 1$ exists, as we know from the NAND latch circuit.

The important concepts in asynchronous sequential circuit analysis are now summarized:

(1) General definitions

(a) A *sequential circuit* is modeled generally as a combinational logic block with external feedback paths.

(b) Sequential circuit inputs and outputs that interface with external circuits, switches, or indicators, are referred to as *primary*. Inputs and outputs that are involved in the feedback paths are referred to as *secondary*.

(c) The propagation delay that is normally present in combinational logic is separated from the combinational logic block of the general model and consolidated into *delay elements* that are embedded in the feedback paths. This is done to simplify the analysis procedure. Each delay element represents the *lumped delay* of one path in the combinational logic. All delay elements are assumed to be of equal value to simplify the analysis, even though this may not actually be the case.

(d) The *state* of a sequential circuit is defined by the output values of the delay elements. A circuit with three delay elements, for example, whose values are '1', '1', and '0', respectively, occupies the state '110'. The maximum number of states a sequential circuit may have is 2^n, where 'n' is the number of delay elements present.

(e) A delay element is said to be *excited* when its input and output values are not equal. After the lumped delay, 'T', has expired, the delay element input and output become equal, and the delay element is said to be *quiescent*. The delay element's output is also a secondary input, and its changing may cause a new condition of excitation. It is possible for a delay element to be in a constant cycle of excitations.

(f) A sequential circuit is defined to be *unstable* whenever one or more delay elements are excited. When all delay elements are quiescent, the circuit is *stable*. Changes of state occur as a result of instability, and progress such that a state of stability is sought. It is possible that stability will never be found for a particular primary input value, in which case oscillator action is observed.

(2) Techniques

(a) The *excitation equations* describe algebraically how the delay elements can be excited by various combinations of primary and secondary input values. One excitation equation is needed for each feedback path. An excitation equation is determined by writing the logic relationship between a secondary output and the primary and secondary inputs.

(b) An *excitation table* is a map that graphically relates primary and secondary inputs to secondary outputs. Generally, primary inputs are shown on the top address row and secondary inputs on the left address column. Each map entry depicts the excitation (secondary output values) resulting from the input values shown in the row and column addresses. In contrast to Karnaugh maps, each entry of an excitation table may have as many bits as there are secondary outputs. The stability of a state is found by comparing the excitation (the binary digits in the entry) to the current secondary inputs (the values of the left address column). Equality of these quantities denotes stability, whereas inequality in any digit position denotes instability. After the delay period 'T', the excitation of an unstable state becomes the new secondary input, and a new map entry is examined.

(c) By expressing stable states with open circles and unstable states with solid circles, and adding pointers to show the direction of state change, a *transition table* is produced.

(d) A *state table* results when binary digit labels for states are replaced by alphabetic letters. This step greatly simplifies the construction of a flow graph. *Output equations* describe the primary outputs as a function of the primary and secondary inputs. In some cases, the primary and secondary outputs are identical. An *output table* is similar to an excitation table in structure, but shows only the primary output values occurring for each state.

(e) To concisely and graphically express the operation of a sequential circuit, a *flow graph* is constructed, based on the state table and output table. Each state that a sequential circuit may have is shown, along with all permissible transitions that may lead to, or exit from, each state. A transition is denoted by a pointer. Each pointer is labeled to reflect primary input values that exist when the transition occurs, and primary output values that are produced.

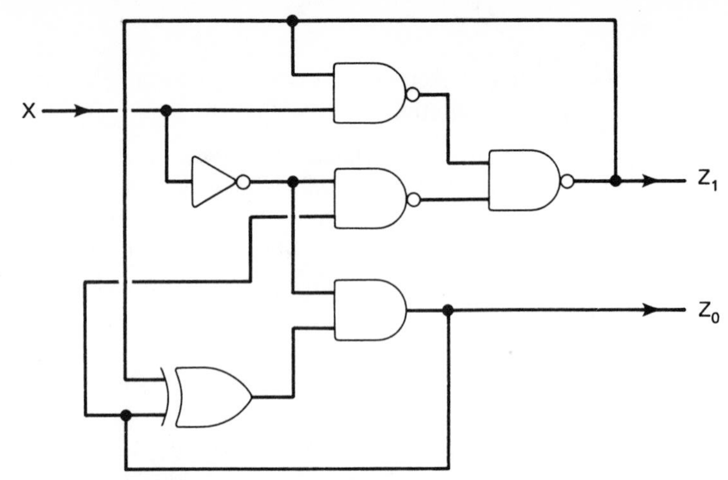

FIGURE 7-15 *Sequential circuit to be analyzed.*

Although the NAND latch is an extremely simple circuit, not requiring the detailed analysis that we have applied to it, the example served well to illustrate the basic ideas of sequential circuit analysis. We will now examine a somewhat more complex circuit whose operation is not easily revealed by just studying the logic schematic. This circuit is illustrated in Figure 7-15.

To begin the analysis, we must identify the primary and secondary inputs and the primary and secondary outputs. The circuit is redrawn in Figure 7-16 to conform with the general model. From this figure, we see that there are two outputs, and that they are both primary and secondary. Also, one primary input is found. Of course, there are always as many secondary inputs as there are secondary outputs. By breaking each feedback path and labeling secondary inputs with lower case letters, two excitation equations are found:

$$Q_0 = \bar{X}(\bar{q}_1 q_0 + q_1 \bar{q}_0) \tag{7-3}$$

$$Q_1 = X q_1 + \bar{X} q_0 \tag{7-4}$$

FIGURE 7-16 *Sequential circuit of Figure 7-15 redrawn to conform with the general model.*

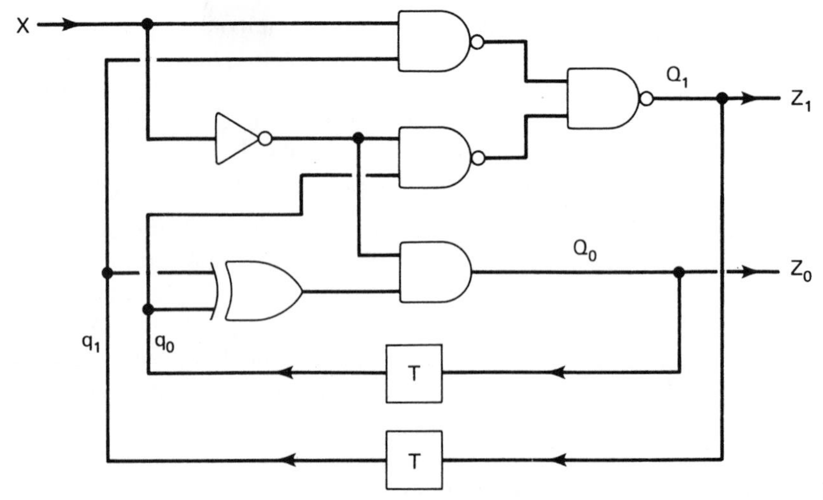

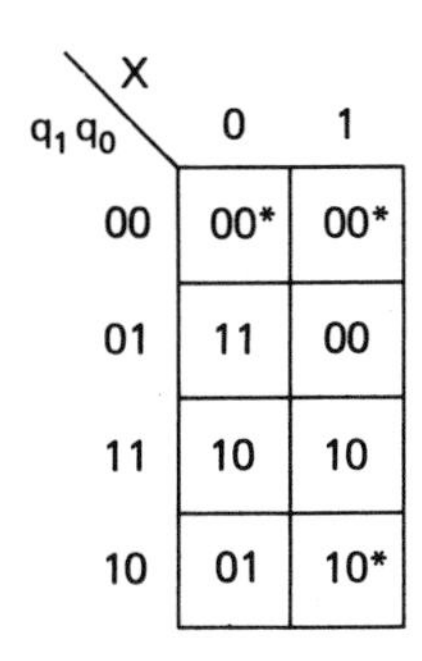

$q_1 q_0$ \ X	0	1
00	00*	00*
01	11	00
11	10	10
10	01	10*

(a) Excitation Table. States Denoted by "*" are Stable Because in Each Case, the Secondary Inputs and Outputs are Equal.

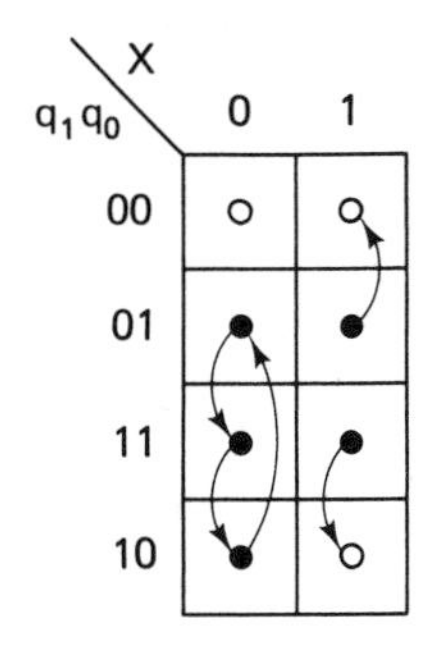

(b) Transition Table.

FIGURE 7-17 *The circuit of Figure 7-15 is analyzed.*

In contrast to the NAND latch circuit, there are, in this case, two secondary inputs and one primary input. An excitation table and transition table corresponding to these equations are shown in Figure 7-17. In the excitation table, we see that three stable states exist [for $(q_1, q_0, X) = (0, 0, 0)$, $(0, 0, 1)$, and $(1, 0, 1)$]. Recall that these three states are stable because, in each case, the excitation resulting from the inputs exactly equals the present state. Thus, the circuit is not inclined to change any of these states. The transition table reveals one three-state closed cycle, containing *no* stable states. Within this loop, the outputs will appear to oscillate with a period of $3T$. Also, note that if the '00' state is ever entered, the circuit will never change state again. Thus, we can think of '00' as a ***trap state***. To summarize our findings, a state table and flow graph are constructed, and are shown in Figure 7-18. Note that since the primary outputs are exactly equal to the secondary outputs, there is no need for additional output equations; the state values are the outputs.

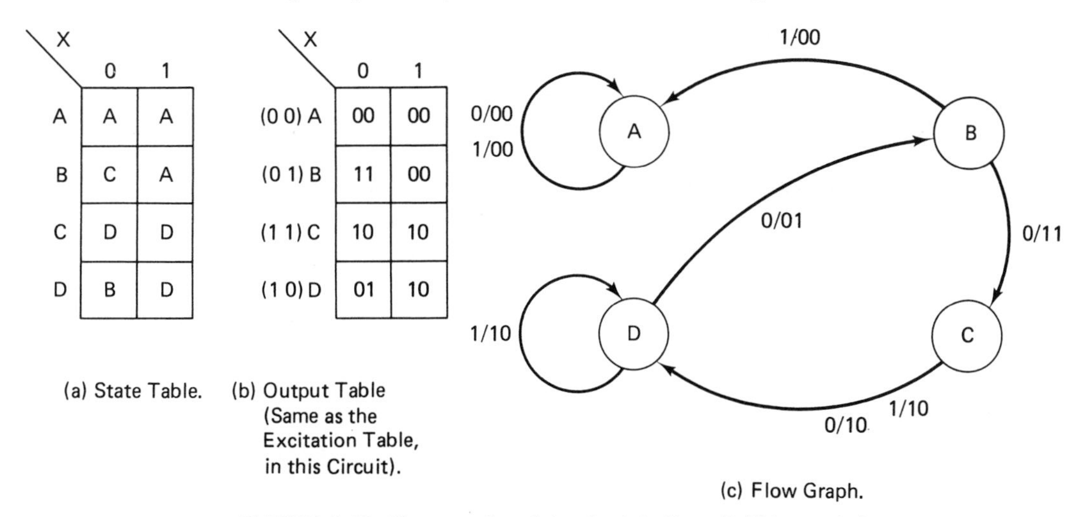

X	0	1
A	A	A
B	C	A
C	D	D
D	B	D

(a) State Table.

X	0	1
(0 0) A	00	00
(0 1) B	11	00
(1 1) C	10	10
(1 0) D	01	10

(b) Output Table (Same as the Excitation Table, in this Circuit).

(c) Flow Graph.

FIGURE 7-18 *The operation of the circuit in Figure 7-15 is revealed.*

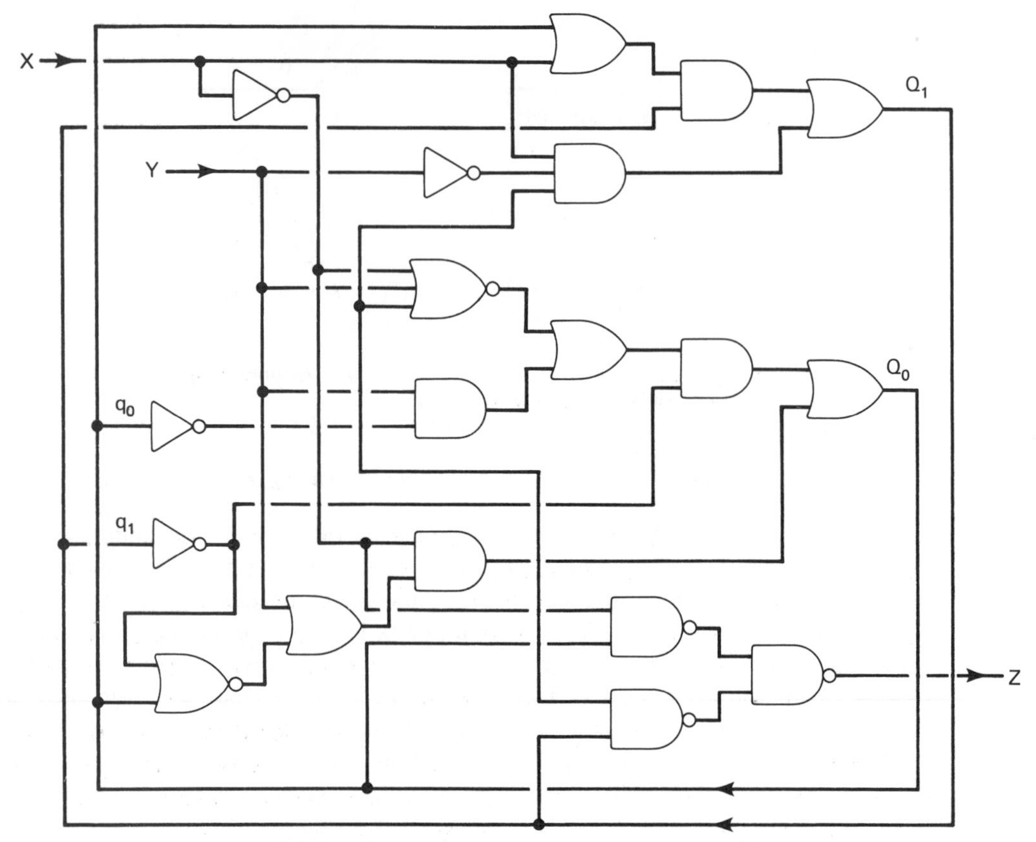

***FIGURE 7-19** Asynchronous sequential circuit to be analyzed.*

7.3.2 Races and Hazards

In our previous discussions, we have assumed that the total propagation delay normally present in the combinational logic block of a sequential circuit is consolidated solely in the delay elements. This assumption made the analysis easier than it would have been otherwise, although it is somewhat unrealistic. At this time, we will examine a sequential circuit that does not behave exactly as our analysis procedure suggests. In fact, we will see that the discrepancy results from the delay being *distributed* among the logic gates rather than being lumped into hypothetical delay elements. The analysis procedure will not be invalidated, but instead, we must simply become aware of the effects that distributed delay can have. The end result of the analysis can then be modified, as needed. In practice, distributed delay is always present in the combinational logic block of a sequential circuit. Consequently, the following discussion has great practical importance.

Consider the sequential circuit shown in Figure 7-19. The excitation and output equations for this circuit are found to be (in MSP form)

$$Q_1 = q_1q_0 + q_1X + \bar{q}_0X\bar{Y} \tag{7-5}$$

$$Q_0 = \bar{q}_1q_0X\bar{Y} + \bar{q}_1\bar{q}_0Y + \bar{X}Y + q_0\bar{q}_0\bar{X} \tag{7-6}$$

$$Z = q_1\bar{q}_0 + q_0\bar{X} \tag{7-7}$$

Figure 7-20 depicts the excitation and transition tables for this circuit. From Figure 7-20 and Eq. 7-7, a state table and an output table can be determined. These are shown in Figure 7-21, along with the flow graph.

At this point, the analysis of the circuit in Figure 7-19 seems complete and did not offer any apparent complications. Unfortunately, however, if the circuit were actually built and tested, it would probably not operate exactly as our analysis suggests. In particular, two-state transitions would be subject to possible error, and are illustrated in Figure 7-22. To understand why the analysis procedure might not tell the whole story, we must ask:

What makes these transitions different from the others?

By studying Figure 7-22, we will note that these two transitions are the only ones in which *both* secondary variables attempt to change at the same time. If our lumped delay model accurately represented the real circuit, there would be no problem. In fact, however, the delay is distributed among the logic gates of the combinational logic block. Thus, it is possible, and likely, that the delay paths for each secondary output are actually different. Consequently, when a change of state such as (Q_1,Q_0) changing from (1,0) to (0,1) is expected, we may actually observe either one of the following:

$$(1,0) \longrightarrow (0,0) \longrightarrow (0,1) \qquad (Q_1 \text{ changes first})$$

or

$$(1,0) \longrightarrow (1,1) \longrightarrow (0,1) \qquad (Q_0 \text{ changes first})$$

(Q_1, Q_0)

q_1q_0 \ XY	00	01	11	10
00	00	01	01	10
01	00	01	00	01
11	10	11	10	10
10	01	01	10	10

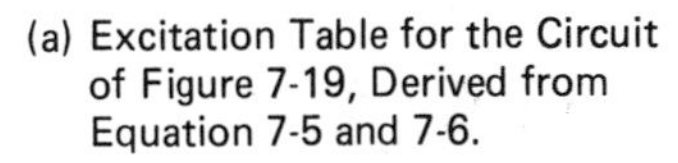
(a) Excitation Table for the Circuit of Figure 7-19, Derived from Equation 7-5 and 7-6.

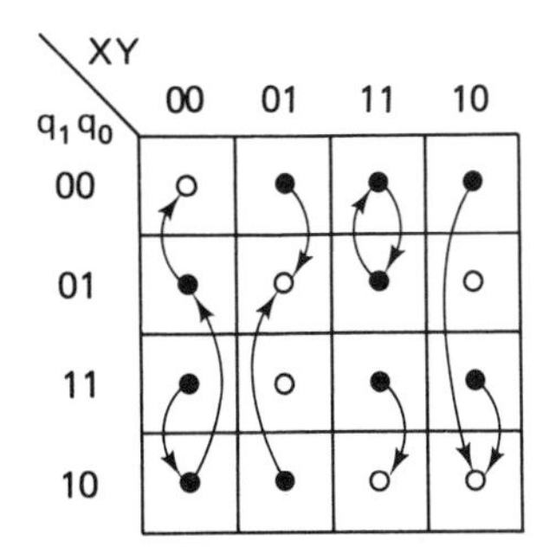

(b) Transition Table Corresponding to the Excitation Table of (a).

FIGURE 7-20

XY	00	01	11	10
(00) A	A	B	B	D
(01) B	A	B	A	B
(11) C	D	C	D	D
(10) D	B	B	D	D

(a) A State Table Follows from the Transition Table of Figure 7-20.

Output

XY	00	01	11	10
(00) A	0	0	0	0
(01) B	1	1	0	0
(11) C	1	1	0	0
(10) D	1	1	1	1

(b) Equation 7-7 is Used to Form an Output Table.

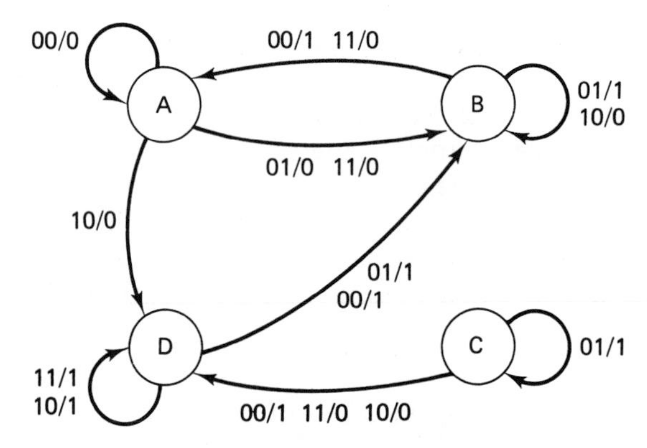

(c) A Flow Graph is Shown for the Circuit of Figure 7-19 as a Final Step in its Analysis.

FIGURE 7-21

FIGURE 7-22 *The transition table is redrawn to illustrate the transitions that are susceptible to error.*

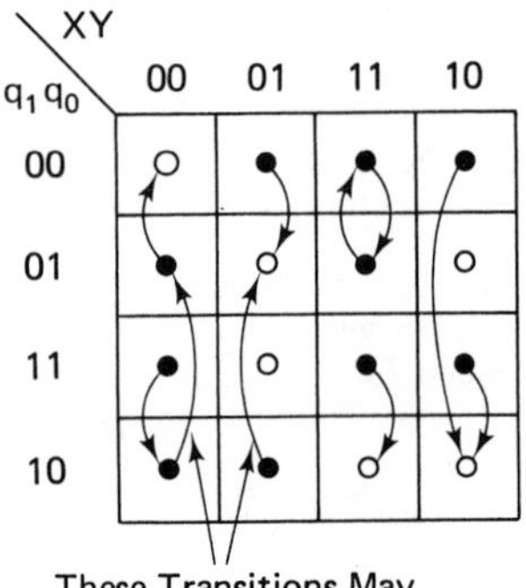

Admittedly, the intermediate state may be of very short duration. Nevertheless, the false secondary inputs resulting from this state could be sufficiently long to interfere with the correct excitation. If this happens, the expected sequence of transitions is not observed. The transition table is redrawn in Figure 7-23 to illustrate the error transitions that are possible. Note that only one of the possibilities will actually happen, and will always happen under the same input and state conditions. Such an error transition is referred to as a ***race***. This term owes its name to the fact that two secondary variables are in a "race" to change, and almost always, one of the variables finds its new value before the other. It is crucial to realize that the intermediate values that occur in the middle of a race may cause a new excitation, which may very well change the expected sequence of transitions. For example, if $(q_1,q_0) = (1,0)$ and $(X, Y) = (1,1)$, the transition table indicates a stable state (Figure 7-23). Now, if the primary inputs become $(X, Y) = (0,1)$, the previously stable state becomes unstable, and the secondary variables are inclined to change. From the transition table, the expected change for (q_1,q_0) is $(1,0) \longrightarrow (0,1)$. Because of the likelihood of a race, however, the actual change will begin as $(1,0) \longrightarrow (0,0) \longrightarrow \ldots$ or $(1,0) \longrightarrow (1,1) \longrightarrow \ldots$. In the first case, the circuit assumes state (0,0) momentarily. Fortunately, since (0,0) leads to (0,1) just as the ideal transition would, the excitation does not change when the momentary state is entered, and (0,1) is correctly reached. In the second case, though, the circuit assumes the state (1,1) momentarily. This time, the momentary state is *stable*, and the excitation that was leading the circuit to (1,0) is removed. Consequently, an improper state is entered and held, and the circuit malfunctions. A race in which the alternative transition possibilities lead to different stable states is referred to as a ***critical race***.

A second example of a race can be seen in Figure 7-23. Assume that the primary inputs are $(X, Y) = (1,0)$ and the secondary inputs are $(q_1,q_0) = (1,0)$. This condition is indicated as a stable state. If the primary inputs are now changed to $(X, Y) = (0,0)$, our focus of attention shifts to the lowest entry in the first column of Figure 7-23. As before, a race is possible. If the secondary variables change as $(1,0) \longrightarrow (0,0) \longrightarrow \ldots$, the proper final state will be entered on the first transition, completely avoiding the (0,1) intermediate state. However, if the secondary variables change as $(1,0) \longrightarrow (1,1) \longrightarrow \ldots$, we see that a two-state oscillation may develop between states (1,0) and (1,1).

FIGURE 7-23 *Possible error transitions resulting from two secondary variables trying unsuccessfully to change at the same time (dashed lines).*

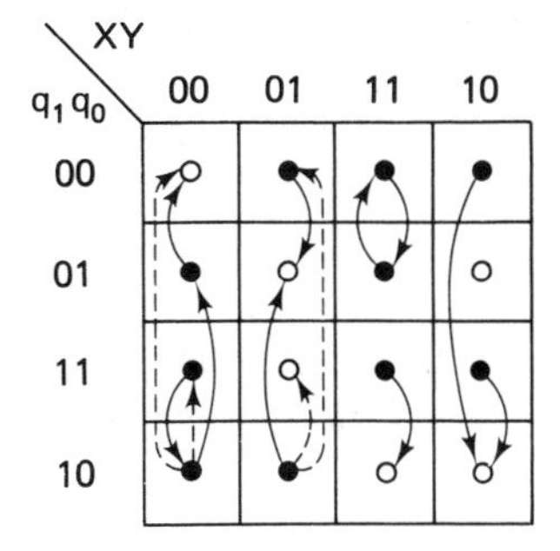

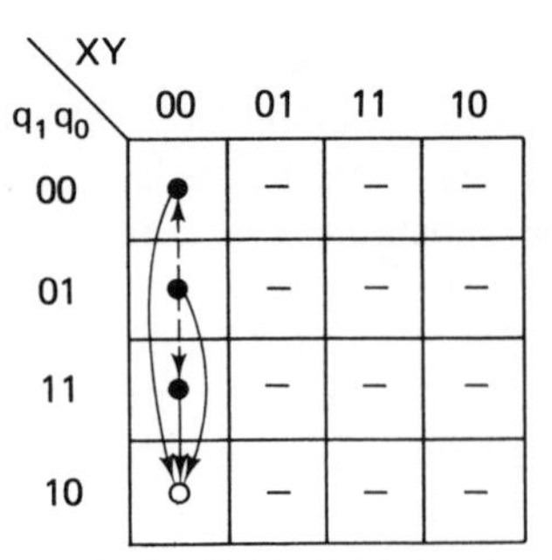

FIGURE 7-24 Example of a noncritical race.

If this happens, it is possible that the proper final state, (0,0), will never be reached. Thus, a critical race exists in the first column.

A ***noncritical race*** occurs when the proper final state is reached, regardless of the sequence of transitions. An example of a noncritical race is shown in Figure 7-24.

Race conditions can also exist in the primary inputs when an attempt is made to change two primary input variables at the same time. Unless the change is perfectly synchronous, false excitations may occur, and improper transitions may then be observed.

Since the primary output values generated by different stable states may be different, it is clearly a danger that the outcome of a critical race may cause improper primary output values to be produced. Perhaps less obvious is the fact that noncritical races may briefly cause improper primary output values to be generated. Even though the possible false primary outputs observed in a noncritical race are of short duration, they could nevertheless have an undesirable effect on succeeding circuitry driven by the primary outputs. By considering the example of a noncritical race in Figure 7-24, and assuming that the output table of Figure 7-21(b) applies, we see that the normally '1' primary output could momentarily become '0' if the race followed the path $(0,1) \longrightarrow (0,0) \longrightarrow (1,0)$. On the other hand, the noncritical race would not produce a false output if the transition $(0,1) \longrightarrow (1,1) \longrightarrow (1,0)$ were observed.

Races result from unequal propagation delay through the combination logic block of the sequential circuit. Another problem closely associated with unequal propagation delays is referred to as a ***hazard***, and affects both primary and secondary outputs. A hazard is possible when two or more gate inputs attempt to change in *opposite directions* at the same time. If "same time" switching were possible, hazards would never be observed. Of course, unequal propagation delays are usually present and can induce hazards. An example of a hazard can be seen in the primary output logic of Figure 7-19. We note that when $X = 0$ and $q_1 = 1$, the output, 'Z', should be '1' regardless of 'q_0'. However, 'q_0' and '$\bar{q}_0$' are not likely to switch at exactly the same time *if the variable changes*. Rather, at least one gate delay will separate the falling edge of 'q_0' and the rising edge of '$\bar{q}_0$'. Thus, momentarily, both 'q_0' and '$\bar{q}_0$' will be '0'. It is possible, then, that a false '0' one gate delay in duration will be observed on the output. This is illustrated in Figure 7-25(a). Hazards are corrected by the addition of redundant terms to the MSP equation describing the logic function. Doing so ensures that the output will be '1' when simultaneous input changes of one variable occur. Determining the redundant term involves a study of the Karnaugh map that describes the logic function. In general, redundant enclosures should be added to the Karnaugh map so that no gaps exist between adjacent essential prime implicants. Figure 7-25(c) illustrates this idea for the output function described by Eq. 7-7. A corrected primary output circuit is drawn in Figure 7-25(d).

Problems due to races and hazards can often be minimized or avoided by careful design. Because races are caused by two or more secondary variables changing simultaneously, the designer should take care to assign states in such a manner that few or no double variable transitions are present. Of course, as the number of secondary variables increase, it becomes much more difficult to do this. The race and hazard problems can also be solved by making the entire circuit *synchronous*, which is the topic of Section 7.4.

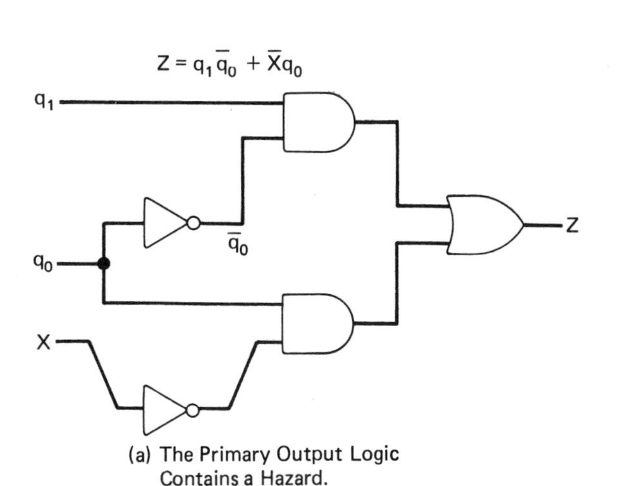

(a) The Primary Output Logic Contains a Hazard.

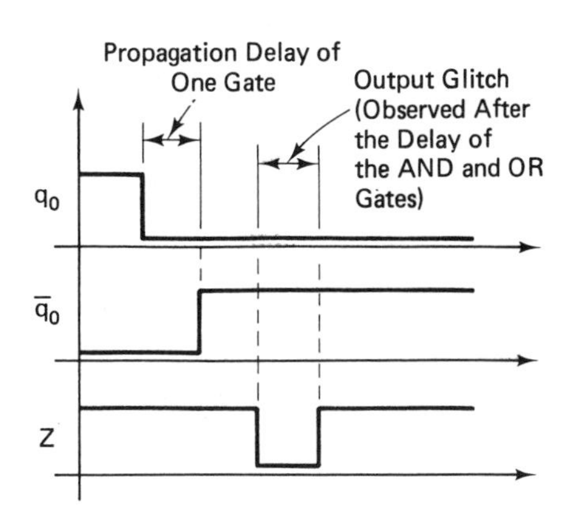

(b) The Hazard in (a) is Demonstrated (Assume that X = 0 and q_1 = 1).

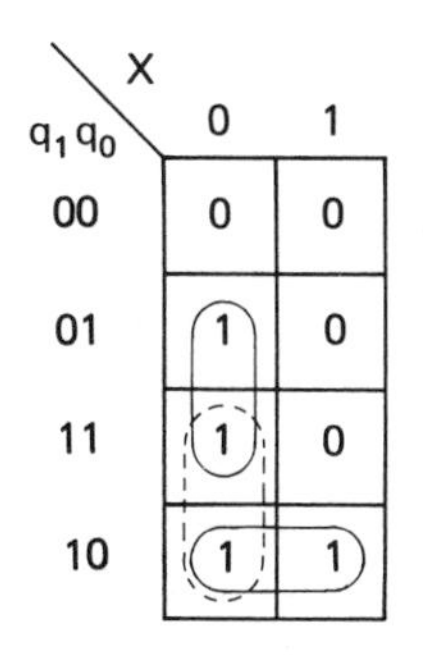

(c) A Redundant Enclosure is Added to Guard Against the Hazard (Dashed Line).

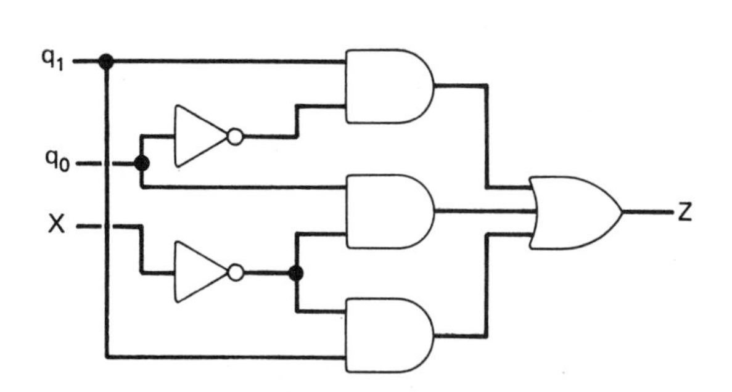

(d) The Primary Output Logic with an Added Gate and Inverter in Accordance with the Redundant Enclosure of (c).

FIGURE 7-25

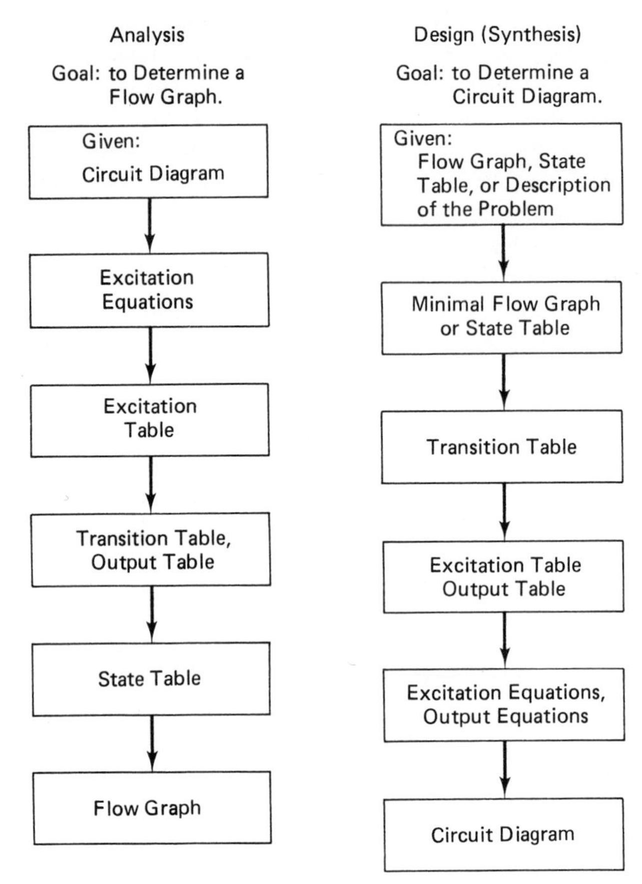

FIGURE 7-26 Comparison of the needed steps for analysis and design.

7.3.3 Design

The design procedure for sequential circuits is basically a reversal of the analysis procedure, as was the case for programmed counters. Since the transition possibilities can be so much more varied in sequential circuits (due to the presence of primary inputs), their flow graphs and state tables can be much more complex. We will see that in many cases, a great amount of effort will be devoted to finding the *simplest possible* flow graph or state table for a design, before progressing to the transition table, excitation table, and excitation equations. This, of course, is necessary and reasonable if a minimal circuit is desired. Figure 7-26 compares the basic procedures for the analysis and design of sequential circuits.

Toggle flip-flop design: To illustrate the design procedure, we will synthesize a minimal circuit for a toggle flip-flop. Rather than the usual T-type flip-flop having '*T*' and '*CK*' inputs, the device we are to design will have only one input, '*CK*', and is to *always* complement its output on the falling edge of '*CK*'. The '*T*' input is eliminated to simplify this first design example. Figure 7-27(a) shows a timing diagram for the desired toggle operation. From this diagram, we see that one primary input ('*CK*') and one primary output ('*Q*') are required. A flow graph for this circuit can be derived from the timing diagram of Figure 7-27(a), and is shown in Figure 7-27(b). Note that when a falling edge occurs, the output does not change until a transition to the next state is complete. Selecting the output in this way eliminates any special combinational logic for the primary output: the primary output is '1' whenever the circuit is in states '*C*' or '*D*', and is '0' otherwise. The primary output will be identical to one of the secondary outputs.

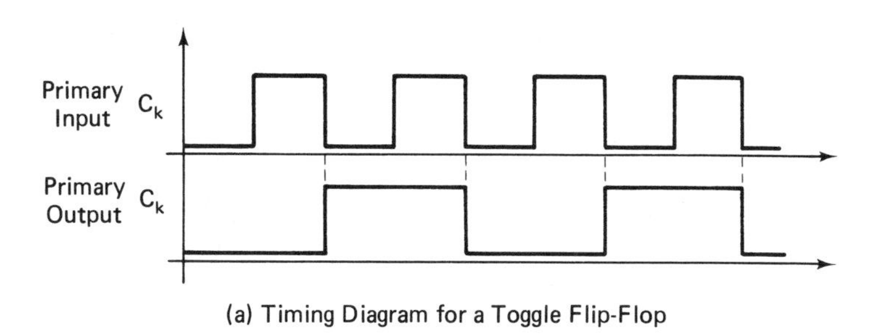

(a) Timing Diagram for a Toggle Flip-Flop

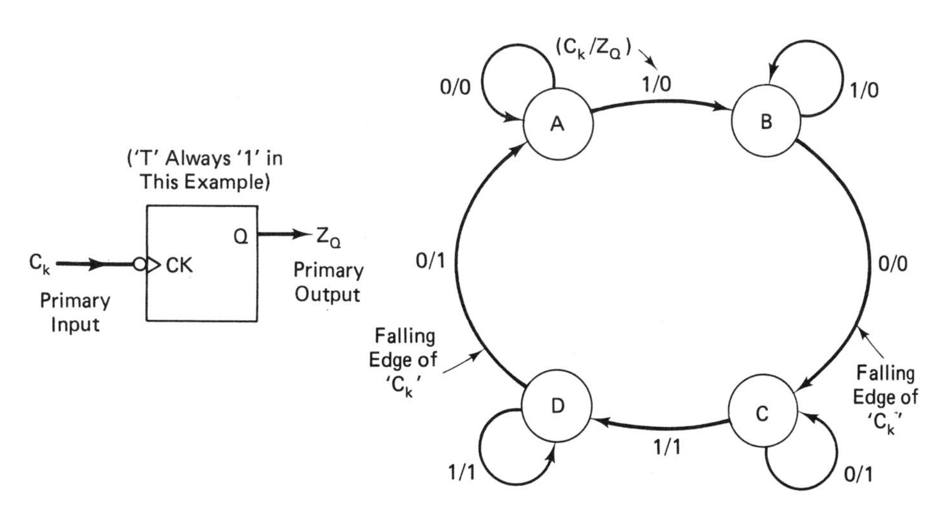

(b) Symbol and Flow Graph for a Toggle Flip-Flop. Two States Would Be Inadequate in the Flow Graph, and Would Permit Oscillation When C_k = 1.

FIGURE 7-27

A state table is shown in Figure 7-28(a) that represents the flow graph of the toggle flip-flop. In this state table, the primary output is shown within the entry. This is useful in that the determination of an output table is somewhat simplified. Figure 7-28(b), (c), and (d) show the transition table, excitation table, and output table, all of which may be obtained from the state table. The assignment of secondary variable values is made in the transition table in such a way that all race possibilities are eliminated. This cannot always be done, but the assignment of secondary variables should always be carefully planned to minimize the possibility of races.

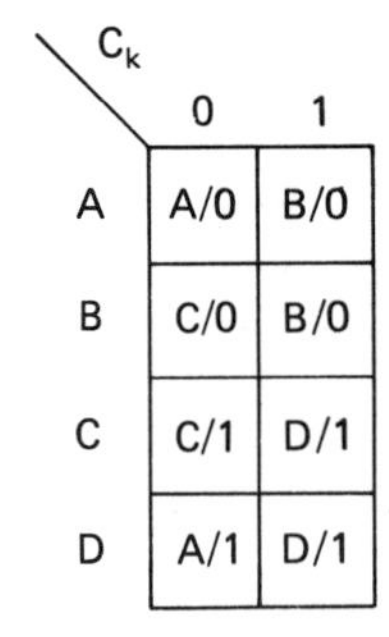

C_k	0	1
A	A/0	B/0
B	C/0	B/0
C	C/1	D/1
D	A/1	D/1

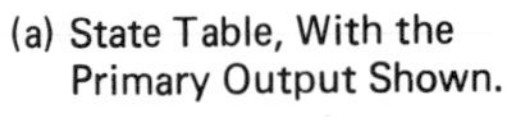

(a) State Table, With the Primary Output Shown.

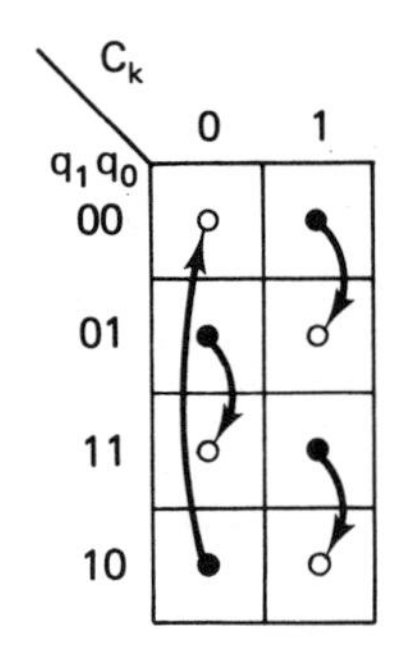

(b) Transition Table. State Assignments, Such as C = 11, are Made in the Transition Table.

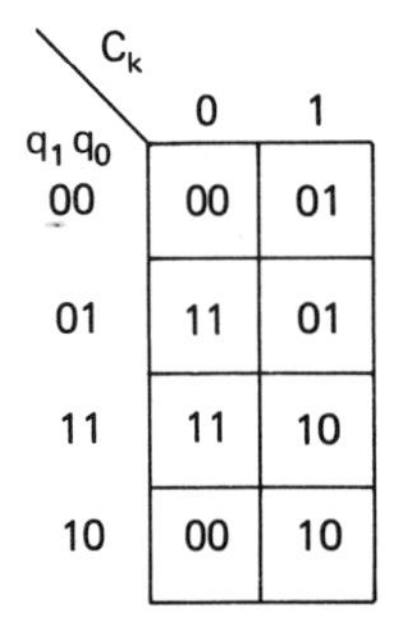

q_1q_0 \ C_k	0	1
00	00	01
01	11	01
11	11	10
10	00	10

(c) Excitation Table.

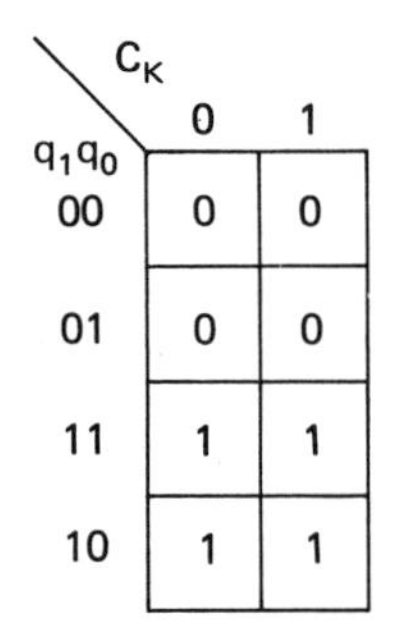

q_1q_0 \ C_K	0	1
00	0	0
01	0	0
11	1	1
10	1	1

(d) Output Table.

FIGURE 7-28

The excitation and output equations are derived in Figure 7-29. Finally, a circuit diagram is produced in Figure 7-30. The circuit diagram does not look familiar, based on the toggle flip-flop circuits that were studied in Chapter 5. In part, this is because of the way in which the circuit is drawn. By first converting the schematic of Figure 7-30 to one in which only NAND gates are shown (by applying DeMorgan's law), and then redrawing the circuit, we obtain the form shown in Figure 7-31. Now, the circuit seems recognizable as a master-slave device.

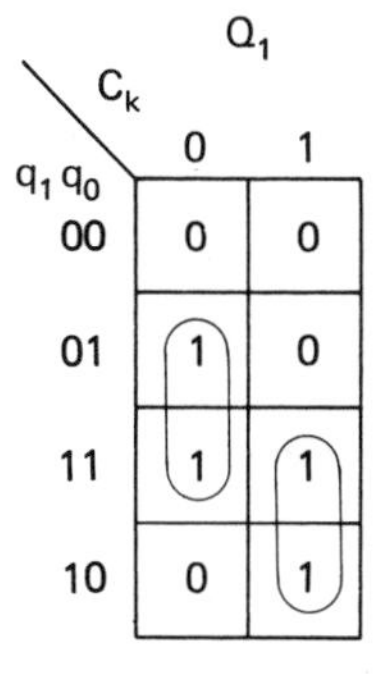

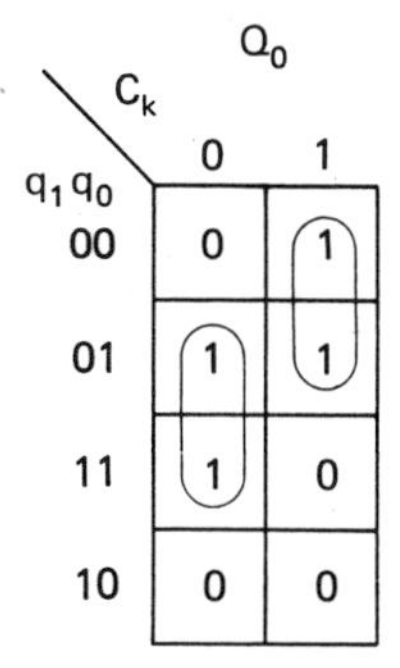

$Q_1 = q_0\overline{C}_k + q_1 C_k$

$$Q_0 = q_0\overline{C}_k + \overline{q}_1 C_k$$
$$= q_0\overline{C}_k + \overline{(q_1 + \overline{C}_k)}$$

(a) The Excitation Equations are Derived.

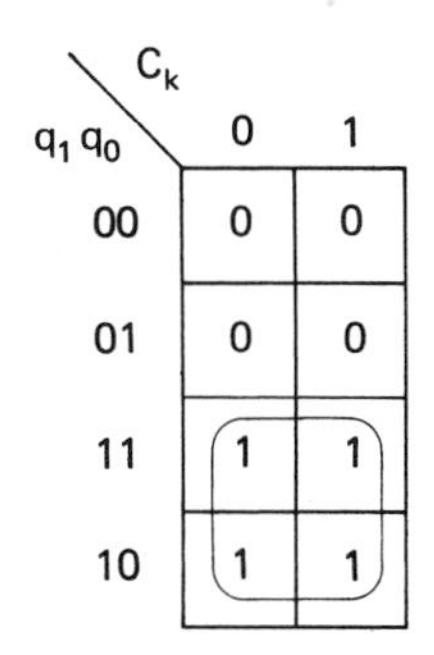

(b) The Output Equation is Derived.

FIGURE 7-29

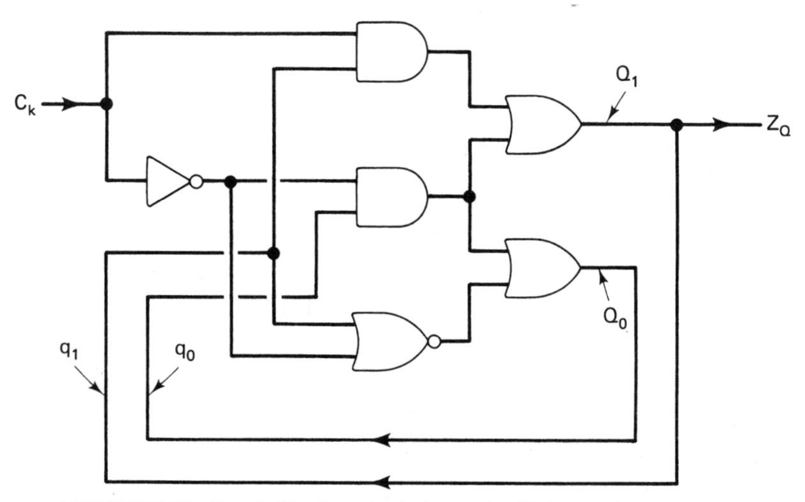

FIGURE 7-30 *Toggle flip-flop circuit design (no 'T' input, always toggles).*

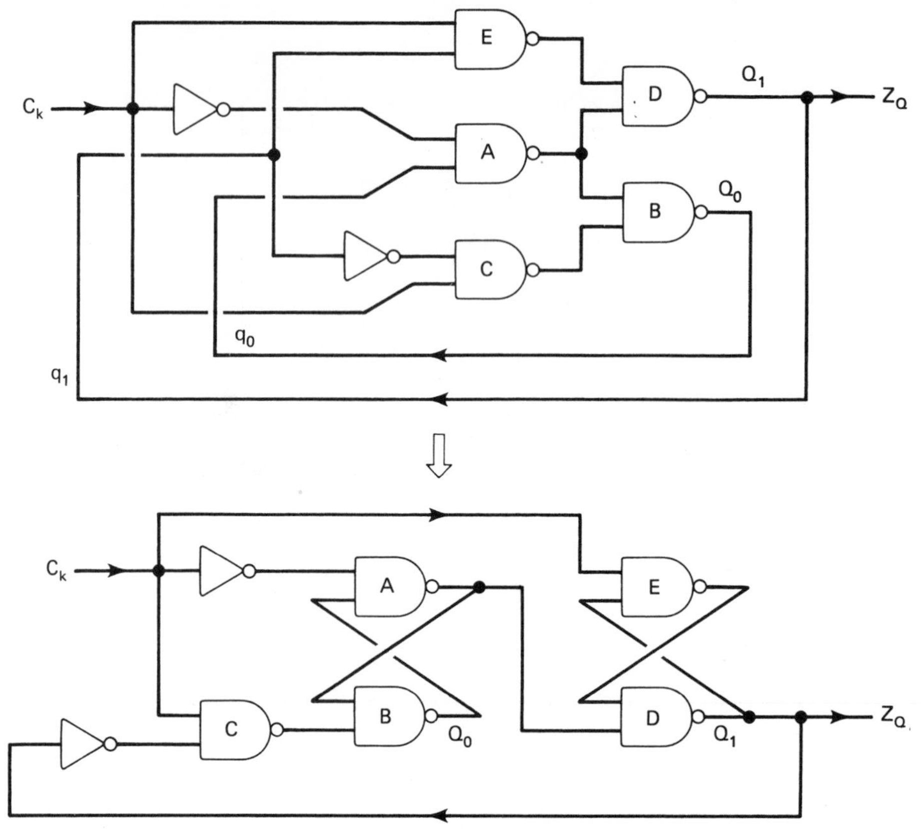

FIGURE 7-31 The schematic of Figure 7-30 is translated to one using only NAND gates (above), and then redrawn (below). Master-slave operation is now evident.

7.4 SYNCHRONOUS SEQUENTIAL CIRCUITS

7.4.1 The Relationship Between Synchronous and Asynchronous Circuits

The most serious problems associated with asynchronous sequential circuits involve the risk of races and hazards. With many three- and four-state circuits, the excitation table can usually be assigned state values in such a way that the probability of races is minimized. Often, however, sequential circuits with five or more states are necessary, and then races are more difficult to deal with. When many states are needed, it is almost impossible to assign excitation values to states such that all transitions show only one secondary variable change, while still providing reasonable circuit economy. Furthermore, when two or more *input* variables are present, we must ensure that each input variable changes simultaneously with respect to other input variables. For example, if two inputs (X, Y) attempt to change as $(0,0) \longrightarrow (1,1)$, a one- or two-gate delay lag in 'Y' that causes the input transition $(0,0) \longrightarrow (1,0) \longrightarrow (1,1)$ could easily send an asynchronous circuit into an improper state.

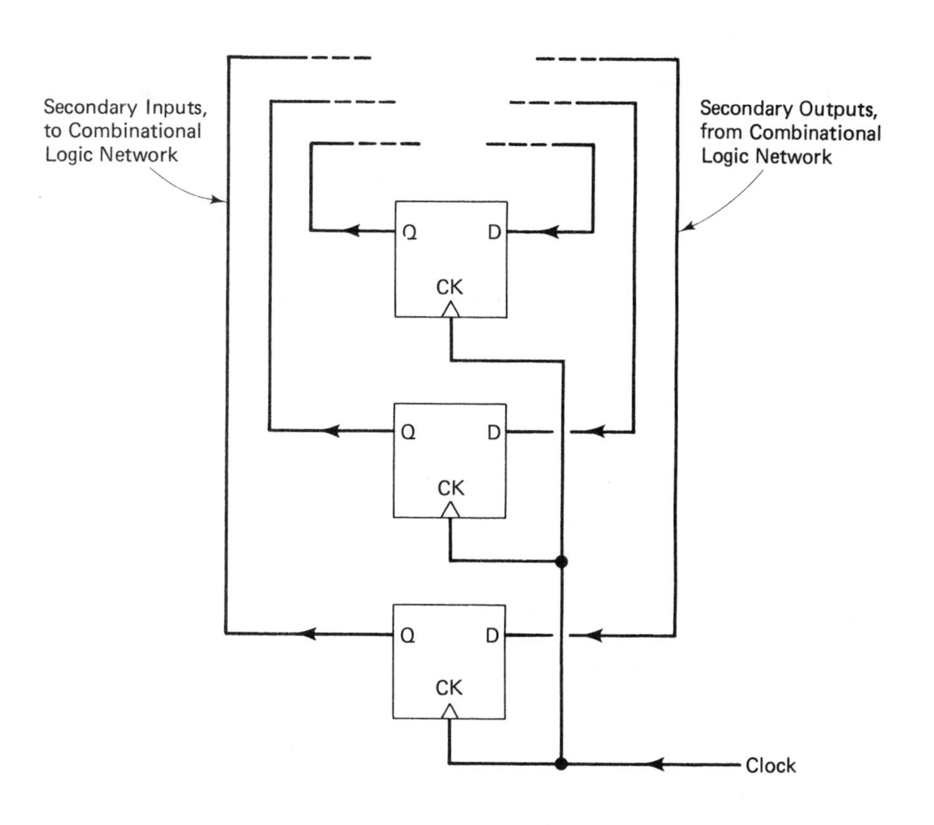

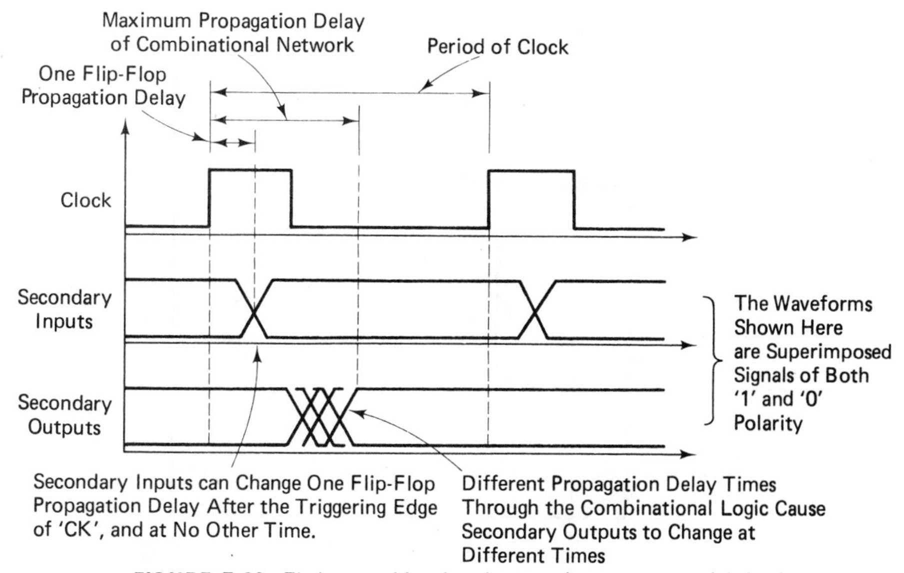

FIGURE 7-32 *Timing considerations in a synchronous sequential circuit.*

The race problem is further aggravated by the fact that if a critical race is biased in the desired direction in one circuit, another circuit of the same design but built with different components could resolve the race differently. This is so because of varying propagation delay, and slightly different electrical properties, between the logic gates.

Race and hazard problems are solved completely by placing a clocked flip-flop in each feedback path of the sequential circuit. Refer to Figure 7-4 at the beginning of Chapter 7 to see this illustrated in the general model of a sequential circuit. The result is a synchronous sequential circuit. We must inquire:

> Why are race and hazard problems eliminated in synchronous sequential circuits?

Referring to Figure 7-32, we see that because the flip-flops are triggered by a common clock, all secondary variables assume their next-state conditions at the same time (assuming each flip-flop has equal propagation delay). Thus, since it is impossible for two or more secondary variables to change independently of one another, false excitations cannot occur. Also, note that even if false excitations *did* occur, no harm would come since spurious changes in the secondary outputs are not passed through the clocked flip-flops. As long as the period of the clock is *longer* than the maximum possible propagation delay between secondary input and secondary output, we can

FIGURE 7-33 *A shift register is connected to a synchronous sequential circuit to store output values generated over the last five clock periods.*

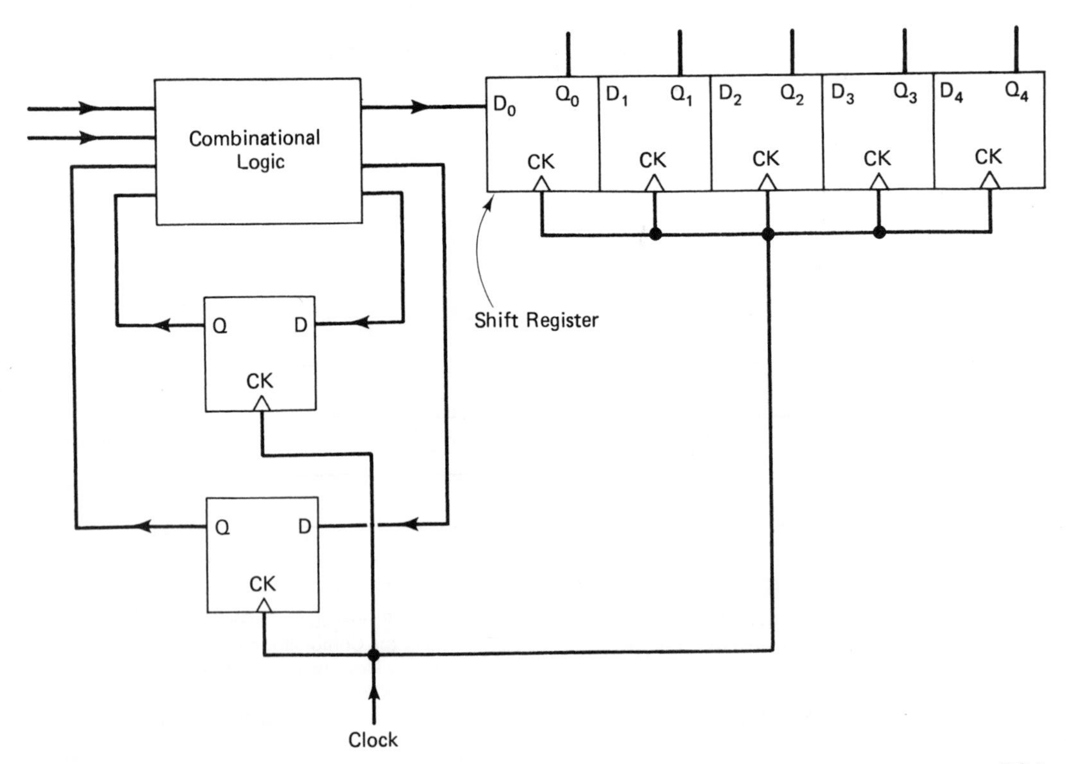

guarantee that no races will occur, regardless of the number of secondary variables that change.

By eliminating the chance of races, synchronous circuits give the designer considerably more freedom in state assignments than would be possible with asynchronous circuits. A further advantage of synchronous circuits is that they can be perfectly coordinated with external circuit functions that are triggered by the same clock signal. The succession of outputs from a synchronous sequential circuit could, for instance, be stored in a shift register, as depicted in Figure 7-33. The ability to be synchronized with other circuit components is extremely important in large-scale digital design and in computer systems.

7.4.2 A Synchronous Circuit Example

A circuit has been designed to recognize a particular five-digit binary sequence whose bits appear in order, MSB first, on a single data line. In addition to the data line, a clock line is present. Each rising edge of the clock signal signifies that valid data exists on the data line. Figure 7-34 illustrates this idea. Note in Figure 7-34(b) that at least six consecutive '0's separate the end of one transmission from the beginning of the next. By using many detector circuits connected to the same data and clock lines, each responsive to a different binary sequence, it would be possible to select one of many different devices by sending an appropriate code. A serial paging scheme such

FIGURE 7-34

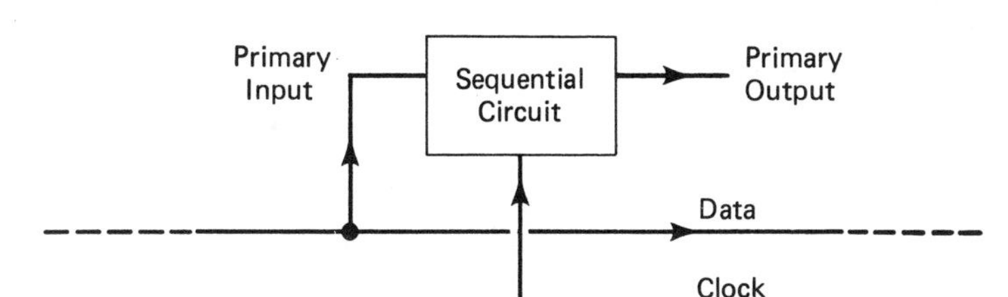

(a) The Block Diagram for a Sequence Detector. An Output is Generated Whenever a Particular Six-Digit Binary Sequence is Recognized on the Data Line.

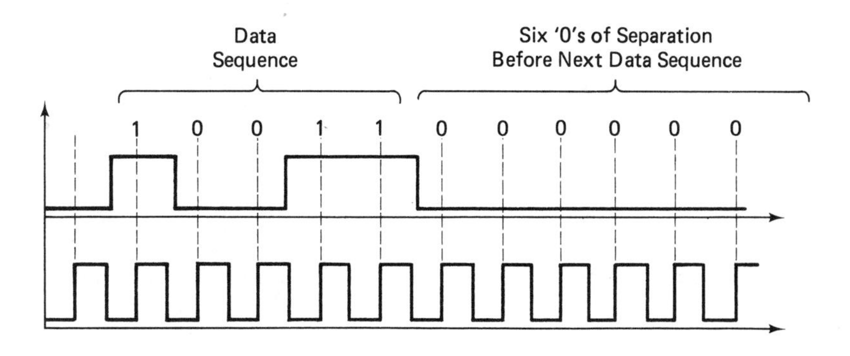

(b) Timing Diagram, Illustrating the Timing on the Data and Clock Lines.

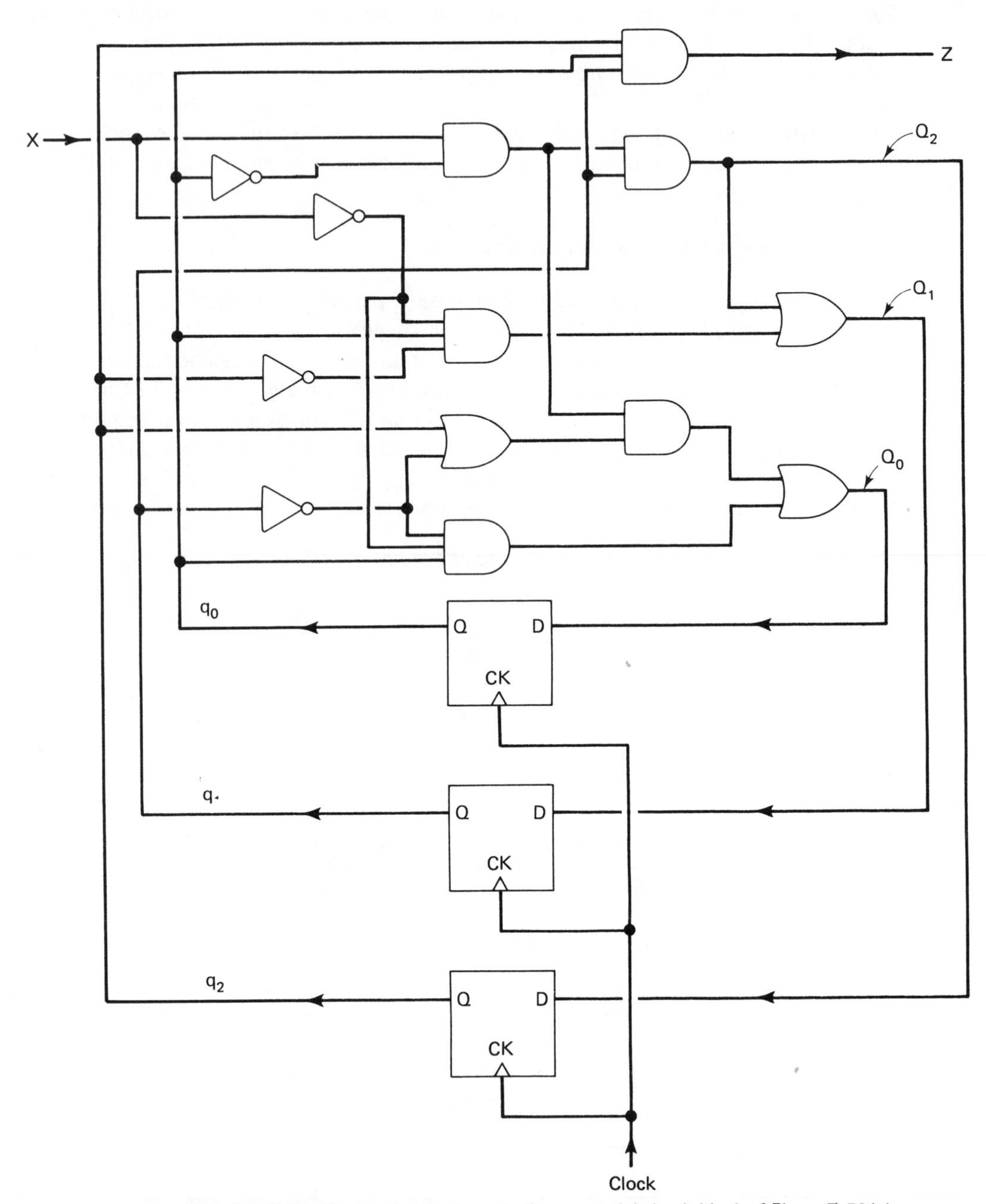

FIGURE 7-35 *Design that implements the sequential circuit block of Figure 7-50(a).*

as this would offer a great savings in cable when compared to the parallel method, in which the full multidigit code is sent at one time in parallel wires.

A circuit schematic is shown in Figure 7-35 that implements the sequential circuit block shown in Figure 7-34(a). This circuit has been designed to respond to the code '1 0 0 1 1', and generates an output pulse after the correct five-digit sequence has been entered. The output pulse is applied for exactly one clock period. Our problem is to redesign this circuit so that it will respond to the code '1 1 0 0 1'. To solve this problem we will first analyze the given circuit to verify that it functions properly. Then, we will modify the flow graph slightly to reflect the new code, and design a new circuit based on the modified flow graph.

Analysis: As we have previously noted, the needed steps for synchronous circuit analysis are virtually identical to those for asynchronous circuit analysis. Actually, even less work is needed: since races are not possible in synchronous circuits, the transition table (which was used primarily to indicate race possibilities) is unnecessary. A state table can be determined directly from the excitation table.

The excitation equations for the circuit of Figure 7-35 are found to be (in MSP form)

$$
\begin{aligned}
Q_2 &= q_1\bar{q}_0X \\
Q_1 &= q_1\bar{q}_0X + \bar{q}_2q_0\bar{X} \\
Q_0 &= \bar{q}_1\bar{q}_0X + q_2\bar{q}_0X + \bar{q}_1q_0\bar{X}
\end{aligned}
$$

From these equations, an excitation table is found and is shown in Figure 7-36(a). The corresponding state table is depicted in Figure 7-36(b), and a flow graph in Figure 7-36(c). The output equation is found to be

$$Z = q_2q_1q_0$$

A primary output of '1' occurs whenever the state '*F*' is reached. By entering the flow graph in any state and applying three consecutive '0's to '*X*' (the minimum number of '0's between transmissions), we will enter the initial state, '*A*'. Applying only the correct code, in synchronization with the clock signal, will cause the circuit to enter state '*F*'. When in state '*F*', a primary output of '1' is generated, while at all other times, the primary output is '0'.

Design: We would now like to alter the circuit in such a way that it responds to a different five-digit code. The basic structure of the flow graph shown in Figure 7-36(c) can remain the same, since the overall circuit function is identical. The only change necessary is in the input sequence that leads to state '*F*'. A modified flow graph for the new detector circuit is shown in Figure 7-37(a). The desired input sequence for this new circuit is '1 1 0 0 1'. A state table corresponding to this flow graph is shown in Figure 7-37(b), and the needed excitation table is shown in Figure 7-37(c). Note that states '*G*' and '*H*' are assumed to have "don't care" transitions to simplify the excitation equations. We will examine the consequences of this assumption when the design is complete.

$q_2q_1q_0$ \ X	0	1
000	000	001
001	011	000
011	010	000
010	000	110
110	000	111
111	000	000
101	001	000
100	000	001

$(Q_2Q_1Q_0)$

(a) Excitation Table for the Circuit of Figure 7-35. Note that the Secondary Inputs are Separated from the Primary Inputs.

\ X	0	1
A	A	B
B	C	A
C	D	A
D	A	E
E	A	F
F	A/1	A/1
G	B	A
H	A	B

Next State

Present Primary Output

(b) A State Table is Shown Based on the Excitation Table of (a) and the Output Equation. To Simplify the Appearance of this Table, only Primary Outputs of '1' are Shown.

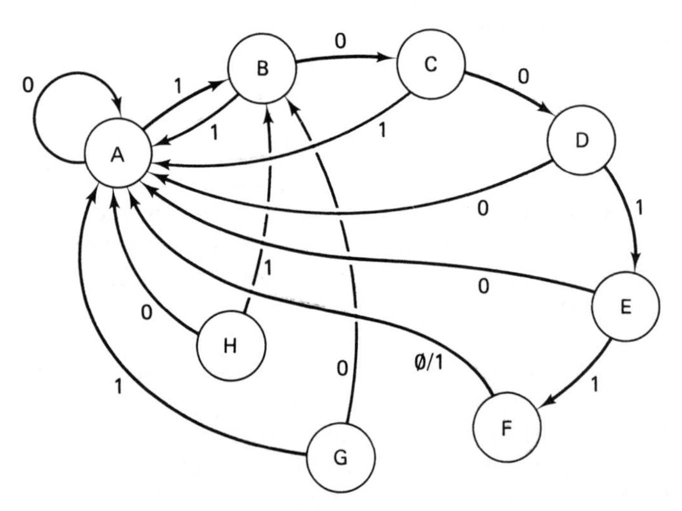

(c) The Flow Graph Derived from the State Table of (b).

FIGURE 7-36

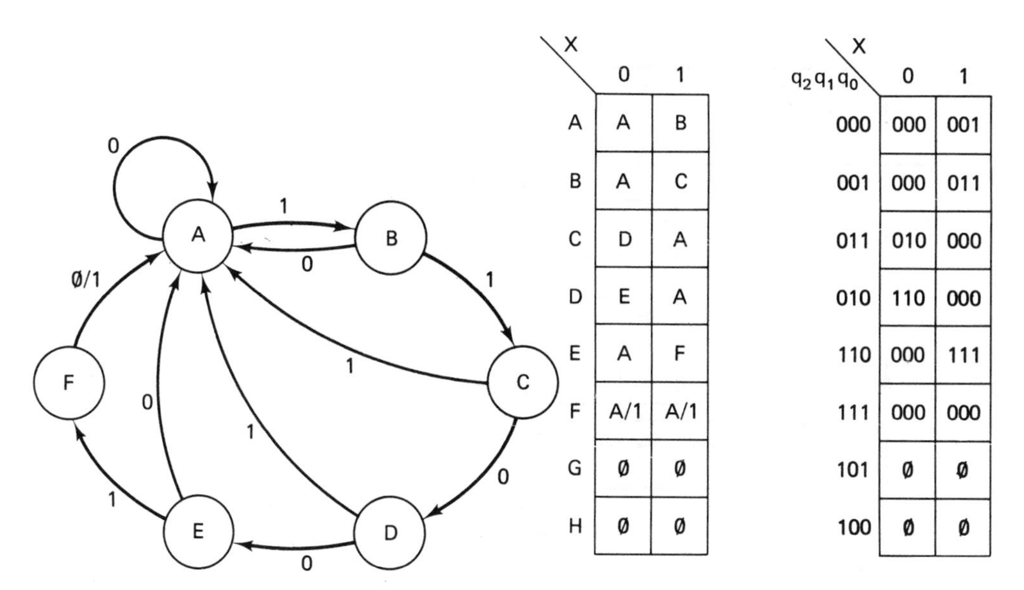

	0	1
A	A	B
B	A	C
C	D	A
D	E	A
E	A	F
F	A/1	A/1
G	Ø	Ø
H	Ø	Ø

$q_2q_1q_0$ \ X	0	1
000	000	001
001	000	011
011	010	000
010	110	000
110	000	111
111	000	000
101	Ø	Ø
100	Ø	Ø

(a) Flow Graph for a Detector Circuit that Responds to '1 1 0 0 1'. (b) State Table Corresponding to the Flow Graph of (a). (c) Excitation Table Derived from (b).

FIGURE 7-37

The excitation equations are determined in Figure 7-38. Note that the secondary and primary inputs need not be separated here. Indeed, if we hope to use Karnaugh map techniques, they must be separated to form a four-by-four map. A two-by-eight Karnaugh map is not practical because there are three-digit addresses in the vertical column that are one bit different, but are *not* adjacent. Thus, the geometric advantage offered by Karnaugh map techniques would be lost in a two-by-eight map. Finally, a circuit schematic based on the excitation equations is shown in Figure 7-39(a). The flow graph corresponding to this circuit is shown in Figure 7-39(b), with the transitions resulting from the initial "don't care" assumption.

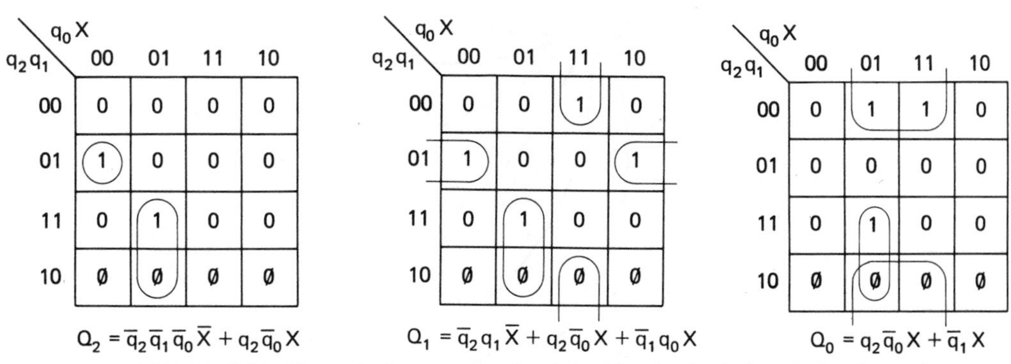

q_2q_1 \ q_0X	00	01	11	10
00	0	0	0	0
01	1	0	0	0
11	0	1	0	0
10	Ø	Ø	Ø	Ø

$Q_2 = \bar{q}_2 q_1 \bar{q}_0 \bar{X} + q_2 \bar{q}_0 X$

q_2q_1 \ q_0X	00	01	11	10
00	0	0	1	0
01	1	0	0	1
11	0	1	0	0
10	Ø	Ø	Ø	Ø

$Q_1 = \bar{q}_2 q_1 \bar{X} + q_2 \bar{q}_0 X + \bar{q}_1 q_0 X$

q_2q_1 \ q_0X	00	01	11	10
00	0	1	1	0
01	0	0	0	0
11	0	1	0	0
10	Ø	Ø	Ø	Ø

$Q_0 = q_2 \bar{q}_0 X + \bar{q}_1 X$

FIGURE 7-38 *The excitation equations are derived for the circuit described in Figure 7-37.*

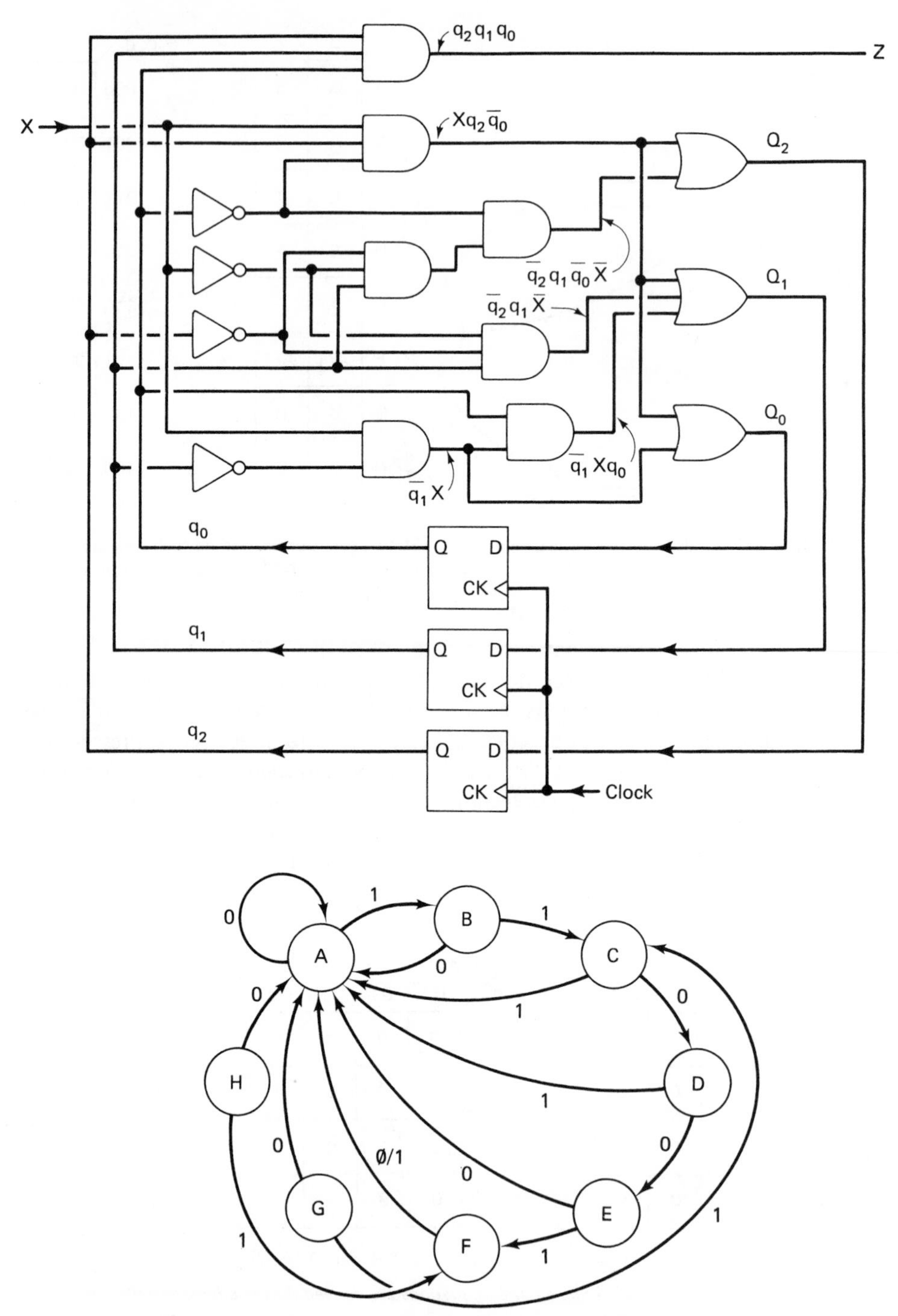

FIGURE 7-39 Design resulting from the excitation equations of Figure 7-38. A flow chart for this circuit shows that the "don't care" assumptions do not cause any trap states.

7.5 APPLICATIONS

7.5.1 Serial Addition

The addition of two multidigit binary numbers is easily implemented using simple combinational logic. Since the basic logic operation is the same for each digit of the addition, a standard one-bit adder block can be designed. Then by cascading several of these standard blocks, a multidigit addition circuit is realized. This was illustrated earlier, in Chapter 4, Figure 4-21. Several important features should be noted about the combinational logic adder. In particular:

1. The sum output of any given stage cannot be determined until the carry digit of the previous stage is valid. Thus, valid sum digits appear in sequence from LSB to MSB at a rate determined by the propagation delay of each stage.
2. Identical logic blocks are used for each stage.

By considering these features for a moment, we realize that a *single* combinational logic block could be used for multidigit addition if

(a) A means existed to feed the addend and augend digits to the single adder block individually, LSB first.

(b) There were a memory device to retain the carry generated by the previous digit's addition.

(c) There were a memory register to retain the sum digits that are produced.

Such an addition circuit is actually possible and is referred to as a ***serial adder***. Requirements (a) and (c) are satisfied using shift registers, while (b) is satisfied with a simple flip-flop. Figure 7-40 illustrates the block diagram for a serial adder circuit. In this problem, we will apply the techniques of synchronous sequential circuit design to arrive at a minimal circuit for the "Sequential Adder Logic" of Figure 7-40.

Our design procedure begins in the usual manner with a flow graph. Given that the carry from the previous digit's addition is known, the needed sum and carry for the current digit can be computed with simple combinational logic. Thus, the flow graph should consist of just two states—one state corresponding to "no carry" and the other corresponding to "carry." Two primary inputs are necessary and offer values for the current addend and augend bits. A single primary output is needed for the sum. Figure 7-41 depicts a flow graph that satisfies these needs. A state table, excitation table, and output table derived from this flow graph are shown in Figure 7-42. The excitation equations and output equations are derived in Figure 7-43. Note that the results are identical to the combinational logic equations for addition developed in Chapter 4 (Eqs. 4-4 and 4-6). This is not surprising and supports our intuitive notion that the flip-flop needed in the sequential adder only serves to retain the previous carry bit. Finally, a circuit schematic based on the excitation and output equations is shown in Figure 7-44.

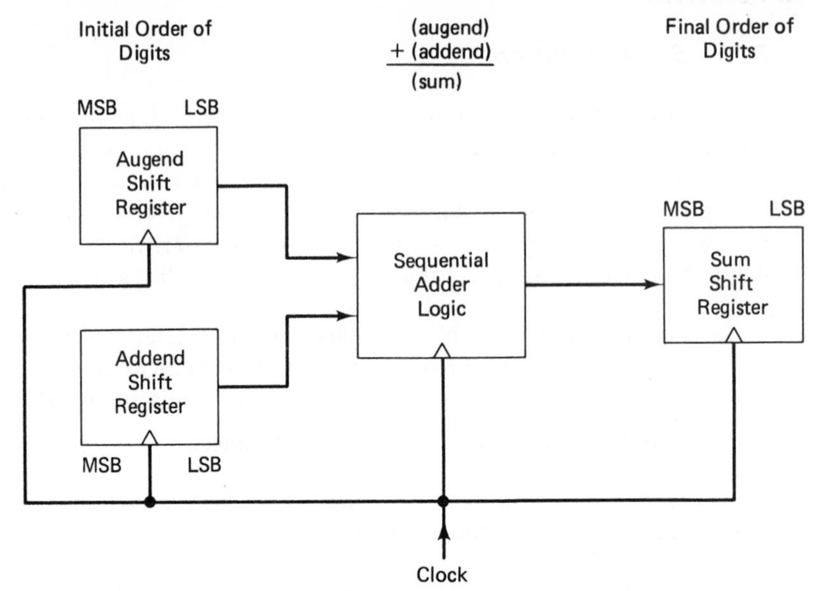

FIGURE 7-40 *Serial adder. If each shift register is 'n' bits long, 'n' clock pulses are required to compute the sum.*

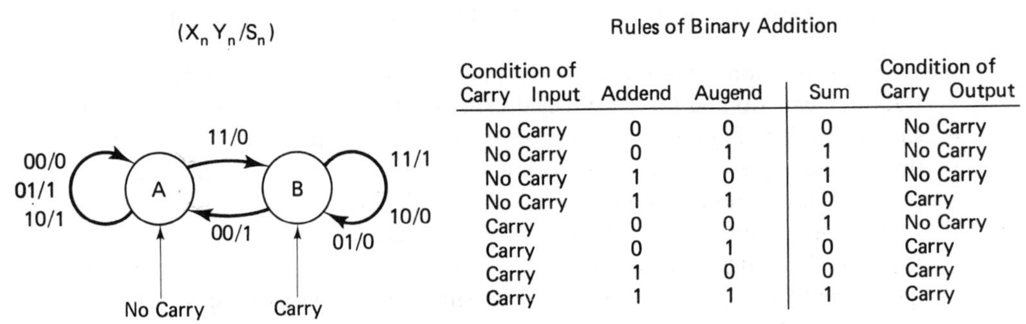

Rules of Binary Addition

Condition of Carry Input	Addend	Augend	Sum	Condition of Carry Output
No Carry	0	0	0	No Carry
No Carry	0	1	1	No Carry
No Carry	1	0	1	No Carry
No Carry	1	1	0	Carry
Carry	0	0	1	No Carry
Carry	0	1	0	Carry
Carry	1	0	0	Carry
Carry	1	1	1	Carry

FIGURE 7-41 *The rules of binary addition, and a corresponding flow graph.*

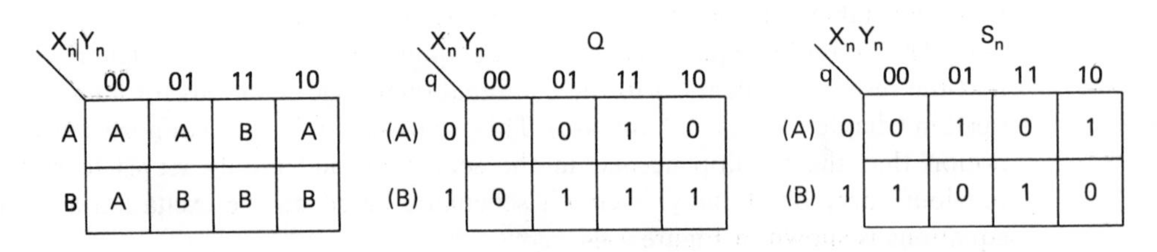

$X_n Y_n$	00	01	11	10
A	A	A	B	A
B	A	B	B	B

(a) State Table.

Q: q \ $X_n Y_n$	00	01	11	10
(A) 0	0	0	1	0
(B) 1	0	1	1	1

(b) Excitation Table.

S_n: q \ $X_n Y_n$	00	01	11	10
(A) 0	0	1	0	1
(B) 1	1	0	1	0

(c) Output Table.

FIGURE 7-42 *Design tables for the sequential adder flow graph of Figure 7-57.*

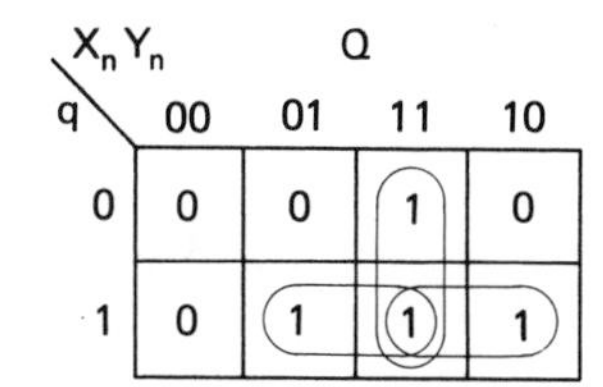

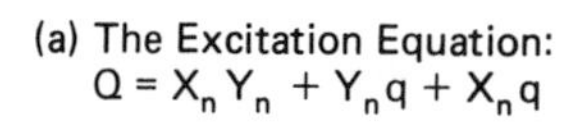

(a) The Excitation Equation:
$Q = X_nY_n + Y_nq + X_nq$

q \ X_nY_n	00	01	11	10
0	0	1	0	1
1	1	0	1	0

(b) The Output Equation:

$$S_n = q\overline{X}_n\overline{Y}_n + qX_nY_n + \overline{q}\overline{X}_nY_n + \overline{q}X_n\overline{Y}_n$$
$$= q \oplus X_n \oplus Y_n$$

FIGURE 7-43 *The excitation and output equations.*

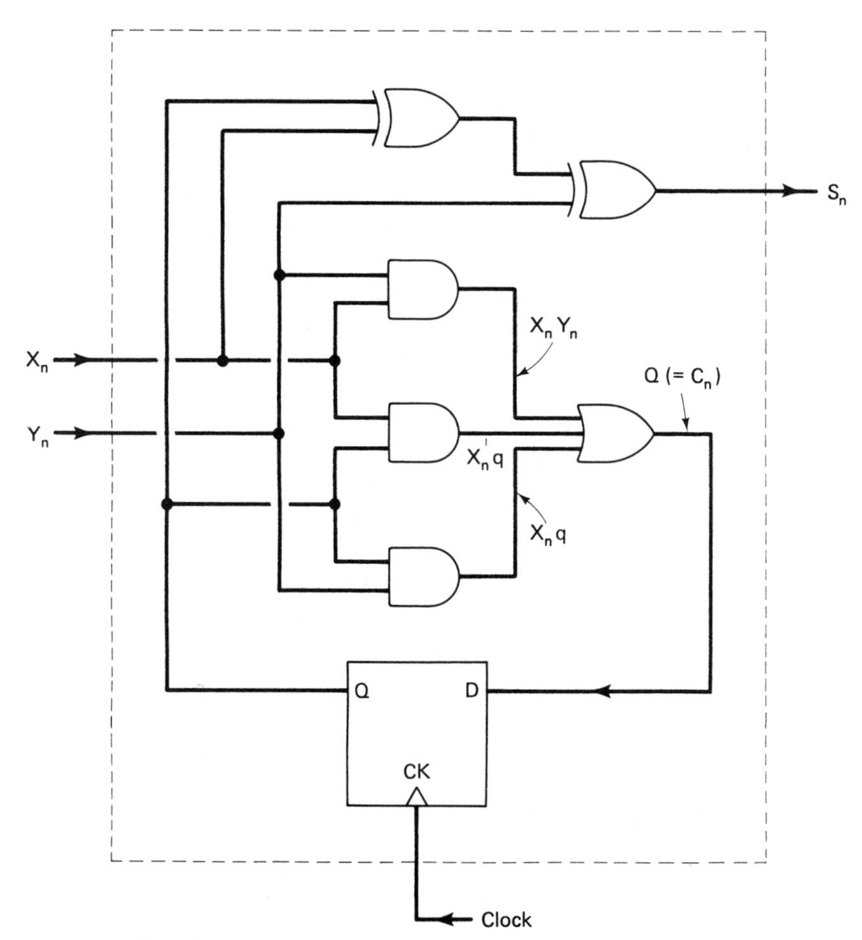

FIGURE 7-44 *Circuit schematic for the sequential adder logic. The dashed line defines the block element shown in Figure 7-56.*

SUMMARY

Digital logic circuits can be classified into two basic catagories. Combinational circuits comprise the first catagory, and include all circuits whose operations can be completely described by a truth table. Sequential circuits, however, do not exhibit a fixed relationship between input values and output values. Instead, the outputs of a sequential circuit depend upon both the circuit's present internal state, which may vary, and the external inputs. Thus, a second catagory of logic circuits is needed to account for the different properties attributed to sequential circuits.

The ability of a sequential circuit to retain an internal state that is independent of external inputs is often referred to as "memory." Flip-flops are simple sequential circuits used primarily because of their memory feature. Because of their simplicity, special techniques are not necessary to understand and apply flip-flops. More complex sequential circuits, however, defy intuition, and require an orderly procedure for analysis and design.

Asynchronous devices operate independetly of any clock signal and can change state in immediate response to a change in inputs. Synchronous devices, on the other hand, cannot change state until the triggering edge of a clock signal is applied, regardless of activity on the inputs.

The general model of a sequential circuit depicts a pure combinational logic block with a series of feedback connections. All sequential circuits have some form of feedback and, thus, can be generally represented as combinational logic with added feedback paths. Asynchronous devices possess direct feedback connections, while synchronous devices have clocked flip-flops embedded in the feedback paths.

Sequential circuit inputs and outputs that interface external devices, switches, indicator lamps, and so on, are primary variables. Inputs and outputs involved in feedback are secondary variables. Excitation equations describe the secondary outputs as a function of primary and secondary inputs. One excitation equation is needed for each secondary output. An excitation table permits us to see how the excitation values will induce the circuit to change state. Whenever the secondary inputs are unequal to the secondary outputs, the sequential circuit is said to be unstable, and is inclined to change state. Asynchronous circuits change state when the lumped propagation delay (represented by delay elements) has expired. Synchronous devices, however, can change only when the triggering edge of a clock pulse occurs. As long as secondary outputs and inputs are equal, one for one, the circuit is stable and will not change state. The transition table graphically indicates whether primary and secondary input conditions result in a stable condition or not, and shows the direction of change when instability is present.

When two or more secondary outputs are induced to change simultaneously in an asynchronous device, a race may develop in which one variable changes before the others. Intermediate false excitations that result from a race may cause improper states to be entered. Races are not possible in synchronous devices since a level of isolation is introduced (by clocked flip-flops) between secondary outputs and inputs.

The flow graph describes in the simplest way the overall operation of a sequential circuit. A state table is prepared from the transition table to assist in the construction of a flow graph.

The most important aspect of sequential circuit design involves translating our understanding of a problem into a suitable flow graph and state table. Assumptions that are made in this step can dramatically affect the simplicity of the resulting circuit design. Once a minimal state table is obtained, the basic steps of analysis are followed in reverse order to yield a workable circuit. The greatest possible care should be exercised in deriving and minimizing the state table.

NEW TERMS

Primary Input
Secondary Input
Primary Output
Secondary Output
State
Lumped Delay
Asynchronous Sequential Circuit
Synchronous Sequential Circuit
Excitation Equations
Output Equations
Excitation Table
Transition Table
State Table
Output Table
Trap State
Unstable State
Excited Delay Element
Quiescent Delay Element
Race
Hazard
Critical Race
Noncritical Race
Redundant Enclosure
Equivalent States
Serial Addition
Feedback Path
Delay Element
Stable State

PROBLEMS

7-1 Classify each of the block diagrams in Figure P7-1 as being a synchronous sequential circuit, an asynchronous sequential circuit, or a combinational logic circuit. Indicate which of the sequential circuits can be analyzed using the techniques presented in

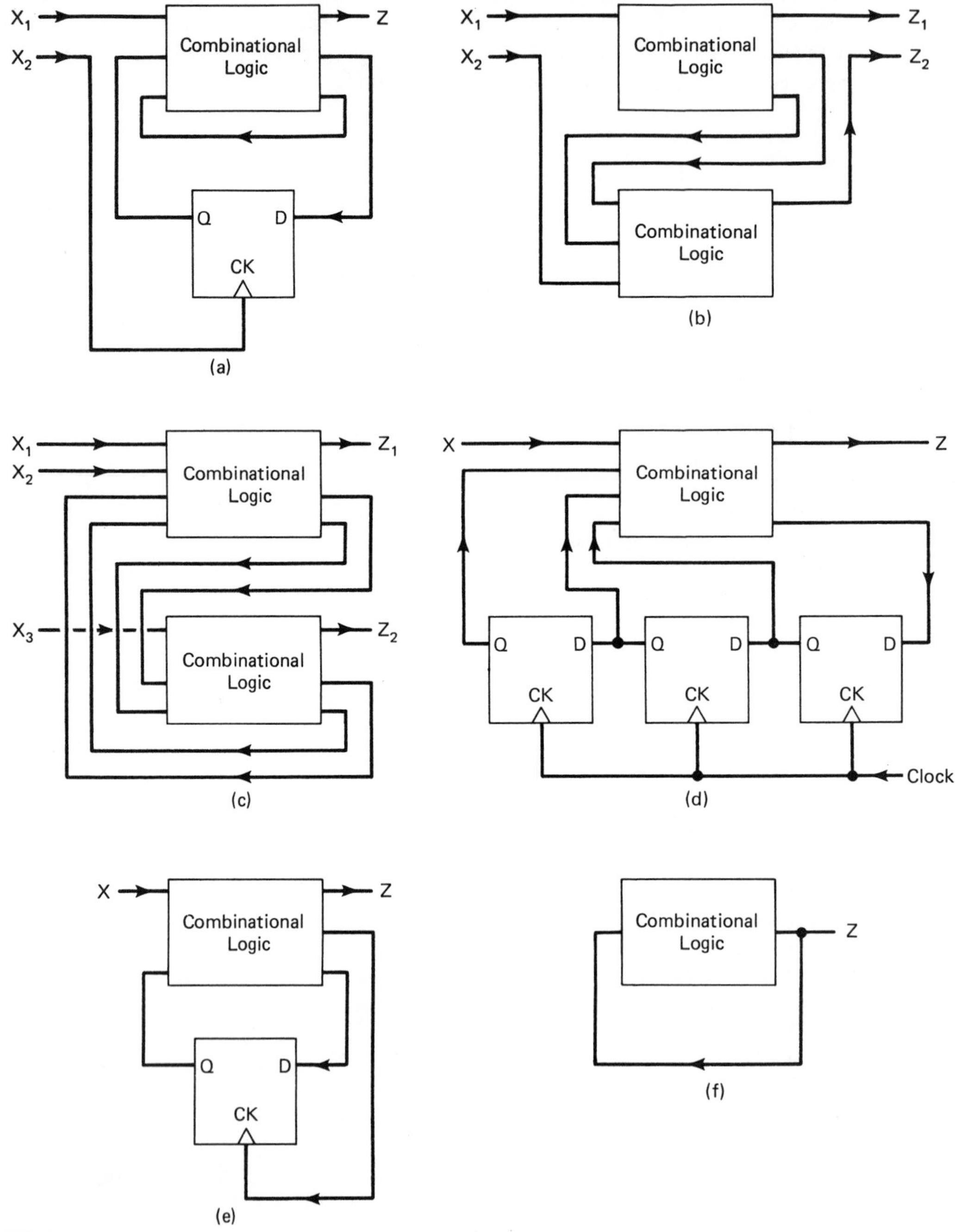

FIGURE P7-1

7-2 Redraw the three-stage shift register of Figure P7-2 to conform with the general model of a synchronous sequential circuit.

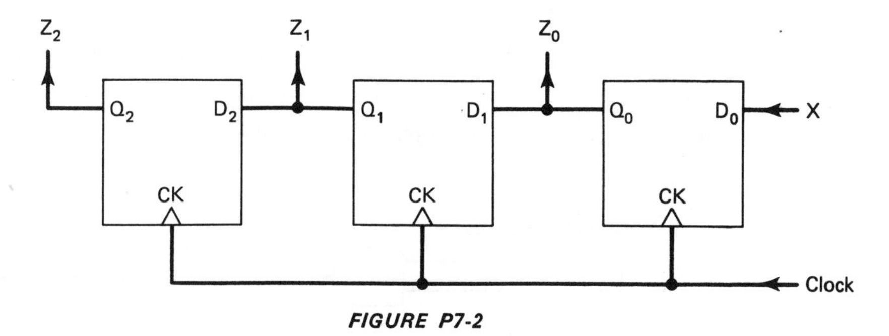

FIGURE P7-2

7-3 Should a ripple counter be considered as a synchronous or asynchronous sequential circuit? Why? (Assume that the clock input is a primary input.)

7-4 Consider a synchronous sequential circuit with one primary input, one primary output, and two feedback paths.

(a) How many states does the circuit have?

(b) In Figure 7-6(a), we saw how only one flow graph is possible for a synchronous counter (the flow graph is a closed loop), since no primary inputs are present. By adding only one primary input, the flow graph of Figure 7-6(b) demonstrates that many more transitions can exist. How many different flow graphs are possible for a sequential circuit with one primary input and two feedback paths?

7-5 Apply the analysis technique for asynchronous sequential circuits to the NOR latch of Figure P7-5. Be certain to

(a) Redraw the NOR latch to conform with the general model of a sequential circuit.

(b) Derive the excitation equation.

(c) Determine the excitation table.

(d) Show a transition table.

(e) Derive the output equation.

(f) Determine the output table.

(g) Determine the state table.

(h) Draw a flow graph.

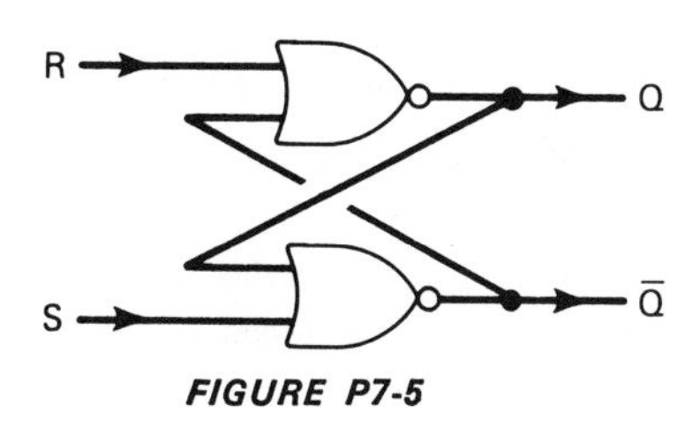

FIGURE P7-5

7-6 Analyze the sequential circuit shown in Figure P7-6.

(a) How many states does this circuit have? How many of these states are stable for at least one value of 'X'?

(b) How many stable conditions (open circles) does the transition table contain?

(c) Are there any trap states in the flow graph for this circuit?

(d) In which state is an output of $Z = 1$ observed?

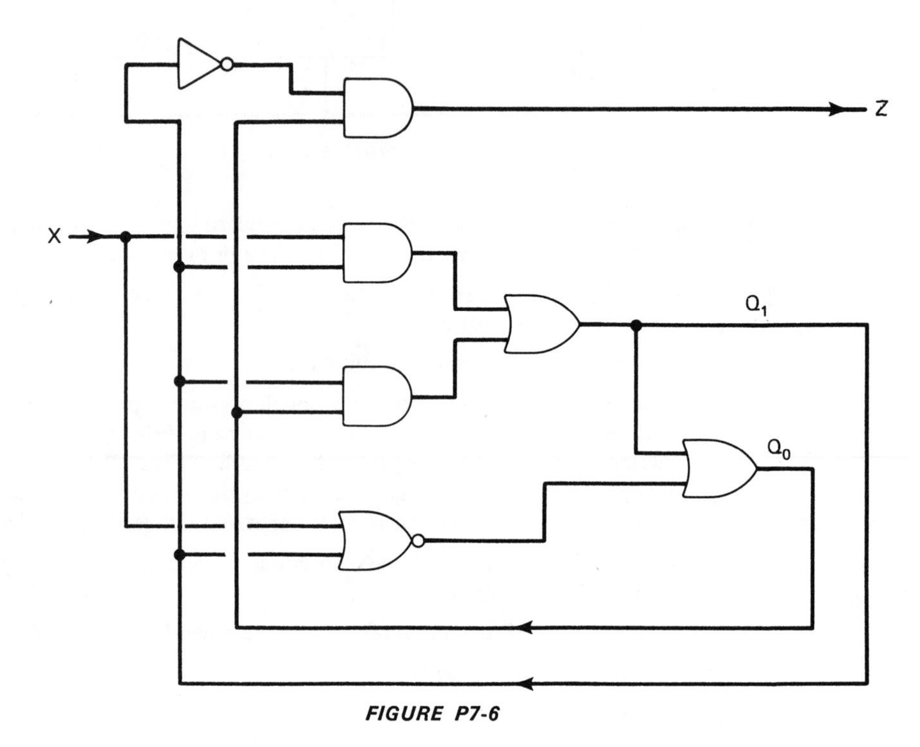

FIGURE P7-6

7-7 Compare asynchronous and synchronous sequential circuits on each of the following issues:

(a) Races

(b) Hazards

(c) Speed of operation

(d) Troubleshooting

(e) Ease of design

(f) Amount of circuit hardware needed (gates, flip-flops, number of connections)

7-8 **(a)** Use asynchronous sequential circuit techniques to analyze the ring oscillator shown in Figure P7-8.

(b) Add one more inverter to the loop of Figure P7-8. Use a transition table to show that the circuit will no longer oscillate.

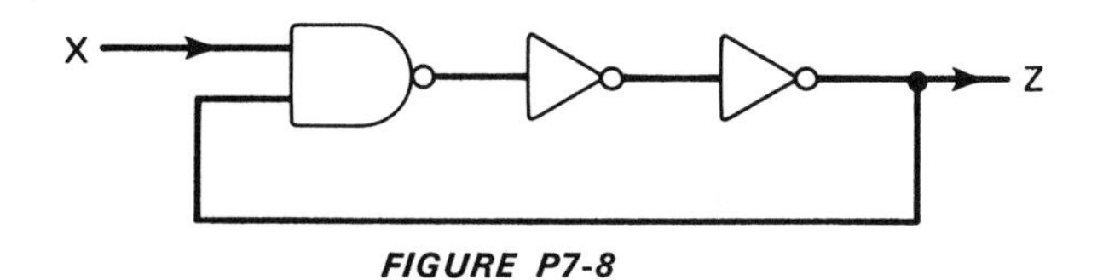

FIGURE P7-8

7-9 Can serial addition be done with the MSB coming first? If not, why? If so, how?

7-10 A serial adder design was discussed in Section 7.5.1. Following the same basic plan, design a serial subtracter circuit. Assume that any type of flip-flops and two input gates are available.

7-11 Determine a flow graph for the circuit shown in Figure P7-11. Note that the flip-flops are *T-type*. Be certain to show all steps in the analysis.

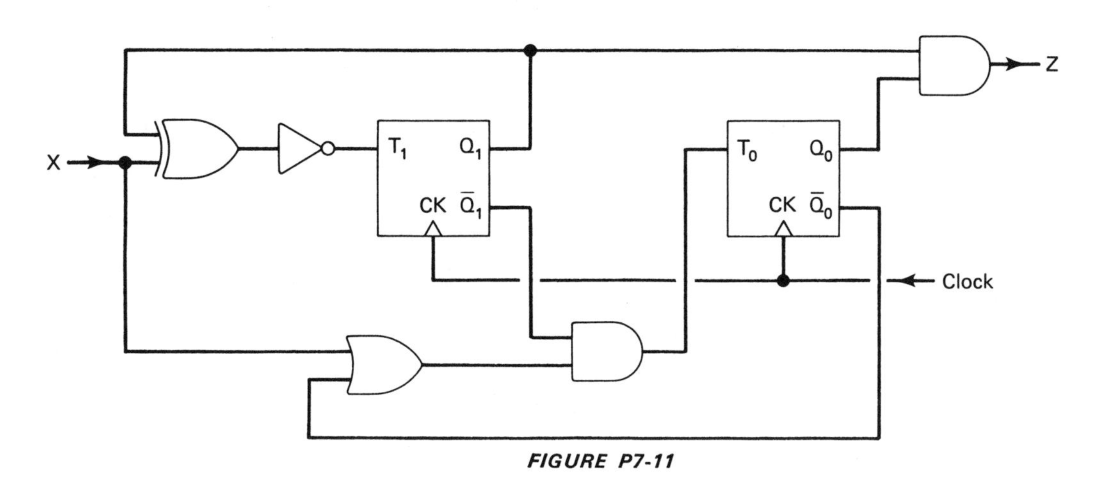

FIGURE P7-11

7-12 Analyze the synchronous sequential circuit whose excitation equations are

$$Q_1 = q_1\bar{X} + \bar{q}_1 X$$
$$Q_0 = \bar{q}_1\bar{q}_0 + q_1 q_0$$

7-13 The asynchronous sequential circuit shown in Figure P7-13 contains two races.

(a) Determine the transition table for this circuit and identify the races. Specify whether the races are critical or noncritical.

(b) It is decided to solve the race problem in this circuit by placing two edge-triggered D flip-flops in the feedback path. This action causes the circuit to become *synchronous*. Assuming that all gate propagation delays are 10 ns, that the flip-flop propagation delays are both 20 ns, and that the flip-flop setup and hold times are 10 ns, what is the maximum possible clock frequency for the circuit? (*Hint:* Determine the maximum propagation delay through the feedback loop.)

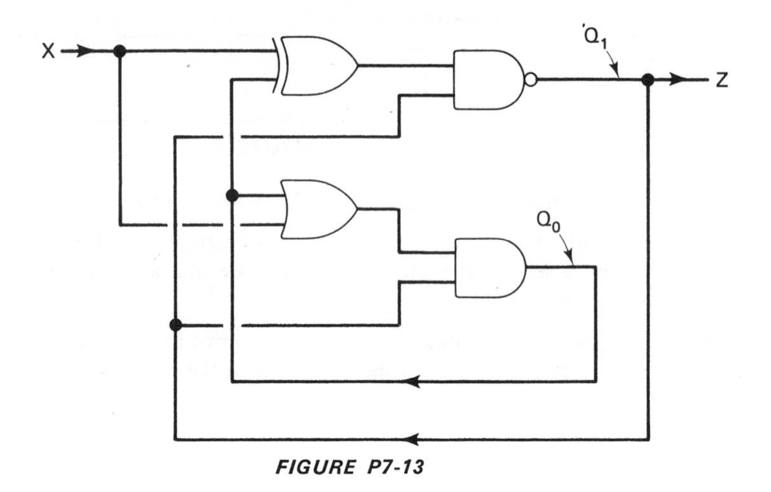

FIGURE P7-13

7-14 An asynchronous sequential circuit is described by the following excitation and output equations:

$$Q_1 = q_0\bar{X}\bar{Y} + q_1 Y + XY$$
$$Q_0 = \bar{q}_1 Y + \bar{q}_1 q_0 \bar{X} + q_0 XY + q_1 \bar{q}_0 \bar{X}\bar{Y}$$
$$Z = \bar{q}_1 \bar{q}_0 + q_0 Y$$

(a) Determine whether any races exist in this circuit. If any are found, classify them as either critical or noncritical races.

(b) Determine whether the output equation contains any hazards. Correct all that are found.

7-15 A synchronous sequential circuit is depicted with its flow graph in Figure P7-15(a). Determine the flow graph that would result if the primary and secondary variables were interchanged, as shown in Figure P7-15(b).

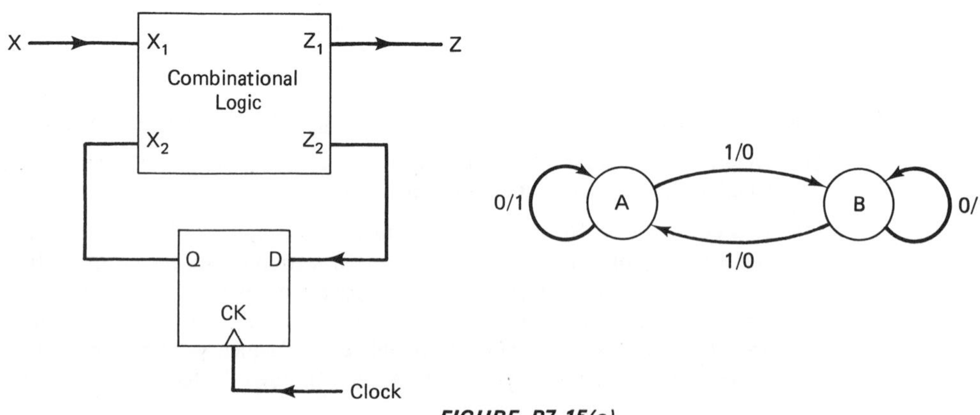

FIGURE P7-15(a)

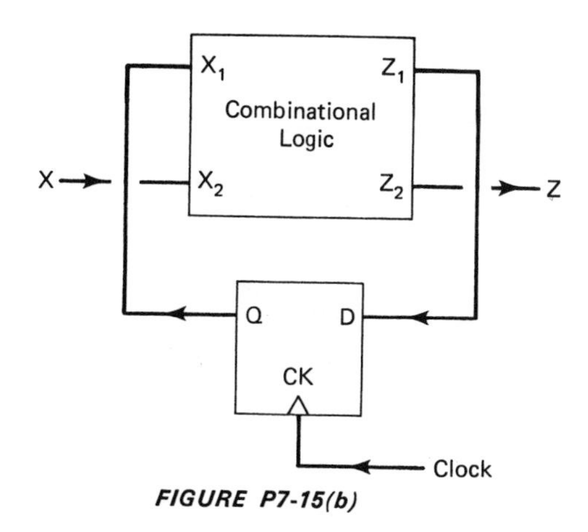

FIGURE P7-15(b)

7-16 An eleven-bit shift register and two synchronous sequential circuits are connected in series as shown in Figure P7-16(a). The sequential circuits are defined by the flow graphs shown in Figure P7-16(b). Determine the output sequence appearing on 'Z_2' if the shift register is initially loaded with (MSB) '0 1 1 0 1 0 0 0 1 0 0' and eleven clock pulses are applied. Assume that both sequential circuits are initialized to state '*A*' and that '*CK*' is initially '0'.

FIGURE P7-16(a)

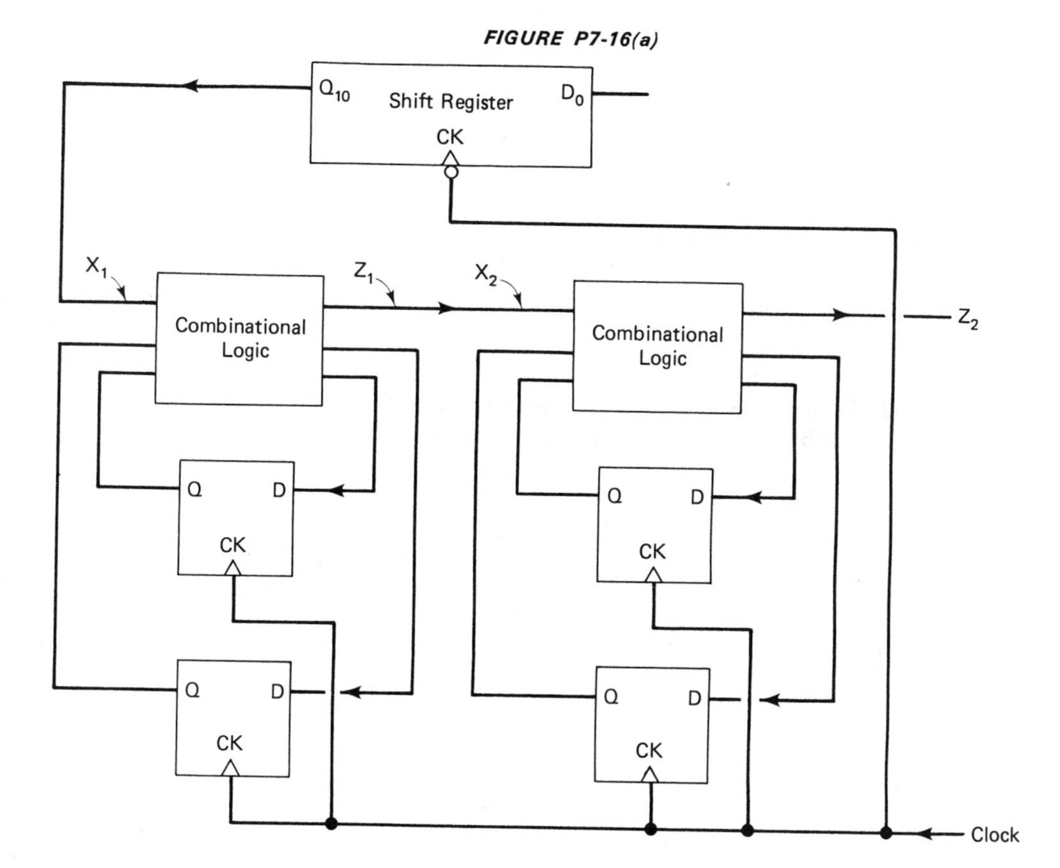

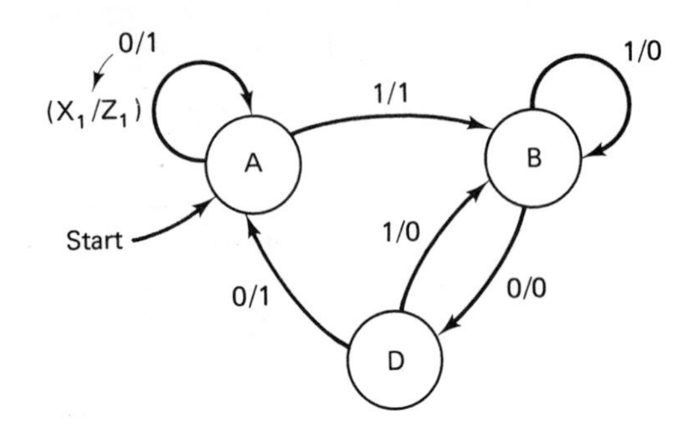

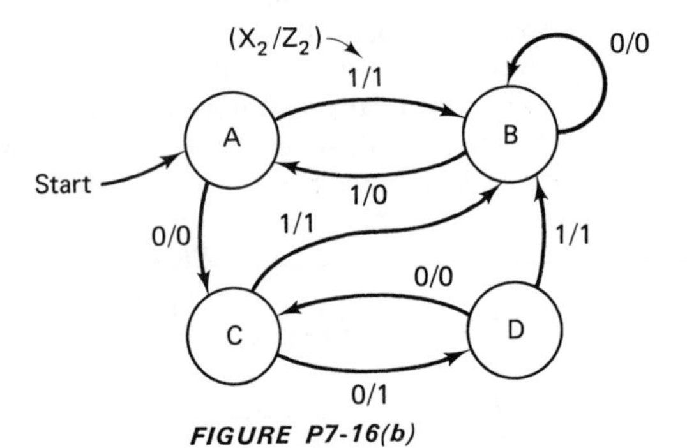

FIGURE P7-16(b)

7-17 **(a)** Can hazards in the excitation logic of an asynchronous sequential circuit cause improper states to be entered? Why?

(b) Investigate the toggle flip-flop design shown in Figure 7-30 for hazards in the excitation logic. If any are found, explain where they occur and how to correct them.

7-18 Design a synchronous sequential circuit using 'T' flip-flops and a minimum amount of combinational logic to implement the flow graph shown in Figure P7-18.

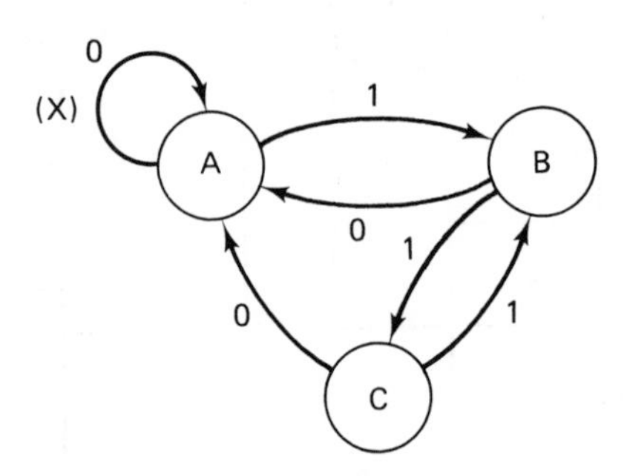

FIGURE P7-18

7-19 Use synchronous sequential circuit techniques to design a four-state synchronous up/down counter. One primary input, 'X', determines the direction of the count sequence: for $X = 0$, an up count should be observed, while for $X = 1$, a down count should occur. D- or T-type flip-flops may be used in the design; however, the choice should be made to minimize the amount of combinational logic necessary. Assume that any type of logic gates may be used.

7-20 A highway crosses railroad tracks at an intersection. To warn motorists of an oncoming train, signals are installed to flash lights and ring a bell before a train crosses the highway. In this problem, a sequential circuit must be designed to control the warning signals. The output, 'Z', of the circuit should be '1' when a train is approaching and while the train is crossing the highway, and '0' when the train is safely past the intersection. Two sensors, 'X' and 'Y', are installed on the tracks as shown in Figure P7-20.

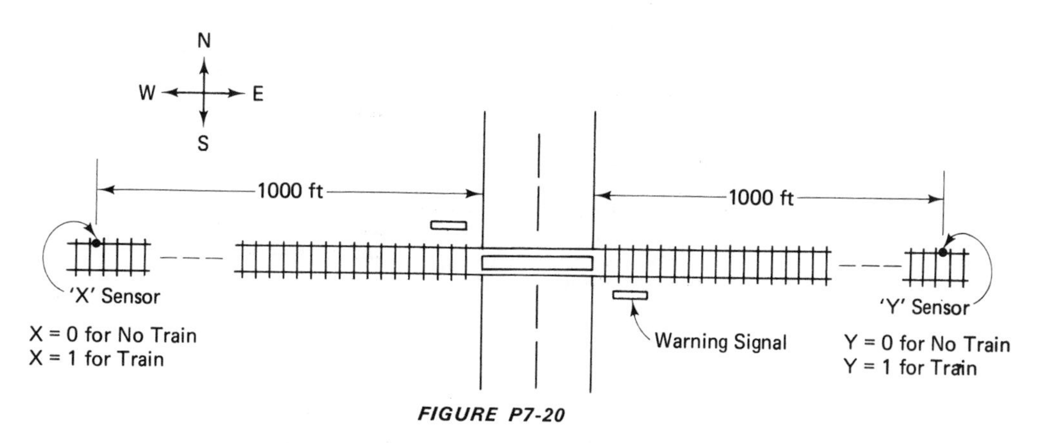

FIGURE P7-20

For west-to-east travel on the track, the following sequence of changes is possible:

$$\begin{array}{l} \text{initial} \\ \text{condition} \\ \text{of } (X,\ Y) \end{array}\ 00 \longrightarrow 10 \longrightarrow \begin{bmatrix} 00 & \text{(short train)} \\ 11 & \text{(long train)} \end{bmatrix} \longrightarrow 01 \longrightarrow 00$$

$$Z = 0 \qquad Z = 1 \qquad\qquad Z = 1 \qquad\qquad Z = 1 \qquad Z = 0$$

(a) Should the sequential circuit design for this signal controller by synchronous or asynchronous? Why?

(b) Determine the input sequence for east-to-west travel.

(c) Using the input sequence for west-to-east travel given above, and the input sequence for east-to-west travel determined in part (b), design one circuit that operates as a signal controller for trains moving in either direction. (*Hint:* The design can be reduced to four states.)

(d) Are there any race problems in your design? If so, can they be solved be making the circuit synchronous?

7-21 An eight-bit serial-to-parallel translator is shown in Figure P7-21. For each eight bits of data that are serially received, a parity bit is provided. Thus, the shift register shown in the figure must be nine bits wide. The parity bit is always selected so that the nine bits of data that are transmitted have an *even* number of '1' digits (even parity). Determine a flow graph for a serial parity check circuit to be connected at the shift register's input, as shown in the figure. Implement this flow graph using one edge-triggered 'D' flip-flop. Assume that a "reset" signal is applied to the parity check circuit preceding each transmission. Would the use of a different type of flip-flop make this circuit simpler?

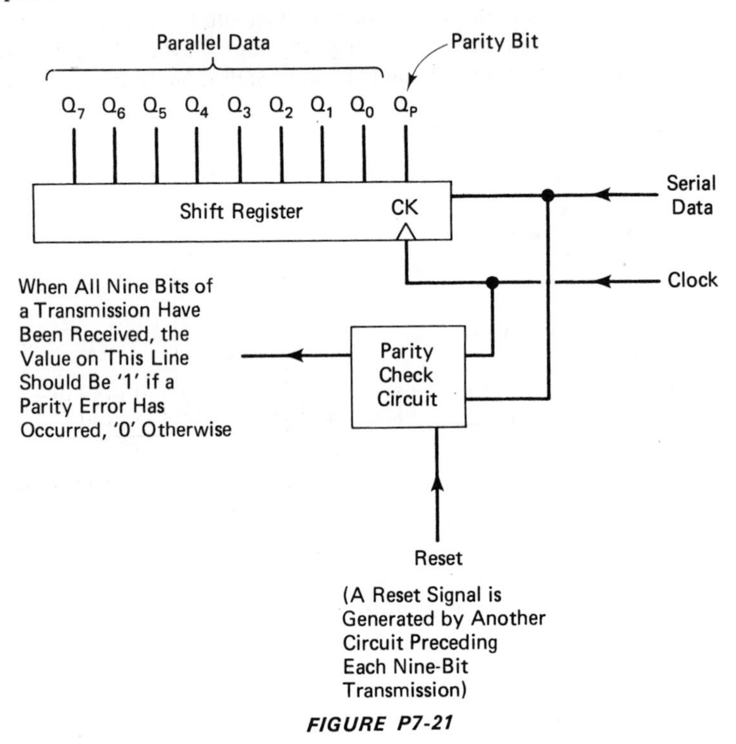

FIGURE P7-21

8 DIGITAL CIRCUIT FAULT ANALYSIS

8.1 MOTIVATION

The previous chapters in this book have presented a collection of basic topics concerning digital circuits. By studying these chapters, we have become familiar with some of the principles that govern digital electronic devices. These principles can be applied in a variety of ways to provide an understanding of existing circuits (analysis), or to allow the design of new ones (synthesis). As we study more advanced topics in the field of digital electronics, our abilities in these areas will become enhanced. However great our abilities may be with pencil and paper techniques, though, we may meet with bitter disappointment in the digital laboratory when a circuit which we construct does not operate. The failure of a circuit to operate might be our fault (miswiring, for example), or it might be caused by factors beyond our control (such as a defective component). In either case, we must be prepared to deal with this frustration in a sensible and efficient manner. Specifically, this means

1. Realizing that a circuit fault is present (as opposed to operator's error).
2. Classifying the fault as either static (constant) or dynamic (transient).
3. Identifying the cause.
4. Taking corrective action.

The intent of this chapter is to offer a sound, general approach to the problem of fault analysis. Really, it is impossible to do anything more: the causes of fault in a digital circuit can be so varied and complex that no specific universal method can be given that will always unravel the fault. Instead, we must rely almost exclusively on patient, clear thinking, and seeing cause/effect relationships. We are most fortunate in this regard, since our subject matter (digital electronics) normally demands patient, clear thinking by its nature, more so than many other subjects in the field of electronics. Thus, it will be less of a task for us to redirect our thinking to fault analysis than it might have been if our subject were different.

The value of studying fault analysis is almost self-evident; no digital circuit can have much worth if it cannot be made to work properly. It is a rare case when someone is never called upon to make their designs operate in the digital laboratory, or to revive an ailing digital circuit. There is another benefit gained by studying fault analysis, perhaps less obvious than the first. Tracing faulty digital circuits gives us firsthand experience in seeing cause/effect relationships in digital devices. This experience often provides a much better understanding of the circuit than could ever have been obtained by a purely "on-paper" analysis. Furthermore, we can often learn of ways in which digital circuits can be designed to facilitate fault detection. For example, incorporating the ability to single-step a circuit through its various states would be an invaluable design feature in terms of fault detection. Finally, few things are more satisfying than seeing our creative designs come alive in a working model. It is an unlikely case that the translation of a design into a working model can come about without the correction of some problems, however small.

8.2 DEFINITION OF FAULT

As we have seen in previous chapters, digital circuits can be classified into two basic catagories:

1. Combinational logic circuits.
2. Sequential circuits.

When a digital circuit fails to operate properly, we can say that it is failing to provide the correct output for a given input or sequence of inputs. Certainly, a circuit has failed if it produces no output or incorrect output all the time. A digital circuit must also be regarded as faulty, however, if it produces only one erroneous output bit in 10,000. Thus, to be functioning properly, a digital circuit must generate the correct output exactly, at all times. "Fault" can be more precisely defined for each classification of digital circuit:

For combinational logic circuits,

> a *fault* exists when a combinational logic circuit does not operate in accordance with its truth table.

For sequential logic circuits,

> a *fault* exists when a sequential logic circuit does not operate in accordance with its flow graph* (or state table).

Since a sequential logic circuit can be transformed into a combinational logic circuit by breaking the feedback paths, the analysis of combinational faults has even greater scope than the study of purely combinational logic devices.

* Recall that a sequential circuit is capable of cycling through a finite number of states. A flow graph describes these states and shows the permissible pattern of flow from state to state.

Deciding that a circuit is free of faults requires more than a simple one-shot test of the device. A combinational circuit, for example, must succeed on consecutive tests of its truth table. If it fails once every one hundred passes through its truth table, it is faulty. It is reasonable to ask:

> How much should a circuit be cycled through a test procedure
> before we decide that it is fault-free?

This is a hard question to answer, and depends upon many factors. If a simple circuit is built in a digital laboratory to demonstrate a counter, for example, it is safe to say that it is fault-free if it provides the correct count sequence once and begins to repeat. However, if that same circuit were to be installed in a spacecraft, we would have to be much more certain that it it fault-free, since repairs after launch are impossible. In this case, operating the counter at high speed for 100 hours, under varying conditions of temperature and power supply voltage, might reveal problems with its construction or design that would normally not have caused malfunction. Having succeeded with such a rigorous test, we could be much more certain that the circuit is fault-free. It is an interesting fact that if a circuit is going to fail during its useful life, that the failure will probably occur very early in this period. Thus, if a circuit survives a prolonged test without failing, it is quite likely that it will operate properly for the remainder of its design life. Fortunately, many of the circuit faults that we encounter make themselves known almost immediately. In any event, when it is clear that a fault exists, we must inquire:

> Why is this fault occurring?
> How did it come about?
> What can be done to correct it?

8.3 CLASSIFICATION OF FAULTS

There are many possible causes of fault in a digital circuit. In general, though, faults can be classified into one of two possible catagories, regardless of cause. These catagories are

1. Static faults.
2. Dynamic faults.

8.3.1 Static Faults

Static faults are characterized by an erroneous output that is constant for a given input condition, or for the duration of a circuit state. It is possible to observe a static fault for both combinational and sequential circuits. To illustrate the concept of a static fault in combinational logic, consider the truth table of Figure 8-1(a). The input column is as usual, that is, a natural binary count sequence. Note that two output columns are present, however. One column defines the desired function and the other shows the observed fault function. Under the column labeled 'Z_f', we see that one

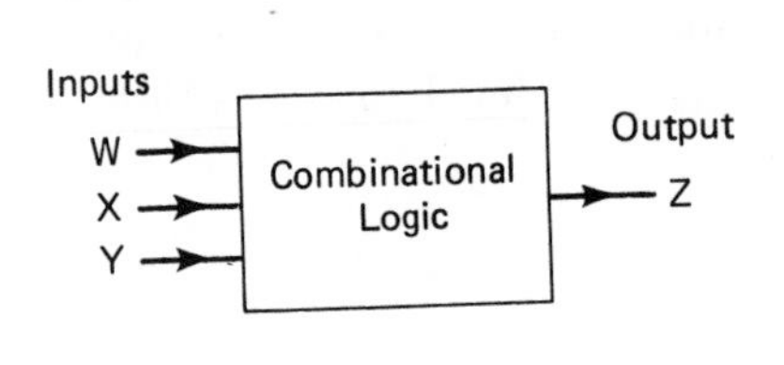

Desired Output → Z; Faulty Output → Z_f

W	X	Y	Z	Z_f
0	0	0	1	1
0	0	1	0	0
0	1	0	1	1
0	1	1	1	1
1	0	0	1	0*
1	0	1	0	0
1	1	0	0	0
1	1	1	1	1

(a) A Static Fault is Illustrated for Combinational Logic. '*' Denotes the Fault Condition.

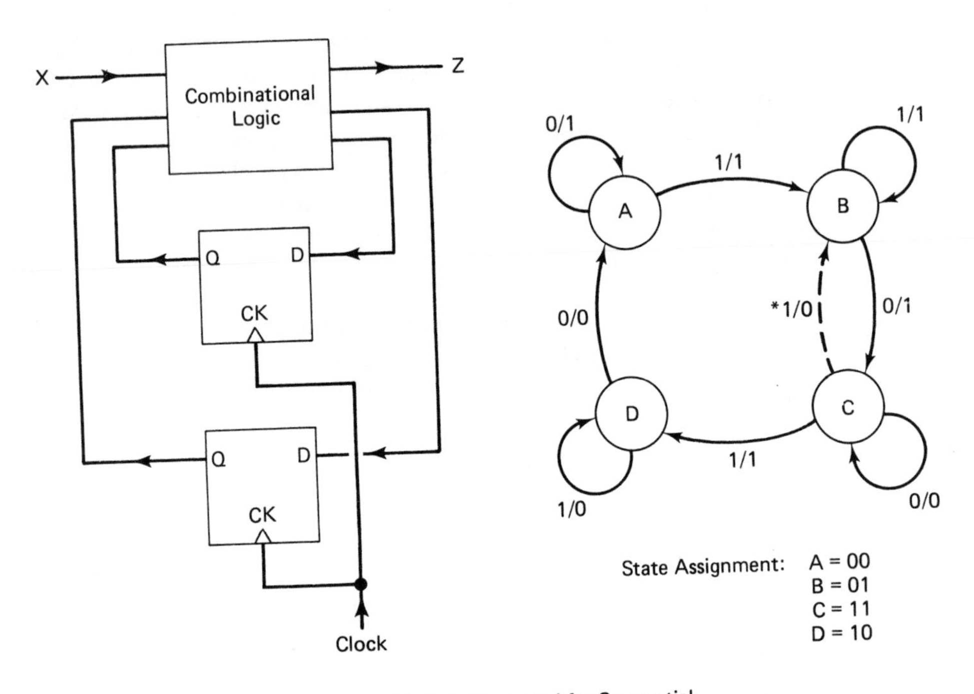

(b) A Static Fault is Illustrated for Sequential Logic. '*' Denotes the Fault Condition.

FIGURE 8-1

output condition is faulty (for $WXY = 100$). We note also that the faulty output is maintained as long as the corresponding input is held constant. This fact identifies the problem as a static fault.

Since combinational logic with a static fault could be incorporated in a sequential circuit, it is possible that sequential circuits can also be afflicted with static fault problems. (Note that a static fault in the combinational logic section of a sequential circuit may give the sequential circuit a static fault or a dynamic fault!) Figure 8-1(b) illustrates this possibility for a synchronous sequential circuit. In this case, the static fault causes the sequential circuit to follow an undesired state transition, with an erroneous primary output.

Static faults are generally not too difficult to isolate since the fault condition is held constant. Now, with the circuit halted in the faulty state, it is possible to trace the problem *backward* toward the inputs. When we find a contradiction (such as the output of an AND gate being '0' while all its inputs are '1') the fault will have been isolated. This will be discussed with more detail in a later section. Many static faults that we find in circuits that are built in the digital laboratory, or while prototyping, are due to our wiring errors, not to defective logic gates. This fact can often be a clue to finding the problem.

8.3.2 Dynamic Faults

Dynamic faults are characterized by an erroneous output that exists for a very short period of time. These transient errors can be extremely difficult to find, requiring the use of a high-speed storage oscilloscope or other special equipment. Hazards and races are examples of dynamic faults, discussed for the first time in Chapter 7.

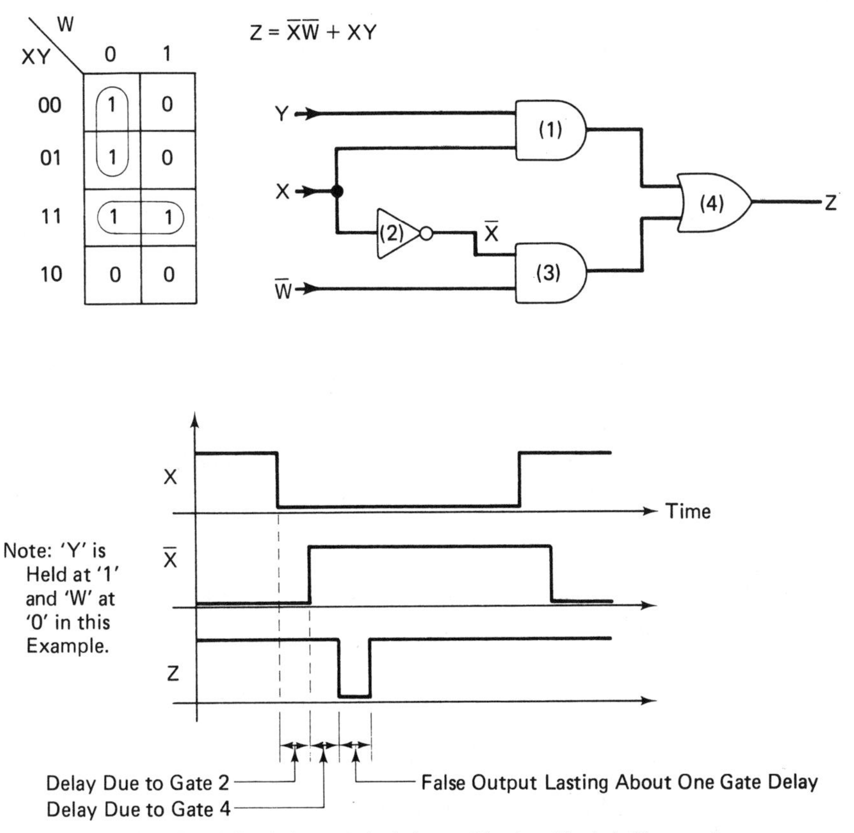

FIGURE 8-2 A dynamic fault in combinational logic is illustrated.

Although the fault condition in a dynamic fault lasts only a very short time, it can have long-lasting and serious consequences. For example, a transient error pulse could increment a counter improperly and cause a false reading.

As with static faults, dynamic faults can exist in both combinational and sequential circuits. To illustrate the concept of a dynamic fault in combinational logic, consider the function

$$Z = \bar{X}\bar{W} + XY$$

This function is implemented in Figure 8-2. In this example, we see that an erroneous output ($Z = 0$, lasting one gate delay) is produced whenever 'X' changes from '1' to '0' while $W = 0$ and $Y = 1$. This is due to the propagation delay imposed by the inverter. Note that no error occurs when 'X' changes from '0' to '1'. Recall that this type of fault is referred to as a *hazard*.

Dynamic faults can drastically affect the operation of asynchronous sequential circuits by causing false transitions, oscillations, and erroneous outputs. Usually, these problems are due to *races*, particularly where the propagation delays through two or more feedback paths are unequal. Refer to Chapter 7, Section 7.3.2 for an example of asynchronous sequential circuit race problems.

8.4 DETECTION AND CORRECTION OF FAULTS

The most difficult aspect of fault analysis is discovering exactly where the fault has occurred. Unfortunately, the symptoms that are first observed can be very misleading, particularly when a single fault causes a multitude of malfunctions. Once the fault is located, however, correcting it is often trivial—perhaps changing an integrated circuit chip or moving a wire. In order to efficiently locate the cause of a malfunction, it is necessary to

(a) Be aware of the possible kinds of fault that are likely.

(b) Have some orderly, logical procedure for locating the problem.

Both of these factors are developed primarily through experience. Nevertheless, we can present some very useful guidance by summarizing the most probable kinds of fault that are observed, and presenting a sound general approach to discover their location.

8.4.1 Summary of Probable Fault Conditions

(A) Static faults

1. Stuck inputs or outputs—occasionally, integrated circuit (IC) chips malfunction in such a way that an input or output appears to be fixed in one logic state ('0' or '1'), regardless of the condition of incoming signals. This could be caused by a defective transistor on the IC chip, or it may be the result of internal shorts to a power supply line. ***Stuck inputs*** or ***outputs*** are due to faults

XY	Z	Z_f
00	0	1
01	0	1
10	0	1
11	1	1

(a) A Single Output Stuck at '1' Could Cause this Fault

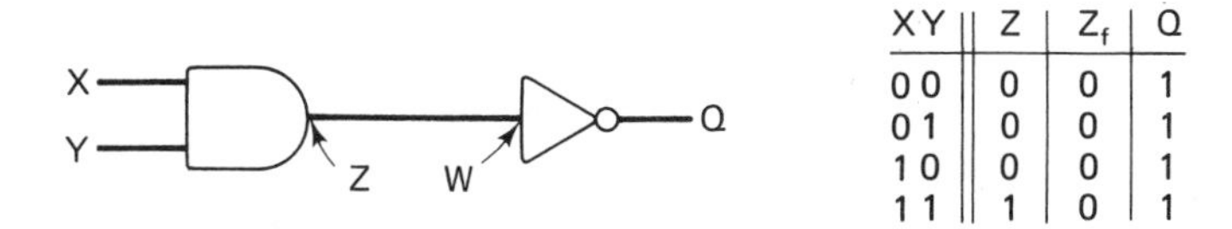

XY	Z	Z_f	Q
00	0	0	1
01	0	0	1
10	0	0	1
11	1	0	1

(b) This Fault Could be Caused by 'X' or 'Y' Stuck at '0', 'Z' Stuck at '0', or Input 'W' Internally Shorted to '0'.

FIGURE 8-3

that are within the hermetically sealed IC package, and thus they cannot be repaired. In such cases, the IC component must be replaced. Refer to Figure 8-3.

2. Bridge faults—most digital circuits are assembled on fiberglass cards containing printed circuit or wire-wrap connections. Such cards usually support a multitude of interconnected IC packages. It is possible that adjacent terminals or printed circuit leads can become accidentally shorted together. When this happens, a ***bridge fault*** has occurred, and the logic values of the shorted lines can be affected. When two bridged lines would normally have the same value (either both '0' or both '1'), no fault will be noticeable. However, when one line is being driven to logic '1' and the other to logic '0', a conflict occurs. Several effects are possible, depending upon the type of electronic logic that is being used (TTL, CMOS, ECL, etc.). When driven in opposite directions, a bridged line can do one of the following:

   (a) Assume a valid logic '1' voltage.

   (b) Assume a valid logic '0' voltage.

   (c) Assume a voltage in the threshold region between '0' and '1'.

   In any event, the output drive circuit on most logic gates is sufficiently strong so that a maintained short between two opposing signals may cause one or both outputs to burn out. In the case of TTL devices, a bridge between two opposing signals results in a valid logic '0' voltage. This is illustrated in Figure 8-4(a). In Figure 8-4(b), we see that this effect causes an inadvertent logic AND between the two bridged lines. While bridge faults can sometimes be easily seen on a circuit board, hairline defects are possible and can be very difficult to locate. Extreme caution should be used to avoid dropping small strands of wire on circuit boards.

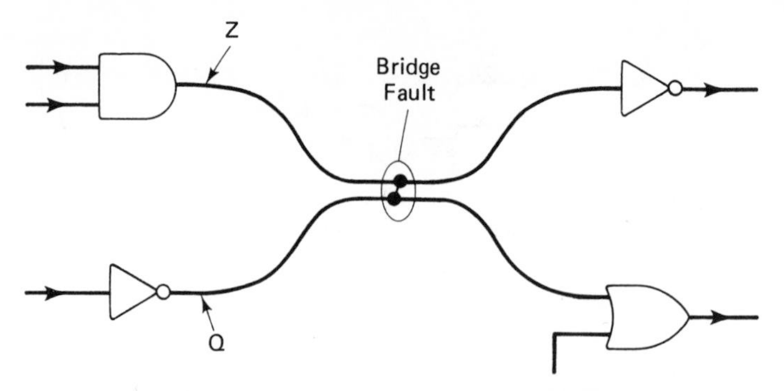

(a) For TTL Devices, Only When Both Outputs are '1' Will the Value On the Bridged Line be '1'.

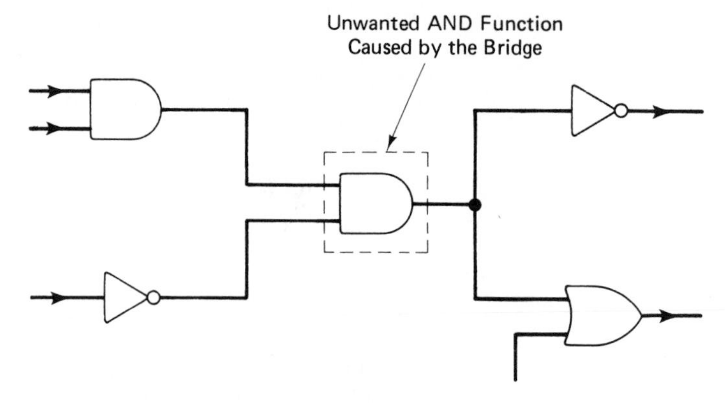

(b) The Result of a Bridge Between Two TTL Outputs Introduces an Effective Logic AND of the Signals Since the Bridged Lines Will Jointly Have a Value of '1' Only if Both Outputs are '1'.

FIGURE 8-4

3. Broken or missing conductors—manufacturing defects or wiring errors can result in an open circuit where there should be a solid connection between two points. Thus, signals driven onto this line at the source never reach their destination. Refer to Figure 8-5. The symptoms often resemble those of a stuck input, since, of course, the signal that should be driving a gate input remains constant. As with bridge faults, open circuits may be found by a careful examination of the circuit board, or by using an ohmmeter to check the resistance between two points. (A good wire conductor should have almost zero resistance.)

FIGURE 8-5 *Broken conductor.*

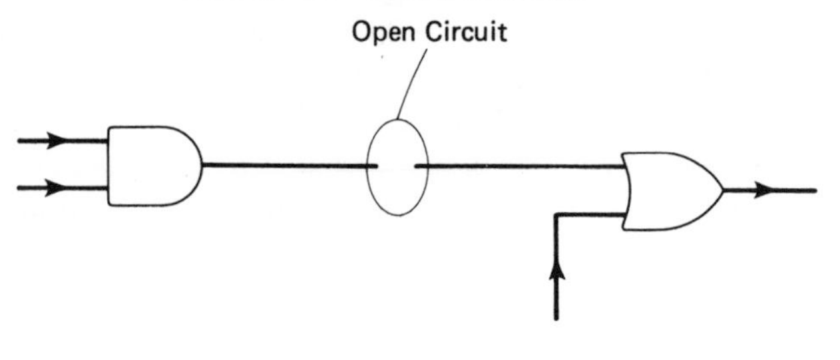

4. Lack of power—every IC device must have a correct and steady supply of voltage at its power supply inputs. Whenever one or more IC's lack power, a malfunction will occur. Normally, if one IC on a board does not have power, it will not be damaged by signals fed to it by other devices. Assuming that there are no other problems, the effect of an IC being unpowered is the same as if the IC were removed from the circuit, in many cases. Thus, the symptoms will be similar to those of stuck inputs or outputs. Depending on the logic family, unconnected gate inputs may assume either logic '1' or logic '0'. Lack of power is a very probable fault and should be one of the first tests made on a defective circuit board. Realize that a more serious defect could have blown a fuse, resulting in lack of power to the entire circuit board.
5. Improperly inserted components—since IC packages are quite symmetrical, it is possible to misinterpret the registration notch and occasionally insert an IC backwards. When this happens, the power supply pins are usually reversed, and the internal circuitry fails to operate. In fact, with the power supply polarity reversed, substantial heat will often develop within the IC in a short time, thus, destroying it. In addition, each signal line may appear to be open-circuited, or shorted to one of the power supply lines.
6. Incorrect components—because integrated circuit packages are often similar in appearance and the labels often hard to read, it is easy to mistake one type of device for another. Clearly, such an error will lead to circuit malfunction. Also, other previously good components connected to the incorrect device may be damaged. All comments relative to individual integrated circuit components can also be applied to entire circuit boards that are interchangeable plug-in modules.
7. Bad electrical contact—while a device may be properly inserted and offer no immediate sign of trouble, it may not be making a good connection at all points because of bent pins, a defective connector, or corrosion. This problem will usually give the symptom of an open circuit.
8. Improper power supply voltage—of course, a circuit cannot be expected to operate correctly if the power supply voltage is improper. Overvoltage is likely to damage components, while undervoltage probably will not.

(B) Dynamic faults

1. Hazards—varying propagation delays within a combinational logic circuit can cause momentary false outputs when the input values are changed. Such faults are referred to as *hazards*, and can sometimes be very difficult to correct. Hazards should be prepared for and corrected at the design stage of development since they are not due to a basic component fault, but to the natural and expected property of propagation delay. Whenever an MSP equation is directly implemented, hazards are easily eliminated by adding redundant terms to the equation. The redundant terms take effect, when needed, to override false '0's in a transition between two of the original MSP terms (refer to Figure

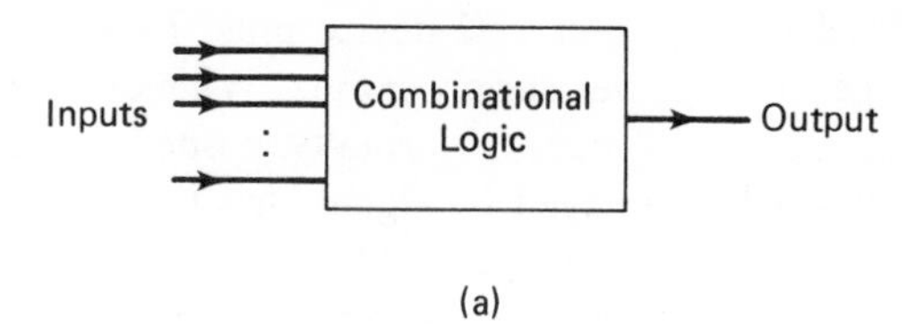

(a)

Ideal Transitions	Actual Transitions
1 → 1	1 → 0* → 1
0 → 0	0 → 1* → 0
1 → 0	1 → 0* → 1* → 0
0 → 1	0 → 1* → 0* → 1

(b)

FIGURE 8-6 *The effect of a hazard. '*' denotes a false output value that persists for a very short period (several gate delays at most).*

7-25 in Chapter 7). Often, however, it is possible to factor an MSP equation so that a more economical construction is possible. In such cases, hazards can become very difficult to find and correct. Thus, when a hazard exists, it is often necessary to make the construction much more complex in order that the hazard be eliminated. Figure 8-6 illustrates the different kinds of hazards that are possible.

2. Races—whenever two or more secondary variables in a sequential circuit attempt to change at the same time, a race condition is possible. In a realistic circuit, it is unlikely that two or more variables will change at *exactly* the same time. Therefore, we can expect them to change at slightly different times. This may result in momentary false excitations and transition to an improper state. Refer to Section 7.3.2 in Chapter 7 for further explanation of races. Races will occur primarily in asynchronous sequential circuits, or in synchronous sequential circuits where the propagation delay of "identical" components differs by more than one gate delay. If a race problem cannot be corrected by a new state assignment, then clocked flip-flops should be added to the feedback paths to make the circuit synchronous.

3. Crosstalk—whenever any two signal lines are in close proximity for any length (more than several centimeters), it is possible that a signal in one line (printed circuit trace or wire) will be electromagnetically coupled into the adjacent line, even though no connection exists between the two. If the coupling or the signal is strong, the voltage induced in the adjacent line can cause a false signal to be present. This fault condition is known as ***crosstalk***. The simple and obvious solution is to separate the two lines; even a small amount will do since the coupling is inversly proportional to the square of the separation distance. Holding the line length and separation constant, coupling is also proportional

to the signal's frequency components: as the frequency increases, so does the crosstalk. If a pulse train is coupled into an adjacent line, the crosstalk can be reduced by increasing the rise and fall times of the signal *without* changing the basic frequency of the pulse train. Very short rise and fall times (found in high-speed logic families) introduce substantial high-frequency components into the signal, which may produce crosstalk. It is generally hard to change the rise time and fall time. The only case where it is easy is when a high-speed integrated circuit can be replaced by a lower-speed device with the same logic function and the same pin connections.

Placing two adjacent lines as close to a ground plane (a sheet of copper at 0 V) as they are to each other will often reduce crosstalk considerably. Also, placing a ground wire between the signal lines has the same effect. Crosstalk is most likely to be found on printed circuit boards, wire-wrap backplanes, and in cables. Crosstalk should be anticipated at the time of design and printed circuit layout.

At no time should the source or coupled line be loaded with external capacitance. Larger current must then be driven into (and out of) the line to charge (and discharge) the added capacitance. This increased current flow will probably *increase* the crosstalk. Refer to Figure 1-38 for an example of crosstalk. Remember also that once a noise spike successfully passes through one gate, it is amplified by that gate and then becomes part of the signal. Thus, in digital systems, it is important to stop noise at the source. Refer to Figure 8-7.

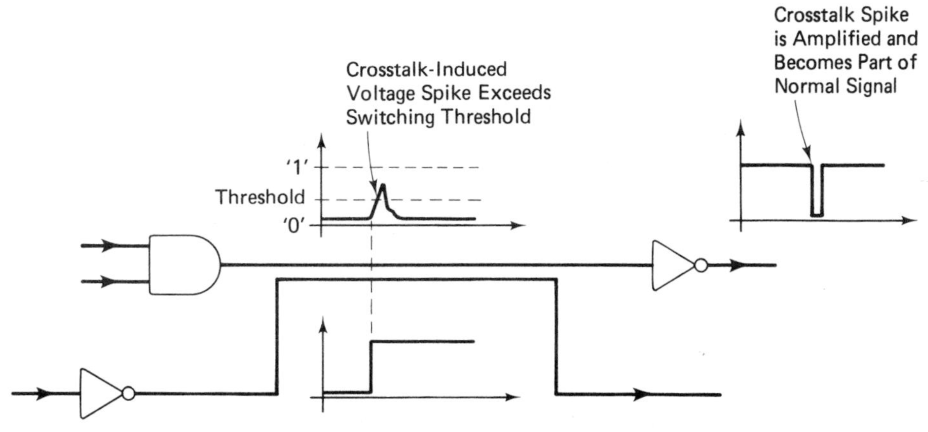

FIGURE 8-7 *Once noise voltages exceed the switching threshold, they are amplified by successive gates and become part of the signal.*

Crosstalk in dense cables is often reduced by using twisted pair wire. In this case, a ground line surrounds the signal line. The twisting of the wires has the effect of canceling any induced voltages due to crosstalk or electromagnetic noise: each twist reverses the induced voltage polarity with respect to the previous twist, resulting in a zero, or very small, net induced voltage over the entire length of the line. The use of a differential receiver also improves

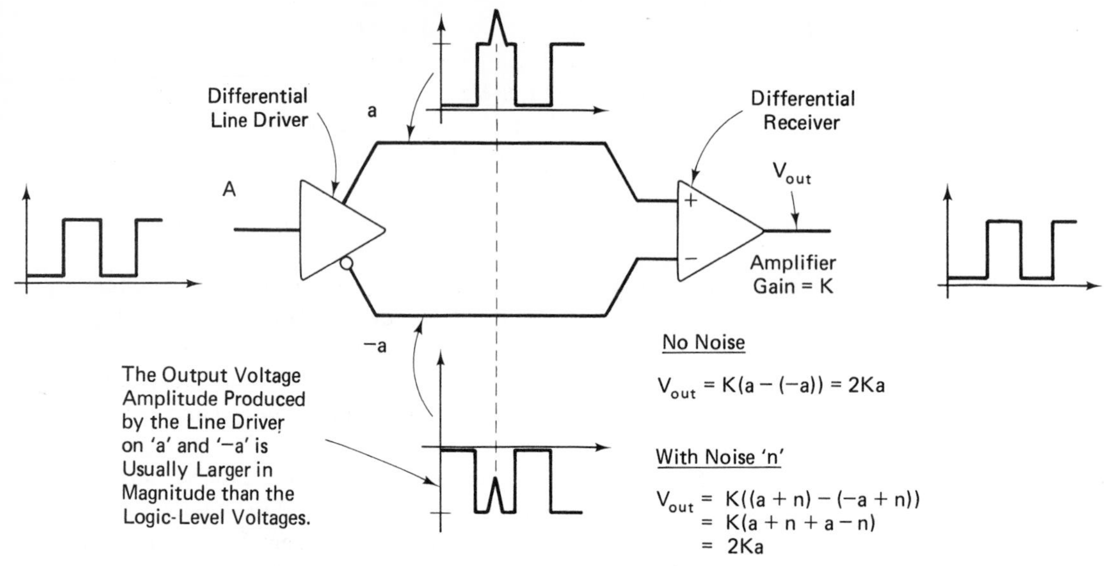

FIGURE 8-8 Common-mode-induced noise on two differential lines is subtracted in a differential receiver, leaving an undisturbed signal.

FIGURE 8-9 Schmitt trigger.

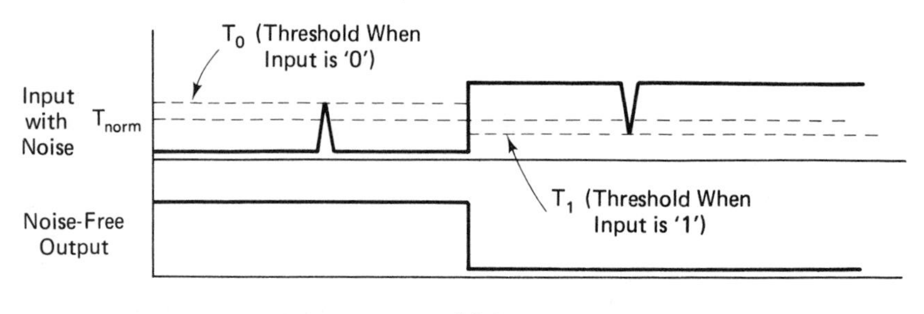

(a) Operation

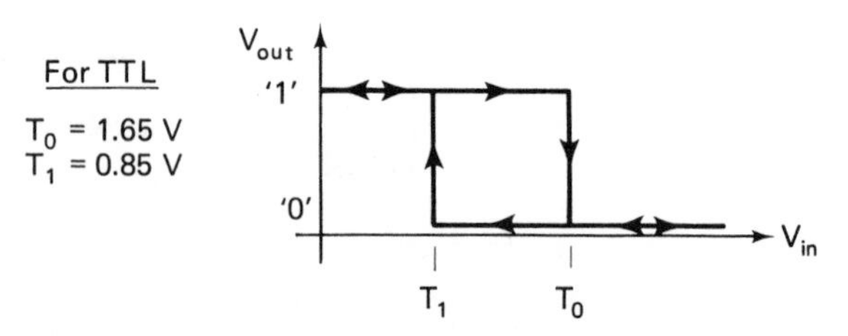

(b) Transfer Characteristic

(c) Gate Symbols

noise immunity on a cable channel. A ***differential line driver*** sends both a normal *and* an inverted signal over the channel. Induced noise that affects both signals equally is subtracted by the ***differential receiver***. This is illustrated in Figure 8-8.

Induced noise due to crosstalk and electromagnetic interference can also be handled by the use of a variable threshold logic gate, often referred to as a ***Schmitt trigger*** gate. This device is useful to eliminate noise voltage that are slightly greater in amplitude than the threshold of normal logic gates. Refer to Figure 8-9.

4. Coupling through the power supply—rapid variations in load due to periodic signals and pulses, as shown in Figure 8-10, usually cannot be immediately compensated for by the voltage regulator in the power supply. Consequently, the power supply voltage may appear to have a small signal voltage superimposed on it. If the voltage is large enough, logic gates can be caused to malfunction. To ease this problem, large electrolytic capacitors are typically placed just after the voltage regulator between the power and ground lines.

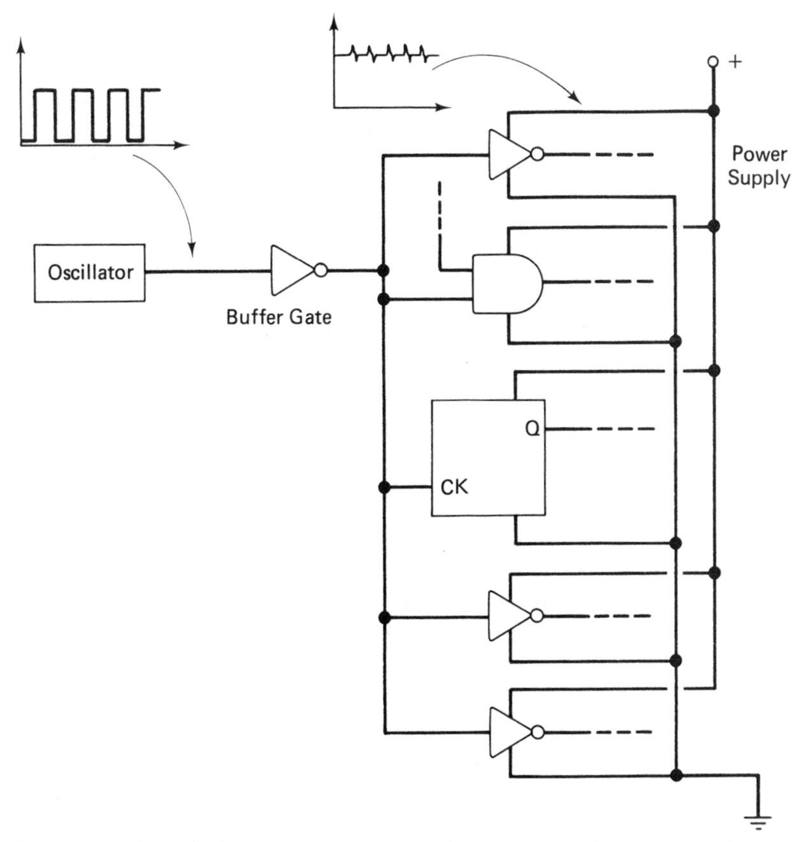

FIGURE 8-10 *Many devices switching simultaneously cause the power supply load to change noticeably, producing a slight variation in the power supply voltage.*

Even that, however, does not always completely eliminate power supply noise; the electrolytic capacitors alone just cannot respond fast enough to absorb all of the small-signal component (this is due to the poor behavior of electrolytic capacitors at high frequencies). Ceramic disc capacitors are one of several types of capacitors made that have especially good high-frequency response, and are used effectively in a power supply system to reduce high-frequency noise. While disc capacitors have a much higher frequency response, they cannot be made to have very large (greater than 0.5 μF) capacitance values. Generally, large electrolytic capacitors (25 μF or more) are used at the power supply source, while small disc capacitors (0.1 μF or less) are distributed throughout the logic card. Often, a single disc capacitor is used in the vicinity of each IC package. Figure 8-11 illustrates this idea. When care is taken to properly decouple the power supply lines with capacitors, transient faults due to power supply noise are unlikely.

5. Reflections—When digital signals must be passed through long lengths of wire, electrical effects that could be ignored for short lengths must now be accounted for. Perhaps the most important property is the characteristic impedance of the source, transmission line, and receiver. Unless these impedances are exactly matched, signal reflections may occur at the interface between source and medium, and receiver and medium, as shown in Figure 8-12. When signal reflections occur, some of the signal energy remains within the transmission line and creates an "echo" effect. The reflections can superimpose on the later signal values to cause severe distortion. Fifty- to 75-Ω transmission lines with matched driver and receiver circuits are commonly used to minimize the reflection problem. Normally, reflections will not concern us unless we are dealing with transmissions lines longer than a few feet, or very high frequency signals (rise and fall times less than 1 ns). Data supplied by the manufacturer of a logic family usually provide information and specific recommendations for the transmission of signals over long-length lines.

6. Transient IC malfunctions—with the advent of medium- and large-scale integration digital circuits, many logic devices can be made available on a single IC chip. At least two kinds of serious faults can occasionally result from the high density of semiconductor components. First, it is possible that two or more independent devices on one chip can interact with each other. This could be due to a manufacturing defect, to excessively high operating temperature, to an external short involving one of several devices on the chip, and for other reasons. Second, defects within a single device, such as a flip-flop, can cause it to malfunction only on occasion, and not always. Occasional spurious errors due to IC malfunctions are extremely difficult to identify and correct. Often, it is only when all other possible sources of the problem (such as crosstalk and power supply coupling) have been ruled out that transient IC errors can be inferred.

7. Unequal propagation delays—while a synchronous circuit design may be perfectly correct on paper, it may fail to operate synchronously in practice,

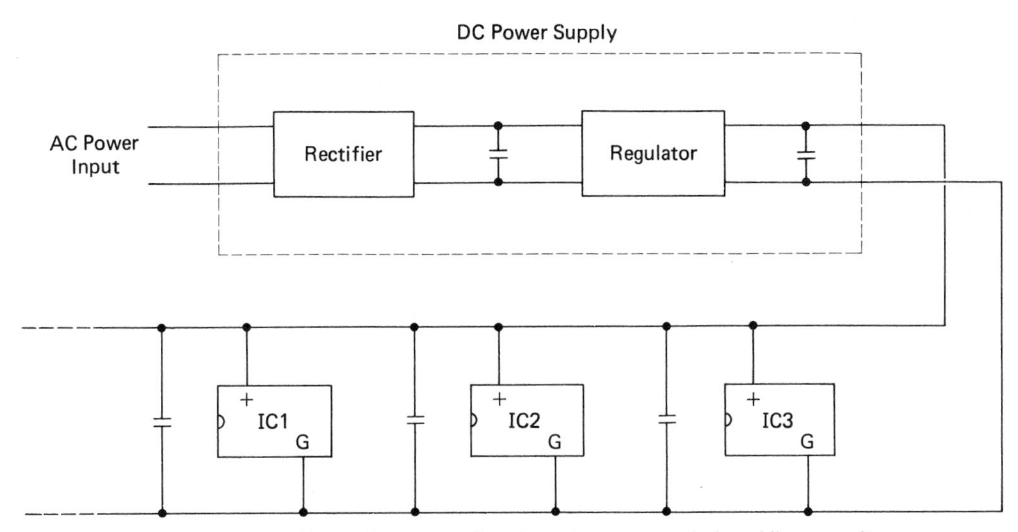

FIGURE 8-11 Reducing coupling through power supply by adding capacitors.

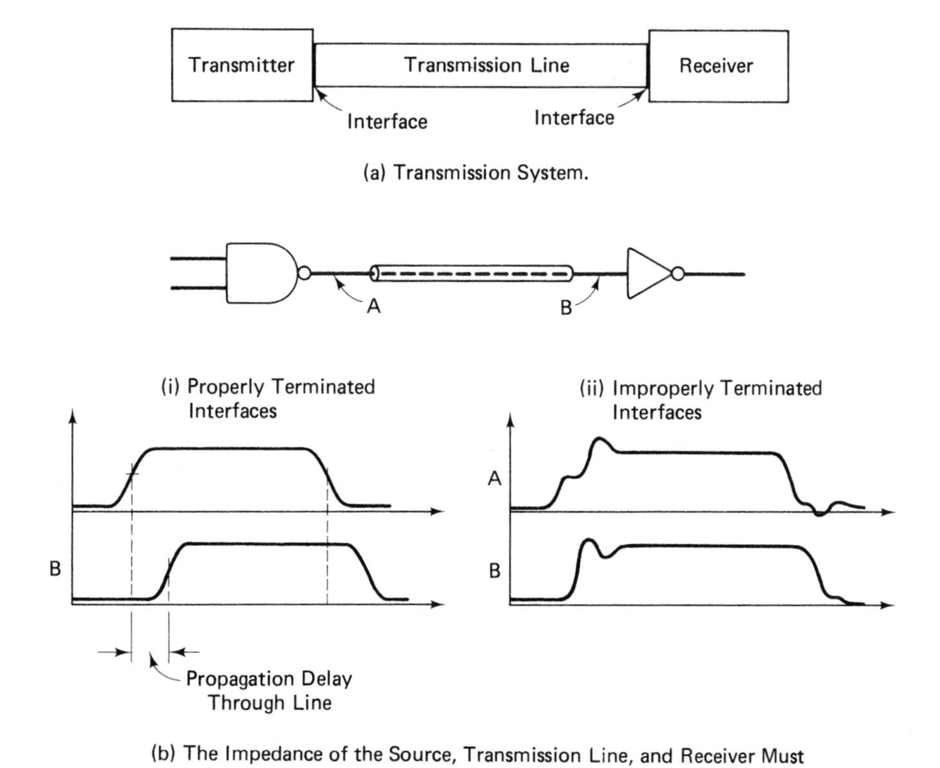

FIGURE 8-12

producing symptoms of hazards and races. This problem can occur if supposedly identical flip-flops, or other devices, have *unequal* propagation delay. Propagation delay is a function of the logic family circuit design, manufacturing variations, and other factors. Being certain to use devices just from one family does not guarantee that all propagation delays are equal for the same IC function. This is especially true if the same type of component is obtained from different manufacturers. For example, in making an eight-bit synchronous counter using two "identical" four-bit counter stages made by different manufacturers, or made at different times, it is possible that one of the four-bit stages will be slightly faster. This, of course, can cause false states to occur, which can result in a false output from a detector gate. Also, unequal delays through printed circuit wiring and backplanes can cause signal skew. Perhaps the most important timing considerations relate to the clock line, which is the synchronizing strobe for most clocked devices.

8. Setup-hold violations—clocked devices such as edge-triggered flip-flops often require a setup and hold time on their inputs. Should these times be violated, the device will fail to operate correctly.
9. Power-up initialization failure—many digital circuits require that particular initializing action be taken when power is first applied. Should this fail to occur, undetermined and possibly damaging consequences may result. When failure to initialize can cause serious side effects, the initializing block of the circuit should be very carefully designed, possibly having redundant components.

8.4.2 Identifying the Cause

Malfunctions in a digital circuit are sometimes very easy to find, and often can be extremely difficult to find. The ease with which faults are located depends upon many factors. Some of the most important are:

(a) Our understanding of how the circuit is suppose to operate.
(b) The type of test equipment that is available.
(c) The number of test points that are available.
(d) Knowing any prior conditions that may have brought about the malfunction.
(e) The ease with which circuit components can be isolated or removed.

Perhaps the best general advice that can be given comes from (a) above: *be certain that the circuit's proper operation is carefully and completely understood.* This may involve hours of careful study of a circuit schematic and a maintenance manual, and possibly the measurement of signals on a different, working version of the faulty circuit. Remember that only when we can recognize the difference between a correct signal and an incorrect one will we be able to distinguish a fault condition.

Experience has shown that the amount of time needed to find the cause of a fault in a circuit tends to increase as more sophisticated test equipment becomes available—

this is not said in jest. With sophisticated test equipment, we are less likely to give careful thought to the problem, and instead lean too much on the equipment itself. This, unfortunately, is self-defeating. We must be very conscious of the tendency, and guard against it. Sophisticated test equipment used in conjunction with a carefully thought out plan can be a powerful approach in the detection of circuit fault. One *hour* of study followed by ten *minutes* of measurement can often take the place of one *minute* of study followed by ten *hours* of measurement.

Some digital circuit boards may have very limited access, so that the number of test points are minimal. This places an additional restriction on our ability to locate faults. Remember, though, that the output of a logic gate can be measured at any point on a line to which it is connected, not just at the logic gate's IC package output pin. Thus, while the output pin of an IC may be inaccessable, the line to which it is connected may lead to a terminal point or gate input that can be measured. Some circuits are provided with special test points for this purpose.

It is extremely helpful to know about any abnormal conditions that may have preceded circuit failure. Knowing a possible cause can very quickly lead to the defect. For example, if a connecting cable were accidently caught in the access door of a device, it is possible that one or more internal conductors could have been severed while the outer insulation remains intact. Thus, the cable is faulty, even though there is no obvious indication of this. Knowing that the cable may have been damaged just before the equipment failed strongly implies the fault. Some factors that may precede failure are listed:

1. Unusual excess heat.
2. Power failure or power line surge.
3. Electrical storms.
4. Static electricity discharge.
5. Unusually high moisture or water damage.
6. Unusually high shock or vibration.
7. Metal fragments or wire strands falling through vent holes.
8. Unusually high stress on cables or connectors.
9. Nearby high-power electrical machinery switching on or off.
10. Unusually large amounts of dust or airborne particles.

When many logic components are connected in a circuit, the effect of a single, faulty component can propagate a fault signal through many successive stages, or can cause good components to change state at improper times. Thus, it is often necessary to trace the faulty signal backward through correctly operating components until a logical inconsistency is found. This may mean, eventually, that we must isolate one or more suspect components before reaching a final decision. For example, consider the faulty combinational logic circuit of Figure 8-13. With the input $X = 1$, $Y = 0, Z = 1$, and $W = 1$, measurements reveal the intermediate logic values shown

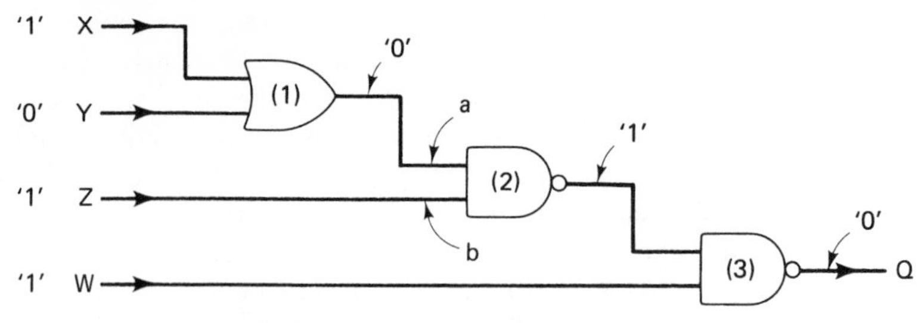

FIGURE 8-13 Faulty combinational logic circuit.

in the figure. It is seen that a logic error appears at the output of gate (1), since a '1' and '0' input to an OR gate should produce a '1' output. We cannot, as yet, assume that the OR gate is at fault, however. If input 'a' to gate (2), or the printed circuit line, were shorted to logic '0' (zero volts in TTL) the output of (1) would be pulled to '0', regardless of whether gate (1) were good or bad. Before being able to decide where the fault is located, we must isolate the output of (1) from the input of (2) and make a new measurement of (1)'s output. If the IC's are socketed, removal of (2) would be easiest. Otherwise, the PC or wire-wrapped trace must be carefully cut. Note also that a solder bridge to ground on the PC trace connecting gates (1) and (2) could also cause the problem. Thus, it is possible that *neither gate* is defective.

8.4.3 Automated Testing

A very direct way of testing the operation of a digital circuit is to compare its response for *each possible* input condition to an identical circuit that is known to be fault-free. If the responses always match for all input conditions, then the circuit under test must be operating properly. This technique is particularly well suited for assembly-line testing since the procedure can be automated easily. The circuit board under test is always electrically isolated from the reference board by buffer gates. This ensures that the reference board is protected against unexpected shorts in the board under test. Sophisticated computer-driven testers using this technique can also either suggest or identify the fault by printing test results on a computer terminal.

8.5 CASE STUDIES

The following case studies describe particular examples of faulty digital circuits. Many of these examples are drawn from actual laboratory experience, while others represent assumed conditions that are, nevertheless, likely to be experienced by some. Each case study is made up of two parts. First, the basic circuit and its fault symptoms are presented. Then, the fault condition is analyzed. The reader is encouraged to study the first section (basic circuit and symptoms), and then think about the problem for a few minutes before reading the analysis. To reduce the temptation to look ahead, the problem descriptions for all case studies are separated from the analyses. A page reference to the analysis is given at the end of each problem description.

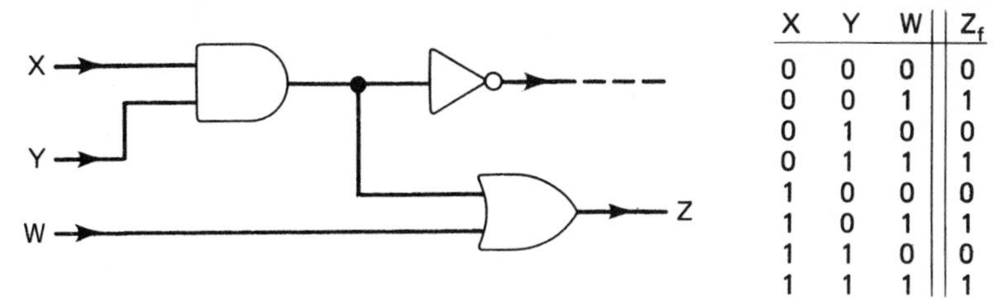

X	Y	W	Z_f
0	0	0	0
0	0	1	1
0	1	0	0
0	1	1	1
1	0	0	0
1	0	1	1
1	1	0	0
1	1	1	1

FIGURE 8-14 *Faulty combinational logic circuit. 'Z_f' is the measured fault function.*

8.5.1 Case Study 1

The combinational logic circuit shown in Figure 8-14 is found to malfunction. A truth table is determined experimentally for this faulty circuit, and is also shown in the figure. Is this a static or dynamic fault? What possible fault could cause this malfunction? The analysis begins on page 377.

8.5.2 Case Study 2

The combinational logic circuit of Figure 8-15 is found to malfunction, as shown in the truth table by fault function 'Z_f'. We note that for the condition $XYZ = 110$, the output oscillates with a period of $4T$, where 'T' is the propagation delay of one gate. The circuit is constructed with standard TTL devices, so that a logic '0' can sink about 1.6 mA. The power supply voltage is correct, and no other circuits are nearby to cause coupling problems. Why should the output oscillate for only one combination of input values? The analysis begins on page 378.

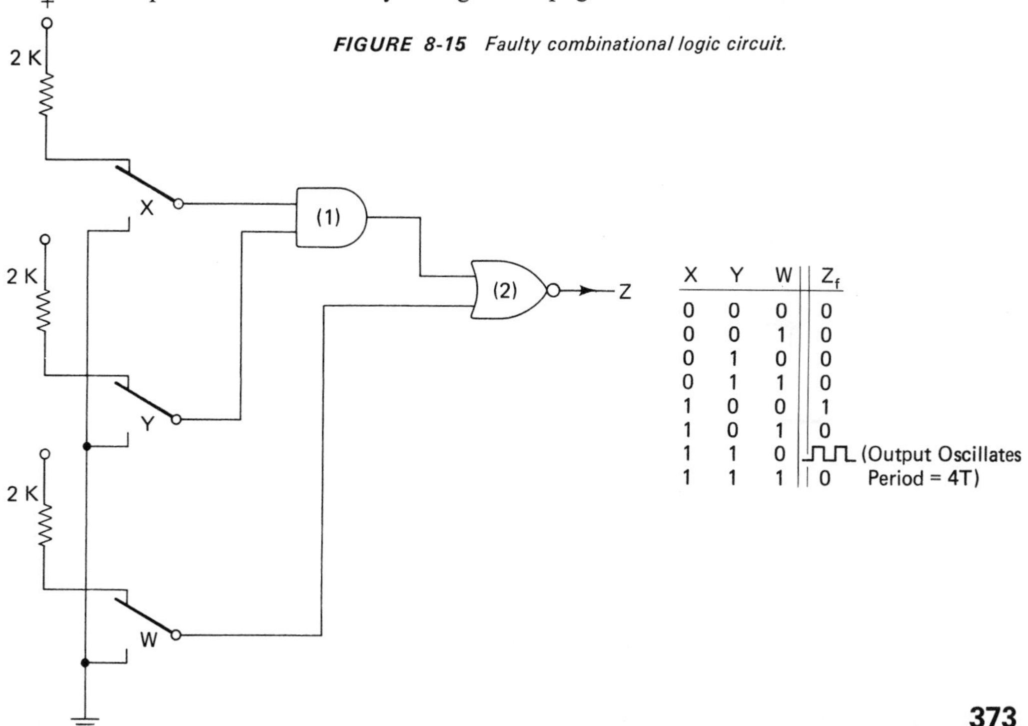

X	Y	W	Z_f
0	0	0	0
0	0	1	0
0	1	0	0
0	1	1	0
1	0	0	1
1	0	1	0
1	1	0	(Output Oscillates Period = 4T)
1	1	1	0

FIGURE 8-15 *Faulty combinational logic circuit.*

8.5.3 Case Study 3

An eight-bit natural binary up counter is constructed using two four-bit synchronous counter modules. An eight-input NOR gate is connected to the 'Q' outputs of this counter to detect the output condition '00000000'. The NOR output, in turn, is connected to the clock input of a J-K-type flip-flop (wired to toggle), so that each time the eight-bit counter completes a full 256-state cycle, the J-K device will complement. When the circuit is operating, a narrow, unwanted pulse (glitch) is being produced by the NOR whenever the counter changes from '00001111' to '00010000', causing the J-K flip-flop to complement at the wrong time. Refer to Figure 8-16. The circuit is being operated at speeds well below the maximum specified by the manufacturer of the components. The analysis begins on page 379.

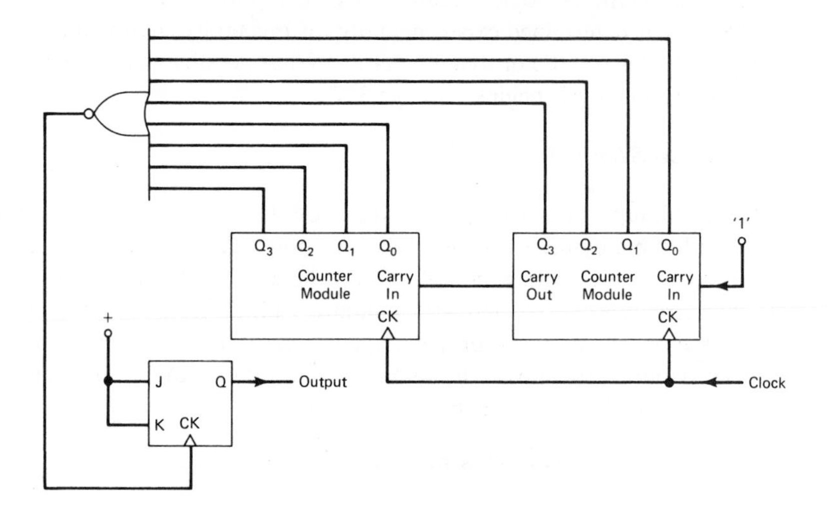

(a) Circuit Connections for the Synchronous Counter, Detector, and Flip-Flop of Case Study 3.

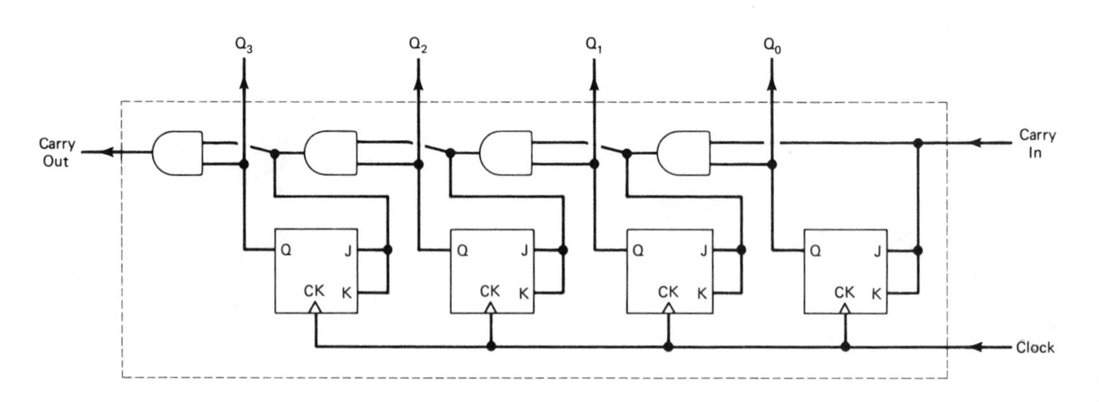

(b) Internal Circuit Design of a Typical Four-Bit Counter Module.

FIGURE 8-16

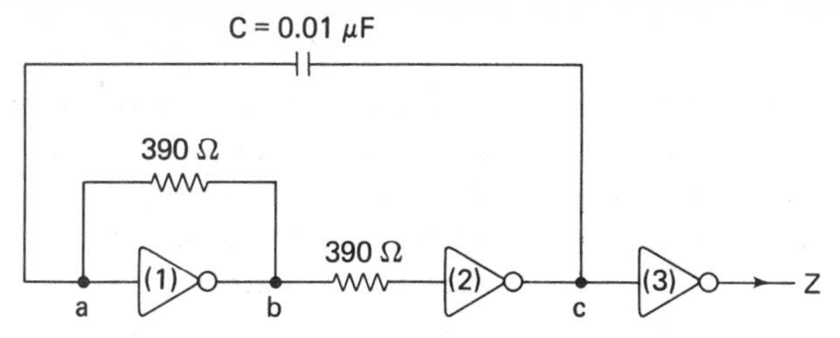

FIGURE 8-17 Simple oscillator circuit.

8.5.4 Case Study 4

A simple oscillator is constructed using TTL gates, as shown in Figure 8-17. This type of circuit has been found to work well over a wide range of component values, and has been used in several earlier applications. However, this particular module, when built, did not produce an oscillating output at 'Z'. Instead, 'Z' was found to be '0' at all times. Voltage measurements at points (a), (b), and (c) in the figure showed that the circuit, was, indeed, oscillating at the desired frequency; the correct charge/discharge waveforms could be seen. Inverter gates (1), (2), and (3) were tested and found to be operating perfectly. Furthermore, all wiring and connections were tested and found to be good. Why is no output seen on 'Z'? The analysis begins on page 379.

8.5.5 Case Study 5

A high-speed digital circuit has several outputs that are to be transmitted to a location 5 m (15 ft) away. D-type flip-flops are used as a temporary storage register, and interface directly to the cable, as shown in Figure 8-18. The high-speed digital circuit's outputs are tested at points (a), (b), and (c), and are found to be correct.

FIGURE 8-18 Cable interface between two circuits.

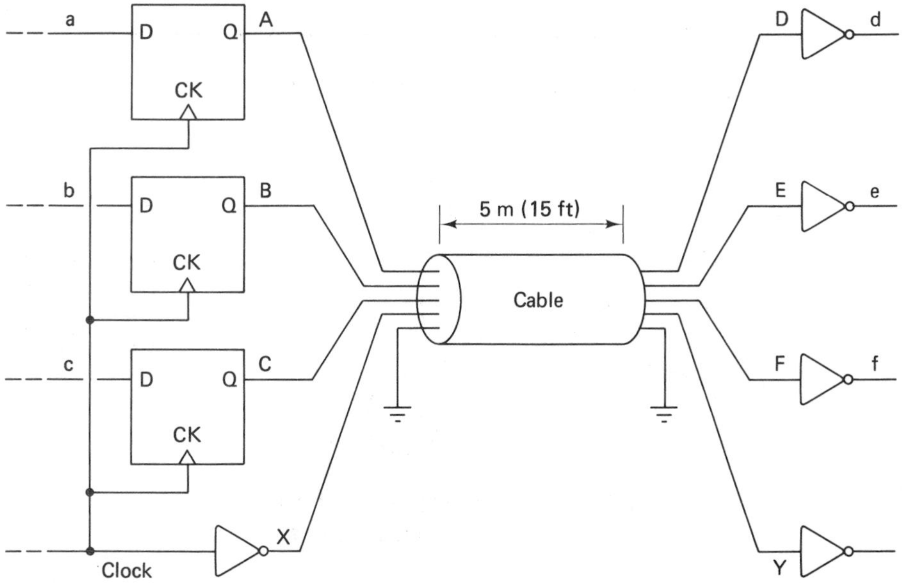

Unfortunately, however, occasional erroneous signals are being received at points (d), (e), and (f). The erroneous signals appear to have very rough rising and falling edges, and steady, incorrect states on occasion, between clock pulses. All inverters and D flip-flops are tested and found to be working properly. The cable is intact, the connectors good, and electromagnetic interference is not present. All cable conductors are twisted pairs, with one conductor of each pair grounded. The power supply voltage is correct, and is properly supplied to each IC. What could be causing the problem? The analysis begins on page 380.

8.5.6 *Case Study 6*

The sequential circuit shown in Figure 8-19(a) produces the erroneous flow graph shown in Figure 8-19(b). All logic components (flip-flops and NAND gates) are tested

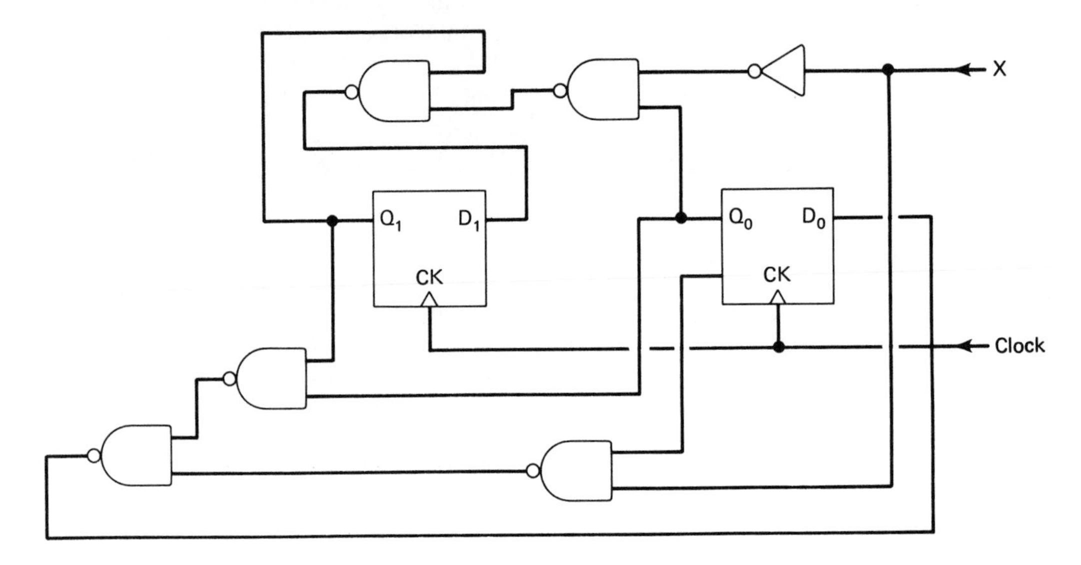

(a) Logic Schematic of a Faulty Circuit.

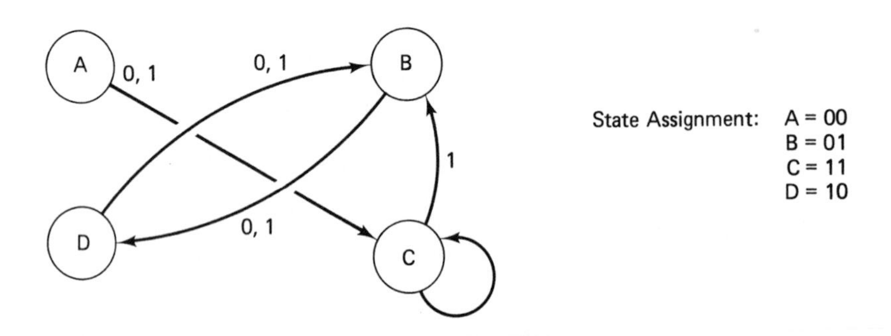

(b) Flow Graph Resulting from the Fault.

FIGURE 8-19

and found to be operating properly. This circuit is constructed on a socketed printed circuit board, and it is suspected that a fault must exist in the printed circuit wiring. The power supply voltage is correct, and was found to be properly applied to each IC. The logic family used is one in which a logic '0' output shorted to a logic '1' output produces a logic '0' result. The analysis begins on page 382.

8.6 CASE STUDY SOLUTIONS

8.6.1 Case Study 1 Analysis

First, we will derive the Boolean equation describing a working circuit of this type and compare the fault-free function to the erroneous one. This is shown in Figure 8-20. Scanning the truth table from top to bottom, we find that the first and only erroneous output comes when $XYW = 110$. At this time, the actual output is '0', while the correct output should be '1'. Referring to the circuit diagram, we see that a false output of $Z = 0$ implies that both inputs to OR gate (3) are '0'. A measurement is made and indicates that inputs (3)a and (3)b are '0'. In a working circuit, however, (3)a should be '1' under these input conditions. Further measurement indicates that for *all* input values, (3)a is '0'. The most likely causes for this condition are

1. Input (3)a is stuck at '0' (internally shorted to logic '0', usually zero volts).
2. Output (1) is stuck at '0'.
3. Input (2) is stuck at '0'.
4. The line connecting (1), (2), and (3) is bridged to logic '0' (zero volts).

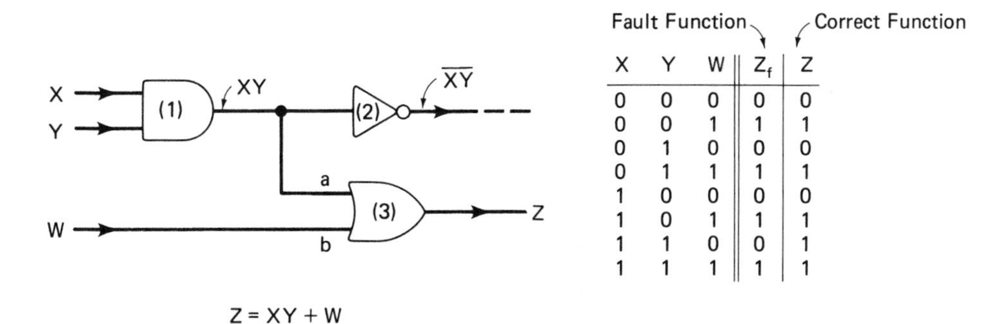

X	Y	W	Z_f	Z
0	0	0	0	0
0	0	1	1	1
0	1	0	0	0
0	1	1	1	1
1	0	0	0	0
1	0	1	1	1
1	1	0	0	1
1	1	1	1	1

$Z = XY + W$

FIGURE 8-20 *The erroneous function 'Z_f', is compared to the correction function, 'Z'.*

To identify the actual cause, the following steps are taken:

(a) Action: Gate (3) is removed from the circuit (we assume that the IC's are socketed), *with the power off.*
Result: the fault remains—the input node to (3)a is '0' under all input conditions.

(b) Action: Gate (2) is removed.
Result: The fault remains.

(c) Action: Gate (1) is removed, and the resistance between the line and power supply bus is measured (with the power off).
Result: Infinite resistance is measured, indicating that the line is electrically isolated from all other nodes in the circuit. Measurable resistance (less than 100 MΩ and greater than 50 Ω) would indicate a bridge to another gate input or output, while very low resistance (less than 5 Ω) would indicate a bridge to the power supply bus (either +, −, or ground).

(d) Action: Gate (1) is replaced with a new circuit.
Result: The fault condition disappears. Thus, the original gate (1) was faulty, having an output that was stuck at '0'.

8.6.2 Case Study 2 Analysis

The oscillating output for $XYW = 110$ might normally imply a faulty logic gate. In this case, however, two unique features cast doubt on that implication. First, the period of oscillation is $4T$ (four gate delays). Parasitic coupling within a gate could cause an oscillating output, but its period would almost certainly be less than four gate delays. Second, if the oscillation were due to an IC failure, it is more likely that the oscillation would be produced for a number of different input conditions, not just one. If we assume that the gates are good, then there must be something wrong with the connections, and a bridge fault is suspected. Now, if a bridge fault is the problem and it causes oscillation, then an *undesired feedback path* must have been created by the bridge. The only conditions under which oscillation can occur is when the feedback loop contains an *odd* number of inversions (an even number of inversions produces a latch). Since this circuit contains only one inverting gate, namely the NOR, we can conclude that one side of the bridge must be on the NOR output. With this conclusion, a careful physical examination of the circuit board might reveal the problem. Let us assume that there were no obvious physical defects, and continue with the analysis.

The Boolean equation for a fault-free circuit of this type is found to be

$$Z = \overline{XY + W} = (\bar{X} + \bar{Y})\bar{W}$$

Figure 8-21 compares the truth table for this function to the fault function of Figure 8-15. The first error that is observed is for $XYW = 000$. With this input, the output is '0' instead of '1'. Assuming that a bridge fault is the problem, and deducing that the NOR gate is one side of the bridge, then we can say that the NOR output is being pulled to '0' by either 'X', 'Y', 'W', or the AND output, all of which are '0'. Let us look at the error condition $XYW = 110$, where we find an oscillating output. Measurement shows that the 'X' input of the AND gate is also oscillating, while 'Y' and 'W' are steady. Thus, the bridge must be between the 'X' input and the NOR output. The fact that there are *two* gates between the bridge (providing a total of two gate propagation delays) explains exactly why the period of the oscillation is $4T$ (two delays for the low-to-high transition and two delays for the high-to-low transition, totaling four delays). It is possible that bridge faults can occur between three or more nodes in a circuit, making the analysis considerably more complex. In such cases, though, careful analysis can usually lead to the problem.

X	Y	W	Z_f	Z
0	0	0	0	1
0	0	1	0	0
0	1	0	0	1
0	1	1	0	0
1	0	0	1	1
1	0	1	0	0
1	1	0	⎍⎍⎍	0
1	1	1	0	0

FIGURE 8-21 *The proper function, 'Z', is compared to the fault function, 'Z_f'.*

8.6.3 Case Study 3 Analysis

By studying the circuit and symptoms carefully, we note that the entire circuit and its components are "basically" working correctly: the counter is progressing properly through all 256 states, the NOR is detecting the '00000000' condition, and the J-K flip-flop is complementing whenever a zero state is reached. The fault seems to be due to a transient '00000000' state appearing at the NOR inputs when progressing from state '00001111' to '00010000'. Thus, it is implied that the synchronous counter is really *not* synchronous. Upon examining the counter modules, we find that they are made by different manufacturers, although they are logically equivalent. The observed fault can be explained if the high-order module ($Q_7 - Q_4$) is faster than the low-order one. When the modules are replaced by identical components (made by the same manufacturer at the same time), the problem dissappears. We can conclude that the fault was due to a race, caused by unequal propagation delays between the two counter modules. Could the problem have been solved by interchanging the original two modules? If not, why? If so, would it have been a good idea?

8.6.4 Case Study 4 Analysis

This circuit confronts us with a very perplexing problem. Why, when all components and wiring are correct, does the circuit fail to produce an output on '*Z*'? This is especially puzzling when we can see evidence of oscillation at points (a), (b), and (c). The two 390-Ω resistors are removed from the circuit and measured. It is found that the resistor connecting the output of gate (1) to the input of gate (2) is really 11 Ω instead of 390 Ω—this resistor was internally defective. Upon replacing the defective component, the circuit operated properly and an oscillating output was produced.

We must still answer the question:

> Why did we see oscillation when the defective resistor was in place?

First, the function of the resistor connection between gates (1) and (2) is not at all clear unless a detailed analysis is made concerning the internal circuit design of TTL gates. When this is done, the need for the component is seen—it enhances the feedback conditions needed for oscillation to occur. In the absence of this component, the circuit is on the brink of oscillation, but the conditions do not quite guarantee that the circuit will actually oscillate. Now, the small amount of capacitance contributed by the *oscilloscope probe* when points (a), (b), and (c) were measured was enough to push the circuit over the brink and cause it to oscillate! When the probe was removed, the oscillation stopped. Since gate (3) is out of the feedback loop, measuring the output on '*Z*' could not have any effect on the oscillation. This case study emphatically demonstrates that the presence of a measuring instrument (such as the oscilloscope) can, on occasion, dramatically affect the circuit we are trying to measure. This fact must always be considered when apparently contradictory measurements are found.

8.6.5 Case Study 5 Analysis

In the absence of component failure, we can assume that the cable or connectors must be having an undesirable electrical effect on the digital signals. This certainly could explain the rough rising and falling edges at (d), (e), and (f). A sustained error, however, after the triggering edge of the clock pulse implies a malfunction of the 'D' flip-flops. Since the flip-flops have been tested and found to be operating properly, and since the clock signal triggering the flip-flops is accurate and glitch-free, we seem to be faced with a contradiction. To learn more about the problem and look for a solution, we take the following steps:

(a) Action: First, the cable is disconnected from the transmitter, and we measure the signals at '*A*', '*B*', '*C*', and '*X*'.
Result: The signals are error-free and undistorted.

(b) Action: The cable is disconnected from the receiver and reconnected to the transmitter. The signals are measured at '*D*', '*E*', '*F*', and '*Y*'.
Result: The signals appear to be quite distorted, as shown in Figure 8-22. We note that the error occurs whenever the flip-flop tries to change from '0' to '1'.

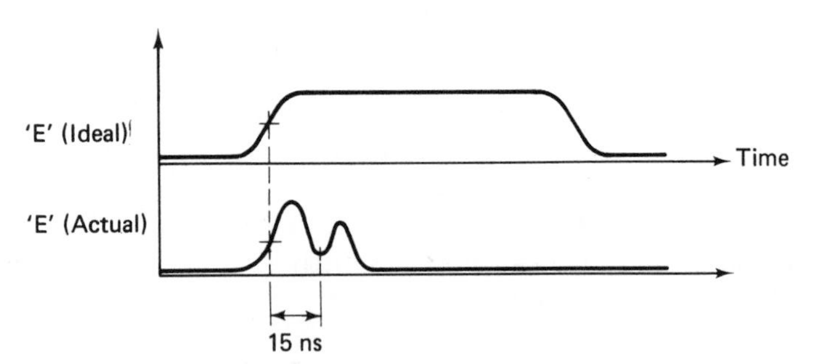

FIGURE 8-22 *The distortion and sustained error found at the receiver end, as seen on line 'E' (typical).*

A study of the 'D' flip-flop's *internal logic design* reveals the problem; refer to Figure 8-23. We see that the design consists of a synchronizing and lock-out circuit for the '*CK*' and '*D*' inputs, followed by a NAND latch. Ringing and reflections in the cable caused the output signal voltage to drop temporarily below the gate threshold voltage after the flip-flop had switched from '0' to '1'. The ringing and reflections are

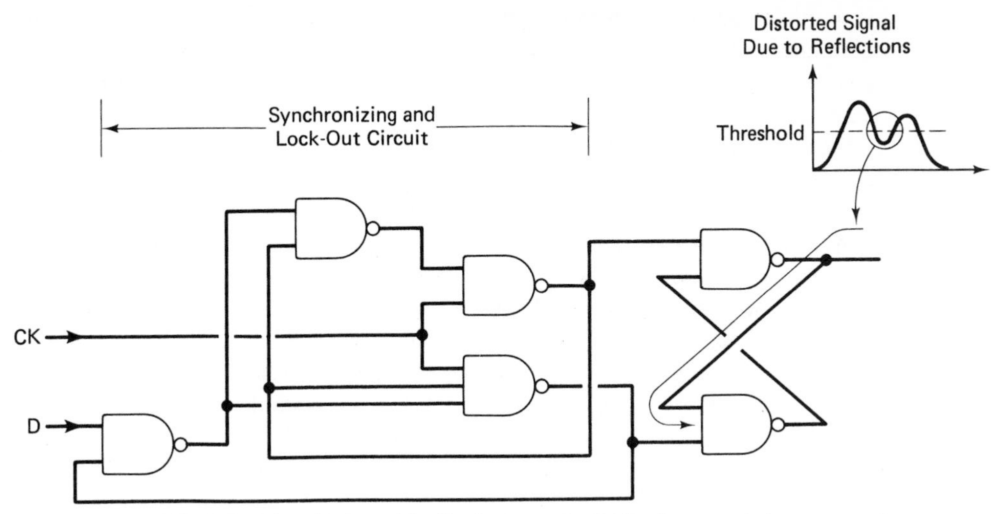

FIGURE 8-23 Undesired feedback causes the 'D' flip-flop to switch unexpectedly. (Set and Clear direct inputs are not shown.)

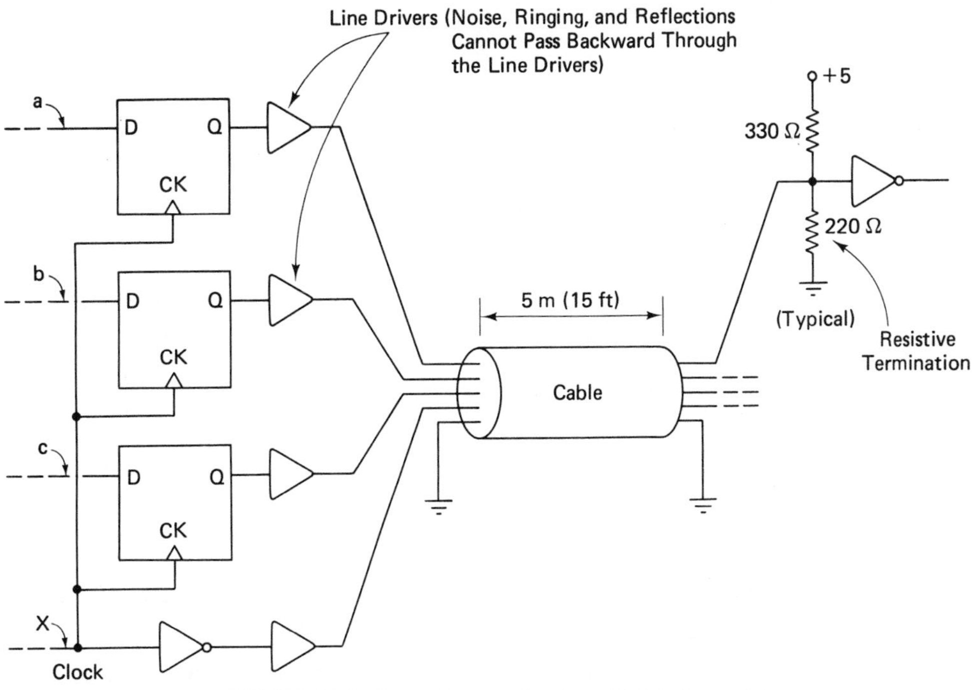

FIGURE 8-24 Corrective steps for the cable interface problem.

due to unwanted energy storage in the transmission line. The transient drop in voltage on the line was *fed backwards* into the flip-flop, causing it to change state again! This second change of state returned the flip-flop to the '0' state, and a sustained error occurred.

The corrective actions that are called for involve two basic steps. First, the ringing and reflection should be reduced as much as possible by making the transmission line match the electrical characteristics of the driving and receiving gates. This means that the impedance of the source, transmission line, and receiver must be approximately equal. By taking this step, we provide that a minimum amount of signal energy is trapped in the line, and electrical "echos" are minimized. Second, large signal variations on the line should not be allowed to be fed back to the flip-flop. This should prevent false switching. These corrective actions are achieved by terminating the line at the receiver, and by adding line driver gates, as shown in Figure 8-24.

8.6.6 Case Study 6 Analysis

To begin, we must determine the correct flow graph. Then, by comparing the correct flow graph to the erroneous one, we should find some clues to the cause of the problem. The excitation equations are determined from Figure 8-19(a).

$$D_1 = \overline{\overline{\bar{X}q_0}\,q_1} = \bar{X}q_0 + \bar{q}_1$$
$$D_0 = \overline{\overline{q_1q_0}\;\overline{\bar{q}_0X}} = q_1q_0 + \bar{q}_0X$$

From these equations, we determine an excitation table, state table, and flow graph, as shown in Figure 8-25. Now, what we must do is find and explain any mismatches between the correct and erroneous flow graphs. We will take the following steps:

(a) Action: The first error we notice is the 'A' to 'C' transition (it should be 'A' to 'D') when $X = 0$. Referring to Figure 8-26, we measure all internal inputs and outputs when in state A and $X = 0$.
Result: $Q_1 = 0$, $Q_0 = 0$, $X = 0$

Outputs:	Q_1	Q_0	1c	3c	$\bar{Q}_0$	X	Q_1	5c	6c	Q_0	$\bar{X}$	4c	2c
Inputs:	1a	1b	2a	2b	3a	3b	4a	4b	5a	5b	6c	D_1	D_0
Correct Value	0	0	1	1	1	0	0	1	1	0	1	1	0
Actual Measured Value	0	0	1	0	1	0	0	1	1	0	1	1	1
Errors				*									*

From this table, we see that an incorrect '0' on input (2)b is causing an incorrect '1' on 'D_0', and thus causing a false transition. Since the logic gates are known to be good, we can conclude that (2)b is being pulled to '0' by either an output line that is '0', or a short to ground.

(b) Action: All logic components are removed, and the printed circuit wiring is tested with an ohmmeter. First the resistance between (2)b and ground (logic

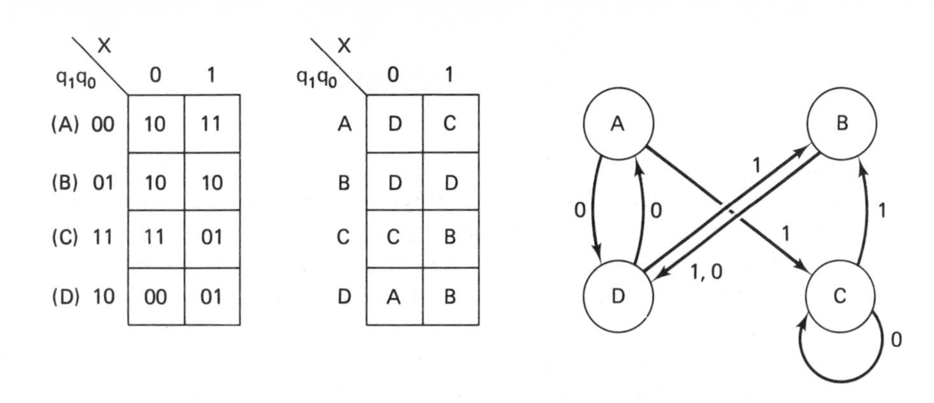

q_1q_0 \ X	0	1
(A) 00	10	11
(B) 01	10	10
(C) 11	11	01
(D) 10	00	01

q_1q_0 \ X	0	1
A	D	C
B	D	D
C	C	B
D	A	B

(a) Excitation Table. (b) State Table. (c) Flow Graph.

FIGURE 8-25 *Excitation table, state table, and flow graph for the circuit being analyzed.*

'0') is measured, then between (2)b and all '0' output points in the previous table.

Result: There is no connection between (2)b and ground; however, a short is found between (2)b and (1)b. Careful examination of the printed circuit board reveals a nearly invisible strand of wire between output (3)c, which is connected to (2)b, and input (1)b.

(c) Action: The bridge is removed, the components are reinstalled, and the circuit is tested.

Result: The circuit functions properly.

Note that we measured the circuit's logic conditions for only one state and one input value. We selected this state and input value so that we could obtain a static inconsistency between the correct excitation and the faulty excitation. This inconsistency was then pursued in order that the error's cause could be isolated.

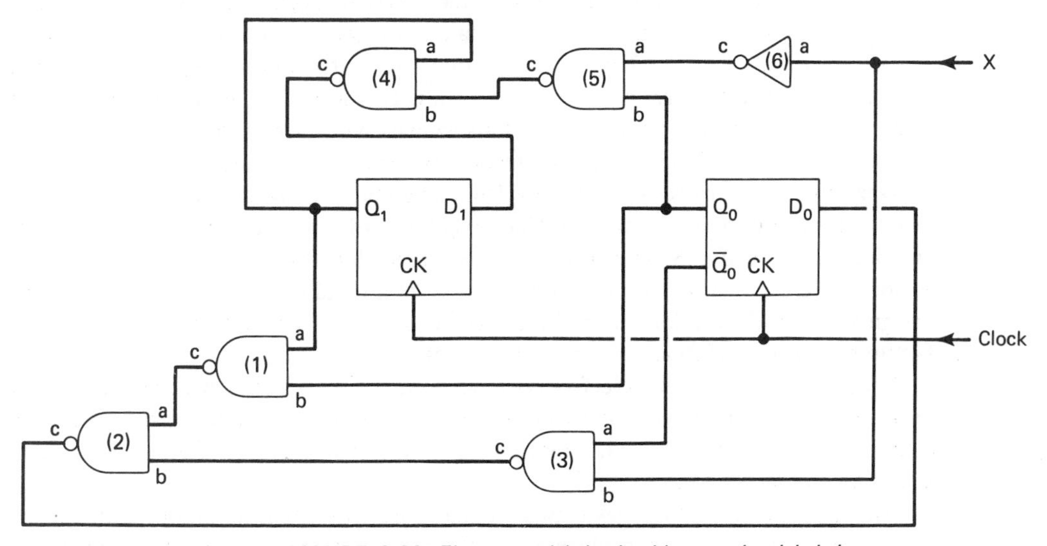

FIGURE 8-26 *The sequential circuit with test points labeled.*

SUMMARY

The analysis of faults in a digital circuit involves four basic steps: realizing that a fault is present, classifying the fault, identifying the cause, and taking corrective action. The large variety of faults that may be found in a digital circuit or system make it impossible to provide a fixed procedure that always reveals the problem. Instead, each case must be considered individually. As a good private investigator would, we must be aware of every detail of a circuit's symptoms, and be careful, clear, and patient in our analysis. At first, suspect the simplest possible faults (such as lack of power, miswiring, or missing components). Only after all simple explanations have been exhausted should more complex problems be considered (such as stuck outputs, bridge faults, or transient IC malfunctions). Be certain that the source of the problem has been found before taking corrective steps (for example, do not replace a blown fuse until the cause of this condition has been determined).

In general, a digital circuit is faulty when it does not respond to inputs in accordance with its design. The fault may be insigificant, causing no serious consequences, or it may result in catastophic failure of a large system. The fault may be transient, occurring unpredictably and occasionally. On the other hand, the fault may be constant, and cause a system to stop functioning completely.

While we may not find a "textbook" sequential circuit in practical digital designs, examples of such circuits serve well to illustrate the basic approach to fault diagnosis. First, an inconsistency is found in the flow graph (or state table) between the circuit's designed response and its actual response. This inconsistency is then pursued by making measurements on preceding stages to determine the direct cause of the inconsistency. If the fault is transient, then periodic excitation should be provided (by a pulse generator, for example), and the transient output viewed on an oscilloscope. Again, the preceding stages should be measured.

Stuck outputs or inputs cause device terminals to appear always '1' or always '0', regardless of logic conditions. Bridge faults are caused by undesired connections between two signal lines. This could be due to printed circuit defects, miswiring, or internal IC shorts. Broken or missing conductors are, in a sense, the opposite of bridge faults in that a conductive path fails to exist where one should exist. This may also be due to printed circuit defects, miswiring, or internal IC defects.

Hazards and races result primarily from unequal propagation delay within electronic logic components. Symptoms that reveal hazards and races are usually transient and difficult to verify. We can sometimes deduce such problems when all other explanations have been ruled out. The prevention of races and hazards should occur at the design stage, and may sometimes be corrected by using identical (in terms of manufacturer, and date of manufacture) components.

Crosstalk results from electromagnetic coupling of signals between physically near conductors or devices, and causes undesired signal energy to be driven onto a transmission path. Differential receivers and Schmitt triggers are frequently used to guard against crosstalk.

The rapid, synchronous switching of many devices (such as flip-flops driven by a common clock signal) can place a rapidly varying load on the power supply, and may cause interference and false switching in other components. The use of small capacitors with good high-frequency response (such as disc capacitors), distributed among the digital components, greatly alleviates this problem.

When high-speed logic components are used, reflections and ringing can become very severe. Thus, care must be taken to properly terminate the interfaces between source, transmission line, and receiver. Failure to do so may result in false edges appearing on the receiver's output.

NEW TERMS

Combinational Logic Fault
Sequential Logic Fault
Static Fault
Dynamic Fault
Stuck Output
Stuck Input
Bridge Fault
Crosstalk
Differential Line Driver
Differential Line Receiver
Schmitt Trigger
Decoupling Capacitor

PROBLEMS

8-1 Is it possible for a fault to exist in a circuit when there are no symptoms of malfunction? If not, why? If so, explain and give an example.

8-2 Can a fault be both static and dynamic at the same time? If not, why? If so, explain and give an example.

8-3 Classify each of the following faults as (1) combinational or sequential, and (2) static or dynamic. Then, suggest what the problem may be.

(a) A four-bit synchronous natural binary up counter is built using four flip-flops and appropriate logic gates. Unfortunately, the count sequence is incomplete, with the output progressing as follows:

(the MSB is always '0')

0 0 0 0
0 0 0 1
0 0 1 0
0 0 1 1
0 1 0 0
0 1 0 1
0 1 1 0
0 1 1 1
0 0 0 0
0 0 0 1
.
.
.

(b) A binary counter and a pushbutton are appropriately connected so that whenever the pushbutton is pressed, the counter advances by one state. However, the counter occasionally and unpredictably advances by itself one or more states, without the button being pushed.

(c) A binary-to-BCD decoder produces incorrect output whenever the input '10110' occurs.

8-4 A simple combinational logic circuit is shown in Figure P8-4(a). Three fault functions are shown in Figure P8-4(b). For each fault function, determine all possible inputs or outputs. Label the gate inputs and outputs as follows: For gate 1, the inputs are I_{1a} and I_{1b}, and the output is 0_1. Follow this convention for all gates. Assume that there is a *single* stuck input or output fault in each case.

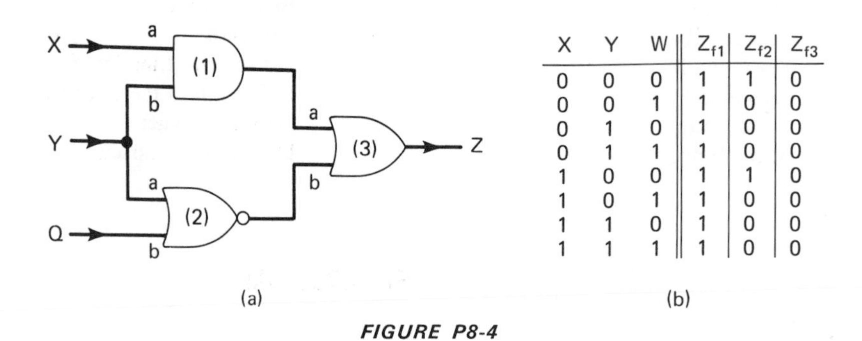

X	Y	W	Z_{f1}	Z_{f2}	Z_{f3}
0	0	0	1	1	0
0	0	1	1	0	0
0	1	0	1	0	0
0	1	1	1	0	0
1	0	0	1	1	0
1	0	1	1	0	0
1	1	0	1	0	0
1	1	1	1	0	0

FIGURE P8-4

8-5 The combinational logic circuit shown in figure P8-5(a) contains a single bridge fault. For each of the fault functions shown in Figure P8-5(b), determine at least one possible location for the bridge fault. Label the inputs and outputs as in Problem P8-4.

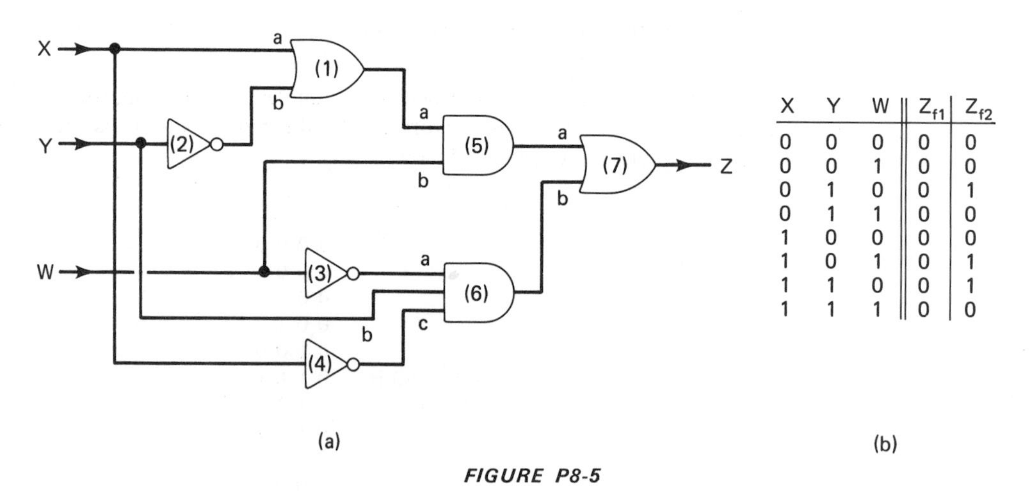

X	Y	W	Z_{f1}	Z_{f2}
0	0	0	0	0
0	0	1	0	0
0	1	0	0	1
0	1	1	0	0
1	0	0	0	0
1	0	1	0	1
1	1	0	0	1
1	1	1	0	0

FIGURE P8-5

8-6 A natural binary up counter is designed using the "detect and steer" method to follow the count sequence 0 0 0, 0 0 1, 0 1 0, 0 1 1, 1 0 0, (repeats). The circuit diagram is shown in Figure P8-6. When the circuit was built in the laboratory, however, a single wiring error was made. What is the wiring error if the actual count sequence is

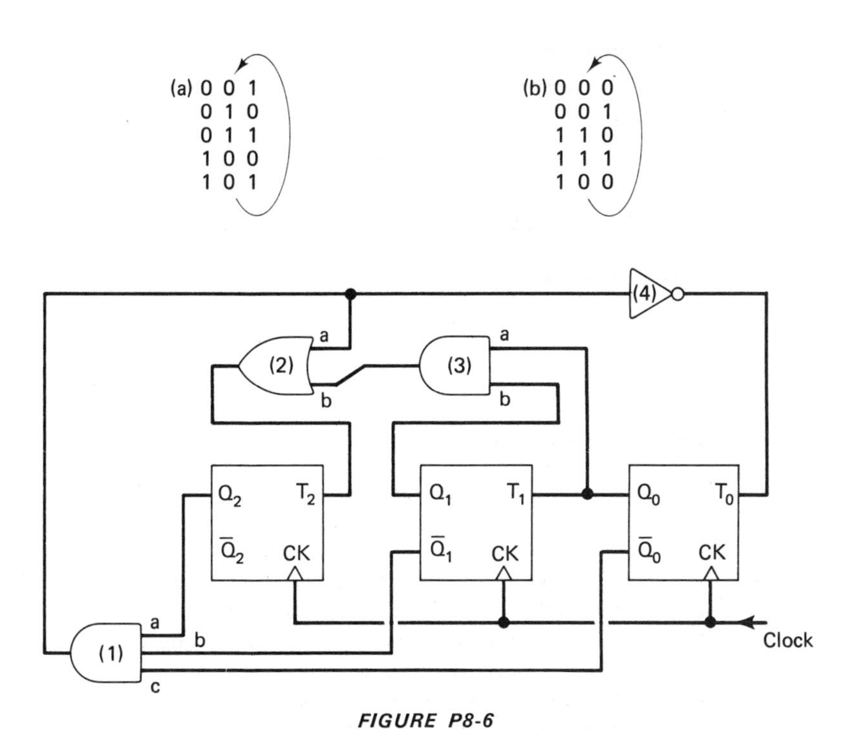

FIGURE P8-6

8-7 **(a)** The NAND latch shown in Figure P8-7 is a simple digital circuit with *four* nodes: How many bridge faults are possible between these four nodes?

(b) How many nodes are present in the counter circuit of Figure P8-6?

(c) How many single bridge faults are possible between the nodes of the counter circuit in Figure P8-6? [*Hint:* Find a general formula for the number of single bridge faults, based on your answer to part (a). Then, apply it to the number of nodes found in part (b).]

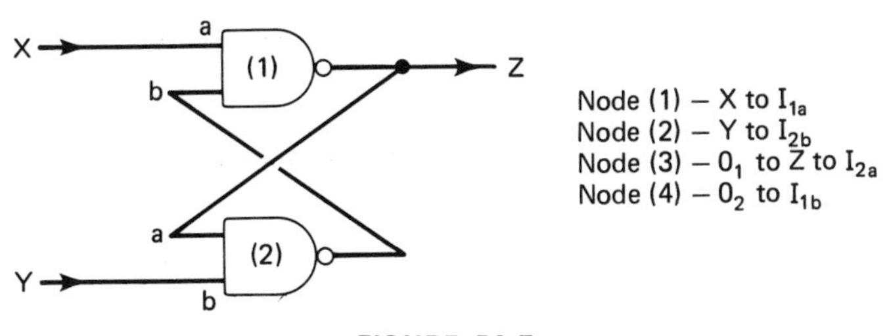

FIGURE P8-7

8-8 Investigate the circuit of Figure P8-8 for hazards. If any are found, correct them by adding redundant terms. Factor the result to obtain a minimal gate implementation. Can factoring alone eliminate or introduce a hazard?

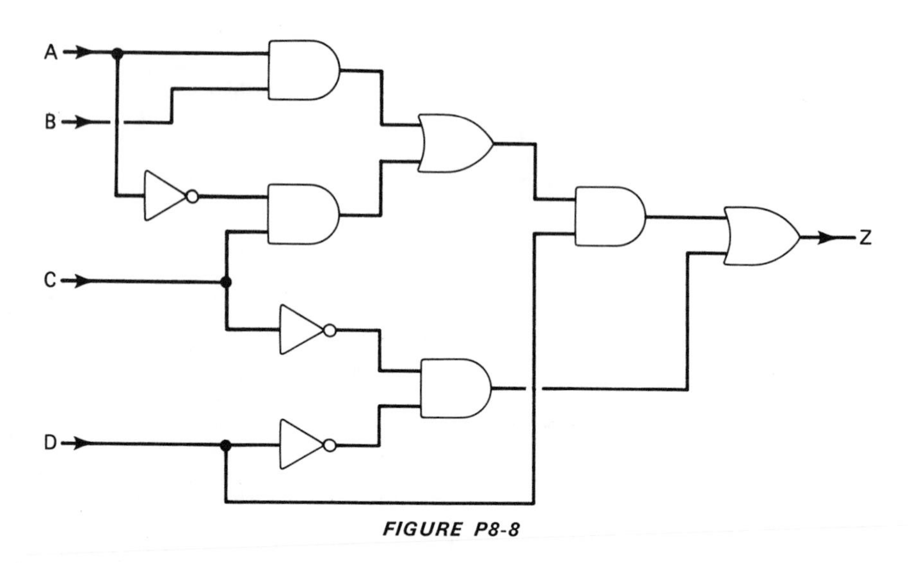

FIGURE P8-8

8-9 The inputs to a hazard-free combinational logic circuit are derived from a 1-m-long ribbon cable, as shown in Figure P8-9(a). This cable is accidently caught in an access door, but is neither severed nor damaged in any obvious way. However, the '*Z*' output of the logic is now incorrect for some input values, as shown in Figure P8-9(b). Assuming that one of the four conductors in the cable is broken and that an unconnected gate input assumes logic '1', determine which cable wire is defective.

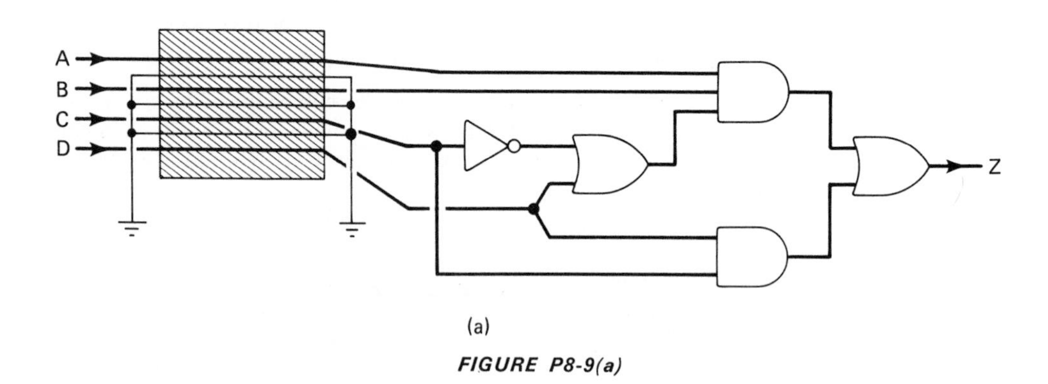

(a)

FIGURE P8-9(a)

A	B	C	D	Z_f
0	0	0	0	0
0	0	0	1	0
0	0	1	0	1
0	0	1	1	1
0	1	0	0	0
0	1	0	1	0
0	1	1	0	1
0	1	1	1	1
1	0	0	0	0
1	0	0	1	0
1	0	1	0	1
1	0	1	1	1
1	1	0	0	1
1	1	0	1	1
1	1	1	0	1
1	1	1	1	1

(b)

FIGURE P8-9(b)

8-10 A high speed four-bit up-down counter is connected, supposedly, as shown in Figure P8-10. Because of a *single* wiring error, however, the circuit malfunctions: once cleared, all 'Q' outputs synchronously *complement* with each clock pulse, producing the output sequence 0 0 0 0, 1 1 1 1, 0 0 0 0, 1 1 1 1, and so on. Name two different single wiring errors that could cause this problem.

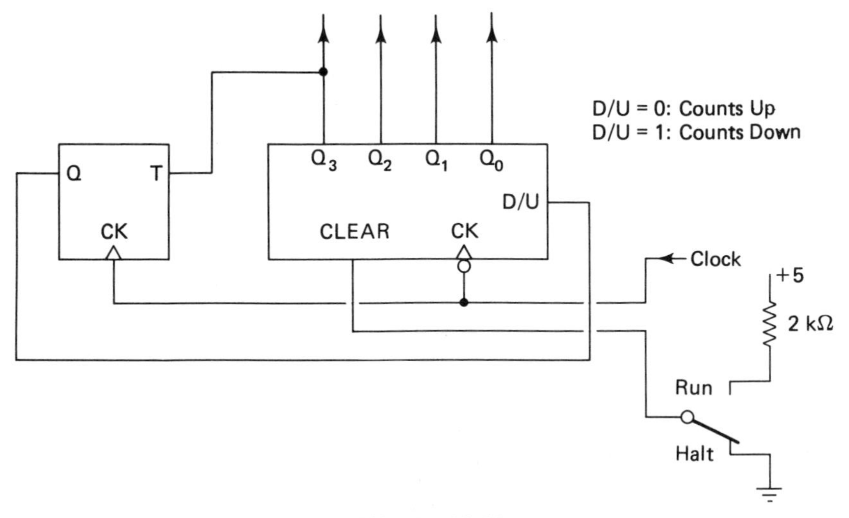

FIGURE P8-10

8-11 If two TTL outputs are shorted together, the logic value on the bridged line is '0' if either of the outputs is '0'. We have seen that this produces an equivalent logic AND between the two bridged lines. Now, let us assume that two shorted outputs produce a '1' if either of the outputs is '1'. What equivalent logic function results?

8-12 The sequential circuit of Figure P8-12(a) was constructed, tested, and found to malfunction. A flowgraph for the faulty circuit was determined and is shown in Figure P8-12(b). A single fault was found, and when corrected, the circuit worked properly. Of the following fault possibilities, which one(s) might be the cause of malfunction in this sequential circuit, and where might it(they) be located?:

1. Input stuck at '0'.
2. Bridge fault.
3. Input stuck at '1'.
4. Open circuit.
5. Signal line shorted to logic '1' (power supply '+').
6. Missing wire.

Assume that an unconnected gate or flip-flop input is logic '1'.

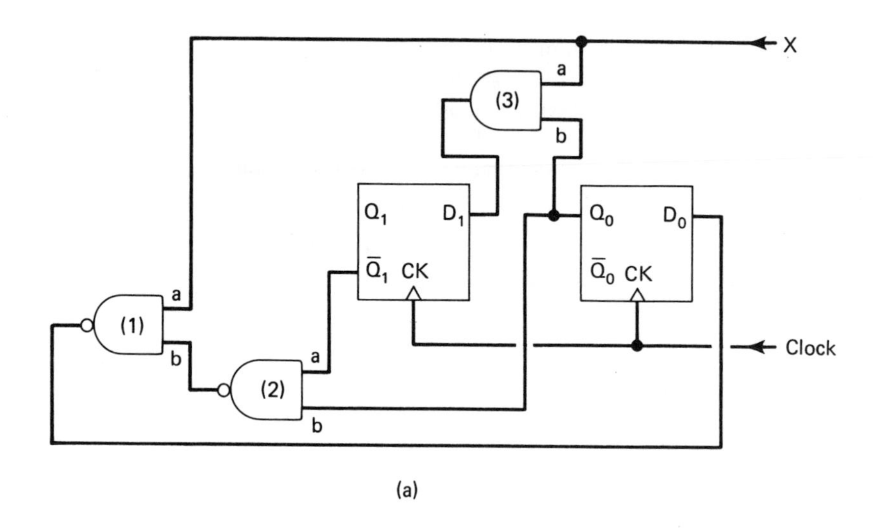

(a)

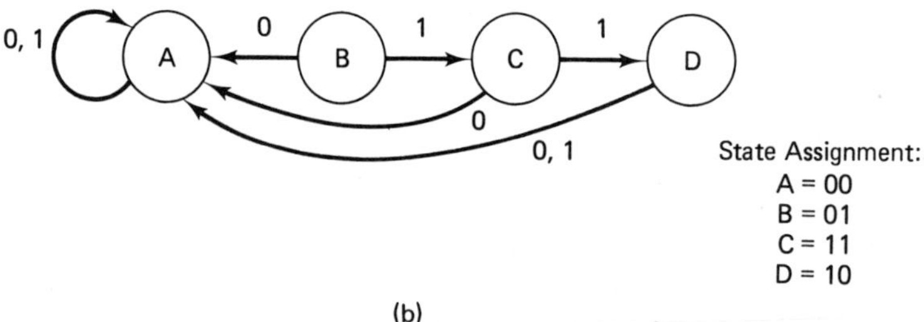

(b)

FIGURE P8-12

9 ANALOG-DIGITAL CONVERSION

9.1 MOTIVATION

9.1.1 Real Signals

Traditional electronic equpiment has been the servant of mankind for many years, and is familiar to us all. Without it, life would be more difficult in numerous ways: there would be no telephones, no television, no radio communication, no stereo high fidelity, no public address systems, and more. It is very interesting to realize that a quiet revolution is under way concerning the internal operation of these devices. It is a revolution because the basic, internal processing of signals is beginning to be handled in a radically different manner from earlier techniques. It is a quiet revolution because the changes in the input/output characteristics resulting from these new techniques are often barely apparent. This is the revolution of *digital signal processing*.

A *signal* can be defined in several ways, as follows*:

1. A visual, audible, or other indication used to convey information.
2. The intelligence, message, or effect to be conveyed over a communications system.
3. A signal wave; the physical embodiment of a message.

We are constantly surrounded by signals to which we must respond appropriately, such as a ringing telephone or the color lamps of a traffic light. It is important to realize, however, that audio and visual signals are handled most effectively only when they have been converted into corresponding *electrical signals*. Even nature has followed this rule by providing that sensory signals within the brains of animals and men are conveyed electrically through the nerves. Thus, we see the importance and necessity of electrical equipment in applications requiring signal processing.

* According to the "IEEE Standard Dictionary of Electrical and Electronics Terms," published by the Institute for Electrical and Electronics Engineers, New York, 1977.

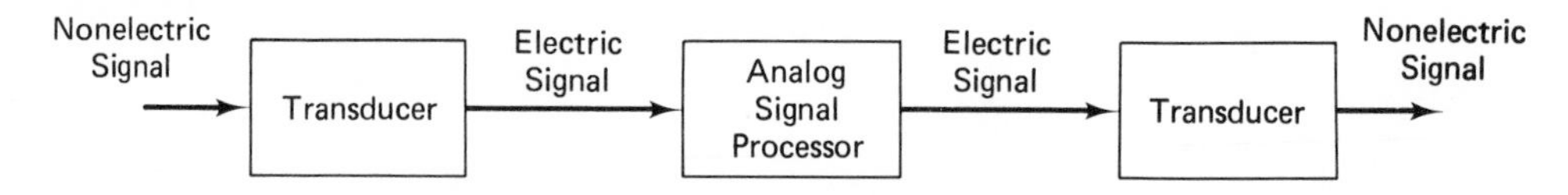

(a) General Analog Signal Processing System.

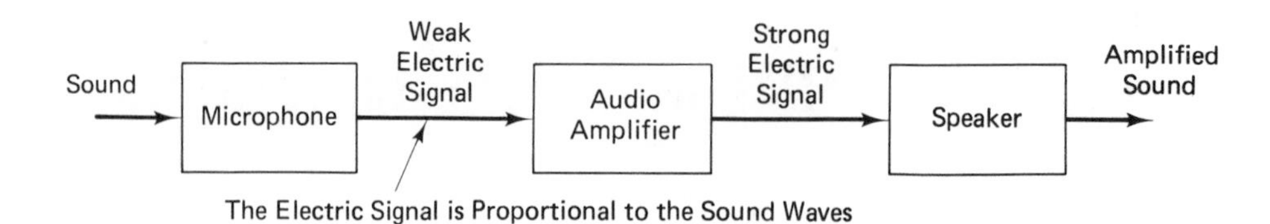

(b) Public Address System; an Example of a Specific Analog Signal Processing System.

FIGURE 9-1

9.1.2 Comparison of Analog and Digital Processing

A *transducer* is a device that serves to translate one type of signal into another type of signal. In the world of electronics, a transducer is generally understood to mean a device that interconverts electrical and nonelectrical signals. A microphone, for example, is a transducer that converts an acoustic signal into an electrical one. The majority of signals produced by transducers are *analog* in nature, that is, the voltage value of the input (or output) signal is exactly proportional to the nonelectrical signal that the transducer produces (or measures). Now, until recently, almost all signal processing (such as amplification) has been carried out on electronic devices that handle purely analog signals. For instance, an audio amplifier accepts a weak analog signal at its input and produces a much stronger one at its output. The block diagram of a general analog signal processing system is shown in Figure 9-1(a). In Figure 9-1(b), a public address system is illustrated as a particular example. While we cannot do anything about the analog nature of real-world signals, such as sound, we *can* make a change in the way in which signals are processed. In a digital signal processing system, the processor block operates only on *binary-coded numbers*. Now, if we are to apply digital processing systems to real-world signals, which are analog

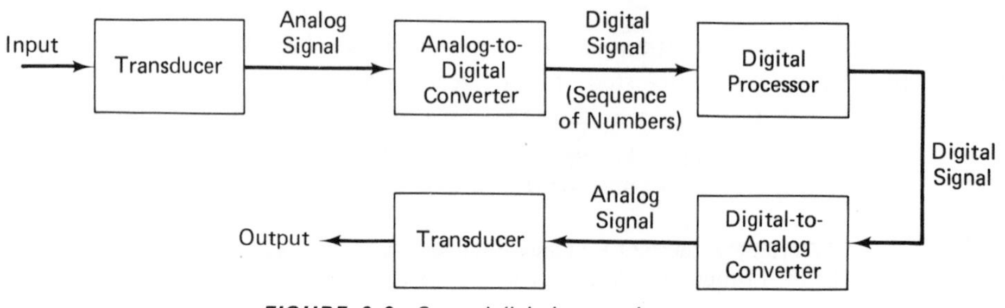

FIGURE 9-2 *General digital processing system.*

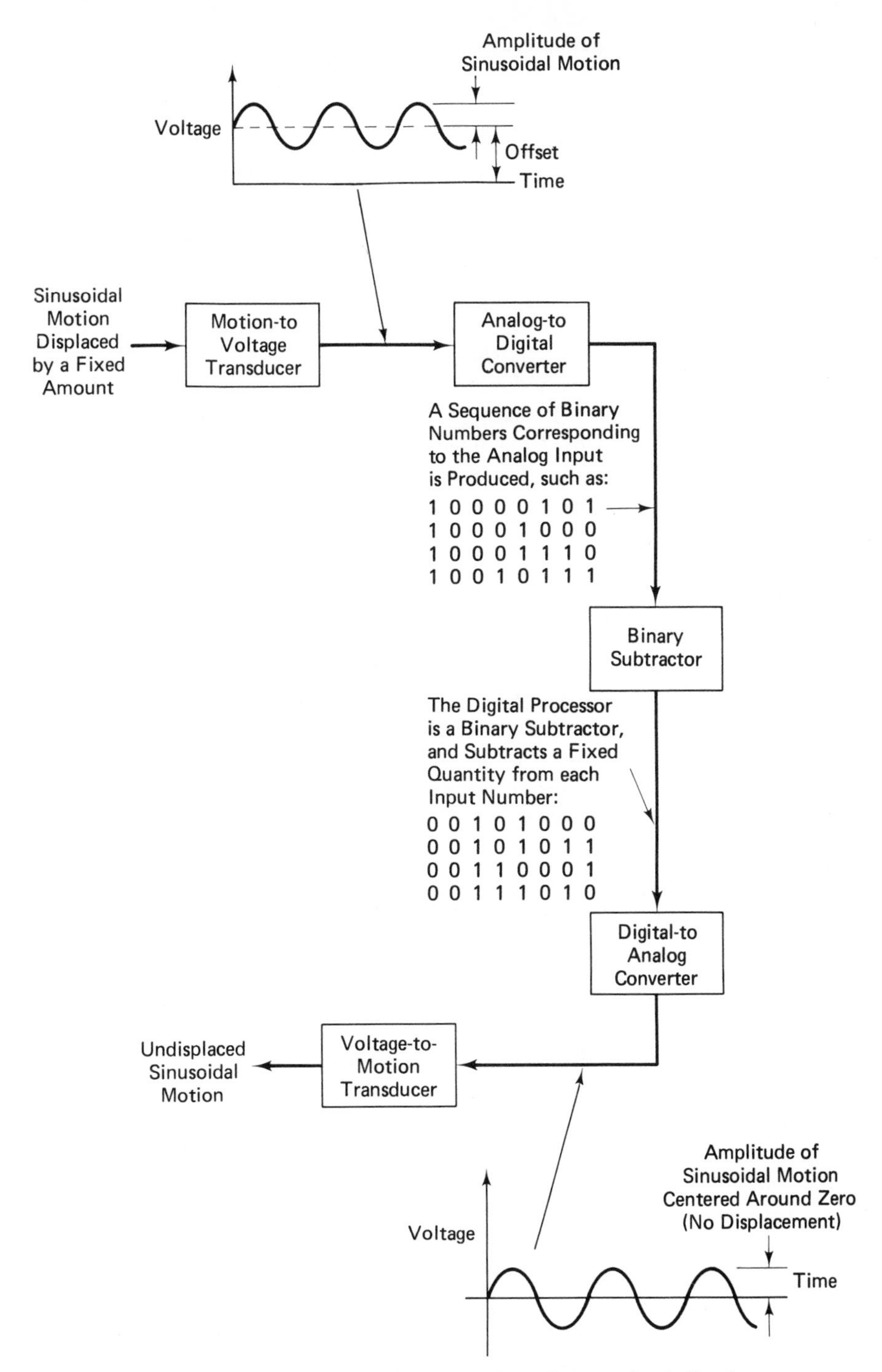

FIGURE 9-3 *Digital processing system that subtracts a fixed offset from an input signal. For the three periods of the sine wave shown, at least thirty eight-bit binary numbers will be produced by the A/D converter.*

in nature, some means must exist to interconvert analog signals and corresponding binary-coded numbers. Fortunately, such a means does exist, and is neither complex nor particularly expensive. ***Analog-to-digital*** converters translate analog signals over a specified range into binary-coded numbers. ***Digital-to-analog*** converters, on the other hand, produce an analog voltage (or current) output from a binary-coded number input. The block diagram for a general digital signal processing system is shown in Figure 9-2. A specific digital processing system is shown in Figure 9-3. In this example, the average value of an analog signal is to be changed. Note the remarkable flexibility that is possible with the processing block once the signal is in a numerical (digital) form. All numerical operations that are allowed by mathematics may be applied to the signal when it is expressed as a sequence of numbers. Operations that would be extremely difficult or impossible to apply to a purely analog signal are now at our disposal. Of course, there are limitations and disadvantages to digital processing systems, perhaps the biggest one being that digital processing systems currently cannot process information at the same maximum speed as an equivalent analog processing system. As semiconductor technology advances, however, the limitations of speed and cost will be greatly offset. In any event, the continual replacement of analog devices with digital processing systems is silent but convincing evidence of the power and advantages behind this approach.

In this chapter, we will focus on the basic properties of analog-to-digital and digital-to-analog converters, and illustrate the important principles governing their operation. Chapter 10, to follow, brings to light the ideas behind a programmable digital processor (computer), and demonstrates how a fixed piece of digital hardware can be made to perform remarkably different operations. We now begin one of the most fascinating studies in the field of electronics, and will look at concepts that can be applied to the benefit of us all for many years to come.

9.2 DIGITAL CODING OF ANALOG SIGNALS

9.2.1 Quantization in Amplitude and Time

An analog signal can be thought of as a continuous, smoothly varying value of voltage or current that changes with respect to time. Even sharp edges of an analog signal (such as the rising and falling edges of a square wave) are continuous, since the signal value cannot change from one level to another without going through *all* intermediate levels. Figure 9-4 illustrates this idea. By studying Figure 9-4, we note that an analog signal is continuous in *amplitude* and also in *time*. Said another way, for every instant of time from the signal's beginning to its end, an exact, corresponding value of amplitude can be found. This means that an analog signal can be resolved into infinitely small increments on the time axis and on the amplitude axis. Now, if we are to have any reasonable hope of translating the analog signal into a digital (numeric) one, we must limit the resolution on the time and amplitude axes. Simply put, we cannot precisely express an analog value *numerically* with anything less than an infinite number of fractional digits, and, of course, such a numerical quantity cannot be processed by a finite amount of digital hardware. What we must do is to

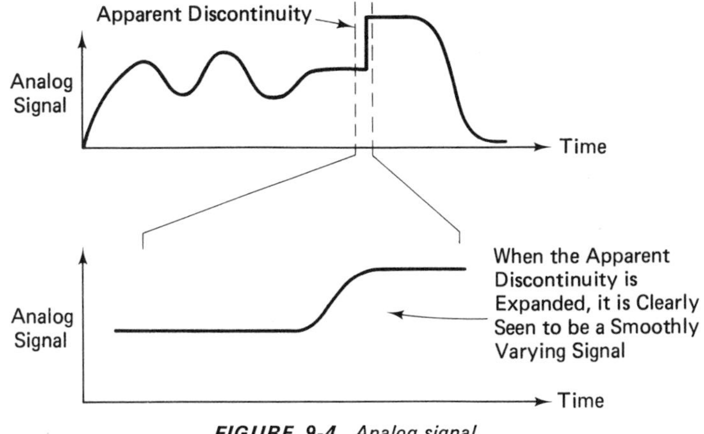

FIGURE 9-4 Analog signal.

subdivide the amplitude and time axes into a number of *equally spaced* intervals, and thereby limit our resolution of measurement to the width of one of these intervals. Subdividing an axis in this manner *quantizes* the variable on that axis. The variable can then assume only one of a fixed number of values. Figure 9-5 illustrates this idea. Note from Figure 9-5 that selecting the width of the interval is very important: too small an interval would produce unreasonably large numbers (too many fractional parts to the right of the decimal point), while an excessively large interval would unfaithfully represent the original signal. Note also that the selection of the interval size on the time axis is independent of the vertical interval; it is possible to have large time intervals and small amplitude intervals, or vice versa.

Sampling: There are two separate issues relating to the quantization of an analog signal. The first is ***sampling***, and refers to subdividing the time axis into a finite number of intervals. In this case, the analog signal can be thought of as being periodically "sampled" to obtain a new signal value. The interval width is a measure of time, such as 1 μs. As the sampling rate increases, so also does the quantity of numbers that are produced to represent the signal. Perhaps the best single example of sampling is the motion picture. By carefully examining a strip of movie film, we see that each frame is a sample of the camera's image at successive moments in time. From one frame to the next the image changes only slightly, yet when the film is viewed on a projector, the original signal (namely, the scene that was being filmed) is smoothly and faithfully reproduced. The sampling rate (frame rate) of a standard motion picture is twenty-four frames per second, corresponding to an interval width of 41.67 ms. We must now ask:

> How are we to know what sampling rate to choose in order to faithfully reproduce the original signal?

To answer this question, we must know how fast the *fastest* part of the signal is expected to be. The sampling rate must be chosen so that only small changes occur in the signal from one sample point to the next. For example, we could not expect to

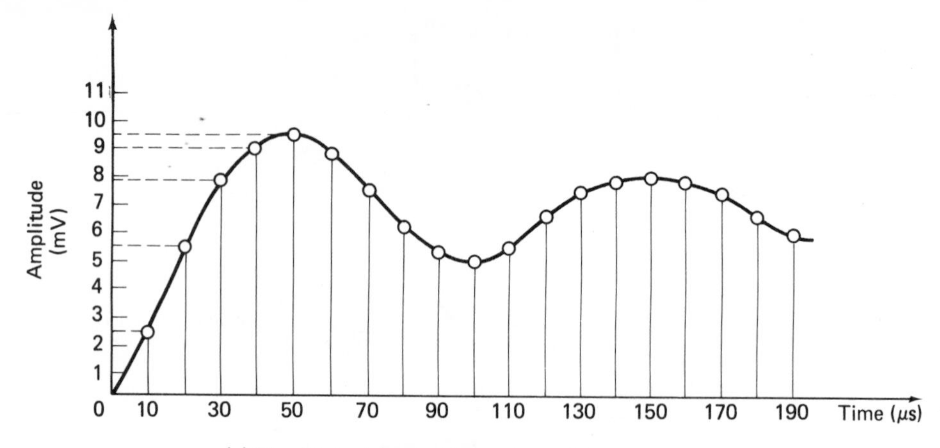

(a) The Original Analog Signal. Signal Values at the Times of Sampling Are Circled.

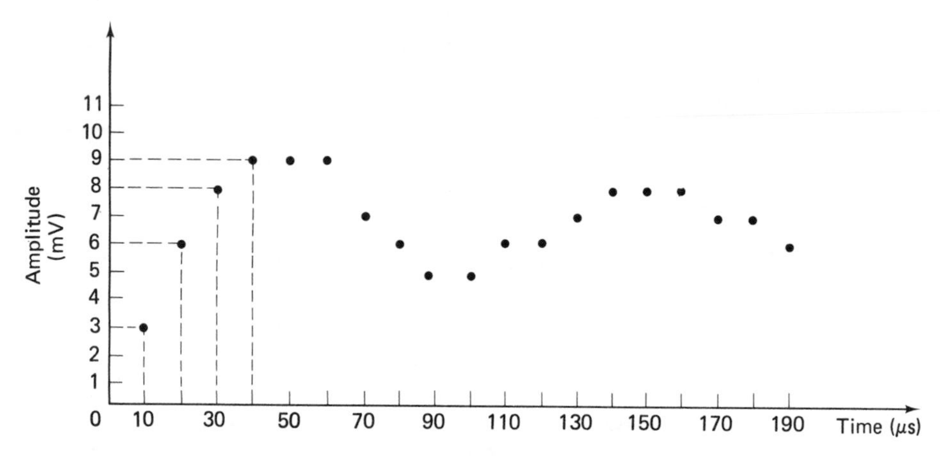

(b) The Analog Signal is Quantized in Time and Amplitude. Note that the Quantized Amplitudes are Rounded to the Nearest Whole Value.

Time	0	10	20	30	40	50	60	70	80	90	100	110	120	130	140	150	160	170	180	190
Quantized Amplitude	0	3	6	8	9	9	9	7	6	5	5	6	6	7	8	8	8	7	7	6

(c) The Sequence of Numbers Represent the Original Analog Signal.

FIGURE 9-5

see the wing motion of a honeybee in flight if we use a twenty-four frame per second motion picture camera; the wings will flap many times between one sample point and the next. For the most part, however, twenty-four frames per second is completely adequate for almost all scenes that are shown in a movie theater. That is, very few entertainers will move faster than that which can be recorded at the standard frame rate.

It has been proven mathematically that an analog signal can be perfectly reproduced from its sample points if the sampling rate is *twice as fast* as the highest-frequency component of the signal. This minimum sampling frequency is known as the ***Nyquist*** rate. To perfectly reproduce a 1000-Hz sine wave, for example, we must sample it at a minimum rate of 2000 Hz. Because of electronic limitations and component imperfections, this theoretical limit cannot be reached. A good general rule is to sample the analog signal at five to ten times the rate of its maximum frequency component. Practically, then, a 1000-Hz sine wave should be sampled about 10,000 times per second. Refer to Figure 9-6(a). Note that the maximum frequency component of a 1000-Hz *square wave* is many times greater than its fundamental frequency of 1000 Hz. If we wish to retain square wave precision to four harmonics, the highest-frequency component would be 8000 Hz, and the sampling rate should then be about 80,000 Hz. To understand this intuitively, think of what the sampling rate should be to retain every detail of the rising and falling edges of the square wave! Sampling a square wave is illustrated in Figure 9-6(b).

Another factor limiting our ability to reach the Nyquist rate relates to the fact that the amplitude must also be quantized. This makes it even more difficult to faithfully reproduce the original signal. The mathematical theory leading to the derivation of the Nyquist rate assumed that the amplitudes would be exactly represented. Of course, in practice this is not possible.

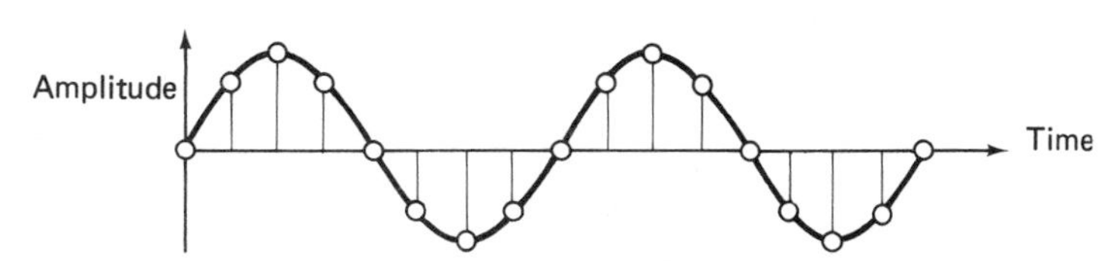

(a) A Sine Wave is Sampled Eight Times Per Cycle. If the Sine Wave Frequency is 1000 Hz, this Sampling Rate is 8000 Hz.

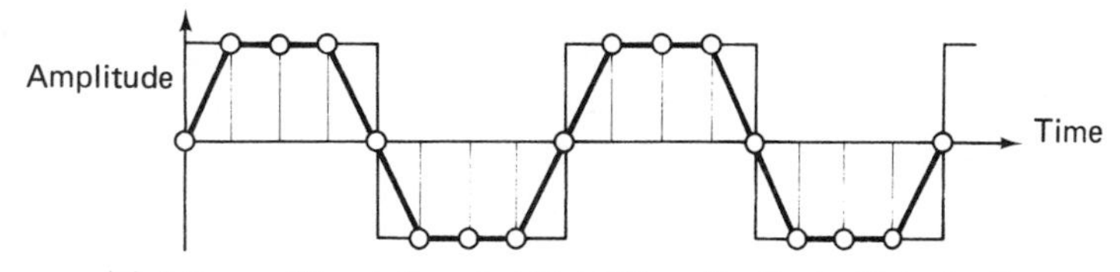

(b) A Square Wave is Sampled Eight Times Per Cycle. Note the Error Between the Original Square Wave (Light Line) and the Signal that is Reproduced from the Samples (Dark Line).

FIGURE 9-6

In order to quantize the time axis of measurement, we need only to generate pulses periodically. Each pulse is a command to measure the amplitude and store the result. The digital hardware required to produce the required pulse train is extremely simple, amounting to only an oscillator circuit and a counter (to divide the oscillator frequency to an appropriate value, namely the sampling rate). A much more challenging and difficult problem is to construct hardware that quantizes the signal amplitudes.

Amplitude quantization: The problem presented by amplitude quantization is really quite easy to understand. Simply stated, the maximum range of an analog signal must be divided into a fixed number of small, equal intervals. Each interval is then associated with a fixed binary number. An analog-to-digital (A/D) converter performs such a function and is illustrated in Figure 9-7. From Figure 9-7(a), we see that two factors must be specified for an A/D converter:

1. The maximum range of analog input voltage.
2. The output resolution (determined by the number of output bits).

Given a fixed input voltage range, we can say that the more binary digits of output that are available, the more resolution that is possible, and the more precisely we can encode the analog signal. ***Resolution*** is a measure of the smallest change in analog input that can be discriminated by an A/D converter. As an example, consider an eight-bit A/D converter that spans an input range of 10 V (0 to +10). With eight bits, the input range is divided into 255 equal intervals,* each of which is 39.22 mV (= 10 V/255). The resolution of this device, then, is 39.22 mV. A ten-bit A/D converter would offer considerably more resolution by dividing the 10-V range into 1023 equal intervals. In this case, the resolution would be 9.78 mV.

The resolution of an A/D converter cannot be arbitrarily high, unfortunately. A fourteen-bit A/D converter, for example, is currently one of the highest-resolution converters available. In this case, a range of 10 V would be divided into 16,383 equal intervals, providing a resolution of 0.61 mV. This corresponds to a dynamic range of 84 dB.† At this level, unavoidable noise voltages generated within the analog circuitry of the A/D converter are almost equal in magnitude to the smallest increment of measurement. When noise voltages exceed the resolution voltage, signal information is lost and is not recoverable. In most cases, between eight and twelve bits of digital output provides adequate resolution.

A digital-to-analog (D/A) converter provides for the translation of a binary coded signal into a corresponding analog signal. This is illustrated in block form in Figure 9-8. Note that the transfer function in Figure 9-8(b) shows that a fixed value of output voltage is produced for each binary input value. Resolution is defined for the D/A converter in the same manner as with the A/D converter. In this case, though, we are concerned with the amount of analog output voltage change for a change of one least significant bit in the input. Figure 9-9 illustrates the effect of varying degrees of resolution.

* In general, the resolution of an 'n'-bit A/D with a voltage range of 'X' is $X/(2^n - 1)$.

† 84 dB = $20 \log 2^{14}$.

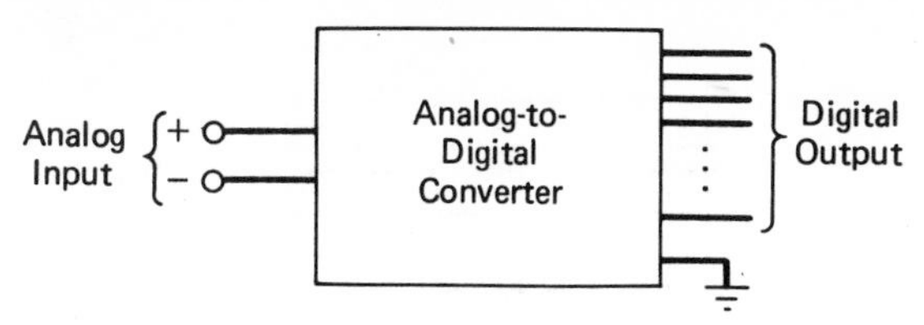

(a) General Block Diagram Form of an A/D Converter.

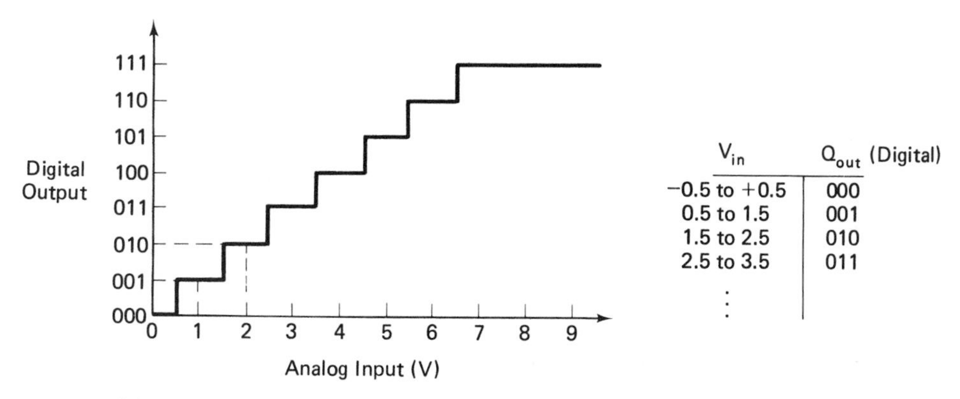

V_{in}	Q_{out} (Digital)
−0.5 to +0.5	000
0.5 to 1.5	001
1.5 to 2.5	010
2.5 to 3.5	011
⋮	

(b) The Input Vs. Output Function of a Three-BitA/D Converter is Illustrated. Sampling Circuitry is Not Shown by this Diagram.

FIGURE 9-7

FIGURE 9-8

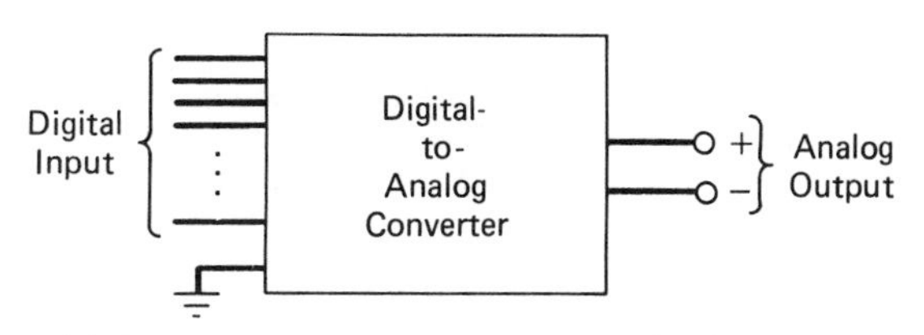

(a) General Block Diagram Form of a D/A Converter.

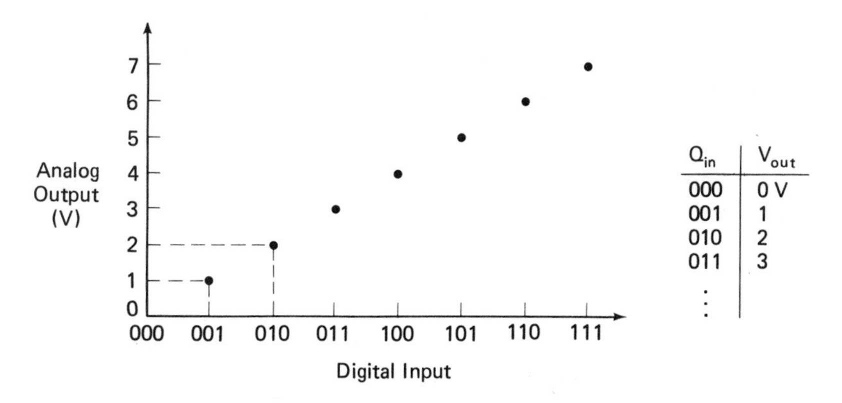

Q_{in}	V_{out}
000	0 V
001	1
010	2
011	3
⋮	

(b) Input Vs. Output Function of a Three-Bit D/A Converter.

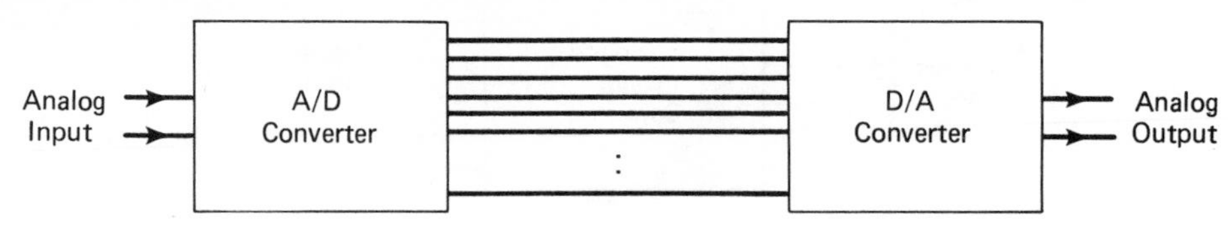

(a) Direct Connection of an A/D and a D/A Converter, with no Digital Processing.

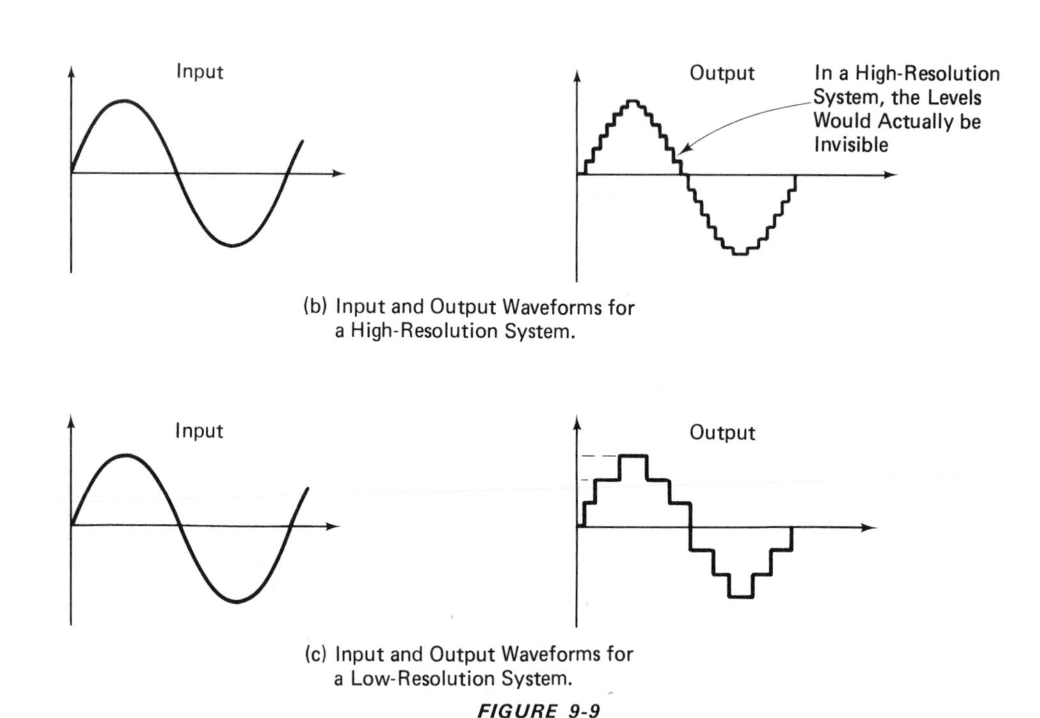

(b) Input and Output Waveforms for a High-Resolution System.

(c) Input and Output Waveforms for a Low-Resolution System.

FIGURE 9-9

In the A/D and D/A devices that have been presented, we note that the binary code used is of the *natural binary* type. That is, an increasing-valued analog signal corresponds to increasingly larger binary numbers. Of course, the binary code that is used could be arbitrarily chosen, as long as each binary number corresponds uniquely to one analog voltage. There are several good reasons, however, for choosing the natural binary sequence. First, as we shall see in an upcoming section, the circuitry needed to accomplish A/D and D/A conversion is greatly simplified if a natural binary code is used. Second, any digital processing that might be done to the numerically coded signal will be much more direct and less time consuming when a natural binary code is applied to the analog signal. Perhaps if we were cryptographers, other considerations might be necessary!

9.2.2 Precision Versus Accuracy

The ability of a circuit or device to function in exactly the same manner over time, or under different conditions, relates to its ***precision.*** The calibration of a circuit or device with respect to a known standard relates to its ***accuracy***. For example, a clock that runs exactly one minute fast per day every day is precise but inaccurate. A clock that is always within 1 minute of the true time but whose minute hand does not always point to the exact minute mark on the dial is accurate (within the one minute range), but imprecise. In another sense, the precision of a measurement is related to the number of numerical digits that are produced, while the accuracy of the measurement tells us how many of the digits can be believed. Of course, it is pointless to have more digits of precision than the basic accuracy of the device can support. The precision of an electronic circuit is dependent on its stability with respect

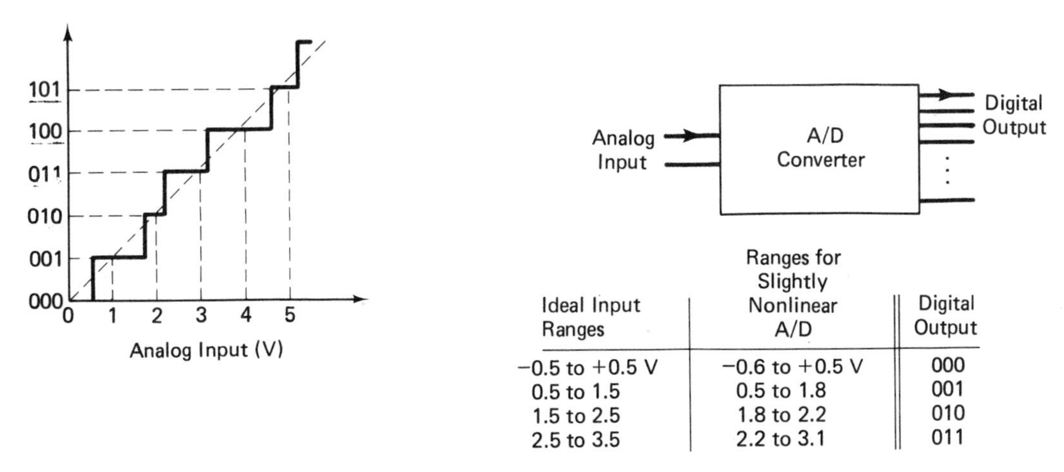

Ideal Input Ranges	Ranges for Slightly Nonlinear A/D	Digital Output
−0.5 to +0.5 V	−0.6 to +0.5 V	000
0.5 to 1.5	0.5 to 1.8	001
1.5 to 2.5	1.8 to 2.2	010
2.5 to 3.5	2.2 to 3.1	011

(a) The Effect of Nonlinearity on an A/D is Shown.

Digital Input	Ideal Output	Actual Output
000	0 V	0 V
001	1	0.8
010	2	2.4
011	3	2.9
100	4	4.6

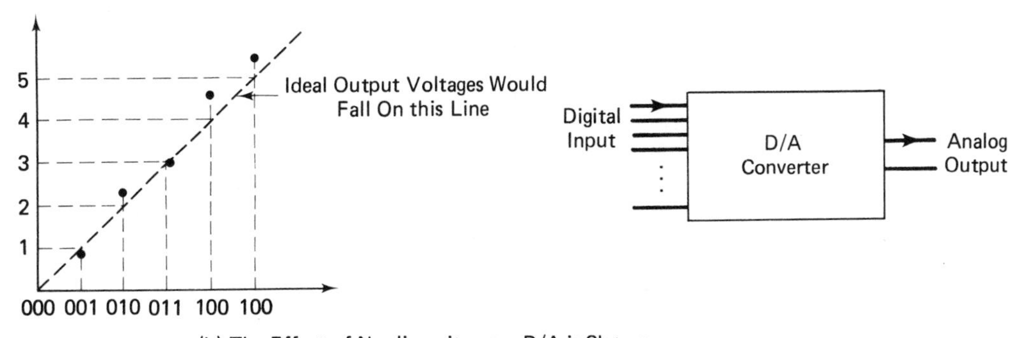

(b) The Effect of Nonlinearity on a D/A is Shown.

FIGURE 9-10

to time, temperature, and also upon random factors such as noise voltages generated within the circuit's components. For a given input value, we would want the circuit's output to remain constant under these varying conditions. Such qualities are particularly important in analog–digital interface circuits. In an A/D converter, for example, it would be very bad to imply more precision than was possible by providing a digital output with an excessively large number of bits.

The ***linearity*** of an A/D or D/A converter is an important measure of its accuracy, and tells us how close the converter's output is to the ideal input versus output characteristic. Figure 9-10 illustrates the concept of linearity for both an A/D and D/A converter. In order to compute a numerical measure of linearity for an A/D or D/A converter, we take the maximum deviation from the ideal characteristic as a fraction of the full-scale range. For example, let us say that an eight-bit D/A spans a 10-V range. Each increment will be about 39 mV. If the output will always be within 2 mV of its ideal value, this D/A is said to be linear to within 0.02% of its full-scale range. Note that the D/A output could be extremely linear in the geometric sense without being accurate: in this case, the slope of the actual output curve would not match the slope of the ideal output, as shown in Figure 9-11. Now, when we compute the linearity of a D/A converter, we are not measuring the consistency of the output slope (which would be precision only), but we are measuring the *deviation* from the ideal characteristic. Since the ideal characteristic is truly "ideal," it is both accurate and precise. For this reason, linearity can be thought of as a joint measure of precision and accuracy.

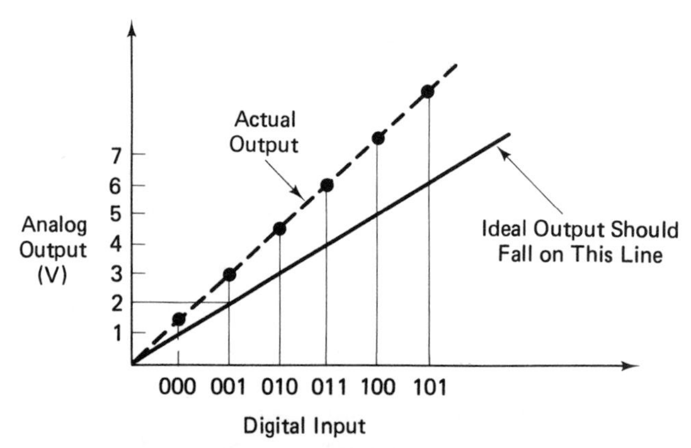

FIGURE 9-11 *Geometrically linear output that is inaccurate.*

Often, the output of an A/D converter is said to be accurate to within $\pm\frac{1}{2}$ LSB (least significant bit). This simply means that the digital output will accurately represent the analog input within $\pm\frac{1}{2}$ the value of the least significant bit. If a four-bit A/D converter were encoding voltage between 0 and +15 V, for example, the least significant bit would represent increments of 1 V. An accuracy of $\pm\frac{1}{2}$ LSB would mean that this A/D produces digital output that truly represents the analog input to within ±0.5 V.

9.3 DIGITAL-TO-ANALOG CONVERSION

Digital-to-analog conversion is somewhat less complex than analog-to-digital conversion, and for this reason, it will be discussed first. Furthermore, several of the A/D techniques that we will study use a D/A converter as one of the components. To begin, we must discuss the function and use of *operational amplifiers*, which are frequently applied in the circuitry of D/A converters.

9.3.1 Operational Amplifiers

The ***operational amplifier*** (abbreviated "op amp") is a fairly complex linear amplifier, usually integrated on a single chip of silicon. It is almost always used in a *feedback* configuration to provide amplification, filtering, and precision nonlinear functions such as rectification. In D/A converters, its primary purpose is that of *current summation*. In this capacity, the operational amplifier is used with a resistive network to add currents that each represent the weight of a binary digit. We shall examine the details of such a circuit within the next few pages.

The operational amplifier itself is represented in block form as shown in Figure 9-12. Remember that this basic block consists of a complex transistor circuit. The equation that describes the output voltage as a function of input voltage is shown also.

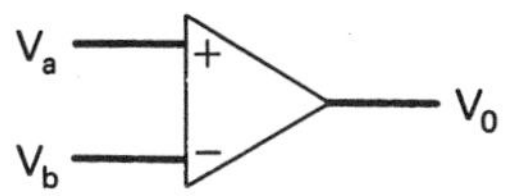

$V_0 = K(V_a - V_b)$ 'K' is the Amplifier Gain

FIGURE 9-12 Basic block diagram and equation of an operational amplifier.

The operational amplifier can be used in a very special way as a result of several important electrical characteristics. They are:

1. The gain, 'K', is extremely high (typically $K = 100{,}000$).
2. The input impedance is extremely high (typically $Z_{in} = 5\ M\Omega$), meaning that very little current will flow into the input terminal.
3. The output impedance is extremely low (typically $150\ \Omega$), meaning that the output terminal can drive a fairly heavy load.

Usually, the frequency response of an operational amplifier will range from DC to 50 kHz.

Judging from the very high gain, we would not expect to use an op amp as a "straight-through" amplifier; a gain of 100,000 or more is usually not needed. Instead, the op amp is usually incorporated in a negative feedback configuration to provide amplification with a lower but very stable gain. Such a circuit is illustrated in Figure 9-13. Note that the output (V_0) is connected to the inverting input ('—'). As 'V_{in}'

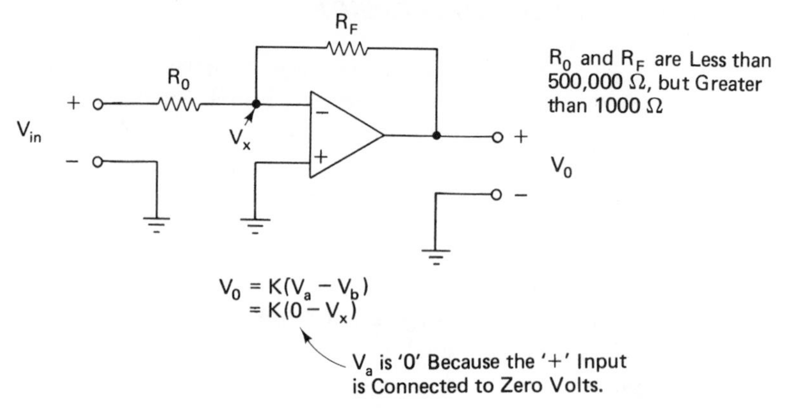

FIGURE 9-13 *Operational amplifier used in a feedback configuration to provide low-gain, stable amplification.*

increases, 'V_x' begins to increase. But, as 'V_x' increases, 'V_0' *decreases*, owing to the fact that 'V_x' is connected to the inverting input. Now, as 'V_0' falls, it pulls the voltage at 'V_x' back down, via 'R_F'. Whenever a change in output voltage is fed back to the inputs to oppose any further change, the circuit is considered to have ***negative feedback***.

It is now desirable to obtain an equation that relates the output voltage, 'V_o', to the input voltage, 'V_{in}'; this will tell us whether the overall circuit functions as an amplifier, and if so, what its gain will be. Fortunately, the equation is not difficult to derive. Consider the internal currents illustrated in Figure 9-14. The very high input impedance that is a characteristic of operational amplifiers tells us that negligible current flows into the '−' or '+' inputs. In this circuit, the '+' input is grounded, so we need to be concerned only with the condition of the '−' input. Since practically no current flows into the '−' input, we can consider 'I_{in}' to equal 'I_F'. Using Ohm's law, we can write 'I_{in}' and 'I_F' in terms of 'V_{in}', 'V_x', 'R_0', and 'R_F':

$$I_{in} = \frac{V_{in} - V_x}{R_0} \tag{9-1}$$

$$I_F = \frac{V_x - V_0}{R_F} \tag{9-2}$$

Since $I_{in} = I_F$, we can write

$$\frac{V_{in} - V_x}{R_0} = \frac{V_x - V_0}{R_F} \tag{9-3}$$

In order to obtain an equation only in terms of 'V_{in}', 'V_0', and the resistor values, 'V_x' must be eliminated. We can do this by using the equation that describes the operational amplifier's function:

$$\begin{aligned} V_0 &= K(V_+ - V_-) \\ &= K(0 - V_x) \\ &= -KV_x \end{aligned} \tag{9-4}$$

Thus, $V_x = (1/K)V_0$. Replacing every occurrence of 'V_x' with '$-(1/K)V_0$', we obtain

$$\frac{V_{\text{in}} - [-(1/K)V_0]}{R_0} = \frac{[-(1/K)V_0] - V_0}{R_F} \tag{9-5}$$

Now, if we assume that 'K' is a very large number, such as 100,000, we can be sure that '$1/K$' is an extremely small number. Thus, '$-(1/K)V_0$' becomes negligible with respect to 'V_{in}' and 'V_0', and we can write

$$\frac{V_{\text{in}} - 0}{R_0} = \frac{0 - V_0}{R_F} \tag{9-6}$$

This is the same as saying that

$$\frac{V_{\text{in}}}{R_0} = -\frac{V_0}{R_F} \tag{9-7}$$

The "gain" of any amplifier circuit is expressed as the ratio of output to input voltages. Therefore, we can rewrite Eq. 9-7 as a ratio of 'V_0' and 'V_{in}' to determine the gain. Doing so, we obtain

$$\frac{V_0}{V_{\text{in}}} = -\frac{R_F}{R_0} \tag{9-8}$$

This is a very important result. It tells us that the gain of the amplifier circuit is a function of *only* the two resistors 'R_F' and 'R_0', and *is not related to the gain of the op amp itself.* Thus, we can obtain any gain that is necessary, within limits, by selecting the two resistors properly. Furthermore, by using a potentiometer (variable resistor) for 'R_F', the circuit's gain can be varied by simply turning the shaft. Note that the amplifier's output is inverted with respect to the input (shown by the negative sign in Eq. 9-8). It is now important for us to see how an amplifier circuit such as this can be modified to become a *summation circuit.*

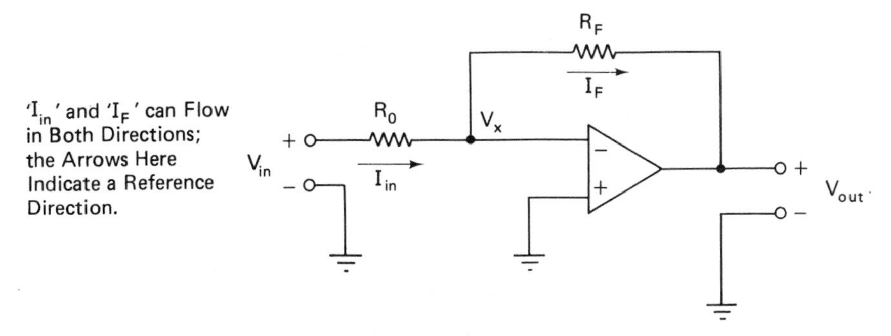

FIGURE 9-14 *The significant currents in a feedback amplifier.*

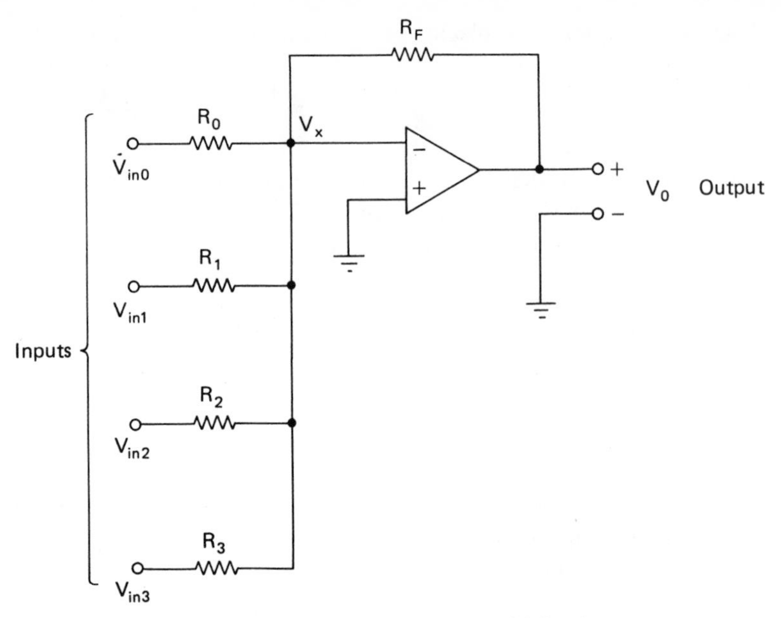

FIGURE 9-15 *Summation circuit with four inputs.*

9.3.2 Summation Circuits

Consider the circuit drawn in Figure 9-15. It appears to be similar to the amplifier circuit of the previous section, except that in this case, there are *four* inputs instead of one. The circuit's output is proportional to the *weighted sum* of the input voltages, with the weighting factor for each input being determined by the corresponding input resistor. These statements can easily be proved by carrying out an analysis similar to the analysis for the single-input case. In Figure 9-15, we see that the '—' node of the amplifier is supplied with current from four inputs, the totality of which is drawn away through the feedback path.

Thus, we can write

$$I_{\text{in}0} + I_{\text{in}1} + I_{\text{in}2} + I_{\text{in}3} = I_F \tag{9-9}$$

As was done for the single-input amplifier, each current variable in Eq. 9-9 can be replaced by an equivalent factor in terms of the voltages and resistances by applying Ohm's law. Doing so, we obtain

$$\frac{V_{\text{in}0} - V_x}{R_0} + \frac{V_{\text{in}1} - V_x}{R_1} + \frac{V_{\text{in}2} - V_x}{R_2} + \frac{V_{\text{in}3} - V_x}{R_3} = \frac{V_x - V_0}{R_F} \tag{9-10}$$

Since 'V_x' can be expressed as '$-(1/K)V_0$' (refer to Eq. 9-4), we can see that for large values of 'K', 'V_x' is approximately zero volts in comparison to the input and output voltages. Thus, Eq. 9-10 can be rewritten as

$$\frac{V_{\text{in}0}}{R_0} + \frac{V_{\text{in}1}}{R_1} + \frac{V_{\text{in}2}}{R_2} + \frac{V_{\text{in}3}}{R_3} = -\frac{V_0}{R_F} \tag{9-11}$$

Finally, we will rewrite this to show the output as a function of the inputs.

$$-V_0 = \frac{R_F}{R_3}V_{in3} + \frac{R_F}{R_2}V_{in2} + \frac{R_F}{R_1}V_{in1} + \frac{R_F}{R_0}V_{in0} \tag{9-12}$$

With this equation, we can see that the output voltage is, indeed, the weighted sum of the input voltages. The weighting factor for each input is determined by the ratio of the feedback resistor to the input resistor in that branch. Again, the output is inverted with respect to the input, shown by the negative sign in Eq. 9-12. Clearly, any number of inputs are possible by simply connecting additional input resistors.

9.3.3 Digital-to-Analog Conversion Circuit

The importance of the summation circuit is realized when we recall how binary numbers are converted into decimal numbers (Chapter 4). A four-digit binary number, '$B_3B_2B_1B_0$', can be translated into the equivalent decimal value, 'D', as follows:

$$D = (B_3 \cdot 2^3) + (B_2 \cdot 2^2) + (B_1 \cdot 2^1) + (B_0 \cdot 2^0) \tag{9-13}$$

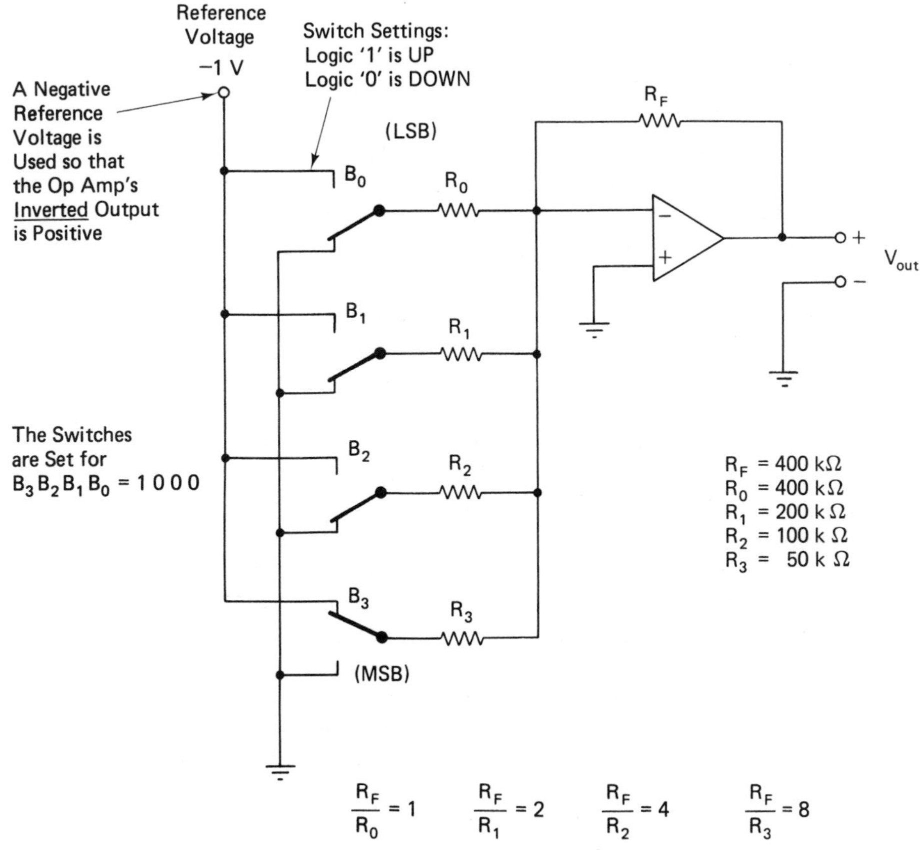

FIGURE 9-16 *Simple four-bit D/A converter utilizing switches.*

Now, note the similarity of this equation to the summation circuit function in Eq. 9-12. With the exception of the inversion, they are identical in form: the resistor ratios in Eq. 9-12 correspond to the powers of '2' in Eq. 9-13, while the input voltages of Eq. 9-12 match the location of the binary digits in Eq. 9-13. With these ideas in mind, we illustrate a simple D/A converter circuit in Figure 9-16. By setting the switches in this circuit to correspond with the digits of the binary number, a voltage at 'V_{out}' is produced that is exactly proportional to the binary input. In order to make all output voltages positive, the reference voltage should be chosen to be a negative value (recall that the summation circuit has an inverted output). In most D/A converters, the switches are implemented with transistors, so that the circuit is fully electronic.

The simple D/A converter of Figure 9-16 can also be used as an amplifier with digitally controlled gain. If the fixed reference voltage is replaced with a signal source, then the output will be an amplified version of the source, whose gain is determined by the switch setting!

9.3.4 Ideal Characteristics Versus Actual Operation

The construction of a D/A converter as shown in Figure 9-16 is not complex, yet it has some serious limitations. The most important factor to be considered is the exponential series of the resistor values in the network. A resistance error of 10% in the least significant bit would cause a 10% error in the basic increment of the analog output. For example, consider a four-bit D/A with a -1 V reference. The basic increment of the analog output is 1 V, and its range is from 0 to 15 V. A 10% tolerance in the LSB resistor means that the output error contributed by the LSB branch would be about ± 0.1 V, which is 10% of the basic 1-V interval. An error of 10% in the MSB resistor, however, would cause an error of about ± 0.8 V, which is 80% of one basic interval in the output. Consequently, the resistor values must be extremely accurate to ensure linear operation of the D/A converter.

Another problem that is of concern to us is the effect of a very wide range of input resistor values on the operational amplifier's operation. The largest input resistor cannot be so large that it approaches the input impedance of the op amp (within a factor of ten). If it does, our earlier assumptions in the analysis about no current flowing into the '$-$' input of the op amp is invalid. On the other hand, we cannot make the smallest resistor so small that it approaches the value of the op amp's output impedance (again, within a factor of ten). A typical op amp with an input impedance of 5 MΩ and an output impedance of 150 Ω limits the input resistor range to between 500 kΩ and 2 kΩ, providing for about eight bits of conversion.

Note that when transistors are used in the place of switches in Figure 9-16, additional caution must be exercised in the selection of resistors. A switching transistor in the conduction mode does not have zero resistance, but rather, contributes some finite, small resistance that can add to the value of the weighting resistor. Similarly, a transistor is not a perfect open circuit when in the cutoff mode. Again, the use of exponentially weighted resistors is not advised.

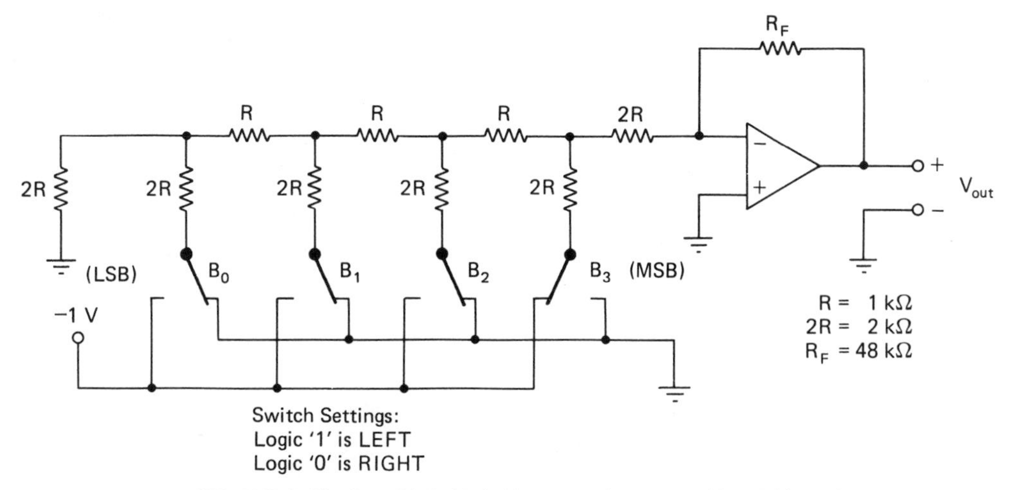

FIGURE 9-17 Four-bit R-2R ladder network connected in a D/A configuration.

9.3.5 R-2R Ladder Networks

To overcome the problems and limitations of exponentially weighted resistors in a D/A converter, a special resistive divider, known as the ***R-2R ladder network***, is commonly used. This circuit is illustrated in Figure 9-17, and derives its name from the fact that only two different values of resistance are used, one being twice the value of the other. The feedback resistor in the op-amp circuit is selected to obtain the desired gain. Although this network might look perplexing at first, it is really very simple and easy to understand. The analysis will rely on the idea of ***superposition***. That is, the combined effect of several different inputs can be considered as the sum of each of these inputs acting separately. In this particular case, the inputs are 'B_3' to 'B_0', and we shall note the effect on the ladder when only one of 'B_3' to 'B_0' is '1' and the others are '0'. Consider the R-2R network shown in Figure 9-18. First, note that the equivalent resistance to the right of 'Y' is exactly '$2R$'. This is easily checked by redrawing these resistors as

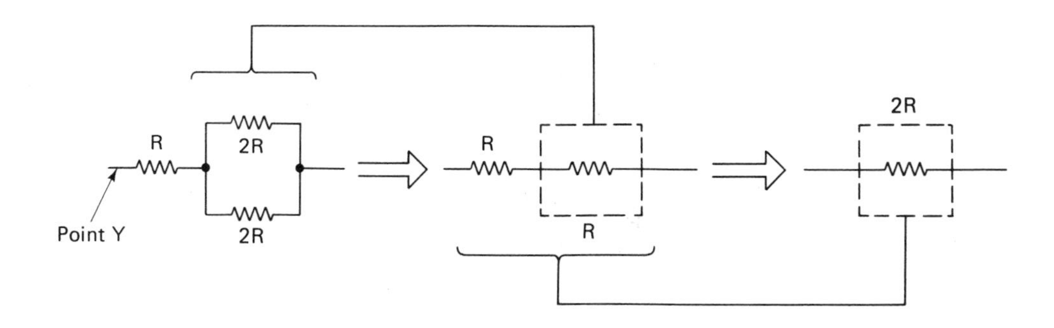

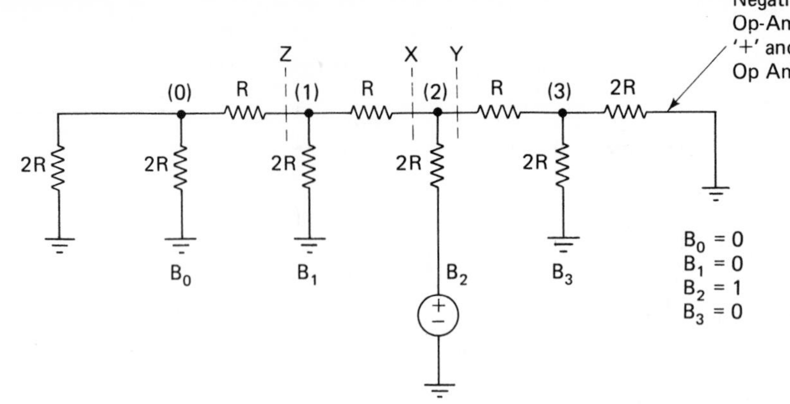

FIGURE 9-18 *R-2R network redrawn for analysis.*

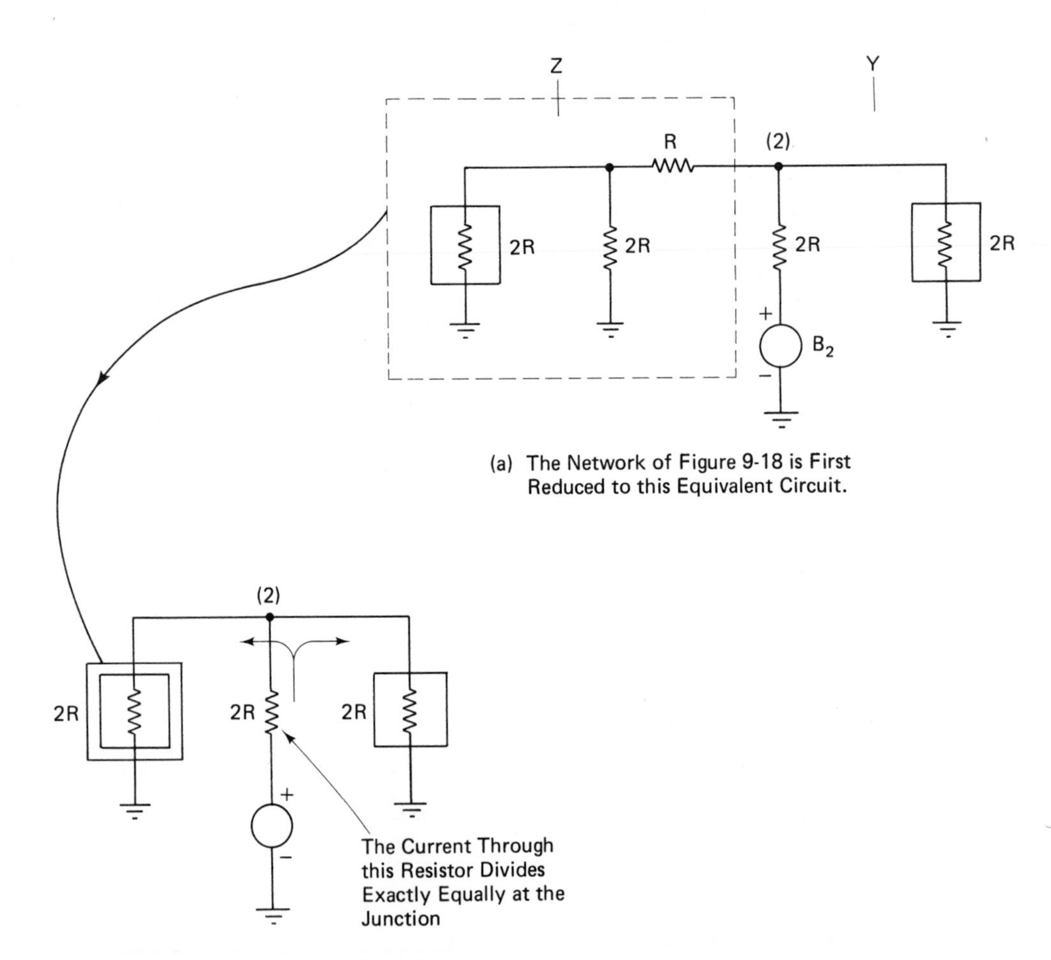

FIGURE 9-19

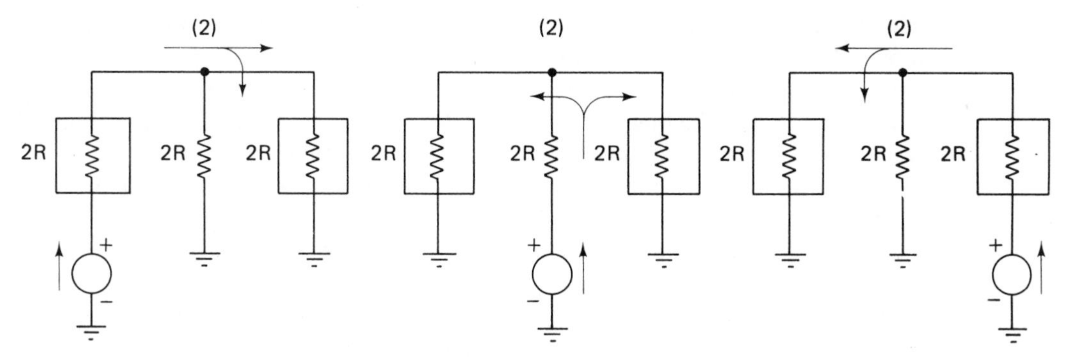

FIGURE 9-20 The current entering node (2) always divides equally no matter what its source branch.

An identical argument can apply to the circuit elements to the left of 'Z'. Now, let us replace each of these groups of three resistors with a single equivalent resistor, as shown in Figure 9-19(a). Reduction of the circuit to equivalent resistances can continue until the circuit of Figure 9-19(b) is obtained. The important fact to realize in this reduction is that the resistance seen to the right of 'Y' and to the left of 'X' in the figure are both equal to '$2R$'. This would be true no matter how extensive the network might be. The importance of this finding is that any current entering a node, such as node (2) in Figures 9-18 and 9-19, is divided exactly equally to the right and to the left. This holds true *no matter which branch the current originates on*. Refer to Figure 9-20. The "equal division" effect is the key to understanding the exponential behavior provided by the R-2R ladder network; this behavior allows the ladder network to be used as a D/A converter.

The four-input ladder network is redrawn in Figure 9-21 to illustrate how a current induced at one end of the network is successively divided by two in each stage. If a current originates at an input closer to the end, it will have been divided less when it

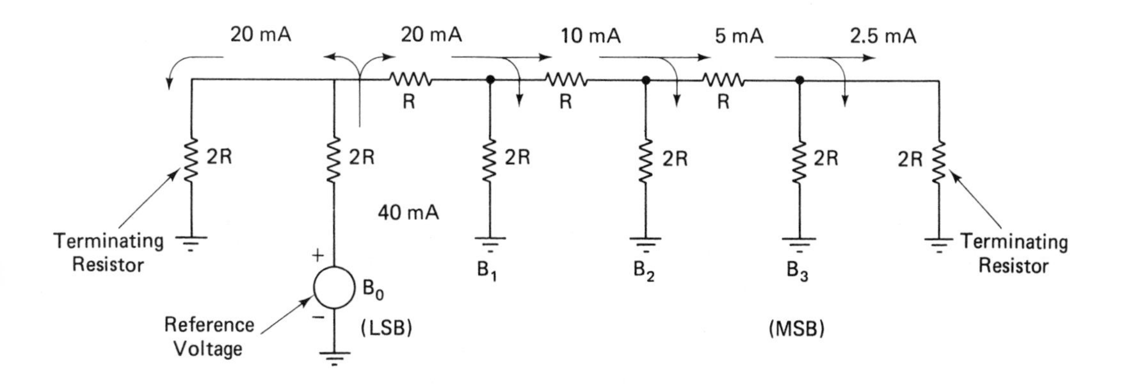

$B_3B_2B_1B_0 = 0001$

FIGURE 9-21 A current is shown to be divided into two equal parts by each stage of the ladder network.

reaches the end, and will, consequently, have more weight. Using the idea of superposition, and assuming that the source voltage for the digits 'B_3' to 'B_0' is identical, we can see that the current flowing through the terminal branch of the ladder network is the weighted sum of the binary input digits. An operational amplifier is added only to amplify the current flow in the terminal branch to a useful level.

The R-2R ladder network overcomes most of the limitations imposed by the exponential resistor network discussed previously. Since only two values of resistor are needed, it is considerably easier to match their tolerances. When the resistors are integrated on a single chip of silicon, their *relative* values will almost always be identical, even though their absolute resistance may vary. Remember that since the resistor *ratios* are the important factors in the R-2R ladder, absolute accuracy is not critically important. For this reason, when the resistors change in equal proportion with temperature, no errors will be seen on the D/A output since the ratios remain identical.

Because the resistance of the network as viewed from the op amp's '—' input is always the same regardless of how many stages are present in the ladder network, we can select a value of 'R' so that the input and output impedances of the op amp are never approached. Thus, loading the op amp will never be a problem.

Finally, the effect of transistor switches is perfectly balanced since the resistance that each transistor faces is identical no matter which branch it controls.

9.4 ANALOG-TO-DIGITAL CONVERSION

9.4.1 General Considerations

The basic intent of an analog-to-digital converter is to translate a value of voltage or current into a corresponding binary number. In contrast to the D/A converter which we discussed in the last section, the A/D converter can be implemented in many different ways, not just one or two. Which method is chosen depends upon many different factors, including:

1. Conversion speed.
2. Accuracy.
3. Cost.
4. Stability.

Before we begin our discussion of A/D techniques, however, it would be a good idea to dispel some common misconceptions about analog-to-digital conversion, and about digital signal processing in general.

Accuracy: First, digital signal processing techniques are not necessarily more accurate or precise than are analog signal processing techniques. It depends upon the A/D converter and on the extent of processing to be performed. An A/D converter is, itself, an analog circuit primarily, and is therefore subject to all the limitations of any analog circuit. If a great amount of signal processing is to take place, then, perhaps, digital signal processing would be more effective. It is considerably more difficult in analog circuits to control noise that is inevitably added to the signal during processing. The noise increases with each processing stage that the signal must pass through. However, when sufficient hardware or processing time is applied to a digital signal (which is a sequence of numbers), processing noise due to roundoff or truncation can be reduced to any level desired.

Speed and consistency: The conversion process between analog and digital signals is, unfortunately, quite time consuming. Thus, we can say almost absolutely that the digital processing of *analog signals* is slower than processing the analog signals as such. Analog systems are faster and generally less expensive for small applications than for analog–digital systems. Yet, digital signal processing can offer almost unlimited processing capability with minimal noise, and such processing is unaffected by changes in temperature, aging, or other enviornmental factors. In addition, extremely complex mathematical operations are possible when a signal is expressed digitally.

Two basic approaches are used in the design of A/D converters: one approach relies on *feedback*, while the other approach does not. We shall discuss the most important techniques of each approach.

9.4.2 Open-Loop Methods (Methods Without Feedback)

Flash converter: We begin by looking at the simplest possible A/D, although it also happens to use the fastest and most expensive technique. The circuit is shown in Figure 9-22(a) and is based on the ***differential comparator***, which is depicted in Figure 9-22(b). Although the comparator is drawn like an op amp, its output is either logic '0' or logic '1', depending upon which of its two inputs has the higher voltage. A small amount of hysteresis is built into the comparator to resolve any problems that might occur if both inputs were of equal voltage. Now, the primary feature noted in the flash converter circuit is the *resistive divider*. At each node of the divider, a comparison voltage is available. Since all resistors are of equal value, the voltage levels available at the nodes are *equally* divided between the reference voltage and ground. The intent of the circuit is to compare the analog input voltage with each of the node voltages. The output of the circuit is illustrated for a typical input voltage. Pure combinational logic follows to produce a binary encoded output. The design for a three-bit A/D is shown in Figure 9-23. We can see that for additional digits of precision in a flash converter, much more circuitry is needed. In fact, the number of circuit components at least doubles with each added digit of precision. Thus, only when extremely high speed or high precision A/D conversion is needed should a flash converter be used. Typical conversion time for such circuits is 100 ns or less.

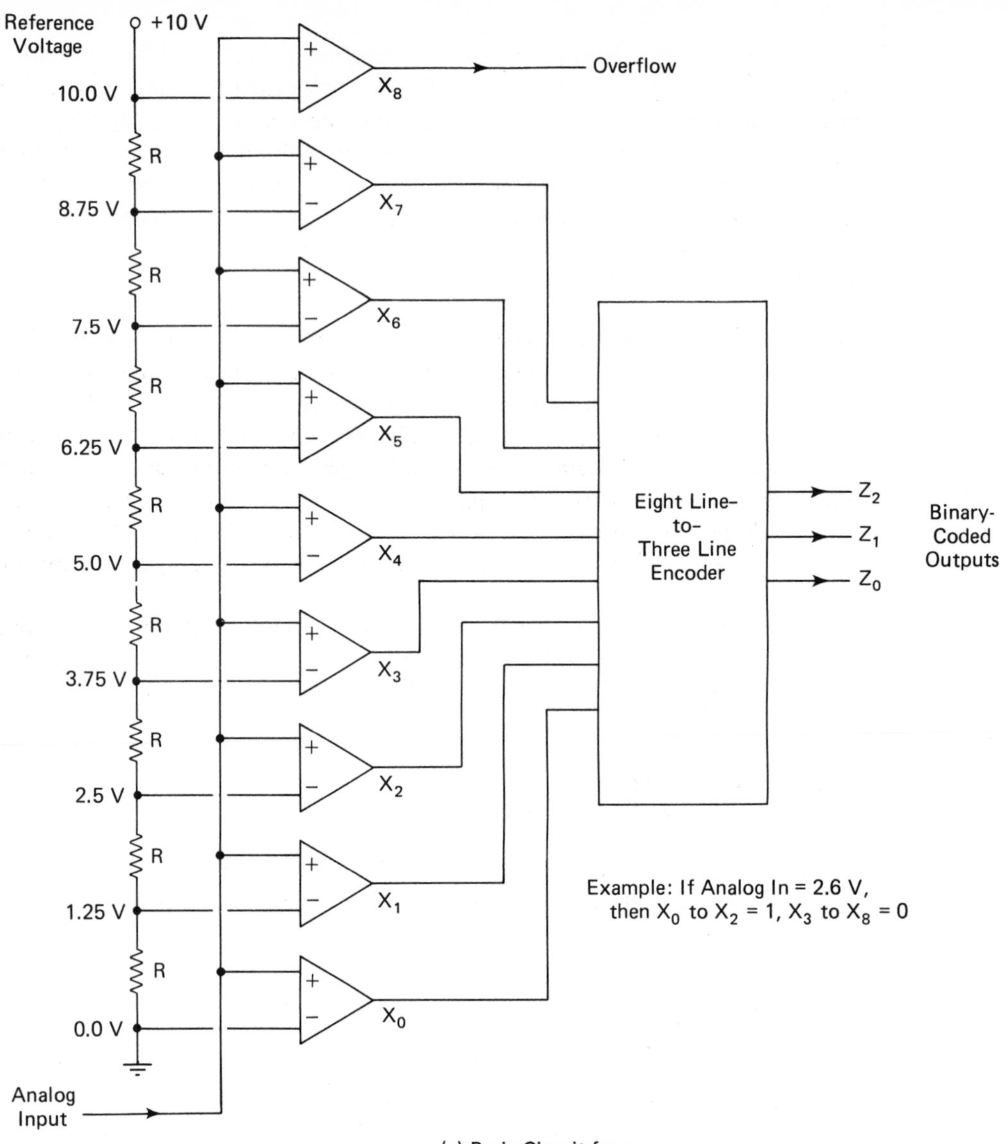

(a) Basic Circuit for a Flash A/D Converter.

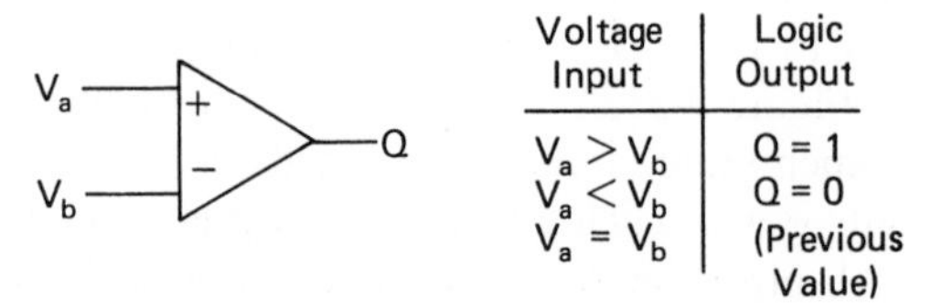

Voltage Input	Logic Output
$V_a > V_b$	Q = 1
$V_a < V_b$	Q = 0
$V_a = V_b$	(Previous Value)

(b) The Function of a Comparator. Hysteresis is Present in the Device to Resolve the Condition Where $V_a = V_b$.

FIGURE 9-22

Input Voltage	X_8	X_7	X_6	X_5	X_4	X_3	X_2	X_1	X_0	Z_2	Z_1	Z_0	U_F	O_F	
Less Than 0 V	0	0	0	0	0	0	0	0	0	0	0	0	1	0	(Underflow)
0 to 1.25	0	0	0	0	0	0	0	0	1	0	0	0	0	0	
1.25 to 2.5	0	0	0	0	0	0	0	1	1	0	0	1	0	0	
2.5 to 3.75	0	0	0	0	0	0	1	1	1	0	1	0	0	0	
3.75 to 5.0	0	0	0	0	0	1	1	1	1	0	1	1	0	0	
5.0 to 6.25	0	0	0	0	1	1	1	1	1	1	0	0	0	0	
6.25 to 7.5	0	0	0	1	1	1	1	1	1	1	0	1	0	0	
7.5 to 8.75	0	0	1	1	1	1	1	1	1	1	1	0	0	0	
8.75 to 10.0	0	1	1	1	1	1	1	1	1	1	1	1	0	0	
Greater Than 10 V	1	1	1	1	1	1	1	1	1	Ø	Ø	Ø	0	1	(Overflow)

$$Z_0 = \overline{X}_8 X_7 + \overline{X}_6 X_5 + \overline{X}_4 X_3 + \overline{X}_2 X_1$$
$$Z_1 = \overline{X}_8 X_7 + \overline{X}_7 X_6 + \overline{X}_4 X_3 + \overline{X}_3 X_2$$
$$Z_2 = \overline{X}_8 X_7 + \overline{X}_7 X_6 + \overline{X}_6 X_5 + \overline{X}_5 X_4$$
$$U_F = \overline{X}_0$$
$$O_F = X_8$$

FIGURE 9-23 *Truth table and design equations for a Flash A/D converter.*

Time-window converter: A simple and inexpensive circuit with very wide range can easily be constructed when high speed is not necessary. This circuit is called the ***time-window converter*** and is based on the monostable multivibrator (one-shot). In this type of A/D, the analog input variable controls the *period* of a monostable multivibrator, whose output is used to gate the clock input of a counter. Thus, the count value present in the counter at the end of the monostable's period is proportional to the analog input. This approach is particularly useful when the analog input is the position of a potentiometer shaft; we would then have direct control over the monostable's period by varying the resistance (shaft angle) of the potentiometer. The time-window converter is illustrated in Figure 9-24. Note that the monostable's period can never be longer than the sample period. If the analog input is electrical in nature, the potentiometer can be replaced by a controlled resistance device such as an FET transistor. With a fairly high speed system clock frequency of 20 MHz, a ten-bit A/D converter designed using the time-window technique would provide a sample period of about 55 μs (approximately 18 kHz), which is too slow for any analog signals with frequency components greater than 1.8 kHz. When speed is not important, however, the time-window converter is the most efficient design possible.

Slope converters: One useful approach to A/D conversion involves the comparison of the analog input to a constantly changing (sweeping) reference voltage. In this case, the time between when the reference begins its sweep and when it becomes equal to the analog input is a measure of the analog voltage. The basic circuit diagram for a slope converter is shown in Figure 9-25(a), and the associated waveforms are shown in Figure 9-25(b). As in the time-window converter, the slope converter translates a voltage level into a measure of time, which can be counted with a binary counter. In both cases, however, the circuits are subject to error because of variations in the timing components with temperature, and due to aging. The "timing components" referred to here are the resistor and capacitor components.

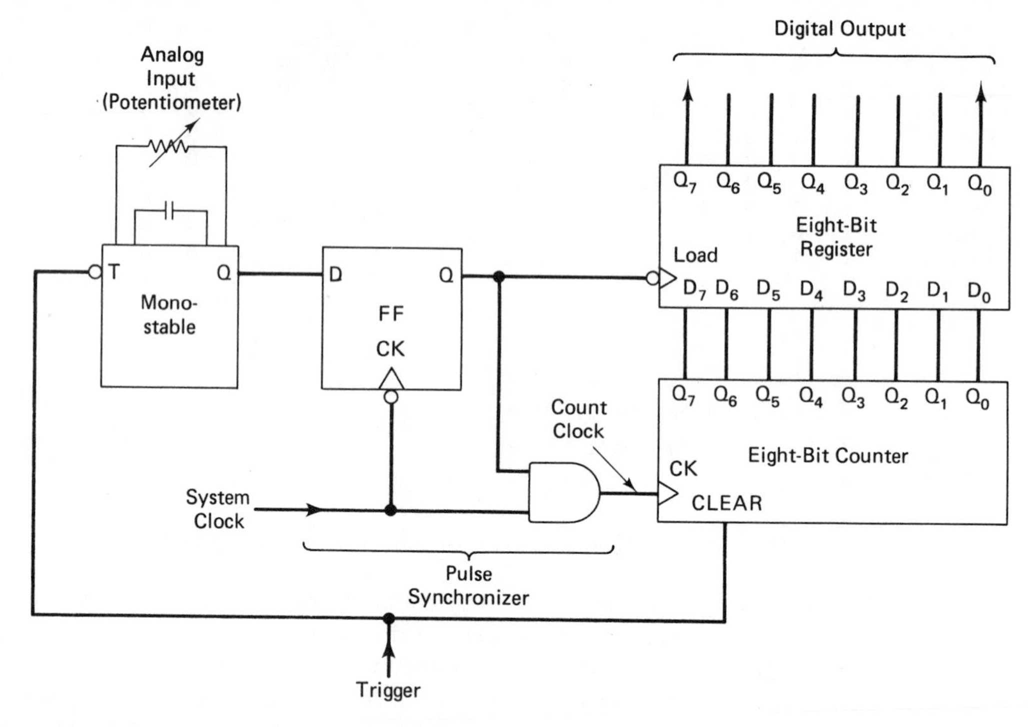

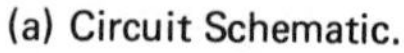

(a) Circuit Schematic.

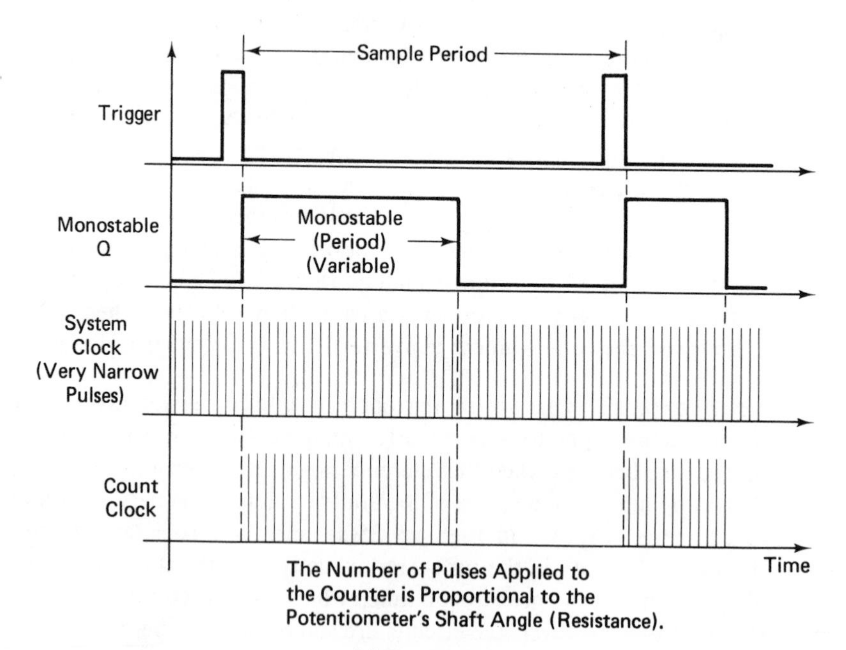

The Number of Pulses Applied to the Counter is Proportional to the Potentiometer's Shaft Angle (Resistance).

(b) Timing Diagram.

FIGURE 9-24

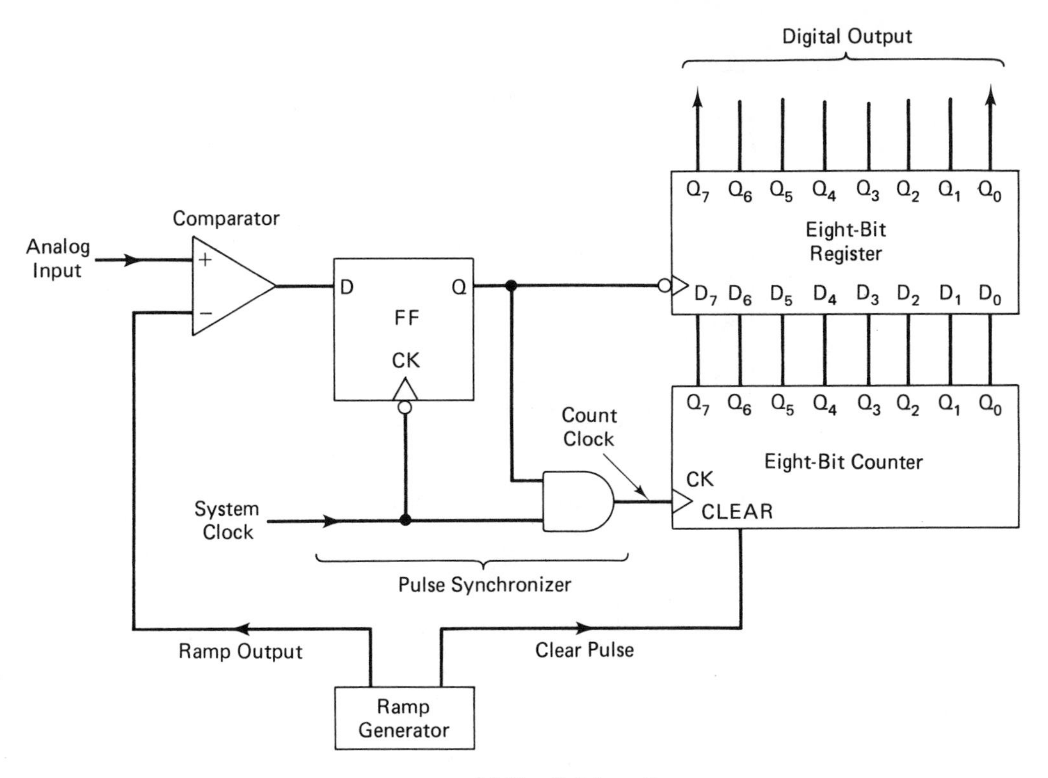

(a) Circuit Schematic.

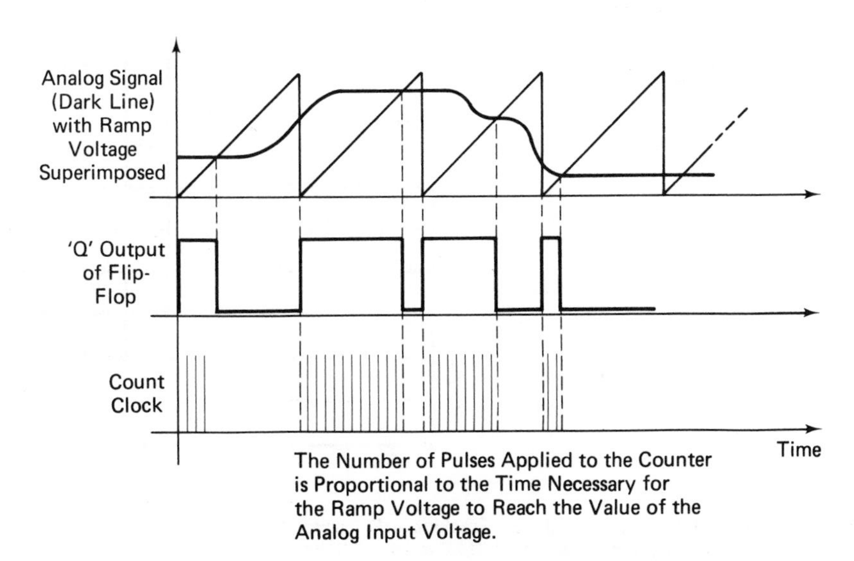

(b) Timing Diagram.

FIGURE 9-25

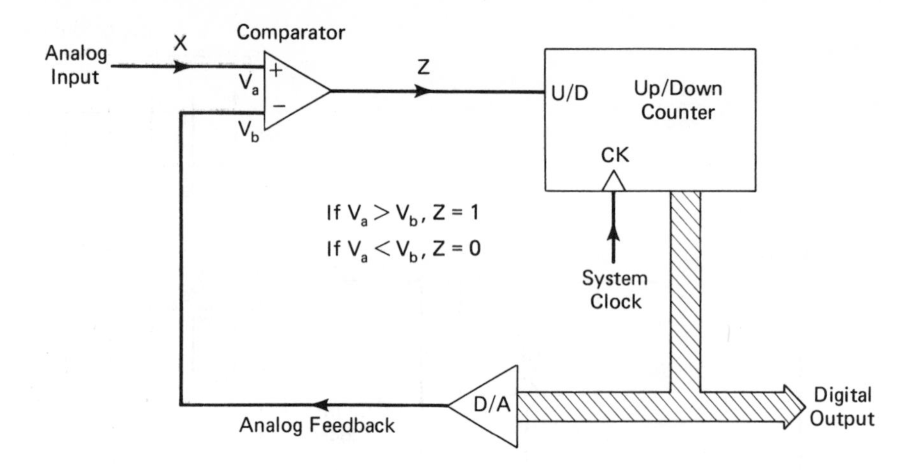

(a) Basic Block Diagram for a Tracking A/D Converter.

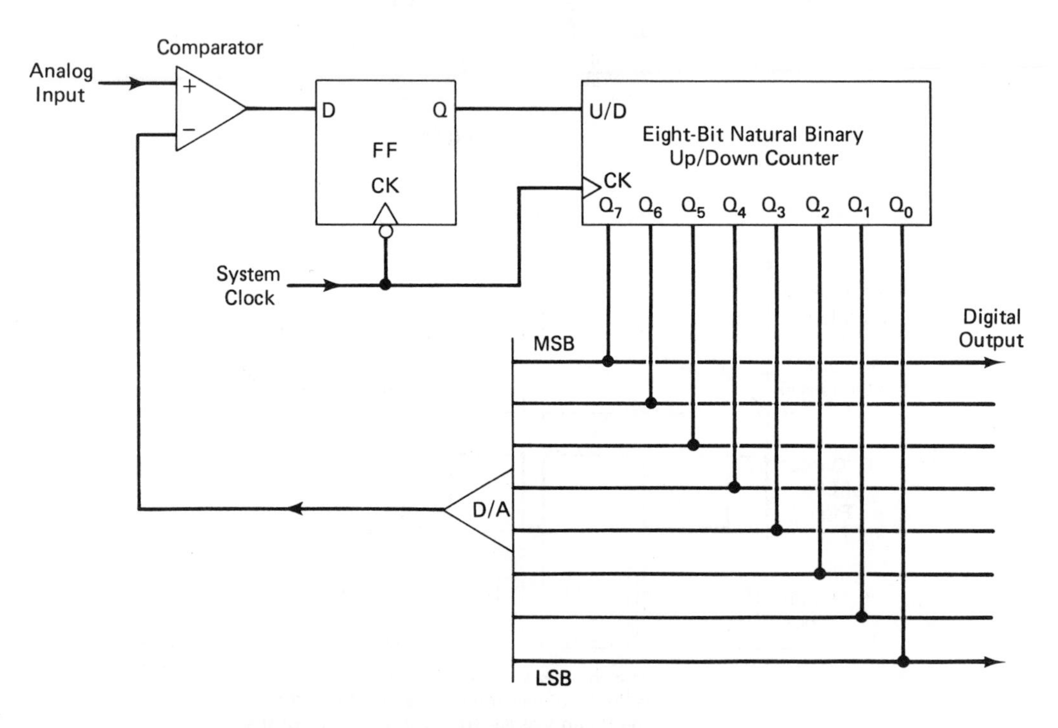

(b) The Details of an Eight-Bit Tracking A/D Converter. The 'D' Flip-Flop is Necessary to Ensure that *'U/D'* does not Change at the Same Time as the Counter's *'CK'* Input.

FIGURE 9-26

Advanced techniques using the slope approach utilize two ramps—one to measure a fixed, accurate reference voltage, and the other to measure the signal source. By first measuring the reference voltage, followed by measurement of the signal source, the timing components of the ramp generator are calibrated on *every conversion* against the accurate voltage reference. Thus, variations in the timing components over time are compensated. This technique is known as *dual-slope A/D* conversion.

9.4.3 Closed-Loop Methods (Methods With Feedback)

Tracking A/D converter: Closed-loop A/D methods are distinguished from open-loop methods by the presence of negative feedback. In closed-loop circuits, the present condition of the output is fed back to the input side of the circuit and compared with the present input value. The result of this comparison causes corrective action to be taken so that the output value more closely matches the input value. This is the basic idea behind the op amp circuits we discussed earlier. Now, if the input of a circuit is *analog* and its output is *digital*, then for feedback to be present, a digital-to-analog converter is needed in order that the comparison of the two signals be possible. Thus, in closed-loop A/D converter circuits, one of the components must be a D/A converter. Perhaps the simplest closed-loop A/D circuit is the ***tracking A/D converter***, which is illustrated in block form in Figure 9-26(a). The actual circuit schematic is shown in Figure 9-26(b). First study the block form and note that there are only three basic components. The binary counter produces a digital output that is the primary output of the circuit and which also drives a D/A. The analog output of the D/A is compared with the circuits analog input. If the input is greater than the feedback signal, the counter is caused to count *up*. As soon as the feedback signal becomes greater than the input, the counter reverses direction and counts down. This operation is illustrated in Figure 9-27. We see that as long as the analog input changes slowly, the tracking A/D will be within one LSB of the correct value. However, when the analog input changes rapidly, the tracking A/D cannot keep up with the change,

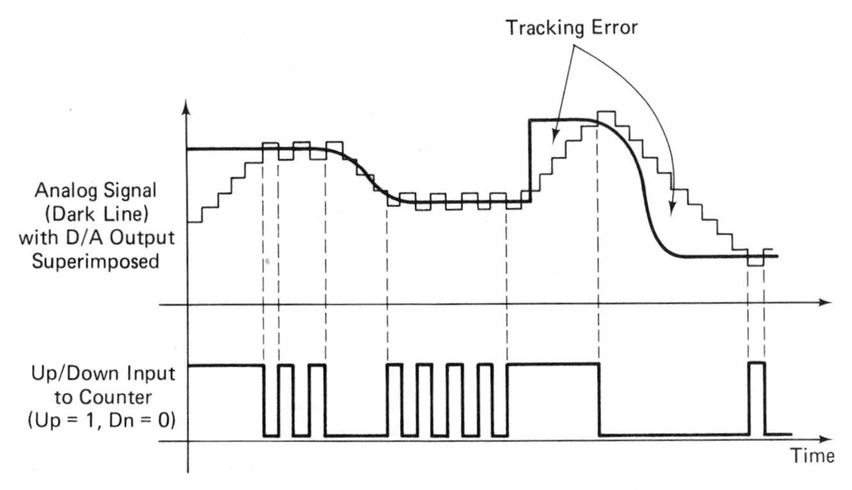

FIGURE 9-27 *Waveforms associated with a tracking A/D converter.*

and error occurs. If we are using a tracking A/D whose output is eight bits and whose input range is 10 V, the resolution of this device is about 39 mV. Now, with a 1-MHz clock signal driving the counter, the maximum rate of change of the feedback signal will be ± 39 mV/μs. Thus, any analog input signal that does not change faster than ± 39 mV/μs can be tracked accurately, within one LSB. A sawtooth waveform is illustrated in Figure 9-28. As an example, we would like to compute the maximum frequency this sawtooth wave can have if it is to be accurately tracked by the A/D.

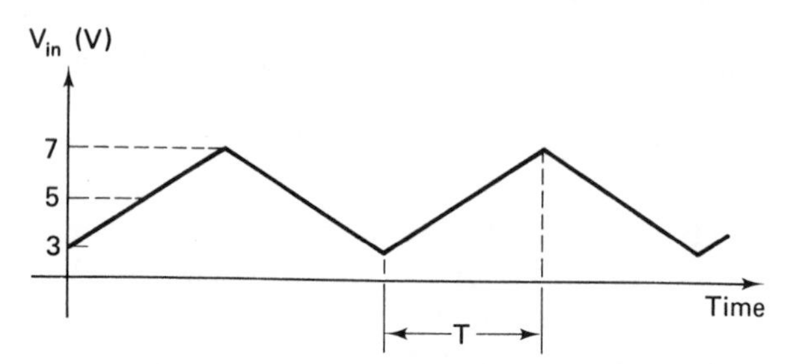

FIGURE 9-28 *Sawtooth wave to be analyzed.*

We see that the wave has an amplitude of 2 V centered around a +5-V level. We see that the slope (which is always the same, but changes sign) of the sawtooth wave is $4/T$. Now, what should 'T' be so that

$$\frac{4\text{ V}}{T} \leq \frac{39\text{ mV}}{1\ \mu\text{s}}?$$

Since $4\text{ V}/T = 4000\text{ mV}/T$, we can say that

$$\frac{4000\text{ mV}}{T} \leq \frac{39\text{ mV}}{\mu\text{s}}$$

Thus,

$$T \geq \frac{4000}{39}\ \mu\text{s} = 103\ \mu\text{s}$$

The period of the sawtooth is '$2T$'. Thus, its period should be greater than or equal to 206 μs. This corresponds to a frequency of 4854 Hz.

The tracking A/D has the advantage of being very simple in design and easy to construct. Unfotunately, however, the time needed for it to stabilize at a new conversion value is directly proportional to the rate at which the analog signal changes. For slowly varying signals (small maximum slopes), there is no problem. For more typical signals that can change quickly at times, the tracking A/D converter is inadequate.

Successive approximation converters: A tracking converter can make only one incremental step per clock pulse. Thus, an eight-bit tracking converter, for example, may take anywhere from 1 to 255 clock pulses in order to reach a new conversion value. With this characteristic the conversion time is directly proportional to

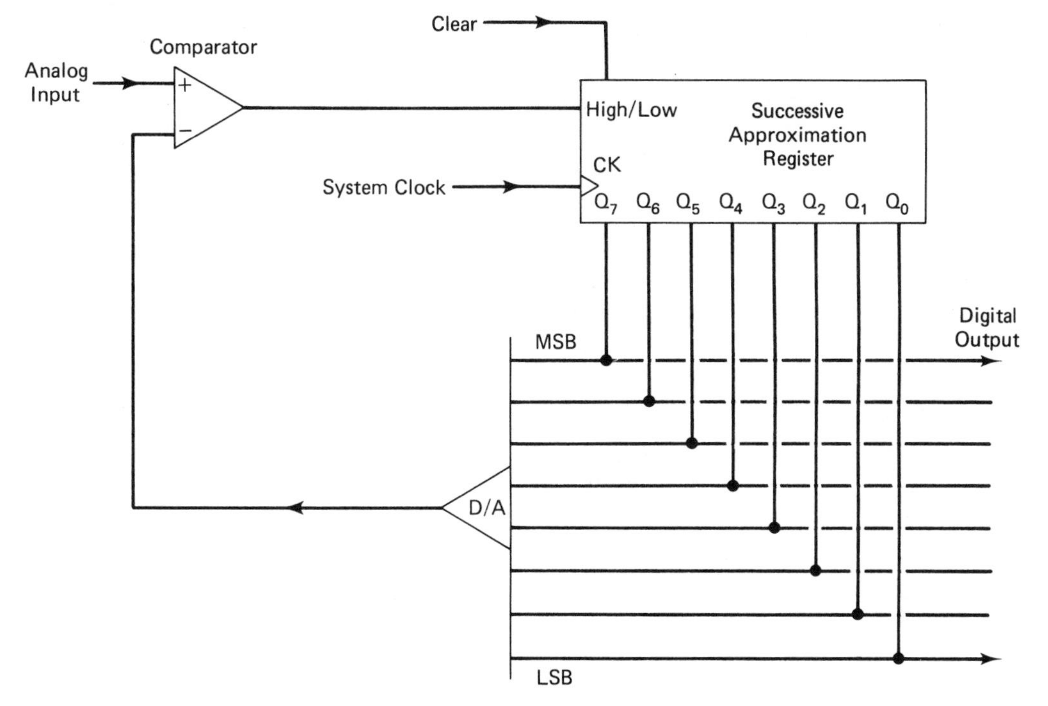

FIGURE 9-29 Block diagram for an eight-bit successive approximation A/D.

the rate of change of the analog signal input, and if any part of the analog signal changes more than one incremental step per clock pulse, substantial error is introduced. The ***successive approximation converter*** is a compromise between the best and worst feature of the tracking converter. For an 'n'-bit successive approximation A/D, exactly '$n + 1$' clock pulses are required for a full conversion cycle, no matter what the analog input may be. An eight-bit converter would, for example, always require nine clock pulses to obtain a valid digital output. The block diagram for an eight-bit converter is shown in Figure 9-29.

The conversion cycle for an eight-bit successive approximation device involves the following sequence of steps:

Bit 7 (MSB):

1. Set the MSB to '1' and all other bits to '0'.
2. Compare the analog input to the D/A output.
   (a) If the analog input is greater than the D/A feedback value, then '10000000' is less than the correct digital representation. Leave the MSB at '1' and go on to the next lower significant bit.
   (b) If the analog input is less than the D/A feedback value, then '10000000' is greater than the correct digital representation. Reset the MSB to '0' and go on to the next lower significant bit.

Bit 6:

3. Set bit 6 to '1', leaving all other bits unchanged.
4. Compare the analog input to the D/A output.
   (a) If the analog input is greater than the D/A feedback value, then 'X1000000' is less than the correct digital representation. Leave bit 6 at '1' and go on to the next lower significant bit.
   (b) If the analog input is less than the D/A feedback value, then 'X1000000' is greater than the correct digital representation. Reset bit 6 to '0' and go on to the next lower significant bit.

FIGURE 9-30

Correct Digital Representation	Successive Approximation Register Output at Different Stages in the Conversion		
1 0 1 1 0 1 0 0	1 0 0 0 0 0 0 0	1	(Initial Output)
	1 1 0 0 0 0 0 0	0	This Column Indicates the Comparator Output
	1 0 1 0 0 0 0 0	1	
	1 0 1 1 0 0 0 0	1	
	1 0 1 1 1 0 0 0	0	
	1 0 1 1 0 1 0 0	1	
	1 0 1 1 0 1 1 0	0	
	1 0 1 1 0 1 0 1	0	
	1 0 1 1 0 1 0 0		

(a) Successive Approximation Conversion Sequence for a Typical Analog Input.

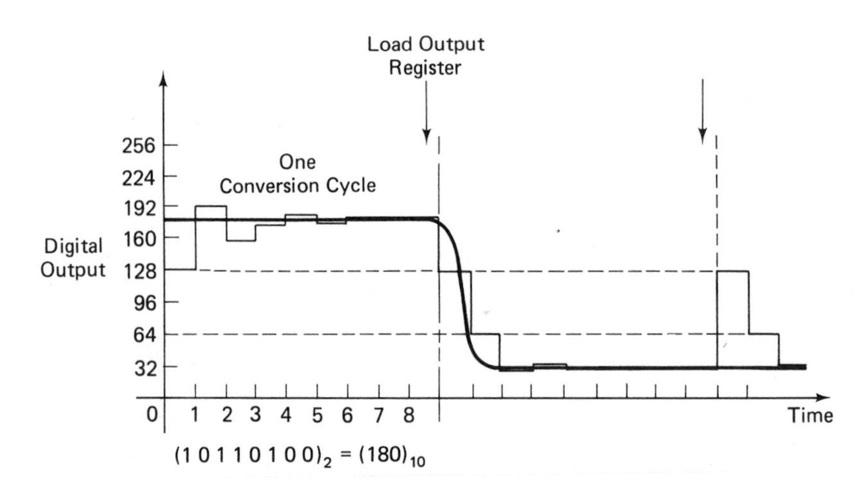

(b) The D/A Output Voltage is Seen to Become Successively Closer to the Actual Analog Input Voltage. Eight Clock Pulses are Always Required to Establish an Accurate Output, Regardless of What the Analog Input May Be. One Additional Clock Pulse is Used to Load the Output Register, and Reinitialize the Circuit.

Bit 5:

5. Set bit 5 to '1', leaving all other bits unchanged.
6. Compare the analog input to the D/A output.
   (a) If the analog input is greater than the D/A feedback value, then 'XX100000' is less than the correct digital representation. Leave bit 5 at '1' and go on to the next lower significant bit.
   (b) If the analog input is less than the D/A feedback value, then 'XX100000' is greater than the correct digital representation. Reset bit 5 to '0' and go on to the next lower significant bit.

Bit 4:

etc.

Figure 9-30(a) illustrates a typical conversion sequence, and Figure 9-30(b) shows the associated waveforms. Figure 9-31 graphically compares the speed of an eight-bit tracking A/D to an eight-bit successive approximation A/D. Given the same clock rate, we see that the tracking circuit is faster only for small changes in the input from one clock pulse to the next. In general, the successive approximation device is much more versatile, and is superior in almost all cases.

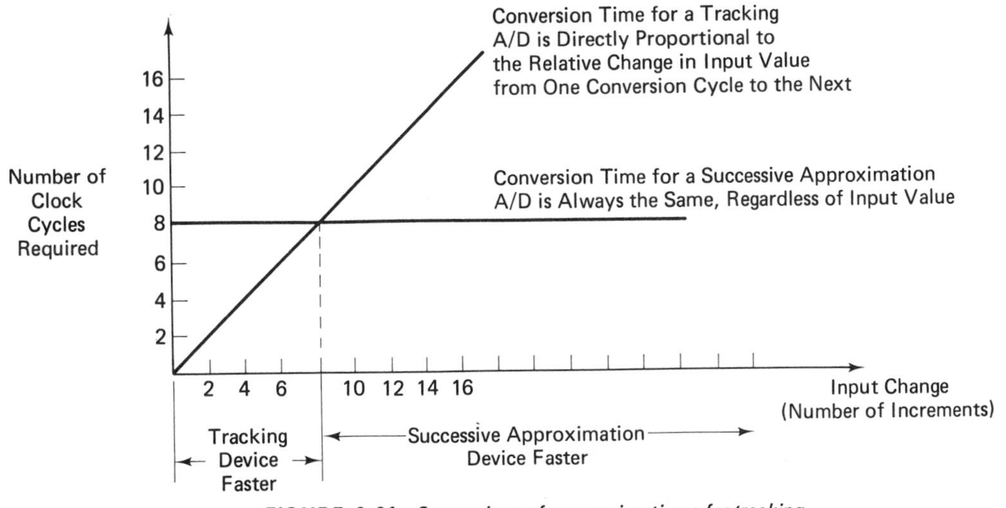

FIGURE 9-31 Comparison of conversion times for tracking and successive approximation A/D devices.

9.5 APPLICATIONS

9.5.1 Digital Readout of Rotational Motion

The shaft of a metal lathe is made to turn at different rates, depending upon the kind of metal and type of cut to be machined. In many cases, an exact indication of the shaft speed is very helpful to the operator and can prevent errors in workmanship. In this problem, we wish to obtain a four-digit numerical readout (in decimal) of the lathe's shaft speed, in revolutions per minute (RPM). This output should be updated once every second.

A sensor is connected to the shaft so that rotational motion between 0 and 5000 RPM can be converted into an electrical signal. One type of sensor that is possible, and perhaps the first to cross our minds, consists of a small DC generator whose output voltage is directly proportional to the rotation rate of its shaft. This electrical output can then be converted into a binary number using an A/D converter. Aside

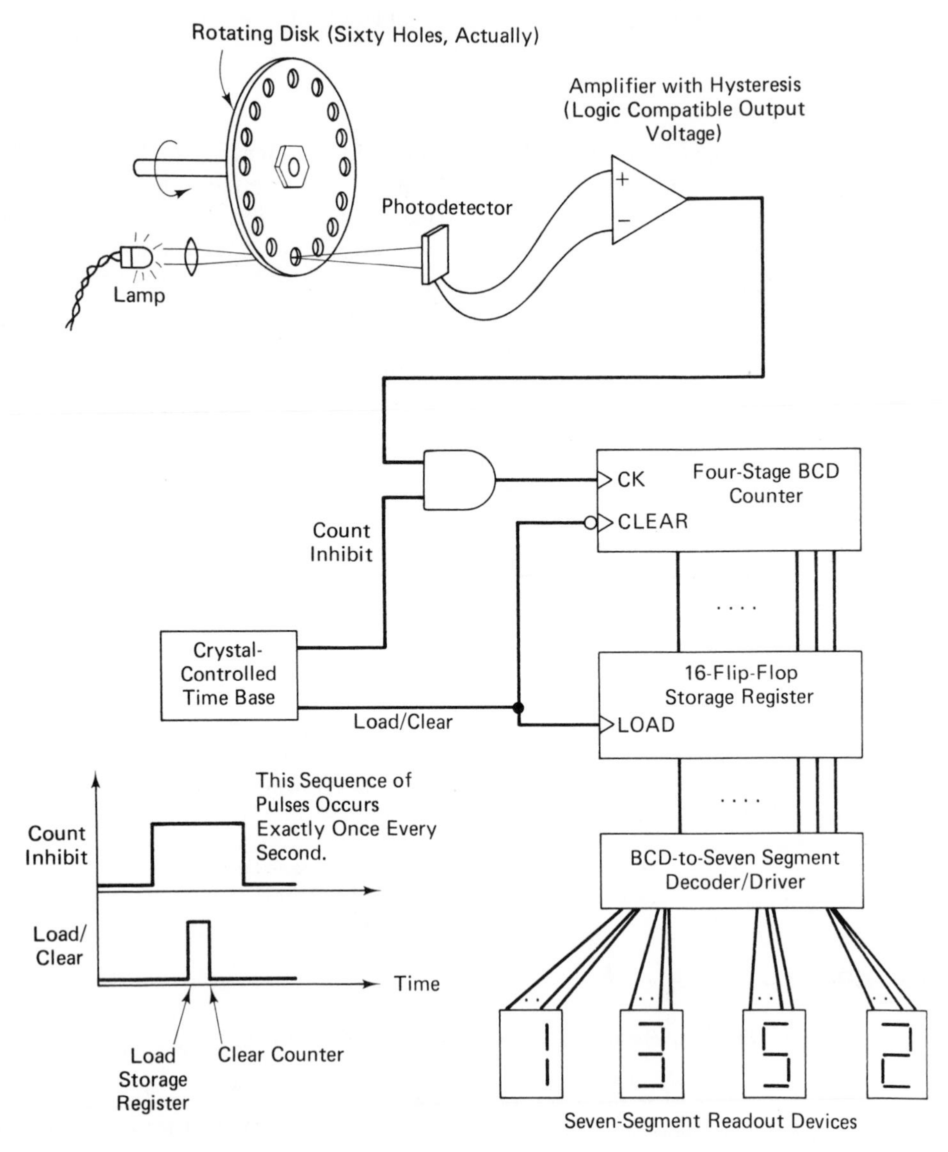

FIGURE 9-32 Block diagram design for digital shaft-speed measurement.

from the need for rather complex components using this approach (linear DC generator, an A/D converter with 13 bits of precision to provide enough levels to encode 0 to 5000), this method requires the conversion of a natural binary number (A/D output) into a *BCD* number suitable for driving a numerical display device. A far simpler approach is possible by realizing that

1. The conversion speed is not critical.
2. Different types of sensors are possible other than the linear DC generator.

The best approach to this problem is to use an *optoelectronic sensor*. With such a sensor, a small disc is directly connected to the shaft and turns with the shaft. One or more holes are present on the outer rim of this disc to permit the passage of light to the optoelectronic sensor only when the hole and light beam are aligned. As the shaft turns, the light beam is interrupted whenever a hole passes. The optoelectronic sensor will produce an electrical output that varies with the amount of light incident upon it, and thus a pulsating electrical signal will be produced as the shaft turns. By counting the number of times the light beam is interrupted in 1 s, we will have a measure of the rate at which the shaft is turning. In the simplest case, we could have a single hole in the disc's rim and count the interruptions in the beam over 1 minute; this will give a direct reading in revolutions per minute. We would like to update the output *once per second*, however, which is sixty times faster than updating it once per minute. Thus, we can place *sixty holes* in the outer rim of the disc to obtain the correct count rate. To solve the BCD output problem, we need only use a *BCD counter* as the counting device; its output is directly compatible with seven-segment decoder circuits, which will produce the desired decimal readout on seven-segment display devices. No binary-to-BCD conversion is necessary. Finally, a crystal-controlled time base is necessary to provide an extremely accurate 1-s timing window so that the counter may be cleared and restarted every second. The counter's outputs are saved in an external register so that the display is held constant over the 1-s period. Figure 9-32 depicts a block diagram for this system.

9.5.2 Digital Sound Recording

Audio tape recording is used frequently by both amateurs and professionals to preserve musical programs and discussion, for playback at a later time. Although the techniques of audio recording have progressed to a very sophisticated state, the recordings nevertheless are subject to unavoidable noise and distortion problems. These flaws result from the basic nature of the magnetic tape, and also from the *analog method* traditionally used to record information. In particular, the usual recording method causes the polarity of microscopic magnetic particles impregnated in the tape to become modulated in proportion to the input signal. Upon playback, the moving tape is passed across a sensitive electromagnet (the playback head), and the variations in polarity that exist on the tape from the recording process are converted back into a proportional electrical signal. Unfortunately, the changing polarity of magnetic particles on the tape cannot be made *exactly* proportional to the signal, and, thus, an amount of random variation (hiss) and distortion affect the playback signal.

Advances in magnetic tape and recording and playback electronics have greatly minimized these effects, yet in the professional studio where perfect reproduction is sought, an improved, noise-free and distortion-free method would be warmly accepted.

Digital magnetic recording can offer virtually flawless reproduction, overcoming most of the earlier defects present in analog systems. Since only two basic levels ever need to be recorded on a digital tape ('0' or '1'), the flux levels are completely in one direction or the other, and small random variations added to the basic flux level will never change a '0' to a '1', or vice versa. On the other hand, much more tape must be used to record an equivalent amount of information in a digital recording.

In this applications problem, we propose to discuss how an analog source signal, such as a musical program, might be recorded on a tape *digitally*, hopefully avoiding the problems of hiss and distortion that are inevitable in analog recording. The block diagrams for an elementary record and playback system are shown in Figure 9-33.

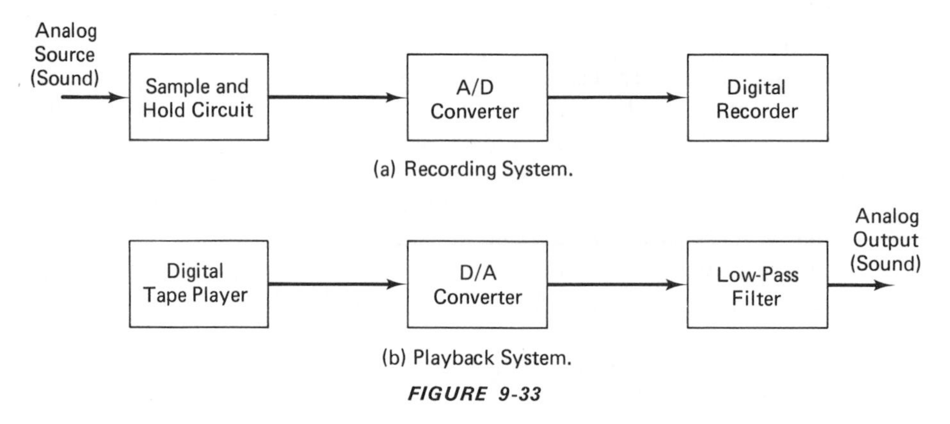

(a) Recording System.

(b) Playback System.

FIGURE 9-33

Recording system: To begin, the analog signal must be sampled at a regular, periodic rate, and sufficiently fast to follow the fastest expected variations in the signal. The human ear generally will not respond to signals higher in frequency than 15,000 Hz. Thus, we will assume that the sampling system need not track a signal higher in frequency than this. With this upper frequency limit, the recording system will be able to faithfully reproduce any signal whose frequency components are 15,000 Hz or lower, which includes most musical signals. Now, the question of sample rate must be addressed. Theoretically, it is not necessary to sample the signal any more than 30,000 times per second (twice the maximum frequency component of the signal) in order to completely reproduce it at a later time. Because of limitations on the components, however, this limit can never be reached. A safe choice for the sampling rate is ten times the maximum expected frequency. Thus, we will sample the signal 150,000 times per second, which is about once every 6.5 μs. The "sample and hold" block in Figure 9-33 will not only sample the input signal, but will hold it constant until the next sample is taken. This feature provides that a steady signal be applied to the A/D even though the actual input may change slightly in 6.5 μs.

The A/D converter has no more than 6.5 μs in which to complete a conversion. Such high speed will necessitate a successive approximation converter. Ten bits of

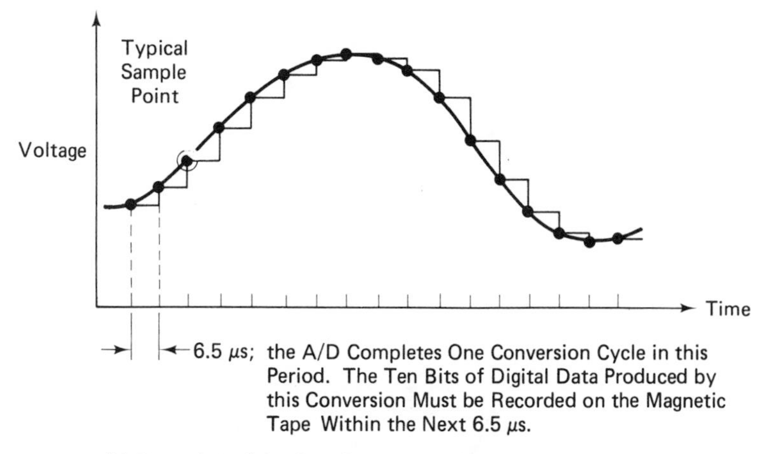

(a) Operation of the Sample and Hold Circuit. The Analog Input (Dark Line) is Shown with the Sample and Hold Output Superimposed.

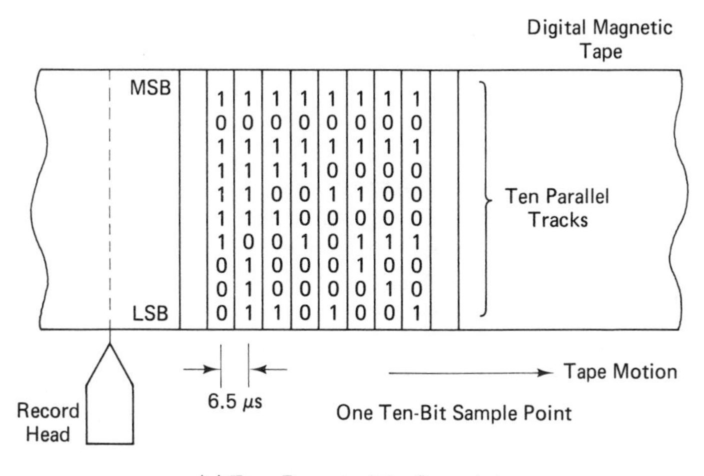

(b) Tape Format of the Recorded Digital Data.

FIGURE 9-34

resolution should be adequate to encode digitally the signal for reproduction that is free of quantization noise. The digital tape recorder will be capable of recording ten parallel tracks of digital data (one track per bit of output data), at a rate of 150,000 samples per second. Figure 9-34 summarizes the requirements of the recording system.

Playback system: Upon playback of the tape, the digital tape player provides a sequence of ten-bit words, with a new word coming every 6.5 μs. The D/A converter must make a complete conversion at least this fast. A D/A output, no matter what its resolution, will be quantized in nature, as shown in Figure 9-35(a). Prior to producing an audible analog signal from the quantized version, the signal should be filtered

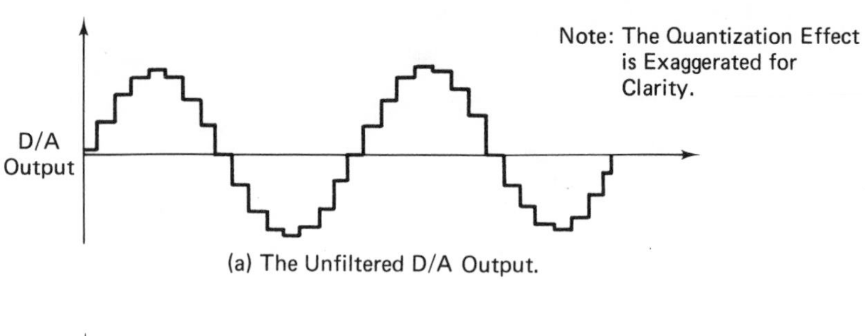

(a) The Unfiltered D/A Output.

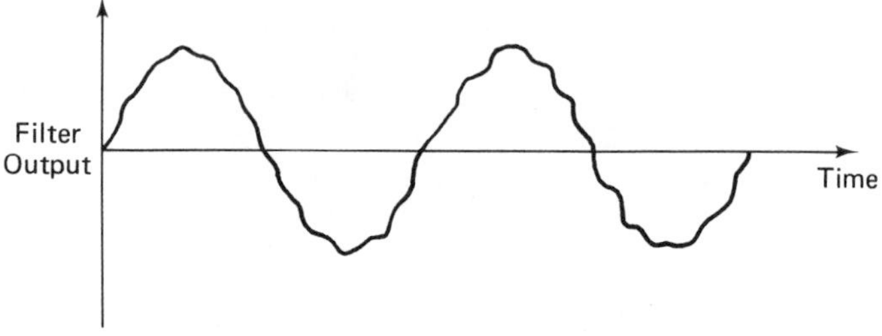

(b) The D/A Output with Filtering.

FIGURE 9-35

to smooth the discontinuities. This filter reduces or eliminates quantization noise that might otherwise be heard. The result is shown in Figure 9-35(b).

Practical factors and limitations: Digital recording is not without technological problems. Occasionally, "dropouts" will occur, at which time the digital data are lost due to magnetic tape imperfections. This fault of the tape is permanent and cannot be corrected. Most tape that will be used for digital recording must be fully tested for dropouts prior to use, thus increasing the tape price considerably. The digital data can be encoded prior to recording using a special error correcting technique. This will improve reliability and reduce tape cost.

One method used to protect the recording against small and infrequent errors (several faulty bits per second, out of thousands of bits) involves *encoding* the data in a special way that allows for the detection and correction of playback errors. This offers some protection against dropouts and tape deterioration. Dropouts do not affect analog recordings in such a severe way because the information density in an analog recording is not nearly as great.

The width of the recording path (ten tracks) is far greater than that which would be found in an analog recorder. This requires more complex electronics in the digital recorder, and wider, more expensive tape. For studio purposes, however, such additional expense may be tolerated for the great improvement in performance.

Signal enhancement and alteration: Once the analog signal is in digital form, many mathematical techniques can be applied to it that allow sophisticated filtering, tonal emphasis, and reduction of backround noise present in the original signal. For example, digital techniques are available that can eliminate audience noise (such as coughs) from a "live" recording. Such sophisticated signal manipulation is virtually impossible using purely analog techniques, and offers almost unlimited means by which the original signal can be improved or changed. Perhaps, a soloist's single mistake in a live performance can be judiciously corrected in a later playback and editing session!

In summary, digital recording can offer flawless reproduction of originally analog signals, but at a price. At the moment, the expense involved in such a system would interest only professional recording studios and scientific laboratories. In the near future, though, the cost should become sufficiently low to make digital audio recording widely available to many people.

SUMMARY

An analog signal is the representation of "change" as a proportional electrical variation. A digital signal, on the other hand, is the representation of "change" as a sequence of varying numbers. Digital signal processing applies numerical techniques to digital signals to modify the original signal so that a desired result is achieved. In order that analog signals may be processed digitally, analog-to-digital converters are necessary to bring about a translation from one signal type to the other. Similarly, for the processed digital signal to be represented as an analog variable, digital-to-analog converters are necessary. While the digital processing of analog signals is slower than processing the analog signals as such, many powerful digital techniques based in mathematics can be applied to yield results that would otherwise be extremely difficult or impossible to obtain.

In order that an analog signal be represented as a sequence of numbers, it must be quantized both in time and in amplitude. This involves the subdivision of the range over which the analog signal can vary into a number of equally spaced segments. Each segment is referred to as a sample, and its location on the time axis is the sample time. The duration of time between successive samples is generally constant, and is referred to as the sample period. Each sample also has an amplitude that represents the function's value at the sample time. This amplitude is quantized into one of a multitude of discrete levels. Thus, each sample is defined by two numbers (usually binary numbers): a sample time and an amplitude. The more finely a given range of time (or amplitude) is divided, the larger the number of sample points (or discrete levels) that are necessary, and the more binary digits that must be used to uniquely identify a particular sample point. If the resolution is not sufficiently fine, the original analog signal cannot be faithfully represented or reproduced.

Digital-to-analog converters rely on a resistive network to exponentially weight binary digits, each of which is represented as a voltage. A logic '1' voltage will always have the same value, yet its position in the binary number determines its weight, and consequently its effect on the value of the analog output. An R-2R ladder network is superior to the straight exponential resistor network because of its well-balanced effect on input voltages. Operational amplifiers are utilized to amplify the resistive network's low output voltage to a more useful level.

Analog-to-digital conversion can be accomplished in many ways. Open-loop systems, such as the flash converter, the time-window converter, and the slope converter, possess no internal feedback, and are generally simple in structure. Closed-loop systems, such as the tracking converter and successive approximation converter, utilize feedback, and therefore require an internal D/A in order that the digital output be converted into an analog feedback value for direct comparison with the analog input voltage. The successive approximation converter is the most effective A/D device when cost and speed are considered jointly. The flash converter, however, is the fastest of all A/D devices, yet it requires a considerable amount of power and is quite expensive.

NEW TERMS

Transducer
Analog Signal
Digital Signal
Digital Signal Processing
Analog-to-Digital Converter
Digital-to-Analog Converter
Quantization
Amplitude Quantization
Time Quantization
Nyquist Rate
Sampling
Resolution
Linearity
Precision
Accuracy
Operational Amplifier
Current Summation
Negative Feedback
R-2R Ladder Network
Superposition
Flash Converter
Differential Comparator
Time-Window Converter
Slope Converter
Tracking Converter
Successive Approximation Converter
Optoelectronic Sensor
Digital Magnetic Recording
Sample and Hold
Conversion Time
Dropout

PROBLEMS

9-1 **(a)** How many output *levels* are possible in a two-bit D/A converter?

(b) If this two-bit D/A spanned the range 0 to 3 V, what would be the width of each of the steps? (In other words, what is its resolution?)

9-2 What is the resolution of an eight-bit A/D that spans the range

(a) 0 to 10 V?

(b) -5 to $+5$ V?

9-3 What output voltage would be produced by a D/A whose output range is 0 to 10 V and whose input number is

(a) 1 0 (assume a two-bit D/A)?

(b) 0 1 1 0 (assume a four-bit D/A)?

(c) 1 0 1 1 1 1 0 0 (assume an eight-bit D/A)?

9-4 What binary number would be produced by an eight-bit A/D converter whose input range is 0 to 10 V and whose input voltage is

(a) 39.3 mV?

(b) 4.48 V?

(c) 7.81 V?

9-5 How many bits of precision are needed for an A/D that is to have a basic step of about 7.33 mV over a 15-V range?

9-6 The basic step of a nine-bit D/A converter is 10.3 mV. If '00000000' represents 0 V, what output is produced if the input is

(a) 1 0 1 1 0 1 1 1 1?

(b) 0 0 1 1 1 0 0 1 0?

9-7 A three-bit A/D converter is connected to a two-bit D/A converter in either one of two ways:

(a) With the MSB's in common, as shown in Figure P9-7(a).

(b) With the LSB's in common, as shown in Figure P9-7(b).

If the input range of the A/D and the output range of the D/A are both 0 to 7 V, sketch the D/A output for *each* case if the A/D input is as shown in Figure P9-7(c).

FIGURE P9-7

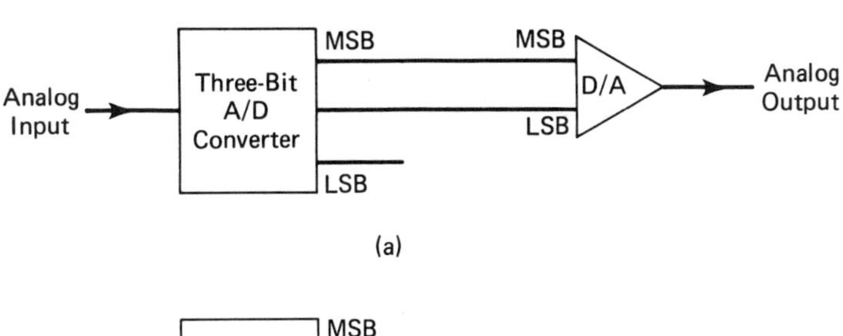

(a)

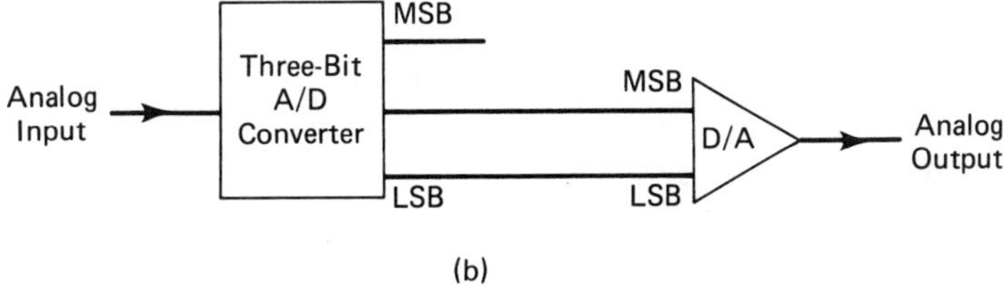

(b)

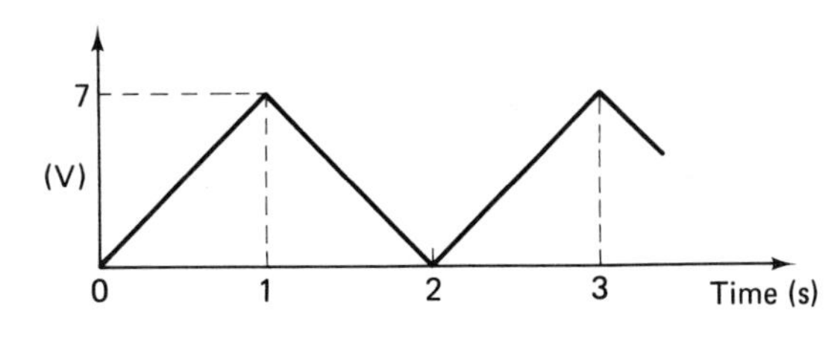

(c)

9-8 Is it possible to have a transducer that produces digital signals directly? If so, give an example. If not, explain why.

9-9 A five-bit D/A converter is available. Assuming that '00000' corresponds to an output of +10 V and that the D/A is connected for −0.1 V per increment, what output voltage will be produced for '11111'?

9-10 A nine-bit A/D converter is used to digitize an analog signal whose values range from −5 to +8 V.

(a) How many quantization levels are available with this A/D?

(b) What is its resolution in volts per increment?

(c) What binary number will be produced when the analog input is zero volts?

9-11 A student working in the digital laboratory connects the output of a four-bit natural binary up counter to a four-bit D/A. Unfortunately, this student accidently *reverses* the MSB and LSB, so that the order of all four bits are reversed. Refer to Figure P9-11. With the D/A converter having a 0- to 15-V range, write a table that gives the output voltage produced for each count value. Sketch the analog output produced as the counter progresses through its sequence.

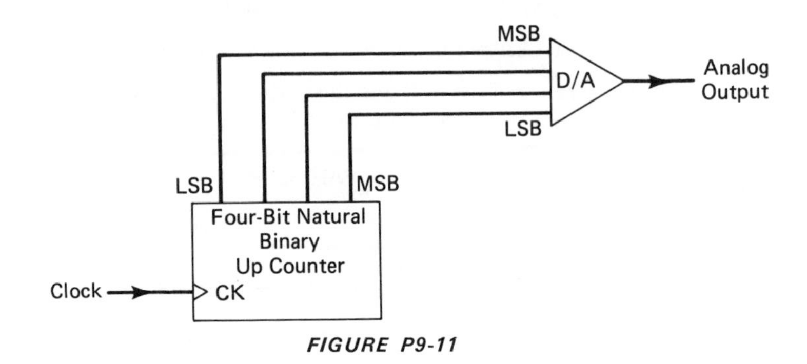

FIGURE P9-11

9-12 If a ten-bit D/A spans a range of 0 to 10 V and is always within 1 mV of its ideal output, what is its linearity as a percent of full-scale range?

9-13 Determine the minimum clock frequency for an eight-bit time-window A/D converter so that a 1000-Hz sawtooth wave that spans the full A/D input range can be sampled twenty times per second.

9-14 Why is the pulse synchronizer needed in the time-window A/D converter? Refer to Figure 9-24.

9-15 Design the control logic for a four-bit successive approximation A/D converter.

9-16 Using an eight-bit D/A converter, an *RC* low-pass filter, and any digital components you wish, design a simple circuit whose output approximates the waveform shown in Figure P9-16. Assume that the output swing of the D/A is 0 to 10 V and that an oscillator is available whose frequency you specify.

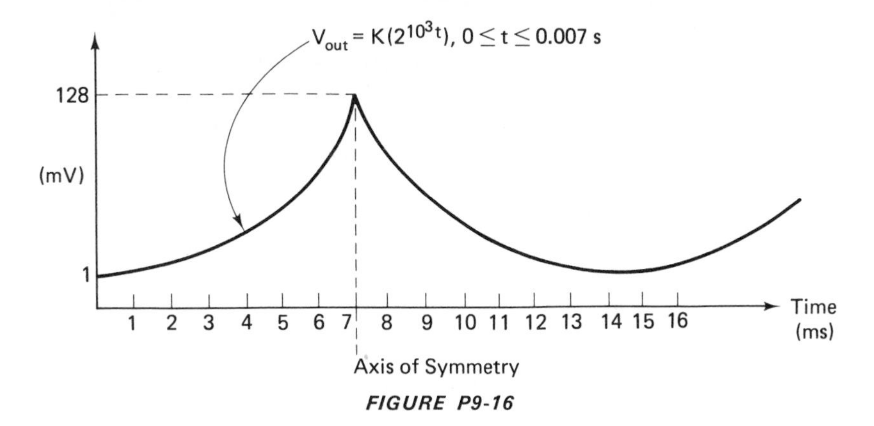

FIGURE P9-16

9-17 An audio tone is to be digitized and sent as a sequence of binary numbers over a digital communications channel. If this tone is a sine wave and varies between 1000 and 2000 Hz, at what minimum rate (the Nyquist rate) should the signal be sampled to ensure that it can be faithfully reproduced? What would be a more practical choice?

9-18 A tracking A/D converter is shown in Figure 9-26(b). Study this circuit carefully, and answer the following questions about it. The system clock frequency is 1 MHz, and the D/A resolution is 1 mV per increment.

(a) Assume that a '00000000' D/A input produces an analog output of zero volts. What voltage is produced for '11111111'? Thus, what is the maximum swing on the analog voltage output of the D/A?

(b) Why is the clock input of the 'D' flip-flop inverted with respect to the clock input of the counter?

(c) Sketch the analog output waveform of the D/A that would result if the analog input was the pulse shown in Figure P9-18(a).

(d) Sketch the analog output waveform that would be observed if the analog input were a constant 281.5 mV.

(e) Could the sawtooth waveform of Figure P9-18(b) be faithfully digitized by this tracking A/D?

(f) At what frequency of the sawtooth wave in Figure P9-18(b) would the tracking A/D no longer be able to follow?

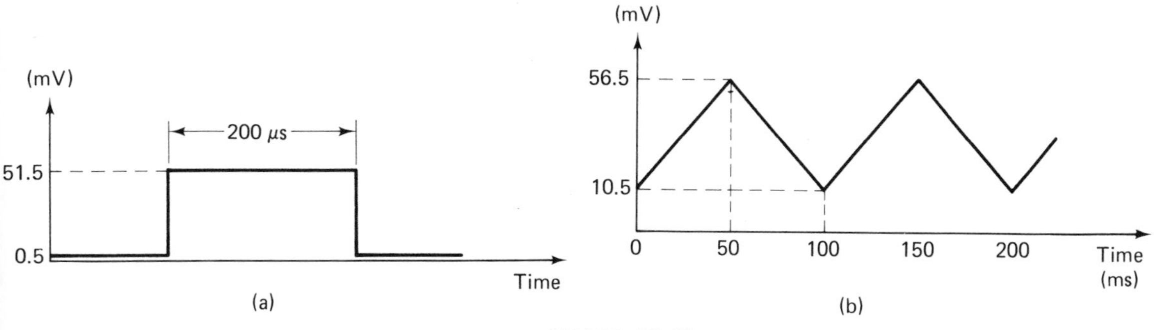

FIGURE P9-18

9-19 **(a)** A sine wave with a 5-V peak amplitude is the input to the tracking A/D converter of Problem 9-18. Assuming that the system clock frequency of the A/D circuit is 1 MHz, what is the highest-frequency sine wave that can be accurately tracked?

(b) Although no special circuitry for sampling is built into a tracking A/D converter, what is the effective sampling rate in such a circuit, assuming that the analog input is being accurately tracked?

9-20 Design a digital bicycle speedometer that reads from 0 to 30.0 miles/hour (three BCD digits) and is updated once per second. Assume that the bicycle wheel is 26 in. in diameter and that there are 5280 ft/mile. Specify any kind of transducer that you wish, and assume that it can be mechanically coupled to the wheel in some way.

9-21 A temperature sensor is the transducer used to measure the temperature of an automobile engine. Its resistance varies between 1000 and 5000 Ω and decreases in direct proportion to *increasing* temperature. Specifically, at 100°F, the resistance is 5000 Ω, and at 300°F the resistance is 1000 Ω. Design a circuit that produces a three-digit decimal readout of engine temperature. The output should be updated once every 5 s. (*Hint:* Consider using a slope converter.)

10 COMPUTERS AND MICROPROCESSORS

10.1 INTRODUCTION

10.1.1 Limitations of Fixed-Wired Digital Circuits

Digital circuit designers have available a collection of standard digital building blocks that can be assembled to form an infinite variety of devices. The complexity of these standard blocks ranges from individual logic gates, through counter modules, shift registers, and adders, to sophisticated functions requiring thousands of transistors. The common factor among all of these blocks is that they are integrated on a single chip of silicon and are available as single components. As advances in semiconductor technology occur, it is discovered how more and more transistors can be produced and connected on a single chip. Realizing that this trend is bound to continue, we must ask the questions

Why is this happening?
and
How will it affect the world of digital electronics?

Most of the progress that is measured in free-enterprise nations is fueled by the profit motive. The electronics industry is no exception. In answer to the "Why?" question, a simple explanation can be provided: it is more economical in the long run to produce large-scale integrated circuits containing thousands of preconnected transistors than it is to produce smaller blocks of individual components that must be connected externally. In other words, when we take into account the time needed to mount, solder, and test a circuit module with many small components, it proves to be much more expensive than integrating all the components and interconnections into a single device that can be manufactured as a whole, at one time, and by one basic process. Furthermore, many new desirable physical characteristics of large-scale integrated (LSI) circuits arise: very small size, low power consumption, and low heat dissipation. The economics of current semiconductor processes is one major limit of circuit density per single chip since the yield of working circuits per batch becomes greatly reduced as circuit density increases. Ultimately, physical limits will be reached that are insurmountable by technological advances.

The answer to the question "How will it affect digital electronics?" is much harder to answer because the effects are many and varied. First, while more and more transistors are integrated on a single chip, so also must be their connections; designing with randomly selected components will no longer be feasible. To take advantage of the economy of LSI circuitry, each electronic circuit producer must, it would seem, have their own LSI design and manufacturing facility. Such a facility is extremely expensive, however, and would not pay for itself unless the company produced thousands upon thousands of each type of circuit. With the uncertainties of the marketplace and the intense competition, very little guarantee is possible for the sale of a massive production of a very specialized digital circuit design. Second, the semiconductor industry is changing so rapidly that expensive LSI facilities that might otherwise be justified could become outdated in a matter of one year or less. So, we see a dilemma: while LSI techniques offer potentially enormous savings in manufacturing costs, the nature of LSI demands that the transistor interconnects (wiring) for a design be as fixed as the transistors themselves, thus allowing for only very specialized circuits. One solution would be to produce a general-purpose circuit whose function is *electrically variable*, and can be specified in such a way that a wide variety of digital circuit needs are satisfied. Such a device is possible, and has actually been available for over twenty years as a large, complex, multicomponent system. It is known as a ***digital computer***. The revolutionary advance has been its implementation as a single LSI circuit, and in this format it is referred to as a ***microcomputer***.

It is the aim of this chapter to present the fundamental concepts of digital computers, to discuss their embodiment as microcomputers, and to reveal the power of *programmed logic*. We cannot hope to obtain a complete working knowledge of computers in the small space of one chapter. Yet, we can convey the primary meaning and method of programmed devices, and make this a springboard for further study in the fascinating field of digital computers.

10.1.2 Stored Program Concepts

"How can a single, unchanging circuit be made to perform an almost infinite variety of different functions?" This ability is possible because most digital electronic functions can be divided into a sequence of small, very simple steps, in much the same way that the preparation of a specific food is defined step by step in a recipe. If a digital device is designed with a carefully chosen "internal dictionary" of basic logical operations, *any* desired function can be produced by specifying a particular sequence of the basic operations. In the same manner, we can form a written sentence of any meaning by selecting a proper sequence of letters and spaces. The "internal dictionary" of a digital computer is referred to as its ***instruction set***, and defines a catalog of simple logical operations that can be called upon to undertake specific actions. The sequence of instructions that is selected to accomplish a certain task is known as a ***program***. Thus, a computer effectively follows a recipe (its program) to obtain desired results, whereas a nonprogrammed digital device has been wired to produce one specific result, and nothing more.

"With this powerful advantage of programming flexibility, why have digital computers not overwhelmed the fixed-wired digital circuits from the beginning?" Until recently, there were two basic advantages that fixed-wired digital circuits had over programmed devices: higher speed and lower cost. With the advent of microcomputers, however, the cost barrier has been broken, and in many applications, microcomputer-controlled functions are sufficiently fast to satisfy most needs. Fixed-wired devices still hold the speed record and are expected to do so for some time to come. The enormous flexibility offered by digital computers paves the way for electrical machines that are self-adaptable to their environment, and, in a sense, "intelligent."

A computer's program is a set of basic commands that is coded as a sequence of multibit binary numbers. This sequence of binary numbers is stored in an electronic memory. Simply by altering the stored sequence of numbers, the computer's program can be changed. We will begin the central material of this chapter by studying electronic memories. Following this, a *programmed sequence generator* is discussed to illustrate by example the difference between fixed-wired devices and programmed devices. Then, we will design a simple digital computer to demonstrate the working principles of programmed logic. Finally, microcomputers will be examined and their application as programmable block elements will be illustrated.

10.2 ELECTRONIC MEMORIES

10.2.1 Definition and Classification of Electronic Storage Devices

All programmable devices rely on a sequence of binary-coded instructions to define the course of action that they take. Once the desired sequence of instructions has been determined by the human operator, these instructions must be placed in a storage medium that can be freely accessed by the device. Such a storage medium is referred to as an ***electronic memory***.

We can classify electronic memories into two general catagories:

1. ***Random access memory*** (RAM).
2. ***Serial access memory*** (SAM).

Most modern RAM devices consist of transistor arrays, and are characterized by the fact that the time needed to retrieve data from any location within the memory is approximately *equal*. Thus, we would say that the ***access time*** for RAM devices is independent of the data's location within the device. On the other hand, SAM devices frequently consist of a movable physical medium, such as magnetic tape or punched paper tape, and require that data be read from the medium only in the sequence in which they were recorded. Consequently, the time needed to gain access to a particular memory location in a SAM device depends upon how much intervening data must be read before the desired memory location is reached. For example, in a spool of tape, the access time to data that is wound deeply in the spool will be much longer than that for data that is located closer to the outer rim.

Figure 10-1(a) illustrates a random access memory in block form, with the access time defined in Figure 10-1(b). The data recorded in the memory can be illustrated by a truth table as shown in Figure 10-1(c). We see that there may be any number of address inputs, denoted by 'N' in Figure 10-1(a), and any number of data outputs, denoted by 'M'. With the address inputs considered as a whole to be a binary number, we also see that 2^{N+1} address combinations are possible. With each address retaining an $(M + 1)$-bit data word, the memory has exactly $2^{N+1} \times (M + 1)$ flip-flop cells.

Note in Figure 10-1(a) that an input labeled 'R/W' is present. This is the ***read/write*** input, and determines whether data is to be read from or deposited into the memory. Although some memory devices have separate connections for input data and output data, most semiconductor RAM's are constructed so that a single set of wires can serve as both data inputs and data outputs, the sense of flow being determined by the 'R/W' input. This is the case in Figure 10-1(a), where the data lines are shown to be *bidirectional.* The electrical nature of bidirectional logic lines is discussed in Section 10.3.

Serial access memories are, as a group, more complex than are random access memories. This is so because of the electromechanical construction that is often required to handle the storage medium. In a magnetic tape system, for example, very great care must be taken to avoid damaging or breaking the tape, entailing a complex mechanical design. Although the access time and complexity of SAM's are much greater than that of RAM's, the storage capacity of a SAM device can be hundreds or thousands of times as great. In addition, the per-bit cost is substantially lower. Furthermore, SAM devices using a magnetic medium (such as magnetic tape) are usually ***nonvolatile***. That is, data is retained without the constant application of power. Semiconductor read/write RAM's however, almost always require a constant flow of current to maintain the individual flip-flop cells in a memory state. Whenever power is removed from ***volatile*** RAM, data is lost. Nonvolatile RAM is possible, but is based on magnetic core storage technology, which is an older but advanced technology that is rapidly being displaced by semiconductor devices.

Serial access devices find their greatest use in ***bulk-storage*** applications. This type of use involves the storage of large amounts of numerical data or instruction codes in an economical form, where random access is not needed. This nonvolatility characteristic is particularly valuable during equipment power-down conditions when we wish to maintain data for future use. Semiconductor-based RAM devices, on the other hand, are especially useful when high-speed nonserial access is required, as would be the case in the main memory of a computer.

Bubble memories and charge-coupled device (CCD) memories are nonmechanical serial access circuits that are becoming more widely used, particularly in small computer systems. In these devices, data is stored in recirculating shift registers and typically has an access time of several hundred microseconds (depending on the size of the loop). Bubble memories utilize small, movable magnetic domains as the storage medium and are nonvolatile. Charge-coupled device memories, however, use electrical charge as the medium and are volatile.

Our primary concern regarding memories in this chapter will be with random access devices, because of their applicability as main memory in computers. Consequently, in all of the following definitions and discussion, we assume RAM's to be the subject.

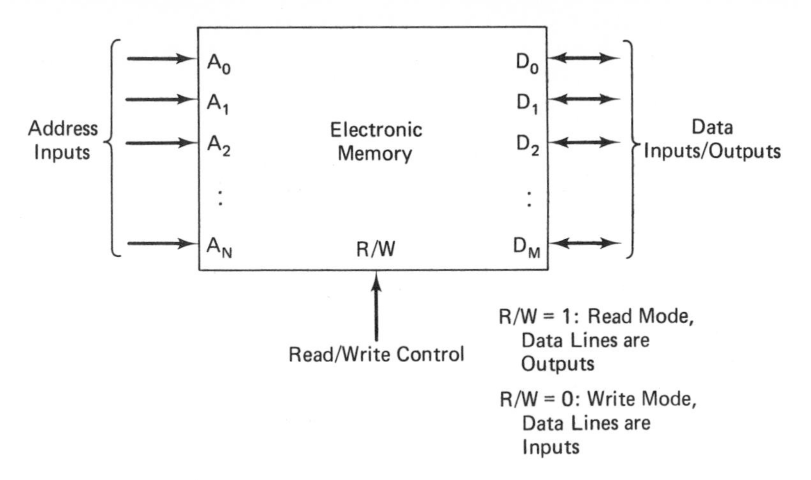

(a) Block Form of an Electronic Memory.

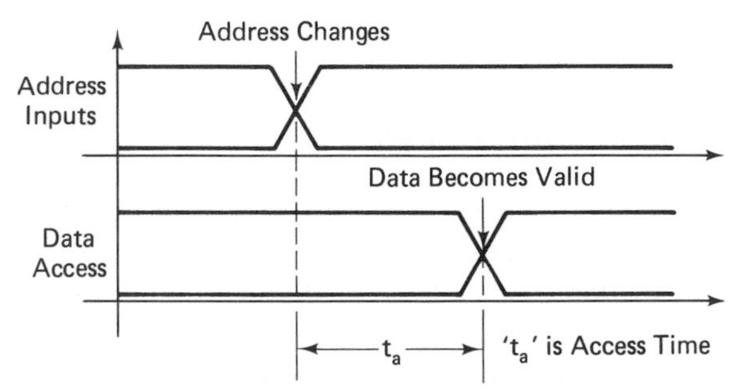

(b) The "Access Time" (t_a) is Shown to be the Time Needed Between the Application of a New Address and the Availability of Valid Data. For RAM Devices, 't_a' is About Equal for Any Address. In SAM Devices, 't_a' Depends on Where the Data is Stored in the Medium.

A_2	A_1	A_0	D_1	D_0
0	0	0	0	1
0	0	1	1	1
0	1	0	0	0
0	1	1	0	0
1	0	0	1	0
1	0	1	1	1
1	1	0	1	0
1	1	1	1	0

(c) The Contents of a Particular Memory (Eight Locations of Two Bits Each) are Shown by this Truth Table. The Data can be Changed at Any Time.

FIGURE 10-1

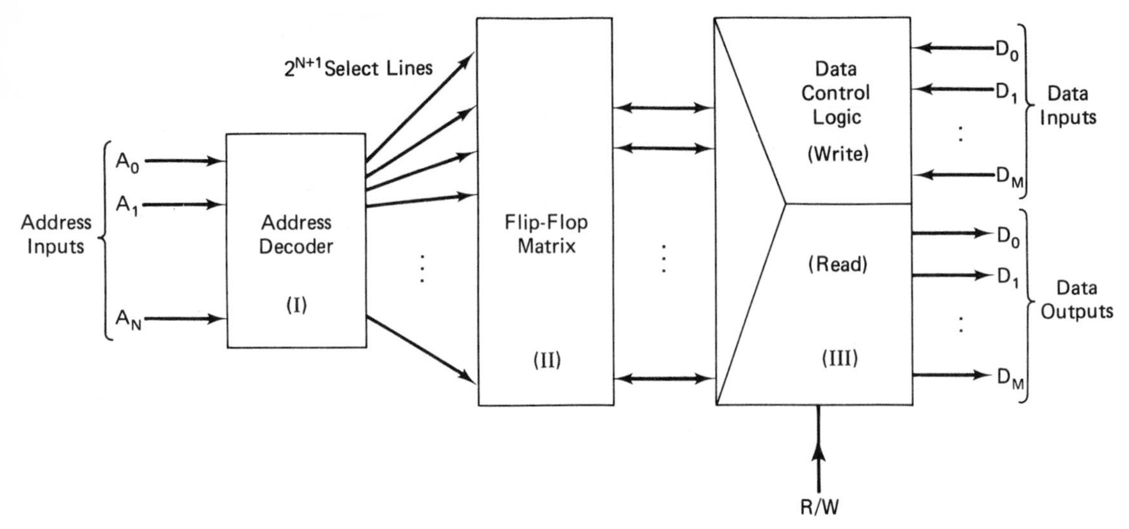

FIGURE 10-2 Functional blocks of a random access memory.

10.2.2 Simple RAM Design

It is our purpose in this section to illustrate the important functional issues of RAM circuits with a very simple example. In particular, we will design a small RAM using standard logic gates. Figure 10-2 depicts the three basic block functions of a RAM circuit, labeled I, II, and III. In Figure 10-2, the data outputs and inputs are assumed to be on separate, unshared wires. The ***address decoder***, (I), causes one of its 2^{N+1} outputs to be asserted in direct correspondence to a particular binary address input. This single asserted output from (I) enables a row of 'M' flip-flop cells (II), each cell of which is the receptacle for one bit of data. For a read operation, the 'M' outputs from this single row of enabled cells are transmitted to the Data Control Logic, (III), where they are driven onto the 'M' data output lines. For a write operation, 'M' data inputs enter at (III), and are passed to the enabled row of 'M' flip-flop cells. These cells are then appropriately set or reset to match the incoming data. The condition of the 'R/W' input determines whether a read or write operation will occur.

As a specific example, a four address, one-bit flip-flop memory is illustrated in Figure 10-3. Note that the basic flip-flop cell is shown in Figure 10-3(a), and is drawn only as a block element in Figure 10-3(b). Also note that the key gates in Figure 10-3 are labeled as I, II, III, to correspond with the functional blocks of Figure 10-2. This circuit is referred to as a *four-word by one-bit* random access memory. Other configurations are possible also, such as a 256-word by 8-bit RAM. In that case though, 256×8 total memory cells are needed.

In practical semiconductor RAM circuits, special electrical designs permit very simple and efficient flip-flop cells. In fact, it would be incorrect to represent actual RAM's using logic schematics since they are usually designed on the transistor level. The block diagram of a typical semiconductor RAM is illustrated in Figure 10-4. In this case, the circuit is 4096 words by one bit.

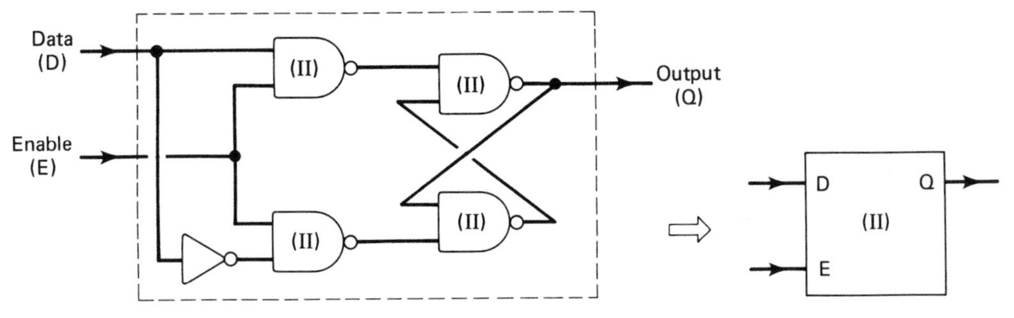

(a) The Construction of a Single Flip-Flop Memory Cell, and its Schematic Equivalent.

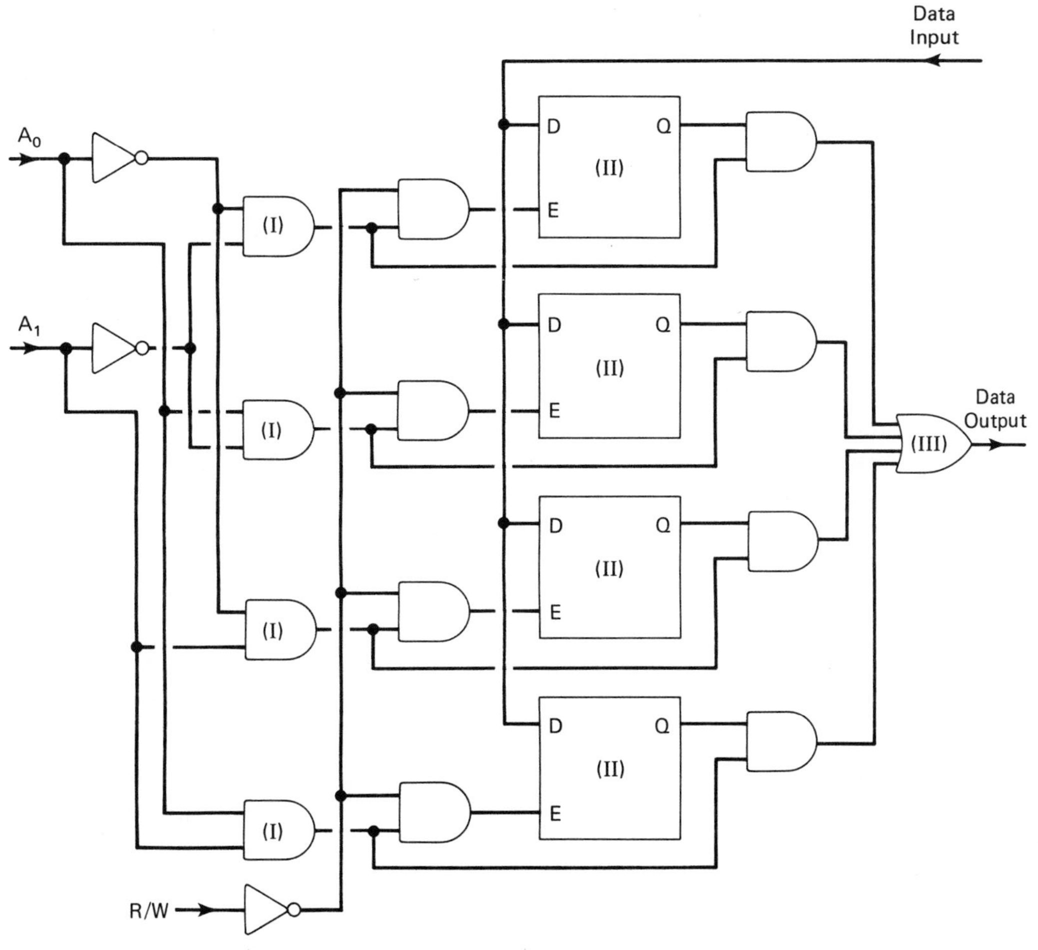

(b) The Logic Schematic for a Four-Address by One-Bit Flip-Flop Memory. The Components Labelled I, II, III, Corresponds with the Blocks of Figure 10-2.

FIGURE 10-3

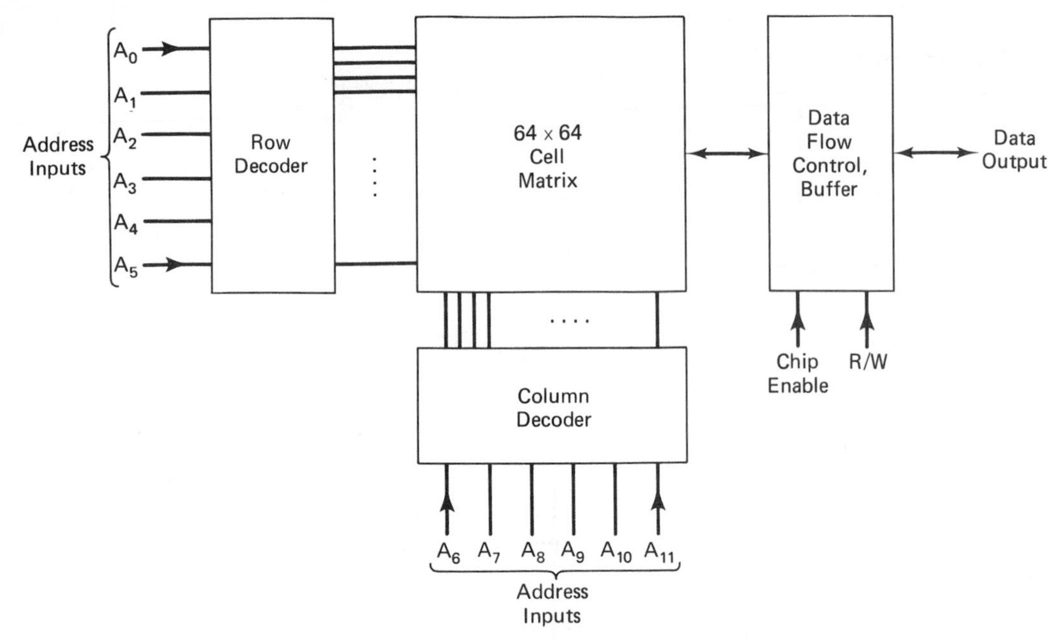

FIGURE 10-4 The basic block diagram of an integrated circuit RAM. In this case, the capacity is 4096 by one bit.

10.2.3 RAM Timing Characteristics

Studying a random access memory data sheet reveals many detailed timing characteristics that must be considered to ensure proper operation. In this section, some of the more important properties are discussed, hopefully to make the data sheets more comprehensible.

There are two basic timing problems that concern us when dealing with RAM circuits:

1. *Read cycle* timing.
2. *Write cycle* timing.

Read cycle timing is the least complex, and that is where we begin.

The ***cycle time*** of a RAM is a measure of the minimum duration between successive read or write operations. Often, the cycle time is longer for write operations, owing to the fact that memory data is not merely read, but altered. Involving a read cycle, two basic time values are important:

1. The *access time*—how much time must elapse between the application of address inputs and the presence of valid data on the data outputs.
2. The ***read cycle time***—how much time must elapse between successive read cycles.

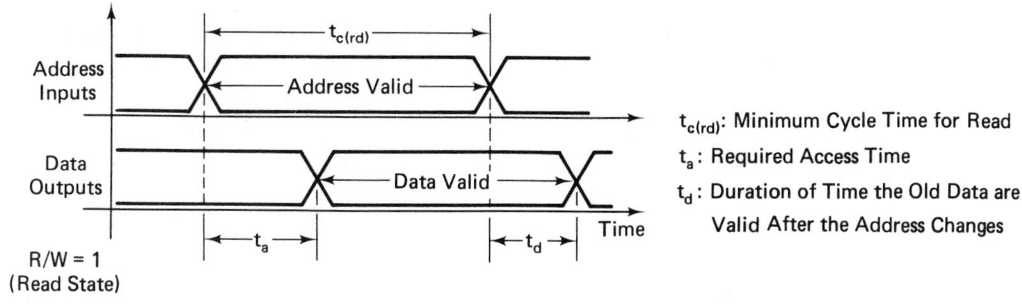

FIGURE 10-5 Read-cycle timing diagram and terms.

This information is summarized, along with additional timing parameters, in Figure 10-5. The data sheet value for '$t_{c(rd)}$' is a minimum value; values greater than this are certainly permissible. '$t_{c(rd)}$' is related to the maximum frequency at which the memory can be read. Note that when multiple data or address lines are referred to in single waveforms, unchanging signal values are denoted by simultaneous '1' and '0' levels, while a changing signal is shown by both levels changing state (crossed lines).

Several additional points must be noted in Figure 10-5. First, different manufacturers will use different symbols to represent the same parameter. Thus, a table of definitions for a particular circuit must always be referred to before timing diagrams can be understood. Second, a manufacturer will often supply additional control inputs, such as "chip-select," to permit the enabling of a single RAM chip in a large memory bank. The chip-select input, for example, would then appear in the timing diagram.

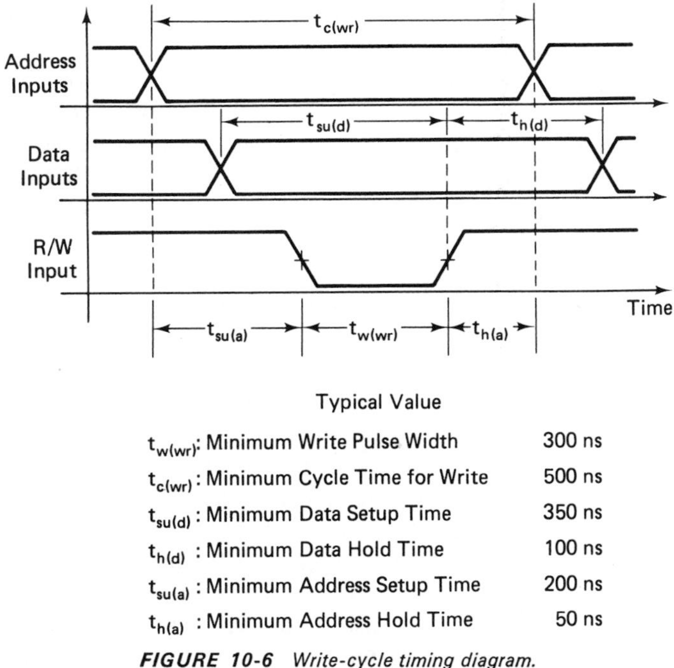

FIGURE 10-6 Write-cycle timing diagram.

Write-cycle timing is somewhat more complex than that for a read cycle only because an additional input, 'R/W', is involved. Normally, 'R/W' is in the '1' state, enabling read operations. When the address lines are stable with a new address, and the data lines steady with valid data, the 'R/W' input is pulsed low for a specified time to enable the writing of data into the memory. It is important that both address *and* data are stable *before* 'R/W' is asserted low, to ensure that the correct memory location is loaded with the correct data. Figure 10-6 summarizes write-cycle timing. The most important feature to remember from the write-cycle timing diagram is that the write pulse must occur *within* the period when both address and data are stable. Generally, computer systems are designed so conservatively that these minimum memory times are only approached, but never reached. Thus, system design will not be as complex as it may seem at this time.

10.2.4 Read-Only Memories

Most semiconductor read/write RAM's consist of an array of flip-flop memory cells which require a constant power source to maintain data. Should power be removed, voluntarily or not, any stored data will be lost. Recall that such memory circuits are referred to as being volatile. In many applications, it is desirable to have random access memories that are nonvolatile, and thus will retain data in a power-down condition. If nonvolatile read/write RAM is desired, battery backup must be provided for semiconductor circuits, or older magnetic core RAM's must be used (magnetic cores do not require power to maintain data). Another alternative is to use ***read-only memory*** (ROM), which is a form of random access memory that can be encoded permanently with data, and is unchangeable after manufacture. While ROM is nonvolatile, it also cannot be written with new data, and therefore its applications are more limited than those for read/write RAM. A typical use for ROM is in a computer system whose function is dedicated to one basic task; the system will probably contain a substantial amount of ROM with instruction codes permanently recorded.

The earliest technique for ROM operation was based on the ***fusible link*** principle. In this case, a matrix of very fine wires (fusible links) is utilized as the storage medium. Data is recorded by selectively burning open the wires for logic '1's, and leaving wires intact for logic '0's. Once a fusible link has been burned open, it cannot be closed and is, therefore, permanent. More sophisticated ROM techniques utilize a special mask applied to a microcircuit transistor matrix at the time of manufacture. The mask represents a data pattern that is to be recorded.

Electrically programmable read-only memories (EPROM's) are very sophisticated devices in which the recorded data is nonvolatile, but alterable by using a special technique. One very popular approach involves the recording of data in insulated gate MOS transistors. By applying a high-voltage pulse to the insulated gate of such a transistor, charge can be transferred to the gate (because of controlled insulation breakdown at high voltage), and trapped (due to the insulated gate). Thus, if a transistor has a charged gate, it will appear "on"; otherwise it will be "off." The charge dissipates in a natural way at a very slow rate (tens of years). However, when the semiconductor matrix of transistors comprising the EPROM is irradiated with ultraviolet light, the material is partially ionized, and the stored charge can escape, thereby erasing the data. EPROM chip packages are provided with a quartz window so that erasure is possible. Electrically erasible versions exist also, but are less common.

10.2.5 Static and Dynamic Memory Circuits

Semiconductor RAM circuits are available in two basic forms, classified as ***static*** and ***dynamic***. Our four-word by one-bit design of Figure 10-3 is an example of a ***static memory***. In this case, power is applied constantly to all flip-flop elements to retain stored data. *Dynamic memory* circuits rely on semiconductor capacitors for bit storage, and not on flip-flop arrays. Microscopic capacitors represent data by being charged for logic '1', and discharged for logic '0'. With such a system, power need not be constantly applied to a capacitor for it to retain its charge (and hence its data). Since charge eventually dissipates in any capacitor, a dynamic memory requires periodic ***refreshing*** to counteract the slow discharge. The chief advantages of dynamic memories over static ones are:

1. Smaller basic storage cell size (less area per bit on the silicon chip), resulting in much greater storage capacity per device.
2. Lower power requirement per bit.
3. Lower heat dissipation per bit.
4. Lower cost per bit.

Offsetting these advantages is the requirement for additional circuitry to enable refreshing. In many cases, the refresh circuitry is incorporated on the memory chip itself or on interrelated LSI chips. Nevertheless, the added complexity makes reliability, system debugging, and system timing more difficult.

10.3 THREE-STATE DEVICES

In Section 10.2.2, it was stated that the same wires could be used to transmit data to a RAM during a write operation, and to receive data from a RAM during a read operation. We will now investigate how this is possible from an electrical point of view.

In any scheme to transmit digital information from one point to another over a set of wires, we generally designate one end of the wires as the *transmitter* and the other end as the *receiver*. The transmitter end is connected to logic gate *outputs* that drive current into (or out of) the wires so that valid logic level voltages are established. The receiving end is connected to one or more logic gate inputs, but no more than the fanout of the driver gates permits. Receiver gates are passive since they do not attempt to force the line into a particular state. Thus, receiver gates can be located at any point along the line, as well as at the end. A parallel array of wires used to transmit a multidigit binary word from one point to another is frequently referred to as a ***bus***. A unidirectional bus has a distinct driving end, and is illustrated in Figure 10-7.

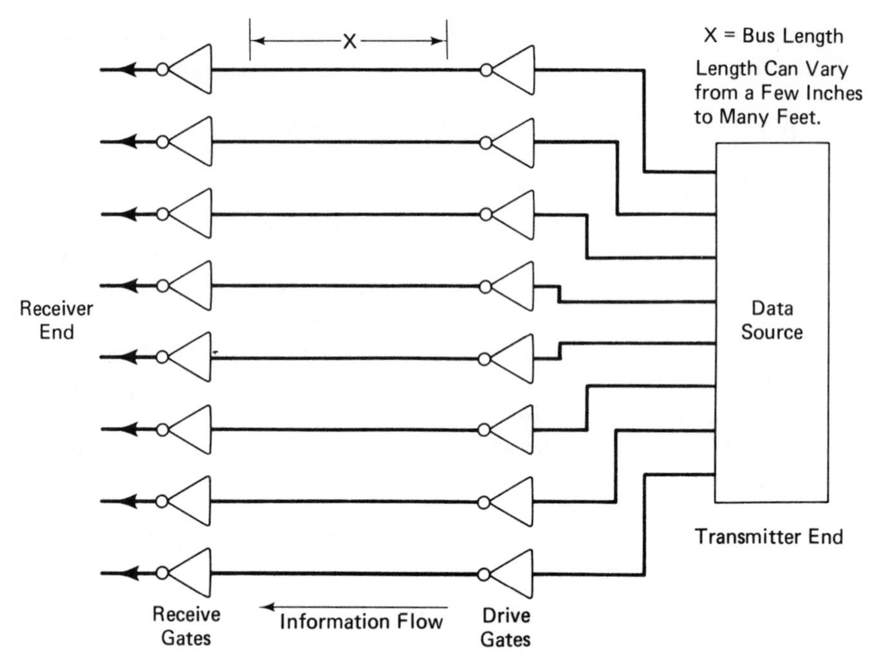

FIGURE 10-7 Unidirectional data bus.

In order to transmit information in the opposite direction on a bus, it is necessary to reverse the location of the transmitter. Yet, unless we physically remove the driver gates from one end and install them at the other end, there appears to be a conflict: How can driver gates be placed at two or more points on a bus without mutual interference? Recall from Chapter 8 that if two normal gate outputs are connected to the same point, an electrical conflict arises when the outputs attempt to go to opposite logic states. For TTL circuits, the logic '0' state dominates. Now, if we wish to successfully drive a bus from multiple points, we must be able to *electrically deenergize* all driver gates on a line *except* for the one that is to control the bus. Special gate circuits are built with this deenergizing feature, and are referred to as ***three-state drivers***. Figure 10-8 illustrates a three-state driver and its schematic symbol. With the control input, '*C*', asserted (= '1'), the driver behaves normally, which in the case of Figure 10-8 means that it acts as an inverter. However, with the control input deasserted, the output is turned off, and neither sources or sinks current from the line to which it is connected. Thus, with the control input deasserted, the driver gate electrically "disappears" from the system.

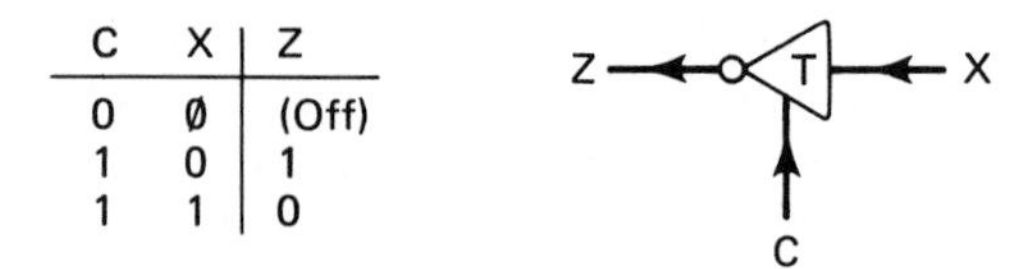

C	X	Z
0	Ø	(Off)
1	0	1
1	1	0

FIGURE 10-8 Schematic symbol and truth table for a three-state driver gate. In this case, it functions as an inverter.

A bus system equipped with three-state drivers and capable of transmitting information in two directions is referred to as a ***bidirectional bus***. Although a bidirectional bus can transmit information in two directions, it must transmit in only one direction at a time. A ***unidirectional*** bus transmits in one direction all the time. Figure 10-9 illustrates the connections required for a four-bit bidirectional bus.

The bidirectional bus concept is an extremely important one that is used on a transistor level in designing memory circuits, and also on a systems level in connecting computer components. While we will not be designing transistor microcircuits, we will be involved with three-state devices in specifying digital computer hardware configurations.

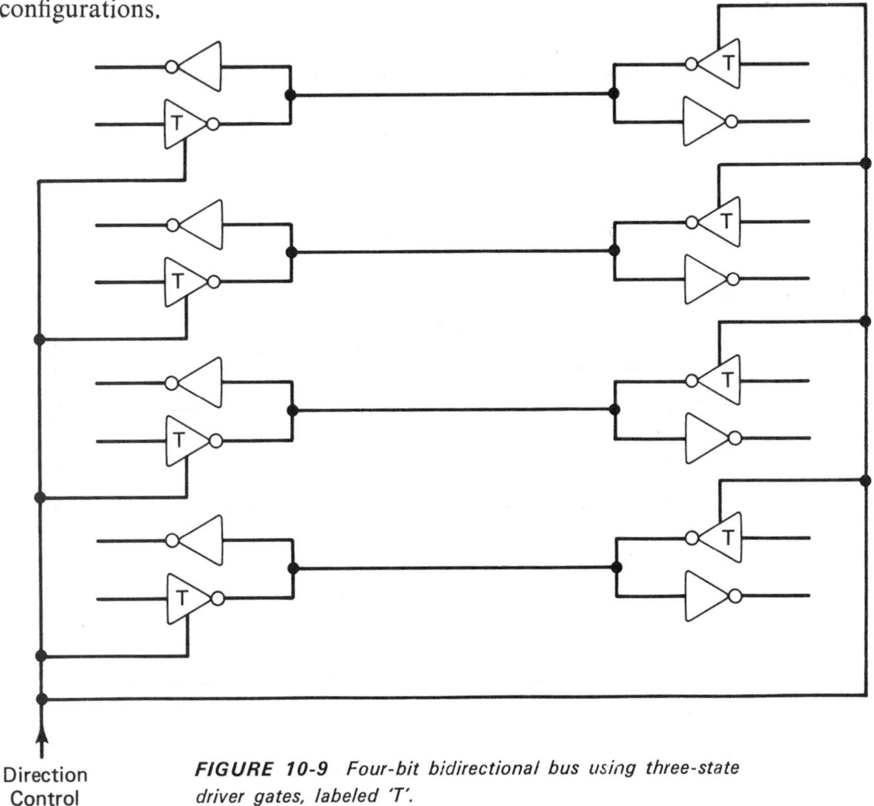

FIGURE 10-9 Four-bit bidirectional bus using three-state driver gates, labeled 'T'.

10.4 PROGRAMMED SEQUENCE GENERATOR

10.4.1 The Basic Concept and an Example

Figure 10-10(a) illustrates the circuit diagram for a three-bit natural binary up counter, a circuit that is familiar to us all. This circuit is redrawn in Figure 10-10(b) to conform with the general model of a sequential circuit. Now, we must ask:

Can the operation of a binary counter, such as this, be obtained by replacing the combinational logic block in Figure 10-10(b) with a random access memory?

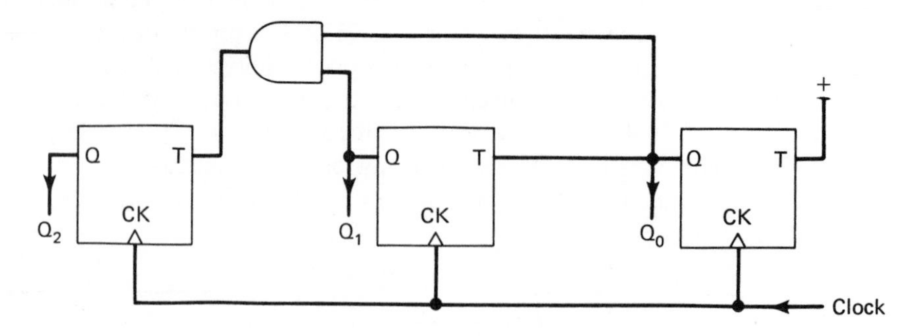

(a) Natural Binary Up Counter.

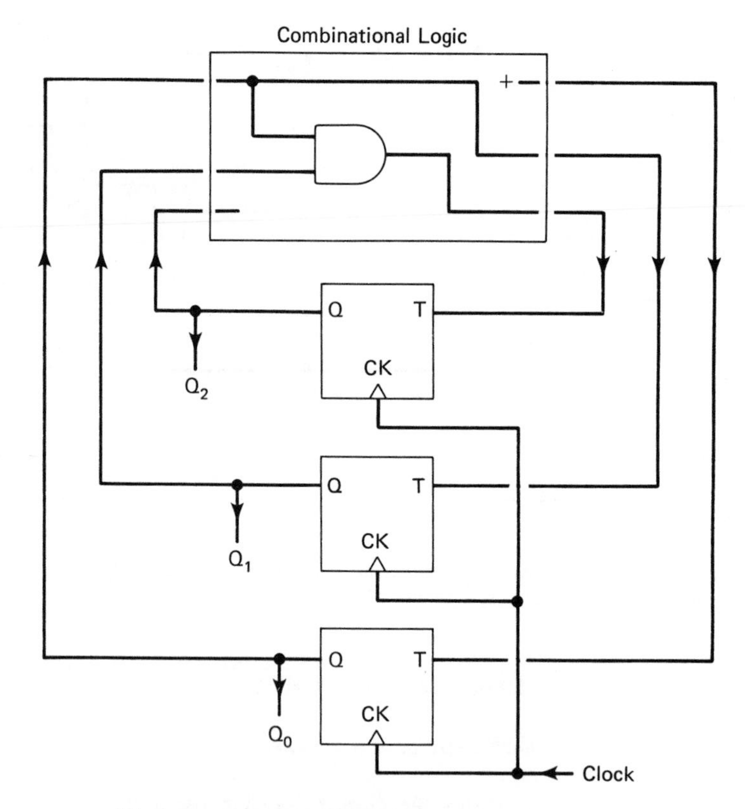

(b) The Counter of (a) Redrawn to Conform with the General Model of a Sequential Circuit.

FIGURE 10-10

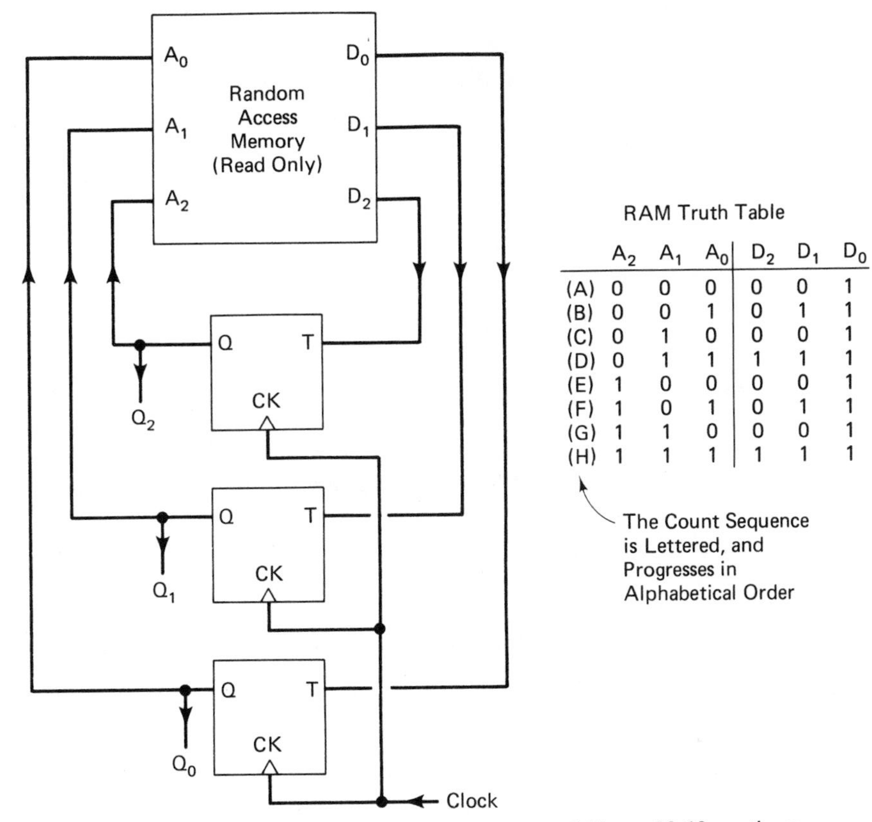

	A_2	A_1	A_0	D_2	D_1	D_0
(A)	0	0	0	0	0	1
(B)	0	0	1	0	1	1
(C)	0	1	0	0	0	1
(D)	0	1	1	1	1	1
(E)	1	0	0	0	0	1
(F)	1	0	1	0	1	1
(G)	1	1	0	0	0	1
(H)	1	1	1	1	1	1

FIGURE 10-11 *The natural binary upcounter of Figure 10-10 continues to function normally when its combinational logic is replaced with a random access memory, containing the data that are shown.*

In short, the answer is "yes," but only if the RAM is preloaded with the proper binary data. It is easy to understand why this is possible.

The function of any combinational logic circuit can be explicitly stated by a truth table. Of course, this also applies to the combinational logic block in the counter circuit of Figure 10-10(b). Now, recall that we can also express the contents of a random access memory using a truth table. We need only to list all possible address inputs, in natural binary order, and observe the data outputs that are produced for each address entry. It would seem, then, that we should be able to *replace* the combinational logic of Figure 10-10(b), on a one-to-one basis, with a RAM whose truth table is *identical.* This is, in fact, possible, and would result in an equivalent circuit, as illustrated in Figure 10-11. One difference that should be noted is the maximum possible speed of operation. While the propagation delay of one level of standard logic gates might be about 10 ns, the access time of a typical RAM is between 200 and 500 ns. Thus, the maximum frequency of operation for the RAM counter is considerably lower. We can compute these values exactly, as follows:

1. Fixed-wired counter

$$20 \text{ ns (flip-flop propagation delay)} + 10 \text{ ns (flip-flop setup time)} + 10 \text{ ns (max gate propagation delay)} = 40 \text{ ns}$$

$$f_{\max} = \frac{1}{40 \text{ ns}} = 25 \text{ MHz}$$

2. RAM counter

$$20 \text{ ns (flip-flop propagation delay)} + 10 \text{ ns (flip-flop setup time)} + 500 \text{ ns (RAM access time)} = 530 \text{ ns}$$

$$f_{\max} = \frac{1}{530 \text{ ns}} = 1.89 \text{ MHz}$$

A circuit very similar to the RAM counter is shown in Figure 10-12(a). It differs in two ways: (a) D-type flip-flops are used, and (b) the RAM contains different data. This circuit also functions as a counter, but is referred to as a ***programmed sequence generator*** since the sequence that it produces is not a natural binary one and depends upon the data contained in the RAM. The count sequence is easily revealed by considering the 'Q' outputs of the flip-flops to represent the "current state," and the RAM data outputs to represent the "next state." A state transition will not occur until a clock pulse is received. By studying the RAM's truth table, we can determine and list the pattern of state transitions. Remember that the circuit's "next-state" data become the "current state" output after a clock pulse is applied.

We must note two important features of the programmed sequence generator. First, the count sequence need not be natural binary, necessarily. It can be any combination of the eight possible states that is desired. Of course, if a larger RAM and more flip-flops were available, the count sequence could be made longer. Second, any random pattern of numbers recorded in the RAM does not guarantee that a periodic count sequence will be obtained. For example, if one RAM entry happened to equal the address value itself, a trap state would be entered, and no further changes on the outputs would be observed.

10.4.2. Fixed-Wired Counter versus the Programmed Sequence Generator

Carefully compare the fixed-wired counter of Figure 10-10 and the programmed sequence generator of Figure 10-12, and look for some basic differences. In particular, ask yourself

What must be done to change the count sequence in each case?

For the fixed-wired counter, the combinational logic block must be changed to alter the count sequence. This probably involves disconnecting existing gates and wires and replacing them with different ones, certainly a time-consuming process. For the programmed sequence generator, however, we need only to change the *bit pattern* stored in RAM! No hardware or wiring must be replaced. In addition to avoiding the

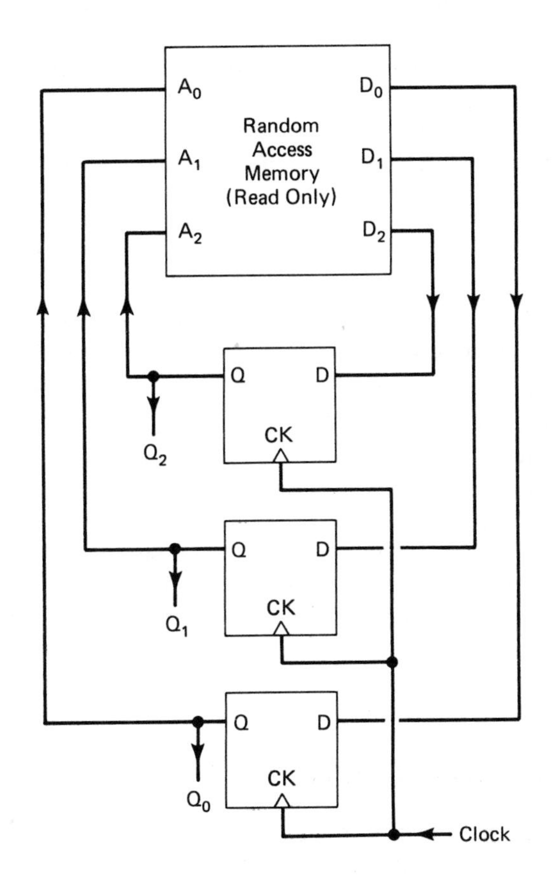

RAM Truth Table

	A_2	A_1	A_0	D_2	D_1	D_0
(A)	0	0	0	0	1	1
(H)	0	0	1	0	0	0
(F)	0	1	0	1	1	0
(B)	0	1	1	1	0	1
(D)	1	0	0	1	1	1
(C)	1	0	1	1	0	0
(G)	1	1	0	0	0	1
(E)	1	1	1	0	1	0

The Count Sequence is Lettered, and Progresses in Alphabetical Order

(a) The Circuit Schematic of a Programmed Sequence Generator.

Current State			Next State		
Q_2	Q_1	Q_0	D_2	D_1	D_0
0	0	0	0	1	1
0	1	1	1	0	1
1	0	1	1	0	0
1	0	0	1	1	1
1	1	1	0	1	0
0	1	0	1	1	0
1	1	0	0	0	1
0	0	1	0	0	0
(Repeats)					

(b) The Count Sequence is Revealed by Studying the Truth Table of (a), and Realizing that "Next-State" Data Become the "Current State" Output After a Clock Pulse is Applied.

FIGURE 10-12

expense of new components, the new RAM data can be entered in a matter of microseconds or less. Thus, the programmed sequence generator is much more flexible than the hard-wired one. Its only disadvantage is the operating speed, which will always be considerably less than the fixed-wired implementation.

In a nutshell, the relationship between the programmed sequence generator and the fixed-wired counter exactly illustrates the difference between a *computer* and *fixed-wired digital hardware*. Specifically:

1. A *computer* is a device whose function is determined by a stored pattern of '1's and '0's in a random access memory. The computer's function can be dramatically changed by altering the stored pattern of bits, and is extremely flexible. However, its speed is limited by the access time of the memory circuits.
2. A *fixed-wired digital circuit* performs one specific function only. Its operation can be substantially changed *only* by redesigning the circuit. It can, however, operate at much higher speeds than an equivalent programmed computer since it does not rely on complex random access memory circuits.

Remember that read/write RAM's are inherently slower than pure combinational logic, given a common semiconductor technology. Thus, improvements in transistor performance will never allow RAM-based designs to overcome the speed advantage of fixed-wired digital hardware.

10.5 DIGITAL COMPUTERS

10.5.1 General Block Diagram and Basic Operation

A *digital computer* is a programmable device capable of operating on numerical data with basic arithmetic and logic processes. Figure 10-13 shows the three functional blocks that comprise a digital computer. It is our objective in this section to investigate the operation of these blocks, and to gain an understanding of how the digital computer can be applied to solve practical problems. To begin, we must ask

How can the simple structure of Figure 10-13 possibly be useful?

FIGURE 10-13 The three basic blocks of a digital computer, and their purposes.

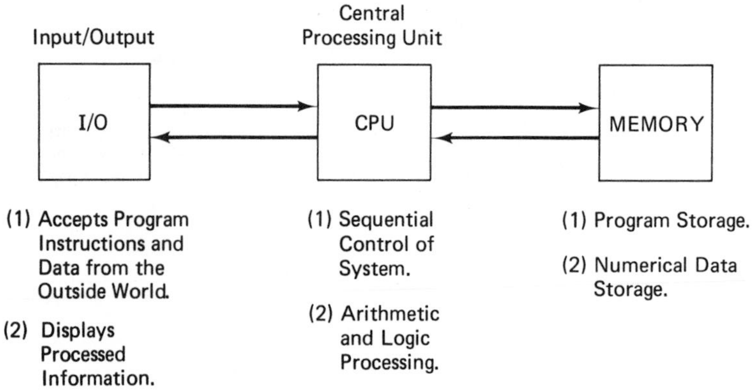

Input/Output
I/O
(1) Accepts Program Instructions and Data from the Outside World.
(2) Displays Processed Information.

Central Processing Unit
CPU
(1) Sequential Control of System.
(2) Arithmetic and Logic Processing.

MEMORY
(1) Program Storage.
(2) Numerical Data Storage.

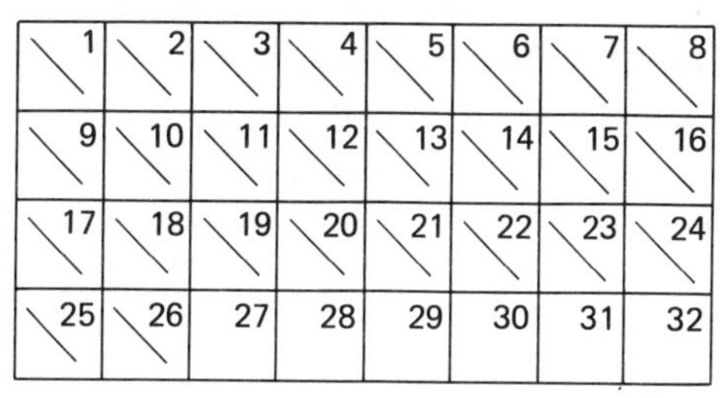

FIGURE 10-14 *Array of numbered, unlocked post office boxes.*

This question is best answered in stages: first, we must understand how the blocks work together, then we can see how they are useful. A simple nonelectronic analogy is used to illustrate the function.

Consider an array of post office boxes, as shown in Figure 10-14. We assume that these boxes are consecutively numbered, and unlocked so that we may gain entrance to any one. In each box resides a 3- by 5-in. file card containing written information. In boxes 1 to 15, the written information is a command that we can follow. In the other boxes are numerical data. Our instructions are to begin at box 1 by removing the card, reading the command, and carrying it out appropriately. When this action is completed, we are to go on to the next successive box (No. 2 in this case), carry out this command, and continue. In such a manner, we will execute a discrete sequence of commands read from successive post office boxes.

What kinds of commands can we expect to find?

Here are some examples:

1. Exchange the data card in box 20 with the data card in box 21.
2. Add the number in box 23 to the number in box 24 and write the result on the card in box 25.
3. Look at a wristwatch and write the time in box 26.
4. Read aloud the value in box 18.
5. Return to box 1 for the next command.
6. Stop.

Note that by the time we get to box 15 (the last "command" box in this example), we should have encountered a command of type 5 or 6 above.

This arrangement can be related to the computer's block diagram (Figure 10-13) as follows:

Post Office Boxes correspond to the computer's *Memory Circuits*
Written Information on the cards corresponds to *Memory Data* (binary numbers)
Our Eyes (for reading the watch) correspond to the *Input Part of the I/O Block*
Our Voice (for reading aloud) corresponds to the *Output Part of the I/O Block*
Our Brain (for understanding the commands) corresponds to the *CPU*

This analogy shows that the computer follows a sequence of *very simple* instructions stored in its memory. Also stored in the memory are numerical data that the computer must process.

Since all the computer's instructions are extremely simple and would certainly not tax the human mind, how can it be useful to us? In one word, it is a matter of "speed." When a computer is implemented with electronic semiconductors, its rate of operation will be *millions* of times faster than the human counterpart. This fact allows the digital computer to perform almost miraculous computational feats, impossible by any person. It is an amplification of mental power far beyond the amplification of mobility offered by the fastest supersonic aircraft, or the amplification of strength offered by the largest machinery.

10.5.2 CPU Functions

The ***Central Processing Unit*** (CPU) controls the flow of information in the computer, interprets coded instructions, and performs arithmetic and logic operations on numeric quantities. In almost all computers, the CPU is a synchronous sequential circuit designed to continuously cycle through a general two-step process:

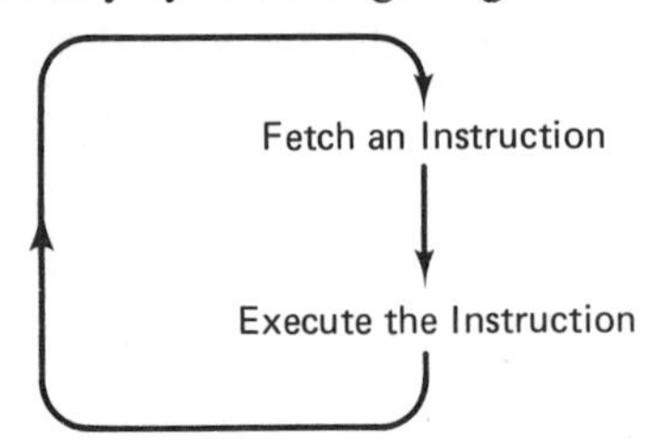

Each of these two basic procedures, ***Fetch*** and ***Execute***, requires a multitude of smaller steps. The sequence of elemental steps required for a Fetch are always the same. However, the sequence of steps required for an Execute are a consequence of the kind of instruction that has been fetched, and, therefore, varies from one instruction type to another. The Fetch and Execute together constitute what is known as an ***instruction cycle***. One complete instruction cycle is necessary to carry out each programmed instruction. The nature of the instruction that is being executed, and the CPU's clock rate, determine the time duration of the instruction. Typical *average* cycle times for modern computers range from about 3 μs or longer, to 20 ns. For a computer with an average cycle time of 200 ns, approximately 5 million typical instructions can be executed per second. Note that some complex instructions can require ten times longer, or more, than the average execution time.

The specific digital implementation of the Fetch/Execute process is referred to as the ***CPU architecture***, and can vary considerably from one computer to another. Typical modern CPU's are designed with a ***bus-structured architecture***. In such a design, registers, counters, combinational logic blocks, and other component parts of the CPU interface to a common internal bus.* This internal bus serves as the

* A CPU could certainly be designed by following the procedure for sequential circuit synthesis, but the resulting design would lack the simplicity of structure desired for debugging ease and reliable operation.

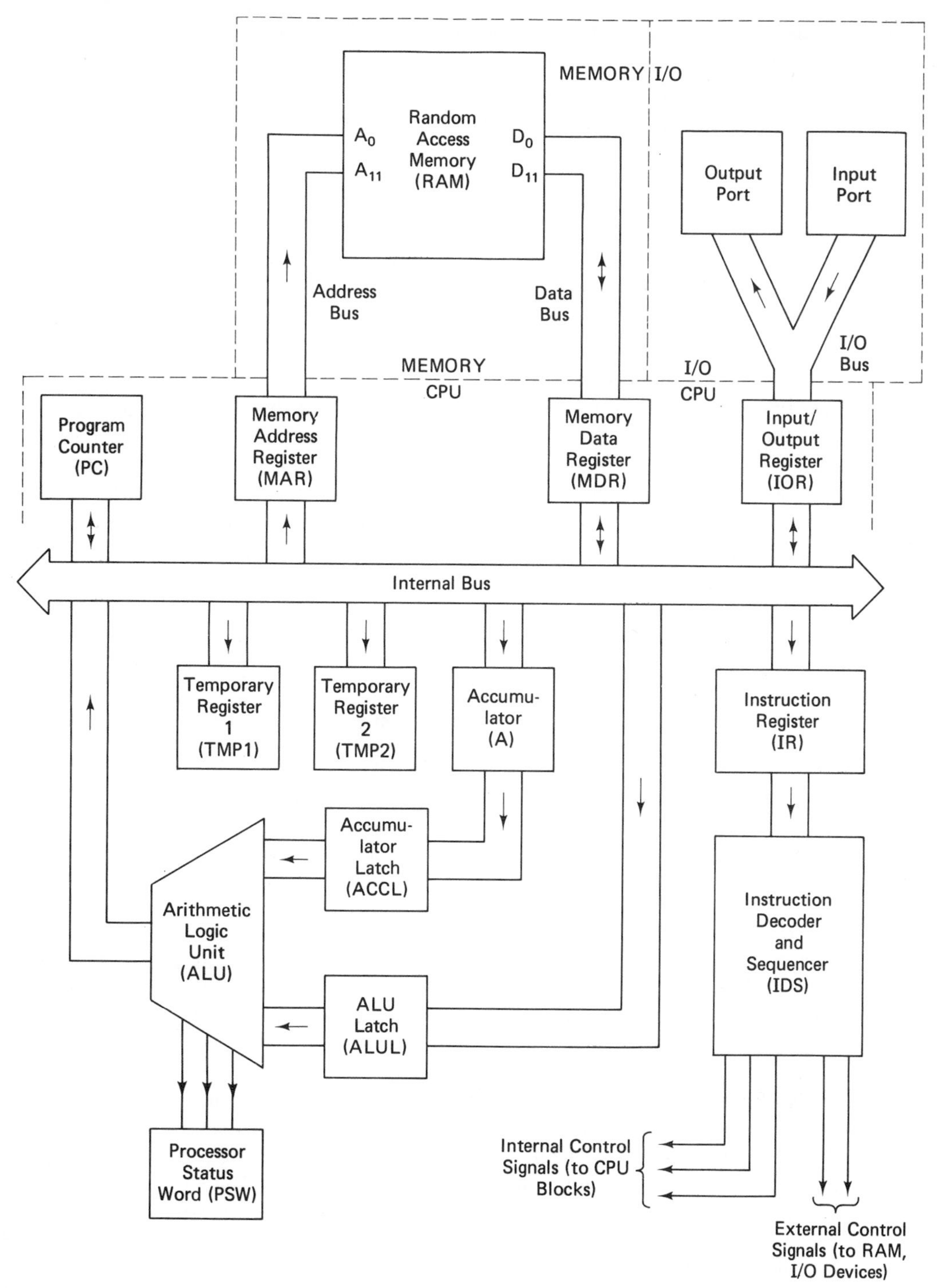

FIGURE 10-15 *Simplified bus-structured architecture, with RAM and I/O ports shown. Arrows indicate whether bus connections are bidirectional or unidirectional.*

primary medium of communication within the CPU, and can transmit information bidirectionally, between any two functional blocks on the bus. Figure 10-15 illustrates the bus-structured architecture of a simplified CPU. Study this figure, and carefully read the following description of the components. In this example architecture, we assume that the width of the data and address paths are identical (for example, both could be sixteen bits wide).

The careful reader will note that there are several busses in Figure 10-15: the internal bus, the address bus, the data bus, and the I/O bus. Our primary interest at this time is the *CPU architecture*, not the ***system architecture***.* Thus, we will focus our attention on the internal bus.

Recall that a "bus" is simply a parallel array of wires used to transmit a multibit binary number between two points. A basic rule that must always be followed in a bus-structured system is that only one source can drive information onto the bus at one time. Therefore, all CPU components that are capable of driving the bus must have three-state outputs; they must be capable of being "turned off" (high impedance outputs). Both the Fetch and Execute processes are defined by specifying a sequence of data transmissions or operations involving component blocks of the CPU. In many cases, the internal bus is used as the medium of transmission. As an example, the Fetch process is defined by the following sequence of events:

1.	PC → MAR	(transmission on internal bus)
2.	MAR → Address Bus	(transmission on external bus)
3.	Enable Memory	(assertion of control signal)
4.	Data Bus → MDR	(transmission on external bus)
5.	MDR → IR	(transmission on internal bus)
6.	Increment PC	(increment of program counter)

We shall discuss this in more detail shortly. First, it is important to understand the function of each CPU block.

PC: In the post office box analogy, we saw how the operator would pull cards from successive boxes, starting at box 1. This function is electronically accomplished in the CPU by the ***Program Counter*** (PC), which is simply a parallel-loadable natural binary counter. After each coded instruction is copied into the CPU, the program counter is pulsed so that its output relates to the next successive memory address.

MAR: The ***Memory Address Register*** (MAR) is a parallel array of flip-flops that serves to store momentarily the address of a memory word that we wish to access. The MAR input is derived from the internal bus, thus allowing it to be loaded from any one of a number of different sources. Its output drives the CPU's address bus, which is externally distributed to the system's memory elements (RAM, ROM).

* System architecture treats the CPU as a single block, and defines the relationship between CPU, memory, and I/O elements.

MDR: The ***Memory Data Register*** (MDR) is a parallel array of flip-flops that serves to hold temporarily the data that is passed to, or returned from, the system's memory elements. Unlike the MAR, the MDR is a *bidirectional* register, capable of holding information that is moving from CPU to memory, or from memory to CPU. The CPU side of the MDR is connected to the internal bus, while the memory side is connected to the system's data bus.

IR: The ***Instruction Register*** (IR) is a parallel array of flip-flops that serves to hold an instruction code that is returned from memory. The IR holds this code for the duration of the instruction execution, while other elements such as the MDR may change.

IDS: The output of the IR drives the ***Instruction Decoder and Sequencer***, which decodes the instruction code, and thereby causes appropriate action to be taken so that the instruction is executed properly. The instruction decoder and sequencer is one of the most complex blocks of the CPU. As well as controlling execution, it is responsible for the Fetch sequence, which always precedes the execution phase. In this architecture the Fetch sequence is always the same, and causes the instruction's operation code to be transmitted to the IR. The instruction decoder has a number of control lines connected to *each* CPU block shown in Figure 10-15. These control lines are not shown in the drawing, however, to avoid obscuring the main ideas.

A: The ***Accumulator*** (A) is, like many other CPU registers, a parallel array of flip-flops. It plays a very important part in all arithmetic and logic processes, being both the source and destination for data in many operations.

ACCL: The ***Accumulator Latch*** is identical to the accumulator, but only provides a temporary buffer for data that are being processed. It is present to avoid race problems when the accumulator is used as both the source and destination in arithmetic operations. Generally, arithmetic operations require two sources of data (for example, in addition, there must be an "addend" and an "augend").

ALUL: The accumulator latch is the source of one, and the ***ALU Latch***, identical in structure to the accumulator latch, is the source of the other.

ALU: The data from the accumulator latch and the ALU latch are combined with the desired function (addition, subtraction, logic AND, etc.) in the ***Arithmetic and Logic Unit*** (ALU). The ALU itself is a large and complex combinational logic block of the CPU responsible for all arithmetic and logic operations. It derives its data inputs from two sources (ACCL and ALUL), and drives the result onto the internal bus.

PSW: To reveal certain features about the result of an ALU operation, the ***Processor Status Word*** (PSW) is appropriately affected. The PSW should be thought of as a group of independently settable flip-flops that are set or reset according to the last ALU result produced. At least three PSW flip-flops are used to denote whether the ALU result

1. Is positive or negative.
2. Caused a carry, or not.
3. Is zero or nonzero.

Other possible (but not required) PSW flip-flops can indicate whether the ALU result

4. Is even or odd parity.
5. Caused a carry from the fourth to the fifth bit location (used in BCD arithmetic).

Additional information, not related to the ALU, is sometimes also included in the PSW. The PSW bits are used primarily for *conditional branch* instructions, in which a specific PSW bit is tested for being '0' or '1', the result of which affects selection of the next instruction.

IOR: Serving a purpose very similar to the MDR, the ***Input/Output Register*** (IOR) is used to temporarily hold data that is being passed between the CPU and input or output devices.* The IOR is a bidirectional register and drives an external I/O bus. In some CPU's the I/O bus and the data bus are shared.

TMP: At least one ***Temporary Register*** (TMP) is present in the CPU to preserve intermediate results and act as a source for data. Some TMP's may be available for general programming purposes, while others may be used only internally by the CPU and not available for general use. In this example CPU (Figure 10-15), two TMP's are shown, both of which are user-accessible.

Fetch process: Now let us review in more detail the sequence of steps, given earlier, for the Fetch process. The objective here is to load the instruction register with an instruction code from memory. The program counter contains the address of the desired instruction code. First, the contents of the program counter are transmitted to the memory address register:

$$\text{(1) PC} \longrightarrow \text{MAR}$$

At this time, the MAR now contains an exact copy of the PC, and is next driven onto the address bus:

$$\text{(2) MAR} \longrightarrow \text{Address Bus}$$

Random access memory circuits, which are also connected to the address bus, now

* An input device could be a bank of switches or a keyboard. An output device could be an array of LED lamps or a printing terminal.

see a valid address signal, consisting of an n-bit binary number (typically, 'n' is between 16 and 24). The next step provides that the memory circuits are placed in the "read" mode, and enabled, initiating the memory access. A special control signal causes this to occur:

(3) Enable Memory Circuits

After the access time is satisfied (typically between 100 and 500 ns), the binary values present on the data bus represent the data at the specified memory address (which in this Fetch example is an instruction code). Thus, we now load these values into the MDR:

(4) Data Bus $\longrightarrow$ MDR

The MDR data is then copied into the instruction register, completing the Fetch:

(5) MDR $\longrightarrow$ IR

Before any more CPU activity can take place, the PC must be incremented so that its value denotes the address of the next instruction, or data related to the current instruction:

(6) Increment PC

At this time, the CPU now interprets the instruction code present in the IR, and initiates a new sequence of steps (the Execute process) to carry out the instruction properly. Of course, the exact sequence of steps will be different for each type of instruction. As an example, we can examine a simple ADD instruction.

Execution process: Most arithmetic operations require two ***operands*** (an operand is a number to be operated on). Addition is no exception. By studying the CPU block diagram in Figure 10-15, we see that the ALU obtains numbers from two sources: the ALU latch and the accumulator latch. Thus, before actually enabling the ALU block for addition, we must preload these two registers (ALUL and ACCL) with the desired operand. In this simple architecture, the accumulator is always the destination for the ALU result. When the ADD command is specified, two arguments must be specified also. A typical ADD instruction in our simplified CPU architecture will look like this:

ADD A, T1

Its meaning is

$$A + T1 \longrightarrow A$$

That is, add the contents of temporary register 1 to the accumulator, and place the result back in the accumulator (thus destroying the original value in A). Other possibilities are

ADD T1, T2	$T1 + T2 \longrightarrow A$
ADD A, A	$A + A \longrightarrow A$

We must now ask:

> What is the exact sequence of CPU operations (transmissions on the internal bus, etc.) that are required to execute an ADD instruction?

For "ADD A,T1" the following steps are taken to fulfill the execute process. First, the two operands are positioned to drive the ALU inputs.

$$(1)\ \text{TMP1} \longrightarrow \text{ALUL}$$
$$(2)\quad \text{A} \longrightarrow \text{ACCL}$$

The ALU is set up for *addition* by the instruction decoder and sequencer, and the addition is performed. When the ALU result is ready, it is driven onto the internal bus and received by the accumulator. This occurs as one step.

$$(3)\ \text{ALU} \longrightarrow \text{A}$$

These three steps constitute the execution of "ADD A,T1".

What we have done in these last few lines is to examine the sequence of steps that the IDS is designed to take, following the Fetch of a typical ADD instruction. Remember that each instruction will cause a different sequence of steps to be taken. In many computers, the IDS is permanently wired, and defines the instruction set of the CPU. A typical computer may have between 50 and 500 different instructions, depending upon its level of sophistication. Note that a more sophisticated CPU will also have a more diverse architecture than the type used in our example. Clearly, we can see that the IDS will be a complex element of the CPU. Unfortunately, space does not permit our discussing how the IDS operates. This subject should be found in a good book on computer fundamentals.

10.5.3 Simple Instruction Set

Goals: Based on the CPU architecture of Figure 10-15, we will now do the following:

1. Define a basic instruction set, and show examples of the sequence of register transmissions and operations needed to implement a typical instruction (Section 10.5.3).
2. Demonstrate how the instruction set of (1) can be applied in a practical computer program (Section 10.5.4).

First, it must be emphasized that we normally would not have the capability, or desire, to define our own instruction set. In a typical computer, the instruction set is determined by the computer's designer, and is wired into the machine. As programmers, we must live with these predetermined instructions. An instruction set is designed in this section only for illustrative purposes.

Types of instructions: Computer instructions can be sorted into five basic groups (refer back to the post office box example):

1. Data Transfer.
2. Arithmetic and Logic.
3. Input/Output.
4. Branch.
5. Control.

Each of these groups perform a different basic function. ***Data Transfer*** operations move data between registers or memory in the computer. Data transfer operations *do not* change source data, but merely copy it from one place to another. ***Arithmetic and Logic*** operations apply well-defined rules (addition, subtraction, etc.) to binary data, thereby altering it in a specific way. Although source data are not necessarily changed, the result usually replaces one of the source operands. ***Branch*** operations cause the sequential execution of instructions to be interrupted, and are use to provide ***program loops*** (which allow a group of instructions to be repeatedly executed), or to skip particular instructions. ***Input/Output*** operations allow data exchange between the computer and external devices. ***Control*** operations affect the function of the CPU by invoking or disabling particular options, and by causing instruction execution to stop.

Instruction mnemonics: Each computer instruction is represented by an instruction mnemonic (silent 'm'), which is a simple three- or four-letter abbreviation of the instruction's function. One or two operands are usually included. Here are some examples:

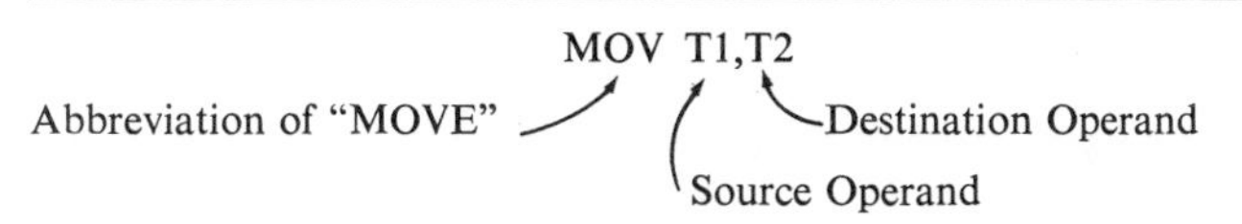

Type of Instruction: Data Transfer
Operation: Copy the contents of temporary register 1 into temporary register 2.

AND A,T1

Type of Instruction: Arithmetic and Logic
Operation: Logic AND the contents of the accumulator with the binary value in temporary register 1, placing the result in the accumulator. All logic instructions, such as this, operate independently on corresponding bits of each register, and never generate carries.

SUB A,A

Type of Instruction: Arithmetic and Logic
Operation: The contents of the accumulator are subtracted from itself, yielding zero. This is a convenient way to clear the accumulator.

JMP 1E3

Type of Instruction: Branch

Operation: Load the program counter with the hexadecimal-coded binary number '1E3'.

MOV T1,IOR

Type of Instruction: Output

Operation: Copy temporary register 1 into the Input/Output register.

HLT

Type of Instruction: Control

Operation: Halt (stop executing instructions).

A computer program is simply a carefully chosen sequence of instruction mnemonics. Realize, though, that when the program is loaded into the computer's memory, each mnemonic must be coded as an appropriate binary number that can be processed by the Instruction Decoder and Sequencer.

Some instructions require numerical information as well as an operation code to be complete, such as "JMP 1E3". In such cases, two or more successive words in memory may be necessary. Other instructions, such as "HLT", where no operands are present, can be coded in a single word. The act of translating instruction mnemonics into appropriate executable binary code is referred to as *assembly*. When a program is "assembled," a sequence of binary numbers is determined that can be processed by the CPU. In a twelve-bit machine, an instruction like "JMP 1E3" would likely assemble into two twelve-bit words, the first being code for the jump operation and the second being the jump address (in this case, '1E3').

Addressing modes: Most instruction sets allow several alternative ways in which the operands can be specified, there being usually at least three. These different ways are referred to as *addressing modes*, and are defined as follows:*

(a) Immediate mode—the operand is specified exactly, and is part of the instruction's assembled code.
Example: MOV#15B,A
Operation: the hexadecimal-coded binary number '15B' is copied into the accumulator.

(b) Direct mode—the *location* of the operand is specified.
Example: MOV T1,A
Operation: The number residing in temporary register 1 is copied into the accumulator. Note that the location of the data is specified, not the data itself.

(c) Indirect mode—the *location of the address* of the operand is specified.
Example: MOV (T1),A

* The mnemonic examples and their binary codes are not associated with any particular CPU, and serve only as examples.

Operation: The number residing in temporary register 1 is used as a *memory address* to denote the location of the operand in memory. The appropriate memory address is accessed, and the data is then copied into the accumulator. Parentheses are used around 'T1' to indicate indirect addressing.

Instruction set: We will now define a simple instruction set representative of one on a practical computer. This instruction set is based on the architecture of Figure 10-15, which we have discussed previously, and assumes twelve-bit-wide data and address paths. Note that with a twelve-bit-wide address bus, the address space in this example is 4096 words ($= 2^{12}$).

Each binary instruction code in our instruction set will be twelve bits wide, and will possess certain *bit fields* that are always used to specify a given component of the instruction. These fields are illustrated and defined in Figure 10-16. Many instructions require a full specification of all bit fields, while some do not. For example, a "HALT" instruction has no operands, and thus only possesses an operation code. In such cases, the unused bits are not interpreted. Table 10-1 defines the instruction set and the appropriate binary instruction codes. Note that the actual binary codes chosen are only related to the design of the Instruction Decoder and Sequencer, and are otherwise arbitrary. In Table 10-1, "a" always refers to the first operand, and has three possible addressing modes (immediate, direct, and indirect), while "b" always refers to the second operand and has two possible addressing modes (direct and indirect). "#d" indicates immediate-mode data, while "#a" indicates an immediate-mode address. All immediate-mode instructions will assemble into *two* twelve-bit words, one for the instruction code, and one for the binary data or address information. In some cases,

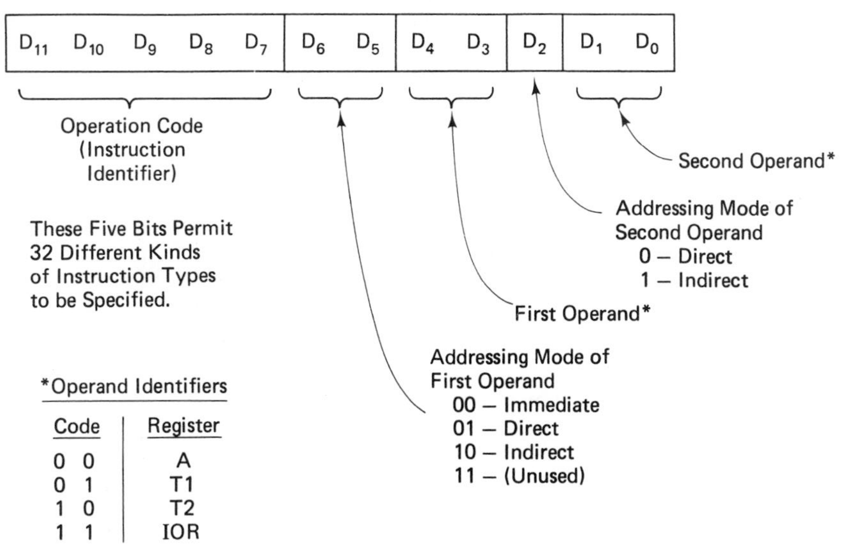

FIGURE 10-16 *Format of a twelve-bit instruction code.*

TABLE 10-1

Instruction set definition.

Nature of Instruction	*Mnemonic*	*Function*	*Operation Code*	*Example*
(1) Data Transfer	MOV a,b	a ⟶ b (Copy the value specified by 'a' into 'b'.)	0 0 0 0 0	MOV #5C,(A)
(2) Addition	ADD a,b	a + b ⟶ A (Add the values specified by 'a' and 'b', place the result in the accumulator.)[a]	0 0 0 0 1	ADD T1,T2
(3) Subtraction	SUB a,b	a − b ⟶ A (Subtract the values specified by 'a' and 'b', place the result in the accumulator.)	0 0 0 1 0	SUB T1,T2
(4) Comparison	CMP a	A − a (Subtract the value specified by 'a' from the accumulator, do not save the result.)	0 0 0 1 1	CMP T1
(5) Logic AND	AND a,b	a · b ⟶ A (Logic AND the values specified by 'a' and 'b', placing the result in the accumulator.)	0 0 1 0 0	AND A,(T1)
(6) Logic OR	OR a,b	a + b ⟶ A (Logic OR the values specified by 'a' and 'b', placing the result in the accumulator.)[a]	0 0 1 0 1	OR (T1),(T2)
(7) Logic '1's Complement	COM b	$\bar{b}$ ⟶ b (Complement the value specified by 'b', and replace the source with the result.)	0 0 1 1 0	COM A
(8) Logic XOR	XOR a,b	a ⊕ b ⟶ A (Logic Exclusive-OR the values specified by 'a' and 'b', placing the result in the accumulator.)	0 0 1 1 1	XOR T1,A
(9) Rotate Left	ROL b	'b ⟶ b (The value specified by 'b' is shifted left, with end-around carry.)	0 1 0 0 0	ROL T1

[a] Note that although '+' is used to represent the ADD and the OR operations, the actual function is different.

TABLE 10-1

(continued)

Nature of Instruction	*Mnemonic*	*Function*	*Operation Code*	*Example*
(10) Rotate Right	ROR b	b' ⟶ b (The value specified by 'b' is shifted right, with end-around carry.)	0 1 0 0 1	ROR T1
(11) Increment	INC b	b + 1 ⟶ b ('1' is added to the value specified by 'b'.)	0 1 0 1 0	INC A
(12) Decrement	DEC b	b − 1 ⟶ b ('1' is subtracted from the value specified by 'b'.)	0 1 0 1 1	DEC A
(13) Branch, Unconditional	JMP #a	#a ⟶ PC (The program counter is loaded with a new value, #a.)	0 1 1 0 0	JMP 4C5
(14) Branch, Conditional	JZ #a	(If the last ALU operation produced a zero result, then #a ⟶ PC. Otherwise, the PC is left unchanged, and the next successive instruction is executed.)	0 1 1 0 1	JZ 203
(15) Branch, Conditional	JNZ #a	(Same as 13, except on the nonzero result.)	0 1 1 1 0	JNZ E1A
(16) Branch, Conditional	JC #a	(Same as 13, except on the carry result.)	0 1 1 1 1	JC 015
(17) Branch	JNC #a	(Same as 13, except on the no carry result.)	1 0 0 0 0	JNC 98C
(18) Control	HLT	Stop executing instructions.	1 0 0 0 1	HLT

the "a" or "b" operands *alone* will be specified when an instruction requires only one operand. Whether "a" or "b" is chosen depends on whether immediate-mode addressing is possible for the instruction.

Table 10-2 illustrates how several of the instructions in Table 10-1 are actually implemented on our example architecture (Figure 10-15). Note that the Fetch is the same in each case, and is not shown; we see only the Execute process. It is important to realize that the circuitry for the ALU and IDS is quite complex and exceeds the scope of our present discussion. Also, note that the desire for a simplified design in these stages would greatly reduce the flexibility of our instruction set.

TABLE 10-2

Examples showing the implementation of instructions on the architecture of Figure 10-15.

Assembled Code	*Mnemonic*	*Register Transmissions and Operations*
0 0 0 0 0 0 0 0 0 1 0 0 0 0 0 1 0 1 0 1 1 0 1 1	MOV #15B,(A)	
	Obtain immediate data from memory	PC → MAR MAR → Address Bus Enable Memory for Read Data Bus → MDR Increment PC
	Store immediate data in memory address specified by 'A'	A → MAR MAR → Address Bus Enable Memory for Write MDR → Data Bus
0 0 0 1 0 0 1 0 1 0 1 0	SUB T1,T2	T1 → A A → ACCL T2 → ALUL
	ALU set up for subtraction by instruction code in IR.	ALU in Subtract Mode ALU → A
0 1 0 0 1 0 0 0 0 0 0 0	ROR A	A → ALUL
	ALU requires only one operand, ACCL unchanged.	ALU in Rotate Mode ALU → A
0 1 1 0 0 0 0 0 0 0 0 0 0 1 0 0 0 1 0 1 1 1 0 0	JMP 45C	PC → MAR MAR → Address Bus
	Obtain immediate address from memory	Enable Memory for Read Data Bus → MDR
	Perform the branch	MDR → PC

10.5.4 Some Practical Programs

Block transfer program: In practical computer systems, it is frequently necessary to transfer an array of data from one area in memory to another. This example depicts how such a transfer may be accomplished using the instruction set defined in the last section.

A first step in any problem solving adventure is to *clearly understand* precise what the problem is, and what we need to know about it. This seemingly obvious step is often left incomplete, leading to incorrect solutions. In this example, the problem is:

> Transfer an array of successive words in RAM from one location to a new one. The source array of data is defined by a *starting address*, and a *length*. The destination is defined by a *starting address*, the length being identical. We assume that the source and destination addresses are such that no overlap occurs.

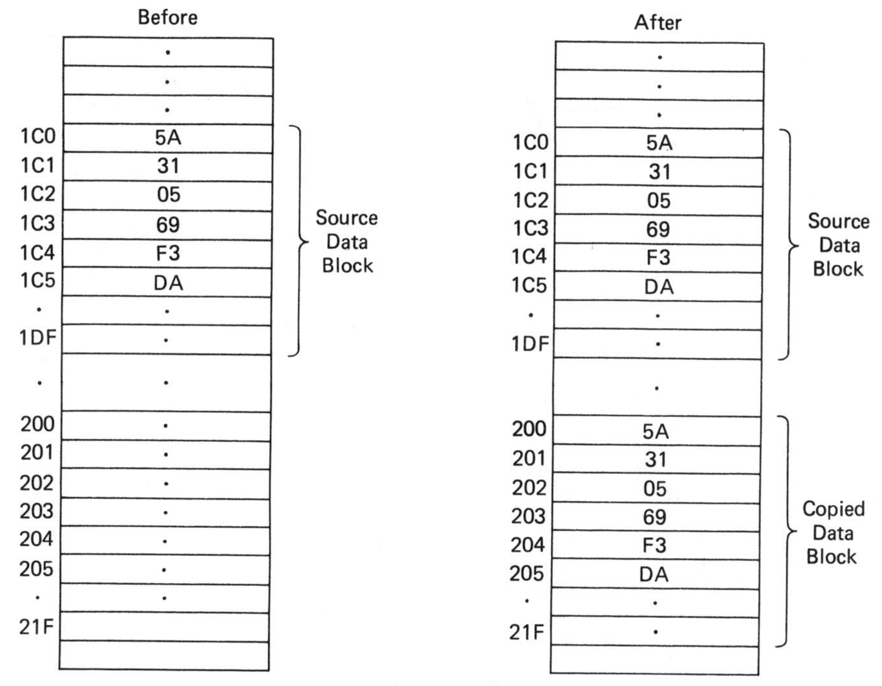

FIGURE 10-17 Images of RAM before and after the block transfer occurs. Arbitrary numerical data and addresses are shown.

Figure 10-17 illustrates the problem. Since no single instruction in our instruction set will accomplish the transfer, we must write a program, using existing instructions, to achieve the desired result. In some cases, knowing which instructions to use, and in which order, may be obvious. Generally, however, it is not, and a systematic approach must be used.

An ***algorithm*** describes an orderly, step-by-step procedure that outlines the solution to a problem. Typically, an algorithm is represented by a ***flowchart***, as illustrated in Figure 10-18 for our block transfer example. In this figure, we see that different shape blocks represent different operations:

1. Circles are used for "start" or "stop."
2. Rectangles are used for processing.
3. Diamonds are used for decisions.

Also, we note that instruction mnemonics are not shown. This is intentional, and should be done to make the flowchart understandable by any programmer, regardless of whether they know our computer's instruction set. Remember, the flowchart describes an algorithm as a general procedure. The procedure is made specific when the computer program is written.

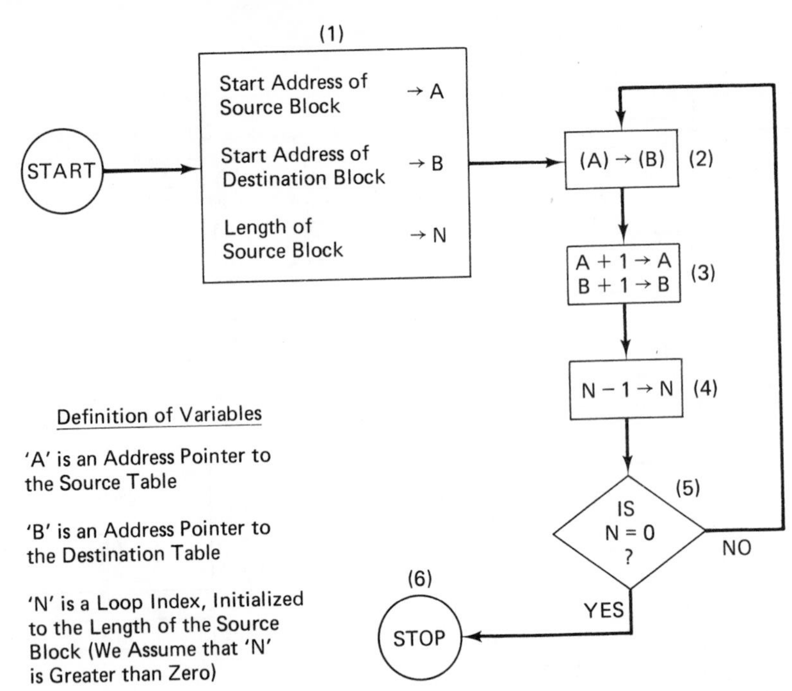

FIGURE 10-18 *Flowchart for the algorithm that copies a block of data from one place in RAM to another. Parentheses indicate indirect addressing.*

The block transfer problem is solved by doing a word-by-word transfer from source area to destination area until all words in the source block have been copied. '*A*' and '*B*' are general-purpose variables in Figure 10-18, and are used to represent the current source and destination word addresses. '*N*' is a general-purpose variable used to account for the number of transfers that have been made. If '*N*' is initialized to the length of the source table, then decrementing it with each pass will cause it to read zero when the transfer is complete. '*N*' should be at least one, initially.

The program solution to the algorithm of Figure 10-18 is shown in Figure 10-19. Remember, this program relies on the architecture and instruction defined in the last section. Study Figure 10-19 carefully.

FIGURE 10-19 *Program solution to the block transfer problem.*

Source Block

Start Address: 1C0
Length: 1F

Destination Block

Start Address: 200

Register Assignment

Flow Chart	Program
A	T1
B	T2
N	A

```
Block         Mnemonic             Comment
(1)           MOV #1C0, T1         ;init. start adr. of source
              MOV #200, T2         ;init. start adr. of destination
              MOV #01F, A          ;init. length
(2) — LINE4:  MOV (T1), (T2)       ;move one word of data
(3)           INC T1               ;increment the source pointer
              INC T2               ;increment the destination pointer
(4) —         DEC A                ;decrement the loop index
(5) —         JNZ LINE4            ;if the loop index is not zero, branch back
(6) —         HLT                  ;otherwise, stop
```

(Block numbers: Corresponding Flow Chart Block)

Before this problem could be entered and executed on a digital computer, it must be assembled (translated into the appropriate sequence of binary instruction codes). The assembled program is shown in Figure 10-20. Carefully review this code to be certain that you understand how it was obtained. Note that the label "LINE4" is *arbitrary*, and is used only to denote which instruction should be returned to when the branch is executed. The label could also have been "LOOP, MORE, WXYZ," and so on. When the program is assembled, however, the label becomes the ***absolute address*** of the jump's destination. Refer to Figure 10-20.

Assembly Reference Chart

Instruction Op-Code Table

Instruction	Op-Code
MOV a, b	0 0 0 0 0
ADD a, b	0 0 0 0 1
SUB a, b	0 0 0 1 0
CMP a	0 0 0 1 1
AND a, b	0 0 1 0 0
OR a, b	0 0 1 0 1
COM b	0 0 1 1 0
XOR a, b	0 0 1 1 1
ROL b	0 1 0 0 0
ROR b	0 1 0 0 1
INC b	0 1 0 1 0
DEC b	0 1 0 1 1
JMP #a	0 1 1 0 0
JZ #a	0 1 1 0 1
JNZ #a	0 1 1 1 0
JC #a	0 1 1 1 1
JNC #a	1 0 0 0 0
HLT	1 0 0 0 1

Hexadecimal-Coded Memory Address / Binary Program Code

Address	Binary Program Code	
100	0 0 0 0 0 0 0 Ø Ø 0 0 1	MOV #1C0, T1
101	0 0 0 1 1 1 0 0 0 0 0 0	
102	0 0 0 0 0 0 0 Ø Ø 0 1 0	MOV #200, T2
103	0 0 1 0 0 0 0 0 0 0 0 0	
104	0 0 0 0 0 0 0 Ø Ø 0 0 0	MOV #01F, A
105	0 0 0 0 0 0 0 1 1 1 1 1	
106	0 0 0 0 0 1 0 0 1 1 1 0	MOV (T1), (T2)
107	0 1 0 1 0 Ø Ø Ø Ø 0 0 1	INC T1
108	0 1 0 1 0 Ø Ø Ø Ø 0 1 0	INC T2
109	0 1 0 1 1 Ø Ø Ø Ø 0 0 0	DEC A
10A	0 1 1 1 0 Ø Ø Ø Ø Ø Ø Ø	JNZ 106
10B	0 0 0 1 0 0 0 0 0 1 1 0	
10C	1 0 0 0 1 Ø Ø Ø Ø Ø Ø Ø	HLT
10D		
10E	:	
10F		
:		
1C0	(Source Table Begins)	Source
:		
1DF	(Source Table Ends)	
:		
200	(Destination Table Begins)	Destination
:		
21F	(Destination Table Ends)	

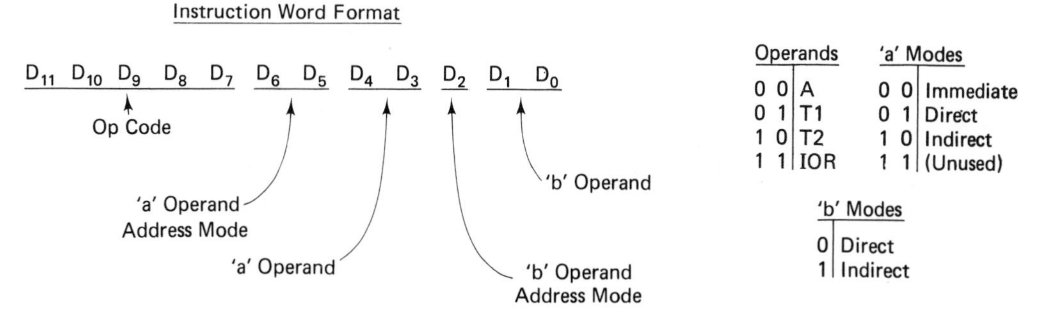

FIGURE 10-20 *The program of Figure 10-19 is assembled into executable code.*

Time delay program: Most computers execute instructions so fast that it is sometimes necessary to intentionally slow them down so that compatibility with other devices is possible. For example, output data may be provided too fast for external circuits, or human operators, to accept it. Although it might be a simple matter to slow down or stop the computer's clock temporarily, it is a better practice to either:

1. Cause the computer to enter a "wait" loop until an external signal is received, or
2. Cause the computer to enter a "wait" loop for a predetermined amount of time.

In this programming example, we show how the second of these two approaches is implemented. Specifically, we will write a program that is totally inert (does not alter any data in memory) and provides only time delay. The technique that is employed involves loading one CPU register with a predetermined value (the time constant), and decrementing it to zero. The flowchart of Figure 10-21(a) illustrates the algorithm. An appropriate program is shown in Figure 10-21(b).

Computation of the exact time delay depends on

1. The computer's clock rate.
2. The number of clock pulses required for each instruction.
3. The value of the time constant.

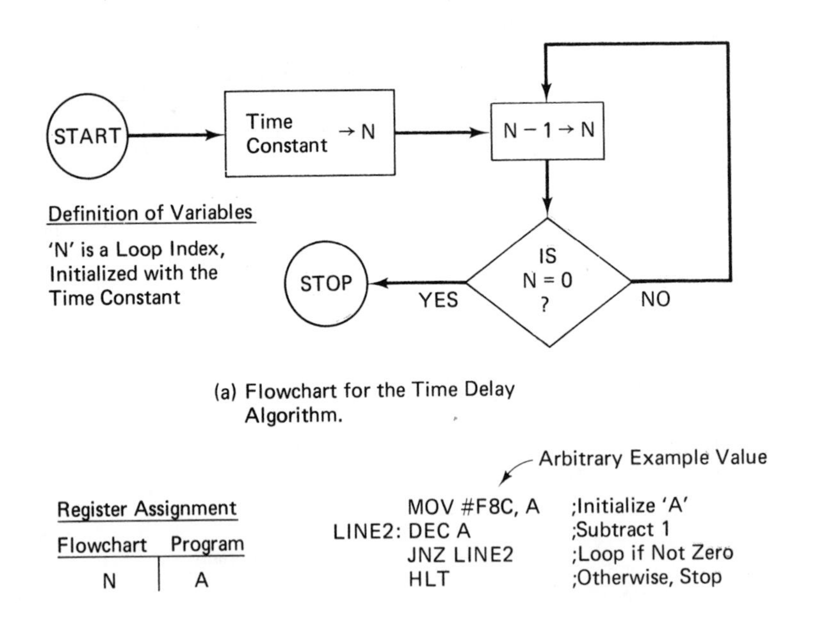

(b) The Corresponding Program.

FIGURE 10-21

We assume the following:

(a) $\tau = 1\ \mu s$ (the period of the CPU clock)

(b) n (the number of clock pulses per instruction)
= 12 for MOV #,A
= 8 for DEC
= 12 for JNZ

(c) The program's time constant is $(F8C)_{16}$:

$$(F8C)_{16} = (3980)_{10}$$

With this information, we can compute the total time delay:

```
        MOV #F8C,A (12)
LINE2:  DEC A      (8)     total delay = 1 μs [12 + 3980(8 + 12)]
        JNZ LINE2  (12)                = 79.6 ms = 0.0796 s
        HLT
```

Although a substantial relative delay in the program flow has been obtained, it nevertheless amounts to only a fraction of 1 s. While the time constant can be made somewhat larger, the twelve-bit limit for CPU register 'A' has almost been reached. Problem 10-20 addresses the concept of nested loops, and describes how delays of more than 1 s are possible.

Table-search program: This will be the most complex of our example programs, and illustrates how a table of data is searched for a specific number. This number may occur exactly once, many times, or not at all. When a match is found, its memory address is to be output via the I/O register in the CPU. Figure 10-22 illustrates how RAM is divided into program and data areas.

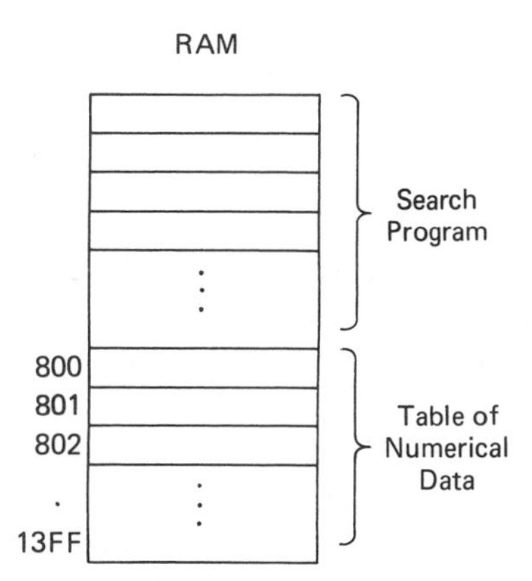

FIGURE 10-22 *Random Access Memory contains two distinct areas: the program space and the data space.*

Our basic procedure will be to individually examine each entry in the table and compare it with the value being sought. If the test entry can be subtracted from the value being sought to produce a *zero* result, these values are equal. For any other result, they are not equal. Figure 10-23 depicts the flowchart for this algorithm.

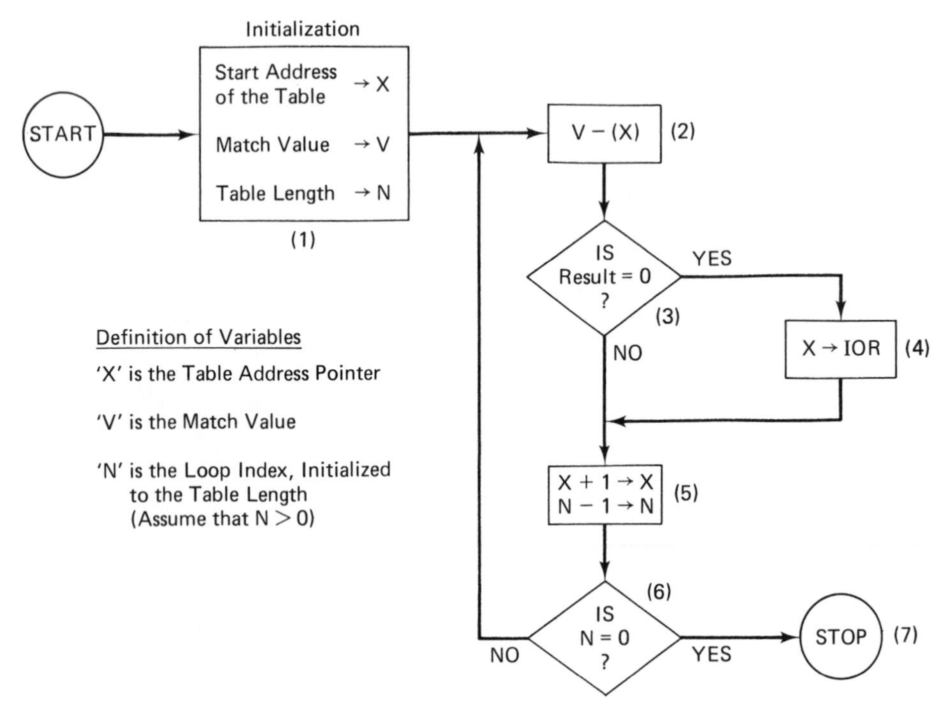

FIGURE 10-23 *Flowchart for the table-search algorithm.*

To write a computer program for the table-search algorithm, we need only to write instructions for each numbered block in Figure 10-23, and then list them all in sequence. The result is shown in Figure 10-24, and should be carefully studied.

There are several points to make about this program. First, 'CMP' and 'SUB' instructions perform the same arithmetic function (subtraction), but CMP *does not* return the result to the accumulator, thus preserving whatever the first operand may be. Since we do not wish to reload 'A' (the match value) after every test, 'CMP' is the better choice. Second, note that all instructions that use the ALU affect the Processor Status Word (PSW). These instructions include ADD, SUB, CMP, INC, and DEC. Conditional jumps are performed based on the *current value* of the PSW at the time the conditional jump is executed. In this program example notice the sequence of instructions in blocks (5) and (6): INC T1, DEC T2, JNZ MORE. Both INC T1 and DEC T2 affect the PSW, but since DEC T2 is last, its result *only* is what the JNZ abides. Simply think of the PSW as another register that some instructions can change and others cannot. Conditional jumps just look at the PSW, and branch according to its contents.

Register Assignment	
Flowchart	Program
V (Match Value)	A
X (Start Adr.)	T1
N (Table Length)	T2

```
              Mnemonic             Comment

     (1) {      MOV #40B, A        ;load search value
                MOV #800, T1       ;load table start adr.
                MOV #BFF, T2       ;load table length
     (2)— MORE: CMP (T1)           ;compare table and search values
     (3)———     JNZ SKIP           ;skip the next inst. if unequal
     (4)———     MOV T1, IOR        ;otherwise, output match adr.
     (5) { SKIP: INC T1            ;increment the table pointer
                DEC T2             ;decrement the loop index
     (6)—       JNZ MORE           ;branch back if table not completely searched
     (7)—       HLT                ;otherwise, stop
```

FIGURE 10-24 *Program that implements the table-search algorithm.*

10.6 MICROPROCESSORS

10.6.1 Definition of a Microprocessor and Typical Applications

In the last section, we have seen how the digital computer can be applied to simple numerical processing tasks. What could not be illustrated, however, is how remarkably well suited it is to performing extremely complex and tedious numerical processing. In fact, the flexibility of the computer's instruction set and its speed of operation allow it to be applied to almost any kind of intelligent activity. Prior to 1971, the cost of a small computer system was no less than $10,000, and thus unaffordable for many possible applications. With the advent of large-scale integrated circuits, thousands of microscopic transistors could be constructed on a single chip of silicon whose area is less than half that of a dime. This development led directly to the development of a one-chip digital CPU. Such a circuit is referred to as a ***microprocessor***, and was introduced for the first time in 1971 by the Intel Corporation. A microprocessor in combination with memory and I/O circuits constitutes a microcomputer system. A microprocessor is characterized by these unique features:

1. It is a single-circuit CPU.
2. It is inexpensive to purchase and operate.
3. It is very small.
4. It requires very little power.

Figure 10-25 shows a typical microprocessor integrated circuit. Some of the consequences of its unique features are:

1. Single-circuit CPU—the cost of assembling a CPU from individual components contributed greatly to the expense of a typical pre-1971 computer. An LSI central processing unit means that the assembly cost is eliminated, and replaced by a modest fabrication charge. Also, reliability is greatly increased, maintenence is simplified, size is reduced, and power consumption is minimal.

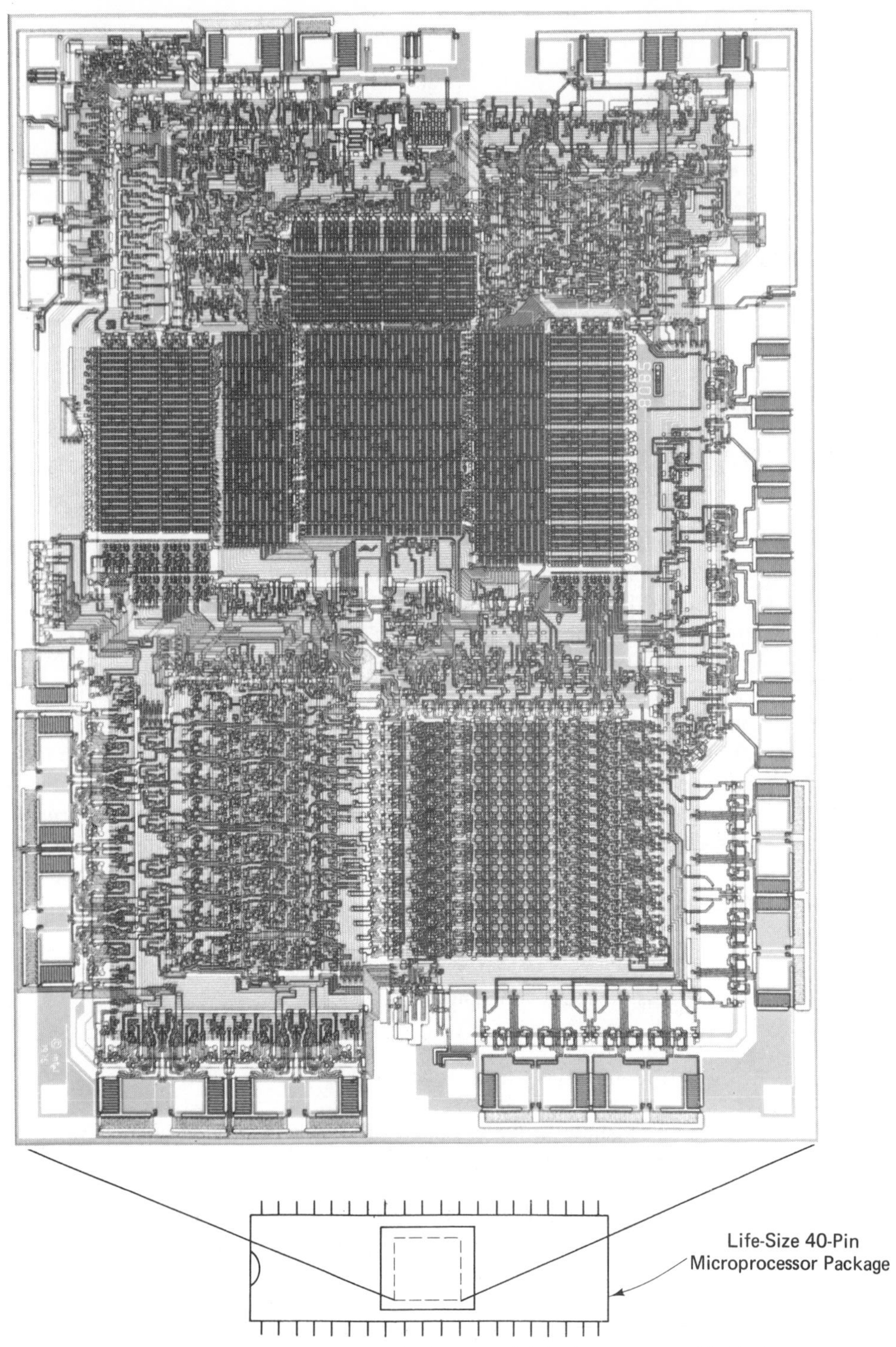

FIGURE 10-25 *Magnified photograph of an Intel 8085 microprocessor chip (top), and a full-size drawing of the package into which it is mounted.*

2. Inexpensive to purchase and operate—with the cost of a microcomputer-based system being several factors of ten less than pre-1971 computers, applications that have previously been impractical or impossible are now economically viable. For example, the microprocessor-based control of automobile carburetors to minimize pollutants is now standard on many American-made cars. A typical microprocessor chip costs between $10 and $25, yet with peripheral circuits and equipment, the cost of a personal computer system may be $500 or more.
3. Small size—the electronic calculator demonstrates how microelectronics, centered on the microprocessor CPU, has affected portable instruments. Small size does not only imply portability, but also volume-sensitive applications. An "intelligent" pacemaker, for example, might utilize a microprocessor for regulating heartbeat according to a complex variety of existing conditions.
4. Low power drain—in most applications where small size is useful, so also is low power drain, particularly in portable devices. In some cases, power drain alone is of great concern while size is not. For example, a weather monitoring buoy in midocean may utilize a microprocessor to control data logging and transmission, and yet remain unattended for several months or longer.

Note that a *microcomputer* is a term that refers to a system containing a microprocessor, memory circuits, and I/O devices. Single-chip microcomputers containing all three basic system components in one IC are available and are constantly becoming more popular in small applications.

10.6.2 Comparison of Large Computers, Minicomputers, and Microcomputers

The first microprocessors were developed when LSI techniques were in their infancy. In those days, the limitation on the maximum number of transistors per chip placed a considerable limitation on the microprocessor's architecture. The benefits of many instructions and a wide variety of addressing modes enjoyed by larger machines had to be sacrificed, and the system clock was limited to less than 750 kHz. Also, data paths were only four to eight bits wide. The effects of improved technology almost ten years later have overcome most of the early limitations on microprocessors. We now find very sophisticated architectures, system clock rates of 2 to 8 MHz, and data paths sixteen to thirty-two bits wide. At the same time, though, improvements in technology have also benefitted non-microprocessor-based computers. Thus, we are prompted to ask:

> What are the basic differences between microprocessor-based systems and large, multicomponent CPU based systems, given that both are technologically up-to-date?

Functionally, there are five important issues; physically there are three. These are illustrated in Table 10-3.

TABLE 10-3

Functional and physical comparison of small and large computer systems.

Issue	*Microprocessor-Based Systems*	*Mainframe Computer Systems*
Functionally		
(1) Clock speed	Slower	Faster
(2) Memory size	Smaller	Larger
(3) Word size (data paths)	Smaller	Larger
(4) Instruction set	More limited	Richer
(5) I/O handling	More basic	More sophisticated
Physically		
(1) Cost	Much less expensive	Very expensive
(2) Size	Small, often portable	Large, requires a large room with air conditioning
(3) Power consumption	Little, sometimes battery operated	Considerable
Applications	Dedicated real-time activities, control (word processing typewriters, carbureror control, portable calculators, TV games, automatic pricing scales, automatic bank tellers, personal computers)	Business data processing, scientific and engineering computations, mathematical modeling, time-shared systems, data-base control (payroll checks, satellite orbit calculation, mechanical stress modeling, electronic library catalog and search, telephone switching)

Remember that all terms in Table 10-3 are relative to each other. Improvements in technology constantly upgrade performance in both categories. In fact, current micro-computer systems match or exceed the performance of many large machines of the 1960's.

Minicomputers fall somewhere between the two extremes of microcomputers and large mainframe computers. Their CPU's are not microprocessor-based (however, considerable LSI is used), yet they are small and less expensive than mainframe types, and find applications both in dedicated real-time activities and in data processing or scientific calculations. They are truely a middle ground in performance, size, and cost.

10.7 COMPUTER SYSTEMS

10.7.1 Hardware Components

Before leaving this introductory chapter on computers, it must be emphasized that a practical computer system is much more than just a CPU with memory circuits. While the CPU is the heart of the system, many other expensive and complex compo-nents are needed to make it useful. In addition, a sophisticated software library,

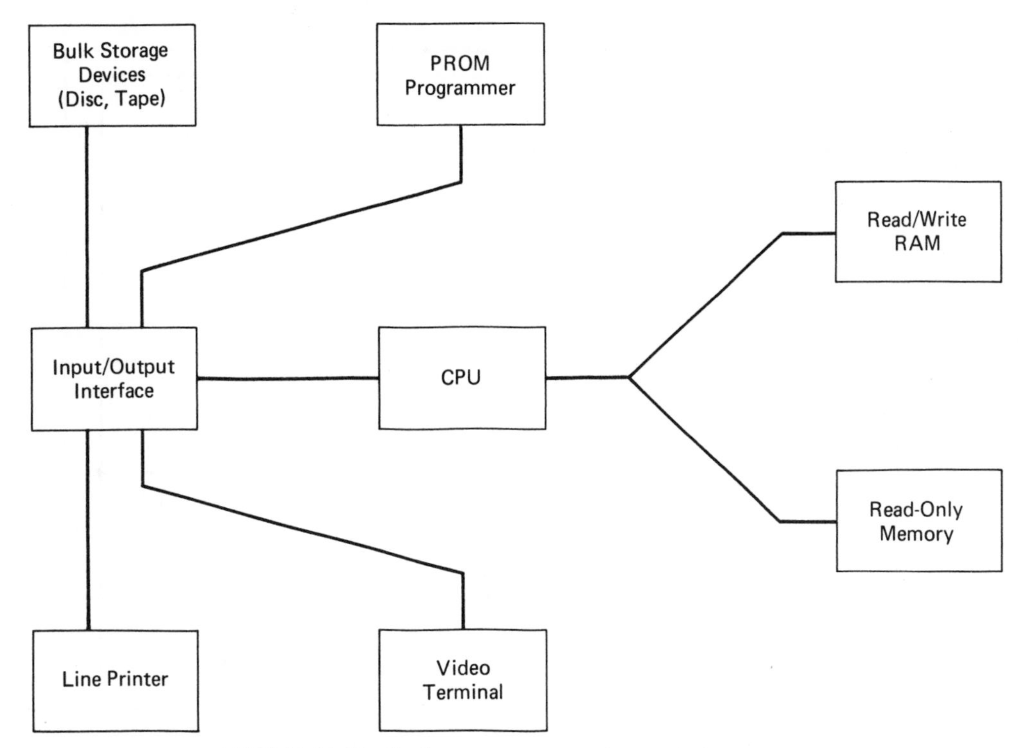

FIGURE 10-26 Hardware components of a practical computer system.

requiring many thousands of man-hours for its design, must be available. Figure 10-26 illustrates the basic hardware components of a practical computer system. Some of the components in this figure we have not discussed, or have mentioned only briefly. A summary discussion of these components follows:

1. Read-only memory (ROM)—a form of RAM in which binary data are permanently or semipermanently recorded. ROM does not require power to retain data, nor can be retained data be changed by CPU write cycles. ROM is used to store an initializing command sequence for use on power-up, and a monitor program that provides utility functions is commonly required when using the system.
2. Video terminal—it is equipped with a typewriter-like keyboard and a television display, and permits communication with the computer using alphanumeric symbols (A, B, C, . . . , 0, 1, 2, . . . , #, $, *, (, +, . . .), rather than just octal- or hexadecimal-coded numbers. Responses by the CPU are displayed on a video screen (CRT).
3. Line printer—a high-speed alphanumeric printing device that provides paper copies of programs and numerical data. Speeds range from 30 to over 8000 characters per second.

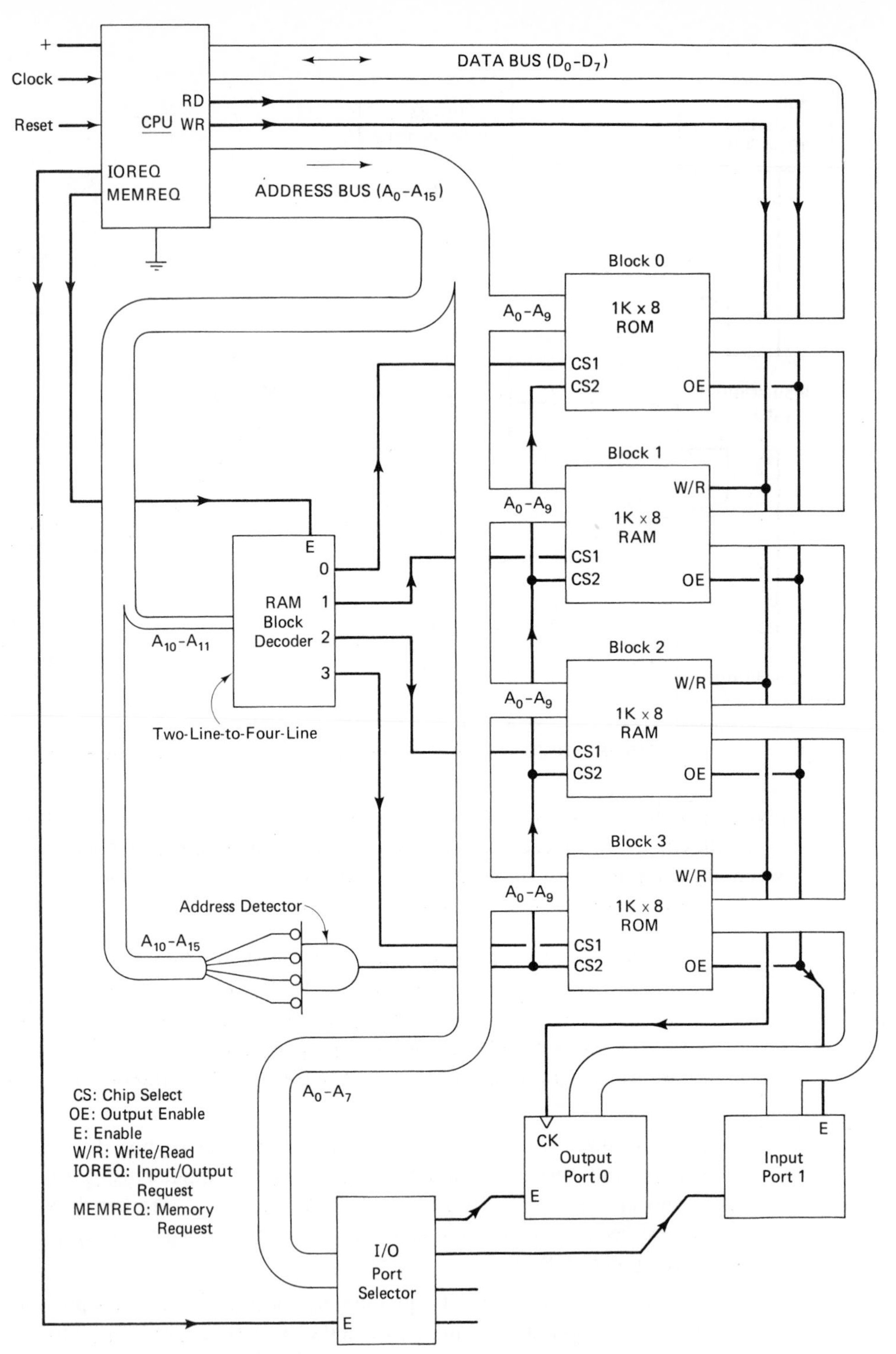

FIGURE 10-27 *Typical microprocessor-based hardware system. Only the most important connections are shown.*

4. Bulk storage—typically, a magnetic-medium serial access storage device capable of retaining 10 to 1000 times the capacity of the RAM memory. The access time of a bulk storage device is considerably longer than RAM (milliseconds vs. microseconds). It is used primarily for long-term nonvolatile storage of programs and numerical data. Punched paper tape is also used occasionally instead of magnetic tape or disc devices.
5. PROM programmer—special circuitry that writes data into programmable read-only memories (PROM's) using high-voltage pulses.

It is fortunate that LSI techiniques have been applied to many other system components besides the CPU chip; the hardware designer's work has been greatly simplified. In particular, compatible RAM, ROM, and interface circuits are usually provided by the microprocessor's manufacturer. The system architecture for a typical microcomputer is generally quite straightfoward. In short, the designer must specify a microprocessor CPU, compatible memory circuits, and I/O interface circuits. Most manufacturers follow a common electrical interface convention which defines voltage levels and timing. Thus, many microcomputer interface circuits are compatible with most I/O devices (terminals, printers, bulk storage). A manufacturer's responsibility in providing compatible components ends with the interface circuit.

The connections for a typical microprocessor-based system are shown in Figure 10-27. In this example system, we assume eight-bit-wide data, and sixteen-bit-wide address paths. Although we cannot, in the alloted space, discuss the details of such a system, we can get an idea of what is involved.

10.7.2 System Software

Productive work on a computer system requires much more than just good hardware. It requires, in addition, a software library consisting of different kinds of utility programs that:

1. Permit alphanumeric text to be entered via a video terminal or similar device.
2. Allow program mnemonics to be automatically translated into an appropriate sequence of binary numbers.
3. Make possible organizational control of the bulk-storage medium so that programs and data are stored in an orderly manner.
4. Permit programs to be single-stepped (single instruction executed) for the purposes of supervised execution.
5. Permit diverse programs and routines to be linked into a common main program.
6. Provide translation of high-level symbolic computer languages.

A ***text editor*** program accomplishes item 1 above, and is used chiefly for program mnemonic entry and alteration. A *file* is created (or reopened if it already exists) when the text editor is invoked. Alphanumeric data are entered, such as:

```
LOOP:_MOV_(T1),A ←
--------ADD_T1,T2 ←
--------INC_T1 ←
--------JNZ_LOOP ←
--------HLT ←
```

Note:
(_) represents "space,"
(←) represents "carriage return, line feed," and both are normally invisible on the character display.

A sequence of instruction mnemonics in this form is referred to as the *source program*, each character of which is handled as a coded binary number. A file, then, consists simply of a sequence of coded binary numbers that represent alphanumeric characters. The text editor program manipulates these characters, as commanded by the human operator. When an editing session is concluded, the file is copied from RAM onto a bulk storage medium, where it can be saved in a power-off condition for future use. The text editor's greatest feature permits fully electronic manipulation of text material, including text that is not necessarily assembly language mnemonics (such as a business letter). Corrections and revisions are made by simply issuing commands to the text editor program via the terminal's keyboard.

An ***assembler*** program translates a sequence of instruction mnemonics existing in a file into appropriate binary code that can be executed by the CPU. The resulting sequence of executable binary numbers is known as an ***object program***. The assembler satisfies item (2) in the list of needed utility programs, and relies upon tables and programmed procedures that tell it which binary codes are needed for each mnemonic. It saves much time over hand assembly, and avoids the coding errors that usually result when carrying out the assembly manually.

The purpose of an ***operating system*** is to facilitate file manipulation, and provide an organized basis for storage and control using bulk storage devices. It is a complex program that is responsible for the coordination of system activity, and accomplishes item 3.

Normally, a great amount of time is spent by the programmer in perfecting programs. Item 4 describes a means by which program debugging is simplified. A ***real-time debug*** program provides information about an operator's program while it is being executed. Specifically, current values of CPU registers and I/O devices are available. The real-time debug program executes concurrently with the operator's program to provide these features. Also, single-step instruction execution is possible. Some special-purpose hardware is necessary, in addition to the debug program, to facilitate these features.

A ***linker-loader*** program, described by item 5, consolidates a variety of smaller programs by loading them into RAM from stored files in a nonoverlapping manner, and adjusting all internal references (jump addresses, data locations) appropriately. Also, program symbols that are not restricted to one particular component program (global symbols) are made to be properly related to all component programs.

In some cases, high-level symbolic computer languages, such as FORTRAN, PL/1, APL, BASIC, and COBOL, are available on a computer system. Such languages provide much greater flexibility than assembly language by allowing

1. Complex mathematical operations to be stated in algebraic form.
2. Simple, symbolic commands to cause a variety of complex responses.
3. English-language-like communication with the computer.

Compiler programs translate high-level-language commands into an appropriate sequence of assembly language commands that can be executed by the CPU. Compilers are themselves very long and complex programs, requiring many hundreds of man-hours to design and perfect.

SUMMARY

A digital computer is a special type of digital device whose operation is dependent upon a stored sequence of binary-coded instructions. These coded instructions constitute a program, and are typically stored in a random access memory circuit. Numerical data to be operated on by the program is also stored in the memory. Each instruction performs only a very basic operation that could easily be done manually by a human being. The computer, however, can execute instructions millions of times faster than a human counterpart, and consequently is capable of performing almost miraculous feats of computation. We know that AND, OR, and INVERT functions can, in combination, implement *any* digital circuit. In the same sense, the instruction set of a computer is chosen so that any desired complex operation can be obtained by executing a sequence of simple commands.

Programmed logic is executed on a digital computer, and is distinguished from wired logic in one crucial way: flexibility. The function of a programmed device is changed merely by altering a stored sequence of binary numbers (the instruction codes). Changing the function of wired logic, however, requires changes in the physical configuration of the circuit.

A random access memory is a digital circuit capable of storing a multitude of binary numbers, and reports the stored data in response to the application of address inputs. The access time of this data is largely independent of where it is stored within the device. A read/write input controls the directional flow of data, and permits new information to be stored in the device. Read-only memory is a form of random access memory in which the recorded data is nonvolatile and permanent. In some read-only devices, data is electrically alterable by applying a special, complex technique, the stored data being semipermanent in this case. The cycle time of a memory circuit defines the minimum permitted time between successive read or write operations. Typically, the cycle time is longer for write operations.

Three-state devices contain special output circuits that enable the output to be completely turned off in one case, or turned on to drive '0' or '1' values onto a line in the other case. An additional input controls the turn-off mode. Whenever bus-structured circuits are designed, three-state circuits are needed at the bus interface so that multiple transmitters can be wired to the same line without mutual interference.

CPU architecture describes the component parts of the CPU and shows how they are interconnected. System architecture treats the CPU as a single component and defines its relationship to memory devices and I/O circuits.

Every computer instruction consists of a sequence of small basic steps that involve data transfer between internal CPU blocks, or the actuation of memory and I/O circuits. The number of steps required depends on the type of instruction that is being carried out. All instructions are divided into two basic processes: Fetch and Execute. The Fetch process initiates every instruction and causes the operation code for that instruction to be loaded into the CPU's instruction register. Execution proceeds only after the op code has been fetched and interpreted.

Computer instructions are represented by mnemonics, which are easily recognizable groups of letters and numbers. All mnemonics, however, must be converted into an appropriate sequence of binary numbers that can be executed by a computer. This conversion process is known as assembly.

An algorithm defines a step-by-step procedure that results in the solution to a problem. This procedure usually involves the repetition of a group of steps many times. Algorithms are graphically described by flowcharts, and ideally bear no relationship to a specific computer's instruction set. Once the flowchart is determined, a corresponding computer program is written using the assembly-language mnemonics of one specific computer. Finally, the mnemonics are assembled into executable binary code.

A practical computer system requires a variety of hardware devices and system programs. In addition to CPU, memory, and interface circuits, the practical system contains a video terminal, a line printer, and a bulk storage device such as a disc or magnetic tape. System programs include an operating system program for file manipulation and program control, a text editor program for entry of instruction mnemonics via a terminal, and an assembler program to translate mnemonics into executable code. Additional hardware and software devices are also available for special-purpose applications.

NEW TERMS

Digital Computer
Microcomputer
Microprocessor
Programmed Logic
Instruction Set
Program
Electronic Memory
Access Time
Random Access Memory (RAM)
Serial Access Memory (SAM)
Read/Write Input
Unidirectional Bus

Central Processing Unit (CPU)
CPU Architecture vs.
System Architecture
Program Counter (PC)
Memory Address Register (MAR)
Memory Data Register (MDR)
Input/Output Register (IOR)
Instruction Register (IR)
Instruction Decoder and
Sequencer (IDS)
Accumulator (A), Accumulator
Latch (ACCL)

Bidirectional Bus
Volatile, Nonvolatile Memory
Bulk Storage
Address Decoder
Read Cycle
Cycle Time
Address Setup Time
Data Setup Time
Address Hold Time
Data Hold Time
Read-Only Memory (ROM)
Fusible Link
Three-State Output
Electrically Programmable
  Read-Only Memory (EPROM)
Programmed Sequence Generator
Fixed-Wired Counter, RAM Counter
Instruction Code
Fetch Process
Execute Process
Instruction Cycle
Computer Architecture
Bus
Bus-Structured Architecture
Assembler
Linker-Loader
Operating System
Real-Time Debug
Arithmetic and Logic Unit (ALU),
  ALU Latch (ALUL)
Processor Status Word (PSW)
Output Port
Input Port
Address Bus
Data Bus
Branch Instruction
Data Transfer Instruction
Program Loop
Instruction Mnemonic
Addressing Mode
  Immediate Mode
  Direct Mode
  Indirect Mode
Operand
Operation Code
Algorithm
Flowchart
Absolute Address
Video Terminal
Line Printer
PROM Programmer
Text Editor
Compiler
Static Memory
Dynamic Memory
Refreshing

PROBLEMS

10-1 **(a)** Determine the total number of bits stored in a RAM which has
(i) Four address inputs and ten data outputs.
(ii) Ten address inputs and four data outputs.
(iii) Twelve address inputs and eight data outputs.

(b) Determine all possible combinations of address inputs and data outputs for a RAM that stores 512 bits of data.

10-2 What might happen if the read/write input on the RAM in Figure 10-1

(a) Is asserted to write before the minimum address setup time?

(b) Is deasserted from write before the data setup time?

(c) Is asserted for too short a time?

10-3 What might happen if the address inputs on a RAM are changed too soon after a write pulse? What parameter is violated? (*Hint:* Refer to Figure 10-6.)

10-4 Identify each of the following memory types as volatile, nonvolatile, or either:

(a) Static RAM (semiconductor) **(b)** Static RAM (magnetic core)

(c) Dynamic RAM (semiconductor) **(d)** ROM

(e) PROM **(f)** EPROM

(g) SAM

10-5 A RAM circuit has eight address inputs and eight data outputs. If the entire circuit requires 0.5 watts, determine the average power required per bit.

10-6 Compute the access time of the four-word by one-bit RAM design of Figure 10-3. Assume that all gate delays are 10 ns.

10-7 Fusible link ROM's are programmed for '1's by burning open microscopic wires, and for '0's by leaving them intact. Once a wire is burned open, it is permanent and cannot be changed. Given the data pattern in Figure P10-7(a) for a four-word by four-bit ROM, determine which of the patterns in Figure P10-7(b) can be obtained from (a) without purchasing a new device.

```
1 0 1 1        1 0 1 0      1 1 1 1      1 0 1 1
0 1 1 1        0 1 0 0      0 1 1 1      0 1 1 1
0 0 1 0        1 1 0 1      1 0 1 0      0 0 1 1
1 0 1 0        0 0 1 0      1 0 1 1      1 1 1 1

                 (i)          (ii)         (iii)

  (a)                         (b)
```

FIGURE P10-7

10-8 Describe electrically each of the three possible states in a three-state driver.

10-9 **(a)** Design a digital circuit using three-state drivers to multiplex four independent logic signals onto a single unidirectional line. The source signal is selected with a two-bit binary input, labeled 'A_1A_0'.

(b) Modify the circuit of part (a) so that one of two source lines are selected for transmission, or one of two receiver gates are enabled (there are four states in total, selected by 'A_1A_0'). This makes the bus bidirectional. If this bidirectional multiplexer is to be coordinated with a similar circuit at the other end of the line, what additional signal lines must be present in the bus besides the one data line and ground?

10-10 **(a)** Determine the RAM contents necessary for the programmed sequence generator of Figure 10-12 to progress through this sequence:

Q_2	Q_1	Q_0
0	0	1
1	0	0
1	1	0
1	1	1
	(repeats)	

(b) What must the RAM contents be if 'T'-type flip-flops are used in part (a)?

10-11 Would we be primarily concerned with system architecture, CPU architecture, or both, if we were designing a

(a) Microprocessor chip?

(b) Microcomputer chip?

10-12 For the instructions depicted in Figure P10-12, answer the following questions:

(a) Which utilize the ALU?

(b) Which are I/O instructions?

(c) Which are fetched from memory?

(d) Which are two-word instructions?

(e) Which affect the PSW?

(f) Which use the PSW in their execution?

(g) Which alter the accumulator?

(h) Which utilize indirect addressing?

(i) Which cause the PC to be parallel-loaded with a new address?

(j) Which change the contents of RAM?

(k) Which have no effect at all?

(l) Which clear the accumulator?

```
(1) MOV T1, A              (8) JNZ #3AA
(2) MOV #03C, (T2)         (9) HLT
(3) JMP #4FC              (10) COM T2
(4) AND T1, T2            (11) MOV T1, T1
(5) INC (A)               (12) JC #105
(6) MOV IOR, A            (13) ADD #000, T2
(7) SUB A, A              (14) XOR T1, T1
```

FIGURE P10-12

10-13 Explain the difference between the “carry” and “zero” bits of the PSW.

10-14 Explain what is incorrect about each of the following instructions, relative to the instruction set defined in Table 10-1:

MOV #100,#3A5	AND T1,T2,A
JMP T1	XOR T3,T1
JNZ #10FA	COM A,T1
MOV #6B1,PSW	SUB T1,#310

10-15 The CPU architecture of Figure 10-15 contains thirteen basic blocks (PC, MAR, MDR, etc.). Which of these blocks

(a) Are always involved in the Fetch process?

(b) Are parallel loadable registers?

(c) Is the implied destination of many ALU operations?

(d) Are involved in the conditional branch operations?

(e) Is the most complex, from a digital circuit viewpoint?

(f) Are always involved in the execute process?

10-16 Define the *exact* sequence of steps the IDS must take for the execution of these instructions:

MOV (T1),(T2)	JNZ #30C
ADD A,A	AND (T1),A
CMP (A)	MOV T1,T2
XOR A,T1	INC A

(*Hint:* Refer to Table 10-2.)

10-17 Classify each of the instructions in Table 10-1 as one of the following:

1. Data Transfer.
2. Arithmetic and Logic.
3. Branch.
4. I/O.
5. Control.

Which type affects the PSW? Which type uses the PSW?

10-18 Can the Instruction Decoder and Sequencer distinguish instruction codes from numerical data? Why?

10-19 Assume that the IR in Figure 10-15 had its 'D_7' bit stuck at '0' all the time. How would the block transfer program in Figure 10-19 be interpreted by the IDS?

10-20 (a) Determine the exact delay offered by the time delay program shown in Figure P10-20. Assume that the period of the computer's clock is 1 μs and that the number of states per instruction are as shown in the figure.

(b) Write a general flowchart for this program. Do not use assembly-language mnemonics! Be sure to show the register assignment table.

(c) Assemble the program shown in Figure P10-20 into a sequence of executable codes, in accordance with the procedure shown in Figure 10-20.

```
                                Number of States
                                Per Instruction
        MOV #FC0, T1     (12)
LOOP2:  MOV #000, T2     (12)
LOOP1:  DEC T2           (8)
        JNZ LOOP1        (12)
        DEC T1           (8)
        JNZ LOOP2        (12)
        HLT
```

FIGURE P10-20

```
200   0 0 0 0 0 0 0 0 0 0 0 1
201   0 0 0 1 0 0 0 0 0 0 0 0
202   0 0 0 0 0 0 0 0 0 0 0 0
203   0 0 0 0 0 0 0 0 0 0 0 0
204   0 0 0 0 0 0 1 0 0 1 0 1
205   0 1 0 1 1 0 0 0 0 0 0 1
206   0 1 1 1 0 0 0 0 0 0 0 0
207   0 0 1 0 0 0 0 0 0 1 0 0
208   1 0 0 0 1 0 0 0 0 0 0 0
```

FIGURE P10-21

10-21 The binary code shown in Figure P10-21 is found in memory.

(a) Disassemble this code in accordance with the instruction set of Table 10-1 and the bit format in Figure 10-16. (In other words, determine the corresponding mnemonics.)

(b) Write a flowchart for this program.

(c) What does the program do to RAM? Over what range?

10-22 (a) Design a flowchart for an algorithm that places a ten-element list in numerical order. Assume that all entries are positive binary integers.

(b) Write a program for this algorithm using the instruction set of Table 10-1.

10-23 (a) Design a flowchart for an algorithm that selects the largest number in a table of ten entries. Assume that all entries are positive binary integers.

(b) Write a program for this algorithm using the instruction set of Table 10-1.

10-24 It can be noted that fifteen op codes are unassigned in the example instruction set of Table 10-1. Design five new instructions that you feel would be a useful supplement to the instruction set. Be sure to specify the exact sequence of register transmissions that are needed, and to conform with the bit-field format used in all other instructions.

10-25 The twelve bits of the output port in Figure 10-15 are to replace the twelve output lines in each of the hardware circuits shown in Figure P10-25. For each one, write a program that causes the output port to exactly simulate the operation of the circuit.

FIGURE P10-25

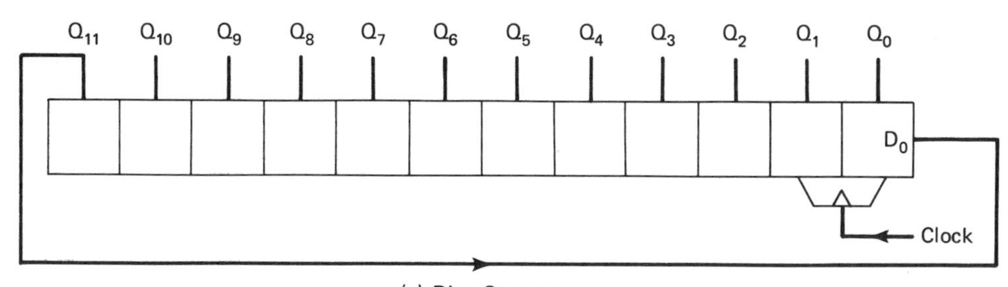

(a) Ring Counter.

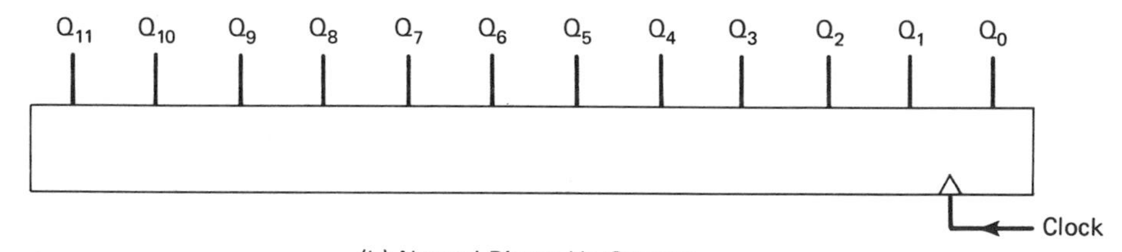

(b) Natural Binary Up Counter.

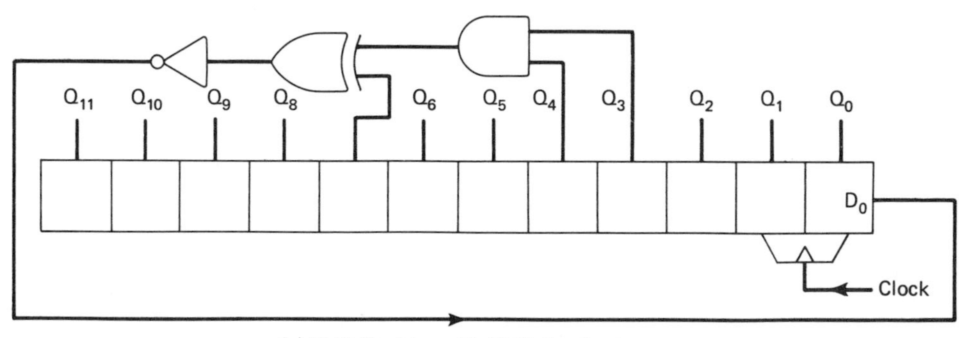

(c) Shift Register with XOR Feedback.

APPENDIX

TECHNOLOGY REFERENCES

The primary thrust of this book is to convey the fundamental concepts behind digital electronic devices. Specific mention of digital circuitry (gate packaging, pin connections, circuit types, etc.) has been avoided intentionally to leave the student open to a variety of technologies, and to not create a bias in one direction or another. Also, with the rapid changes occurring in the circuit designs that implement logic functions, circuitry that is used at this writing may be obsolete when these words are read. For those students that are interested in learning about specific circuitry, a list of some major semiconductor manufacturers and their addresses is provided. A variety of data books and applications manuals are available from manufacturers, usually at a nominal price. Writing to a manufacturer, or calling a local distributor, should reveal exactly how data books can be obtained, and how much they cost.

Advanced Micro Devices, Inc.
901 Thompson Place
Sunnyvale, California 94086

American Microsystems, Inc.
3800 Homestead Road
Santa Clara, California 95051

Fairchild Camera and Instrument Corp.
P.O. Box 880A
Mountain View, California 94042

Harris Semiconductors, Inc.
P.O. Box 883
Melbourne, Florida 32901

Intel Corporation
3065 Bowers Ave.
Santa Clara, California 95051

Mostek Corporation
1215 W. Crosby Road
Carrollton, Texas 75006

Motorola Semiconductor Products, Inc.
3501 Ed Bluestein Blvd.
Austin, Texas 78721 (CMOS)
or
Box 20912
Phoenix, Arizona 85036
(microcomputers)

National Semiconductor Corp.
2900 Semiconductor Drive
Santa Clara, California 95051

NEC Microcomputers, Inc.
173 Worcester Street
Wellesley, Massachusetts 02181

RCA Solid State Division
Box 3200
Somerville, N.J. 08876

Texas Instruments, Inc.
6000 Denton Drive
P.O. Box 5012
Dallas, Texas 75222

Zilog Corporation
10460 Bubb Road
Cupertino, California 95014

TABLE OF ACRONYMS

AC—Alternating Current
A/D—Analog-to-Digital
ADC—Analog-to-Digital Converter
ALU—Arithmetic and Logic Unit
ANSI—American National Standards Institute
ASCII—American Standard Code for Information Interchange
BCD—Binary-Coded Decimal
CCD—Charge-Coupled Device
CML—Current Mode Logic
CMOS—Complementary Metal Oxide Semiconductor
CPU—Central Processing Unit
CRC—Cyclic Redundancy Check
CRT—Cathode Ray Tube
D/A—Digital-to-Analog
DAC—Digital-to-Analog Converter
DC—Direct Current
DIP—Dual In-Line Package
DMA—Direct Memory Access
DMUX—Demultiplexer
DTL—Diode-Transistor Logic
DVM—Digital Voltmeter
EBCDIC—Extended Binary Coded Decimal Interchange Code
ECL—Emitter Coupled Logic
EPROM—Electrically Programmable Read-Only Memory
FET—Field Effect Transistor
FIFO—First-In First-Out
HTL—High-Threshold Logic
IC—Integrated Circuit
IDS—Instruction Decoder and Sequencer
I^2L (or IIL)—Integrated Injection Logic
I/O—Input/Output
IOR—Input/Output Register
IR—Instruction Register
LCD—Liquid Crystal Display
LED—Light-Emitting Diode
LIFO—Last-In First-Out
LSB—Least Significant Bit
LSI—Large-Scale Integration
MAR—Memory Address Register
MDR—Memory Data Register
MODEM—Modulator/Demodulator
MOS—Metal Oxide Semiconductor
MPS—Minimum Product of Sums
MPU—Microprocessor Unit
MSB—Most Significant Bit
MSI—Medium-Scale Integration
MSP—Minimum Sum of Products
MUX—Multiplexer
n-MOS—n-channel MOS
OCR—Optical Character Recognition
PC—Printed Circuit
PC—Program Counter
PLA—Programmable Logic Array
PLL—Phase-Locked Loop
p-MOS—p-channel MOS
PROM—Programmable Read-Only Memory
PSW—Processor Status Word
RAM—Random Access Memory
ROM—Read-Only Memory

R/L—Right/Left
RTL—Resistor-Transistor Logic
R/W—Read/Write
SAM—Serial Access Memory
SOS—Silicon on Sapphire
SPDT—Single-Pole Double-Throw
SPST—Single-Pole Single-Throw
SSI—Small-Scale Integration
TTL (or T^2L)—Transistor-Transistor Logic
TTY—Teletypwriter
UART—Universal Asynchronous Receiver/Transmitter
USART—Universal Synchronous/Asynchronous Receiver/Transmitter
VCO—Voltage-Controlled Oscillator

GLOSSARY

GLOSSARY

All new terms found at the end of each chapter are compiled and defined in this glossary. Refer to the Table of Acronyms for any abbreviations that are unfamiliar.

Absolute Address: a memory address that is given exactly, and not computed or inferred from other information.

Access Time: the time needed to gain access to stored information in an electronic memory.

Accumulator: a parallel array of flip-flops in the CPU of a computer that serves as the source and destination of most arithmetic operations.

Accuracy: the degree to which a device is calibrated against a known standard.

Active Edge: the triggering edge that causes an edge-sensitive device to sample its input(s).

Active Level: the logic state in which a device is enabled.

Addend: in arithmetic addition, the quantity added to the augend.

$$(\text{augend}) + (\text{addend}) = (\text{sum})$$

Address Bus: a parallel array of conductors capable of transmitting address information from the CPU of a computer to external elements such as memory and I/O circuitry. It is a unidirectional bus. (*see* "Data Bus")

Address Decoder: an 'n' line–to–'2^n' line decoder used in semiconductor memories to select one specific storage cell.

Addressing Mode: (1) General: the means by which data in a computer instruction is specified. (2) Immediate mode: the numerical data is given exactly as part of the instruction code. (3) Direct mode: the address of the data is given. (4) Indirect mode: the location of the address of the data is given.

Adjacent Cells: (1) Graphically: Karnaugh map cells that have a common border. (2) Numerically: Karnaugh map cells whose addresses differ by only one bit.

Algorithm: a step-by-step procedure that outlines the sequence of actions necessary to solve a problem.

Analog Information Processing: the operations on analog signals that cause information translation, information manipulation, or information storage. (*see* "Digital Information Processing")

Analog Signal: a voltage or current value that is a proportional representation of some measured or calculated parameter. An analog signal smoothly spans a range of values. (*see* "Digital Signal")

Analog-to-Digital Converter: a circuit that translates an analog signal, spanning a smooth range of voltage (or current) into a digital signal suitable for digital processing. (*see* "Digital-to-Analog Converter")

AND: a logic function which is defined as follows:

Truth Table

X	Y	Z
0	0	0
0	1	0
1	0	0
1	1	1

Algebraically

$$Z = X \cdot Y$$

Schematically

AND Precedence: a convention in which AND operations are performed before OR operations in the absence of parentheses.

Argument: the values upon which a function operates.

Arithmetic and Logic Unit (ALU): a complex combinational logic block in the CPU of a computer responsible for all arithmetic and logic operations.

Assembler: a computer program that translates binary-coded alphanumeric symbols that represent instruction mnemonics into executable instruction codes.

Associative Law: a Boolean property which states that the location of parentheses in a compound OR or AND function does not matter:

$$(X + Y) + Z = X + (Y + Z)$$
$$(XY)Z = X(YZ)$$

Astable Multivibrator: an oscillator circuit whose output is a binary signal. Typically, an astable produces a pulse train or a square wave.

Asynchronous Device: a circuit or module that *does not* operate in response to or in synchronization with a clock signal. (*see* "Synchronous Device")

Asynchronous Sequential Circuit: a sequential circuit possessing no flip-flops, and whose operation is not controlled by a clock signal. (*see* "Synchronous Sequential Circuit")

Augend: in arithmetic addition, the quantity to which the addend is added:

$$(\text{augend}) + (\text{addend}) = (\text{sum})$$

Base: *same as* "Radix"

BCD Code: "Binary-Coded-Decimal" code is a modification of the decimal number system different only in that the decimal digits are independently coded as four-bit binary numbers. Example: $(138)_{10} = (0001\ 0011\ 1000)_{BCD}$.

BCD Counter: a ten-state natural binary counter module whose sequence begins at 0000 and ends at 1001, cascadable to any length to produce a multidigit binary-coded-decimal count sequence.

Bidirectional Bus: a parallel array of conductors capable of transmitting information in two directions, requiring three-state driver gates at each end. (*see* "Unidirectional Bus")

Binary Counter: an interconnection of flip-flops whose outputs progress through a natural binary sequence when a periodic signal is applied to its clock input.

Binary Digital Signal: a digital signal having exactly two discrete levels. (*see* "Digital Signal")

Binary Number: a numerical quantity expressed as a weighted combination of two symbols, '0' and '1'. Examples: $(9)_{10} = (1001)_2$, $(31)_{10} = (11111)_2$.

Binary Variable: a letter of the alphabet used to represent generally a binary signal which could have a value of either '0' or '1'.

Bistable Multivibrator: an alternative name for a flip-flop.

Bit: a *BI*nary digi*T* (either '0' or '1').

Boolean Algebra: a set of rules and properties by which symbolic models of two two-state logic circuits can be manipulated.

Boolean Constant: a specific, unchanging logic value of either '0' or '1'.

Boolean Function: a well-defined relationship between Boolean variables, specified by either a Boolean equation or a truth table.

Boolean Variable: *same as* "Binary Variable"

Borrow Digit: a digit produced from the MSB stage in subtraction when the subtrahend is larger than the minuend.

Branch Instruction: a type of computer instruction which causes transfer to a nonsuccessive memory address for the next instruction.

Branch State: a counter state which is not part of the cyclic count sequence. A branch state is possible only for a non-maximal-length count sequence.

Bridge Fault: a static fault condition that exists when two logic outputs are inadvertantly shorted together.

Buffer Gate: a gate whose output can drive substantially more inputs that a standard gate, thereby providing increased fanout.

Bulk Storage: the storage of large quantities of binary information on an inexpensive medium, usually magnetic tape or disc.

Bus-Structured Architecture: a computer architecture in which the component register and logic blocks all interface to one or more common internal busses.

Byte: an eight-bit binary quantity.

Carry Digit: a digit produced from the MSB stage in arithmetic addition when the sum cannot be expressed with the alloted number of digits.

Carry Input: in cascadable modules, an input accepted from a previous stage.

Carry Output: In cascadable modules, an output intended for a successive stage, which, when asserted, indicates overflow in the current stage.

Checksum: a modulus sum of a sequence of binary words, useful for error detection.

Clear Direct: an asynchronous input found on flip-flops, registers, and counters that causes the internal state of such devices to become '0' (or all '0's).

Clocked D Latch: an S-R latch with $S = D$, and $R = \bar{S}$. When enabled, 'Q' follows 'D'. When disabled, 'Q' is held steady at its last correct value.

Clocked S-R Latch: a type of latch that contains "set," "reset," and "enable" (or clock) inputs, and requires an enable signal to load data.

Combinational Logic: memoryless logic circuitry that can be unambiguously specified by a truth table, and in which no feedback exists. (*see* "Sequential Logic")

Combinational Logic Fault: a fault condition in a combinational logic circuit shown by the desired truth table and measured truth table being unequal for one or more entries. (*see* "Sequential Logic Fault")

Commutative Law: a Boolean property which states that the order of two variables connected by AND or OR functions does not matter:

$$X + Y = Y + X$$
$$XY = YX$$

Comparator: (1) A digital circuit or module used to compare the values of two binary numbers. The output of such a circuit consists of three lines, labeled "greater than," "equal," and "less than." (2) *See* "Differential Comparator."

Compiler: a computer program that translates high-level-language commands into machine-executable code, often producing intermediate-level assembly-language mnemonics.

Complement: the value obtained by reversing the state of a binary digit. (*see* "One's Complement" and "Two's Complement")

Computer Architecture: the specific interconnection of registers and logic that defines the design of a central processing unit. (*see* "System Architecture")

Conversion Time: the minimum time needed by an A/D or D/A converter to make one conversion.

Count Sequence: the repeating sequence of binary numbers observed on the outputs of a counter circuit.

CPU Architecture: *same as* "Computer Architecture"

Critical Race: *see* "Race"

Crosstalk: a dynamic fault condition that exists when two parallel conductors are physically too close, and become electromagnetically coupled.

Crystal Oscillator: an astable circuit whose frequency-determining element is a quartz crystal, characterized by frequency stability over time and over variations in temperature.

Current Mode: a system of electronic logic in which signal currents (not voltages) are measured and related to the logic conditions of '0' and '1'. (*see* "Voltage Mode")

Current Summation: the addition of currents in a resistor network to arrive at a sum current, which is then amplified to a useful level with an operational amplifier circuit.

Cycle Time: the minimum time between successive read or write cycles in an electronic memory.

D Flip-Flop: a one-bit clocked memory element that is represented, and behaves, as follows:

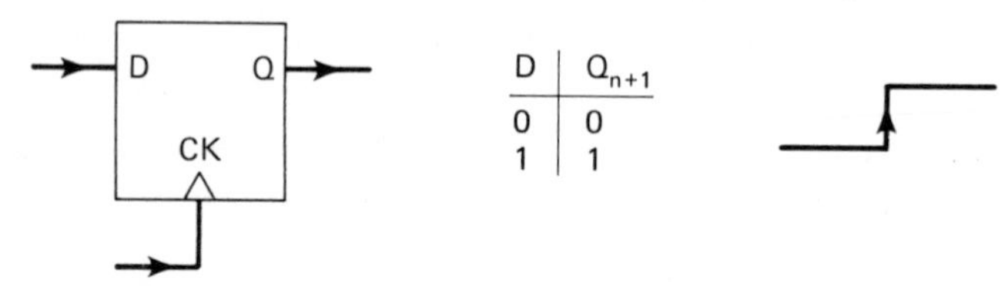

D	Q_{n+1}
0	0
1	1

Data Bus: a parallel array of conductors capable of transmitting and receiving data between the CPU of a computer and external elements, such as memory and I/O circuitry. It is often a bidirectional bus. (*see* "Address Bus")

Data Transfer Instruction: a type of computer instruction that causes numerical data to be moved between internal CPU registers, or between CPU registers and memory.

DC Offset: an unchanging voltage value added to a signal, causing the '0' state of a digital to be offset from zero volts.

Decoupling Capacitor: a capacitor connected locally across the power supply pins of an IC to prevent high-frequency variations in load from affecting nearby IC's. Usually, a decoupling capacitor is of the disc variety, and between 0.01 and 0.1 μF.

Delay Element: a block representing the lumped delay of a combinational network, found in the feedback path of an asynchronous sequential circuit.

DeMorgan's Law: the Boolean property that states:

$$\overline{XY} = \bar{X} + \bar{Y}$$
$$\overline{X + Y} = \bar{X}\bar{Y}$$

Demultiplexer: a logic circuit in which a single input is gated onto one of a multitude of output lines. (*see* "Multiplexer")

Detect and Steer: a method of counter design in which the last correct state in the sequence is sensed by a detector gate, which in turn causes the component flip-flops to be steered back to an earlier state.

Difference: the result of subtraction. (*see* "Minuend" or "Subtrahend")

Differential Comparator: an analog integrated circuit used to compare the magnitudes of two input voltages, producing a logic-compatible output whose value depends upon which input is greater.

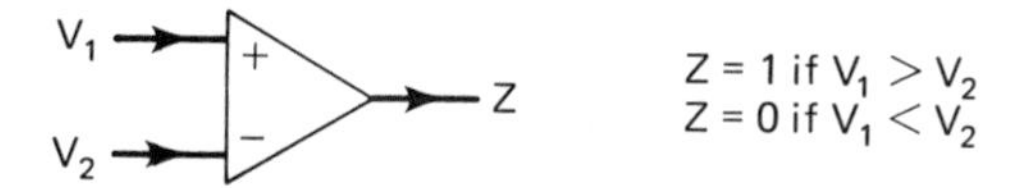

Differential Line Driver: an interface circuit used to convert a single logic signal into a complementary pair of signals for noise-resistant transmission over long cables.

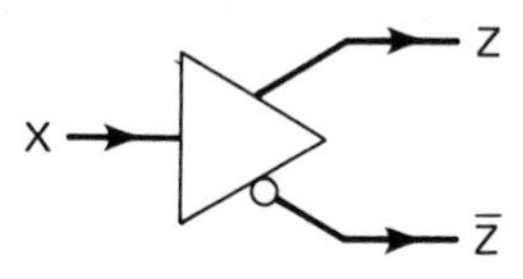

Differential Line Receiver: an interface circuit that converts a complementary signal pair from a cable into a single logic output.

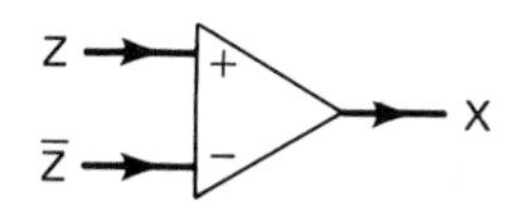

Digital Computer: a digital device whose function and operation depends upon a stored sequence of binary numbers.

Digital Information Processing: the operations on digital signals that cause information translation, information manipulation, and information storage. (*see* "Analog Information Processing")

Digital Magnetic Recording: the means by which digital information is stored on a moving magnetic medium, such as magnetic tape and disc.

Digital Signal: (1) A waveform whose value at any point in time will be precisely one of a group of select discrete levels. (*see* "Binary Digital Signal") (2) A sequence of binary numbers that represents a measured or calculated parameter. A digital signal depicts discrete quantities with a well-defined precision.

Digital Signal Processing: the processing of digitally coded signals on computers, or with special-purpose digital hardware.

Digital-to-Analog Converter (D/A): a circuit that transforms binary numbers into corresponding proportional voltages or currents. (*see* "Analog-to-Digital Converter")

Distributive Law: a Boolean property that shows how factored AND, OR expressions can be expanded:

$$X(Y + Z) = XY + XZ$$

$$X + (YZ) = (X + Y)(X + Z)$$

Division: (1) Restoring division: a shift-and-subtract technique for binary division. (2) Nonrestoring division: a division method similar to (1), but where a comparison between divisor and partial dividend is not necessary.

Don't Care State: a particular output or state condition that can be either '0' or '1' without affecting the overall function of a logic circuit, denoted by '∅'.

Dropout: the inadvertant absence of signal on a digital magnetic recording medium due to microscopic irregularities in the medium.

Dual Postulate: a new postulate obtained by complementing every constant, and the functional operation, of a given postulate.

Dynamic Fault: a transient fault condition that exists only momentarily, between state or input changes. (*see* "Static Fault")

Dynamic Memory: a type of memory design utilizing a semiconductor capacitor matrix, and requiring periodic refreshing to maintain data. (*see* "Static Memory")

Edge Triggered Device: a synchronous device whose state may change only on the rising or falling edge (but not both) of a clock signal. Also referred to as a "clocked" device. (*see* "Level-Sensitive Device")

Electrically Programmable Read-Only Memory (EPROM): a type of random access memory whose data are semipermanently recorded, and alterable only by using a special technique.

Electromagnetic Coupling: the condition in which voltage or current variations in one wire are induced in an adjacent unconnected wire by electric and magnetic fields.

Electronic Memory: (1) A digital electronic circuit capable of storing multibit binary numbers, each of which is located at a unique address within the memory circuit. (2) Digital circuitry in combination with electromechanical components capable of storing multibit binary numbers on a movable magnetic medium.

Enable Input: an input usually found on counter and register modules that, when asserted, permits normal operations to occur. When deasserted, normal operation is disabled.

Enclosure: an encirclement on a Karnaugh map that denotes an implicant.

Equation: an algebraic statement that shows the equality of two expressions.

Equivalent Function: a logic function that is identical to another function, but implemented with different logic gate connections.

Equivalent States: two separate states in a sequential circuit that lead to the same new state under identical input conditions and with identical outputs.

Essential Prime Implicant: a prime implicant that encloses one or more '1'-cells, and cannot be enclosed by any other prime implicants.

Even Parity: the condition where a binary number has an even number of '1' digits. Example: 01111011—even parity, 01100010—odd parity. (*see* "Odd Parity")

Excitation Equations: (1) Counters: a Boolean equation which describes how a flip-flop within a counter is driven by the counter's 'Q' outputs. (2) Sequential circuits: a Boolean equation which describes how a flip-flop or delay element is driven by a sequential circuit's primary or secondary inputs.

Excitation Table: a tabular representation of the excitation equations showing how flip-flops or delay elements are driven for all possible input combinations.

Excited Delay Element: a delay element whose input and output states are unequal. (*see* "Quiescent Delay Element")

Execute Process: the second phase of an instruction cycle during which the instruction code (or, alternatively, "op code") is interpreted.

Expression: (1) Simple: an algebraic statement in which only one Boolean function is involved. Example: $Z = X + Y$. (2) Compound: an algebraic statement in which two or more Boolean functions are involved. Example: $Z = XY + \bar{Y}Q$.

Fall Time: the duration of time in a '1' to '0' transition for a signal to change from 90% to 10% of its logic '1' value.

Fanout: the number of inputs that can be driven by a single output. TTL circuits typically have a fanout of ten.

Feedback Path: a connection between an output and input of the same circuit.

Fetch Process: the first phase of an instruction cycle during which an instruction code (or, alternatively, "op code") is fetched from memory. (*see* "Execute Process")

Fixed-Wired Counter: a counter whose connections are permanently wired, and whose count sequence is unalterable.

Flash Converter: a type of A/D converter utilizing a resistive divider (for the analog input), a comparator for each quantization level, and a binary encoder.

Flip-Flop: a one-bit memory element possessing between two and five inputs to control the entry of data.

Flow Chart: a graphical means of expressing an algorithm. Circles are used to denote "start" or "stop," rectangles are used for "processing," and diamonds are used for "decisions."

Flow Graph: the description of a sequential circuit's operation by the use of interconnected circles, each of which represents a circuit state, to show the permitted pattern of transitions between states.

Frequency: the repetition rate of a cyclic signal, expressed in hertz (Hz) (1 Hz is equivalent to 1 cycle per second). Frequency 'f' is related to period 'T' by

$$f = 1/T$$

Frequency Divider: a counter whose clock input is derived from a periodic signal, and whose 'Q' output frequencies are submultiples of the clock frequency.

Full Adder: a one-bit cascadable adder circuit accepting three inputs (C_{n-1}, X_n, Y_n) and producing two outputs (C_n, S_n). (*see* "Half Adder")

Functional Result: the result obtained by assigning fixed values to Boolean variables and evaluating the expression.

Fusible Link: a fine wire that, when intact, represents logic '0', and when burned open, represents logic '1'. Fusible-link matrices are used in some ROM circuits.

Gray Code: a binary code in which the difference between successive values is exactly one bit. Example: 000, 001, 011, 010, 110, 111, 101, 100.

Ground Return: a connection that completes the electrical circuit in logic gate circuits.

Half Adder: a simplified one-bit adder module used as the LSB stage of a multibit adder circuit, where the carry input is assumed to be '0'.

Hazard: a condition that exists in combinational logic when two or more inputs change simultaneously, but because of unequal path propagation delays in the logic, cause improper excitations and momentary false outputs to exist.

Hexadecimal Code: a simplified means of representing binary numbers in a radix-sixteen number system.

Hold Time: (1) General: in synchronous devices, an amount of time after the application of a clock signal when data input(s) must remain stable. (2) Address: the minimum time that address inputs to a RAM must be stable after read or write operations. (3) Data: the minimum time that the data inputs to a RAM must be stable after applying a "write" pulse. (*see* "Setup Time")

Hysteresis: a condition in which the threshold between two discrete states varies, and is determined by the state that is currently occupied. Hysteresis operates so as to allow the widest possible analog variation within a discrete state prior to switching. Having switched to the next state, the threshold of that new state widens to include points that were previously stable in the last state. (*see* "Schmitt Trigger")

Implicant: a product term in a sum-of-products expression. (*see* "Enclosure," "Prime Implicant," and "Essential Prime Implicant")

Implied AND: a convention in which the AND symbol, '$\cdot$', is eliminated for simplicity. Example: $X \cdot Y \cdot Z = XYZ$.

Information Manipulation: altering or combining information in accordance with a well-defined procedure.

Information Translation: a change in the relative format or nature of an informational quantity by a transducer. For example, the conversion of sound waves into corresponding electrical variations by a microphone.

Information Storage: preserving information for later use.

Input/Output Register (IOR): a parallel array of flip-flops in the CPU of a computer used to hold data that is being received from, or transmitted to, the input/output bus.

Input Port: the input interface between the CPU and external digital hardware in a computer system, consisting of a parallel-readable array of flip-flops (external to the CPU).

Instruction Code: the binary-encoded form of a computer instruction that can be interpreted by the Instruction Decoder and Sequencer in the CPU.

Instruction Cycle: the detailed sequence of steps required to carry out one instruction in a computer.

Instruction Decoder and Sequencer (IDS): a complex sequential logic block in the CPU of a computer that determines the sequence of register transmissions needed to execute an instruction.

Instruction Mnemonic: a three- or four-letter abbreviation that represents a computer instruction. Examples: MOV for move, JMP for branch, SUB for subtract. All mnemonics must be assembled into binary code before execution on a computer.

Instruction Register (IR): a parallel array of flip-flops in the CPU of a computer used to hold the instruction code after the fetch process is complete.

Instruction Set: the set of basic operations that a computer can execute, specified by assembly-language mnemonics and coded as binary numbers.

Intermediate Variable: a binary variable that is neither the primary input or output of a digital circuit, but indicates only a value internal to the circuit.

Inversion Circle: a small circle appended to a logical device to indicate that an incoming or outgoing signal is complemented.

INVERT: a logic function which is defined as follows:

Truth Table

X	Z
0	1
1	0

Algebraically

$Z = \overline{X}$

Schematically

X —▷o— Z

J-K Flip-Flop: a one-bit memory element that is represented, and behaves, as follows:

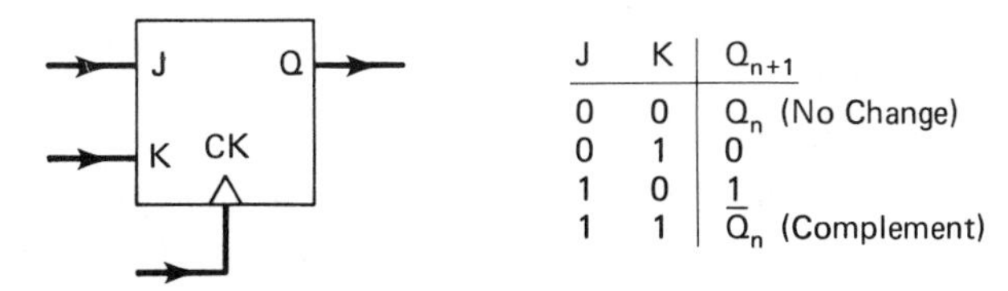

J	K	Q_{n+1}
0	0	Q_n (No Change)
0	1	0
1	0	1
1	1	$\overline{Q}_n$ (Complement)

Karnaugh Map: a matrix of cells for expressing a logic function, and used to obtain graphically a minimum algebraic expression for the function.

Latch: the simplest type of memory element, possessing asynchronous "set" and "reset" inputs.

Level-Sensitive Device: a device whose state may change only when its Enable input is asserted. (*see* "Edge-Triggered Device")

Line Printer: an output device used in computer systems to print program and data information on paper at a high rate (one line in parallel at one time).

Linearity: the degree to which the output of an A/D converter is exactly proportional to changes in its digital input.

Linker-Loader: a computer program that permits a multitude of small, diverse computer programs to be joined into one large program and loaded into memory.

Logic Circuit: an electronic circuit that operates on digital signals in accordance with a logic function.

Logic Family: a group of logic gates, modules, and components built around a standardized integrated circuit transistor configuration. Example families: TTL—Transistor-Transistor Logic, CMOS—Complementary Metal-Oxide Semiconductor logic, ECL—Emitter-Coupled Logic.

Logic Function: a well-defined relationship between inputs and outputs of a logic circuit. A logic function predicts the output of a logic circuit given a specific set in input conditions.

Logic Gate: an electronic circuit that implements a specific logic function, and can be treated as a simple, cascadable block element.

Logic State: one of two possible conditions a logical quantity may have, represented numerically as '0' or '1' and electrically as 'low' or 'high' voltage levels. (*see* "Bit")

LSB: an acronym for *L*east *S*ignificant *B*it, which is the digit of a binary number weighted least heavily.

Lumped Delay: the delay of a combinational network consolidated into one block and shown in the feedback path of an asynchronous sequential circuit, for the purpose of analysis. (*see* "Delay Element")

Map Address: in a Karnaugh map, the row and column numbers that identify a particular cell.

Master-Slave Flip-Flop: a clocked flip-flop consisting of two serial S-R latches. Input data is transferred to the master when the clock is asserted, and from the master to slave when the clock is deasserted.

Maximal Length Sequence: a binary count sequence in which all possible states are represented for a given number of bits.

Memory Address Register (MAR): a parallel array of flip-flops in the CPU of a computer, used to hold the address of the RAM cell currently being accessed. The MAR outputs drive the address bus.

Memory Data Register (MDR): a parallel array of flip-flops in the CPU of a computer used to hold data that is being received from or transmitted to RAM.

Merger Diagram: a graph that simplifies finding mergers from a merger table.

Merger Table: a table revealing how equivalent and unassigned states may be combined. (*see* "Merger Diagram")

Microcomputer: a computer system utilizing a microprocessor CPU.

Microprocessor: the central processing unit (CPU) of a computer, implemented as a single integrated circuit containing 5000 or more transistors.

Minimal Circuit: a logic circuit that has been implemented with the minimum number of logic gates.

Minimal Equation: a Boolean equation containing the fewest number of sum and product terms.

Minimization: generally thought of as the act of reducing the number of terms in a Boolean equation, or the number of gates in a logic circuit.

Minimum Sum of Products (MSP): a sum-of-products expression that contains the fewest possible terms.

Minuend: in subtraction, the quantity from which the subtrahend is subtracted.

$$\text{(minuend)} - \text{(subtrahend)} = \text{(difference)}$$

Modulus (often abbreviated "mod"): the maximum number of representable states in a device which has a specific number of digits. A four-digit binary counter is mod-sixteen, for example.

Modulus Property: a property which states that the modulus of an N-bit number can be determined by adding that number, its one's complement, and one:

$$q + \bar{q} + 1 = \text{modulus}$$

Monostable Multivibrator: a special type of flip-flop possessing one momentary state and one stable state. The period of the momentary state is determined by external resistor and capacitor timing components.

MSB: an acronym for Most Significant Bit, which is the digit of a binary number weighted most heavily.

Multiple Output Circuit: a single logic circuit that has two or more primary outputs.

Multiplexer: a logic circuit in which one of a multitude of input lines is gated to a single output line. (*see* "Demultiplexer")

NAND: a logic function which is defined as follows:

Truth Table

X	Y	Z
0	0	1
0	1	1
1	0	1
1	1	0

Algebraically

$$Z = \overline{X}\,\overline{Y}$$

Schematically

X, Y → NAND gate → Z

Natural Binary Sequence: a progression of binary numbers in which successive entries correspond to successive numerical quantities:

Number:	0, 1, 10, 11, 100, 101, 110, 111, 1000, . . .
Value:	0 1 2 3 4 5 6 7 8

Negative Feedback: a circuit connection in which a proportion of the output signal is returned to and subtracted from the input to bring about a reduction in output.

Negative Logic: the convention in electronic logic by which logic '0' values are represented by a more positive voltage than logic '1' values. Thus, the assertion level for logic '0' is 'high'. Negative logic is rarely used in modern systems. (*see* "Positive Logic")

Nine's Complement: the value obtained by subtracting a decimal digit from nine, or a decimal number from a succession of nine's.

Example: the nine's complement of 251 is 999 − 251 = 748.

Noise Margin: the maximum noise voltage that can be added to (or subtracted from) a digital signal before passing a threshold.

Noncritical Race: *see* "Race"

Nonvolatile Memory: a type of electronic memory that does not require power to retain stored data. Example: read-only memory. (*see* "Volatile Memory")

NOR: a logic function which is defined as follows:

Truth Table

X	Y	Z
0	0	1
0	1	0
1	0	0
1	1	0

Algebraically

$Z = \overline{X + Y}$

Schematically

X, Y → NOR gate → Z

Number Circle: an adaptation of the number line reflecting the modulus nature of all digital hardware.

Number Line: a graphical description of a sequence of numbers, consisting of a line with zero located at the center, and positive and negative integers equally spaced to the right and left, respectively.

Number System: a scheme whereby counted or measured quantities are symbolically expressed. The decimal number system uses ten symbols, 0, 1, 2, 3, 4, 5, 6, 7, 8, 9, while the binary number system uses only two: 0 and 1.

Nyquist Rate: the minimum theoretical rate at which an analog signal can be sampled to ensure faithful representation, equal to twice the frequency of the signal's highest frequency component.

Octal Code: a simplified means of representing binary numbers in a radix-eight number system.

Odd Parity: the condition where a binary number has an odd number of '1' digits. Example: 10110101—odd parity, 11001010—even parity. (*see* "Even Parity")

One's Complement: the value obtained by reversing the logic state of each bit in a given binary number. The one's complement of '0110111' is '1001000'. (*see* "Two's Complement")

One Shot: *same as* "Monostable Multivibrator"

Operand: data to be operated on by a computer instruction. Most arithmetic operations require two operands. (*see* "Operation Code")

Operating System: a program that supervises and coordinates system activity.

Operation Code: that part of a computer instruction that specifies the operation to be performed on the operands.

Operational Amplifier: a high-gain differential amplifier used in feedback circuits for linear amplification, current summation, and a variety of nonlinear applications.

Optoelectronic Sensor: a transducer whose electrical properties (resistance or voltage output) change under varying intensities of light.

OR: a logic function which is defined as follows:

Truth Table

X	Y	Z
0	0	0
0	1	1
1	0	1
1	1	1

Algebraically

$$Z = X + Y$$

Schematically

X, Y → Z

Output Equation: an equation describing the logic function of a sequential circuit's primary output.

Output Port: the output interface between the CPU and external digital hardware, consisting of a parallel-loadable array of flip-flops (external to the CPU). (*see* "Input Port")

Output Skew: a transient condition where related outputs which should change synchronously actually change at slightly different times, due to cascaded or unequal propagation delays. For example, output skew exists on the outputs of a ripple counter.

Output Table: a tabular representation of the output equation(s), principally in sequential circuits.

Overflow: a condition which occurs in an arithmetic operation when the result is greater than that which can be expressed with the permissible number of bits. (*see* "Underflow")

Parallel Load Input: an input, often synchronous, that causes a multibit counter, register, or other device to be loaded with data from parallel input lines.

Parallel-to-Serial Converter: a register whose memory elements can be simultaneously loaded with parallel data, and then caused to shift in one direction until all bits have been transmitted from the last stage. (*see* "Serial-to-Parallel Converter")

Parity Bit: a bit appended to a binary number so that the resulting '$n + 1$' bit number conforms with one of the two parity conventions. (*see* "Even Parity" and "Odd Parity")

Period: the duration between congruent sections of a cyclic waveform. Period, T, is related to frequency, 'f', by $T = 1/f$. (*see* "Frequency")

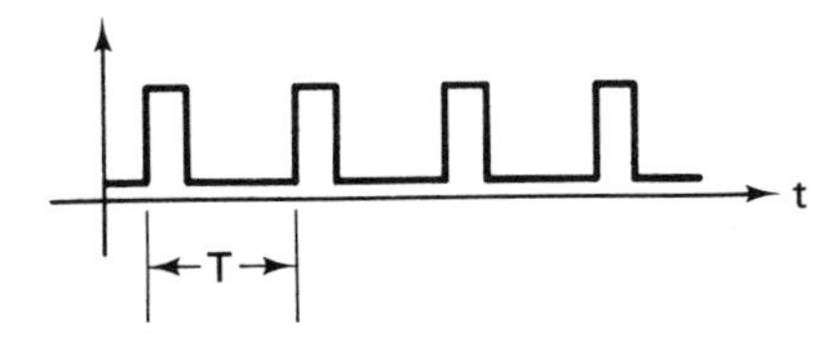

Periodic Digital Signal: a digital signal that regularly repeats a specific sequence of values.

Positive Logic: the convention in electronic logic by which logic '1' values are represented by a more positive voltage than logic '0' values. Thus, the assertion level for logic '1' is 'high'. Most modern systems employ a positive logic convention. (*see* "Negative Logic")

Postulate: an initial assertion or presupposition upon which a mathematical system is based.

Precision: (1) The degree to which small incremental changes can be represented (as with the number of significant digits in a number). (2) The ability of a device to operate in a consistent manner over time, and over changes in environment. (*see* "Accuracy")

Preset Input: *same as* "Set Direct"

Primary Input: an input to a sequential circuit derived from external sources, such as switches and sensors. (*see* "Secondary Input")

Primary Output: an output of a sequential circuit that is transmitted to external devices, such as indicator lamps and actuators. (*see* "Secondary Output")

Prime Implicant: an implicant that cannot be fully enclosed by another implicant on a Karnaugh map.

Processor Status Word (PSW): a collection of individual bits each of which represent and store a different possible outcome of an ALU operation. Example, PSW bits: zero bit, carry bit, sign bit.

Program: a sequence of instructions in a computer that exactly define a specific activity, and stored in memory as a sequence of binary numbers.

Program Counter (PC): a natural binary up counter in the CPU of a computer whose output indicates the memory address of the next instruction.

Program Loop: a sequence of computer instructions that are executed many times. A branch instruction is required at the end of the sequence to close the loop.

Programmed Counter: a counter that utilizes special combinational logic to cycle through a nonnatural binary sequence.

Programmed Logic: an orderly, mechanized plan of events specified by a stored sequence of binary numbers.

Programmed Sequence Generator: a circuit using a RAM and D-type flip-flops that progresses through a sequence of states determined by the stored binary data.

PROM Programmer: a special circuit used to write data into PROM and EPROM devices.

Proof: a demonstration that a property is true by showing that no contradictions exist for all possible conditions of the component variables.

Propagation Delay: the intrinsic delay existing in a logic gate or device, beginning with the application of an input and ending with the device's response. Propagation delay is due to unavoidable parasitic capacitance or inductance in an electronic circuit. NAND gate propagation delays in TTL are typically 10 ns.

Property: an equation whose left side contains a variable in combination with constants or other variables, and whose right side is logically equivalent but utilizing a different (often simpler) arrangement of variables and constants.

Pulse Synchronizer: a logic circuit that synchronizes the AND of a gate signal with a pulse train to ensure that no pulses are truncated.

Pulse-Triggered Device: another term for a master-slave device.

Quantization: (1) Time: the subdivision of the time axis of an analog signal into a sequence of discrete equally spaced sample points. (2) Amplitude: the subdivision of a smooth range of voltage or current values into a finite number of encodable levels.

Quiescent Delay Element: a delay element whose input and output states are equal. (*see* "Excited Delay Element")

R-2R Ladder: a resistor network constructed using only two values of resistance, one of which is twice the value of the other, and applied in digital-to-analog converter circuits where currents must be exponentially weighted and summed.

Race: a condition that exists in sequential circuits when simultaneous changes in flip-flop states cause temporary false excitations, resulting from output skew. A noncritical race is one where the correct final state is reached regardless of intermediate false excitations. A critical race leads to an erroneous final state, or produces false outputs.

Radix: a quantity that expresses the number of unique symbols used in a number system. Same as "Base."

Radix Circle: a number circle divided into as many segments as the radix of the chosen number system.

Random Access Memory (RAM): a type of electronic memory in which the access time to data is independent of the data's address. (*see* "Serial Access Memory")

RC Oscillator: an astable circuit whose frequency determining elements are a resistor and capacitor.

Read Cycle: the sequence of signal changes needed to constitute a read operation in an electronic memory. The "read-cycle time" is the minimum duration between successive read cycles. (*see* "Write Cycle")

Read-Only Memory (ROM): a type of RAM whose data is permanently recorded.

Read/Write Input: the input found on an electronic memory that determines whether data will be read from or written into the addressed cell.

Real-Time Debug: a computer program that facilitates the debugging of other programs by providing for single instruction execution, breakpoints, and various other features.

Redundant Enclosure: the addition of a nonessential implicant on a Karnaugh map to eliminate a hazard.

Refreshing: the act of recharging capacitor memory cells in a dynamic memory so that the stored information is not lost due to dissipation of charge.

Register: a parallel array of flip-flops used to store a binary word.

Relay: an electromechanical switch in which the contact positions can be changed by applying an electrical signal.

Reset Input: *see* "Clear Direct"

Resolution: the capacity to distinguish difference between closely spaced values.

Ring Oscillator: an astable circuit created by connecting an odd number of inverter gates in a ring network.

Ripple Counter: a counter circuit obtained with T flip-flops by connecting the 'Q' output of one stage to the 'CK' input of the next stage (all 'T' inputs are connected to '1'). (*see* "Synchronous Counter")

Rise-Time: the duration of time in a '0' to '1' transition for a signal to change from 10% to 90% of its logic '1' value.

Sample and Hold: a device used in analog-to-digital converter circuits to sample the analog input, then store and hold the sampled value while conversion takes place.

Sampling: quantization of the time axis of an analog signal.

Schmitt Trigger: a logic gate with greater noise immunity on its inputs due to dynamically variable thresholds. The threshold function is described by a hysteresis curve:

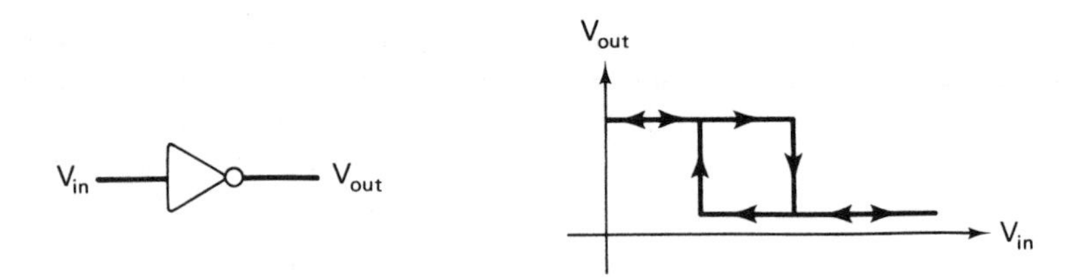

Secondary Input: an input to a sequential circuit derived from internal feedback paths. (*see* "Primary Input")

Secondary Output: an output of a sequential circuit that is used for internal feedback.

Self-Complementing Code: a binary code in which the one's complement of a BCD quantity equals the nine's complement of the corresponding decimal digit.

Sequential Logic: logic circuitry with memory, whose operation is a function of both the circuit's past history and its present inputs. Sequential logic is characterized by the presence of feedback paths.

Sequential Logic Fault: a fault condition in a sequential circuit shown by the desired flow graph being unequal to the measured flow graph for one or more state transitions. (*see* "Combinational Logic Fault")

Serial Access Memory (SAM): a type of electronic memory in which the data is stored serially in a medium, and where the access time depends upon where the data is located in the device or on the medium. (*see* "Random Access Memory")

Serial Addition: addition that is performed bit by bit on two serial-format binary numbers, from LSB to MSB, using a single flip-flop to store carries from one iteration to the next.

Serial-to-Parallel Converter: a register that can accept data in the first stage, shift it to later stages upon the receipt of clock pulses, and present the accumulated data on parallel outputs.

Series-Parallel Network: a switch network that can be considered as a combination of series and parallel switch connections only.

Set Direct: an asynchronous input, found on flip-flops, registers, and counters, that causes the internal state of such devices to become '1' (or all '1's). (*see* "Clear Direct")

Setup Time: (1) General: in synchronous devices, an amount of time preceding the application of a clock signal wherein the data input(s) must remain stable. (2) Address: the minimum time that address inputs to a RAM must be stable prior to reading or writing data. (3) Data: the minimum time that data inputs to a RAM must be stable prior to applying a "write" pulse. (*see* "Hold Time")

Seven-Segment Display: a display device containing seven independently controllable light bars (usually LED's) arranged as follows:

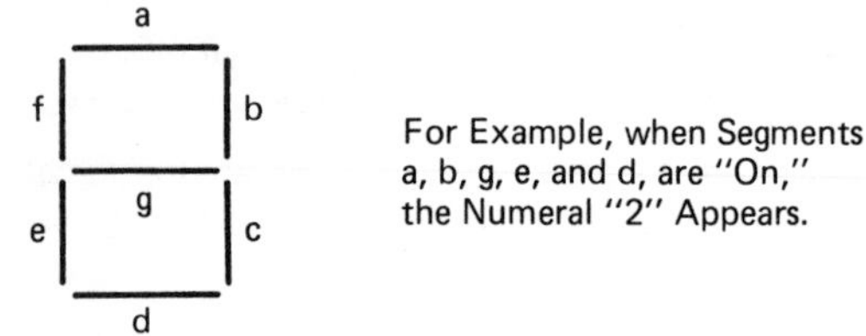

Shift Register: a succession of interconnected flip-flops in which data can be translated to the left, or right, upon the application of a clock pulse. The direction of shift is determined by an 'R/L' input, or is only in one direction.

Sign Bit: a bit appended to a binary number for the purpose of indicating the sign of that number. Typically, the sign bit is the MSB.

Sign-Magnitude Representation: identifying positive and negative numbers by a magnitude and a sign.

Simplification: the act of reducing a Boolean equation to (a) a single value when all variables are exactly specified, or (b) to an expression with fewer terms, using algebraic variables only.

Single-Pole Double-Throw (SPDT) Switch: an electrical switch schematically represented as

Single-Pole Single-Throw (SPST) Switch: an electrical switch schematically represented as

Slope Converter: a type of analog-to-digital converter in which a comparator measures a constant analog input voltage against increasing reference voltage. The time at which the comparator's output switches is a measure of the analog input, and is used to gate a binary counter.

Square Wave: a pulse train with equal '0' and '1' periods.

Stable State: a condition in a sequential circuit where the input states of the delay elements (or flip-flops) are equal to their corresponding output states. (*see* "Unstable State")

State: (1) The condition of a digital signal, being either logic '1' or logic '0'. (2) A particular pattern of '1's and '0's observed on the outputs of a multibit device.

State Detector: a multi-input gate, usually an AND gate, connected to be responsive to one particular state of a counter or sequential circuit.

State Table: a concise and abstract version of the transition table, and a tabular representation of a flow graph.

Static Fault: an output fault that will persist when the input is held steady with a particular excitation.

Static Memory: a type of memory design using a flip-flop matrix. (*see* "Dynamic Memory")

Steering Logic: the added gates that are present in a natural binary counter to cause the component flip-flops to reenter a previous state, responsive only when a "detect" signal is asserted from a detector gate. (*see* "Detect and Steer")

Stuck Input: a static fault in which a logic device input appears to be always in one state (either '0' or '1'), regardless of actual input conditions.

Stuck Output: a static fault in which a logic device output is always in one state (either '0' or '1') regardless of the correct output.

Subtrahend: in subtraction, the quantity subtracted from the minuend.

$$(\text{minuend}) - (\text{subtrahend}) = (\text{difference})$$

Successive Approximation Converter: a type of analog-to-digital converter employing a D/A device in a feedback loop, and a shift register with a special logic design to arrive at the digital equivalent of the analog input. More complex but more efficient than a tracking converter.

Sum: the result of addition. (*see* "Addend" or "Augend")

Sum of Products (SP): a Boolean expression that contains no parentheses. (*see* "Minimum Sum of Products")

Superposition: a procedure in which a composite signal is arrived at by the summation of component signals.

Switch Bounce: irregular electrical contact exhibited when a mechanical switch is closed or opened, usually lasting less than 10 ms.

Synchronous Counter: a counter circuit in which all flip-flops are triggered simultaneously, allowing all output bits to change at the same instant. (*see* "Ripple Counter")

Synchronous Device: a circuit or module whose operation depends on, and is synchronized with, a clock signal. (*see* "Asynchronous Device")

Synchronous Sequential Circuit: a sequential circuit with clocked flip-flops embedded in the feedback paths, and whose state transitions are controlled by a clock signal. (*see* "Asynchronous Sequential Circuit")

System Architecture: the specific interconnection of CPU, I/O, and memory components that defines a computer system. (*see* "Computer Architecture")

T Flip-Flop: a one-bit memory element that is represented, and behaves, as follows:

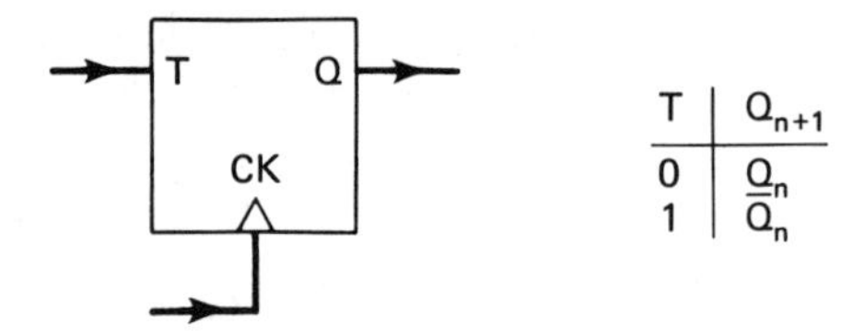

T	Q_{n+1}
0	Q_n
1	$\overline{Q}_n$

Ten's Complement: the value obtained by adding one to the nine's complement of a number.

Text Editor: a computer program that permits alphanumeric text to be entered, electronically manipulated, and stored in a computer system.

Three-State Output: a type of logic output where a third, zero-current condition is possible, as well as logic '0' and logic '1' conditions. The third state is the "turned off" state, and permits multiple outputs to be connected in a bus network without mutual interference. A three-state inverter is schematically drawn as follows:

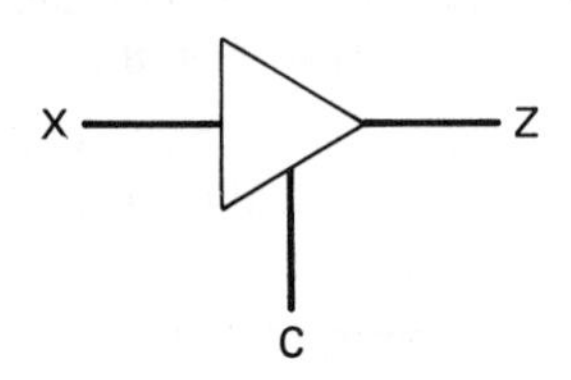

Threshold: the borderline between a valid logic-state voltage and a transition zone.

Time-Division Multiplexing: the sharing of a single transmission channel (wire) by two or more signals at specific and periodic times. No two signals are ever on line at the same time.

Time-Window Converter: a type of analog-to-digital converter in which the analog input controls the time constant of a monostable, and whose output gates a clock signal. The total number of pulses passed is accumulated in a counter and is proportional to the analog value.

Timing Diagram: a waveform drawing showing the time relationship between two or more digital signals.

Tracking Converter: a type of analog-to-digital converter employing a D/A device in a feedback loop with an up/down counter, to track the analog input.

Transducer: a device that converts one type of signal (such as an electrical one) into another type of signal (such as a mechanical one).

Transition Table: a table particularly useful in asynchronous sequential circuit analysis to show the permissible patterns of change between states, and to reveal race conditions. It is derived from the excitation table during analysis, and from the state table during design.

Transition Zone: an intermediate range of voltage between the defined voltage limits of logic '0' and logic '1'. For standard TTL circuitry, these limits are:

Logic '0'	0 to 0.8 V
Transition zone	0.8 to 2.0 V
Logic '1'	2.0 to 5.0 V

Transmittance: the state of conduction through a switch network. A transmittance of '1' indicates that current can flow, whereas '0' shows that it cannot.

Trap State: an undesirable internal state in flip-flop circuits where no further changes can occur until a master reset is applied.

Truth Table: a table that describes logic functions in which an exact output value is shown for every possible input combination.

Two's Complement: the value obtained by adding one to the one's complement of a given binary number. The two's complement of '0101100' is '1010100'.

Two's-Complement Subtraction: a method of subtraction in which the "difference" is obtained by adding the minuend and the two's complement of the subtrahend.

Underflow: a condition which occurs in an arithmetic operation when the result is smaller than can be expressed with the permissible number of bits.

Unidirectional Bus: a parallel array of conductors transmitting information in one direction only. (*see* "Bidirectional Bus")

Unique Function: a Boolean function that can be described by only one MSP equation. A "nonunique" function is represented by two or more MSP equations.

Universal Counter Stage: a multibit synchronous up/down counter module, usually four to eight bits long and parallel-loadable, that can be cascaded to any desired length.

Universal Gate: a logic gate that can be connected with copies of itself to obtain any desired logic function. NAND and NOR gates are examples of universal gates.

Unstable State: a condition in a sequential circuit where the input state of one or more delay elements (or flip-flops) is unequal to the corresponding output states. (*see* "Stable State")

Video Terminal: an interface device between the computer and a human operator consisting of a typewriter-like keyboard and a television display capable of displaying alphanumeric characters.

Volatile Memory: a type of electronic memory requiring a constant supply of power to retain stored data. Example: most semiconductor read/write RAM circuits are volatile. (*see* "Nonvolatile Memory")

Voltage Mode: a system of logic in which signal voltages (not currents) are measured and related to the logic conditions of '0' and '1'. (*see* "Current Mode")

Weighted Code: a numeric code in which the digit positions are weighted with different values.

Word: a multibit binary quantity.

Write Cycle: the sequence of signal changes needed to constitute a write operation in an electronic memory. The "write cycle time" is the minimum duration between successive write cycles. (*see* "Read Cycle")

XOR: exclusive OR, a logic function which is defined as follows:

Truth Table

X	Y	Z
0	0	0
0	1	1
1	0	1
1	1	0

Algebraically

$$Z = X \oplus Y$$

Schematically

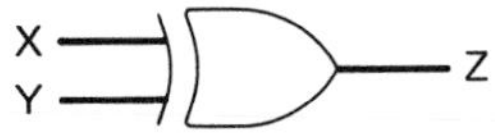

X
Y
Z

INDEX

INDEX

D

E

F

G

H

I

J

K

L

M

N

O

P

Q

R

S

T

U

V

W

X